ANNUAL REVIEW OF
EARTH AND
PLANETARY SCIENCES

ANNUAL REVIEW OF EARTH AND PLANETARY SCIENCES

VOLUME 16, 1988

GEORGE W. WETHERILL, *Editor*
Carnegie Institution of Washington

ARDEN L. ALBEE, *Associate Editor*
California Institute of Technology

FRANCIS G. STEHLI, *Associate Editor*
DOSECC Science Advisory Committee

ANNUAL REVIEWS INC   4139 EL CAMINO WAY   P.O. BOX 10139   PALO ALTO, CALIFORNIA 94303-0897

ANNUAL REVIEWS INC.
Palo Alto, California, USA

International Standard Serial Number: 0084-6597
International Standard Book Number: 0-8243-2016-6
Library of Congress Catalog Card Number: 72-82137

Annual Review and publication titles are registered trademarks of Annual Reviews Inc.

Annual Reviews Inc. and the Editors of its publications assume no responsibility for the statements expressed by the contributors to this *Review*.

TYPESET BY AUP TYPESETTERS (GLASGOW) LTD., SCOTLAND
PRINTED AND BOUND IN THE UNITED STATES OF AMERICA

Annual Review of Earth and Planetary Sciences
Volume 16, 1988

CONTENTS

ERRATUM

Volume 15 (1987)

In "Architecture of Continental Rifts With Special Reference to East
Africa" by B. R. Rosendahl, the top panel of Figure 4 (p. 464) should
appear as follows:

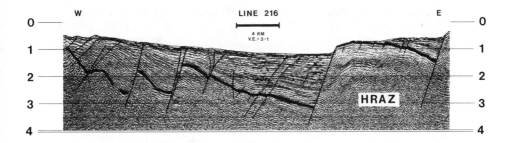

SOME RELATED ARTICLES IN OTHER *ANNUAL REVIEWS*

From the *Annual Review of Astronomy and Astrophysics*, Volume 25 (1987):

Star Formation in Molecular Clouds: Observation and Theory, Frank H. Shu, Fred C. Adams, and Susana Lizano

Comets and Their Composition, Hyron Spinrad

The Local Interstellar Medium, Donald P. Cox and Ronald J. Reynolds

The IRAS View of the Galaxy and the Solar System, C. A. Beichman

From the *Annual Review of Biochemistry*, Volume 56 (1987):

Transfer RNA Modification, Glenn R. Björk, Johanna U. Ericson, Claes E. D. Gustafsson, Tord G. Hagervall, Yvonne H. Jönsson, and P. Mikael Wikström

From the *Annual Review of Ecology and Systematics*, Volume 18 (1987):

Architecture of Tropical Plants, P. B. Tomlinson

Molecular versus Morphological Approaches to Systematics, David Hillis

Size-Based Demography of Vertebrates, J. R. Sauer and N. A. Slade

Patterns of Tropical Vertebrate Frugivore Diversity, Theodore H. Fleming, Randall Breitwisch, and George H. Whitesides

Tropical Alpine Plant Ecology, Alan P. Smith and Truman P. Young

Tropical Limnology, William M. Lewis, Jr.

Evolution of Animal Mitochondrial DNA: Relevance for Population Biology and Systematics, C. Moritz, T. E. Dowling, and W. M. Brown

Speciation in the Deep Sea, George D. F. Wilson and Robert R. Hessler

Stable Isotopes in Ecosystem Studies, Bruce J. Peterson and Brian Fry

The Ecology of Indeterminate Growth in Animals, Kenneth P. Sebens

From the *Annual Review of Fluid Mechanics*, Volume 20 (1988):

Fractals in Fluid Mechanics, D. L. Turcotte

Fluid Models of Geological Hotspots, John A. Whitehead

Remote Sensing of the Sea Surface, O. M. Phillips

(*continued*)

Surf-Zone Dynamics, J. A. Battjes

Sand Transport on the Continental Shelf, K. R. Dyer and R. L. Soulsby

From the *Annual Review of Materials Science*, Volume 17 (1987):

Synthesis of Diamond Under Metastable Conditions, R. C. DeVries

From the *Annual Review of Microbiology*, Volume 41 (1987):

Rapid Evolution of RNA Viruses, D. A. Steinhauer and J. J. Holland

Enzymatic "Combustion": The Microbial Degradation of Lignin, T. Kent Kirk and Roberta L. Farrell

Genetic Research With Photosynthetic Bacteria, Pablo A. Scolnik and Barry L. Marrs

From the *Annual Review of Nuclear and Particle Science*, Volume 37 (1987):

Nuclear Techniques for Subsurface Geology, D. V. Ellis, J. S. Schweitzer, and J. J. Ullo

ANNUAL REVIEWS INC. is a nonprofit scientific publisher established to promote the advancement of the sciences. Beginning in 1932 with the *Annual Review of Biochemistry*, the Company has pursued as its principal function the publication of high quality, reasonably priced *Annual Review* volumes. The volumes are organized by Editors and Editorial Committees who invite qualified authors to contribute critical articles reviewing significant developments within each major discipline. The Editor-in-Chief invites those interested in serving as future Editorial Committee members to communicate directly with him. Annual Reviews Inc. is administered by a Board of Directors, whose members serve without compensation.

ANNUAL REVIEWS OF		SPECIAL PUBLICATIONS
Anthropology	Materials Science	
Astronomy and Astrophysics	Medicine	Annual Reviews Reprints:
Biochemistry	Microbiology	Cell Membranes, 1975–1977
Biophysics and Biophysical Chemistry	Neuroscience	Immunology, 1977–1979
Cell Biology	Nuclear and Particle Science	
Computer Science	Nutrition	
Earth and Planetary Sciences	Pharmacology and Toxicology	Excitement and Fascination
Ecology and Systematics	Physical Chemistry	of Science, Vols. 1 and 2
Energy	Physiology	
Entomology	Phytopathology	Intelligence and Affectivity,
Fluid Mechanics	Plant Physiology	by Jean Piaget
Genetics	Psychology	
Immunology	Public Health	Telescopes for the 1980s
	Sociology	

A detachable order form/envelope is bound into the back of this volume.

Robert P. Sharp

Ann. Rev. Earth Planet. Sci. 1988. 16: 1–19

EARTH SCIENCE FIELD WORK: ROLE AND STATUS

Robert P. Sharp

Division of Geological and Planetary Sciences, California Institute of Technology, Pasadena, California 91125

Introduction

This essay contends that the outstanding contributions currently being made to the earth sciences by theoretical and laboratory endeavors increase rather than decrease the need for sound field observations. The presentation is strongly prejudiced in favor of field studies and, accordingly, invites critical examination by skeptical minds.

Many earth science problems being investigated have their source in field studies. Samples of materials worthy of analysis by sophisticated laboratory techniques and apparatuses are selected on the basis of field studies. Furthermore, the field is where the results of theory and laboratory experimentation are tested for conformity to nature and the truth. Field investigators of all types are as sorely needed now as at any time in the past. Their role in the earth sciences merits respect and recognition.

A statement supporting field studies may seem to be championing the obvious, but Francis Pettijohn's (1984) outspoken memoirs amply demonstrate that field activities have not always been respected. Many earth scientists would profess regard for, if not devotion to, field activities, and the US Geological Survey, the greatest assemblage of earth science talent ever, is strongly field oriented. Yet many field geologists feel their discipline is on the decline, that their efforts and products are looked down upon, that time spent in field work is seen as less productive than time spent in the laboratory or before a computer, that greater and more spectacular advances are made by laboratory experimenters or theoreticians than by field workers, and that field geologists may be an endangered species.

An unusually large number of exciting earth science developments have recently come from experimental and theoretical work. Much of the

1

0084–6597/88/0515–0001$02.00

glamor in the earth sciences now seems to be associated with such endeavors, stimulated and supported by rapid developments in instrumentation and generous research grants. But close inspection shows that many of these fruitful advances have depended upon complementary and cooperative field investigations.

Field and laboratory work are mutually supportive, not adversative. Most field geologists need laboratory analyses to support their field work, and most laboratory investigators depend upon field data to bring reality to their studies. Few field geologists now make a career solely from field mapping, as Tom Dibblee (Steller 1986) has done so spectacularly. By the same token, few theoreticians or experimentalists proceed without reliance upon field data.

Field work may seem at times routine, unproductive, or even boring, but the same can be said of much laboratory work. One can labor long in either arena without rewarding results. Still, the chance of turning up something exciting is as great in the field as in the laboratory.

Most broadly, the term *geology* encompasses the study of planet Earth (Gary et al 1974, p. 293). In this sense, it includes the disciplines of geophysics, geochemistry, and geobiology (paleontology), as well as other subdisciplines involving the solid earth. A common modern practice, however, is to use the term *earth science* as embracing all these disciplines, including geology. This usage is followed here.

The terms *field geology* and *field work* are used in their broadest sense to include the observation, study, and investigation of natural materials, features, phenomena, and processes in their natural setting by any of a wide spectrum of procedures, techniques, and instruments. This broad concept of field work is developed further in a following section. A more classical concept is based on a geologist walking out contacts between rock units and transferring data to a base map. Geological mapping, however, is only one kind of field work.

Historical Perspective

My candidate for the greatest North American geologist would be G. K. Gilbert (Pyne 1980). Gilbert was above all a field geologist. His two monumental contributions, the *Lake Bonneville* monograph (Gilbert 1890) and the *Report on Geology of the Henry Mountains* (Utah) (Gilbert 1877), are products of field work. This is not surprising, because during the 50 years of his professional career, 1869–1918, geological studies were primarily field oriented, and western United States, where Gilbert did the bulk of his work, was geologically unexplored.

In some respects, Gilbert was before his time in setting up flume experiments to study the behavior of alluvial wastes derived from California gold placers (Gilbert 1914). From this study he formulated some of the basic laws governing fluvial transport of coarse rock debris. Gilbert (1893) was even more adventuresome in peering through an astronomical telescope at the moon's surface. He was one of the first to look at lunar features through the eyes of a field geologist. As a result, he postulated an impact origin for lunar craters, anticipating by many years conclusions drawn by Ralph Baldwin (1949) and Harold Urey (1951). Gilbert even resorted to throwing steel ball bearings into soft mud to simulate impact features. Through such activities, he pointed the way for modern earth scientists who combine field activity and laboratory experiments (Baker & Pyne 1978).

Outside his administrative duties for the US Geological Survey, Gilbert probably spent 80 to 85% of his research time and effort on field work and in preparing field data for publication. For many earth scientists today, laboratory effort is more likely to predominate. Modern instrumentation, techniques, and data-processing facilities make laboratory work especially productive and rewarding. As a result, fewer present-day earth scientists who devote the major part of their time and effort to field activities are regarded as outstanding by their peers, in contrast to Gilbert's time.

Types of Field Work

One's concept of field work naturally depends on special needs and interests. Near one end of the spectrum is the classical field geologist who wants to lay hands upon the earth's rocks, minerals, and fossils in their natural setting and to observe natural processes in action on the earth's surface. Such a person uses mostly eyes, feet, training, and experience in conducting field work and is likely to end up making a map showing relationships between geological units. Near the other end of the spectrum are those who employ highly sophisticated instruments, apparatuses, and techniques to learn about the physical properties, behavior, and relationships of masses composing the earth that cannot be ascertained by direct visual observation. The objective of such studies often lies within the earth rather than on its surface. Between these extremes is a wide variety of tasks, such as collecting specimens, measuring stream velocity, operating tiltmeters, and detailing stratigraphic sections.

Most exploratory geophysical procedures are forms of remote sensing, which does not at first thought seem like field work. Seismology, magnetometry, and gravimetry are examples. When Vening-Meinesz (1948) boarded a Netherlands submarine planning to operate a gravity meter over ocean basins, he actually embarked on a field program. If a magnetometer is used by someone walking over the ground, its measurements would probably be accepted as a product of field work. Why not regard measurements made by the same magnetometer flown over the area in an aircraft as the product of field work? If a geologist charters a plane for one day to fly over an area being mapped, has that day been devoted to something other than field work? Most of us would think not. The same geologist may make liberal use of aerial photos for mapping without thinking of them as a product of remote sensing.

A wide array of geophysical techniques is used for probing the earth, and if one, such as magnetometry, qualifies as field work, why not similarly regard the others? Geophysical techniques are used to observe and record characteristics of the earth in its natural state, whether the instruments are carried by hand, car, boat, submarine, airplane, spacecraft, or donkey. In most instances, field data not available in any other way are gathered by techniques of remote sensing. The fact that a remote-sensing technique, such as shuttle imaging radar in North Africa (Elachi et al 1982), benefits from subsequent ground studies (McCauley et al 1982) does not make it any less a form of field work. To map mantle tomography by use of earthquake waves, without ever leaving the laboratory, is in the broadest sense also a type of field work.

Remote sensing has the virtue of providing integrated views of large-scale relationships. Its needs have stimulated the development of techniques, instruments, and procedures that make possible more effective scientific observation of our Earth. It seems high time to recognize that remote sensors are engaged in an important form of field activity.

Essentially everyone accepts geological field mapping as classical field work. Two easily identified types of mapping are exploratory and directed. Exploratory mapping is carried on primarily to discover what exists within areas of unknown terrane. It is widely practiced in government and industry, and by some individuals. Exploratory mapping raises more questions than it answers, because by reconnaissance it turns up many new and unexpected findings. Directed mapping is normally conducted to solve specific problems or to support other field activity. It is normally more detailed but of more limited scope, both geographically and intellectually, than exploratory mapping, and it is designed to produce answers. To a purist, geological mapping of either type is the most basic form of field work.

Direct observations of geological phenomena in action, such as floods, surging glaciers, and volcanic eruptions, are productive and exciting— even hazardous—types of field work. Underground mapping in mines is a specialized activity largely of commercial interests. Although field geologists tend to rely on natural exposures, modern earth-moving tools, such as backhoes, are now often employed by Quaternary geologists, in place of shovels, to make artificial exposures at critical sites.

A widely exercised type of field activity is what might be called "show-and-tell." This is a procedure in which persons who have conducted field studies of relationships in a specific location show them to others and present explanations and interpretations. Show-and-tell field trips, usually sponsored by organizations, are an effective way of sharing knowledge.

Field work involves the observation, visual or instrumental, of natural materials and processes of the earth in their natural setting. Sometimes the only workable means are by the techniques of remote sensing. Defined this broadly, field work is done far more extensively by earth scientists than they or others may realize.

Field Work as Related to Research

Field activity introduces reality into earth science research. Nature can be a harsh critic, destroying elegant theoretical models with a few hard, cold facts. It makes sense to base such models on as much salient, sound field data as possible. No matter how attractive, a theoretical product remains incomplete until shown to be compatible with field relationships. Successful theoreticians respect and value field data and do not hesitate to enlist the cooperation of field workers in obtaining more. Essentially the same can be said for laboratory experimentation.

Much modern analytical equipment is so productive that one has to guard against letting the satisfaction of doing laboratory procedures overwhelm the significance of the analyses. Time and resources can be wasted analyzing specimens that do not merit the effort and expense. Since the earth sciences deal with complex, messy systems, a large number of relatively imprecise data are often more useful than a few highly precise values. There is little point in measuring the width of a city street to fractions of a millimeter with a micrometer. Field work is relatively inexpensive compared to most laboratory procedures, so it makes sense to invest in a thorough field study before launching an expensive analytical program. Contrariwise, in some situations, a few blind laboratory analyses may be required to establish the need for detailed field work.

A good field map is a necessity for many earth science research projects (US Geological Survey 1987, Reinhardt & Miller 1987). Such a map may have already been made, or it can be custom made, perhaps by someone other than the principal investigator. The problem being investigated may even have been identified in the first place by mapping. Geological knowledge of the area will help any research program, no matter how specialized or localized, to avoid later surprises.

Not all earth scientists need be adept at field mapping, but awareness of its value is desirable, as has been demonstrated many times. Consider the Heart Mountain overthrust of Wyoming, which for many years after

its initial description (Hewitt 1920) was regarded as a rooted thrust produced by compression in the crust. Years of detailed field work by William
G. Pierce (1941, 1957, 1960, 1963, 1973, 1979) were required to demonstrate that the Heart Mountain structure is a detachment thrust, essentially a huge slide, originating in the northeast corner of Yellowstone Park.
The field relationships are striking, but a program of devoted mapping
over a large region was required to demonstrate their true meaning.

Some geological settings allow us to observe and monitor experiments
being conducted by nature at real scales in natural environments. Tracking
natural experiments has yielded good results for rapidly acting processes
and agents, such as glaciers, rivers, volcanism, wind, tectonism, and shoreline activities. Working with such situations requires patience, sometimes
for years, before useful results are obtained, and even then it can be
frustrating. Nature has a habit of changing parameters and variables
indiscriminately at inopportune moments. Because of their duration and
attending uncertainties, many natural experiments are not suitable subjects
for graduate student research; they lie more in the domain of established
professionals with a stable base.

Pedagogical Value of Field Work

The first geological field experience of many people is likely to be a show-
and-tell field trip (Figure 1). Such trips are effective for stimulating interest
among nonprofessionals and attracting the attention of other scientists to
the opportunities and satisfactions of working on geological problems.

Show-and-tell can generate interest in the earth sciences without a large
investment of time and effort on the part of nongeologists. Something seen
in its natural state arouses more enthusiasm than it does through the
medium of written or photographic representation, and it is certainly better
retained in memory. Through stimulation by show-and-tell experiences,
physicists, chemists, and planetary scientists, among others, have actually
become professional earth scientists. It is a rewarding experience to have
a radio astronomer wax enthusiastic about geological field phenomena
after participating in some show-and-tell experiences. This has happened
more than once. The show-and-tell procedure sows seeds widely, and one
never knows what may sprout from some unexpected spot.

As a purely pedagogical procedure, show-and-tell has limitations in not
demanding enough from the audience (students). Its educational function
can be enhanced by presenting problems and challenges that require audience participation. The emphasis can be on show, with students providing
the tell.

Educational insitutions could use field experiences more effectively in elementary geology courses by designing self-guided field excursions to replace some of the indoor exercises of the usual physical geology laboratory. This works better in some environments than others, but even the polished slabs of granite and gneiss in the local village bank front, the glacial erratic in the quad fronting the library, the cut behind the bookstore, or the gully and stream through the arboretum can, with imagination and thought, be used as a basis for field observation and interpretation by students. Send students into the field with written descriptive guides and thought-provoking questions. Don't lead them by the hand. It is also probably more effective to send them back into the field alone, with a corrected exercise in hand, than to conduct them en masse for a review of relationships.

Participation in a full-fledged field project is a still more effective means of generating a commitment to the earth sciences, on the part not only of students but of scientists from other disciplines. The British seem to practice this with particular success. Take, for example, Gerald Seligman's (1941) Jungfraujoch research program in glaciology, which enlisted the talents of Nobel laureate-to-be Max Perutz (Perutz & Seligman 1939), among others. Similar results attended a program of field research on the Austerdalsbrea in Norway, which led to major glaciological contributions from British physicists John Nye (1952, 1960, 1963) of Bristol and John Glen (1955, 1956) of Birmingham.

Geological field mapping may be regarded by the uninitiated as a simple-minded task of putting lines on maps. In truth, it is a first-class pedagogical discipline requiring keen observation, synthesis, and interpretation. Learning to arrive at workable conclusions, often on the basis of insufficient evidence, is part of the art of doing both geology and field mapping. Field mapping demands decisions; otherwise the map remains blank. Any reasonably intelligent person can be trained to do geological mapping, but as with many other pursuits, really good field mappers seem to be born, not made. The knack is not given to everyone, and those who have it deserve to be nurtured and respected.

Nature is a perverse ego humbler, and she exercises that trait freely in field geology. She delights in throwing spitball curves that send the overconfident neophyte, and often the hardened, experienced field mapper, back to the dugout muttering to themselves.

Except in the simplest areas, mapping involves a steady flux of surprises and complications. It is a detective game, solving ancient crimes committed by nature, with the clues now obscured. The experience is good for students. One of the main goals of a college education should be to establish a discipline of self-education. Few places are better suited to do that than the field. Nature is a stern teacher.

Figure 1 (above and facing)   Sharp and students during a field excursion east of the Sierra Nevada in the early 1980s devoted to inspection of glacial features.

Special Projects Related to Field Work

At least two long-range communal projects could be initiated by field geologists for the benefit of future generations of earth scientists.

THE CENTURY PROJECT   A central committee of retired field workers could be formed under the auspices of one of our national earth science organizations. This committee would design, select, and implement a number of field experimental and observational projects aimed at determining the rates at which geological changes take place over a period of 100 years.

Initiators of the program would obviously derive no direct benefit from it, hence the emphasis on older, retired people, who presumably have arrived, at least intellectually, and can afford to be generous with their time and effort. Raising funds for the project would be the responsibility of the central committee, which would also be charged with seeing that monitoring was maintained. The committee would have to be a continuing body with a slowly changing membership, managed by the sponsoring organization.

Suggestions for experiments should be solicited nationwide, and a selection of the most desirable, that could be adequately financed, would be made by the committee or a panel operating under its auspices. The central committee would assign responsibility for installation and periodic monitoring of each experiment, probably most often to the originators of the accepted proposals. For each experiment, the responsible entity would be charged with seeing that the baton of maintenance and monitoring was passed along to assure continuity to the completion of the project. Periodic reviews of each experiment by the central committee should be made to assure continuity.

Reports could be prepared periodically by experimenters, and the central committee could issue summaries through publication or other means, with ample credit to individuals. In 100 years, a final report on each experiment would be published under auspices of the central committee with full credit to all who participated.

Monitoring already going on in fields such as volcanology, glaciology, and neotectonics would not be duplicated. The types of experiments considered, all in natural settings, might include sedimentation rates for deposits in various environments, weathering and erosion rates, slow mass-movements, ground-ice wasting, dune migration, clay and caliche accumulation in soils, groundwater deposition, and shoreline modifications. There is no dearth of things that might be done. The task would be to select feasible experiments, likely to produce meaningful results, that could be maintained and monitored at reasonable cost and effort over a century.

A CENTER FOR FIELD TRIP GUIDEBOOKS   Unfortunately, much excellent field information is never published except in road logs and guides for field trips associated with a one-time event. The amount of valuable information packed away in such media is huge, but it is not easily accessible.

Field trip guides are issued by a number of widely dispersed sponsoring organizations; only a limited number are part of established serial publications. Guidebooks are commonly printed in small numbers, as special publications, and rapidly become unavailable. Few libraries have more than a sampling of field guides, and those they do have are a headache because of classification.

The Guidebooks Committee of the Geosciences Information Society has compiled a splendid catalog of geological field trip guidebooks of North America, up to 1980, and a limited listing of library holdings. A fourth edition of their catalog has recently been published by the American Geological Institute (1986). This is a commendable accomplishment, but more is needed.

Someone with library skills, organizational sense, and entrepreneurial spirit could render a major service to the earth sciences, hopefully at a profit, by establishing a Center for Field Trip Guidebooks. The Center should search out and assemble single copies of every field guide that can be located. Frequently, updated lists of these holdings should be distributed.

Through arrangements with authors and publishers, permission should be obtained allowing the Center to reproduce and sell copies of the guides at a price designed to support the Center's operation and provide a profit. If some tax-exempt organization undertook the service, it could be operated more economically on a no-gain basis and initial outlays might be underwritten by grants.

Personal Experiences

Field geology can be done almost anywhere at any time, with no more equipment than a notebook and pencil, by anyone with reasonable training and experience. Furthermore, it is usually worth doing. The following personal experiences are offered to illustrate the point.

In the summer of 1937, while finishing field work on a PhD thesis on geology of the Ruby–East Humboldt Mountains of northeastern Nevada (Sharp 1939), I began getting letters from Ian Campbell at the California Institute of Technology exploring possibilities of my joining a geological boat expedition through the Grand Canyon in October and November. The purpose of the expedition was to study exposures of the Archean igneous-metamorphic complex within the inner gorges, which are hard to reach except from river level. The expedition was to be jointly sponsored by Caltech and the Carnegie Institution.

That arrangement came about because John C. Merriam, famed vertebrate paleontologist at the University of California (Berkeley), had become Director of the Carnegie Institution of Washington. He wished to see the geology of the Grand Canyon thoroughly studied. Aside from early explorations, largely by Powell (1875), Gilbert (1875), Dutton (1882), and Walcott (1883, 1890, 1894, 1895), modern work had focused primarily on the Paleozoic (McKee 1934, 1938, Wheeler & Kerr 1936) and Proterozoic rocks (Van Gundy 1934, Hinds 1935). The only published product of comprehensive field mapping in the canyon was Levi Noble's (1914) excellent bulletin on the Shinumo quadrangle, which lies west of that part of the national park normally visited by tourists.

Solomon-like, Merriam apportioned the Paleozoic section to E. D. McKee, then ranger naturalist of the National Park Service, later of the US Geological Survey; the Proterozoic sedimentary and volcanic rocks to N. E. A. Hinds of the University of California (Berkeley); and the Archean complex to Ian Campbell and John H. Maxson of Caltech. Merriam supported these investigators with financial grants and facilitated publication of results in Carnegie Institution monographs (McKee 1934, 1938, McKee & Resser 1945, Hinds 1935).

Thus, I found myself in earliest October 1937 at Lee's Ferry on the Colorado River in the company of three experienced boatmen and three senior geologists—Campbell, Maxson, and J. T. Stark of Northwestern University. We were to board three wooden Stone-Galloway river boats for a two-month voyage of 280 miles through the Grand Canyon into Lake Mead, then filling behind Boulder (Hoover) Dam. McKee later joined the party at the foot of the old Bass Trail, partway along our route.

In 1937, the Colorado River was not a tourist's run; probably less than 100 people had made the trip through the canyon. Only one professional geologist other than Powell was known to have preceded us on such a voyage: Raymond C. Moore (1925), Professor of Geology at the University of Kansas and Kansas State Geologist, was a member of the 1923 Birdseye expedition (LaRue 1925) sponsored by the US Geological Survey to locate and evaluate dam sites.

Our expedition was truly exploratory with a promise of scientific discovery. I was very much the junior member of the group, and why Campbell took me rather than an experienced igneous-metamorphic petrologist, I'll never know. We traveled as fast as conditions, mostly rapids, permitted in reaches through Paleozoic and Proterozoic rocks and more slowly within the inner-gorge Archean exposures.

I asked Campbell if he had any specific geological chores in mind for me, and he decided I should keep track of pegmatite bodies in the Archean terrane. This and other activities did not fully occupy my time, so I cast around for something to do on my own. The Paleozoic belonged to McKee, the Proterozoic to Hinds, and the Archean to Campbell and Maxson, so I had to find something in between to avoid stepping on toes. Two great uncomformities are spectacularly exposed in the canyon walls: the younger one at the base of the Paleozoic beds where they rest upon truncated Proterozoic strata or the Archean complex, and the older one separating Proterozoic beds from truncated Archean rocks.

These appeared to fill the bill nicely. The near-horizontal pre-Paleozoic surface is exposed in cross section in the canyon walls along the wandering course of the river and its tributaries, so one gets a reasonable view of its three-dimensional relief. The pre-Proterozoic surface is preserved only in tilted fault blocks scattered throughout the region, so it is not so completely exposed. The study had to be improvised without a literature search or preconceived ideas. I had a Brunton compass, geological hammer, pencil, and notebook. Modern US Geological Survey topographic maps were available for only two or three quadrangles along the entire extent of the canyon, but river profile maps from the Birdseye work were helpful.

My improvised project meant some inconvenience for others of the expedition. Most of their work was conducted from daytime stops along the river, rather than from established camps. When a stop was made, if at all feasible I took off up a tributary canyon to get a look at one of my unconformities. As a result, when the rest of the party was prepared to move on, Sharp was often 500 to 1000 feet above on the Cambrian-Archean contact and out of touch. Nonetheless, Campbell was remarkably patient and supportive in letting me pursue my project.

The unconformities were produced by uplift and subsequent terrestrial erosion and weathering over hundreds of millions of years. Both of them retained remnants of the regolithic mantle produced on and within the underlying rocks. This weathering and erosion had occurred on a landscape devoid of vegetative cover under oxidizing conditions. The ancient erosion surfaces were each subsequently slowly invaded by a shallow sea, and the regolith had clearly been reworked into the basal layers of the initial marine sediments (Sharp 1940a). Striking sea cliffs were cut into residual knobs on the pre-Paleozoic surface by waves of the encroaching Cambrian sea. Looking at cross-section exposures of such cliffs in the canyon walls, one can almost hear the roar of the Cambrian surf hurling itself against the cliff and retreating to gather strength for its next attack. The knobs, with their sea cliffs flanked by outward-thinning tongues of coarse, reworked debris, were eventually submerged by the ever-deepening water, and ultimately buried by finer seafloor deposits.

In one locality, a cuesta on the pre-Paleozoic surface, created by sub-aerial erosion of tilted layers of Proterozoic quartzite, had been undercut by waves, generating a huge rockslide that spread outward over the seafloor across earlier beds of fine sediment. The slide contorted these beds into convolutions and whorls and created a tongue of coarse, angular quartzite breccia (Sharp 1940b).

The opportunity to work with these relationships was a stimulating scientific adventure for a neophyte geologist. The Grand Canyon voyage alone would have been a great adventure through magnificent scenery. Thanks to the ease with which one can do field geology, it was an enriching intellectual experience for me.

In late 1940 and early 1941, as a young faculty member at the University of Illinois, I received communications from Walter A. Wood, Director of Exploration and Field Research at the American Geographical Society in New York City, telling of explorations he and others had been conducting for years in the remote Mt. Steele and Wolf Creek (later renamed Steele Creek) area in the ice-bound St. Elias Range along the Yukon-Alaska border. Wood, trained as a geodesist in Switzerland, was using modern Swiss techniques to set up a triangulation network in this unsurveyed region.

Would I be interested in joining the expedition as a geologist, doing whatever I wished in the way of field work? My early wanderings in California's Sierra Nevada and mapping of glacial features in the Ruby–East Humboldt Range of Nevada (Sharp 1938) had whetted my interest in glaciers. The invitation to become familiar with active ice bodies was irresistible, and the summer's experience led to subsequent glaciological work extending over twenty years.

Although the Mt. Steele project allowed greater preparation than the Grand Canyon exercise, I was working entirely by myself with minimal equipment; the only special items were a Swedish increment borer for coring trees and a small ROTC tripod and plane table for mapping. No base maps were available, but Wood's triangulation network proved useful. The only paths were game trails made by Dall sheep and grizzly bears. The bears repeatedly devastated our caches of canned goods, except for an early version of the Army C-ration, which they disdainfully rejected.

It was a productive and educational summer of new experiences with glaciers (Sharp 1947), bedrock (Sharp 1943), frozen ground and ground ice (Sharp 1942c), debris flows (Sharp 1942a), and patterned ground (Sharp 1942b). A lot can be learned in two and a half months doing field work in a virgin area. I didn't have the background, however, to realize that the stagnant condition of the lower few kilometers of Wolf Creek (Steele) glacier showed it had undergone an earlier episode of surging. Twenty-five years later, in 1965–66, the glacier surged again, rejuvenating its stagnant lower reach (Post 1969, p. 230).

During World War II, as part of an Army–Air Force intelligence unit dealing with arctic, desert, and tropic environments, I was in the Aleutians during the summer of 1945 working on survival problems for the Eleventh Air Force. In the course of this duty, a visit to Shemya, the third western-most island of the Aleutian chain, brought me to a firm-minded general who said, "I don't believe all this junk you fellows write about survival in this area. Let's put you over on Agattu for a few days and see how you do." So a small speedboat deposited me on the east coast of Agattu with a sleeping bag and the meager bailout kit of a single-seater fighter pilot. They kindly gave me a cup of coffee before putting me ashore.

Even in summer, the Aleutians are cool and damp. I had matches, but the Aleutian tundra provides little fuel, and shoreline driftwood is meager and wet. A small cooking fire was possible, but not a warming fire. I would have appreciated the pilot's parachute, but that was not available because I had not bailed out. Some shelter from light rains and mist was obtained beneath a stream-cut bank capped by tundra.

Food proved to be no problem. The bailout kit contained a line and fishhooks. Within the first day, I caught enough small arctic char out of a stream to feed myself and a friendly gray fox for days. Sea urchins were plentiful along shore, and their roe is extremely nutritious, but it proved too rich for my stomach without blander foods. Chitons and small mussels were abundant, and edible plants could be found within the complex of tundra vegetation.

Once settled in and tired of watching puffins and whales, I explored coastal cliffs in search of a cave for shelter. Instead, I soon found extensive, modestly inclined exposures of thinly bedded, light-colored, siliceous sedimentary rocks some 600 meters thick (Sharp 1946). I was flabbergasted. The Aleutians are known to be an island arc, a chain of volcanic cones. What were sedimentary rocks like these doing on Agattu?

A literature search later revealed that other small accumulations of sedimentary rocks were known in the islands, and more may have been discovered since. The Agattu rocks proved to be porcellanites, presumably formed by silicification of fine-grained pyroclastics, so they were at home in a volcanic province, although they reflected an unusual environment of accumulation.

I was elated by the discovery. This was the one occasion during the war when I could do a bit of geology, albeit solely with notebook and pencil. An isotopic geochemist might have found Agattu uninteresting and forbidding. To a field geologist, it was fascinating.

When plans were formulated for experiments to be carried aboard spacecraft Mariners 3 and 4, during a flyby of Mars in 1965, Caltech physicist R. B. Leighton, designer and principal experimenter for the television camera, realized that pictures of the martian surface needed to be studied by a geologist, so he drafted Bruce Murray of Caltech's geology division. Murray recognized that the pictures would show primarily landscape features, so he drafted geomorphologist Sharp, and we three constituted the TV scientific team. Craters and faults proved to be the two primary forms shown on the 22 photos taken by Mariner 4. (Mariner 3 had shroud troubles and never got out of Earth orbit.)

This early work led to further participation in the Mariner 6 and 7 flybys and the Mariner 9 orbiter, which produced many good photos of martian surface features. Extensive experience in field work on Earth was an invaluable background for interpreting the martian terrains.

I have been fortunate to participate as a colleague in modern geochemical stable isotope research (Epstein & Sharp 1959) and in planetary science projects (Sharp 1973), largely because as a classical, old-time field geologist, I could bring something useful to the investigations.

Conclusion

Are field geologists an endangered species? Only if they themselves think so and behave accordingly.

As the earth sciences use more sophisticated laboratory procedures and theoretical models, the need for good field data increases rather than decreases. The relationship is symbiotic: The field and laboratory need each other, and as one prospers, the other benefits. Not everyone need be a field geologist, but the earth sciences will advance more effectively if workers use and have respect for field data and those who produce it. The earth sciences will always need people who are skilled at making field observations.

Literature Cited

American Geological Institute. 1986. *Union List of Geologic Field Trip Guidebooks of North America.* 200 pp. 4th ed.

Baker, V. C., Pyne, S. 1978. G. K. Gilbert and modern geomorphology. *Am. J. Sci.* 278: 97–123

Baldwin, R. B. 1949. *The Face of the Moon.* Chicago: Univ. Chicago Press. 238 pp.

Dutton, C. E. 1882. Tertiary history of the Grand Canyon district. *US Geol. Surv. Monogr. 2.* 264 pp.

Elachi, C., Brown, W. E., Cimino, J. B., Dixon, T., Evans, D. L., et al. 1982. Shuttle imaging radar experiment. *Science* 218: 996–1003

Epstein, S., Sharp, R. P. 1959. Oxygen isotope variations in the Malaspina and Saskatchewan glaciers. *J. Geol.* 67: 88–102

Gary, M., McAfee, R., Wolf, C. L., eds. 1974. *Glossary of Geology.* Washington, DC: Am. Geol. Inst. 805 pp.

Gilbert, G. K. 1875. Report on the geology of portions of Nevada, Utah, California, and Arizona examined in the years 1871 and 1872. *US Geogr. Geol. Surv. W. of the 100th Meridian Rep.* 3: 17–187

Gilbert, G. K. 1877. *Report on the Geology of the Henry Mountains.* Washington, DC: US Geogr. Geol. Surv. Rocky Mtn. Reg. 2nd ed. (1880). 160 pp., 170 pp.

Gilbert, G. K. 1890. Lake Bonneville. *US Geol. Surv. Monogr. 1.* 438 pp.

Gilbert, G. K. 1893. The moon's face; a study of the origin of its features. *Philos. Soc. Washington Bull.* 12: 241–92

Gilbert, G. K. 1914. The transportation of

debris by running water. *US Geol. Surv. Prof. Pap. 86*. 263 pp.

Glen, J. W. 1955. The creep of polycrystalline ice. *Proc. R. Soc. London Ser. A* 228: 519–38

Glen, J. W. 1956. Measurement of the deformation of ice in a tunnel at the foot of an ice fall. *J. Glaciol.* 2: 735–45

Hewitt, D. F. 1920. The Heart Mountain overthrust, Wyoming. *J. Geol.* 28: 536–57

Hinds, N. E. A. 1935. Ep-Archean and Ep-Algonkian intervals in western North America. *Carnegie Inst. Washington Publ. 463*. 52 pp.

LaRue, E. C. 1925. Water power and flood control of Colorado River below Green River, Utah. *US Geol. Surv. Water Supply Pap. 556*. 171 pp.

McCauley, J. F., Schaber, G. G., Breed, C. S., Grolier, M. J., Haynes, C. V., et al. 1982. Subsurface valleys and geoarcheology of the eastern Sahara revealed by shuttle radar. *Science* 218: 1004–20

McKee, E. D. 1934. The Coconino sandstone—its history and origin. *Carnegie Inst. Washington Publ. 440, 7*: 77–115

McKee, E. D. 1938. The environment and history of the Toroweap and Kaibab formations of northern Arizona and southern Utah. *Carnegie Inst. Washington Publ. 492*. 268 pp.

McKee, E. D., Resser, C. E. 1945. Cambrian history of the Grand Canyon region. *Carnegie Inst. Washington Publ. 563*, pp. 3–168

Moore, R. C. 1925. Geologic report on the inner gorge of the Grand Canyon of Colorado River. *US Geol. Surv. Water Supply Pap. 556*, pp. 125–71

Noble, L. F. 1914. The Shinumo quadrangle, Grand Canyon district, Arizona. *US Geol. Surv. Bull. 549*. 100 pp.

Nye, J. F. 1952. The mechanics of glacier flow. *J. Glaciol.* 2: 82–93

Nye, J. F. 1960. The response of glaciers and ice sheets to seasonal and climatic changes. *Proc. R. Soc. London Ser. A* 256: 559–84

Nye, J. F. 1963. On the theory of the advance and retreat of glaciers. *Geophys. J. R. Astron. Soc.* 7: 431–56

Perutz, M. F., Seligman, G. 1939. A crystallographic investigation of glacier structure and the mechanism of glacier flow. *Proc. R. Soc. London Ser. A* 172: 335–60

Pettijohn, F. J. 1984. *Memoirs of an Unrepentant Field Geologist: A Candid Profile of Some Geologists and Their Science.* Chicago: Univ. Chicago Press. 260 pp.

Pierce, W. G. 1941. Heart Mountain and South Fork thrusts, Park County, Wyoming. *Am. Assoc. Pet. Geol. Bull.* 25: 2021–45

Pierce, W. G. 1957. Heart Mountain and South Fork detachment thrusts of Wyoming. *Am. Assoc. Pet. Geol. Bull.* 41: 591–626

Pierce, W. G. 1960. The "break-away" point of the Heart Mountain detachment fault in northwestern Wyoming. *US Geol. Surv. Prof. Pap. 400-B*, pp. 236–37

Pierce, W. G. 1963. Reef Creek detachment fault, northwestern Wyoming. *Geol. Soc. Am. Bull.* 74: 1225–36

Pierce, W. G. 1973. Crandall Conglomerate, an unusual stream deposit and its relation to Heart Mountain faulting. *Geol. Soc. Am. Bull.* 84: 2631–44

Pierce, W. G. 1979. Clastic dikes of Heart Mountain fault breccia, northwestern Wyoming, and their significance. *US Geol. Surv. Prof. Pap. 1133*. 25 pp.

Post, A. 1969. Distribution of surging glaciers in western North America. *J. Glaciol.* 8: 229–40

Powell, J. W. 1875. *Exploration of the Colorado River and Its Canyons.* New York: Dover. 400 pp.

Pyne, S. J. 1980. *Grove Karl Gilbert: A Great Engine of Research.* Austin: Univ. Tex. Press. 306 pp.

Reinhardt, J., Miller, D. M. 1987. Cogeomap: a new era in cooperative geologic mapping. *US Geol. Surv. Circ. 1003*. 12 pp.

Seligman, G. 1941. The structure of a temperate glacier. *Geogr. J.* 97: 295–315

Sharp, R. P. 1938. Pleistocene glaciation in the Ruby–East Humboldt Range, northeastern Nevada. *J. Geomorph.* 1: 296–323

Sharp, R. P. 1939. Basin-range structure of the Ruby–East Humboldt Range, northeastern Nevada. *Geol. Soc. Am. Bull.* 50: 881–919

Sharp, R. P. 1940a. Ep-Archean and Ep-Algonkian erosion surfaces, Grand Canyon, Arizona. *Geol. Soc. Am. Bull.* 51: 1235–69

Sharp, R. P. 1940b. A Cambrian slide breccia, Grand Canyon, Arizona. *Am. J. Sci.* 238: 668–72

Sharp, R. P. 1942a. Mudflow levees. *J. Geomorph.* 5: 222–27

Sharp, R. P. 1942b. Soil structures in the St. Elias Range, Yukon Territory. *J. Geomorph.* 5: 274–301

Sharp, R. P. 1942c. Ground-ice mounds in tundra. *Geogr. Rev.* 32: 417–23

Sharp, R. P. 1943. Geology of the Wolf Creek area, St. Elias Range, Yukon Territory, Canada. *Geol. Soc. Am. Bull.* 54: 625–49

Sharp, R. P. 1946. Note on the geology of Agattu, an Aleutian Island. *J. Geol.* 54: 193–99

Sharp, R. P. 1947. The Wolf Creek glaciers,

St. Elias Range, Yukon Territory. *Geogr. Rev.* 37: 26–52

Sharp, R. P. 1973. Mars: fretted and chaotic terrains. *J. Geophys. Res.* 78: 4073–83

Stellar, D. L. 1986. Thomas Wilson Dibblee, Jr. *Calif. Geol.* 39: 202–6

Urey, H. C. 1951. The origin and development of the earth and other terrestrial planets. *Geochim. Cosmochim. Acta* 1: 209–77

US Geological Survey. 1987. National Geologic Mapping Program. *US Geol. Surv. Circ. 1020.* 29 pp.

Van Gundy, C. E. 1934. Some observations on the Unkar group of the Grand Canyon Algonkian. *Grand Canyon Nat. Notes* 9: 338–49

Vening-Meinesz, F. A. 1948. *Gravity Expeditions at Sea 1923–1938.* Delft: Publ.

Neth. Geod. Comm. 4. 233 pp.

Walcott, C. D. 1883. Pre-Carboniferous strata in the Grand Canyon of the Colorado, Arizona. *Am. J. Sci.* 26: 437–42

Walcott, C. D. 1890. Study of a line of displacement in the Grand Canyon of the Colorado, in northern Arizona. *Geol. Soc. Am. Bull.* 1: 49–64

Walcott, C. D. 1894. Pre-Cambrian igneous rocks of the Unkar terrane, Grand Canyon of the Colorado, Arizona. *US Geol. Surv. Ann. Rep. 14,* pt. 2, pp. 497–524

Walcott, C. D. 1895. Algonkian rocks of the Grand Canyon of the Colorado. *J. Geol.* 3: 312–30

Wheeler, R. B., Kerr, A. R. 1936. Preliminary report on the Tonto Group of the Grand Canyon, Arizona. *Grand Canyon Nat. Hist. Assoc. Bull.* 5: 1–16

Photo of near-life-size caricature of Bob Sharp in the field, complete with staff and Filson jacket, as created in 1986 by Sharon Martens, emphasizing a characteristic gesture being imitated by the roadrunner, one of Bob's favorite birds.

Ann. Rev. Earth Planet. Sci. 1988. 16: 21–51

PHASE RELATIONS OF PERALUMINOUS GRANITIC ROCKS AND THEIR PETROGENETIC IMPLICATIONS[1]

E-an Zen

US Geological Survey, Reston, Virginia 22092

INTRODUCTION

Peraluminous granitic rocks are magmatic rocks that contain quartz, potassic feldspar and/or sodic plagioclase, and one or more aluminum-rich minerals; thus these rocks contain more aluminum than could be accommodated by feldspar in a CIPW normative calculation and show normative corundum. Peraluminous granites are defined by their chemical nature, but their recognition is often based on petrography; this discrepancy thus may lead to inconsistency. Although peraluminous granitic rocks are commonly true granite (Streckeisen 1973), they range from tonalite to granite. Of special interest are "strongly peraluminous" (as defined below) granite, granodiorite, or, in rare instances, trondhjemite; these rocks contain aluminous minerals (almandinic garnet, cordierite, tourmaline, muscovite) that are thought to be stable in silicate magmas only if these magmas are themselves peraluminous.

The origin of peraluminous granite and related intrusive rocks is of much current interest. Many of these form shallow-level intrusions, and they can reach enormous sizes; examples are the batholiths in the Lachlan Fold Belt in southeastern Australia (White & Chappell 1983), the Cornubian batholith in England (Exley & Stone 1966), the South Mountain batholith in Nova Scotia (McKenzie & Clarke 1975), the Manaslu batholith in the Himalayas (LeFort 1981), much of the Idaho batholith

---

(Hyndman 1983), and the tin-bearing plutons of Thailand and Malaysia (Hutchinson 1983). The chemistry and mineralogy of peraluminous granitic rocks may indicate their tectonic setting and the nature of the source rocks of the magmas, as well as the processes under which these magmas evolved. Peraluminous granitic rocks may host economic metal deposits, especially of "incompatible elements" such as tin and tungsten, and may be petrogenetically related to skarn, vein, pegmatite, or disseminated deposits of other lithophile elements (U, Li, B, Ta, Nb); understanding their setting and origin is thus of practical interest.

In 1981, the *Canadian Mineralogist* produced a special issue (Clarke 1981a) on the petrogenesis and significance of peraluminous granitic rocks. Barker (1981) also summarized ideas on the origin of peraluminous granites. Three major modes of genesis of peraluminous granitic rocks have been accepted: partial melting, especially of peraluminous sedimentary rocks; differentiation of metaluminous magma, whatever its origin, with or without assimilation; and late-stage but super-solidus metasomatic loss of alkali through vapor phase transport. There are many examples of each process, and certainly not all peraluminous granitic rocks had the same origin. This review uses the AFM mineral relations of peraluminous granites to probe their origin and to highlight what we still need to learn.

DEFINITIONS

A granitic rock is said to be peraluminous if the molar ratio of Al_2O_3 to the sum of CaO, K_2O, and Na_2O [the aluminum saturation index, or ASI (Zen 1986); often also referred to as A/CNK (e.g. Clarke 1981b)] exceeds unity:

$$ASI = Al_2O_3/(CaO + K_2O + Na_2O) > 1$$

(Shand 1949, Chappell & White 1974, Miller & Bradfish 1980, Barker 1981, Clarke 1981a, Speer 1984). If the ASI calculation is first corrected for CaO in apatite, then the definition for "peraluminosity" is equivalent to stating that the CIPW normative minerals of the rock include corundum; if the norm instead contains wollastonite or diopside, then it is metaluminous. These definitions are entirely descriptive and say nothing about the actual minerals or the petrogenesis.

"Strongly peraluminous granitic rocks," a petrographic subset of peraluminous granitic rocks, require the presence of cordierite, muscovite, almandine, or the aluminum silicate minerals, as well as possibly tourmaline and topaz. The existence of these minerals in equilibrium magmatic assemblages generally connotes the peraluminosity of the magma; however, because minerals in most coarse-grained intrusive rocks are at least partly cumulate in origin (Zen 1986, Clemens & Wall 1981, O'Hara 1968), the bulk rock chemistry should not be directly equated with the composition of a former melt. Another caution is that subsolidus alteration of granitic rocks tends to deplete the alkalis, so chemical analyses should always be accompanied by petrographic data (preferably including modes) so as to allow reasonable estimates of this error as well as of errors in the analysis.

INDICATOR MINERALS OF PERALUMINOUS GRANITIC ROCKS

Two groups of minerals found in peraluminous granitic rocks can be identified: those minerals that could have equilibrated with a peraluminous silicate melt but do not require peraluminous melts for their stable precipitation, and those that could precipitate out of a silicate melt only if the melt itself was peraluminous.

Minerals That Do Not Imply Peraluminous Magmas

Among mafic minerals that could coexist with both metaluminous and at least mildly peraluminous magmas are biotite, hornblende, and sphene. The ideal formula of biotite is $H_2K(Fe, Mg)_3AlSi_3O_{12}$ (ASI = 1). Actual igneous biotite deviates from this formula substantially and, if we ignore the Ti component, averages to about $H_2K_{0.9}(Fe, Mg)_{2.75}Al_{1.4}Si_{2.85}O_{12}$ (Zen 1986; cf. Foster 1960, Clemens & Wall 1984, Monier & Robert 1986a, Lee et al 1981, Hammarstrom 1982). Biotite from granitic rock has molar $Mg/(Fe^{2+} + Mg)$ ratios (the mg number) around 0.3–0.5; it typically shows a persistent A-site alkali deficiency but excess octahedral and tetrahedral Al. The ASI is around 1.3–1.4 for igneous biotite coexisting with hornblende but is > 1.4 for igneous biotite coexisting with muscovite and other peraluminous minerals, reflecting increased alumina activity in the magma (Table 1). The ASI of biotite is readily determined by microprobe analysis and is a useful guide to the peraluminosity of the rock.

Igneous hornblende has ASI about 0.3–0.5 and can coexist with melts having ASI as high as 1.1–1.2 (Table 1; Zen 1986). Because of the high ASI of igneous biotite, a rock whose modal proportion of hornblende to biotite is 1 : 3 or less should have ASI > 1. For this reason, many hornblende-biotite granites, granodiorites, and even some tonalites are mildly peraluminous.

Magmatic sphene is rare in strongly peraluminous granitic rocks but is common in mildly peraluminous rocks. The stability of magmatic sphene is related to both the bulk composition and the redox potential of the magma, as is discussed in what follows. Much research remains to be done on the phase relations and mineral chemistry of sphene in peraluminous granites.

Minerals That Do Imply Strongly Peraluminous Magmas

This category (Table 1) includes almandine-rich garnet, cordierite, the aluminum silicate polymorphs (mainly andalusite and sillimanite), muscovite, topaz, and tourmaline; the crystallization of the last two minerals requires, respectively, fluorine and boron.

GARNET   Magmatic garnets of different compositions occur in intrusive rocks. Those in shallow-level peraluminous granite are largely almandine-spessartine solid solutions having minor pyrope component (Clarke 1981b); however, almandine- and grossular-rich but spessartine-poor gar-

Table 1   ASI of nonquartzofeldspathic minerals in peraluminous granites

Mineral	Ideal formula	Ideal ASI	ASI range
A. Some subaluminous minerals			
Edenite	$H_2NaCa_2(Fe, Mg)_5AlSi_7O_{24}$	0.2	
Common hornblende	$H_2Ca_2(Fe, Mg)_4Al_2Si_7O_{24}$	0.5	0.3–0.5[a]
Pargasite	$H_2NaCa_2(Fe, Mg)_4Al_3Si_6O_{24}$	0.6	
Garnet	$Ca_3Al_2Si_3O_{12}$	0.33	0.33
Sphene	$CaTiSiO_5$	0	0
Apatite	$Ca_5(PO_4)_3(F, OH, Cl)$	0	0
B. Some "neutral" minerals			
Plagioclase	$NaAlSi_3O_8-CaAl_2Si_2O_8$	1	1
K-feldspar	$KAlSi_3O_8$	1	1
Ilmenite	$FeTiO_3$	Indeterminate	Indeterminate
Magnetite	Fe_3O_4	Indeterminate	Indeterminate
C. Some peraluminous minerals			
Biotite	$H_2K[Fe, Mg]_3AlSi_3O_{12}$	1	1.3–1.5[b]
Celadonite	$H_2KMgFe^{3+}Si_4O_{12}$	0	
Leucophyllite	$H_2KMgAlSi_4O_{12}$	1	2–2.5[b]
Muscovite	$H_2KAl_3Si_3O_{12}$	3	
Cordierite	$(Fe, Mg)_2Al_4Si_5O_{18}$	Infinite	Infinite
Fe, Mg, Mn garnet	$(Fe, Mg, Mn)_3Al_2Si_3O_{12}$	Infinite	Infinite
Tourmaline	$H_4Na(Fe, Mg)_3Al_6B_3Si_6O_{31}$	6	~6
Andalusite/ sillimanite	Al_2SiO_5	Infinite	Infinite
Topaz	$Al_2SiO_4(OH, F)_2$	Infinite	Infinite

[a] Hornblende in igneous rocks typically is a complex solid solution that may be described, simplistically, as among edenite, common hornblende, and pargasite. For justification of the range of values of ASI as shown, see Zen (1986).

[b] See text for actual compositions of these minerals in peraluminous granitic rocks.

net can form in high-pressure peraluminous granitic rocks (Anderson & Rowley 1981; J. M. Hammarstrom & E-an Zen, unpublished data). The mole fraction of spessartine ranges up to 0.3, but many garnets are not particularly Mn rich (Miller & Stoddard 1981, Allan & Clarke 1981, Phillips et al 1981) and the mg numbers are commonly 0.1–0.2 (see Table 2).

CORDIERITE   Cordierite occurs in peraluminous granite and rarely in trondhjemite (Table 2). Its occurrence as phenocrysts in peraluminous volcanic rocks proves that it could be in equilibrium with a granitic magma. The mg number of magmatic cordierite is as low as 0.3 but is mostly around 0.5 (Thompson 1976, Clemens & Wall 1981, 1984, Wyborn et al 1981, Phillips et al 1981, Speer 1981, Fang & He 1985, Maillet & Clarke 1985).

ALUMINUM SILICATE POLYMORPHS   Andalusite and sillimanite are the principal polymorphs of magmatic aluminum silicate in granitic rocks, indicating the generally low-pressure conditions of crystallization of peraluminous granitic magmas. These minerals are discussed here along with muscovite. There seems to be no reason why kyanite cannot form in high-pressure and relatively dry granitic magmas; indeed, Huang & Wyllie (1981) reported kyanite from their melting experiments on the Harney Peak Granite.

MUSCOVITE, TOURMALINE, TOPAZ, AND ANDALUSITE IN PERALUMINOUS GRANITIC ROCKS

A persistent petrogenetic problem with peraluminous granitic rocks has been the correct interpretation of possibly magmatic muscovite. Euhedral muscovite phenocrysts in volcanic and dike rocks (see below) demonstrate the reality of magmatic muscovite, but this demonstration is more tenuous for plutonic rocks. The problem is both to determine firmly that a particular muscovite equilibrated with a silicate magma and to use such muscovite to interpret the conditions of crystallization of the magma. The problem, then, devolves to the criteria for magmatic muscovite in intrusive rocks and to the conflict between experimental data on muscovite stability and geological inferences on the depth of emplacement of plutons containing it.

Criteria for Magmatic Muscovite

Miller et al (1981) reviewed the data on chemical differences between muscovite that appears to be magmatic and muscovite that appears to be subsolidus reaction products. To do so, they first established the following textural criteria for magmatic muscovite: coarse grain size comparable to that of obviously magmatic phases; clean crystal termination, ideally showing subhedral to euhedral forms; crystals not enclosed by or not raggedly enclosing another mineral, such as feldspar, from which muscovite might have formed by subsolidus alteration; and occurrence in rocks having clean, unaltered igneous texture.

The criteria individually can be challenged. For example, coarse grain size might actually be favored in subsolidus rather than magmatic crystallization; certainly not all magmatic minerals have the same grain size ranges. Clean termination could reflect surface energy relations and does not prove the magmatic origin of a mineral. Inclusion relations (e.g. of hornblende, zircon, and apatite in feldspar) have been used to indicate earlier magmatic crystallization of the enclosed mineral. Finally, how does one ever prove that a given set of textural relations is primary? Some might use the mere presence of muscovite to "prove" subsequent alteration!

Nevertheless, there has to be a first step, and the criteria of Miller et al (1981) are probably a near-consensus view of the state of the art. The presence of euhedral contact of muscovite against magmatic-looking biotite, especially euhedral butt-end parallel growth, may remove part of the textural ambiguity (Zen 1985a). Moreover, muscovite of supposedly magmatic origin is consistently richer in Ti, Na, and Al, but poorer in Mg and Si, than muscovite of hydrothermal origin (Miller et al 1981). These authors also showed that the microprobe data of muscovite computed in terms of only ferrous iron can give seriously misleading values of site occupancy. Indeed, future studies may show that the ferric/ferrous relations of magmatic muscovite in appropriate assemblages can provide information on the redox conditions during crystallization, as biotite has been shown to do (Wones & Eugster 1965).

The chemical and textural criteria of Miller et al (1981) are supported by Speer (1984), who on the basis of additional data, especially those of Monier et al (1984), concluded that the higher Ti content of magmatic muscovite is real. This conclusion is fully supported by the data and inferences of Anderson & Rowley (1981). All workers agree that magmatic muscovite is never close to the ideal end-member formula, H_2KAl_2 $(AlSi_3)O_{12}$ (the "Mu" component), but has 20–25 mole percent of other components that can be described (Zen 1985a) as leucophyllite ("Lp," $H_2KMgAlSi_4O_{12}$) and ferrimuscovite ["Fu," $H_2KAlFe^{3+}(AlSi_3)O_{12}$] within the reciprocal system Mu-Lp-Fu-Ce [celadonite, $H_2K(Mg, Fe^{2+})$ $Fe^{3+}Si_4O_{12}$; see Figure 1]. Other substitutions in muscovite include Ti [end member assumed to be $H_2KTiMgAlSi_3O_{12}$ (Miller et al 1981, Monier & Robert 1986a)], Na (paragonitic), and possibly trioctahedral substitutions (Monier & Robert 1986b). Muscovites having at least ~0.6 wt%

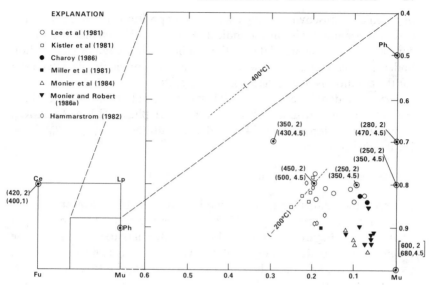

Figure 1 Compositions of selected analyses of magmatic muscovite found in peraluminous granites, calculated in terms of the reciprocal components Mu (muscovite), Lp (leucophyllite), and Fu (ferrimuscovite). [The fourth component is Ce (celadonite; see text). Ph (phengite) is halfway between Mu and Lp.] The ferric/ferrous iron ratio is taken to be 3 unless separate determinations are given in the report. The titanium component is removed prior to calculation, assuming that the substitution scheme is $[Ti, Mg] = 2Al^{VI}$ into end-member muscovite. The segregation of the data of Monier et al and of Monier & Robert is unexplained but could reflect an inconsistent set of criteria for magmatic muscovite (cf. Zen 1985a). The experimental data on the thermal stability relations refer to compositions indicated by the bull's-eyes (mainly from Velde 1965, 1972) and are given by paired temperature (°C) and pressure (kbar) between parentheses. For end-member muscovite, the values of, or interpolated from, Chatterjee & Johannes (1974) are given between square brackets. The two short-dashed lines are approximate predicted temperature lowerings (given between parentheses) due to increase in Si content in the mica, as given by Massonne & Schreyer (1983). The thermal stability of end-member celadonite at two pressures is from Wise & Eugster (1964).

TiO_2 are candidates for magmatic origin (see also Miller & Stoddard 1981, Kistler et al 1981, Charoy 1986, Monier & Robert 1986a, Lee et al 1981, Zen 1986, Table 6; Hammarstrom 1982, Table 11). One reason that Ti is a useful indicator of magmatic origin may be that it does not readily undergo subsolidus exchange or oxidation reactions as do Mg, Fe, and the alkali cations.

The combined use of textural and chemical criteria inevitably involves elements of circular reasoning. However, the successful separation of the chemical data that conform to textural contrasts suggests a real discrimination, and, lacking evidence to the contrary, I accept the existence of magmatic muscovite having chemical compositions removed from the end-member muscovite in the indicated manner. The challenge is to reconcile the experimental data for the stability of muscovite solid solution under magmatic conditions with the oftentimes conflicting field evidence for the conditions of emplacement of the plutons.

The most definitive experimental hydrothermal studies of the phase relations of muscovite are on the end-member phase, $H_2KAl_2(AlSi_3)O_{12}$ (Chatterjee & Johannes 1974). Under conditions of $P_{total} = P_{H_2O}$, the breakdown reaction

muscovite + quartz = sanidine + Al-silicate + H_2O

intersects the solidus for either the haplogranite (Tuttle & Bowen 1958) or a peraluminous granite (Wyllie 1977) at a pressure of 3–4 kbar, or a depth of 10–14 km (Figure 2). This conclusion contradicts the evidence of late magmatic muscovite in many shallow-level ($\sim$ 1–2 kbar) intrusions. If we assume that the field data showing a shallow depth of intrusion for many plutons are correct, then three explanations are possible. First, other components in the magma, not considered in the experiments, can materially lower the solidus temperature of the real melt so that the intersection with the breakdown curve for muscovite would be lowered. Second, other dissolved components in magmatic muscovite could enhance its thermal stability to accomplish the same end. Third, the muscovite might be a relict of an earlier stage of the magmatic history. We now examine these options in turn.

Other Components in the Magma and the Enigma of Tourmaline and Topaz

Additional components might affect the solidus temperature of a granitic melt and significantly change the minimum pressure for muscovite stability. Major rock-forming components antecedent to haplogranite in Bowen's reaction series, especially CaO and MgO, raise rather than lower the solidus (Figure 2; Tuttle & Bowen 1958). The An component raises the haplogranite solidus by several tens of degrees (Johannes 1984; cf. Wyllie 1977), and the thermal stability of muscovite + quartz is decreased by the addition of the Ab component and the phase albite (Chatterjee & Flux 1986, Thompson 1974). H_2O activity less than unity raises the temperature of the solidus and at the same time decreases the thermal stability of muscovite. All these effects increase the minimum pressure for the intersection of the solidus and the muscovite stability limit. In addition, the experimental data of Huang & Wyllie (1973, 1981), of Piwinskii & Wyllie (1970), and of Clemens & Wall (1981) showed that the com-

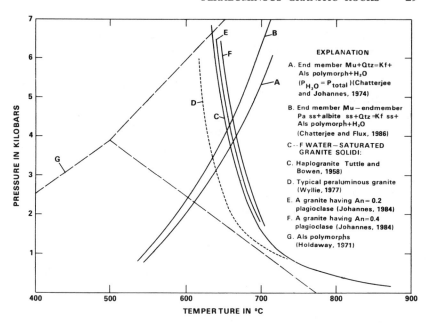

Figure 2 Experimental phase equilibrium data that bear on the stability of magmatic muscovite. Abbreviations used: Mu, muscovite; Qtz, quartz; Kf, potassic feldspar; Als, aluminum silicate polymorph; Pa, paragonite; ss, solid solution; An, anorthite component. Note that none of the solidi intersects the stability field of andalusite.

positional departures of these peraluminous granites (ASI between 1.02 and 1.4) from haplogranite do not significantly lower the solidus temperature.

On the other hand, components that preferentially partition into the melt and are not antecedent to granite in Bowen's reaction series could lower the solidus below that for haplogranite. Especially interesting among these are Cs, Rb, Li, possibly Fe and Al itself, and the fugitive components that might escape into the surroundings through a late vapor phase but that are known to occur in some late and highly evolved granites. Charoy (1986) showed that biotite and muscovite from the Carnmenellis pluton of the Cornubian batholith are highly enriched in Cs, Rb, and Li, so presumably the magma was also enriched in these elements. Glyuk et al (1977) and Glyuk & Trufanova (1977) reported that these elements drastically lower the solidus of hydrous granite melts: 0.5 wt% Cs and Rb as fluoride and chloride apparently suffice to lower the solidus to about 450°C at 1 kbar, but whether this effect is caused by the cations or the anions is not clear. These cationic components are commonly not analyzed for and are vulnerable to removal from the system through vapor-phase transport.

Among the fugitive components, B and F are special suspects. Chorlton & Martin (1978), Pichavant (1981), and Pichavant & Manning (1984) studied the influence of B on the solidus temperature of synthetic haplo-granite at 1 kbar. If the partition constant (as weight percent) for B_2O_3 between melt and aqueous solution near the liquidus, 0.33 (Pichavant 1981), applies to solidus temperatures, then about 2 and 6 wt% of B_2O_3 in the melt would lower the solidus temperature by 60 and 130°C, respectively (Pichavant 1981). These temperature drops imply that the minimum pressure for the intersection of the solidus and the muscovite-albite-quartz stability curve would be lowered by about 1 and 2 kbar, respectively.

However, these B_2O_3 contents are not realistic for natural granites. The lowering refers to the solidus temperature of the final dregs of magma (M. Pichavant, written communication, 1987) and is petrologically irrelevant. Moreover, Pichavant & Manning (1984) estimated, on the basis of the boron content of the peraluminous granitic glass from Macusani (Herrera et al 1984), that the natural limit for boron is about 1 wt% B_2O_3. Thus, the final-dregs solidus temperature of natural boron-rich granitic magma should be lowered by about 30°C, corresponding to a drop of the minimum pressure of only about 0.5 kbar. This small effect is corroborated by the experimental studies on tourmaline-bearing granites (Pichavant & Manning 1984, Benard et al 1985). The data show that for tourmaline in the dravite-schorlite series (ideal compositions $H_4NaMg_3Al_6B_3Si_6O_{31}$ and $H_4NaFe_3Al_6B_3Si_6O_{31}$, respectively), its presence lowers the solidus temperatures only by 10–15°C for the leucogranite Mar 2 from the Massif Central and by less than 5°C for the leucogranite Him 3 from the Manaslu pluton in the Himalayas (see also LeFort 1981). The rocks used in the experiments do not contain biotite, but glass obtained from melting the Manaslu granite, upon recrystallization at 2 kbar, did produce biotite as well as tourmaline. The solidus temperature drops are not enough to account for the low-pressure environment of consolidation of these plu-tons.

The boron content of muscovite that might coexist with tourmaline is unknown (see analyses in Charoy 1986, p. 602); if this content is enough to stabilize muscovite to higher temperatures, then the minimum pressure might be significantly lowered. Exley & Stone (1966), quoting Brammall & Harwood (1932; but not verifiable from the latter reference), gave the boron content of a biotite from a porphyritic granite of Dartmoor (part of the Cornubian batholith) as a negligible 179 ppm, but the mineral assemblage of the rock is not known. Given these uncertainties, can we say anything about the efficacy of tourmaline in buffering the activity of B_2O_3 in a granitic melt?

Biotite and tourmaline do coexist in peraluminous magmas. Apart from the unreversed experimental runs from Him 3 cited above (Benard et al 1985), these two minerals occur together in the Macusani ignimbrite (Herrera et al 1984) and in the Cornubian batholith of southwest England (Brammall & Harwood 1925, Power 1968, Charoy 1986; reviewed by Exley & Stone 1966). Textural relations described by these authors [e.g. tourmaline inclusions in biotite similar to inclusions of zircon, apatite, and monazite (Exley & Stone 1966, p. 169; E-an Zen, unpublished data)] suggest that some tourmaline and some biotite, as well as muscovite, feldspar, and quartz, did coexist during the magmatic stage. We can therefore write a buffering reaction for the magma using the compositions of igneous biotite and muscovite discussed earlier:

$$2.2H_2K_{0.9}(Fe, Mg)_{2.75}Al_{1.4}Si_{2.85}O_{12} + 12.25H_2K(Fe, Mg)_{0.2}Fe^{3+}_{0.2}Al_{2.4}Si_{3.2}O_{12}$$

  biotite   muscovite

$$+ 3.65NaAlSi_3O_8 + 8.17SiO_2 + 5.475B_2O_3$$

  albite   quartz

$$= 14.23KAlSi_3O_8 + 3.65H_4Na(Mg, Fe)_3Al_6B_3Si_6O_{31} + 7.15H_2O + 1.225O$$

  sanidine   tourmaline   oxygen

Thus, in a two-mica granite the fugacity of B_2O_3 would be controlled by tourmaline, the solidus temperature would not be notably affected by boron (Benard et al 1985), and the minimum pressure for stable muscovite would not be much affected. The mg numbers of presumed magmatic tourmaline and biotite are similar [0.65–0.8 (Charoy 1986)], and so the thermodynamic variance of the above reaction is probably not materially affected by the mg numbers of the minerals. The reaction predicts that the formation of tourmaline from biotite would be accompanied by potassic feldspar; this process might account for some of the late-magmatic potassic feldspar megacrysts typical of the Cornubian granites, especially those intergrown with tourmaline (see Brammall & Harwood 1925, Power 1968, Exley & Stone 1966, Vernon 1986).

Even though the formation of muscovite in two-mica granites cannot be explained by the effect of boron, in the absence of biotite the fugacity of B_2O_3 could increase, allowing muscovite-tourmaline-bearing late leucogranites (including many "elvans" of Cornubian terminology) to form at substantially lowered solidus temperatures. This conclusion does not conflict with the experimental data of Benard et al (1985), which showed little change in the solidus temperature of leucogranite by tourmaline, because in the experiments all boron came from the tourmaline.

The other components that could significantly lower the solidus temperature are the halogen elements, especially fluorine. Topaz is the major fluorine-bearing phase stable in peraluminous magma that, together with the fluorine-bearing micas, could buffer the fugacity of this component. Liquidus phase relations of fluorine-bearing granite were studied by Manning (1981), and those involving topaz were studied by Barton (1982). Barton (1982) concluded that for the reaction

muscovite (F, OH solid solution) + quartz = sanidine + topaz,

the thermal stability of muscovite + quartz is enhanced by as much as 50°C when the fugacity of fluorine is buffered by the assemblage topaz + andalusite + vapor. Manning & Pichavant (1983) inferred a similar effect on biotite; the buffer anorthite + fluorite + sillimanite + quartz has a comparable effect (Munoz & Ludington 1977). Because at 1-kbar total pressure each weight percent of fluorine lowers the solidus temperature by about 20°C, the combined effect would be to lower the minimum pressure for fluorine-saturated magmatic muscovite (Pichavant & Manning 1984, their Figure 4), possibly to values as low as 1 kbar. The relatively high fluorine content of muscovite from eastern Nevada (Lee et al 1981), as much as 6 mol% in the OH site, could reflect stabilization of muscovite by this component; the partitioning of fluorine in muscovite and coexisting biotite suggests exchange equilibrium, presumably established under magmatic conditions (see Munoz & Ludington 1977). However, topaz occurs mainly in late-stage greisen-type rocks and is rare in most two-mica granites (see, however, Kortemeier & Burt 1986); whether, in the absence of the buffer assemblages, dissolved fluorine has a comparably large effect on the stability of muscovite in main-stage peraluminous granite remains to be shown.

An explanation for low-pressure magmatic muscovite by fluorine dissolved in the magma could also be used for magmatic andalusite. Apparently magmatic andalusite is well known [e.g. the South Mountain batholith (Clarke et al 1976), the Cornubian batholith (Exley & Stone 1966), and the Macusani ignimbrite (Herrera et al 1984)]. However, neither the solidus of the haplogranite (Tuttle & Bowen 1958) nor those of natural peraluminous leucogranites (Huang & Wyllie 1973, 1981, Clemens & Wall 1981) intersect the stability field of andalusite (Holdaway 1971; see Figure 2). The presence of magmatic andalusite requires a lowering of the solidus temperature by additional components because the minor departure of magmatic andalusite composition from that of the end member is inadequate to explain its stability.

Other Components in Natural Muscovite

Magmatic muscovite is always a complex and fairly extensive solid solution. The effects of this solid solution on its thermal stability have been investigated (Figure 1; all stability data are in the presence of quartz).

Massonne & Schreyer (1983) studied the compositions within the system $K_2O\text{-}MgO\text{-}Al_2O_3\text{-}SiO_2\text{-}H_2O$, Wise & Eugster (1964) the end-member celadonite, and Velde (1965) the compositions that lie on different trends from end-member muscovite to celadonite, to leucophyllite, and to a composition halfway between these two. Velde (1965) and Wise & Eugster (1964) found that the upper thermal stability limit of the mica, under conditions of $P_{total} = P_{H_2O}$ and under various oxygen fugacity conditions, is considerably lower than for end-member muscovite (Chatterjee & Johannes 1974). Changing the mg number of the run material from 1 to 0.5 affected the results only marginally (Velde 1965). Massonne & Schreyer reported that each increase in the per-formula (11 oxygen atoms, anhydrous) Si content of the mice by 0.1 leads to a decrease of its stability by about 2.5 kbar, or by about 200°C; these results are qualitatively similar to those of Velde. Even though the experimental results are not well constrained and Velde's data on the intermediate composition trend seem inconsistent with those for the other two trends, clearly the solid solution decreases the thermal stability of end-member muscovite in this ternary reciprocal system. The effect of Ti on the stability of muscovite is not known, but the small amount of TiO_2 (<1 wt%, corresponding to <5 mol% of the Ti white mica end-member defined above) is unlikely to enhance the stability by 200–300°C, as would be needed to compensate for the effects of the reciprocal components and to make natural igneous muscovite more stable than the end-member composition. The experimental data indicate that the solid solution enhances the thermal stability of phengite and celadonite toward that of muscovite, not the other way as proposed by Anderson & Rowley (1981). J. L. Anderson (oral communication, 1986) no longer subscribes to the latter suggestion.

Monier & Robert (1986b) experimentally investigated the effect of trioctahedral substitution in muscovite along the Mu-phlogopite and Mu-annite joins. They showed that solid solution toward biotite becomes most significant above 600°C, appropriate to igneous rocks. However, the thermal stability of muscovite is not enhanced by the solid solutions.

Relict Magmatic Muscovite

It is conceivable that muscovite precipitated out of a magma at pressures appropriate to its stability but persisted metastably when the magma congealed at higher levels. This is clearly a heroic explanation; nonetheless, it might have merit in situations where textural features suggest that the muscovite was not among the last phases to precipitate out of the magma. Other hydrous silicate minerals (notably hornblende and biotite) do apparently persist metastably to lower pressures; why not muscovite? The presence of muscovite phenocrysts without signs of breakdown in ignimbrites and other extrusive rhyolitic rocks (Herrera et al 1984, Noble et al 1984, Schleicher & Lippolt 1981) and in dikes that show clear evidence of intrusion from considerable depth and of rapid chilling (Evans & Vance 1987) indicates that muscovite can crystallize out early and can persist metastably at least in a rapidly cooled environment. The involvement of muscovite in flow banding of a granite implies that the mineral was present during emplacement of the crystal-melt mush and was not necessarily a late mineral. However, to address fully the question of metastable muscovite, we need not only data on the thermal stability of muscovite of natural compositions in appropriate assemblages, but also information on the various rates applicable to specific geological instances.

Summary

The presence of magmatic muscovite in plutons that apparently consolidated at pressures significantly less than 3–4 kbar might be due to the metastable persistence of the phase formed at an earlier stage in a deeper level of the crust. It cannot be explained by the known departure of the composition from end-member muscovite. It might also be explained by the high concentrations of Cs, Rb, and Li in the magma (but critical experimental and chemical data are lacking) or by the high values of boron fugacity in the magma (but only for leucogranites without biotite, and preferably without the mafic components to form tourmaline). However, the best explanation seems to be a lowering of the solidus temperature by fluorine (aided by the other compositional effects); this process has an added attraction because fluorine helps to mobilize the alkalis, remove them from the system, and thereby enhance the peraluminosity of the melt (Manning & Pichavant 1983). For granites that contain magmatic muscovite but no topaz or fluorite, however, the presence of adequate magmatic fluorine remains conjectural, especially where the micas themselves are not fluorine rich.

The problem of interpreting magmatic muscovite is discussed in detail because magmatic muscovite is a common mineral in peraluminous granites and our inability to explain its stability relations shows that our understanding of peraluminous granitic rocks is less than adequate. Bridging this conceptual gulf between field and experimental inferences must remain high on the agenda of igneous petrologists.

SOME AFM PHASE RELATIONS IN PERALUMINOUS GRANITIC ROCKS

The phase relations of biotite, muscovite, garnet, cordierite, andalusite/ sillimanite, and (more rarely) orthopyroxene provide useful information on crystallization conditions and on the inferred source terrane for a peraluminous granite. These minerals can be graphically represented in an Al_2O_3-K_2O-FeO-MgO (AKFM) tetrahedron and projected onto the AFM plane through the ubiquitous (and assumed end-member) potassic feldspar (Barker 1961). Within the AFM triangle (Figure 3), the composition of muscovite is represented by $Mu_{0.8}Lp_{0.2}$, biotite by $H_2K_{0.9}(Fe, Mg)_{2.75}Al_{1.4}Si_{2.85}O_{12}$, cordierite by $(Fe, Mg)_2Al_4Si_5O_{18}$, and garnet by $(Fe, Mg)_3Al_2Si_3O_{12}$. Orthopyroxene is taken to be nonaluminous and so plots along the Fe-Mg boundary line.

The AFM projection assumes excess quartz, potassic feldspar, and plagioclase. In general, the system will have an iron oxide or iron-titanium oxide phase and, of course, a silicate melt. Table 2 gives data from the literature of apparently coexisting AFM mineral assemblages from peraluminous ignimbrite and lava, as well as from selected peraluminous plutonic rocks. The data show that (a) the mg numbers of the mafic phases vary in the sequence Gt < Bt $\simeq$ Opx < Crd, comparable to that in metamorphic rocks (Thompson 1976, Fang & He 1985); and (b) the AFM phase assemblages appear to be related to each other by the following overall dehydration reaction:

$$\text{K-feldspar} + \text{garnet} + \text{cordierite} + H_2O = \text{muscovite} + \text{biotite} + \text{quartz}.$$

As the activity of H_2O increases (or as the temperature decreases; the activity of SiO_2 is fixed), the two-mica assemblage is favored. The assemblage Bt-Mu-Gt-Crd has been observed both for the South Mountain batholith and related plutons and for the Cornubian batholith (Table 2); however, here the Mn component might cause garnet to appear in the assemblage (Miller & Stoddard 1981, Allan & Clarke 1981). The presence elsewhere in the same batholith of the assemblage biotite-muscovite-andalusite shows that, in detail, the muscovite-breakdown reaction and the tieline-switch reaction outlined above are separate events.

How these AFM mineral assemblages are chemographically related to the evolving granitic magmas is poorly known, and existing experimental data are of limited use. Different workers (Huang & Wyllie 1973, 1981, Piwinskii & Wyllie 1970, Clemens & Wall 1981, London et al 1986, Benard et al 1985) used different natural rocks for starting materials, but they did not assess the possible effects of added components, nor did they establish the reversal of the equilibria. As a result, we cannot reconstruct accurate liquidus diagrams. Note also that crystallization experiments on totally fused rocks cannot reproduce the crystallization of the parent rocks because the constraints imposed by preexisting phases that must have existed in natural rocks (Bowen 1928) are missing.

36   ZEN

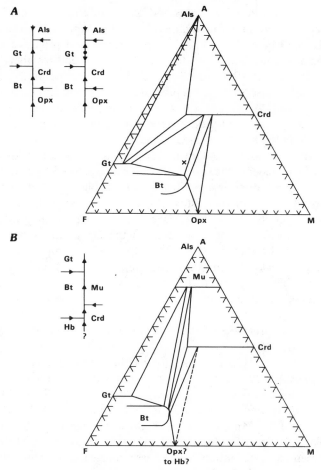

Figure 3 AFM diagram for peraluminous granitic rocks, as projected from potassic feldspar
and in the presence of plagioclase (commonly Ab-rich) and quartz. Muscovite is shown as
a solid solution, about 20% of the way to Lp only, because ferric iron is not included. The
biotite composition is that discussed in the text. In (*A*), biotite is the only stable hydrous
phase, but in (*B*), the system is sufficiently hydrous for muscovite solid solution to be stable.
The mineral compositions of coexistent phases are mainly based on microprobe data in the
literature (see Table 2). Parts (*A*) and (*B*) are related by the overall reaction garnet + cordierite + potassic feldspar + H$_2$O = muscovite + biotite + quartz (see text). Also
shown in (*A*) is the composition of a Strathbogie granite containing the depicted AFM
minerals (×); this sample was used in crystallization experiments (Clemens & Wall 1981)
that showed that the orthopyroxene field on the liquidus lapped onto the garnet-cordierite-
biotite solid assemblage field by a peritectic relation. Possible topologies of crystallization
fields of minerals on the liquidus surfaces are shown and are discussed in the text; two
alternatives are given for (*A*). Abbreviations: Gt, garnet; Crd, cordierite; Mu, muscovite; Bt,
biotite; Als, aluminum silicate polymorphs; Hb, hornblende; Opx, orthopyroxene.

Nevertheless, we can make several observations. First, even with the same melt composition and P-T conditions, a loss of volatiles (for example, through volcanic eruptions) could affect the solid assemblage. Second, in crystallization experiments on the cordierite-biotite-garnet-bearing Strathbogie granite, orthopyroxene appeared, and then disappeared, in the melting interval; the final assemblage was the same as the starting assemblage (Clemens & Wall 1981). On the liquidus surface, therefore, compositions in the garnet-cordierite-biotite field could initially crystallize orthopyroxene but later lose it peritectically. Whether there is a "biotite-out" peritectic reaction that would permit the liquid to evolve into the andalusite-garnet-cordierite field is not known, so both versions of the topology of liquid descent are given in Figure 3A. The common succession of biotite-muscovite rocks, biotite-muscovite-garnet rocks, and leuco-granite (muscovite + garnet ± tourmaline) apparently genetically related to biotite-muscovite granite (e.g. the Cornubian batholith) suggests that biotite indeed can react out peritectically in hydrous peraluminous magmas (Figure 3B; Miller & Stoddard 1981).

Except for the Andean volcanics [the Macusani (Noble et al 1984) and the Morococala and Los Frailes volcanics (Ericksen et al 1985)] and possibly the Rheingraben rocks (Schleicher & Lippolt 1981),[2] peraluminous volcanic rocks contain less hydrous AFM assemblages (Table 2); Friedman & Smith (1958) observed that the magmatic H_2O content of rhyolite is low. More hydrous melts would be expected to quench upon pressure drop and loss of volatiles.

In terms of the AFM diagram, how do metaluminous magmas evolve into peraluminous magmas? Abbott (1981; see also Abbott & Clarke 1979, Speer 1986) suggested that a change in the melting behavior of calcic amphibole, from eutectic to peritectic depending on the mg number of the melt, could extend the stability of hornblende into peraluminous melts. Such peritectic behavior must exist for real rocks; once a metaluminous magma evolves into the peraluminous field, even after hornblende has reacted out, further fractionation will cause the melt to become more peraluminous, possibly resulting in strongly peraluminous, two-mica granite (Zen 1986; see Chayes 1952). However, all these inferences await experimental confirmation.

---

[2] The Permian volcanic rocks from Germany pose a problem because the chemical analyses and petrographic descriptions of Schleicher & Lippolt (1981) show subsolidus alteration and possibly alkali loss and gain. The extreme range of ASI (corrected for apatite), 0.97–1.92, points in the same direction. The barrel-shaped muscovite phenocryst showing magmatic minerals along pried-apart cleavage planes seems truly magmatic, but its low TiO_2 content suggests possible subsolidus chemical changes.

Table 2 Typical nonquartzofeldspathic minerals of peraluminous igneous rocks

Formation	Assemblage[a]	Mg/(Mg+Fe)	References and notes
	Volcanic Rocks		
Violet Town, Victoria, Australia	Bt-Gt-Crd ⎱ Bt-Crd-Gt-Opx ⎰	Gt($\sim$0.25) < Bt($\sim$0.4) $\simeq$ Opx($\sim$0.4) < Crd($\sim$0.58)	Clemens & Wall 1984. Only averages for all samples and both assemblages given
Hawkins Formation	Bt-Gt-Crd-Opx-Herc	Gt(0.25) < Opx(0.47) < Bt(0.55) < Crd(0.63)	Wyborn et al 1981. The Hawkins, Goobarragandra and
Goobarragandra Formation	Bt-Crd-Opx	Bt(0.44) $\sim$ Opx(0.47) < Crd(0.64)	Laidlaw formations are all from the Lachlan Fold Belt,
Laidlaw Formation	Bt-Opx	Bt(0.53) $\sim$ Opx(0.55)	New South Wales, Australia
Macusani ignimbrite, Peru	Bt-Mu-And(Sil)-Tm	Bt(0.3) $\sim$ Tm(0.3)	Noble et al 1984, Herrera et al 1984
Morococala and Los Frailes ignimbrites, Bolivia	Bt-Mu-Crd-And		Ericksen et al 1985
Permian rhyolite, southwest Germany	Bt-Mu		Schleicher & Lippolt 1981
Cerro del Hoyazo, Spain	Bt-Crd-Sil-Gt(?)		Zeck 1970; Gt may be restitic
Jianzhou volcanics, China	Gt-Crd-Opx		Fang & He 1985
	Selected Intrusive Rocks		
South Mountain batholith, Nova Scotia	Bt-Mu-Crd-And ⎫ Bt-Mu-Crd-(Gt) ⎬ Bt-Mu-Crd-Gt ⎭	Bt($\sim$0.3) < Crd($\sim$0.55)	Maillet & Clarke 1985
	Bt-Gt	Gt($\sim$0.15–0.25) < Bt(0.2–0.4)	Allan & Clarke 1981
Musquodoboit batholith, Nova Scotia	Bt-Crd-(Gt-Mu)		Maillet & Clarke 1985
Ellison Lake pluton, Nova Scotia	Bt-Crd-(Mu-Gt) Crd-Mu-Gt		Maillet & Clarke 1985
Kinsman quartzmonzonite, New Hampshire	Bt-Mu-Gt		Barker 1961

Table 2 (*continued*)

Formation	Assemblage[a]	Mg/(Mg + Fe)	References and notes
	Selected Intrusive Rocks		
Clouds Creek, South Carolina	Bt-Crd	Bt(0.3) < Crd(0.5)	Speer 1981
Stumpy Point, North Carolina	Bt-Crd-Gt	Gt(0.1) < Bt(0.25) < Crd(0.3)	Speer 1981
Willard Creek pluton, Nevada	Bt-Mu-Gt		Lee et al 1981; see also Miller & Bradfish 1980
Old Woman–Piute pluton, California	Bt-Mu Bt-Mu-Gt Mu-Gt	Gt(0.05) < Bt(0.22)	Miller & Stoddard 1981
Pine Lakes trondhjemite, Cornucopia Stock, Oregon	Bt-Mu-Crd		Taubeneck 1967; E-an Zen, unpublished data
Strathbogie batholith, New South Wales, Australia	Bt-Crd-Gt	Gt(0.1) < Bt(0.3) < Crd(0.55)	Clemens & Wall 1981, Phillips et al 1981
Southern Brittany, France	Bt-Mu-Gt-Tm		Strong & Hanmer 1981
Cornubian batholith, southwest England	Bt-Mu-Tm-Gt-Crd	Bt(0.26) < Crd(0.36)	Brammall & Harwood 1932
	Bt-Mu-And		Exley & Stone 1966
	Bt-Mu-Crd-And-Tm	Bt(~0.2–0.3) ~ Tm(0.2–0.3)	Charoy 1986
Ryoke Belt, Honshu, Japan	Bt-Mu-Gt		Ishihara 1978
Manaslu batholith, Nepal	Bt-Mu-Sil Mu-Gt		LeFort 1981 Benard et al 1985
SE China:			Fang & He 1985
Darongshan and Shihe granites	Bt-Gt-Crd	Gt(0.2) < Bt(0.4) < Crd(0.6)	
Jiuzhou and Dasi granites	Bt-Gt-Crd-Opx	Gt(0.3) < Crd(0.6)	
Taima granite	Bt-Opx		

[a] Abbreviations: Bt, biotite; Gt, garnet; Crd, cordierite; Opx, orthopyroxene; Herc, hercynite; Mu, muscovite solid solution; And, andalusite; Sil, sillimanite; Tm, tourmaline. Names between parentheses were indicated in reference as minor phase. All rocks contain potassic feldspar.

REDOX RELATIONS OF PERALUMINOUS GRANITIC ROCKS

Most strongly peraluminous granitic rocks are thought to have formed under relatively reducing conditions. The common opaque oxide phase in such rocks is ilmenite (Ishihara 1977, White et al 1986a), whereas magnetite is common for hornblende-biotite-bearing metaluminous to weakly per-aluminous rocks. Ishihara (1977) showed that the Fe^{3+}/Fe_{total} ratio of biotite in his "ilmenite series" tends to indicate more reducing conditions than the nickel-bunsenite buffer according to the calibration of Wones & Eugster (1965); this ratio of biotite in his "magnetite series" tends to be more oxidizing. These observations are consistent with the postulate that strongly peraluminous granites were derived from partial melting of sedimentary rocks (Chappell & White 1974) having inherited organic material; indeed, graphite is an accessory phase in many ilmenite-bearing plutons of Japan (Ishihara 1977).

The relation between ilmenite and magnetite involves more than a simple redox reaction because magnetite-bearing granitic rocks characteristically contain sphene as the primary titanium-carrying magmatic phase, rather than ilmenite. A schematic reaction showing the relation is

$$3FeTiO_3 + 3CaO + 3SiO_2 + O_2 = 3CaTiSiO_5 + Fe_3O_4,$$

$$\text{ilmenite} \qquad \text{quartz} \qquad \text{sphene} \quad \text{magnetite}$$

where CaO and O are components in the magma. Decreasing activity of either component would favor ilmenite. Thus, to use ilmenite-magnetite relations to compare the redox relations in the magma, one must establish both the absence of another adequate host phase for Ti and commensurate activities for CaO (corresponding to that in andesine/oligoclase). The latter requirement disqualifies some highly evolved peraluminous granites.

Dickenson & Hess (1981) showed that the Fe^{2+}/Fe_{total} value in synthetic silicate melts in the system K_2O-FeO-Fe_2O_3-Al_2O_3-SiO_2 depends on the alkali/alumina ratio of the melt: It is at a minimum at ASI ~ 1, increases slowly for ASI < 1, and increases rapidly for ASI > 1. As these experiments were carried out in the open air, the fugacity of oxygen was uniform at ~ 0.2 bar; the Fe^{2+}/Fe_{total} values reflect charge distribution changes due to compositional and structural adjustments in the melt, and not oxygen fugacity. No data exist as to whether these melts would precipitate the same or different iron-bearing solid phases. It seems prudent to suppose that all three factors—redox potential in the melt, structural and com-

positional effects on Fe^{2+}/Fe_{total}, and mass-action properties—could affect
the nature of the iron oxide phases in metaluminous and peraluminous
granitic rocks, so the mineral data must be interpreted with caution.

A usable monitor of the oxygen fugacity for some peraluminous granites is the reaction biotite+almandinic garnet+oxygen = quartz+mus-
covite+magnetite (Zen 1985b). For minerals having typical igneous
compositions, the oxygen fugacity of this assemblage lies between the
nickel-bunsenite and the magnetite-hematite buffers. Muscovite-magnetite-
bearing granitic rocks could be pink because they could coexist with
hematite, but biotite-garnet-bearing granites are not expected to be pink
except through later alteration.

CAN PERALUMINOUS GRANITES YIELD CLUES TO THE SOURCE TERRANE? THE S- AND I-TYPE SYSTEMS OF GRANITE NOMENCLATURE

Study of the batholithic complexes in the Lachlan Fold Belt of New South
Wales in the early 1970s enhanced our appreciation of the implications of
the chemistry, petrography, and geology of different kinds of granitic
plutons. White et al (1974) and Chappell & White (1974) separated these
batholiths into those presumably derived from anatexis of sedimentary
rocks (the S-type granite) and those presumably derived from igneous
sources that have not been significantly altered by weathering (the I-type
granite); the two rock types convey information on processes of granite
genesis as well as provide glimpses of the nature of the hidden source
terranes.

The S- and I-type granites ideally possess the contrasting features shown
in Table 3 (White & Chappell 1977, White et al 1986a). The significance
of each pair of features may be quickly summarized. High peraluminosity
(item 1) implies a highly aluminous source related to weathered material,
where Na and Ca are preferentially removed. The cutoff at ASI = 1.1
rather than 1.0 recognizes the fact that many mildly peraluminous granitic
rocks could be derived by fractionation or melting of originally metaluminous or slightly peraluminous material (see Zen 1986). The emphasis
on alkali ratios (item 2) accents this effect of weathering and the fact that
potassium tends to be fixed in clay minerals. The significance of the range
of SiO_2 (item 3) in comagmatic series is that most sedimentary rocks have
excess quartz; thus the silica contents of the resulting melts are more
restricted than those of melts evolved from differentiation of igneous
source material, which could encompass the entire range of compositions

Table 3 Criteria for I-type versus S-type granitic rocks

I-type	S-type
1. ASI < 1.1; CIPW norm yields diopside or <1% corundum	ASI > 1.1; CIPW norm yields >1% corundum
2. Relatively high Na_2O and high Na/K ratio	Relatively low Na_2O and low Na/K ratio
3. Extended SiO_2 range for rocks within series	Restricted SiO_2 range for rocks within series
4. Low initial Sr ratios	High initial Sr ratios
5. Low $\delta^{18}O$	High $\delta^{18}O$
6. Inclusions are mafic material	Inclusions are peraluminous meta-sedimentary material
7. Opaques are magnetite or ilmenite	Opaques are ilmenite
8. No cordierite phenocrysts	Cordierite phenocrysts

from gabbro to granite. High initial Sr ratios (item 4) for S-type rocks reflect the cumulative enrichment of Rb relative to Sr in crustal material, as compared with mantle material, through geological processes. The high $\delta^{18}O$ (item 5) results from the fact that sedimentary rocks are formed in surface waters at low temperatures (Taylor 1980). The inclusions (item 6), if cognate, reflect the nature of the source terranes. The nature of opaque oxide phases (item 7)—only ilmenite in S-type rocks but including magnetite (typically associated with sphene) in I-type rocks—suggests differences in the redox conditions reflecting the presence of organic material in S-type source rocks. Finally, magmatic cordierite (item 8) is considered definitive of S-type granite (White et al 1986a,b,c).

These lines of evidence, taken in concert, probably do provide cogent argument for the nature of the source terranes; however, when the various kinds of evidence are contradictory, use of the S- and I-type classifications becomes problematic. The following comments, keyed to Table 3, may be added (see also Miller & Bradfish 1980, Miller 1985, 1986, Dickson 1986, Halliday et al 1981):

1. High values of ASI certainly could, and normally probably do, result from anatexis of peraluminous rocks. But what if the source terrane was itself a peraluminous igneous rock? What if the source terrane consisted of immature arkose and graywacke that had largely preserved their igneous compositions (Miller 1986, White et al 1986a)? The geological nature of these source terranes would not be readily illuminated by the S- and I-type criteria.

Differentiation of metaluminous magmas, presumably of I-type origin, could lead to peraluminous rocks having ASI as high as 1.15 and could even yield strongly peraluminous magmas by continued fractionation after hornblende has reacted out (Zen 1986). Fractionation of a metaluminous parent magma is a very inefficient way to obtain peraluminous residual melts, however; large quantities of antecedent rocks would be expected. Differentiation combined with fractionation-assimilation (the "afc process") could lead to ambiguous results because the assimilation of peraluminous material by a metaluminous magma should initially promote precipitation of phases already present (Bowen 1928); however, judicious use of isotopic tracers could reduce the ambiguity (Taylor 1980).

2. Use of alkali content and alkali ratios presupposes that the bulk chemistry reflects a large average sample of the source, that melting extended beyond the minimum-composition range (for which the alkali ratio and alkali content would be nearly buffered by the assemblage), and that there was no late-magmatic or postmagmatic alkali exchange. It implies that the chemical signature was not masked by the geochemical complements to pelitic rocks such as evaporites. The alkali content and ratios can only be used to corroborate an already strong case for a sedimentary protolith.

3. In the last analysis, it is the mineralogy and overall chemistry of the comagmatic rocks, rather than the SiO_2 range, that provide useful information.

4. The correlation of the initial strontium ratios to the peraluminosity of the plutons enabled workers to infer the nature of the source material in the Lachlan Fold Belt (White et al 1986a). Similarly, in the Appalachian Piedmont of Georgia, metaluminous rocks ranging from gabbro to granite show low initial ratios, and two-mica, tourmaline-bearing granites show high initial ratios (Whitney & Wenner 1980). However, Halliday et al (1981) analyzed the expected effects of various processes of magma genesis on the initial strontium ratios and concluded that these effects cannot always yield unequivocal signatures. In support of this doubt, various plutons in the Pioneer batholith of southwest Montana broadly conform to the bulk chemical and mineralogical features of I- and S-type granites, but all have initial strontium values in excess of 0.711, and the two dominant series based on this ratio each extends from quartz diorite and/or tonalite to two-mica granite (Zen 1987, Arth et al 1986).

5. Use of $\delta^{18}O$ can elucidate the nature of the source material, but the situation is more complex than envisioned for the Lachlan Fold Belt batholiths (Halliday et al 1981, Miller 1985, Taylor 1980). Magmas derived from metamorphosed and dehydrated pelitic source material can be expected to show low $\delta^{18}O$ (Halliday et al 1981), mimicking nonsedimentary sources. Normally high $\delta^{18}O$ values of the Atlanta lobe of the Idaho batholith [much of which consists of two-mica granitic rocks

(Hyndman 1983)] have been decreased by interaction with meteoric-hydrothermal circulation, which otherwise left few traces (Criss & Taylor 1983, Fleck & Criss 1985).

6. The restitic inclusions can indicate the nature of the source terrane, but mixed and accidental inclusions can be troublesome. Using the presence of mafic inclusions, White et al (1974) argued against stoping as a general mechanism of emplacement of large batholiths. However, the criteria for discriminating restite clots, mixed magma, mineral segregations within the magma body, and accidental xenoliths remain controversial (Wall et al 1987).

7. The nature of the opaque oxide phase has been discussed earlier. We know too little about the redox reactions in a magma chamber to attribute the presence of accessory oxide phases or of graphite to a sedimentary origin of the source material. Even mantle-derived ultramafic rocks contain graphite (Elliott et al 1982).

8. White et al (1986a) cited phenocrysts of cordierite as the most persuasive evidence for a sedimentary source for S-type granites. As shown in Figure 3, both cordierite and muscovite can appear over wide ranges of bulk aluminosity, depending on the bulk mg number and the H_2O fugacity. Indeed, the cordierite-garnet-biotite-bearing Strathbogie granite (Clemens & Wall 1981) is equivalent to a two-mica $\pm$ cordierite granite with the addition only of H_2O. Oxidation of iron could increase the effective mg number and result in cordierite; high sulfur fugacity could lead to pyrite or pyrrhotite and accomplish the same result. The apatite-corrected ASI of some granites and ignimbrites containing cordierite-bearing and muscovite-bearing assemblages (including some of the rocks that precipitated the debate among Miller, Dickson, and White et al) are compared in Figure 4; the cordierite-bearing rocks are not more peraluminous. (The wide range of ASI values for the two-mica rocks may reflect a greater extent of fluid-phase alteration.) The points made by White et al (1986a) can be restated as follows. Peraluminous magmas derived from differentiation of subaluminous material probably could not remain dry enough to result in a biotite-cordierite-garnet assemblage (as occurs in the Lachlan Fold Belt); such highly differentiated magmas also would probably be too magnesia poor for cordierite to be stable. Thus it seems reasonable to say that cordierite-bearing peraluminous granites are likely to be S-type, but lack of cordierite does not connote a non-S-type source terrane.

SUMMARY: CLUES TO ORIGINS

To distinguish peraluminous granitic rocks derived through different processes, we must face two familiar questions: What was the source terrane

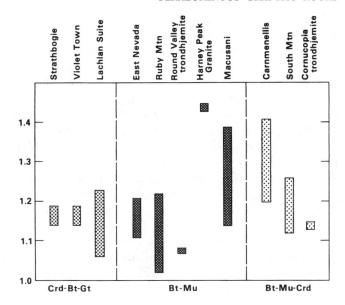

Figure 4 Comparison of the aluminum saturation index (ASI) for plutonic and volcanic rocks bearing different AFM mineral assemblages—in particular, cordierite-bearing and cordierite-free assemblages. All ASI values have been corrected for CaO in apatite. Sources of data: Strathbogie pluton (Phillips et al 1981, Clemens & Wall 1981); Violet Town Volcanics (Clemens & Wall 1984); Lachlan volcanic suite (Wyborn et al 1981); East Nevada pluton (Lee et al 1981); Ruby Mountain granite (Kistler et al 1981); Round Valley trondhjemite (E-an Zen & J. M. Hammarstrom, unpublished data); Harney Peak Granite (Huang & Wyllie 1981); Macusani volcanics (Noble et al 1984); Carnmenellis granite of the Cornubian batholith (Charoy 1986); South Mountain batholith (McKenzie & Clarke 1975); Cornucopia trondhjemite (Taubeneck 1967).

for a given pluton? How did the pluton become what it is, and how did it get here? Although the geochemical, isotopic, mineralogic, and petrographic criteria of Chappell & White (1974), as modified by later discussions and reviewed earlier, are intimately related to their model of restite-melt unmixing (see Wall et al 1987), these criteria do address important genetic questions and furnish reasonable starting points for this summary.

How does one identify a peraluminous plutonic body as having resulted from alkali loss to the surroundings? Alkali transport depends heavily on a volatile-rich phase and thus can be expected to be effective mainly during late-magmatic stages. If the rock became peraluminous only through this process, then the earlier petrogenetic history and even chemical and mineralogical features should betray its origin (Halliday et al 1981). Evidence for a high content of volatiles, especially halogens that facilitate alkali loss

(Orville 1963, Manning & Pichavant 1983), can be expected; one might also expect the alkali loss to be uneven in the pluton, to find remnant or relict metaluminous mineral assemblages especially in the cumulate fractions, and the peraluminous assemblages concentrated in later parts. Textural evidence for these relations should be sought. Fluid and glass inclusions in minerals might contain important information. If an ample vapor phase was generated by addition of H_2O from the surroundings, then the rocks ought to show clear evidence of this in their $\delta^{18}O$ signature. Finally, if strongly peraluminous minerals can be shown on textural evidence to have been early or if the magma can be shown to have been relatively dry, then late-stage vapor-phase alkali loss would be an unlikely mechanism for the origin of these peraluminous rocks.

Fractional differentiation is unlikely to produce large volumes of strongly peraluminous granite. Especially for plutons that do not belong to a demonstrably consanguineous series, such an explanation seems unlikely. Nevertheless, sizable bodies of at least mildly peraluminous granite may originate through fractionation without voluminous consanguineous antecendents, if the source material nearly straddles the metaluminous-peraluminous boundary [for example, the Cenozoic Cascades volcanic rocks of northwestern United States (Chayes 1969)].

For large bodies of strongly peraluminous granitic rocks, such as those of southeastern Australia, southwest England, or the Himalayas, peraluminous sources seem necessary. The important question to be addressed is what should be done when indications of the nature of the source rock are in conflict. I propose the following hierarchy based on geochemical and petrological considerations.

The whole-rock initial Sr ratios are important. Low initial Sr ratios may prove little, but if the rock has high initial Sr ratios, a source having long crustal residence (likely an ultimately recycled rock source) seems to be required. The isotopic ratios of feldspar lead (Ayuso 1986) and of neodymium (Farmer & DePaolo 1983) in plutons can provide additional useful clues to the nature of the source rock as averaged over the entire source area, and these should be used along with the initial Sr ratios wherever possible.

High bulk ASI and the presence of minerals that characterize strongly peraluminous rocks can also be useful clues, especially in conjunction with volumetric considerations and with a lack of comagmatic and more mafic, less silicic rocks. This caution is needed because we do not yet have experimental data that quantify possible paths of liquid descent from metaluminous, through weakly peraluminous, to strongly peraluminous magmas. K/Na ratios are not by themselves a reliable indicator of the nature of the source terrane. Likewise, high $\delta^{18}O$ values for the rocks are neither necessary nor sufficient conditions for sedimentary sources. The $\delta^{18}O$ values are subject to too many factors to be a reliable guide and are highly sensitive to postconsolidation alteration processes.

The criteria of Chappell & White (1974), as amended and discussed by various workers (e.g. White & Chappell 1977, White et al 1986a,b, Miller 1986, Dickson 1986, Halliday et al 1981), are useful petrological tools precisely because these criteria do not generally provide consistent answers to the question of sources, and conflicts provide opportunities for better understanding of the details of petrogenesis. Classification, by whatever scheme, should conclude, rather than start, a petrogenetic inquiry.

Regardless of what process gave rise to a peraluminous body, and from what source terrane, questions remain that have beleaguered petrologists since the early part of this century. What are the sources of heat for magma generation? Can we satisfactorily account for the thermal budgets? Is "mantle heat" separable from mantle material in situations where we require the energy to be supplied rapidly enough to cause major melting to occur? Is the assimilation-fractionation-cumulation model (Taylor 1980) adequate to account for the thermal balance, for the mineralogy and petrography, and for the processes of magma derivation and ascent?

The ascent of a peraluminous magma to shallow levels from deeper levels of generation implies a H_2O-undersaturated magma, for otherwise the magma should lose its volatiles upon decompression and be pressure quenched. This constraint applies both to anatexis of sedimentary rocks and to fractionation of metaluminous rocks, for even these latter rocks, upon attaining a stage of differentiation to produce granite, must be both volatile- and solid-rich. Initially dry peraluminous magmas may have behaved as desiccators, sucking up moisture as they rose, thereby lowering the solidus temperatures, promoting the assimilation-fractionation-cumulation process, and permitting the differentiation process to be stretched out. If so, however, the original anatectic melts must have been very hot, enhancing the problem of adequate sources of thermal energy.

ACKNOWLEDGMENTS

I thank J. Lawford Anderson, Robert A. Ayuso, Paul C. Bateman, D. Barrie Clarke, John D. Clemens, Charles V. Guidotti, Jane M. Hammarstrom, Ralph A. Haugerud, J. Julian Hemley, Stephen D. Ludington, Calvin F. Miller, Michel Pichavant, and Priestley Toulmin III for their thoughtful reviews of the manuscript. Clarke and Clemens kindly gave me advance copies of their papers. Discussion of the genesis of granite with these colleagues and with Joe Arth, Fred Barker, Allan White, the late Dave Wones, and Pete Wyllie over the years have helped my education in this fascinating subject. Kathleen Krafft helped to wordcraft the manuscript, and I thank her.

Literature Cited

Abbott, R. N. Jr. 1981. AFM liquidus projections for granitic magmas, with special reference to hornblende, biotite and garnet. *Can. Mineral.* 19: 103–10

Abbott, R. N. Jr., Clarke, D. B. 1979. Hypothetical liquidus relationships in the subsystem Al_2O_3-FeO-MgO projected from quartz, alkali feldspar and plagioclase for $a(H_2O) \leq 1$. *Can. Mineral.* 17: 549–60

Allan, B. D., Clarke, D. B. 1981. Occurrence and origin of garnets in the South Mountain batholith, Nova Scotia. *Can. Mineral.* 19: 19–24

Anderson, J. L., Rowley, M. C. 1981. Synkinematic intrusion of peraluminous and associated metaluminous granitic magmas, Whipple Mountains, California. *Can. Mineral.* 19: 83–101

Arth, J. G., Zen, E-an, Sellers, G., Hammarstrom, J. 1986. High initial Sr isotopic ratios and evidence for magma mixing in the Pioneer batholith of southwest Montana. *J. Geol.* 94: 419–30

Ayuso, R. A. 1986. Lead-isotopic evidence for distinct sources of granite and for distinct basements in the northern Appalachians, Maine. *Geology* 14: 322–25

Barker, F. 1961. Phase relations in cordierite-garnet-bearing Kinsman quartz monzonite and the enclosing schist, Lovewell Mountain quadrangle, New Hampshire. *Am. Mineral.* 46: 1166–76

Barker, F. 1981. Introduction to special issue on granites and rhyolites: a commentary for the nonspecialist. *J. Geophys. Res.* 86B: 10,131–35

Barton, M. D. 1982. The thermodynamic properties of topaz solid solutions and some petrologic applications. *Am. Mineral.* 67: 956–74

Benard, F., Moutou, P., Pichavant, M. 1985. Phase relations of tourmaline leucogranites and the significance of tourmaline in silicic magmas. *J. Geol.* 93: 271–91

Bowen, N. L. 1928. *The Evolution of the Igneous Rocks.* Princeton, NJ: Princeton Univ. Press. 334 pp. (Reissued by Dover Publ., 1956)

Brammall, A., Harwood, H. F. 1925. Tourmalinization in the Dartmoor granite. *Mineral. Mag.* 20: 319–30

Brammall, A., Harwood, H. F. 1932. The Dartmoor granites: their genetic relationships. *Q. J. Geol. Soc. London* 88: 171–237

Chappell, B. W., White, A. J. R. 1974. Two contrasting granite types. *Pac. Geol.* 8: 173–74

Charoy, B. 1986. The genesis of the Cornubian batholith (south-west England): the example of the Carnmenellis pluton. *J. Petrol.* 27: 571–604

Chatterjee, N. D., Flux, S. 1986. Thermodynamic mixing properties of muscovite-paragonite crystalline solutions at high temperatures and pressures, and their geological applications. *J. Petrol.* 27: 677–93

Chatterjee, N. D., Johannes, W. 1974. Thermal stability and standard thermodynamic properties of synthetic $2M_1$-muscovite, $KAl_2[AlSi_3O_{10}(OH)_2]$. *Contrib. Mineral. Petrol.* 48: 89–114

Chayes, F. 1952. The finer-grained calcalkaline granites of New England. *J. Geol.* 60: 207–54

Chayes, F. 1969. The chemical composition of Cenozoic andesite. *Oreg. Dep. Geol. Miner. Ind. Bull.* 65: 1–11

Chorlton, L. B., Martin, R. F. 1978. The effect of boron on the granite solidus. *Can. Mineral.* 16: 239–44

Clarke, D. B. 1981a. Peraluminous granites. *Can. Mineral.* 19: 1–2

Clarke, D. B. 1981b. The mineralogy of peraluminous granites: a review. *Can. Mineral.* 19: 3–17

Clarke, D. B., McKenzie, C. B., Muecke, G. K., Richardson, S. W. 1976. Magmatic andalusite from the South Mountain batholith, Nova Scotia. *Contrib. Mineral. Petrol.* 56: 279–87

Clemens, J. D., Wall, V. J. 1981. Origin and crystallization of some peraluminous (S-type) granitic magmas. *Can. Mineral.* 19: 111–31

Clemens, J. D., Wall, V. J. 1984. Origin and evolution of a peraluminous silicic ignimbrite suite: the Violet Town Volcanics. *Contrib. Mineral. Petrol.* 88: 354–71

Criss, R. E., Taylor, H. P. Jr. 1983. An $^{18}O/^{16}O$ and D/H study of Tertiary hydrothermal systems in the southern half of the Idaho batholith. *Geol. Soc. Am. Bull.* 94: 640–63

Dickenson, M. P., Hess, P. C. 1981. Redox equilibria and the structural role of iron in alumino-silicate melts. *Contrib. Mineral. Petrol.* 78: 352–57

Dickson, W. L. 1986. Comment on "S-type granites and their probable absence in southwestern North America." *Geology* 14: 894–95

Elliott, W. C., Grandstaff, D. E., Ulmer, G. C., Buntin, T., Gold, D. P. 1982. An intrinsic oxygen fugacity study of platinum-carbon associations in layered intrusions. *Econ. Geol.* 77: 1493–1510

Ericksen, G. E., Smith, R. L., Luedke, R. G., Flores, M., Espinosa, A., et al. 1985. Preliminary geochemical study of ash-flow tuffs in the Morococala and Los Frailes volcanic fields, central Bolivian tin belt. *US Geol. Surv. Open-File Rep. 85-258.* 8 pp.

Evans, B. W., Vance, J. A. 1987. Epidote phenocrysts in dacitic dikes, Boulder County, Colorado. *Contrib. Mineral. Petrol.* 96: 178–85

Exley, C. S., Stone, M. 1966. The granitic rocks of south-west England. In *Present Views of Some Aspects of the Geology of Cornwall and Devon*, ed. K. F. G. Hosking, G. J. Shrimpton, pp. 131–84. Penzance, Engl: R. Geol. Soc. Cornwall

Fang, Q.-h., He, S.-y. 1985. Application of cordierite-garnet geothermo-barometer to Darongshan S-type granite. In *The Crust—the Significance of Granites Gneisses in the Lithosphere*, ed. L.-r. Wu, T.-m. Yang, K.-r. Yuan, J. Didier, J. K. Greenberg, et al, pp. 463–78. Athens: Theophrastus Publ.

Farmer, G. L., DePaolo, D. J. 1983. Origin of Mesozoic and Tertiary granite in the western United States and implications for pre-Mesozoic crustal structure, 1, Nd and Sr isotopic studies in the geocline of the northern Great Basin. *J. Geophys. Res.* 88B: 3379–3401

Fleck, R. J., Criss, R. E. 1985. Strontium and oxygen isotopic variations in Mesozoic and Tertiary plutons of central Idaho. *Contrib. Mineral. Petrol.* 90: 291–308

Foster, D. M. 1960. Interpretation of the composition of trioctahedral micas. *US Geol. Surv. Prof. Pap. 354-B.* 49 pp.

Friedman, I., Smith, R. L. 1958. The deuterium content of water in some volcanic glass. *Geochim. Cosmochim. Acta* 15: 218–28

Glyuk, D. S., Trufanova, L. G. 1977. Melting at 1000 kg/cm^2 in a granite-H$_2$O system with the addition of HF, HCl, and Li, Na, and K fluorides, chlorides, and hydroxides. *Geokhimiya* 1977(7): 1003–12. Transl., 1978, in *Geochem. Int.* 1977: 28–36

Glyuk, D. S., Bazaroba, S. B. B., Trufanova, L. G. 1977. Phase relations in system granite-H$_2$O with addition of CsF, CsCl, Cs$_2$CO$_3$, RbCl and Rb$_2$CO$_3$ at 450–550°C and 1000 kg/cm^2. In *Geochemistry of Endogenic Processes*, ed. L. V. Tauson,

pp. 170–75. Irkutsk: Sib. Inst. Geochem. (In Russian)

Halliday, A. N., Stephens, W. E., Harmon, R. S. 1981. Isotopic and chemical constraints on the development of peraluminous Caledonian and Acadian granites. *Can. Mineral.* 19: 205–16

Hammarstrom, J. M. 1982. Chemical and mineralogical variation in the Pioneer batholith, southwest Montana. *US Geol. Surv. Open-File Rep. 82-148.* 178 pp.

Herrera, J. V., Pichavant, M., Esteyries, C. 1984. Le volcanisme ignimbritique peralumineux plio-quaternaire de la région de Macusani, Pérou. *C. R. Acad. Sci. Ser. II* 298: 77–82

Holdaway, M. J. 1971. Stability of andalusite and the aluminum silicate phase diagram. *Am. J. Sci.* 271: 97–131

Huang, W. L., Wyllie, P. J. 1973. Melting relations of muscovite-granite to 35 kbar as a model for fusion of metamorphosed subducted oceanic sediments. *Contrib. Mineral. Petrol.* 42: 1–14

Huang, W. L., Wyllie, P. J. 1981. Phase relationships of S-type granite with H$_2$O to 35 kbar: muscovite granite from Harney Peak, South Dakota. *J. Geophys. Res.* 86B: 10,515–29

Hutchinson, C. S. 1983. Multiple Mesozoic Sn-W-Sb granitoids of southeast Asia. *Geol. Soc. Am. Mem.* 159: 35–60

Hyndman, D. W. 1983. The Idaho batholith and associated plutons, Idaho and western Montana. *Geol. Soc. Am. Mem.* 159: 213–40

Ishihara, S. 1977. The magnetite-series and ilmenite-series granitic rocks. *Min. Geol. (Jpn.)* 27: 293–305

Ishihara, S. 1978. Two-mica granitoids in Japan. *Sci. Terre, Inf. Géol.* 1978(11): 81–86

Johannes, W. 1984. Beginning of melting in the granite system Qz-Or-Ab-An-H$_2$O. *Contrib. Mineral. Petrol.* 86: 264–73

Kistler, R. W., Ghent, E. D., O'Neil, J. R. 1981. Petrogenesis of garnet two-mica granites in the Ruby Mountains, Nevada. *J. Geophys. Res.* 86B: 10,591–10,606

Kortemeier, W. T., Burt, D. M. 1986. Evidence for the magmatic origin of ongonite and topazite dikes in the Flying W Ranch area, Tonto Basin, central Arizona. *Geol. Soc. Am. Abstr. With Programs* 18: 661

Lee, D. E., Kistler, R. W., Friedman, I., van Loenen, R. E. 1981. Two-mica granites of northeastern Nevada. *J. Geophys. Res.* 86B: 10,607–16

LeFort, P. 1981. Manaslu leucogranite: a collision signature of the Himalaya, a model for its genesis and emplacement. *J. Geophys. Res.* 86B: 10,545–68

London, D., Weaver, B. L., Hervig, R. L.

50   ZEN

1986. Liquidus relations of Macusani rhyolite, an analogue for rare-element granite-pegmatite systems. *Geol. Soc. Am. Abstr. With Programs* 18: 675

Maillet, L. A., Clarke, D. B. 1985. Cordierite in the peraluminous granites of the Meguma Zone, Nova Scotia, Canada. *Mineral. Mag.* 49: 695–702

Manning, D. A. C. 1981. The effect of fluorine on liquidus phase relationships in the system Qz-Ab-Or with excess water at 1 kb. *Contrib. Mineral. Petrol.* 76: 206–15

Manning, D. A. C., Pichavant, M. 1983. The role of fluorine and boron in the generation of granitic melts. In *Migmatites, Melting and Metamorphism*, ed. M. P. Atherton, C. D. Gribble, pp. 94–109. Cheshire, Engl: Shiva Publ.

Massonne, H.-J., Schreyer, W. 1983. A new experimental phengite barometer and its application to a Variscan subduction zone at the southern margin of the Rhenohercynicum. *Terra Cognita* 3: 187 (Abstr.)

McKenzie, C. B., Clarke, D. B. 1975. Petrology of the South Mountain batholith, Nova Scotia. *Can. J. Earth Sci.* 12: 1209–18

Miller, C. F. 1985. Are strongly peraluminous magmas derived from pelitic sedimentary sources? *J. Geol.* 93: 673–89

Miller, C. F. 1986. Comment on "S-type granites and their probable absence in southwestern North America." *Geology* 14: 804–5

Miller, C. F., Bradfish, L. J. 1980. An inner Cordilleran belt of muscovite-bearing plutons. *Geology* 8: 412–16

Miller, C. F., Stoddard, E. F. 1981. The role of manganese in the paragenesis of magmatic garnet: an example from the Old Woman–Piute Range, California. *J. Geol.* 89: 233–46

Miller, C. F., Stoddard, E. F., Bradfish, L. J., Dollase, W. A. 1981. Composition of plutonic muscovite: genetic implications. *Can. Mineral.* 19: 25–34

Monier, G., Robert, J.-L. 1986a. Titanium in muscovites from two mica granites: substitutional mechanism and partition with coexisting biotites. *Neues Jahrb. Mineral. Abh.* 153: 147–61

Monier, G., Robert, J.-L. 1986b. Muscovite solid solutions in the system K₂O-MgO-FeO-Al₂O₃-SiO₂-H₂O: an experimental study at 2 kbar P_{H_2O} and comparison with natural Li-free white micas. *Mineral. Mag.* 50: 257–66

Monier, G., Mergoil-Daniel, J., Labernardiere, H. 1984. Générations successives de muscovites et feldspaths potassiques dans les leucogranites du massif de Millevaches (Massif Central français). *Bull. Minéral.* 107: 55–68

Munoz, J. L., Ludington, S. 1977. Fluorine-hydroxyl exchange in synthetic muscovite and its application to muscovite-biotite assemblages. *Am. Mineral.* 62: 304–8

Noble, D. C., Vogel, T. A., Peterson, P. S., Landis, G. P., Grant, N. K., et al. 1984. Rare-element-enriched, S-type ash-flow tuffs containing phenocrysts of muscovite, andalusite, and sillimanite, southeastern Peru. *Geology* 12: 35–39

O'Hara, M. J. 1968. The bearing of phase equilibria studies in synthetic and natural systems on the origin and evolution of basic and ultrabasic rocks. *Earth-Sci. Rev.* 4: 69–133

Orville, P. M. 1963. Alkali ion exchange between vapor and feldspar phases. *Am. J. Sci.* 261: 201–37

Phillips, G. N., Wall, V. J., Clemens, J. D. 1981. Petrology of the Strathbogie batholith: a cordierite-bearing granite. *Can. Mineral.* 19: 47–63

Pichavant, M. 1981. An experimental study of the effect of boron on a water saturated haplogranite at 1 kbar vapour pressure. *Contrib. Mineral. Petrol.* 76: 430–39

Pichavant, M., Manning, D. 1984. Petrogenesis of tourmaline granites and topaz granites; the contribution of experimental data. *Phys. Earth Planet. Inter.* 35: 31–50

Piwinskii, A. J., Wyllie, P. J. 1970. Experimental studies of igneous rock series: felsic body suite from the Needle Point pluton, Wallowa batholith, Oregon. *J. Geol.* 78: 52–76

Power, G. M. 1968. Chemical variation in tourmalines from south-west England. *Mineral. Mag.* 36: 1078–89

Schleicher, H., Lippolt, H. J. 1981. Magmatic muscovite in felsitic parts of rhyolites from southwest Germany. *Contrib. Mineral. Petrol.* 78: 220–24

Shand, S. J. 1949. *Eruptive Rocks*. New York: Wiley. 488 pp.

Speer, J. A. 1981. Petrology of cordierite- and almandine-bearing granitoid plutons of the southern Appalachian piedmont, U.S.A. *Can. Mineral.* 19: 35–46

Speer, J. A. 1984. Micas in igneous rocks. *Mineral. Soc. Am. Rev. Mineral.* 13: 299–356

Speer, J. A. 1986. Evolution of magmatic AFM mineral assemblages in granitoid rocks: the Hbl + Liq = Bt reaction of the Liberty Hill pluton, South Carolina. *Geol. Soc. Am. Abstr. With Programs* 18: 759

Streckeisen, A. L. 1973. Plutonic rocks: classification and nomenclature recommended by the IUGS Subcommittee on the systematics of igneous rocks. *Geotimes* 18(10): 26–30

Strong, D. F., Hanmer, S. K. 1981. The leucogranites of southern Brittany: origin by faulting, frictional heating, fluid flux and fractional melting. *Can. Mineral.* 19: 163–76

Taubeneck, W. H. 1967. Petrology of Cornucopia tonalite unit, Cornucopia Stock, Wallowa Mountains, northeastern Oregon. *Geol. Soc. Am. Spec. Pap. 91.* 56 pp.

Taylor, H. P. Jr. 1980. The effect of assimilation of country rocks by magmas on $^{18}O/^{16}O$ and $^{87}Sr/^{86}Sr$ systematics in igneous rocks. *Earth Planet. Sci. Lett.* 47: 243–54

Thompson, A. B. 1974. Calculation of muscovite-paragonite-alkali feldspar phase relations. *Contrib. Mineral. Petrol.* 44: 173–94

Thompson, A. B. 1976. Mineral reactions in pelitic rocks: II. Calculation of some P-T-X(Fe-Mg) phase relations. *Am. J. Sci.* 276: 425–54

Tuttle, O. F., Bowen, N. L. 1958. Origin of granite in the light of experimental studies in the system $NaAlSi_3O_8$-$KAlSi_3O_8$-H_2O. *Geol. Soc. Am. Mem. 74.* 153 pp.

Velde, B. 1965. Phengite micas: synthesis, stability, and natural occurrence. *Am. J. Sci.* 263: 886–913

Velde, B. 1972. Celadonite mica: solid solution and stability. *Contrib. Mineral. Petrol.* 37: 235–47

Vernon, R. H. 1986. K-feldspar megacrysts in granites—phenocrysts, not porphyroblasts. *Earth-Sci. Rev.* 23: 1–63

Wall, V. J., Clemens, J. D., Clarke, D. B. 1987. Models for granitoid evolution and source compositions. *J. Geol.* In press

White, A. J. R., Chappell, B. W. 1977. Ultrametamorphism and granitoid genesis. *Tectonophysics* 43: 7–22

White, A. J. R., Chappell, B. W. 1983. Granitoid types and their distribution in the Lachlan Fold Belt, southeastern Australia. *Geol. Soc. Am. Mem.* 159: 21–34

White, A. J. R., Chappell, B. W., Cleary, J. R. 1974. Geologic setting and emplacement of some Australian Paleozoic batholiths and implications for intrusive mechanisms. *Pac. Geol.* 8: 159–71

White, A. J. R., Clemens, J. D., Holloway, J. R., Silver, L. T., Chappell, B. W., et al.

1986a. S-type granites and their probable absence in southwestern North America. *Geology* 14: 115–18

White, A. J. R., Clemens, J. D., Holloway, J. R., Silver, L. T., Chappell, B. W., et al. 1986b. Reply to "S-type granites and their probable absence in southwestern North America." *Geology* 14: 805–6

White, A. J. R., Clemens, J. D., Holloway, J. R., Silver, L. T., Chappell, B. W., et al. 1986c. Reply to "S-type granites and their probable absence in southwestern North America." *Geology* 14: 895

Whitney, J. A., Wenner, D. B. 1980. Petrology and structural setting of post-metamorphic granites of Georgia. In *Excursions in Southeastern Geology, Geol. Soc. Am. Ann. Meet., Atlanta,* ed. R. W. Frey, 2: 351–78. Falls Church, Va: Am. Geol. Inst.

Wise, W. S., Eugster, H. P. 1964. Celadonite: synthesis, thermal stability and occurrence. *Am. Mineral.* 49: 1031–83

Wones, D. R., Eugster, H. P. 1965. Stability of biotite: experiment, theory, and application. *Am. Mineral.* 50: 1228–72

Wyborn, D., Chappell, B. W., Johnston, R. M. 1981. Three S-type volcanic suites from the Lachlan Fold Belt, southeast Australia. *J. Geophys. Res.* 86B: 10,335–48

Wyllie, P. J. 1977. Crustal anatexis: an experimental review. *Tectonophysics* 43: 41–71

Zeck, H. P. 1970. An erupted migmatite from Cerro del Hoyazo, SE Spain. *Contrib. Mineral. Petrol.* 26: 225–46

Zen, E-an. 1985a. Muscovite. In *Yearbook of Science and Technology 1985,* pp. 283–87. New York: McGraw-Hill

Zen, E-an. 1985b. An oxygen buffer for some peraluminous granites and metamorphic rocks. *Am. Mineral.* 70: 65–73

Zen, E-an. 1986. Aluminum enrichment in silicate melts by fractional crystallization: some mineralogic and petrographic constraints. *J. Petrol.* 27: 1095–1117

Zen, E-an. 1987. Bedrock geology of the Vipond Park 15-minute, Stine Mountain 7-1/2 minute, and Maurice Mountain 7-1/2 minute quadrangles, Pioneer Mountains, Beaverhead County, Montana. *US Geol. Surv. Bull. 1625.* In press

Ann. Rev. Earth Planet. Sci. 1988. 16: 53–72

CHONDRITIC METEORITES AND THE SOLAR NEBULA[1]

John A. Wood

Harvard-Smithsonian Center for Astrophysics, 60 Garden Street, Cambridge, Massachusetts 02138

OVERVIEW

Chondrites are stony meteorites: $\sim 87\%$ of the meteorites that fall to Earth are chondrites. Their radiometric ages, which cluster around 4.55 Gyr (Dodd 1981, Chap. 6; Tilton 1988), are understood to define the time when the solar system formed. The abundance pattern of nonvolatile elements in chondrites corresponds closely to abundance patterns in the atmospheres of the Sun and many stars in the Galaxy (Cameron 1982), indicating that chondrites are condensed samples of undifferentiated cosmic matter. They are specimens of planetary material in very nearly the same state it assumed when the planets were first formed. In particular, they have never experienced planetary melting and the chemical fractionations that attend igneous processes.

Chondrites have broadly ultramafic compositions (Dodd 1981), consistent with the predominance of Mg, Si, and Fe in the elemental abundance pattern of undifferentiated cosmic material. The chief mineral constituents of chondrites are olivine, pyroxene, and Ni,Fe metal. Their textures are aggregational; they consist of mixtures in various proportions of three particulate components. (*a*) *Chondrules* are more or less spheroidal, ultramafic igneous bodies, roughly a millimeter in dimension. They are very abundant (30–70% by volume in the chondrites that this article discusses) and tend to dominate the textural patterns of their hosts. (*b*) *Refractory inclusions* (RI) are more irregular in shape and range in size from several centimeters down to the limit of microscopic visibility. Most of them contain enhanced concentrations of the most involatile elements (e.g. Al,

---

Ca, Ti), which appear in Si-poor minerals such as Mg-spinel and melilite that would not be stable if the RI had unfractionated compositions. RI are a minority component of chondrites, making up 0–10% by volume of various subtypes of chondrites. (*c*) *Matrix* is porous, fine-grained (submicron to 10 μm) mineral matter that fills the spaces between chondrules and inclusions if these are abundant, or that forms the greater part of the chondrite in some cases (Huss et al 1981, Scott et al 1988). The composition of most matrix is similar to that of the bulk chondrite. All three components are understood to have been formed, or to have been extensively altered from some earlier state, while dispersed in space (the matrix as dust particles), after which they aggregated into chondritic masses.

Stages in the History of Chondritic Material

A complex sequence of events is recorded in the microchemical, micro-isotopic, and textural patterns of chondrites. The following discrete stages can be resolved.

THE INTERSTELLAR PRELUDE   The various atoms that make up chondrites were not created when the solar system was. They had an existence prior to that, most of them presumably in dust grains that were part of a volume of interstellar material which collapsed to form the Sun and a protosolar nebula. Some refractory inclusions and matrix constituents appear to preserve presolar isotopic patterns (R. N. Clayton 1978, Anders 1987).

PROCESSING IN THE SOLAR NEBULA   The infant Sun is understood to have been surrounded for the first 10^5–10^6 yr by a rotating disk of gas and dust, the *solar nebula*. Disks are predicted by astrophysical models of star formation via self-gravitational collapse of interstellar gas and dust (Wood & Morfill 1988), and they have been observed by astronomers to accompany very young stars (Harvey 1985). Interstellar dust that joined the solar nebula instead of accreting directly to the Sun was transformed at high temperatures into the chondrules, inclusions, and matrix dust we see in chondrites.

AGGREGATION OF CHONDRITE COMPONENTS   After the chondrite components were formed, they aggregated into small (~ 100 km) planetesimals (Weidenschilling 1984, 1988). The accreting material was not uniform in character. The nature and degree of nebular processing and of fractionation associated with accretion varied with time and position in the nebula, so different batches of accreting chondritic material had characteristic populations of chondrules and refractory inclusions and differed somewhat from one another in chemical composition (Dodd 1981). Accretion must have followed promptly after the processing events that created chondritic components, or else these populations would have been mingled and lost (Wood 1985). Of the many chondrite varieties probably generated, samples of nine (coded CI, CM, CO, CV, H, L, LL, EH, EL) have fallen to Earth abundantly in recent centuries.

METAMORPHISM   Immediately after their accretion the chondrite planetesimals were heated internally, metamorphically altering the character of virtually all the chondrites that have been studied. The source of the heat is not known. Two forms of alteration have been observed in different chondrites: (a) anhydrous metamorphism at temperatures of 700–1200 K, which equilibrated mineral compositions and recrystallized textures (Van Schmus & Wood 1967, Dodd 1969, McSween et al 1988); and (b) low-temperature hydrotheral metamorphism that altered mafic minerals (especially finely divided matrix grains) into clay minerals (McSween 1979, Zolensky & McSween 1988). [Degrees of metamorphism ("facies") experienced are coded with numbers from 1 to 6: 1 signifies extensive hydrothermal alteration, 2 less of it; 3 stands for minimal and 6 for maximum anhydrous metamorphism. Chondrites are classified by a letter-number combination (e.g. CV3) that expresses their compositional class and degree of metamorphism.]

COLLISIONS AND FURTHER ACCRETION OF PLANETESIMALS   A great many chondritic planetesimals were at large in the early solar system, and collisions were frequent during and after their metamorphic heating. In some cases the planetesimals were disrupted and dispersed, but in others they merged and grew in size, ultimately attaining diameters of hundreds of kilometers. They became what we now call asteroids. The substance of the planetesimals was extensively shattered during their collisional history; many if not most chondrite specimens are breccias (fragmented but rehealed rocks) (Wahl 1952). Veins (fracture surfaces filled with shock-melted glass) are common, and other mineralogical evidence of shock pressures is abundant (Stöffler 1971, 1974, Stöffler et al 1988). Most chondrite breccias consist of fragments of a single chondrite type. Sometimes two meteorite types are mixed, presumably representing two dissimilar planetesimals that collided in space; sometimes a more varied population of fragments is present. Some chondrites are *regolith* breccias—shock-lithified assemblages of fragments and fine dust that were once part of unconsolidated residual debris layers (regoliths) on the surfaces of planetesimals, where they were broken and stirred by continuing impacts and bathed in the space environment (Housen & Wilkening 1982).

EJECTION AND ORBITAL EVOLUTION   The metamorphic heating cycle of chondrites was brief (10^7–10^8 yr). They spent the rest of geologic time buried in cold, inert planetesimals, disturbed only a few times by the shock effects of occasional collisions between these bodies. The collisions tended to break down the asteroids parental to chondrites into smaller and smaller fragments. Initially in near-circular heliocentric orbits between Mars and Jupiter, some of these fragments were perturbed by Jupiter's gravity (applied with a particular resonant frequency) into elliptical orbits that crossed Earth's orbit, and ultimately they were captured by Earth (Greenberg & Chapman 1983, Wetherill 1985). The time chondrites existed as small fragments (less than a few meters) in space before Earth capture was typically a few tens of million years (Crabb & Schultz 1981).

ATMOSPHERIC FLIGHT AND FALL TO EARTH   Meteorites plunge into the Earth's atmosphere at velocities of 12–16 km s^{-1}, and all but the largest are decelerated by air friction to the free-fall velocity by the time they reach the ground. During the early stages of deceleration, atmospheric friction heats and melts their surfaces, creating a bright fireball and a glowing trail (ReVelle 1979), but this hot material is ablated as fast as it is heated; no significant amount of heat has penetrated more deeply than $\sim$ 1 mm into the residual decelerated stony masses that reach the ground. Roughly 4000 chondrites, 1 kg or more in mass, fall to Earth each year (Halliday et al 1984). Only a small percentage of these are recovered.

Scope of This Article

To achieve some depth of discussion in the limited length of this article, I focus on only one major topic: the nebular history of planetary material, as recorded by the chondrites. The chondritic record of presolar history, the planetary history of chondritic material after it accreted, and the later collisional and dynamic vicissitudes of these rocks are not treated.

This means that only a small subset ($\sim$ 8%) of known chondrites figure in this discussion: the most primitive, i.e. least metamorphosed, of them (the chondrites of metamorphic grades 3 and 2). Higher degrees of metamorphism have erased many of the properties that were imparted to chondritic components by the nebula and have replaced them with planetary properties.

The Condensation Sequence

Before nebular processes can be discussed, an important concept of material behavior must be noted. At high temperatures silicates not only melt, but they also vaporize. Vaporization and recondensation played a major role in the high-temperature processing of chondrite components in the nebula. The equilibrium compositions of coexisting solids and vapors as a function of temperature can be calculated approximately from thermodynamic data. This has been done by Grossman (1972; see also Grossman & Larimer 1974). The results are traditionally described in the context of a cooling system that was hot enough at the outset to be totally vaporized: What is the sequence of minerals that condense out of the cooling gas, and at what temperatures do they appear, if equilibrium is maintained? The condensation sequence calculated for a system assumed to contain the cosmic abundances of the chemical elements, at an overall pressure of 10^{-3} atm, is shown in Table 1.

Several aspects of this concept should be underlined. First, the appearance of these minerals during cooling results from reactions between solids and gas as well as from simple condensation. Minerals high in the sequence do not persist at low temperatures; they tend to react with residual vapors to form new, lower temperature minerals. For example, at 1620 K condensed corundum begins to react with Ca and Si vapor to form gehlenitic melilite. When 1513 K is reached, the last of the surviving corundum is used up. Second, it is unfortunate that this list of minerals has come to be called "the condensation sequence," because the concept works just as well in reverse as an "evaporation sequence." Beginning with completely condensed cold cosmic matter, increases of temperature would cause the elements to vaporize selectively, leaving a shrinking solid residue of min-

Table 1 The equilibrium condensation sequence[a]

Temperature[b] (K)	Mineral
1758 (1513)	Corundum, Al_2O_3
1647 (1393)	Perovskite, $CaTiO_3$
1625 (1450)	Melilite, $Ca_2Al_2SiO_7$-$Ca_2MgSi_2O_7$
1513 (1362)	Spinel, $MgAl_2O_4$
1471	Fe,Ni metal
1450	Diopside, $CaMgSi_2O_6$
1444	Forsterite, Mg_2SiO_4
1362	Anorthite, $CaAl_2Si_2O_8$
1349	Enstatite, $MgSiO_3$
< 1000[c]	Alkali-bearing feldspar, $(Na,K)AlSi_3O_8$-$CaAl_2Si_2O_8$
< 1000[c]	Ferrous olivines, pyroxenes; $(Mg,Fe)_2SiO_4$, $(Mg,Fe)SiO_3$
700	Troilite, FeS
405	Magnetite, Fe_3O_4

[a] Major minerals that would condense from a gas having the cosmic proportions of chemical elements and a pressure of 10^{-3} atm (Grossman 1972, Fuchs et al 1973, Dodd 1981).

[b] In most cases, temperature at which condensation or reaction begins in a cooling system. At temperature in parentheses, reaction has completely used up the phase and converted it into other minerals.

[c] Temperatures at which substantial amounts of alkalis are incorporated by plagioclase and Fe^{2+} by mafic minerals.

erals that appear higher and higher in Table 1. Third, though chondrites consist mostly of objects (chondrules) that were melted in the nebula, nothing has been said about melt-vapor equilibrium. This is partly because such systems are thermodynamically intractable, but also because (as Table 1 shows) chondritic matter should vaporize at temperatures lower than its melting range (1500–2000 K). Thus chondrules appear not to be equilibrium systems.

This illustrates the fourth and most important aspect of the condensation sequence—that it applies only under equilibrium conditions. Nature often does something different than this because there is not enough time to achieve the equilibrium condition before a system becomes too cool to be reactive. The condensation sequence provides an indispensable first-order basis for understanding the thermal processing of chondrules and refractory inclusions in the nebula, but allowance should always be made for processes that were controlled by kinetics rather than equilibrium.

PROCESSING OF CHONDRITE COMPONENTS IN THE NEBULA

Chondrules

CHONDRULES AS IGNEOUS DROPLETS   Chondrites consist mostly of chondrules. The significance of the spheroidal shapes and igneous character of chondrules has long been appreciated: They were once dispersed molten droplets (Sorby 1877). An obvious explanation for these would be high-velocity impacts on planetary surfaces, which tend to melt rock and splash droplets of it up (e.g. Dodd 1971), but there are severe difficulties with this mechanism. Taylor et al (1983) have marshaled the evidence against it: the scarcity of "chondrules" on the Moon's extensively impacted surface; the compositional variability of chondrules (see below); the absence of gross fragments of suitable target rock among the meteorites; and the apparent absence of chondrules younger than ~4.4 Gyr. (The lunar surface was heavily bombarded until much later than this.) To these might be added the fact that some chondrules have igneous rims or multiply concentric internal structures that required the addition of successive layers of molten material, an effect very difficult to produce by impact melting.

The alternative is that chondrules were melted by high-temperature events or processes in the solar nebula; they represent a stage of the nebular processing that converted interstellar dust grains into planets. This conclusion has been reached by most workers in the field, but so far the nature and site of the formative process have defied understanding. Chondrules are such an abundant constituent of chondrites (up to 70% by volume) that their formation must be seen as a major, pervasive process that operated in early solar system history, not as some freakish occurrence.

The various chemical classes of primitive chondrites contain more or less distinctive populations of chondrules, differing somewhat from one another in their distributions of chondrule dimensions, morphologies, internal structures, compositions, and mineralogies. The chondrule-forming process must have operated in slightly different ways to produce these systematic differences, but there is widespread agreement that the same basic process (whatever it was) was at work.

TIME SCALE OF COOLING   The rate at which an igneous system cools and crystallizes affects several observable parameters: the size and morphology of crystals, and the extent of chemical zoning in them. Laboratory experiments in which melts of appropriate composition were cooled at various rates have established that cooling rates in the range 100–2000 K hr^{-1} were required to reproduce the textures and chemical zonations in most meteoritic chondrules (Hewins 1983, 1988). Thus chondrules cooled through their crystallization range in the order of an hour.

DISEQUILIBRIUM STATE OF CHONDRULES   The simplest model for the nebular environment in which chondrule processing took place would contain the cosmic proportions of chemical elements (Cameron 1982) and would have temperatures in the range required to melt and crystallize the chondrules [1300–1900 K (Hewins 1983)] and gas pressures near that predicted at the nebula midplane at 3-AU radial distance by astrophysical models [10^{-6}–10^{-5} atm (Wood & Morfill 1988)]. Molten chondrules would be out of equilibrium with this environment for several reasons. First, vapor pressures of Si, Mg, Fe, and other rock-forming elements in the model defined are orders of magnitude lower than vapor pressures in equilibrium with the melt; the chondrules should vaporize away, and experiments indicate they would do so in ~ 1 hr (Hashimoto 1983). Second, the cosmic O/H ratio ($\sim 7 \times 10^{-4}$) defines a redox state that is too reducing to be in equilibrium with the Fe^{2+} content of minerals in most chondrules. Olivine in equilibrium with a cosmic gas should contain only ~ 0.03 mol% Fe_2SiO_4 at 1800 K, ~ 0.07 mol% at 1400 K; fayalite in excess of this should be reduced by the gas to Fe-metal plus $MgSiO_3$. A few chondrules contain olivine almost as Fe-poor as this (Johnson 1986), but the great bulk of them are more ferrous, up to Fa_{47} (Dodd et al 1967).

The discrepancy in vapor pressures appears to exclude the possibility that chondrules condensed from a cooling vapor. The alternative is that they were formed by the melting of some solid precursor material, and the fact that they did not vaporize must mean that the duration of heating and melting was short, less than ~ 1 hr (which is consistent with the cooling time scale inferred from crystal morphologies and zonations). Usually the melting was not complete: A large proportion of chondrules have been found to contain corroded *relict grains* of olivine and pyroxene dissimilar in composition to the igneous minerals surrounding them (e.g. Kracher et al 1984), which must be surviving precursor material. Also, laboratory experiments have shown that the textures of most chondrules cannot be reproduced by cooling charges that have been completely melted, i.e. that lack crystallization nuclei (Hewins 1983).

In many cases the discrepancy in oxidation state probably also results from thermal events too brief to allow melted ferrous precursor material to equilibrate with a more reducing environment. However, this does not appear to be the explanation in all cases. Olivine compositions in the chondrules of C2 chondrites peak very sharply at $\sim Fa_{0.5}$ (Wood 1967, Johnson 1986), a value that would be in equilibrium with an environment in which O/H was 5–20 times greater than the cosmic ratio. This is unlikely to be an example of incomplete reduction by an environment with cosmic O/H: There is no reason why reduction effective enough to bring the C2 olivines to such a uniform composition would stop short of the equilibrium value ($\sim Fa_{0.05}$). It is much more likely that the chondrules reacted with an environment that really had 5–20 times the cosmic ratio of O to H. Also, some chondrules have magnesium cores ($\sim Fa_6$) enclosed in more ferrous ($\sim Fa_{70}$) igneous rims (Kring 1987). In some cases another magnesian rim envelops the ferrous rim. This could be rationalized as resulting from coalescence of droplets that had not reacted at all with their environment, and which retained the compositional differences of their precursor materials; but a more straightforward interpretation is that the redox potential of the local environment fluctuated widely during chondrule formation.

CHEMICAL COMPOSITIONS   Most chondrules in chondrites of a given chemical class are rather similar in bulk composition: They contain a fairly constant 5–20 wt% of normative feldspathic constituents, and most or all of the remainder is oxides that go into mafic minerals (e.g. McSween 1977). Variability of SiO_2 content produces a spectrum of mafic mineral compositions from pure olivine to pure pyroxene, but (except in the E class) olivine-rich chondrules tend to be more abundant. FeO tends to be low, and the mafic minerals are Mg-rich varieties in most chondrules of CM2, CV3, and CO3 chondrites (McSween et al 1983), but in H3, L3, and LL3 (the unequilibrated ordinary chondrites, or UOC) the FeO content and mafic mineral compositions are highly variable (Dodd et al 1967).

A minor proportion of chondrules fall outside this simple pattern of compositions. A few percent have basaltic rather than ultramafic compositions. Less than a percent are highly siliceous ($>60\%$ SiO_2) and consist largely of glass (e.g. Fredriksson & Reid 1965); a few contain SiO_2 minerals (Brigham et al 1982). Rare chondrules are almost pure Fe,Ni metal, troilite, or chromite, or have bulk compositions so unorthodox that they give rise to hitherto unknown minerals [e.g. merrihueite, $(K,Na)_2(Fe,Mg)_5Si_{12}O_{30}$ (Dodd et al 1965)].

Apart from differences in the redox environment, what produced this compositional variability among chondrules? If the starting material was well-mixed, compositionally uniform, interstellar dust, and if formation occurred in the nebula (not on a planet, where igneous fractionation offers an explanation), the only possible mechanism seems to be volatility fractionation during the partial evaporation of chondrules that occurred while they were molten. Chondrules must have experienced some evaporative loss if they were melted in the space environment, but a relationship between chondrule compositions and volatile loss is not obvious. Hashimoto (1983) studied the fractionating effects of evaporation into vacuum of melts with cosmic proportions of Fe, Mg, Si, Ca, and Al oxides; the compositional trends he produced among residues do not correspond closely to trends among chondrule populations, especially in UOC. Surprisingly, chondrules still contain approximately their cosmic complement of Na, the most volatile of major lithophile elements. Grossman & Wasson (1983) have raised this as an objection to the idea of evaporative fractionation of chondrules, concluding that the observed compositional diversity must already have existed in the precursor material which was melted to form chondrules, and that chondrules were molten for too short a time to permit evaporation of Na or anything else.

However, the fact is that chondrules were molten for ~ 1 hr (Hewins 1983, 1988), long enough to permit substantial evaporation to occur (Hashimoto 1983); yet even the smallest chondrules have not been depleted in Na (Grossman & Wasson 1983). Tsuchiyama et al (1981) have shown that the evaporation rate of Na from chondrule melts is retarded by the presence of enhanced levels of O in the environment; as noted above, there is evidence that melting occurred in environments where O/H was greater than the cosmic value. Chondrules may have experienced evaporative fractionation under conditions more complex than those of the simple model nebular environment sketched at the beginning of the DISEQUILIBRIUM subsection, or in Hashimoto's experiments, or in our imaginations.

NONCOSMIC CHEMICAL ENVIRONMENTS   If the nebula contained O-rich zones, how could these have been created? Where temperatures were high enough to melt chondrules, they also would have been high enough to vaporize silicates and oxides. Most of the oxygen in the vaporizing silicates and oxides would have joined the gas phase as H_2O, effectively increasing O/H. Thus O/H was necessarily enhanced in zones of chondrule formation. The effect would have been slight, however, if the zone contained the cosmic proportions of gas and involatiles: In this case only $\sim 1/7$ of the system's O resides in its silicates and oxides when fully condensed, so total vaporization could not enhance O/H in the gas to more than 7/6 of its original value.

In zones where physical fractionation had concentrated involatile particles relative to gas prior to the heating event, on the other hand, even partial vaporization of the particles could drive O/H to values much greater than the cosmic ratio. Silicate particles that remained unvaporized (perhaps only the largest objects or aggregations that were in the original population of particles), now melted into droplets, would find themselves embedded in an oxidizing gas (Wood 1967). As the system cooled and recondensed, of course, O/H would decrease again. Thus vaporization and recondensation in dust-rich regions would have produced major fluctuations of O/H in the gas phase, of the sort that chondrule rims with widely varying olivine compositions (Kring 1987) and other nonequilibrium mineral assemblages in chondrites appear to require.

Particle/gas sorting in the nebula is not an improbable concept. If some of the chondrule precursor particles were relatively coarse (as many of the relict grains preserved in chondrules are), particle/gas fractionation inevitably would have occurred. The most obvious mechanism of fractionation would have been gravitational settling of solids toward the midplane of the nebula (Weidenschilling 1984, 1988), but other modes of concentration are also possible.

Refractory Inclusions

This category of objects in chondrites, also referred to as Ca,Al-rich inclusions (CAI), has been reviewed by Grossman (1980), Kornacki & Wood (1984), and MacPherson & Wark (1988). The topic is hard to come to grips with. The literature of refractory inclusions (RI) is mostly descriptive, consisting in large part of long, densely packed petrographic treatises. The properties of RI are so diverse that it is difficult to generalize about them. No conclusion can be drawn about one subset of RI that is not inconsistent with observations made in other inclusions. It is hard to isolate the scientific issues addressed by RI research, and it is hard even to define the class of objects referred to.

WHAT ARE REFRACTORY INCLUSIONS?   As already noted, they are structural entities in chondrites, most about the same size as chondrules (though a few are as large as several centimeters), that appear to have formed dispersed in space just as the chondrules did. As the name declares, most "refractory inclusions" are enriched in the least volatile elements (such as Ca, Al, Ti, REE), which appear in a characteristic set of refractory minerals (especially spinel, melilite, hibonite, fassaite, perovskite, diopside). However, some objects (amoeboid olivine aggregates) that are neither particularly rich in Ca and Al nor exceptionally refractory are also grouped with RI. Some RI have massive igneous textures; some are clearly aggregates of smaller (10–100 μm) structured entities (Cohen et al 1983); and some contain a second generation of minerals resulting from metasomatic alteration of earlier primary minerals. A refractory inclusion can have an ultramafic composition or it can be igneous, but it cannot be both: Igneous ultramafic objects in chondrites are chondrules, by definition. (This suggests that the distinction between RI and chondrules may be somewhat artificial.)

REFRACTORY INCLUSIONS AS NEBULAR CONDENSATES   Most meteorite researchers did not become aware of RI until 1969, when the fall of the Allende CV3 chondrite provided copious amounts of inclusion-rich material for them to study. At that time Cameron's (1962) first model of the solar nebula was widely accepted: In it the material of the nebula fell together abruptly and was heated by compression of the gas to > 1800 K. This is a high enough temperature to have completely vaporized any interstellar dust in the nebula. As the system radiated its heat and cooled, the metal and metal-oxide vapors would have recondensed. Under equilibrium conditions the minerals expected to form would have been those of the condensation sequence (Table 1; Grossman 1972). The minerals observed in Allende RI correspond strikingly to the highest-temperature minerals in the condensation sequence, which led meteoriticists to conclude that RI are specimens of the first material that condensed from the cooling nebula (e.g. Marvin et al 1970). The inclusions somehow had been spared from reacting further at lower temperatures and being converted to the minerals lower on the sequence.

In more recent years our view of the situation has matured, and the simple picture of condensation from a monotonically cooling nebula is no longer tenable. First, astrophysical models after that of Cameron (1962) conclude that the nebula was not abruptly compressed when it formed and was not heated to such high temperatures that interstellar grains in it were vaporized (Wood & Morfill 1988). High temperatures must have been achieved on a local scale to account for the properties of chondrules and inclusions, but there is no indication that *all* the interstellar dust which joined the nebula was vaporized and recondensed. Second, some RI contain O, Mg, Si, Ca, Sr, Ba, Nd, Sm (R. N. Clayton 1978), and Ti (Niederer et al 1981) with isotopic compositions that are not representative of solar system material. These inclusions must contain components that had special presolar nucleosynthetic histories, in nonrepresentative proportions. If everything had been vaporized in a hot nebula, mixing of the vapors would have wiped out presolar memories like these. Therefore, not all presolar grains were vaporized: Some survived to be incorporated in RI, which they made isotopically anomalous. Third, the genetic associations of minerals in most inclusions (the order in which they seem to have been formed) differ somewhat from the order predicted by the condensation sequence.

REFRACTORY INCLUSIONS AS EVAPORATION RESIDUES   The alternative to condensation is the opposite process, evaporation. Perhaps some aggregates of cold, fully condensed cosmic matter in the nebula were heated so intensely by the postulated transient local heating events that they were largely vaporized: The unvaporized residues, consisting mostly of involatile elements, are RI (Kurat 1970, Chou et al 1976, Hashimoto et al 1979). This model has several advantages over simple condensation. It becomes possible to understand the formation of melted inclusions even if the condensation sequence does not predict the existence of condensed matter at the temperature needed to melt them: Like chondrules, these inclusions were heated to melting for too short a time to allow complete evaporation. It is no longer necessary to worry about how the substance of inclusions got together—how slowly the nebula would have to cool in order for a condensing inclusion to come in contact with enough of the thin nebular gas to obtain all the atoms of condensable elements it contains. [This is a real problem for the condensation model. It would take a condensing 1-mm inclusion ~ 0.3 yr to accumulate its content of Ca from a nebula of cosmic composition at a pressure of 10^{-3} atm. Such a long time scale for initial condensation of the substance of the RI cannot be disproven, but it is inconsistent with the time scales established by petrologic experiments for chondrule cooling (minutes or hours) and the cooling of some igneous RI (hours or days) (Hewins 1983, 1988).]

In spite of these factors favoring evaporation over condensation, however, there is compelling evidence that condensation did in fact play an important role in the formation of RI. The strongest evidence comes from studies of REE abundances. Certain of the inclusions (Type II; Mason & Martin 1977) are greatly depleted in both the most volatile (Eu, Yb) and the most refractory (Er, Lu) rare earth elements, relative to other REE. The volatile depletion could be understood if the Type II inclusions had formed by partial evaporation, but the refractory depletion could not. The only plausible way to deplete the most refractory REE would be to physically remove them from a largely vaporized system (either as earliest condensate or evaporation residue), then condense the remaining less refractory elements into Type II inclusions (Boynton 1978).

It is clear at this point that both evaporation and condensation played a role in the formation of RI. (Among other considerations, if transient heating events evaporated major amounts of material to leave refractory residues, the vaporized matter would have had to condense when the systems cooled; some would condense as dust, but some would inevitably rejoin the residues it had come from.) Condensation and evaporation tend to fractionate the stable isotopes of a given element in different ways; the mass-dependent fractionation patterns of Mg, Si, Ca, and Ti observed in RI appear to result from complex, multistage histories probably involving both processes (R. N. Clayton et al 1985).

NONCOSMIC CHEMICAL ENVIRONMENTS   There is evidence that RI, like chondrules, were thermally processed in gaseous environments of non-cosmic, generally O-enriched composition. A variety of mineralogical and chemical properties of inclusions testify to this (Fegley & Palme 1985, Rubin et al 1988). For example, Fegley & Palme (1985) attribute the depleted levels of Mo and W in a series of inclusions they studied to the tendency of these elements to form volatile oxides under oxidizing conditions. They calculate that O/H in the environment would have to be 30–100 times the cosmic value to account for the observed depletions of Mo and W in RI. [Boynton (1985), however, notes that Ce also tends to form a volatile oxide, and he argues that Ce depletions in most RI are not large enough to support more than a factor of 10–100 enhancement of O/H over the cosmic ratio.]

Chondrules and Refractory Inclusions as Related Objects

The literature of meteoritics says little about a relationship between chondrules and RI. This is partly because individual meteoriticists tend to concentrate on one class of objects and ignore the other; partly because chondrules and inclusions seem bimodal in their properties—very few objects of intermediate nature are known; and partly because such unrelated origins have been pictured for them. Though it is almost never explicitly stated, the basic model that underlies most published meteoritic interpretations is as follows: At first everything was vaporized and homogenized in a hot Cameron (1962) nebula; as this cooled, RI condensed at high temperatures, and mafic dust (similar to, and perhaps identical with, chondrite matrix) at lower temperatures; transient heating events then melted aggregations of these condensates (especially the mafic dust) to form chondrules.

As already noted, however, this simplistic picture is outdated. The nebula was not hot enough where the chondrites formed to vaporize and homogenize everything. RI were not formed by condensation in a monotonically cooling gas of cosmic composition, but rather experienced complex histories of vaporization and condensation in environments of (apparently) noncosmic composition. Chondrules were molten for long enough times that they, too, must have experienced substantial vaporization, again in environments of noncosmic composition. The ways in which RI and chondrules were processed seem to have been very similar. It may turn out to be fruitful to regard both as having been formed by the same process, operating to different extents. Higher temperatures were not necessarily required to form RI, by partial evaporation of cosmic solids, instead of chondrules; longer thermal cycles, allowing more time for the loss of volatile constituents, may have sufficed.

EFFECTS OF A NONCOSMIC ENVIRONMENT   Repeated reference has been made to evidence that chondrules and inclusions were processed in noncosmic environments, almost always environments in which O/H was enhanced by one or more orders of magnitude over the cosmic value. As noted, these environments were probably created in dust-rich regions (e.g. near the nebula midplane) when heating events vaporized much of the dust and released its O into the gas phase. The gas phase also would have been enriched by similar factors in its content of metal vapors, of course. This enriched composition of the gas would have resulted in some major changes in the "rules of the meteorite game," which have been little appreciated:

1. Higher vapor pressures of the elements and volatile compounds cause condensed minerals to be stable at higher temperatures. Enrichment of metal vapors in the gas above the cosmic abundances could increase temperatures in the condensation sequence (Table 1) by as much as 600 K.

2. If the condensation temperatures in Table 1 are increased by $> \sim 200$ K, it changes the situation in a crucial way. Where previously the major igneous minerals diopside, forsterite, anorthite, and enstatite were stable only below temperatures at which significant melting occurs (~ 1500 K), now they are stable above this temperature. Therefore igneous melts in certain compositional ranges are stable against evaporation. Wark (1987) has argued that one class of RI (his type C) condensed as liquids under circumstances similar to these. It becomes possible that liquid condensation played a role in the formation of chondrules. Evaporative fractionation can explain the existence of chondrules in which Si/Mg is less than the cosmic value, since Si is a relatively volatile element; however, it

cannot explain the Si-rich (even silica-normative) chondrules that are also present in chondrites. Condensation could enhance the Si contents of molten droplets.

3. The vapor pressure changes contemplated would not just shift the condensation sequence upward in temperature, but they would also change its structure. For example, Hashimoto et al (1987) have shown that under oxidizing circumstances Ca tends to form volatile hydroxides and behaves as a rather volatile element instead of one of the most refractory. This would have a drastic effect on the structure of the condensation sequence.

OXYGEN ISOTOPES Chondrites, and the individual chondrules and inclusions in them, have stable O-isotope compositions that differ from terrestrial compositions and from each other in ways that cannot be attributed to simple mass-dependent fractionation (R. N. Clayton et al 1977, 1985). The effects are much larger than stable isotope anomalies reported in other elements (Mg, Si, Ca, etc), amounting to deviations of several percent from terrestrial standards. The first-order observation is that chondritic materials seem to have recorded mixing or reaction between two reservoirs, one rich in ^{16}O (apparently the condensed material) and the other poor (the nebular gas). Additional complexities in the O-isotope picture can probably be explained (though not easily) by multi-stage processes of reservoir exchange and mass-dependent fractionation associated with evaporation and condensation (R. N. Clayton, personal communication).

How were these two reservoirs created? A chemical process that fractionates ^{16}O from the other isotopes has been demonstrated in the laboratory by Thiemens & Heidenreich (1983). Associative reactions between O and oxide vapor molecules tend to favor the formation of ^{17}O- and ^{18}O-bearing polyatomic molecules (equally) over molecules containing only ^{16}O. Apparently, excited ^{16}O-only molecules formed by atom-molecule collisions have a smaller chance of stabilizing before they redissociate than other molecules because there are fewer diffuse energy levels associated with the symmetric ^{16}O-only molecule than with asymmetric molecules (Thiemens 1988). It has been suggested that this type of fractionation, operating during nebular gas-phase reactions such as $^{18}O + Si^{16}O$ (g) → $Si^{16}O^{18}O$ (g, asymmetric), might have produced the non-mass-dependent O fractionations observed in chondrites. However, no detailed model relating this mechanism to chondrite-forming processes has been developed. Reactions like the one just cited would appear to yield condensed [e.g. from the reaction product SiO_2 (g)] material that is impoverished rather than enriched in ^{16}O.

The alternative is that the two isotopically different O reservoirs were presolar in origin, like the Mg, Si, Ca, etc, stable isotope anomalies discussed earlier; the nucleosynthetic history of interstellar dust incorporated by the solar system had been different from that of the gas accompanying it, such that the dust had a larger component of ^{16}O than the gas. This interpretation has been most widely discussed in the literature. The O-isotope anomalies are larger than other presolar stable isotope anomalies, presumably because O is the only element that is abundant in two reservoirs (gas and oxide solids) over a wide range of temperatures (Wood 1981). Other elements (Mg, Si, Ca, etc) are wholly vaporized at some temperatures, wholly condensed at others; in either case, isotopic homogenization is facilitated.

THE ISOTOPIC EFFECT OF OXYGEN ENRICHMENTS IN THE NEBULA   If the chondrules and RI reacted chemically with gaseous environments enriched by large and variable factors in O, as has been suggested earlier, this must have profoundly affected the establishment of the ^{16}O mixing lines observed in chondrites. In a system where solids had been concentrated to > 10 times the cosmic abundance before they were in large part evaporated by a heating event, the gas phase would contain more O that had been derived from the solids than O that had been brought into the nebula from the interstellar gas. The isotopic composition of O in the gas phase of such a system would be dominated by the contribution from vaporized solids. Any difference that had originally existed between the isotopic composition of O in the gas and the solid, of the sort that seems required to explain the ^{16}O mixing line observed in primitive chondrites (see the previous paragraph), would be largely washed out. This is, perhaps, the most serious objection to the concept that chondrules and RI were thermally processed in environments where O had been enriched by the evaporation of concentrated dust (E. Anders, personal communication).

On the other hand, this aspect of the oxygen situation may be a source of insight rather than an objection. The problem would really only arise in the case where the solids concentrated and vaporized were fresh, (presumably) ^{16}O-rich interstellar grains. Then the gas would become ^{16}O rich too, and ^{16}O-poor minerals that occur on the mixing lines observed by R. N. Clayton et al (1977, 1985) could not be explained. However, it is possible that most of the solid particles in the nebula at any given time were not "fresh," but were the product of multiple cycles of vaporization and condensation. Such grains could not have retained an ^{16}O-rich character. In the final analysis the bulk nebula contains much more O that came into it from interstellar gas than from interstellar involatile grains, so repeated vaporization and condensation would have the effect of bringing the isotopic composition of O in reprocessed grains to the ^{16}O-poor value of the average gas. Concentration and evaporation of these reprocessed grains would elevate the O content of the local gas without increasing its ^{16}O content importantly. It may be that RI, which display O isotope mixing lines spanning a very wide range of ^{16}O contents, were formed when aggregations of fresh, ^{16}O-rich interstellar grains were introduced into, and reacted with, environments that had been more or less enriched in ^{16}O-poor O by the evaporation of reprocessed grains.

SOURCE OF HEAT THAT PROCESSED CHONDRULES AND INCLUSIONS   Model studies of the solar nebula by astrophysicists have shown that the disk would have evolved in a way that dissipated mechanical energy as heat, but that the temperature attributable to this effect at the radial distance where the chondrites formed was only a few hundred degrees Kelvin, far less than the temperature required to melt chondrules (Wood & Morfill 1988). Some other pervasive high-energy process or series of events, outside the evolutionary studies of astrophysicists, operated to process the nebula's complement of interstellar dust into chondrules and other chondrite components. The relatively short time scale of cooling of chondrules requires that these events were transient and local: If the whole nebula was heated, chondrules in it could not cool in ∼ 1 hr.

Transient nebular heating events that have been suggested are lightning discharges (Whipple 1966, Cameron 1966); energy released by the reconnection of twisted magnetic field lines (Sonett 1979, Levy 1988); the conversion of kinetic energy to heat by infalling interstellar matter when it strikes the nebula surface (Larson 1972, Wood 1986); aerodynamic drag heating experienced by presolar dust aggregates when they fell into the nebula (Wood 1984); and the exothermic reaction in grain aggregates of unstable mixtures of presolar oxides, once triggered in the nebula (D. D. Clayton 1980). There are difficulties with all these models; none has won wide acceptance among meteoriticists. The source of heat that melted the chondrules is still an open question.

Conclusion

We remain profoundly ignorant of how the solar system's oldest rocks were formed. This writer believes it happened in dynamic, energetic, dust-rich zones of the nebula, where local dust/gas ratios were constantly changing and undefined releases of energy caused temperatures to fluctuate through ∼ 1000 K. Mineral material caught in this maelstrom went through multiple cycles of melting, evaporation, recondensation, crystallization, and aggregation, forming in different circumstances all the components of chondrites: chondrules, RI, matrix dust. Newly accreted interstellar grains constantly joined the melee, so at any time there was always a component of condensed material that still retained presolar isotopic anomalies. Periodically, batches of chondritic material were removed from these systems by being rapidly aggregated into small planetesimals.

ACKNOWLEDGMENTS

I am grateful to R. N. Clayton, J. F. Kerridge, A. Hashimoto, D. A. Kring, and G. W. Wetherill for discussions of chondrites and other forms of help. This work was supported in part by NASA Grant NAG 9-28.

Literature Cited

Anders, E. 1987. Local and exotic components of primitive meteorites, and their origin. *Philos. Trans. R. Soc. London Ser. A.* In press

Black, D. C., Matthews, M. S., eds. 1985. *Protostars and Planets II.* Tucson: Univ. Ariz. Press. 1293 pp.

Boynton, W. V. 1978. The chaotic solar nebula: evidence for episodic condensation in several distinct zones. In *Protostars and Planets*, ed. T. Gehrels, pp. 427–38. Tucson: Univ. Ariz. Press. 756 pp.

Boynton, W. V. 1985. Meteoritic evidence concerning conditions in the solar nebula. See Black & Matthews 1985, pp. 772–87

Brigham, C., Murrell, M. T., Burnett, D. S. 1982. SiO₂-rich chondrules in ordinary chondrites. *Conf. Chondrules and Their Origins*, p. 4. Houston: Lunar Planet. Inst. (Abstr.)

Cameron, A. G. W. 1962. The formation of the Sun and planets. *Icarus* 1: 13–69

Cameron, A. G. W. 1966. The accumulation of chondritic material. *Earth Planet. Sci. Lett.* 1: 93–96

Cameron, A. G. W. 1982. Elemental and nuclidic abundances in the solar system. In *Essays in Nuclear Astrophysics*, ed. C. A. Barnes, D. D. Clayton, D. Schramm, pp. 23–43. Cambridge: Cambridge Univ. Press. 562 pp.

Chou, C.-L., Baedecker, P. A., Wasson, J. T. 1976. Allende inclusions: volatile-element distribution and evidence for incomplete volatilization of presolar solids. *Geochim. Cosmochim. Acta* 40: 85–94

Clayton, D. D. 1980. Chemical energy in cold-cloud aggregates: the origin of meteoritic chondrules. *Astrophys. J. Lett.* 239: L37–41

Clayton, R. N. 1978. Isotopic anomalies in the early solar system. *Ann. Rev. Nucl. Part. Sci.* 28: 501–22

Clayton, R. N., Onuma, N., Grossman, L.,

Mayeda, T. K. 1977. Distribution of the presolar component in Allende and other carbonaceous chondrites. *Earth Planet. Sci. Lett.* 34: 209–24

Clayton, R. N., Mayeda, T. K., Molini-Velsko, C. A. 1985. Isotopic variations in solar system material: evaporation and condensation of silicates. See Black & Matthews 1985, pp. 755–71

Cohen, R. E., Kornacki, A. S., Wood, J. A. 1983. Mineralogy and petrology of chondrules and inclusions in the Mokoia CV3 chondrite. *Geochim. Cosmochim. Acta* 47: 1739–57

Crabb, J., Schultz, L. 1981. Cosmic-ray exposure ages of the ordinary chondrites and their significance for parent body stratigraphy. *Geochim. Cosmochim. Acta* 45: 2151–60

Dodd, R. T. 1969. Metamorphism of the ordinary chondrites: a review. *Geochim. Cosmochim. Acta* 33: 161–205

Dodd, R. T. Jr. 1971. The petrology of chondrules in the Sharps meteorite. *Contrib. Mineral. Petrol.* 31: 201–27

Dodd, R. T. Jr. 1981. *Meteorites.* Cambridge: Cambridge Univ. Press. 368 pp.

Dodd, R. T. Jr., Van Schmus, W. R., Marvin, U. B. 1965. Merrihueite, a new alkali-ferromagnesian silicate from the Mezö-Maderas chondrite. *Science* 149: 972–74

Dodd, R. T. Jr., Van Schmus, W. R., Koffman, D. M. 1967. A survey of the unequilibrated ordinary chondrites. *Geochim. Cosmochim. Acta* 31: 921–51

Fegley, B. Jr., Palme, H. 1985. Evidence for oxidizing conditions in the solar nebula from Mo and W depletions in refractory inclusions in carbonaceous chondrites. *Earth Planet. Sci. Lett.* 72: 311–26

Fredriksson, K., Reid, A. M. 1965. A chondrule from the Chainpur meteorite. *Science* 149: 856–60

Fuchs, L. H., Olsen, E., Jensen, K. J. 1973. Mineralogy, mineral-chemistry, and composition of the Murchison (C2) meteorite. *Smithsonian Contrib. Earth Sci. No. 10.* 39 pp.

Greenberg, R., Chapman, C. R. 1983. Asteroids and meteorites: parent bodies and delivered samples. *Icarus* 55: 455–81

Grossman, J. N., Wasson, J. T. 1983. The compositions of chondrules in unequilibrated chondrites: an evaluation of models for the formation of chondrules and their precursor materials. See King 1983, pp. 88–121

Grossman, L. 1972. Condensation in the primitive solar nebula. *Geochim. Cosmochim. Acta* 36: 597–619

Grossman, L. 1980. Refractory inclusions in the Allende meteorite. *Ann. Rev. Earth Planet. Sci.* 8: 559–608

Grossman, L., Larimer, J. W. 1974. Early chemical history of the solar system. *Res. Geophys. Space Phys.* 12: 71–101

Halliday, I., Blackwell, A. T., Griffin, A. A. 1984. The frequency of meteorite falls on the Earth. *Science* 223: 1405–7

Harvey, P. M. 1985. Observational evidence for disks around young stars. See Black & Matthews 1985, pp. 484–92

Hashimoto, A. 1983. Evaporation metamorphism in the early solar nebula— evaporation experiments on the melt FeO-MgO-SiO$_2$-CaO-Al$_2$O$_3$ and chemical fractionations of primitive materials. *Geochem. J.* 17: 111–45

Hashimoto, A., Kumazawa, M., Onuma, N. 1979. Evaporation metamorphism of primitive dust material in the early solar nebula. *Earth Planet. Sci. Lett.* 43: 13–21

Hashimoto, A., Wood, J. A., Weinberg, A. 1987. Experimental determination of Ca(OH)$_2$ vapor pressure as a key to understanding the alteration of Ca,Al-rich inclusions. *Meteoritics.* In press

Hewins, R. H. 1983. Dynamic crystallization experiments as constraints on chondrule genesis. See King 1983, pp. 122–33

Hewins, R. H. 1988. Experimental studies of chondrules. See Kerridge & Matthews 1988. In press

Housen, K. R., Wilkening, L. L. 1982. Regoliths on small bodies in the solar system. *Ann. Rev. Earth Planet. Sci.* 10: 355–76

Huss, G. R., Keil, K., Taylor, G. J. 1981. The matrices of unequilibrated ordinary chondrites: implications for the origin and history of chondrites. *Geochim. Cosmochim. Acta* 45: 33–51

Johnson, M. C. 1986. The solar nebula redox state as recorded by the most reduced chondrules of five primitive chondrites. *Geochim. Cosmochim. Acta* 50: 1497–1502

Kerridge, J. F., Matthews, M. S., eds. 1988.

Meteorites and the Early Solar System. Tucson: Univ. Ariz. Press. In press

King, E. A., ed. 1983. *Chondrules and their Origins.* Houston: Lunar Planet. Inst. 377 pp.

Kornacki, A. S., Wood, J. A. 1984. Petrography and classification of Ca,Al-rich and olivine-rich inclusions in the Allende CV3 chondrite. *Proc. Lunar Planet. Sci. Conf., 14th, J. Geophys. Res.* 89: B573–87 (Suppl.)

Kracher, A., Scott, E. R. D., Keil, K. 1984. Relict and other anomalous grains in chondrules: implications for chondrule formation. *Proc. Lunar Planet. Sci. Conf., 14th, J. Geophys. Res.* 89: B559–66 (Suppl.)

Kring, D. A. 1987. Fe,Ca-rich rims around magnesium chondrules in the Kainsaz (CO3) chondrite. *Lunar Planet. Sci. XVIII,* pp. 517–18. Houston: Lunar Planet. Inst. 1141 pp.

Kurat, G. 1970. Zur Genese der Ca-Al-reichen Einschlüsse im Chondriten von Lancé. *Earth Planet. Sci. Lett.* 9: 225–31

Larson, R. B. 1972. Collapse calculations and their implications for the formation of the solar system. In *l'Origin du Systèm Solaire,* ed. H. Reeves, pp. 142–53. Paris: CNRS. 383 pp.

Levy, E. 1988. Energetics of chondrule formation. See Kerridge & Matthews 1988. In press

MacPherson, G. J., Wark, D. A. 1988. Primitive material surviving in the chondrites: inclusions. See Kerridge & Matthews 1988. In press

Marvin, U. B., Wood, J. A., Dickey, J. S. Jr. 1970. Ca-Al rich phases in the Allende meteorite. *Earth Planet. Sci. Lett.* 7: 346–50

Mason, B., Martin, P. M. 1977. Geochemical differences among components of the Allende meteorite. *Smithsonian Contrib. Earth Sci.* 19: 84–95

McSween, H. Y. Jr. 1977. Chemical and petrographic constraints on the origin of chondrules and inclusions in carbonaceous chondrites. *Geochim. Cosmochim. Acta* 41: 1843–60

McSween, H. Y. Jr. 1979. Are carbonaceous chondrites primitive or processed? A review. *Rev. Geophys. Space Phys.* 17: 1059–78

McSween, H. Y. Jr., Fronabarger, A. K., Driese, S. G. 1983. Ferromagnesian chondrules in carbonaceous chondrites. See King 1983, pp. 195–210

McSween, H. Y. Jr., Sears, D. W. G., Dodd, R. T. 1988. Secondary processing: thermal metamorphism. See Kerridge & Matthews 1988. In press

Niederer, F. R., Papanastassiou, D. A.,

Wasserburg, G. J. 1981. The isotopic composition of titanium in the Allende and Leoville meteorites. *Geochim. Cosmochim. Acta* 45: 1017–31

ReVelle, D. O. 1979. A quasi-simple ablation model for large meteorite entry: theory vs observations. *J. Atmos. Terr. Phys.* 41: 453–73

Rubin, A. E., Fegley, B., Brett, R. 1988. Oxidation state in chondrites. See Kerridge & Matthews 1988. In press

Scott, E. R. D., Peck, J. A., Alexander, C., Barber, D., Mackinnon, I. 1988. Primitive material surviving in chondrites: matrix. See Kerridge & Matthews 1988. In press

Sonett, C. P. 1979. On the origin of chondrules. *Geophys. Res. Lett.* 6: 677–80

Sorby, H. C. 1877. On the structure and origin of meteorites. *Nature* 15: 495–98

Stöffler, D. 1971. Deformation and transformation of rock-forming minerals by natural and experimental shock processes. I. Behavior of minerals under shock compression. *Fortschr. Mineral.* 49: 50–113

Stöffler, D. 1974. Deformation and transformation of rock-forming minerals by natural and experimental shock processes. II. Physical properties of shocked minerals. *Fortschr. Mineral.* 51: 256–89

Stöffler, D., Bischoff, A., Buchwald, V., Rubin, A. E. 1988. Shock effects in meteorites. See Kerridge & Matthews 1988. In press

Taylor, G. J., Scott, E. R. D., Keil, K. 1983. Cosmic setting for chondrule formation. See King 1983, pp. 262–78

Thiemens, M. H. 1988. Heterogeneity in the nebula: evidence from stable isotopes. See Kerridge & Matthews 1988. In press

Thiemens, M. H., Heidenreich, J. E. 1983. The mass independent fractionation of oxygen: a novel isotope effect and its possible cosmochemical implications. *Science* 219: 1073–75

Tilton, G. R. 1988. Age of the solar system. See Kerridge & Matthews 1988. In press

Tsuchiyama, A., Nagahara, H., Kushiro, I. 1981. Volatilization of sodium from silicate melt spheres and its application to the formation of chondrules. *Geochim. Cosmochim. Acta* 45: 1357–67

Van Schmus, W. R., Wood, J. A. 1967. A chemical-petrologic classification for the chondritic meteorites. *Geochim. Cosmochim. Acta* 31: 747–65

Wahl, W. 1952. The brecciated stony meteorites and meteorites containing foreign fragments. *Geochim. Cosmochim. Acta* 2: 91–117

Wark, D. A. 1987. Plagioclase-rich inclusions in carbonaceous chondrite meteorites: liquid condensates? *Geochim. Cosmochim. Acta* 51: 221–42

Weidenschilling, S. J. 1984. Evolution of grains in a turbulent solar nebula. *Icarus* 60: 553–67

Weidenschilling, S. J. 1988. Formation processes and timescale for meteorite parent bodies. See Kerridge & Matthews 1988. In press

Wetherill, G. W. 1985. Asteroidal source of ordinary chondrites. *Meteoritics* 20: 1–22

Whipple, F. L. 1966. Chondrules: suggestion concerning their origin. *Science* 153: 54–56

Wood, J. A. 1967. Olivine and pyroxene compositions in Type II carbonaceous chondrites. *Geochim. Cosmochim. Acta* 31: 2095–2108

Wood, J. A. 1981. The interstellar dust as a precursor of Ca,Al-rich inclusions in carbonaceous chondrites. *Earth Planet. Sci. Lett.* 56: 32–44

Wood, J. A. 1984. On the formation of meteoritic chondrules by aerodynamic drag heating in the solar nebula. *Earth Planet. Sci. Lett.* 70: 11–26

Wood, J. A. 1985. Meteoritic constraints on processes in the solar nebula. See Black & Matthews 1985, pp. 687–702

Wood, J. A. 1986. High temperatures and chondrule formation in a turbulent shear zone beneath the nebula surface. *Lunar Planet. Sci. XVII*, pp. 956–57. Houston: Lunar Planet. Inst. 1014 pp.

Wood, J. A., Morfill, G. E. 1988. A review of solar nebula models. See Kerridge & Matthews 1988. In press

Zolensky, M. E., McSween, H. Y. Jr. 1988. Secondary processes: aqueous alteration. See Kerridge & Matthews 1988. In press

Ann. Rev. Earth Planet. Sci. 1988. 16: 73–99

VOLCANIC WINTERS[1]

Michael R. Rampino

Department of Applied Science, New York University, New York, NY 10003 and NASA Goddard Space Flight Center, Institute for Space Studies, New York, NY 10025

Stephen Self

Department of Geology, University of Texas at Arlington, Arlington, Texas 76019

Richard B. Stothers

NASA Goddard Space Flight Center, Institute for Space Studies, New York, NY 10025

INTRODUCTION: THE VOLCANO/CLIMATE CONNECTION

Accounts of prolonged darkness, often associated with abnormally cold weather and hardship, are common in the myths and legends of many cultures. Egyptian papyruses corroborate the statement in the Book of Exodus in the Bible that "there was a thick darkness in all the land of Egypt for three days." Similar kinds of stories can be found in ancient Sumerian, Greek, and Mayan literature. These have on occasion been used to argue for global catastrophes, such as encounters or collisions with comets, in early historical times. A more reasonable explanation comes from the similarity of these reports to more recent historical accounts of the aftereffects of large volcanic eruptions. Thus, a substantial case has been made for the connection of the Egyptian and Biblical reports of darkness and ash rains at the time of the Exodus with the explosive

---

[1] The US Government has the right to retain a nonexclusive royalty-free license in and to any copyright covering this paper.

eruption of Santorini (Thera) in the Aegean Sea in the second millenium BC (see references in Downey & Tarling 1984, Stanley & Sheng 1986), which also somehow contributed to the demise of Minoan Crete (Marinatos 1939) and had effects as far away as China (Pang & Chou 1985). In more recent times, the inhabitants of interior New Guinea speak of the "time of darkness," a tradition passed down over generations since the seventeenth century AD. Blong (1982) has shown that this darkness was accompanied by large local temperature changes, both hot and cold, and was most likely related to the ash cloud from an eruption of Long Island, one of the active volcanoes off the northeast coast of New Guinea.

Considering the knowledge that we have gained in the past few decades about the mechanisms of volcanic eruptions and the generation of volcanic aerosol clouds in the atmosphere, it is reasonable to ask what the atmospheric effects of the largest eruptions might be. Beyond the local devastation and regional effects, it is known that some historical eruptions had a noticeable impact on climate and agriculture on a hemispheric to global basis. This being the case, much larger eruptions may possibly have caused severe "volcanic winters," perhaps similar to the recently proposed "nuclear winter." These "supereruptions" must therefore be considered in discussions of natural hazards that might have global consequences (Rampino et al 1985, Burke & Francis 1985, Smith 1985).

Modern interest in the problem of the impact of volcanic eruptions on the atmosphere and climate is traditionally traced back to the observations of Benjamin Franklin at the time of the eruption of Laki (Lakagígar) in Iceland in 1783. Franklin (see Lamb 1970, p. 433) described what he termed a "dry fog" in Europe during his stay there as minister to France. As Franklin wrote, the rays of the Sun "were indeed rendered so faint in passing through it that, when collected in the focus of a burning glass, they would scarce kindle brown paper." Franklin connected the dry fog and the reduced solar radiation with the severe winter of 1783/1784 in Europe and eastern North America, and he proposed that the Icelandic eruption of that time was to blame.

The connection between large volcanic eruptions and worldwide perturbations of the optical properties of the atmosphere was established by the classic study of the Krakatoa Commission (Russell & Archibald 1888) in the aftermath of the changes in the atmosphere seen after the Krakatau eruption in 1883 (see also Simkin & Fiske 1983). A number of atmospheric optical phenomena were identified, including noticeable dimming and blurring of celestial objects, unusual blue or green color of the Sun and Moon, enhanced sunrises and sunsets with lavender glows that appeared high over the horizon, Bishop's rings (a complex halo around the Sun produced by diffraction of sunlight by small particles, in which the normal order of colors is reversed, with red on the outside), and also unusually dark lunar eclipses (Flammarion 1884). A link between climate change and volcanic eruptions was also based on later studies that suggested a possible correlation between some eruptions, decreases in solar radiation measured at ground observatories, and short-term coolings of the Earth's surface (Humphreys 1913, Abbot & Fowle 1913; for an early review, see Humphreys 1940); however, in other similar studies the same or other noteworthy eruptions seemed to show no noticeable effect on the global climate (Gentilli 1948, Deirmendjian 1973, Landsberg & Albert 1974, Ellsaesser 1986).

In 1970, a classic study by H. H. Lamb clearly presented the empirical evidence for a volcano/climate connection as understood at that time. Lamb reviewed previous work on the subject, but his most valuable contribution was his tabulation of a chronology of important volcanic eruptions for the period subsequent to AD 1500 and his definition of the volcanic dust veil index (d.v.i.), an estimate of the amount of fine volcanic ash or dust lofted into the upper atmosphere by specific historical eruptions. Lamb concluded that some significant correlations existed between "volcanic eruption years" with high d.v.i. values and climatic cooling, but he stressed that in some cases where the d.v.i. was assessed largely on the evidence of temperature variation, one was in clear danger of arguing in a circle when investigating the possible effects of eruptions on climate. Newhall & Self (1982; see also Simkin et al 1981) attempted to further quantify volcanic "explosivity" in their Volcanic Explosivity Index (VEI), which combined estimates of eruption volume with explosive energy as evidenced by the height of the eruption plume. Hirschboeck (1980) proposed another, but simpler, index based largely on the volume of the eruption. By that time, however, it was clear that the composition of the volcanic ejecta, particularly the amount of sulfur volatiles released, had an importance above and beyond that of the total amount of ash ejected. The geographic location, time of year, and prevailing climatic conditions (e.g. phase in the quasi-biennial oscillation cycle) were also seen to be critical factors in determining the spread and lifetime of volcanic aerosol clouds.

NINETEENTH AND TWENTIETH CENTURY ERUPTIONS: SULFUR IS THE KEY

Although most workers prior to the late 1960s stressed the importance of "volcanic ash" in the stratospheric clouds (e.g. Jacobs 1954, Mitchell 1961), Lamb (1970) and Deirmendjian (1973) both recognized a possible connection between sulfur gases (primarily SO_2 and H_2S) injected into the upper atmosphere by volcanic eruptions and the evidence discovered by Junge et al (1961) for a supposedly permanent layer of sulfate aerosols (consisting largely of small droplets of sulfuric acid) in the stratosphere at around 25 km. It was later shown that the majority of the H_2SO_4 aerosols in the so-called Junge Layer were volcanic in origin (Castleman et al 1974).

A number of studies followed that focused on photochemical reactions and nucleation of sulfuric acid aerosols in the lower stratosphere. It was soon well established that the bulk of volcanic "dust" veils consisted of fine droplets of sulfuric acid [see Turco et al (1982) for a review]. Most of the volcanic ash fell out of the stratosphere in a few months, while the aerosols continued to nucleate and grow, creating the volcanic cloud that spread over wide areas of the globe and persisted for several years (Pollack et al 1976, Cadle et al 1976, 1977, Hunt 1977, Capone et al 1983). In addition, HCl and water vapor that are injected into the stratosphere during an eruption may have significant effects on the ozone concentrations (Hofmann 1987, Pinto et al 1987).

Stratospheric aerosols affect the global radiation budget by absorbing and, more importantly, backscattering incoming solar radiation (although they also absorb some outgoing infrared radiation from the ground). Absorption and backscattering of solar radiation should cause a cooling of the lower atmosphere and the surface. The absorption of infrared radiation should also cause an increase in the stratospheric temperatures (see Turco et al 1982). The volcanic signal expected in hemispheric or zonal surface temperature records in historical times, however, is about the same as the background interannual variations in temperature. Several studies have made use of the method of "compositing" or "superposed epoch analysis," in which the temperature records of several years bracketing a number of different eruptions are superposed in order to strengthen the contrast between the possible volcanic signal and background noise (Figure 1). These studies identified a statistically significant average temperature decrease of about 0.2 to 0.5°C for 1 to 3 years following the times of known nineteenth and twentieth century eruptions (Mitchell 1961, Mass & Schneider 1977, Taylor et al 1980, Self et al 1981, Angell & Korshover 1985, Lough & Fritts 1987). Other statistical studies have come to similar conclusions regarding the magnitude and duration of cooling after volcanic eruptions of the past 100 years (for a review, see Angell & Korshover 1985). A recent superposed epoch analysis of Northern Hemisphere sea-surface temperatures after major eruptions of the last 100 years, however, did not show any consistent response of posteruption cooling (Parker 1985), but sea-surface temperatures are expected to be less responsive to short-term temperature perturbations, and other factors such as the Southern Oscillation/El Niño phenomenon may be masking the volcanic climate signal.

Another approach is to focus on the largest and/or best documented

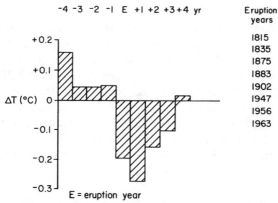

Figure 1 Composite plot of the temperature departure for the Northern Hemisphere in the four years immediately before and after some large nineteenth and twentieth century eruptions (after Self et al 1981).

volcanic perturbations and to reexamine them in detail. Hansen et al (1978) examined the 1963 eruption of Mt. Agung on the Indonesian island of Bali, which occurred at a time when tropospheric and stratospheric temperatures were being routinely measured (Newell 1970, Newell & Weare 1976) and accurate measurements of aerosol optical depth were being made at a number of observatories (Volz 1970). [Optical depth is equal to the negative natural logarithm of the attenuation of incident light, or $\tau = -\ln(I/I_0)$, where I_0 and I are the initial and final light intensity, respectively.] High-altitude aircraft above 20 km could also directly collect stratospheric aerosols for analysis (Mossop 1964).

Hansen et al (1978) calculated the expected effect on the stratospheric and tropospheric temperatures by using the measured time history of aerosol optical depth and a simple one-dimensional radiative/convective climate model. They then compared these calculated temperature changes with the observed temperature perturbations. The theoretical results agreed with the observation that stratospheric temperatures rose by 4 to 8°C in the region from 10°N to 30°S, while surface temperatures in the region from 30°N to 30°S showed a decrease of a few tenths of a degree over a period on the order of a year. Hansen et al (1981) later confirmed, to a large extent, their earlier theoretical results for surface cooling as well as the simpler box-model predictions of Schneider & Mass (1975) and Harshvardhan & Cess (1976). Further studies of the effects of volcanic aerosols on climate using more sophisticated two-dimensional climate models have in general confirmed the empirical evidence for cooling of a few tenths of a degree Celsius following large historic eruptions (Robock 1981, 1984, Chou et al 1984, McCracken & Luther 1984).

A major finding in both the statistical and individual studies of volcanic perturbations of the atmosphere is that relatively small volcanic eruptions (measured by the total volume of magma ejected as pumice and ash) such as that of Mt. Agung in 1963 [estimated volume of ejected magma of only 0.3–0.6 km^3, dense-rock equivalent (DRE); Rampino & Self 1984a] can lead to a relatively dense aerosol cloud, totaling perhaps 10 megatons (Mt) of H_2SO_4 aerosols in the specific case of Agung [at least in the Southern Hemisphere; the Northern Hemisphere stratosphere contained only about one fifth of this mass of aerosols (see Table 1)]. By contrast, the much larger Krakatau eruption in 1883 [~ 10 km^3 (DRE) of ejected magma] produced a cloud only five times more massive (~ 50 Mt of H_2SO_4 aerosols). Obviously a number of factors, both volcanological and meteorological, involving the amount of H_2SO_4 aerosols created in the upper atmosphere and their potential for widespread distribution come into play in determining the impact of any particular volcanic eruption on the climate.

Volcanologists have attempted to measure the amount of sulfur volatiles emitted by past eruptions through analysis of the composition of the solid products (pumice, scoria, lava, or fine ash). One method is to determine the sulfur volatile content in glass inclusions in crystals of minerals that formed within the magma chamber just prior to the eruption; this gives a

Table 1 Estimates of stratospheric aerosols and climatic effects of some volcanic eruptions[a]

Volcano	Latitude	Date	Stratospheric aerosols (Mt)	Northern Hemisphere τ_D	Northern Hemisphere ΔT (°C)
Explosive eruptions					
St. Helens	46°N	May 1980	0.3	<0.01	<0.1
Agung	8°S	March/May 1963	10	<0.05[b]	0.3
El Chichón	17°N	March/April 1982	20	0.15	<0.4
Krakatau	6°S	August 1883	50	0.55	0.3
Tambora	8°S	April 1815	200	1.3	0.5
Rabaul?	4°S	March 536	300	2.5	large?
Toba	3°N	−75,000 yr	1000?	10?	large?
Effusive eruptions					
Laki	64°N	June 1783 to February 1784	~0	Locally high[c]	1.0?
Roza	47°N	−14 Myr	6000?	80?[d]	large?

[a] References: Rampino & Self (1984a) and text. Optical depths are visual, direct beam.
[b] Southern Hemisphere $\tau_D \approx 0.2$.
[c] Aerosols were mostly tropospheric.
[d] If the aerosols were dispersed globally, the average Northern Hemisphere optical depth would have been about 40.

measure of preeruption sulfur. Knowing the volume of magma erupted, one can then estimate the sulfur volatile release by determining the amount of sulfur contained in the erupted ash and taking the difference (Sigurdsson 1982, Devine et al 1984). These methods are currently being refined (Sigurdsson et al 1985).

Since a positive general correlation has been established between the solubility of sulfur and the iron content in a magma, basaltic (mafic) magmas tend to be richer in dissolved sulfur than more silicic magmas. Sulfur release into the atmosphere from a basaltic eruption may be an order of magnitude greater than that of a silicic eruption of similar volume, but this is balanced in climatic impact by the fact that silicic eruptions in general are more explosive and therefore tend to create eruption columns that reach well into the stratosphere (Wilson et al 1978). Some eruptions, however, may be "anomalously" rich in sulfur volatiles with regard to their major element chemistry, such as was the case for the 1982 eruption of El Chichón (Table 1), which is discussed later.

Analyses of volcanic emissions show that sulfur is emitted mostly as sulfur dioxide (SO_2) and also as hydrogen sulfide (H_2S), which is soon oxidized to SO_2. In the stratosphere, the sulfur dioxide reacts with hydroxyl (OH^-) radicals produced by the photodissociation of water vapor. Gaseous sulfuric acid condenses on minute seed particles of dust (possibly volcanic or meteoritic) or on ions or small clusters of molecules. The photochemical reactions may be protracted, with complete conversion of emitted sulfur gases into aerosols taking weeks to months.

The residence time of the aerosols depends upon the dynamics of nucleation and growth of the droplets. After the initial input of sulfur volatiles and the conversion to droplets, the volcanic aerosols typically have a modal diameter of about half a micron [see Turco et al (1982) for a comprehensive review of observations and theory related to stratospheric aerosols]. For historic eruptions of all sizes, from those just capable of stratospheric injection (e.g. Fuego in 1974) to the largest known (e.g. Tambora in 1815), the e-folding time for fallout of the stratospheric aerosols has been observed to be about one year (Stothers 1984a). This means that for significant eruptions (those that create more than 1 Mt of stratospheric sulfuric acid aerosols) the stratospheric aerosol optical depth can be perturbed for several years.

INFORMATION FROM ICE CORES

Reasonably continuous, direct estimates of atmospheric optical depth from astronomical observations of the Sun, Moon, and stars are available only for the time period since 1883. During this period, changes in the atmospheric transparency have been correlated with significant volcanic eruptions (e.g. Pollack et al 1976). A major problem is to establish methods of estimating the amounts of sulfur aerosols created by significant eruptions prior to that time. In an important paper, Hammer et al (1980) presented evidence from the yearly ice layers in deep Greenland ice cores for sharp increases in acidity that coincided with the times of historic eruptions. They used the acid concentrations to estimate the global stratospheric aerosol burden in these volcanic years.

The ice-core method has several limitations, however, including the fact that eruptions in relatively high northern latitudes can produce especially large acidity spikes because of closeness of the volcanoes to Greenland. For example, the years 1963 and 1964 are associated with a noticeably high acidity peak in Greenland ice cores—enough to suggest about 20 Mt of aerosols in the Northern Hemisphere stratosphere if the source were the equatorial Agung eruption. But we know from direct atmospheric observations that less than one fifth of this amount was actually spread in the stratosphere north of the equator from the Agung eruptions in Bali (Volz 1970). The Greenland acidity spike is too high to be the result of Agung aerosols and is almost certainly due to tropospheric transport from smaller nearby eruptions such as the ongoing Surtsey eruptions off Iceland (Cronin 1971). Significantly, Koerner & Fisher (1982) detected no excess ice acidity for the years 1963 and 1964 on Ellesmere Island, Canada, whereas Delmas & Boutron (1980) and Legrand & Delmas (1987) did discern a strong acid signal in Antarctic ice which could be attributed to Agung. Icelandic eruptions in general are overrepresented in Greenland ice cores (Hammer 1984). Globally, however, the effectiveness of transport of acid aerosols to Greenland may also vary with seasonal and year-to-year changes in atmospheric circulation patterns (Hammer et al 1980).

It is worth emphasizing that if an eruption, even a moderately large one, does not inject sufficient sulfur into the atmosphere it will not appear above the noise level in the polar ice acidity record. This means that most eruptions with potential climatic impact are expected to leave a discernible trace in polar ice, although the amount deposited may not be proportional to the actual mass of sulfur produced, for the meteorological reasons just discussed. Probably any explosive eruption bigger than Krakatau's in 1883 can be detected in one or both of the polar ice sheets, because so much magma is erupted that the sulfur release is bound to be fairly large in any case.

One way of correcting for all these problems is by making comparisons with ice cores from several other localities to get a better estimate of the global distribution of the aerosols. Although such a method of calibration holds promise, for example, with the conductivity records of the Quelccaya ice core from the Peruvian Andes (Thompson et al 1986) and the Yukon ice core (Holdsworth et al 1986), it has proven to be difficult even with cores from different parts of the Greenland Ice Sheet (Herron 1982). The accumulation rate of snow in Antarctica is much less than that in Greenland, and hence the yearly ice layers there are generally thinner and more difficult to count and date accurately (Delmas et al 1985). Obviously, the best way of calibrating the ice-core acidity records as a record of global stratospheric aerosols is to identify, if possible, the location of the specific eruptions that produced the acid spikes.

It is worth noting that some periods of enhanced ice-core acidity and microparticle accumulation, lasting for decades and longer, coincide with historic cool intervals such as the Little Ice Age (Hammer et al 1980, Porter 1981, 1986, Thompson & Mosley-Thompson 1981, Thompson et al 1986). It is possible, therefore, that episodes of greater-than-average volcanism may modulate the climate over periods of tens to hundreds of years (for a review, see Bryson & Goodman 1980).

MT. ST. HELENS AND EL CHICHÓN: A STUDY IN CONTRASTS

The spectacular Mt. St. Helens eruption of May 1980 produced a strato-spheric cloud of ash but released a relatively small amount of SO_2 into the Northern Hemisphere stratosphere, with the result that only about 0.3 Mt of sulfuric acid aerosols were produced (Table 1) and no significant climatic effects [aside from a local daytime cooling the next day for stations down-wind of the ash (Robock & Mass 1982)] were detected (Newell & Deepak 1982). In March–April 1982, however, the Mexican volcano El Chichón erupted explosively and sent a huge cloud rich in SO_2 up to about 26 km in the stratosphere. Observations from the ground and by satellite showed that this eruption was having a large impact on the stratosphere, and the spread of the cloud could be tracked accurately.

El Chichón provided the test case for which volcanologists and atmo-spheric scientists had been waiting (see reviews by Rampino & Self 1984b, Hofmann 1987). Like St. Helens, it was a small eruption volumetrically, producing only about 0.3 to 0.4 km³ (DRE) of magma, but it was extremely rich in sulfur (derived perhaps from deposits of $CaSO_4$ beneath the vol-cano); thus, while the volcanic ash contribution to the atmosphere was small, the sulfuric acid aerosol contribution was considerable, about 20 Mt (Table 1). This is enough, theoretically, to lower the surface temperatures in the Northern Hemisphere a few tenths of a degree Celsius, and the year 1982 showed such a cooling (Table 1), although it seems that the cooling began with colder than average weather from January to March—before the El Chichón eruptions. But the summer was cool, with a unique snowfall in Vermont in August. Kelly & Sear (1984) proposed that Northern Hemisphere eruptions can cause cooling within the first 2 to 3 months after an eruption. Longer term effects from El Chichón were predicted by some models (Robock 1984), and Reiter & Jäger (1986) suggested that the cold winter of 1984/1985 was possibly related to lingering aerosols from the eruption.

The severe El Niño event of 1982/1983 added some confusion as to climatic cause and effect. Handler (1984), among others, has suggested that the El Chichón aerosol cloud either triggered the El Niño or led to its intensification—a proposal that has generated a good deal of debate. (It is interesting to note that the 1963 Agung eruption was also followed by an "off-season" El Niño event.)

THE GREATEST HISTORIC ERUPTIONS AND THEIR ATMOSPHERIC EFFECTS

With the Greenland ice-core record of acidity as a guide to notable "eruption years" (Hammer et al 1980, Hammer 1984), it has become possible to attempt to identify the source volcanic eruptions for the acid-rich layers deposited on the polar ice sheets. For more recent historical times it is relatively easy to pinpoint the eruptions that caused acid spikes (for example, Krakatau in 1883 and Tambora in 1815), but even here the situation may be more complicated than it first appears. Mt. Augustine in Alaska also erupted in 1883, and the Mayon eruption in the Philippines in 1814 may have contributed some acids to the ice layers of 1815 and 1816 (Stothers 1984a). Asama in Japan erupted at the same time as Laki during 1783.

In order to identify the extent of aerosol clouds and possible sources of ice-core acidity spikes prior to AD 1500, it has been necessary to search through historical records for evidence both of local eruptions (mostly in the Mediterranean region) and of the atmospheric perturbations caused by aerosol clouds from perhaps distant eruptions. A virtually complete search of the European records prior to AD 630 (Stothers & Rampino 1983a,b) turned up occasional evidence of significant atmospheric disturbances, such as a dim Sun and Moon, unusual atmospheric optical phenomena, and unusually cold weather accompanied by crop failures and famine. Similar work is now being done for the extensive Chinese historical records (Pang & Chou 1984, 1985, Pang et al 1986). It will be useful in this review to summarize the atmospheric and climatic effects occurring in some of the most severe and better known of the historical eruption years.

1816: The Tambora Effect and the Year Without a Summer

The year 1816 has gone down in the annals of climate history as the "Year Without a Summer" and "Eighteen Hundred and Froze to Death" (Stommell & Stommell 1983). In fact the entire decade from 1810 to 1820 was a time of noticeably cool temperatures in the Northern Hemisphere, and this has been correlated by some authors with the low sunspot maximum in 1816 (e.g. Humphreys 1940). The unusual weather in 1816 followed the spectacular April 1815 eruption of Tambora volcano on Sumbawa Island in Indonesia—one of the largest known ash-producing eruptions [150 km^3 of ash and pumice, equal to about 50 km^3 (DRE) of magma] in the last 10,000 years (Stothers 1984a, Self et al 1984).

Ash fallout was noted over an area in excess of 4×10^5 km^2 (and probably fell over an area of more than 10^6 km^2); darkness lasted for up to 2 days at distances of 600 km from the volcano. The eruption rate and the area of ash dispersal both suggest that the eruption column may have reached 50 km into the stratosphere. The volcanic cloud traveled around the world, and within 3 months its optical effects were observed at distant locations in Europe. For example, around the end of June, and later in September, several observers near London reported prolonged and brilliantly colored sunsets and twilights.

The following year (1816) was marked by a persistent dry fog, or dim Sun, as reported in the northeastern United States. The haze was clearly located above the troposphere, since neither surface winds nor rain dispersed it and because the total lunar eclipse of 9–10 June was extremely dark. Stothers (1984a) has derived a time history of the optical properties of the Tambora aerosol cloud (Figure 2) by using indicators of reduced atmospheric transmissivity such as dimming of the Sun (shown by increased naked-eye visibility of sunspots) and dimming of starlight (noted by astronomical observers). The calculated aerosol mass for Tambora (Table 1) is in good agreement with estimates based on the 4-yr-long (1815–1818) acidity enhancement in the Crête, Greenland, ice core (Hammer et al 1980). Fallout from this eruption has probably also been detected in Antarctic ice (Thompson & Mosley-Thompson 1981, Delmas et al 1985, Legrand & Delmas 1987).

The exceptional meteorological conditions spawned by the explosion started with a hot, followed by an "extremely cold," pocket of air directly under the tropospheric ash clouds (at least at Banjuwangi, 400 km from the volcano) and then continued with freezing temperatures in Madras, India, two weeks later (Stothers 1984a). Analogous studies of the effects of the Canadian wildfires of 1950 (Wexler 1950) and the Siberian wildfires of 1915 (Seitz 1986) have shown that surface temperatures in those cases

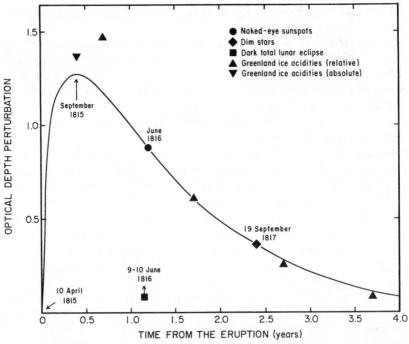

Figure 2 Excess visual optical depth at northern latitudes as a function of time reckoned from the date of the 1815 Tambora eruption. The plotted point for 9–10 June 1816 is only the lower limit to the true value (after Stothers 1984a).

dropped several degrees Celsius in areas that were thickly covered by high-altitude smoke clouds, as a result of the attenuation of the incoming solar flux. On the other hand, low-altitude smoke clouds heated up the boundary layer locally. Since volcanic ash sufficiently resembles sooty smoke, the meteorological analogies probably can be validly made in a qualitative way.

The summer of 1816 in western Europe was cool and exceedingly wet; crop failures (compounded by the aftereffects of the Napoleonic Wars) led to famine, disease, and social unrest, referred to by Post (1977) as "The Last Great Subsistence Crisis in the Western World." Kelly et al (1984) have suggested that an important effect of volcanic aerosol clouds is to produce a marked drop in surface pressure across the midlatitudes of the North Atlantic sector, leading to a southward shift in the track taken by middle-latitude cyclones. A major anomaly would thus be centered over England and would extend over much of western Europe, giving rise to a cold, wet summer. In support of this, Kelly et al have reconstructed pressure anomaly charts of Europe based on the available data; these charts are dominated by negative pressure anomalies over Europe beginning in early 1816. Data from Manley (1974) show that the summer months of 1816 in central England were about 1.5°C cooler than during the summer of 1815. The dismal European summer is credited with having inspired Mary Shelley to write *Frankenstein*, and Lord Byron his poem *Darkness*.

In North America, records of Hudson's Bay Company posts on the eastern side of Hudson Bay show that the summers of 1816 and 1817 were the coldest of any in the modern record (Wilson 1985a,b). Tree-ring data from northern and western Quebec support these observations (Filion et al 1986, Jacoby et al 1987). The distribution and severity of sea ice in Hudson Strait in 1816 suggests prevailing northerly or northwesterly winds, which again supports the idea that these years were marked by the development of strongly meridional atmospheric circulation patterns allowing southward penetrations of Arctic air across eastern North America and western Europe (Catchpole & Faurer 1983). Outbreaks of unusually cold weather during the spring and summer of 1816 in eastern Canada and the eastern United States are well documented (Post 1977, Stommel & Stommel 1983, Hamilton 1986). For example, the summer of 1816 was the coldest in New Haven, Connecticut, for the entire period from 1780 to 1968 (Landsberg & Albert 1974). From late spring through the summer, repeated frosts in New England caused crop failures, resulting in poor harvests and food shortages.

The outbreaks of cold weather and raininess during the summer months of 1816 are seen clearly in a number of other climatic indicators from around the world, from lateness in the grape harvests in France (Stommel & Swallow 1983) to frost damage rings in trees in the western United States (LaMarche & Hirschboeck 1984) and in South Africa (Dunwiddie & LaMarche 1980).

On a zonal to hemispheric basis, the deviation of annual mean temperature is more difficult to assess, since the station coverage in 1816 was very spotty. Using W. Köppen's compilation of temperature data, Stothers (1984a) finds an average deviation for the "Northern Hemisphere" in 1816 of −0.4 to −0.7°C, whereas the value for northern midlatitudes is about −1.0°C. This agrees with other estimates based on less data and somewhat different averaging (Lamb 1970, Rampino & Self 1982, Angell & Korschover 1985).

1783: The Fire Fountains of the Laki Fissure Eruption

Franklin's observations of the "dry fog" produced by the Laki (or Lakagígar) eruption have focused the attention of climatologists and volcanologists on the events surrounding this unusual eruption and its aftermath. The Laki eruption began in June 1783 and lasted for 8 months. The eruptions were not the typical explosive eruptions of the sort that produce great amounts of pumice and ash; Laki was primarily a fissure basalt, lava-flow type, and it erupted about 12.3 km^3 of lava, the bulk of it coming from a 13 km length of fissure during June and July, in the first 50 days of the eruption (Thorarinsson 1969). From eyewitness accounts, it appears that during the first days the eruption was extremely violent, with "enormous" Hawaiian-type lava fountains. Thorarinsson (1969) estimated that about 0.3 km^3 (DRE) of tephra was erupted, mostly during this early violent phase, and fine ash from the eruption fell as far away as northern Europe.

The effects of the eruption in Iceland were disastrous. The toxic volcanic gases and aerosols created a "blue haze" that spread all across the island and led to the destruction of the summer crops. About 75% of the livestock in Iceland died, and the resulting "Blue Haze Famine" claimed 24% of the Icelandic population. The dry fog reported by Franklin was also reported by others in Europe (for example, Gilbert White in England) and was even seen in Asia and North Africa (Holm 1784, Russell & Archibald 1888). Wood (1984) has established from eyewitness reports that much of the haze over Europe lay in the lower troposphere. However, the reported visibility of the haze high up in the Alps and its continued observability throughout Europe for weeks in spite of changing wind directions and rainfall suggest that it extended upward at least into the upper troposphere. The haze in Europe appeared most intense during June and July, precisely the same months that Laki was most active. The eruption ceased by early February 1784, but the dry fog had already largely disappeared by the end of December 1783 (Stothers et al 1986). In three Greenland ice cores, Hammer et al (1980) and Hammer (1984) found the acidity of the 1783 layers to be extraordinarily high, but no excess acidity was found in the 1784 layers, contrary to what one would expect if the stratosphere had been significantly loaded with aerosols. Moreover, the total lunar eclipse of 10 September 1783 was not unusually dark (Maclean 1984). Thus, the eruption column of fine ash and volcanic gases from Laki must have normally reached only up to, at most, the tropopause (8 to 11 km during the Icelandic summer).

In Iceland itself, a prompt and extreme cooling at the surface was observed directly under the ash clouds during the summer of 1783 (Stephensen 1813). Elsewhere, as Sigurdsson (1982) has shown from early Northern Hemisphere temperature records, an abnormal temperature decline began in the autumn of 1783 and reached a minimum in the period from December 1783 to February 1784. This period showed the lowest mean winter temperature in 225 years, which was 4.8°C below the long-term average. The spring, autumn, and winter mean temperatures for 1784 and 1785 were below normal as well. Northern tree-ring data also indicate cold growing seasons during 1783 and 1784 (Oswalt 1957, Filion et al 1986). Theoretically, very fine solid particles and *small* tropospheric aerosols in continuous production, with extensive horizontal dispersion, might have been able to initiate such a cooling (see Hansen et al 1980).

536: The Mystery Cloud

The densest and most persistent dry fog in recorded history was observed during AD 536–537 in Europe and the Middle East (Stothers & Rampino 1983a, Stothers 1984b) and in China (K. D. Pang & H.-h. Chou, in Weisburd 1985). In the Western literature five reliable contemporary descriptions of the atmospheric conditions of 536–537 exist. According to one contemporary writer (probably John of Ephesus), conditions in Mesopotamia (30° to 37°N) were such that "the sun was dark and its darkness lasted for eighteen months; each day it shone for about four hours, and still this light was only a feeble shadow . . . the fruits did not ripen and the wine tasted like sour grapes." The winter in Mesopotamia was exceptionally severe, with freak snowfalls and much hardship. In Italy, a high government official (Cassiodorus Senator, not mentioned by Stothers & Rampino) wrote in the late summer of 536: "The sun . . . seems to have lost its wonted light, and appears of a bluish color. We marvel to see no shadows of our bodies at noon, to feel the mighty vigor of the sun's heat wasted into feebleness, and the phenomena which accompany a transitory eclipse prolonged through almost a whole year. The moon, too, even when its orb is full, is empty of its natural splendor. . . . We have had . . . a spring without mildness and a summer without heat . . . the months which should have been maturing the crops have been chilled by north winds . . . rain is denied . . . the reaper fears new frosts" (Cassiodorus Senator AD 536). Cold and drought finally succeeded in killing off the crops in Italy and Mesopotamia and led to a terrible famine in the immediately following years. These and other accounts of the time read remarkably like Franklin's and others' modern descriptions of the "dim Sun" conditions following known volcanic eruptions.

It is possible to estimate from these historical accounts that such a pronounced reduction in solar brightness would require an excess visual atmospheric optical depth of $\tau_D = 2.5$ (Stothers 1984b). Under these conditions, at maximum altitude, the Sun (and Moon) would have appeared about 10 times fainter than normal, thus accounting for the reported darkening; scattered sunlight would have illuminated the rest of the sky. The mystery cloud first appeared in the Mediterranean region in late March of 536. Observers at 41 to 42°N reported effects lasting for 12 to

15 months, whereas at 30 to 37°N the duration recorded was 18 months. This suggests that the eruption that produced the aerosols was situated somewhere to the south; a possible source is the large eruption of Rabaul (4°S), on the island of New Britain off New Guinea, radiocarbon dated to AD 540±90 (Heming 1974).

Similar atmospheric effects were seen in China during the same years. Pang & Chou (see Weisburd 1985) have recently noted, for example, that the bright star Canopus was not visible when looked for at the equinoxes of 536. They have also documented and reconstructed the distribution of summer snows and frosts, drought, and famine throughout China in the years 536–538. The situation in China clearly paralleled that of Europe and the Middle East; the mystery cloud and the anomalously cold weather seem to have occurred throughout at least a large portion of the Northern Hemisphere. Moreover, a deep Greenland ice core shows a high acidity at roughly this date, originally given as 540±10 (Hammer et al 1980, Herron 1982), but later revised to 516±4 (Hammer 1984) and perhaps still in need of revision.

44 BC: Caesar's Death and the Year of the Failing Sun

Another significant dimming of the Sun is reported in the ancient literature for 44 BC and the subsequent two years. In the Western records (Stothers & Rampino 1983a), Plutarch, writing ca. AD 100, gives the fullest account of the "dim Sun" conditions after the murder of Julius Caesar: "For during all that year its orb rose pale and without radiance, while the heat that came down from it was slight and ineffectual, so that the air in its circulation was dark and heavy owing to the feebleness of the warmth that penetrated it, and the fruits, imperfect and half ripe, withered away and shrivelled up on account of the coldness of the atmosphere." Ovid (ca. AD 10) describes the Moon as "bloody" and Venus as "darkly rusty" in 44 BC, while Calpurnius Siculus (ca. AD 60) alludes to the "bloody" color of the comet of 44 BC (not cited by Stothers & Rampino). One possible cause for these atmospheric conditions was an eruption of Etna dated to the same year. As Vergil relates, "Mighty Etna . . . from its burst furnaces breathes forth flame; and . . . all Sicily moans and trembles, veiling the sky in smoke." Livy and, later, Pliny the Elder independently testified to the exceptional magnitude of this explosive Etnan eruption. Etna's last large eruption had occurred about 77 years earlier. (We now think that Lucan's and Petronius's apparent allusions to an eruption in 50–49 BC actually refer to the 44 BC eruption.)

Atmospheric effects in China were reported in the *Chronicles of the Han Dynasty* (Schove 1951, Pang & Chou 1984). For example, in April–May 43 BC, "It snowed. Frosts killed mulberries." In May–June, "The sun was bluish white and cast no shadow. At high noon there were shadows but dim." In October, "Frosts killed crops, widespread famine. Wheat crops damaged, no harvest in autumn." The historical data and the Greenland ice-core acidity record support the idea of multiple eruption clouds in the years from 44 to 42 BC (Stothers & Rampino 1983a, Pang et al 1986). A strong 3-yr-long acidity peak in two Greenland ice cores has been dated around 50 ± 4 BC (Hammer et al 1980, Hammer 1984; see also Herron 1982), and it probably correlates with the veiled Sun and other peculiar optical phenomena of 44–42 BC.

COMETARY WINTER AND NUCLEAR WINTER

Interest in the aftermath of a proposed collision of an asteroid or comet with the Earth at the time of the Cretaceous-Tertiary mass extinctions (66 Myr BP) has led to scenarios of Sun-blocking dust clouds originating from the huge cratering event, and smoke clouds rising from widespread wildfires [see Alvarez (1986) for a review]. These studies created the impetus for an analysis of the possible atmospheric effects of the sooty smoke from burning cities in the wake of a nuclear war (Crutzen & Birks 1982, Turco et al 1983). The initial "nuclear winter" simulation studies, based on the results of simple one-dimensional radiative/convective climate models, suggested the possibility of drastic temperature decreases of up to 30°C, with subfreezing conditions for weeks to months over large portions of the Northern Hemisphere and effects penetrating into the Southern Hemisphere as well, after a "baseline" nuclear exchange (Turco et al 1983, 1984b). This work prompted several research groups to study the effects of smoke generated by nuclear war, and the results obtained touched off a debate regarding the uncertainties in the amount of smoke that could be lofted to high altitudes and the scale and severity of the climatic effects of nuclear wildfires (for a review, see Colbeck & Harrison 1986).

Two independent study groups (National Research Council 1985, SCOPE 1985) concluded that the climatic effects could be severe, but that there were many uncertainties in the nuclear winter analyses. The most recent studies, using more sophisticated climate models (GCMs), suggest that the cooling would be much less severe though still significant ("nuclear autumn"), with worst case (July) decreases of perhaps 5°C in low latitudes (10–30°N) and 10–15°C at higher northern latitudes, lasting only a few weeks and with considerable unevenness. Most of the reduction in the degree of cooling is related to the probable removal of 75% of the smoke from the atmosphere within the first 30 days, as well as to patchiness in the smoke distribution and to the moderating effects of the oceans in the climate system (Thompson & Schneider 1986, Covey 1987).

VOLCANIC WINTER?

Large Explosive Eruptions

We have thus far discussed the climatic aftereffects of a number of the greatest historical volcanic eruptions. The nuclear winter debate raised a suggestion that the atmospheric aftereffects of these volcanic eruptions might be used as a basis for estimating the severity of cooling from dense smoke clouds (Maddox 1984, Brown & Peczkis 1984). The differing optical properties between volcanic aerosols and black, sooty smoke from urban fires, however, makes such a comparison difficult (Turco et al 1984a, 1985). More importantly, historic eruptions have produced relatively small amounts of aerosols. But perhaps the historic eruptions can be used as a baseline for estimating the possible atmospheric effects of the largest volcanic eruptions in the geologic past—much larger than recent historic events such as Tambora or Krakatau, or even the mystery eruption of 536. One good example is the Toba eruption in Sumatra about 75,000 years ago, which is the best-known late Quaternary "supereruption," with a recent estimate of the volume of erupted pyroclastics being equivalent to more than 2000 km^3 (DRE) of magma (Rose & Chesner 1986).

The Toba ash layer is extraordinarily widespread (Ninkovich et al 1978), and rough calculations, using the same methods as Murrow et al (1980), suggest that a maximum of $\sim 0.8\%$ of the erupted material could be in the form of fine dust $<2~\mu$m in diameter, for a total of about 20,000 Mt of volcanic dust. If only 10% of this dust were injected into the stratosphere, conditions of total darkness could have existed over a large area for weeks to months (see also Kent 1981).

In order to estimate any longer-lasting atmospheric effects, however, it is necessary to calculate the total amount of sulfuric acid aerosols that could have been produced by the Toba eruption. From simply scaling upward from historical eruptions of similar composition magma, Toba is estimated to have been capable of producing between 1000 to 5000 Mt of sulfuric acid aerosols (Rampino et al 1985). For volcanic aerosols, the globally averaged optical depth is $\tau_D = 6.5 \times 10^{-3}~M_D$, where M_D is the global aerosol loading in megatons (Stothers 1984a). For Toba the equivalent global aerosol optical depths are 6 to 33 (Figure 3). Local, regional, and hemispheric effects could have been greater, depending on the spread of the cloud. These values may be compared with the peak aerosol optical depth of about 2 estimated for the AD 536 mystery cloud or the value of about 1 following the 1815 Tambora eruption. The atmospheric aftereffects of a Toba-sized explosive eruption might be comparable to some scenarios of nuclear winter, although the aerosols are expected to have a longer atmospheric residence time than would the nuclear winter smoke.

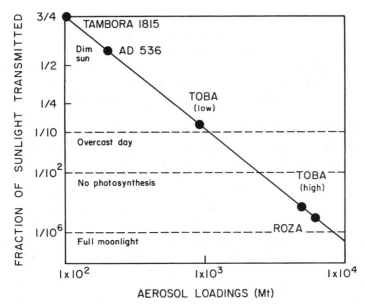

Figure 3 Fraction of sunlight transmitted through stratospheric aerosol and/or fine-ash dust clouds of different masses (theoretical line, after Turco et al 1984b). Points refer to great historic and prehistoric eruptions (see text).

If such an aerosol cloud could form and persist in the stratosphere, the climatic effects would almost certainly be quite severe. It is important to stress, however, that these are "worst-case" situations, made simply by extrapolating a linear increase in mass of aerosols under the assumption that the behavior of very dense aerosol clouds is not qualitatively different from that of the less dense aerosol clouds observed after historical eruptions. Recent work has shown that differences in aerosol nucleation, saturation, and fallout in dense clouds may affect the concentrations and atmospheric lifetimes of the aerosols. Pinto et al (1987) have recently used a one-dimensional radiative/convective model including aerosol microphysical and photochemical processes to examine the conversion of sulfur dioxide to aerosols in the stratosphere after volcanic eruptions. They find that for successively larger injections of SO_2, in the range of 10 to 200 Mt, the processes of condensation and coagulation produce larger particles; these particles have a smaller optical depth and fall out of the stratosphere faster. The rate of SO_2 oxidation may also be limited by conversion of OH to HO_2 radicals, which could limit the formation of aerosols. These results all suggest that the buildup of H_2SO_4 aerosols in the stratosphere might be self-limiting to a degree. However, Pinto et al have not yet modeled the injection of SO_2 burdens of >1000 Mt, accompanied by possibly large amounts of water vapor, as could be the case after "supereruptions" such as Toba. In such cases the dynamics of gas to particle conversion and the e-folding time of aerosols and ash in the stratosphere and troposphere might be quite different from those in the less dense clouds that have been observed and modeled thus far.

Flood-Basalt Eruptions

As mentioned above, basaltic volcanic eruptions may release an order of magnitude more sulfur volatiles than do silicic eruptions of the same volume (Devine et al 1984). Very recent results indicate that episodes of flood-basalt volcanism in the geologic past have involved the outpouring of up to 10^6 km^3 of basaltic magma over peak time periods of less than a million to a few million years (Bellieni et al 1984, Courtillot & Cisowski 1987, White 1987). Individual eruptions seem to have generated tens to hundreds of cubic kilometers of magma in periods of days to weeks. In the past, "quiet" effusive basaltic eruptions were considered unlikely to produce high-altitude aerosol clouds (Lamb 1970). Recent study has shown, however, that even relatively small historic fissure basalt eruptions, such as the 1783 Laki eruption, have produced widespread aerosol clouds. Theoretical plume modeling of such eruptions indicates that at rapid eruption rates the sulfur volatiles are efficiently released and can be carried to high altitudes in convective plumes rising above large fire fountains (Figure 4).

Stothers et al (1986) have recently considered the possible atmospheric impact of the large flood-basalt eruptions in the geologic past. For example, the Roza flow eruption of the Columbia River Basalt Group (about 14 Myr BP) produced some 700 km^3 of basaltic lava in about 7 days (Swanson et al 1975). The estimated eruption rates of 10^4–10^5 m^3 s^{-1} from 1 to 10 km fissure lengths are predicted to have generated Hawaiian-type fire fountains about 1 km in height (Wilson & Head 1981) and stratospheric (>10 km) eruption plumes (Figure 4).

The quantity of atmospheric aerosols produced by such large basalt eruptions can be roughly estimated by scaling from the known amounts of aerosols generated by the largest modern fissure eruptions, such as Laki in 1783. Laki erupted about 12 km^3 of magma and is estimated by various methods to have released about 30 Mt of sulfur (Stothers et al 1986). The sulfur release from the Roza flow eruption is therefore computed to have been (700 km^3/12 km^3) $\times$ 30 Mt $\approx$ 2000 Mt, equivalent to about 6000 Mt of H_2SO_4 aerosols; the corresponding global average optical depth would be about 40. Such a thick aerosol cloud, distributed worldwide, would allow only a small fraction (10^{-5}) of sunlight to reach the Earth's surface

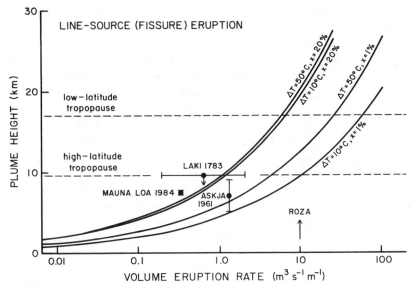

Figure 4 Convective plume height as a function of volume eruption rate per meter length of fissure for a line-source eruption. Theoretical curves are given for two values of the fine-ash content (*x*) and for two values of the temperature drop of the fountain clasts (Δ*T*) appropriate for fire-fountain activity. Predicted plume heights for the Roza flow eruption can be read off the observationally calibrated theoretical curves. The plotted plume height for Laki is an observational upper limit (after Stothers et al 1986).

(Figure 3). In this case, barring efficient self-limiting mechanisms, the cloud's atmospheric effects would be comparable to those in recent nuclear winter models, but more extended in time.

VOLCANISM AND MASS EXTINCTION?

One of the great current debates in geology concerns the cause of mass extinctions. Strong evidence now exists for a comet or asteroid impact at the time of the Late Cretaceous (66 Myr BP) mass extinctions (Alvarez 1986). But recent studies provide evidence that the Deccan Traps flood basalts in India, and perhaps the North Atlantic flood basalts, were erupted at the same time (Courtillot & Cisowski 1987, Officer et al 1987). Episodes of flood-basalt volcanism with peak periods lasting up to a few million years have occurred from time to time in the Earth's history. During the last 250 Myr, there were at least nine major flood-basalt episodes, some involving eruptions in more than one geographic area. When the ages of these flood-basalt episodes are subjected to a formal time-series analysis, they reveal a possible periodicity of roughly 30 Myr (Rampino & Stothers

1986). This is similar to the recent finding of a 26–32 Myr periodicity in the ages of biological mass extinctions (Fischer & Arthur 1977, Raup & Sepkoski 1984) and in the ages of episodes of impact cratering (Rampino & Stothers 1984, Alvarez & Muller 1984). Within the errors of dating, the ages of some of the flood basalts agree very well with the estimated ages of the mass extinctions and impact craters.

What can we infer from this? It may be that some massive outpourings of basalt are triggered by extraterrestrial impacts (Öpik 1958, Urey 1973, Rampino 1987). Or flood basalts might be generated by a quasi-periodic cycle of hotspot activity related to internal mantle dynamics (Loper & Stacey 1983). More speculatively, mass extinctions could be the result both of the aftereffects of a large impact and of related flood-basalt or explosive volcanism. In this case, cometary winters could be succeeded by volcanic winters that prolong the conditions adverse for life.

CONCLUSIONS

As has been shown in a number of studies, some of the largest historic eruptions are associated with atmospheric perturbations that have had a considerable impact on climate and agriculture. Even the greatest of these historic eruptions, however, was small compared with the very large explosive and effusive eruptions that are well known from the geologic record. A simple scaling-up of the effects of historic eruptions suggests that the much larger eruptions could have brought about severe, short-term coolings or "volcanic winters" over considerable portions of the globe. A very large eruption in the near future might have drastic effects on crop yields and could create food-supply crises in many areas, especially those regions of marginal productivity. Eruptions like these constitute a very real "volcanic hazard" in terms of the number of people that would be affected. There is no question that such large eruptions will recur, the only uncertainty lies in where and when.

Could individual "supereruptions" also lead to greatly prolonged climate cooling? The residence time of volcanic aerosols in the stratosphere is only of the order of a few years; therefore, one must invoke some form of positive climatic feedback to extend the cooling. One possibility is that a few cool summers could lead to significantly increased snow and ice cover at high latitudes, such that the increased albedo would further cool the Earth (see, e.g., Bray 1976). Both the Toba and Roza eruptions occurred at times of relatively rapid climate cooling, but no firm causal connection has been established. Although other large volcanic eruptions are not known to coincide with coolings, certain climatic regimes may be more sensitive to volcanic perturbations. In addition, climatic changes themselves may be able to influence the incidence and severity of volcanism (Rampino et al 1979). However, such connections remain elusive.

ACKNOWLEDGMENTS

We thank H. E. Brooks, R. Delmas, C. U. Hammer, G. C. Jacoby, A. A. Lacis, M. Legrand, K. D. Pang, J. P. Pinto, W. I. Rose, H. Sigurdsson, L. Wilson, J. A. Wolff, and C. A. Wood for helpful discussions and correspondence, and the American Philosophical Society for a research grant. The Center for Global Habitability at Columbia University also provided research support.

Literature Cited

Abbot, C. G., Fowle, F. E. 1913. Volcanoes and climate. *Smithson. Misc. Collect.* 60(29): 1–24

Alvarez, W. 1986. Toward a theory of impact crises. *Eos, Trans. Am. Geophys. Union* 67: 649–58

Alvarez, W., Muller, R. A. 1984. Evidence from crater ages for periodic impacts on the earth. *Nature* 308: 718–20

Angell, J. K., Korshover, J. 1985. Surface temperature changes following the six major volcanic episodes between 1780 and 1980. *J. Climatol. Appl. Meteorol.* 24: 937–51

Bellieni, G., Brotzu, P., Comin-Chiaramonti, P., Ernesto, M., Melfi, A., et al. 1984. Flood basalt to rhyolite suites in the southern Paranà plateau (Brazil): palaeomagnetism, petrogenesis, and geodynamic implications. *J. Petrol.* 25: 579–618

Blong, R. J. 1982. *The Time of Darkness.* Seattle: Univ. Wash. Press. 257 pp.

Bray, J. R. 1976. Volcanic triggering of glaciation. *Nature* 260: 414–15

Brown, W. H., Peczkis, J. 1984. Nuclear war—counting the cost. *Nature* 310: 455

Bryson, R. A., Goodman, B. M. 1980. Volcanic activity and climatic changes. *Science* 207: 1041–44

Burke, K., Francis, P. 1985. Climatic effects of volcanic eruptions. *Nature* 314: 136

Cadle, R. D., Kiang, C. S., Louis, J.-F. 1976. The global scale dispersion of the eruption clouds from major volcanic eruptions. *J. Geophys. Res.* 81: 3125–32

Cadle, R. D., Fernald, F. G., Frush, C. L. 1977. Combined use of lidar and numerical diffusion models to estimate the quantity and dispersion of volcanic eruption clouds in the stratosphere: Vulcán Fuego, 1974, and Augustine, 1976. *J. Geophys. Res.* 82: 1783–86

Calpurnius Siculus. ca. AD 60. *Eclogues* 1.80–83

Capone, L. A., Toon, O. B., Whitten, R. C., Turco, R. P., Riegel, C. A., Santhanam, K. 1983. A two-dimensional model simulation of the El Chichón volcanic eruption cloud. *Geophys. Res. Lett.* 10: 1053–56

Cassiodorus Senator. AD 536. *Variae* 12.25

Castleman, A. W. Jr., Munkelwitz, H. R., Manowitz, B. 1974. Isotopic studies of the sulfur component of the stratospheric aerosol layer. *Tellus* 26: 222–34

Catchpole, A. J. W., Faurer, M.-A. 1983. Summer sea ice severity in Hudson Strait, 1751–1870. *Clim. Change* 5: 115–39

Chou, M.-D., Peng, L., Arking, A. 1984. Climate studies with a multilayer energy balance model. Part III: climatic impact of stratospheric volcanic aerosols. *J. Atmos. Sci.* 41: 759–67

Colbeck, I., Harrison, R. M. 1986. The atmospheric effects of nuclear war—a review. *Atmos. Environ.* 20: 1673–81

Courtillot, V. E., Cisowski, S. 1987. The Cretaceous-Tertiary boundary events: external or internal causes? *Eos, Trans. Am. Geophys. Union* 68: 193–200

Covey, C. 1987. Protracted climatic effects of massive smoke injection into the atmosphere. *Nature* 325: 701–3

Cronin, J. F. 1971. Recent volcanism and the stratosphere. *Science* 172: 847–49

Crutzen, P. J., Birks, J. W. 1982. The atmosphere after a nuclear war: twilight at noon. *Ambio* 11: 114–25

Deirmendjian, D. 1973. On volcanic and other particulate turbidity anomalies. *Adv. Geophys.* 16: 267–96

Delmas, R., Boutron, C. 1980. Are the past variations of the stratospheric sulfate burden recorded in central Antarctic snow and ice layers? *J. Geophys. Res.* 85: 5645–49

Delmas, R. J., Legrand, M., Aristarain, A. J., Zanolini, F. 1985. Volcanic deposits in Antarctic ice and snow. *J. Geophys. Res.* 90: 12,901–20

Devine, J. D., Sigurdsson, H., Davis, A. N., Self, S. 1984. Estimates of sulfur and chlorine yield to the atmosphere from volcanic eruptions and potential climatic effects. *J. Geophys. Res.* 89: 6309–25

Downey, W. S., Tarling, D. H. 1984. Archaeomagnetic dating of Santorini volcanic eruptions and fired destruction levels of late Minoan civilization. *Nature* 309: 519–23

Dunwiddie, P. W., LaMarche, V. C. Jr. 1980. A climatically responsive tree-ring record from *Widdringtonia cedarbergensis*, Cape Province, South Africa. *Nature* 286: 796–97

Ellsaesser, H. W. 1986. Comments on "Surface temperature changes following the six major volcanic episodes between 1780 and 1980." *J. Climatol. Appl. Meteorol.* 25: 1184–85

Filion, L., Payette, S., Gauthier, L., Boutin, Y. 1986. Light rings in subarctic conifers as a dendrochronological tool. *Quat. Res.* 26: 272–79

Fischer, A. G., Arthur, M. A. 1977. Secular variations in the pelagic realm. *Soc. Econ. Paleontol. Mineral. Spec. Publ.* 25: 19–50

Flammarion, C. 1884. L'éclipse totale de lune du 4 Octobre. *C. R. Acad. Sci. Paris* 3: 401–8

Gentilli, J. 1948. Present-day volcanicity and climatic change. *Geol. Mag.* 85: 172–75

Hamilton, K. 1986. Early Canadian weather observers and the "year without a summer." *Bull. Am. Meteorol. Soc.* 67: 524–32

Hammer, C. U. 1984. Traces of Icelandic eruptions in the Greenland ice sheet. *Jökull* 34: 51–65

Hammer, C. U., Clausen, H. B., Dansgaard, W. 1980. Greenland ice sheet evidence of post-glacial volcanism and its climatic impact. *Nature* 288: 230–35

Handler, P. 1984. Possible associations of stratospheric aerosols and El Niño type events. *Geophys. Res. Lett.* 11: 1121–24

Hansen, J. E., Wang, W.-C., Lacis, A. A. 1978. Mount Agung eruption provides test of a global climatic perturbation. *Science* 199: 1065–68

Hansen, J. E., Lacis, A. A., Lee, P., Wang, W.-C. 1980. Climatic effects of atmospheric aerosols. *Ann. NY Acad. Sci.* 338: 575–87

Hansen, J., Johnson, D., Lacis, A., Lebedeff, S., Lee, P., et al. 1981. Climate impact of increasing atmospheric carbon dioxide.

Science 213: 957–66

Harshvardhan, Cess, R. D. 1976. Stratospheric aerosols: effect upon atmospheric temperature and global climate. *Tellus* 28: 1–10

Heming, R. F. 1974. Geology and petrology of Rabaul caldera, Papua New Guinea. *Geol. Soc. Am. Bull.* 85: 1253–64

Herron, M. M. 1982. Impurity sources of F^-, Cl^-, NO_3^- and SO_4^{2-} in Greenland and Antarctic precipitation. *J. Geophys. Res.* 87: 3052–60

Hirschboeck, K. K. 1980. A new worldwide chronology of volcanic eruptions. *Palaeogeogr. Palaeoclimatol. Palaeoecol.* 29: 223–41

Hofmann, D. J. 1987. Perturbations of the global atmosphere associated with the El Chichón volcanic eruption of 1982. *Rev. Geophys.* 25: 743–59

Holdsworth, G., Krouse, H. R., Peake, E. 1986. Relationship between volcanic emission peaks and the oxygen isotope signature in an ice core from the Yukon Territory, Canada. *Eos, Trans. Am. Geophys. Union* 67: 883 (Abstr.)

Holm, S. M. 1784. *Vom Erdebrande auf Island im Jahr 1783.* Copenhagen: Proft

Humphreys, W. J. 1913. Volcanic dust and other factors in the production of climatic changes, and their possible relation to ice ages. *Bull. Mt. Weather Obs.* 6: 1–34

Humphreys, W. J. 1940. *Physics of the Air.* New York: McGraw-Hill. 676 pp.

Hunt, B. G. 1977. A simulation of the possible consequences of a volcanic eruption on the general circulation of the atmosphere. *Mon. Weather Rev.* 105: 247–60

Jacobs, L. 1954. Dust clouds in the stratosphere. *Meteorol. Mag.* 83: 115–18

Jacoby, G. C., Ivanciu, I. S., Ulan, L. 1987. Major climatic shift at the beginning of the last century: evidence from tree rings. Preprint (Lamont-Doherty Geol. Obs.)

Junge, C. E., Chagnon, C. W., Manson, J. E. 1961. Stratospheric aerosols. *J. Meteorol.* 18: 81–108

Kelly, P. M., Sear, C. B. 1984. Climatic impact of explosive volcanic eruptions. *Nature* 311: 740–43

Kelly, P. M., Wigley, T. M. L., Jones, P. D. 1984. European pressure maps for 1815–16, the time of the eruption of Tambora. *Clim. Monitor* 13: 76–91

Kent, D. V. 1981. Asteroid extinction hypothesis. *Science* 211: 649–50

Koerner, R. M., Fisher, D. 1982. Acid snow in the Canadian high Arctic. *Nature* 295: 137–40

LaMarche, V. C. Jr., Hirschboeck, K. K. 1984. Frost rings in trees as records of major volcanic eruptions. *Nature* 307: 121–26

Lamb, H. H. 1970. Volcanic dust in the atmosphere; with a chronology and assessment of its meteorological significance. *Philos. Trans. R. Soc. London Ser. A* 266: 425–533

Landsberg, H. E., Albert, J. M. 1974. The summer of 1816 and volcanism. *Weatherwise* 27: 63–66

Legrand, M., Delmas, R. J. 1987. A 220-year continuous record of volcanic H_2SO_4 in the Antarctic ice sheet. *Nature* 327: 671–76

Loper, D. E., Stacey, F. D. 1983. The dynamical and thermal structure of deep mantle plumes. *Phys. Earth Planet. Inter.* 33: 304–17

Lough, J. M., Fritts, H. C. 1987. An assessment of the possible effects of volcanic eruptions on North American climate using tree-ring analysis, 1602 to 1900 A.D. *Clim. Change* 10: 219–39

Maclean, A. D. I. 1984. The cause of dark lunar eclipses. *J. Br. Astron. Assoc.* 94: 263–65

Maddox, J. 1984. From Santorini to Armageddon. *Nature* 307: 107

Manley, G. 1974. Central England temperatures: monthly means 1659 to 1973. *Q. J. R. Meteorol. Soc.* 100: 389–405

Marinatos, S. 1939. The volcanic destruction of Minoan Crete. *Antiquity* 13: 425–39

Mass, C., Schneider, S. H. 1977. Statistical evidence on the influence of sunspots and volcanic dust on long-term temperature records. *J. Atmos. Sci.* 34: 1995–2004

McCracken, M. C., Luther, F. M. 1984. Preliminary estimate of the radiative and climatic effects of the El Chichón eruption. *Geofis. Int. Vol.* 23(3): 385–401

Mitchell, J. M. 1961. Recent secular changes of global temperature. *Ann. NY Acad. Sci.* 95: 235–50

Mossop, S. C. 1964. Volcanic dust collected at an altitude of 20 km. *Nature* 203: 824–27

Murrow, P. J., Rose, W. I. Jr., Self, S. 1980. Determination of the total grain size distribution in a vulcanian eruption column, and its implications to stratospheric aerosol perturbation. *Geophys. Res. Lett.* 7: 893–96

National Research Council. 1985. *The Effects on the Atmosphere of a Major Nuclear Exchange.* Washington, DC: Natl. Acad. Press

Newell, R. E. 1970. Stratospheric temperature change from the Mt. Agung volcanic eruption of 1963. *J. Atmos. Sci.* 27: 977–78

Newell, R. E., Deepak, A., eds. 1982. *Mount St. Helens Eruptions of 1980. Atmospheric Effects and Potential Climatic Impact. NASA SP-458.* Washington, DC: NASA.

118 pp.

Newell, R. E., Weare, B. C. 1976. Factors governing tropospheric mean temperature. *Science* 194: 1413–14

Newhall, C. G., Self, S. 1982. The Volcanic Explosivity Index (VEI): an estimate of explosive magnitude for historical volcanism. *J. Geophys. Res.* 87: 1231–38

Ninkovich, D., Shackleton, N. J., Abdel-Monem, A. A., Obradovich, J. D., Izett, G. 1978. K-Ar age of the late Pleistocene eruption of Toba, north Sumatra. *Nature* 276: 574–77

Officer, C. B., Hallam, A., Drake, C. L., Devine, J. D. 1987. Late Cretaceous and paroxysmal Cretaceous/Tertiary extinctions. *Nature* 326: 143–49

Öpik, E. J. 1958. On the catastrophic effects of collisions with celestial bodies. *Ir. Astron. J.* 5: 34–36

Oswalt, W. H. 1957. Volcanic activity and Alaskan spruce growth in A.D. 1783. *Science* 126: 928–29

Ovid. ca. AD 10. *Metamorphoses* 15.789–90

Pang, K. D., Chou, H.-h. 1984. A correlation between Greenland ice core climatic horizons and ancient Oriental meteorological records. *Eos, Trans. Am. Geophys. Union* 65: 846 (Abstr.)

Pang, K. D., Chou, H.-h. 1985. Three very large volcanic eruptions in antiquity and their effects on the climate of the ancient world. *Eos, Trans. Am. Geophys. Union* 66: 816 (Abstr.)

Pang, K. D., Pieri, D., Chou, H.-h. 1986. Climatic impacts of the 44–42 BC eruptions of Etna, reconstructed from ice core and historical records. *Eos, Trans. Am. Geophys. Union* 67: 880–81 (Abstr.)

Parker, D. E. 1985. Climatic impact of explosive volcanic eruptions. *Meteorol. Mag.* 114: 149–61

Pinto, J. P., Toon, O. B., Turco, R. P. 1987. Self-limiting physical and chemical effects in volcanic eruption clouds. Submitted for publication

Pollack, J. B., Toon, O. B., Sagan, S., Summers, A., Baldwin, B., Van Camp, W. 1976. Volcanic explosions and climatic change: a theoretical assessment. *J. Geophys. Res.* 81: 1071–83

Porter, S. C. 1981. Recent glacier variations and volcanic eruptions. *Nature* 291: 139–42

Porter, S. C. 1986. Pattern and forcings of Northern Hemisphere glacier variations during the last millenium. *Quat. Res.* 26: 27–48

Post, J. D. 1977. *The Last Great Subsistence Crisis in the Western World.* Baltimore: Johns Hopkins Press. 240 pp.

Rampino, M. R. 1987. Impact cratering and flood basalt volcanism. *Nature* 327: 468

Rampino, M. R., Self, S. 1982. Historic eruptions of Tambora (1815), Krakatau (1883), and Agung (1963), their stratospheric aerosols, and climatic impact. *Quat. Res.* 18: 127–43

Rampino, M. R., Self, S. 1984a. Sulphur-rich volcanic eruptions and stratospheric aerosols. *Nature* 310: 677–79

Rampino, M. R., Self, S. 1984b. The atmospheric effects of El Chichón. *Sci. Am.* 250(1): 48–57

Rampino, M. R., Stothers, R. B. 1984. Terrestrial mass extinctions, cometary impacts and the Sun's motion perpendicular to the galactic plane. *Nature* 308: 709–12

Rampino, M. R., Stothers, R. B. 1986. Periodic flood-basalt eruptions, mass extinctions, and comet impacts. *Eos, Trans. Am. Geophys. Union* 67: 1247 (Abstr.)

Rampino, M. R., Self, S., Fairbridge, R. W. 1979. Can rapid climatic change cause volcanic eruptions? *Science* 206: 826–29

Rampino, M. R., Stothers, R. B., Self, S. 1985. Climatic effects of volcanic eruptions. *Nature* 313: 272

Raup, D. M., Sepkoski, J. J. Jr. 1984. Periodicity of extinctions in the geologic past. *Proc. Natl. Acad. Sci. USA* 81: 801–5

Reiter, R., Jäger, H. 1986. Results of 8-year continuous measurements of aerosol profiles in the stratosphere with discussion of the importance of stratospheric aerosols to an estimate of effects on the global climate. *Meteorol. Atmos. Phys.* 35: 19–48

Robock, A. 1981. A latitudinally dependent volcanic dust veil index, and its effect on climate simulations. *J. Volcanol. Geotherm. Res.* 11: 67–80

Robock, A. 1984. Climate model simulations of the effects of the El Chichón eruption. *Geofis. Int. Vol.* 23(3): 403–14

Robock, A., Mass, C. 1982. The Mount St. Helens volcanic eruption of 18 May 1980: large short-term surface temperature effects. *Science* 216: 628–30

Rose, W. I., Chesner, C. A. 1986. Dispersal of the ash in the great Toba eruption, 75,000 yrs B.P. *Geol. Soc. Am. Abstr. With Programs* 18: 733 (Abstr.)

Russell, F. A. R., Archibald, E. D. 1888. On the unusual optical phenomena of the atmosphere, 1883–1886, including twilight effects, coloured suns, moons, etc. In *The Eruption of Krakatoa and Subsequent Phenomena*, ed. G. J. Symons, pp. 151–463. London: Truebner

Schneider, S. H., Mass, C. 1975. Volcanic dust, sunspots, and temperature trends. *Science* 190: 741–46

Schove, D. J. 1951. Sunspots, aurorae and blood rain: the spectrum of time. *Isis* 42: 133–38

SCOPE (Scientific Committee on Problems of the Environment). 1985. *Environmental Consequences of Nuclear War, SCOPE 28.* New York: Wiley

Seitz, R. 1986. Siberian fire as "nuclear winter" guide. *Nature* 323: 116–17

Self, S., Rampino, M. R., Barbera, J. J. 1981. The possible effects of large 19th and 20th century volcanic eruptions on zonal and hemispheric surface temperatures. *J. Volcanol. Geotherm. Res.* 11: 41–60

Self, S., Rampino, M. R., Newton, M. S., Wolff, J. A. 1984. Volcanological study of the great Tambora eruption of 1815. *Geology* 12: 659–63

Sigurdsson, H. 1982. Volcanic pollution and climate: the 1783 Laki eruption. *Eos, Trans. Am. Geophys. Union* 63: 601–2

Sigurdsson, H., Devine, J. D., Davis, A. N. 1985. The petrologic estimation of volcanic degassing. *Jökull* 35: 1–8

Simkin, T., Fiske, R. S. 1983. *Krakatau 1883.* Washington, DC: Smithson. Inst. 464 pp.

Simkin, T., Siebert, L., McClelland, L., Bridge, D., Newhall, C., Latter, J. H. 1981. *Volcanoes of the World.* Stroudsburg, Pa: Hutchinson Ross. 233 pp.

Smith, J. V. 1985. Protection of the human race against natural hazards (asteroids, comets, volcanoes, earthquakes). *Geology* 13: 675–78

Stanley, D. J., Sheng, H. 1986. Volcanic shards from Santorini (Upper Minoan ash) in the Nile Delta, Egypt. *Nature* 320: 733–35

Stephensen, M. 1813. In *Journal of a Tour in Iceland, in the Summer of 1809*, ed. W. J. Hooker, 2: 121–261. London: Longman et al

Stommel, H., Stommel, E. 1983. *Volcano Weather.* Newport, RI: Seven Seas. 177 pp.

Stommel, H., Swallow, J. C. 1983. Do late grape harvests follow large volcanic eruptions? *Bull. Am. Meteorol. Soc.* 64: 794–95

Stothers, R. B. 1984a. The great Tambora eruption and its aftermath. *Science* 224: 1191–98

Stothers, R. B. 1984b. Mystery cloud of AD 536. *Nature* 307: 344–45

Stothers, R. B., Rampino, M. R. 1983a. Volcanic eruptions in the Mediterranean before A.D. 630 from written and archaeological sources. *J. Geophys. Res.* 88: 6357–71

Stothers, R. B., Rampino, M. R. 1983b. Historic volcanism, European dry fogs, and Greenland acid precipitation, 1500 B.C. to A.D. 1500. *Science* 222: 411–13

Stothers, R. B., Wolff, J. A., Self, S., Rampino, M. R. 1986. Basaltic fissure eruptions, plume heights, and atmospheric aerosols. *Geophys. Res. Lett.* 13: 725–28

Swanson, D. A., Wright, T. L., Helz, R. T. 1975. Linear vent systems and estimated rates of magma production and eruption for the Yakima Basalt on the Columbia Plateau. *Am. J. Sci.* 275: 877–905

Taylor, B. L., Gal-Chen, T., Schneider, S. H. 1980. Volcanic eruptions and long-term temperature records: an empirical search for cause and effect. *Q. J. R. Meteorol. Soc.* 106: 175–99

Thompson, L. G., Mosley-Thompson, E. 1981. Temporal variability of microparticle properties in polar ice sheets. *J. Volcanol. Geotherm. Res.* 11: 11–27

Thompson, L. G., Mosley-Thompson, E., Dansgaard, W., Grootes, P. M. 1986. The Little Ice Age as recorded in the stratigraphy of the tropical Quelccaya ice cap. *Science* 234: 361–64

Thompson, S. L., Schneider, S. H. 1986. Nuclear winter reappraised. *Foreign Aff.* 1986 (Summer): 981–1005

Thorarinsson, S. 1969. The Lakagígar eruption of 1783. *Bull. Volcanol.* 33: 910–29

Turco, R. P., Whitten, R. C., Toon, O. B. 1982. Stratospheric aerosols: observation and theory. *Rev. Geophys. Space Phys.* 20: 233–79

Turco, R. P., Toon, O. B., Ackerman, T. P., Pollack, J. B., Sagan, C. 1983. Nuclear winter: global consequences of multiple nuclear explosions. *Science* 222: 1283–92

Turco, R. P., Toon, O. B., Ackerman, T. P., Pollack, J. B., Sagan, C. 1984a. "Nuclear winter" to be taken seriously. *Nature* 311: 307–8

Turco, R. P., Toon, O. B., Ackerman, T. P., Pollack, J. B., Sagan, C. 1984b. The climatic effects of nuclear war. *Sci. Am.* 251(8): 33–43

Turco, R. P., Toon, O. B., Ackerman, T. P., Pollack, J. B., Sagan, C. 1985. Ozone, dust, smoke and humidity in nuclear winter. *Nature* 317: 21–22

Urey, H. C. 1973. Cometary collisions and geological periods. *Nature* 242: 32–33

Volz, F. E. 1970. Atmospheric turbidity after the Agung eruption of 1963 and size distribution of the volcanic aerosol. *J. Geophys. Res.* 75: 5185–93

Weisburd, S. 1985. Excavating words: a geological tool. *Sci. News* 127: 91–94

Wexler, H. 1950. The great smoke pall— September 24–30, 1950. *Weatherwise* 3: 129–42

White, R. S. 1987. When continents rift. *Nature* 327: 191

Wilson, C. 1985a. The Little Ice Age on eastern Hudson/James Bay: the summer weather and climate at Great Whale, Fort George and Eastmain, 1814–1821, as derived from Hudson's Bay Company records. *Syllogeus* 55: 147–90

Wilson, C. 1985b. Daily weather maps for Canada, summers 1816–1818—a pilot study. *Syllogeus* 55: 191–218

Wilson, L., Head, J. W. III. 1981. Ascent and eruption of basaltic magma on the Earth and Moon. *J. Geophys. Res.* 86: 2971–3001

Wilson, L., Sparks, R. S. J., Huang, T. C., Watkins, N. D. 1978. The control of volcanic column heights by eruption energetics and dynamics. *J. Geophys. Res.* 83: 1829–36

Wood, C. A. 1984. Amazing and portentous summer of 1783. *Eos, Trans. Am. Geophys. Union* 65: 410

Ann. Rev. Earth Planet. Sci. 1988. 16: 101–19

MASS WASTING ON CONTINENTAL MARGINS

J. M. Coleman and D. B. Prior

Coastal Studies Institute, School of Geoscience,
Louisiana State University, Baton Rouge, Louisiana 70803

INTRODUCTION

Mass-wasting processes on slopes comprise a variety of natural phenomena, including landsliding, rockfalls, debris flows, mudflows, soil and rock avalanching, and slow deformational creep. The recent eruptions of Mount St. Helens in the United States and Nevado del Ruiz in Colombia caused relatively large, disastrous mass-wasting processes to occur in the subaerial environment. These mudflows were relatively large in comparison with other subaerial types of mass movements, yet they pale in comparison with subaqueous mass-wasting phenomena. Varnes (1975) justifiably claimed, "It is well to remember that the largest of all slope movements on earth appear to have occurred on the bottom of the sea. . . ."

Modern-day subaqueous slope mass-wasting processes have been known since the laying of submarine communication cables in the nineteenth century (Milne 1897, Benest 1899). By the 1930s various authors were unequivocally invoking submarine landsliding to explain anomalous sediment properties, sudden bathymetric changes, and submarine slope morphology (Archanguelsky 1930, Shepard 1932). The most striking evidence of subaqueous slope instability was the breaking of cables south of the Grand Banks (shoals in the western Atlantic, southeast of Newfoundland) after the earthquake of 1929 (Heezen & Ewing 1952). Submarine cables broke in an orderly sequence downslope, away from the earthquake epicenter, leading to the postulation that the breaks were caused by movements of sediment-laden water downslope (Heezen & Ewing 1952) or by the progressive and temporary liquefaction of continental slope sediments moving like a wave down the slope (Terzaghi 1956).

101

TECHNIQUES

The detection and recognition of continental margin mass movement rely on the identification of seafloor morphology and sediment deformation analogous to better-known terrestrial landslide geometry. Continuing improvement of acoustic remote-sensing equipment and acquisition of soil engineering properties by direct sampling and indirect measuring techniques have significantly improved our ability to map and evaluate these features, but problems still remain.

Seafloor profiles are constructed along ship survey lines by monitoring return echoes from the seafloor of emitted high-frequency sound pulses. Multiple profiles are then used to construct contoured bathymetric charts; comparisons of bathymetric charts constructed at differing times are used to detect areas of active seafloor slope instability (Shepard 1932, Coleman & Prior 1983).

A wide variety of high-frequency acoustical sources (pingers, sparkers, boomers, air guns, water guns, etc) are available that detect acoustic variations in sediment properties, each type providing different resolutions and sediment penetration depths. These data are particularly important in assessing the three-dimensional geometry and internal characteristics of subaqueous sediment instability features. In some areas, especially where evaluations of offshore drilling structure placement have been conducted, relatively dense seismic grids exist, and it is relatively easy to evaluate the geometry and characteristics of subaqueous sediment instabilities. Along the vast majority of continental margins, however, evidence for submarine slope failure is available only from widely spaced preliminary reconnaissance survey lines, and it is extremely difficult to obtain specific geometries of the landslide feature.

Side-scan sonar techniques, first developed in the 1940s, now routinely provide detailed images of the seafloor topography along swaths of predetermined widths. A transducer and receiver are towed behind the ship on a transmission cable linked to a multichannel digital or analog recorder on the vessel. The recorded image, referred to as a sonograph, can be considered analogous to an aerial photograph made when the Sun is low and behind the camera, except that a camera records reflected light, whereas the sonograph records reflected acoustic energy. This technique is schematically illustrated in Figure 1.

Low-frequency systems such as the GLORIA (6.5 kHz) can be tuned for a maximum range of 30 km, resulting in an image swath of 60 km. At such low frequencies, resolution is relatively low, but rapid coverage of large areas of the continental margin can be attained (Kenyon et al 1978, Kidd 1982, Field et al 1982).

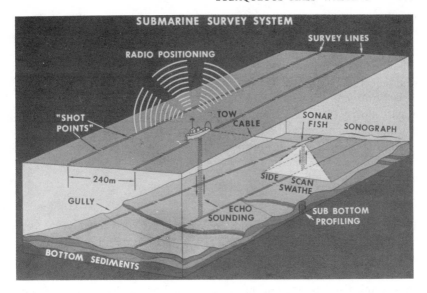

Figure 1 Schematic of modern submarine survey system used to map submarine slope instabilities (after Prior & Coleman 1984).

Medium-range systems, such as the SEAMARC I, operate at frequencies of 27/30 kHz and are commonly used at ranges of 1 to 5 km. Such systems improve considerably the resolution of seafloor morphology, but they require considerably more time to cover large areas of the continental margins. McGregor et al (1982), Ryan (1982), and Piper et al (1985) have documented the usefulness of this system for evaluating and interpreting subaqueous failures.

The highest resolution of seafloor imaging is obtained with shorter range (100–500 m), 100-kHz frequency systems. Seafloor object resolution is outstanding, and morphologic elements as small as a meter can be detected. High-frequency sonar systems, used in conjunction with high-resolution seismic techniques, can provide details of landslide geometry and internal characteristics that are considerably better than those of many subaerial investigations (Prior & Coleman 1982, Field et al 1982).

TYPES OF SUBAQUEOUS MASS WASTING

Subaqueous mass wasting is particularly difficult to classify because deformational mechanisms and processes must usually be inferred from resulting landslide geometry (remotely obtained) and from depositional characteristics observed or measured at some time after the particular event.

Various authors have proposed different classification schemes. Dott (1963) suggested four main categories: submarine falls, slides or slumps, flows, and turbidity flows. Moore (1977) omitted turbidity flows and subdivided the other categories on the basis of the type of material. Middleton & Hampton (1976) proposed the term "sediment gravity flows" and suggested that landslides may evolve into different types of gravity flows. Prior & Coleman (1984) suggested a subdivision of submarine slides and flows based on the morphology and geometry of the features (Figure 2). In this approach, the concept of a process continuum of slides to flows is maintained, but subdivisions of slide and flow phenomena are also indicated.

Falls

The free-fall of rock, mud, or sand-sized particles is restricted to very steep, near-vertical underwater slopes. Dill (1966) recorded sand falls along the walls of submarine canyons. Similarly, Prior & Doyle (1985) identified rocky debris composed of weathered Eocene carbonate rocks within canyons along the US Atlantic margin. The debris resulted from episodic rockfall activity from Eocene outcrops along steep canyon walls. Dating of canyon-floor sediments containing rockfall debris suggests that the last

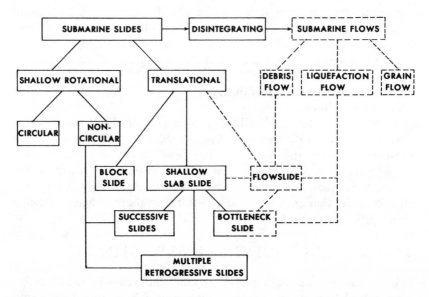

Figure 2 Diagram showing types of sediment movement processes and terminology, as well as gradations between slides (solid boxes) and flows (dashed boxes) (after Prior & Coleman 1984).

discrete period of rockfall activity began about 10,000 yr ago and ended about 6000 yr ago. Other localities where rockfalls may be anticipated are trench walls, fault scarps, and carbonate buildups.

Slides

A common type of subaqueous failure involves movement of a sediment mass along discrete basal shear planes. The volume of sediment involved in sliding is often extremely large; the Agulhas slide off South Africa involved a sediment mass of more than 20,000 km^3 (Dingle 1977), and the Storegga slide off Norway involved more than 5000 km^3 (Bugge 1983). Table 1 lists some examples of large submarine slides and their characteristics.

Three basic subdivisions of slides are possible on the basis of the geometry of basal shear surfaces: translational, multiple retrogressive, and rotational. In translational slides, the basal shear plane is inclined approximately parallel to the surface slope and slabs of soft sediment or partially lithified sediment are displaced downslope. The Straits of Florida slide (Wilhelm & Ewing 1972) and many Mississippi River delta mudslides are typical examples (Prior & Coleman 1982). Multiple retrogressive slides occur where continued failure and interaction of adjacent slides produce multiple features and extension of instability upslope (Moore 1964, Dingle 1977, Prior & Coleman 1982). Rotational slides are typified by displacement of relatively intact blocks over curved or spoon-shaped slip surfaces.

The Norwegian continental margin provides excellent examples of subaqueous slide phenomena (Bugge et al 1978, Bugge 1983). There appears to have been extensive sliding commencing in early or mid-Tertiary times, with the most recent events being associated with the Storegga slide. About 11,000 to 13,000 yr ago, a volume of 3880 km^3 of shelf-edge sediment moved downslope 400 km from the headwall into a water depth

Table 1   Characteristics of large submarine slides[a]

Location	Length (km)	Width (km)	Area (km^2)	Volume (km^3)	Slope
Agulhas, South Africa	750	106	79,488	20,331	
Bassein, Bay of Bengal	108	37	4,000	900	6–1°
Grand Banks, Newfoundland	240	140	27,500	760	4–3°
Rockall, NE Atlantic	160	13.8	2,200	300	2°
Ranger, Baja California	35	8.6	300	20	3°
Kidnappers, New Zealand	45	5.6	250	8	4–1°

[a] From Prior & Coleman (1984).

of 3000 m. The second event involved about 450 km^3 of sediment and dates from 6000 to 7000 yr ago. The entire complex comprises large arcuate head scarps indenting the shelf break and extensive deposits of blocky, acoustically transparent debris, which has been transported considerable distances over very low angles.

Such massive retrogressive slides may also play a major role in the creation and evolution of some types of submarine canyons. In the Gulf of Mexico, Coleman & Prior (1983) documented two large retrogressive failures that resulted in the formation of large (> 14 km wide, > 100 km long) submarine canyons involving sediment thicknesses in excess of 600 m.

Flows

Subaqueous downslope flows of sediment are believed to involve four main processes: debris flow, liquefaction flow, grain flow, and turbidity flow (Middleton & Hampton 1976).

Debris flows are defined as consisting of sediment in which larger clasts are supported by a matrix (a mixture of interstitial fluid and fine sediment that has a finite yield strength). Large submarine debris flows have been described by Hampton (1970, 1972) and Embley (1976). The large debris flow off the Canary Islands is typical (Figure 3; Embley 1976). Beginning with massive slide scars extending over an area of 18,000 km^2, the slide ends with finger-like lobes of deposited material covering an area of the seafloor in excess of 30,000 km^2. The total flow volume exceeds 600 km^3. The downslope debris lobes, in water depths of 4600 to 4800 m and on slopes of about 0.1°, are acoustically unstratified and overlie deep-water pelagic sediments. Cores described by Embley (1976) contained irregularly shaped clasts, folds, and pebbles, believed to be "exotic," derived from shallow-water sources. Dating of sediments indicates that the slide took place between 16,000 and 17,000 yr ago.

Similar debris flows, on a smaller scale, have been documented off the Mississippi River delta (Terzaghi 1956, Coleman & Garrison 1977, Coleman et al 1980). The debris flows begin in water depths of 10–30 m and continue downslope as elongate chutes and channels to water depths of 150 m; at the downslope ends of the chutes are extensive areas of blocky and hummocky topography composed of debris discharged from the chutes and channels.

Liquefaction flows occur when a loosely packed, coarse-grained sediment collapses; the grains temporarily lose contact with each other, and the particle weight is transferred to the pore fluid (Terzaghi 1956). Temporary excess pore-fluid pressures are induced, and downslope gravitational stresses cause sediment to flow. Typical of liquefaction flows are the

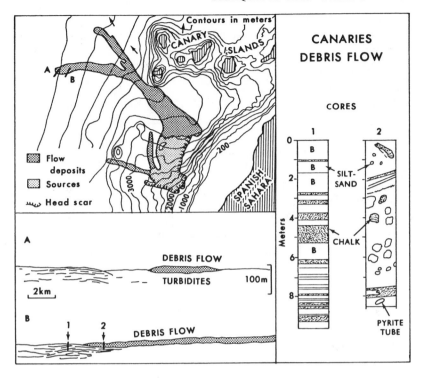

Figure 3   Characteristics of the Canary Islands debris flow (after Embley 1976).

"flowslides" in sand off Zeeland, Holland (Koppejan et al 1948), and liquefaction of shelf sands near the Klamath River delta, California (Field & Hall 1982, Field et al 1982). A liquefaction type of process seems also to be significant in fine silts and clays in the Mississippi River delta region (Prior & Suhayda 1979). Pore pressures are extremely high (approaching geostatic), and the voids contain large amounts of biogenic gas. Following initial shear failure, strain softening of the highly pressured clay/water/gas system causes remolding and strength loss, which accounts for in situ collapse and observed sinking of pipelines vertically into the sediment.

Grain flows comprise downslope sediment movement of loose individual sediment grains and have been observed by Dill (1966). These flows may be associated with submarine weathering processes and local rockfall processes.

Turbidity flows involve a downslope transport of sediment that is supported by an upward component of fluid turbulence within the density current. Menard (1964), Morgenstern (1967), and Hampton (1972) have discussed the conditions whereby slope instability can evolve into turbidity flows. Hampton (1972) cites boundary shear between advancing debris flows and the water column as a mechanism for the production of turbulent dense suspensions or turbidity flows.

MECHANISMS OF CONTINENTAL MARGIN MASS WASTING

Subaqueous slope failure thresholds for the initiation of mass wasting appear to be the result of a complex interaction of numerous variables rather than the product of a single factor. Although analytical methods are available for analysis of subaqueous instabilities, a major uncertainty is still the determination of appropriate sediment parameters before and leading to movement. Figure 4 illustrates some of the major factors that appear to be associated with subaqueous slope instability and shows how they may interact to increase stresses or to lower sediment strength.

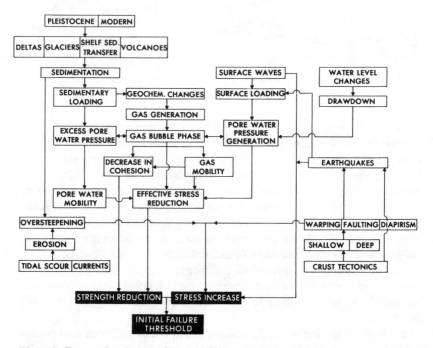

Figure 4 Factors that may combine to initiate submarine sediment and slope instability (after Coleman & Prior 1983).

Crustal Tectonics

In regions experiencing active crustal tectonics, bottom-slope angles can be increased and cyclic loading by earthquakes can cause increased pore-fluid pressures. Moore (1977) has identified particularly large slides on submarine trench walls in which deep-ocean sediments are affected by "oversteepening of slopes as part of tectonic off-scraping." Almagor & Wiseman (1977) have cited tectonic warping as a factor in initiating slides in the eastern Mediterranean.

Earthquakes cause application of horizontal and vertical acceleration stresses, leading to increased pore-fluid pressures and strength reduction. Hampton et al (1978) used an effective stress approach to examine the combined effects of earthquake-related pore pressure changes and direct loading effects of acceleration for several large slides in the Gulf of Alaska. Sediment instabilities described by Field et al (1982) in the Klamath River delta, California, appear to have been caused by a major 1980 earthquake. This earthquake induced liquefaction on slopes of 0.25°, and remolded sediment subsequently covered large areas of the seafloor.

Sedimentary Loading

Rapid sedimentation associated with prograding deltas, glacier margins, and site-specific shelf locations where sediment is concentrated may indicate situations where excess pore-fluid pressures can develop. Terzaghi (1956) described the process by which excess hydrostatic pressures can develop for areas undergoing rapid sedimentation. Increased pore-fluid pressures lead to a reduction in sediment strength and to sediment instability on relatively low slopes. Excessive sedimentary loading also contributes to vertical movement of salt and shale diapirs, which causes local oversteepening and failure of recently deposited sediments. During the last low sea-level stand (15,000 to 18,000 yr ago), rapid sedimentation occurred at the sites of the present continental shelf margins, and many continental margin failures occurred worldwide. Rapid sedimentation associated with prograding deltas results in a wide range of sediment instabilities (Shepard 1932, Coleman & Prior 1982).

Rapid Water-Level Changes

Nearshore and intertidal areas experience sediment slope failure that is related to rapid fluctuations in water level. Koppejan et al (1948), Terzaghi (1956), Wells et al (1980), and Karlsrud & Edgers (1982) have noted that failures in a wide range of settings have accompanied rapid changes in water levels, particularly with rapid drawdown associated with low tides. Seepage pressures and elevated pore water pressures develop with ebbing tide in a manner similar to drawdown, which causes river-bank instability. The Kitimat, British Columbia, failure of a delta-front slope at the head of a fjord occurred about one hour after an unusually low tide (Prior et al 1982).

Cyclic Loading by Waves

In shallow shelf regions, large surface waves apply hydraulic forces to the seafloor (Watkins & Kraft 1978). Henkel (1970) has shown theoretically that waves impose oscillatory motion on unconsolidated sediment sufficient to cause downslope movement, and his methods have been used by many investigators to explain downslope mass movement in water depths generally less than 150 m. In the Mississippi River delta, a series of field experiments (SEASWAB) measured the sediment perturbation and increases in pore-fluid pressures caused by cyclic loading during the passage of large surface waves associated with a tropical depression (Suhayda 1977, Dunlap et al 1979).

Biologically Induced Sediment Failures

In regions of more consolidated sediments and steep slopes, organisms boring into the substrate are believed to sufficiently weaken the sediment to cause failure. Steep sedimentary slopes associated with seaward margins of carbonate buildups can be weakened by intensive boring, and rockfalls result.

Biochemical processes are active in fine-grained sediment that has high organic content and may lead to the generation of pore gasses, particularly methane (Whelan et al 1976). Biogenic methane first saturates the pore fluid and then exists as bubble-phase gas within the pore spaces. The exact effects of pore gas on interparticle cohesion and friction are not known, but this gas is believed to reduce sediment strength substantially.

EXAMPLES OF CONTINENTAL MARGIN MASS WASTING

Gulf of Mexico–Mississippi River Delta

The submarine slopes of the offshore Mississippi River delta are subject to a variety of mass-movement processes; these features impose design constraints for offshore platforms and pipelines. Figure 5 illustrates schematically the relationship of the various types of sediment instabilities in the delta front associated with a prograding distributary channel. Offshore slopes are extremely small, rarely exceeding 1.5° and generally averaging 0.5°.

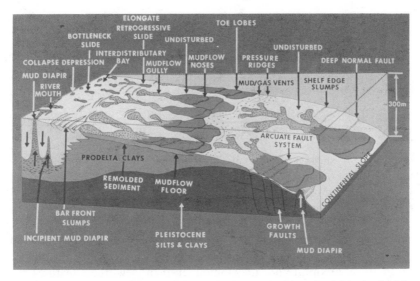

Figure 5   Schematic distribution and morphology of subaqueous landslides in the Mississippi River delta (after Prior & Coleman 1982).

PERIPHERAL ROTATIONAL SLUMPS   Downslope movement of large sediment masses begins high on the upper delta front, near the distributary mouths of the river. Arcuate scarps, ranging in height from 3 to 8 m, display curved or curvilinear plan views and give the bar front a stairstepped appearance in profile view (Figure 5). Tensional crown cracks are often present upslope from the major scarp; the surface of the slump block normally displays extensive hummocky, irregular bottom topography and displaced clasts of sediment. Shear planes are concave upward and generally merge into bedding planes downslope. This type of morphology is indicative of rotational sliding, and failure occurs when stresses exceed sediment strength because of excess pore-fluid pressures.

COLLAPSE DEPRESSIONS AND BOTTLENECK SLIDES   Collapse depressions and bottleneck slides are confined primarily to shallow-water areas of interdistributary bays (Prior & Coleman 1978). Collapse depressions are associated with relatively low slopes of 0.1–0.2° and are comparatively small (50 to 150 m). The depressions are bounded by curved escarpments up to 3 m high, within which the bottom is depressed and filled with irregular blocks of sediment. These features are interpreted to result from subsidence of parts of the seafloor and represent a decrease in the volume of the sediment-gas-water system.

On slightly steeper slopes in the interdistributary bays (0.2–0.4°) are found bottleneck slides. These features are similar to collapse depressions, but the boundary scarps do not form a totally closed perimeter around the instability (Figure 5). They have narrow openings at the downslope margins through which debris is discharged over the surrounding intact slope. Bottleneck slides vary in length from 150 to 600 m.

ELONGATE RETROGRESSIVE SLIDES   Extending radially seaward from each river mouth are major elongate bathymetric gullies, first described by Shepard (1955). Each gully has a clearly recognizable area of rotational instability or shear slumps at its upslope margin and a long, sinuous, narrow chute or channel that links a depressed, hummocky source area upslope to composite overlapping depositional lobes on the seaward end (Figure 5). The formation of elongate chutes of this type is very similar to the morphology associated with subaerial debris flows and some types of subaerial mudflows. The widths of individual gullies range from 20 to 150 m at the narrow points to 350 to 1500 m at the widest points. Lengths of up to 10 to 20 km are not uncommon. In plan view, these features are rarely straight and quite commonly are markedly sinuous, with alternating narrow constrictions or chutes and wider bulbous sections. At the seaward or downslope ends of mudflow gullies there are extensive areas of irregular bottom topography (labeled D in Figure 6) composed of discharged blocky, disturbed debris. The thickness of the discharged debris ranges from 5 to 15 m, but because of the overlapping nature of multiple discharges from the same gully system, overall thicknesses can approach 50 to 60 m. Coleman & Garrison (1977) measured a volume of 11.2×10^6 m^3 of discharged debris covering an area of 877 km^2.

SHELF-EDGE FAILURES   Near the edge of the continental shelf off the modern Mississippi River delta are found large, massive, arcuate-shaped families of shelf-edge failures and deep-seated contemporaneous or growth faults (Figure 5). Shelf-edge slumps give a stairstepped appearance to the edge of the continental shelf and are highly reminiscent of the rotational peripheral slumps described above, except these features are much larger in scale. The slumps generally display arcuate shear planes in plan view that can be traced from a few kilometers to 8–10 km; the shear planes cut into the sediment to a depth of 500 m. One failure, off South Pass, Louisiana, moved approximately 8600 km^2 of shelf and upper-slope sediments downslope a distance in excess of 300 km. The debris accumulated in water depths of 3000 m. Site 616 of Leg 96 of the Deep Sea Drilling Project drilled through this debris and revealed tilted blocks stacked one on top of another for a total thickness of 105 m of displaced material (Bouma et al 1986).

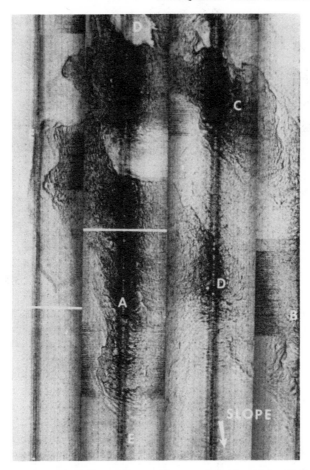

Figure 6 Scale-corrected side-scan sonar mosaic across depositional lobes (A, B, C), different block patterns (D), and pressure ridges (E). Mosaic covers an area of 1500 m × 1000 m (after Prior & Coleman 1982).

Canadian Fjords Debris Flows

Many fjords have histories of submarine slope instability involving slides and debris flows (Terzaghi 1956, Karlsrud & Edgers 1982, Prior et al 1984a,b). Kitimat Fjord, British Columbia, has a recent history of slope failure on the delta front at the head of the fjord, culminating in a major failure in 1975 (Luternauer & Swan 1978, Murty 1979). The debris flow extended a distance of 5 km down the fjord from the delta front to a water depth of 210 m, where bottom slopes range from 0.37 to 0.56°. Figure 7

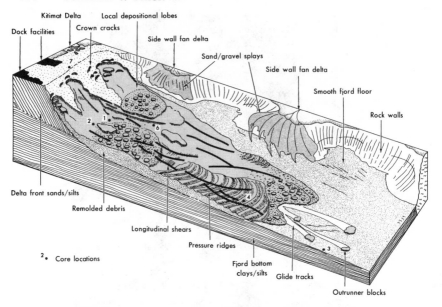

Kitimat Delta   Local depositional lobes
Dock facilities   Crown cracks
Side wall fan delta
Sand/gravel splays
Side wall fan delta
Smooth fjord floor
Rock walls
Delta front sands/silts
Remolded debris
Longitudinal shears
Pressure ridges
²• Core locations
Fjord bottom clays/silts   Glide tracks
Outrunner blocks

Figure 7 Schematic illustration of the morphology of the seafloor near the Kitimat River delta, based on bathymetry, side-scan sonar, and subbottom profile data (after Prior et al 1984b).

illustrates schematically the various types of seafloor morphology resulting from this type of mass movement. Shallow rotational slides combine to form small submarine channels on the upper delta-front slope; these channels deliver sediment to the base of the delta front and the fjord basin. Along the margins of the channels, tensional shearing of intact fjord bottom sediments allows incorporation of marine sediments into the debris flow. The central area of differential sliding and flowage consists of heavily remolded debris, longitudinal shears, irregularly shaped blocks, and large clasts. Near the distal ends of the debris flow, compressional forces have produced arcuate pressure ridges and fold systems and scattered debris blocks. The debris blocks range in size from a few meters to large tabular slabs of sediment 35–50 m across; intriguingly, some of these blocks traveled more than 500 m from the debris lobe, leaving a prominent impression, or glide track, on the seafloor. This event, containing both erosional and depositional elements, covers a total area of about 7.5 km² and involved a total volume of displaced sediment of about 55,000,000 m³.

Mid-Atlantic Continental Margin, USA

The mid-Atlantic continental margin off the east coast of the United States has been the site of numerous process-oriented and geophysical studies

(Knebel 1984, Robb et al 1981, Farre & Ryan 1987). High-resolution geophysical data indicate that the canyons and gullies that crease the continental slope here are highly complex in nature. The floors of the canyon are generally flat, vary in bottom width (average width is 500 m), and have extremely complex sidewalls displaying numerous benches and outcrop scarps. The morphology of the canyons is controlled largely by underlying bedrock characteristics. Intraslope canyons (Prior & Doyle 1985), lineations, joint patterns, subparallel stripes, and localized angular steep scarps are present.

The processes responsible for the formation of the canyons have long been debated; density currents and mass-wasting processes have obviously played a role in the evolution of the canyons. A variety of types of slumping, rockfall processes, biological erosion, undersea discharge of groundwater, and sediment resuspension by currents have played a role in shaping these canyons. One view is that the canyons are the cumulative result of long periods of gradual slope degradation rather than the products of short-term catastrophic processes (Prior & Doyle 1985).

An example of a slide involving relatively consolidated sediment is the Currituck slide, which occurred on the continental margin off Cape Hatteras, North Carolina (Bunn & McGregor 1980, Prior et al 1986). The slide area, about 20 km wide, begins at approximately 500-m water depth and continues for more than 20 km downslope to depths greater than 2000 m. Recent deep-tow seismic data indicate that the slide complex is the result of two separate slides (Figure 8). An upslope headwall encloses a

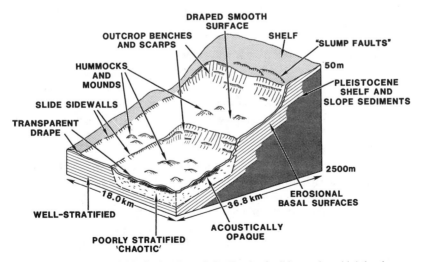

Figure 8 Illustrative schematic drawing of the Currituck slide on the mid-Atlantic continental margin (after Prior et al 1986).

stripped, smooth seafloor area, and a second headwall/sidewall system bounds a trough filled with chaotic and poorly stratified slide debris. Basal shear planes of this translational slide are inclined at 4° and 2°, close to the regional dip of the underlying strata. The total volume of the sediment in the slides is at least 128 km^3. Figure 8 illustrates schematically the morphology of these two slides. Failure of the deeper seated, lowermost slide left the upslope segment of the original continental slope unsupported and a shallower failure ensued, enlarging the entire slide complex.

SCIENTIFIC CONTROVERSIES

The remoteness and inaccessibility of submarine mass-wasting features and processes means that there are many problems and unknown aspects that are subject to debate in the scientific literature. The principal areas of controversy are as follows:

1. *Recognition* Identification of the products of mass wasting relies heavily on interpretation of acoustic imagery and recognition of diagnostic and characteristic morphologies and sediment geometries. Such interpretations are supported by sediment samples, most often limited to shallow drop cores, which penetrate only the near-surface components. Embley's (1976) recognition of large-scale debris flows is an excellent example of the combined use of acoustics and coring, and their successful correlation. In most instances, however, the evidence is ambiguous and limited by the types of acoustic systems used. For example, surface-deployed seismic profilers suffer considerable loss of resolution and penetration in deep water. Beam spreading, or an enlarged seismic "footprint," results in offline echoes, which may give false geometries. Such problems can be addressed only by using deep-towed instruments, and relatively few continental margins have been surveyed with these instruments. Even when high-quality acoustic imagery is available, there is a problem of multiple origins for the same acoustic character. "Chaotic" acoustic returns, for example, are characteristic of debris flows, near-surface gas, and sandy, cemented sediment types, and interpretation requires extensive "ground-truthing" by other techniques.

2. *Movement Initiation* Many examples of submarine mass movements have been recognized only after the fact, perhaps thousands or tens of thousands of years after their activity has ceased. The overall lack of precise age dating and our inability to "hindcast" the geologic and environmental conditions at the time of initiation of movement mean that the actual causes and factors responsible are poorly known and often controversial. While conceptual models exist for the factors likely to be involved, there

are, as yet, few fully quantified analyses with which to refine the models. The lack of precision as to causes means that prediction of the location, magnitude, and timing of future events remains extremely difficult. On-site, specific engineering studies, the most successful approach to evaluation of future risk, rely upon definition of site characteristics and broad conservative estimations of future conditions, such as the magnitude of future storms, earthquakes, etc.

3. *Movement Mechanisms*   Unlike mass-movement processes on land, where there are actual eyewitness accounts, knowledge of how subaqueous landslides move on underwater slopes is extremely poor. Our understanding of movement mechanisms, velocity, thickness, and how movement alters sediment properties with time is based almost exclusively on back-analysis from sediment properties after movement has ceased. This aspect remains perhaps one of the most significant unknowns, resulting in our inability to forecast distances of travel and forces within the moving material, which may damage underwater engineering facilities.

4. *Mass-Wasting Sediments*   In view of the uncertainties about movement mechanisms, the recognition of sediment bodies and layers emplaced by mass-wasting processes is fraught with difficulties. Our perception is limited to the most obvious types, such as sheared and rotated blocks caused by sliding, supported blocks and clasts within remolded matrixes resulting from debris flows, or various turbidite sequences. Problems remain in the correct identification of complex systems where movement and sedimentation dynamics have changed within individual mass-wasting episodes.

Literature Cited

Almagor, G., Wiseman, G. 1977. Analysis of submarine slumping in the continental slope off the southern coast of Israel. *Mar. Geol.* 2: 349–88

Archanguelsky, A. D. 1930. Slides of sediments on the Black Sea bottom and the importance of this phenomenon for geology. *Bull. Soc. Nature, Moscow* 38: 38–80

Benest, H. 1899. Submarine gullies, river outlets and fresh water escapes beneath the sea level. *Geogr. J.* 14: 394–413

Bouma, A. H., Coleman, J. M., Meyers, A. W., et al, eds. 1986. *Initial Reports of the Deep Sea Drilling Project,* Vol. 96. Washington, DC: US Govt. Print. Off.

Bugge, T. 1983. Submarine slides on the Norwegian continental margin, with special emphasis on the Storegga area. *Publ. No. 10,* Cont. Shelf Inst., Trondheim, Norway. 152 pp.

Bugge, T., Reidar, L. L., Rokoengen, R. 1978. Quaternary deposits off More and Trondlag, Norway: seismic profiling. *Publ. No. 99,* Cont. Shelf Inst., Trondheim, Norway. 55 pp.

Bunn, A. R., McGregor, B. A. 1980. Morphology of the North Carolina continental slope, western North Atlantic, shaped by deltaic sedimentation and slumping. *Mar. Geol.* 37: 253–66

Coleman, J. M., Garrison, L. E. 1977. Geological aspects of marine slope stability, northwestern Gulf of Mexico. *Mar. Geotechnol.* 2: 9–44

Coleman, J. M., Prior, D. B. 1982. Deltaic sand bodies. *Am. Assoc. Pet. Geol. Cont. Educ. Course Note Ser. No. 15.* 171 pp.

Coleman, J. M., Prior, D. B. 1983. In *Deltaic Environments of Deposition, Am. Assoc. Pet. Geol. Mem.,* 31: 139–78

Coleman, J. M., Prior, D. B., Garrison, L.

E. 1980. Subaqueous sediment instabilities in the offshore Mississippi River delta. *Rep. No. 80-01*, Bur. Land Manage., New Orleans, La, 48 pp.

Dill, R. F. 1966. Sand flows and sand falls. In *Encyclopedia of Oceanography*, ed. R. W. Fairbridge, pp. 763–65. New York: Reinhold

Dingle, R. V. 1977. The anatomy of a large submarine slump on a sheared continental margin (southeast Africa). *J. Geol. Soc. London* 3: 293–310

Dott, R. H. 1963. Dynamics of subaqueous gravity depositional processes. *Am. Assoc. Pet. Geol. Bull.* 47: 104–28

Dunlap, W. A., Bryant, W. R., Williams, G. N., Suhayda, J. N. 1979. Storm wave effects on deltaic sediments—results of SEASWAB I and II. *Proc. Conf. Port Ocean Eng. Under Arct. Cond., Trondheim, Norway*, 2: 899–920

Embley, R. W. 1976. New evidence for occurrence of debris flow deposits in the deep sea. *Geology* 4: 371–74

Farre, J. A., Ryan, W. B. F. 1987. Surficial geology of the continental margin offshore New Jersey in the vicinity of DSDP Sites 612 and 613. In *Initial Reports of the Deep Sea Drilling Project*, ed. C. W. Poag, A. B. Watts et al, 95: 725–59. Washington, DC: US Govt. Print. Off.

Field, M. E., Hall, R. K. 1982. Sonographs of submarine sediment failure caused by the 1980 earthquake off northern California. *Geomar. Lett.* 2: 135–41

Field, M. E., Gardner, J. V., Jennings, A. E., Edwards, B. D. 1982. Earthquake-induced sediment failures on a 0.25° slope, Klamath River delta, California. *Geology* 10: 542–46

Hampton, M. A. 1970. *Subaqueous debris flow and generation of turbidity currents*. PhD thesis. Stanford Univ., Stanford, Calif.

Hampton, M. A. 1972. The role of subaqueous debris flow in generating turbidity currents. *J. Sediment. Petrol.* 42: 775–93

Hampton, M. A., Bouma, A. H., Sangrey, D. A., Carlson, P. R., Molnia, B. M., Clukey, E. C. 1978. Quantitative study of slope stability in the Gulf of Alaska. *Proc. Offshore Tech. Conf., Houston, Tex.* Pap. 3314, pp. 2307–18

Heezen, B. C., Ewing, M. 1952. Turbidity currents and submarine slumps and the 1929 Grand Banks earthquake. *Am. J. Sci.* 250: 849–73

Henkel, D. J. 1970. The role of waves in causing submarine landslides. *Geotechnique* 20: 75–80

Karlsrud, K., Edgers, L. 1982. Some aspects of submarine slope stability. In *Marine Slides and Other Mass Movements*, ed. S.

Saxov, J. K. Nieuwenhuis, 6: 61–81. New York/London: Plenum

Kenyon, N. H., Belderson, R. H., Stride, A. H. 1978. Channels, canyons and slump folds on the continental margin between southwest Ireland and Spain. *Oceanol. Acta* 1: 369–80

Kidd, R. B. 1982. Long-range sidescan sonar studies of sediment slides and the effects of slope mass sediment movement on abyssal plain sedimentation. In *Marine Slides and Other Mass Movements*, ed. S. Saxov, J. K. Nieuwenhuis, 6: 289–303. New York/London: Plenum

Knebel, H. J. 1984. Sedimentary processes on the Atlantic continental slope of the United States. *Mar. Geol.* 61: 43–74

Koppejan, A. W., van Wamelan, B. M., Weinberg, L. J. H. 1948. Coastal flow slides in the Dutch province of Zeeland. *Proc. Conf. Soil Mech. Found. Eng., 5th*, pp. 89–96

Luternauer, J. L., Swan, D. 1978. Kitimat Submarine Slump deposit(s): a preliminary report, part A. *Geol. Surv. Can. Pap. 78-1A*, pp. 327–32

McGregor, B. A., Stubblefield, W. C., Ryan, W. F. B., Twichell, D. C. 1982. Wilmington submarine canyon: a fluvial-like system. *Geology* 10: 27–30

Menard, H. W. 1964. *Marine Geology of the Pacific*. New York: McGraw-Hill. 271 pp.

Middleton, G. V., Hampton, M. A. 1976. Subaqueous sediment transport and deposition of sediment gravity flows. In *Marine Sediment Transport and Environmental Management*, ed. D. J. Stanley, D. J. P. Swift, pp. 197–218. New York: Wiley

Milne, L. 1897. Suboceanic changes. *Geogr. J.* 10: 129–46, 259–89

Moore, D. G. 1977. Submarine slides. In *Rockslides and Avalanches, Dev. Geotech. Eng.*, ed. B. Voight, 1: 563–604. Amsterdam: Elsevier

Moore, J. G. 1964. Giant submarine landslides on the Hawaiian ridge. *US Geol. Surv. Prof. Paper 501D*, pp. 95–98

Morgenstern, R. N. 1967. Submarine slumping and the initiation of turbidity currents. In *Marine Geotechnique*, ed. A. Richards, pp. 189–220. Urbana: Univ. Ill. Press

Murty, T. S. 1979. Submarine slide-generated water waves in Kitimat Inlet, British Columbia. *J. Geophys. Res.* 84: 7777–79

Piper, D. J. W., Shor, A. N., Farre, J. A., O'Connell, S., Jacobi, R. 1985. Sediment slides and turbidity currents on the Laurentian Fan: sidescan sonar investigations near the epicenter of the 1929 Grand Banks earthquake. *Geology* 13: 538–41

Prior, D. B., Coleman, J. M. 1978. Submarine landslides on the Mississippi River delta-front slope. In *Geoscience and Man*, 19: 41–53. Baton Rouge, La.: State Univ. Press

Prior, D. B., Coleman, J. M. 1982. Active slides and flows in underconsolidated marine sediments on the slopes of the Mississippi delta. In *Marine Slides and Other Mass Movements*, ed. S. Saxov, J. K. Nieuwenhuis, 6: 21–49. New York/London: Plenum

Prior, D. B., Coleman, J. M. 1984. Submarine slope instability. In *Slope Instability*, ed. D. Brunsden, D. B. Prior, pp. 419–55. New York: Wiley

Prior, D. B., Doyle, E. H. 1985. Intra-slope canyon morphology and its modification by rockfall processes, U.S. Atlantic continental margin. *Mar. Geol.* 67: 177–96

Prior, D. B., Suhayda, J. N. 1979. Application of infinite slope analysis to submarine sediment instabilities, Mississippi delta. *Eng. Geol.* 14: 1–10

Prior, D. B., Bornhold, B. D., Coleman, J. M., Bryant, W. R. 1982. Morphology of a submarine slide, Kitimat Arm, British Columbia. *Geology* 10: 588–92

Prior, D. B., Coleman, J. M., Doyle, E. H. 1984a. Antiquity of the continental slope, U.S. middle-Atlantic margin. *Science* 223: 926–28

Prior, D. B., Bornhold, B. D., Johns, M. W. 1984b. Depositional characteristics of a submarine debris flow. *J. Geol.* 92: 707–27

Prior, D. B., Doyle, E. H., Neurauter, T. 1986. The Currituck slide, mid-Atlantic slope—revisited. *Mar. Geol.* 73: 25–45

Robb, J., Hampson, J., Twichell, D. 1981. Geomorphology and sediment stability of a segment of the U.S. continental slope off New Jersey. *Science* 211: 935–37

Ryan, W. B. F. 1982. Imaging of submarine landslides with wide-swath sonar. In *Marine Slides and Other Mass Movements*, ed. S. Saxov, J. K. Nieuwenhuis, 6: 175–88. New York/London: Plenum

Shepard, F. P. 1932. Landslide modifications of submarine valleys. *Trans. Am. Geophys. Union* 13: 226–30

Shepard, F. P. 1955. Delta front valleys bordering the Mississippi distributaries. *Geol. Soc. Am. Bull.* 66: 1489–98

Suhayda, J. N. 1977. Surface waves and bottom sediment response. *Mar. Geotechnol.* 2: 135–46

Terzaghi, K. 1956. Varieties of submarine slope failures. *Proc. Tex. Soil Mech. Eng. Conf., 8th*, pp. 1–41

Varnes, D. J. 1975. Slope movements in the western United States. In *Mass Wasting, Guelph Symp. Geomorphol., 4th*, ed. E. Yatsu, A. J. Ward, F. Adams, pp. 1–17

Watkins, D. J., Kraft, L. M. 1978. Stability of continental shelf and slope off Louisiana and Texas: geotechnical aspects. In *Framework, Facies, and Oil Trapping Characteristics of the Upper Continental Margin*, ed. A. Bouma, G. T. Moore, J. M. Coleman, 7: 267–86. Tulsa, Okla: Am. Assoc. Pet. Geol.

Wells, J. T., Prior, D. B., Coleman, J. M. 1980. Flowslides in muds on extremely low angle tidal flats, northeastern South America. *Geology* 8: 272–75

Whelan, T., Coleman, J. M., Roberts, H. H., Suhayda, J. N. 1976. The occurrence of methane in Recent deltaic sediments and its effect on soil stability. *Int. Assoc. Eng. Geol. Bull.* 14: 55–64

Wilhelm, O., Ewing, M. 1972. Geology and history of the Gulf of Mexico. *Geol. Soc. Am. Bull.* 83: 575–600

Ann. Rev. Earth Planet. Sci. 1988. 16: 121-45

EARTHQUAKE GROUND MOTIONS[1]

Thomas H. Heaton and Stephen H. Hartzell

U.S. Geological Survey, Seismological Laboratory, California Institute of
Technology, Pasadena, California 91125

INTRODUCTION

In this review, we discuss some of the many phenomena that can determine
the nature of strong ground motion. Some of these phenomena have been
inferred from the study of existing ground-motion records, and others
are inferred from theoretical models of earthquake sources and wave
propagation.

Perhaps the most common question posed in engineering seismology is,
what is the size of ground motion at a given distance from an earthquake
of a given magnitude? The answer to this question is surprisingly ambiguous. To illustrate, we show response spectra from horizontal ground
motions recorded at distances near 50 km from shallow, crustal, strike-
slip earthquakes of about magnitude 6.5 in Figure 1*a* (from Heaton et al
1986). The distances actually vary from 36 to 65 km and the magnitudes
from $M_w = 6.1$ to $M_w = 6.6$, and the spectra have been scaled to account
for differences in these parameters. Even so, the largest motion is over 10
times larger than the smallest, and it is obvious that a wide variety of
ground motions have occurred at a distance of about 50 km from magnitude 6.5 strike-slip earthquakes. The same degree of scatter can be seen
in Figures 1*b* and 1*c* (from Heaton & Hartzell 1987), where we show
response spectra (unscaled) from shallow subduction earthquakes between
magnitude 7.0 and 7.5 and observed at distances between 50 and 100 km,
and between 101 and 150 km, respectively. This problem arises in virtually

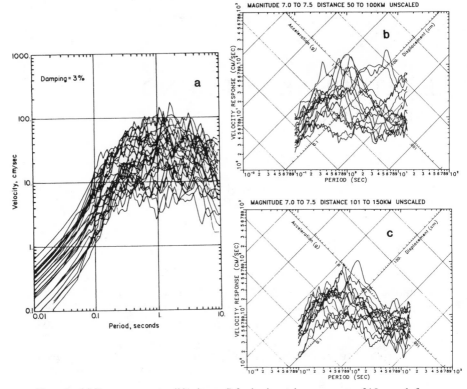

Figure 1   (*a*) Response spectra (3% damped) for horizontal components of 15 records from strike-slip earthquakes that are scaled to a distance of 50 km and a magnitude of 6.5 (from Heaton et al 1986). (*b*) Response spectra (5% damped) of horizontal components (not scaled for distance or magnitude) of 7 records from sites ranging between 50 and 100 km from shallow subduction earthquakes having magnitudes between 7.0 and 7.5 (from Heaton & Hartzell 1987). (*c*) Same as (*b*), except that sites range between 101 and 150 km. These composite plots show the type of scatter that is typical of ground motions recorded at a given distance and magnitude.

any data set that characterizes ground-motion amplitude as a function of distance and magnitude.

What phenomena are responsible for this large scatter, and what other parameters, besides distance and magnitude, can be used to predict near-source ground motions? The answer to this problem divides naturally into two classes of phenomena: those that are related to the rupture process (the source), and those that are related to the propagation of waves between the source and observer (path effect). Since the seismic velocity structure of the Earth is essentially constant in time, we expect that the effects of wave propagation can be included in any attempt to predict near-source ground motion. Predicting the rupture characteristics of future earthquakes may be impractical for the foreseeable future.

EFFECTS DUE TO WAVE PROPAGATION

The *P*- or *S*-waves that are radiated from a simple point dislocation in a homogeneous whole-space are described simply by a radiation pattern and an inverse distance-amplitude decay; the motion at the receiver is described by the time derivative of the dislocation history (ignoring, for the moment, near-field terms). For the sake of simplicity, ground-motion phenomena are sometimes interpreted assuming that our universe is indistinguishable from a homogeneous whole-space. In reality, seismic waves travel through a medium having a free surface, systematic variations (usually increases) of velocity with depth, large-scale lateral variations (mountains and basins), small-scale lateral variations (scatterers), and dramatically different elastic properties at individual observation sites (soil conditions). Even for very simple sources, such as explosions, very complex wave trains result from the propagation of waves through such complex media. When one is confronted with the variety of ground motions that often result from different stations recording the same earthquake, it is easy to despair at the seemingly uninterpretable variation in waveforms. However, careful studies of ground motions from well-recorded earthquakes and increasingly more realistic numerical models of wave propagation in complex structures are steadily improving our ability to understand and predict these complex waveforms. Studies of ground motions in densely instrumented regions have demonstrated that near-source waveforms, although complex, are quite coherent (Spudich & Cranswick 1984, Hanks 1975, Liu & Heaton 1984).

In Figure 1, we showed the large variability of ground motions encountered at similar distances from similar-sized earthquakes. In Figure 2, we show response spectra from several different earthquakes but recorded at the same sites (from Heaton & Hartzell 1987). Clearly, the particular location of the recording site greatly influences the shape of the response spectra. What are some of the factors that control the path effect?

Near-Site Soils

The 19 September 1985 magnitude 8.1 subduction earthquake on the west coast of Michoacán, Mexico, has provided one of the most dramatic examples of the effects of localized site geology. Ground acceleration time histories for a variety of sites are shown in Figure 3 (from Singh et al

124   HEATON & HARTZELL

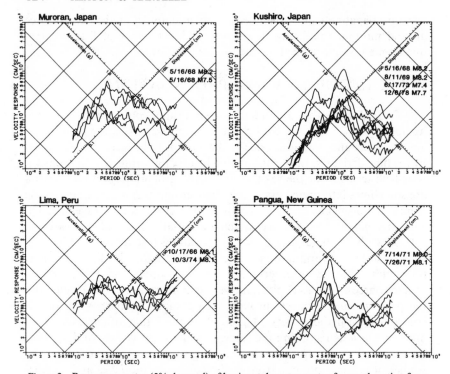

Figure 2 Response spectra (5% damped) of horizontal components of ground motion from several different earthquakes recorded at the same site. Notice that features in the response spectra are similar from one earthquake to another, thus indicating the importance of path effects (from Heaton & Hartzell 1987).

1987). Ground motions in the vicinity of Mexico City are dramatically enhanced at periods of several seconds when compared with records from hard-rock sites closer to the earthquake. This phenomenon was previously noted by Zeevaert (1964), who interpreted it as the excitation of a fundamental shear modal vibration within the upper 50 m of very low shear-velocity (about 60 m s^{-1}) lacustrine deposits underlying parts of Mexico City. Tsai (1969, pp. 152–60) even generated synthetic ground motions that were similar to those that occurred in September 1985.

Similar amplification has been proposed in other regions, such as the soft San Francisco Bay muds (Borcherdt 1970). Differences in ground motions with depth in a 186-m-deep hole in the San Francisco Bay muds were well modeled by assuming plane shear waves in a horizontally stratified medium (Joyner et al 1976).

A recording of the ground motion in Niigata, Japan, from the 12 June

EW component of acceleration, 19 Sep 1985

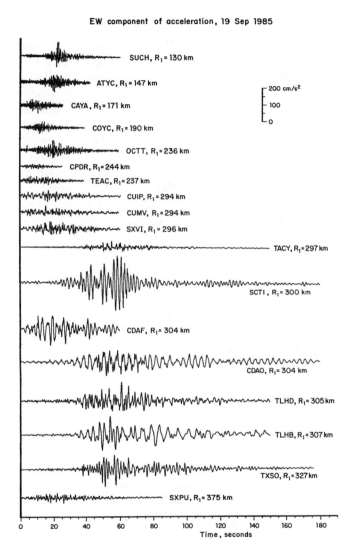

Figure 3   East-west component of acceleration from the 19 September 1985 Michoacán, Mexico, earthquake from selected sites. Records are shown in order of increasing distance from the rupture area. Sites CUIP, CUMV, SXVI, TACY, SCT1, CDAF, CDAO, TLHD, and TLHB are all in Mexico City. Note the variation in ground-motion characteristics for this earthquake (from Singh et al 1987; copyright Seismological Society of America).

1964 magnitude 7.5 earthquake (Figure 4) provides another example of a dramatic site effect. High-frequency ground motions can only be seen in the first 6 s of the record, after which there is an abrupt transition to much longer-period motion (from Ishihara 1985). This site experienced extensive liquefaction during this event, and it is believed that the abrupt transition from short-period to long-period motion represents the onset of liquefaction. Liquefaction occurs during earthquakes when shaking causes materials to lose stiffness or strength, and it is usually associated with saturated, cohesionless soils. This is one of the most striking examples of nonlinear, nonelastic site response. Ishihara et al (1981) documented a less dramatic, but convincing, example of nonelastic site response in which they show an increase in pore fluid pressures (presumably caused by the incremental volume changes in the soil) coincident with strong shaking on a hydraulic-fill island in Tokyo Bay during an earthquake on 25 September 1980. However, the pore fluid did not reach values corresponding to complete liquefaction of the soil.

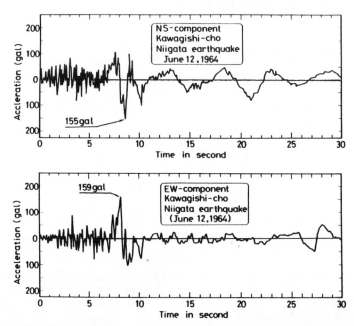

Figure 4   Ground acceleration from a site in Niigata, Japan, that experienced liquefaction. The change in frequency content of the record at about 8 s may be due to a loss of stiffness (liquefaction) in the materials beneath the site (Ishihara 1985; figure from National Academy of Sciences 1985).

Basins

In a study of strong-motion data (principally from the 9 February 1971 San Fernando earthquake), Trifunac (1976) pointed out that larger ground velocities and displacements tend to occur more at sites characterized as "soft soils" than at sites on "hard rock." Curiously, this conclusion does not hold for the higher-frequency motions characterized by ground acceleration. This same tendency has also been observed in other strong-motion data sets (e.g. Joyner & Boore 1982, Kawashima et al 1984). Although this observation is often thought of as a local site effect, careful examination of the ground motions from the 9 February 1971 San Fernando earthquake leads to the conclusion that the excitation of surface waves (both Love and Rayleigh) within the San Fernando and Los Angeles basins is the origin of the large long-period ground motions observed at soil sites (Hanks 1975, Liu & Heaton 1984). In Figure 5 (from Liu & Heaton 1984), we show ground-velocity records versus distance along an epicentral distance profile crossing two major basins, together with a profile of the subsurface geology. It is evident that long-period motions are predominantly surface waves that are excited within each basin, and that these surface waves do not propagate across the Santa Monica Mountains that separate these two basins. Vidale (1987) constructed finite-difference wave-propagation models to simulate the effects of this geologic structure, and he was able to reproduce many of the features seen in these records.

Basin structures may also serve to focus P and S body waves in much the same way that a lens focuses light (Rial 1984). Ihnen & Hadley (1986) have constructed three-dimensional body-wave ray-trace models of the Puget Sound region for the 29 April 1965 Seattle earthquake (magnitude 6.5). This earthquake occurred at a depth of 65 km beneath Puget Sound and created a complex distribution of shaking intensities that can be explained, in part, by the focusing of body waves.

f_{max}

It has long been recognized that the average Fourier amplitude spectra of strong ground acceleration is approximately constant between frequencies of approximately 0.3 and 15 Hz. It seems clear that the spectra fall off at low frequencies because of the finite duration of seismic sources. However, there have been several theories to explain the spectral falloff at high frequencies. Hanks (1982) has defined the term f_{max} to be the frequency at which an acceleration spectrum begins to fall off at high frequencies. We show an example of this spectral falloff in Figure 6 from Anderson & Hough (1984). They point out that the spectral decay at high frequencies seems to fit the functional form e^{-kf} better than it does the form f^{-v}. Their

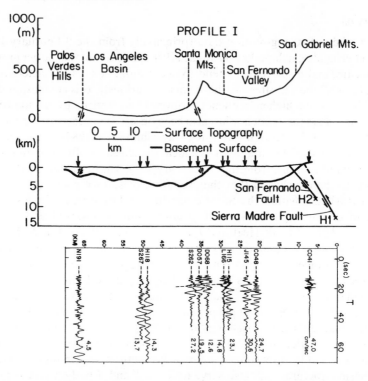

Figure 5 Transverse component of ground velocities measured during the 9 February 1971 San Fernando earthquake. Records are plotted as a function of epicentral distance along a profile running south across the San Fernando Valley and the Los Angeles basin. The corresponding free-surface and basement-surface profiles are shown to the left. The dashed line indicates the possible phase arrival of surface waves. Notice that apparent surface waves seen within the basins do not appear to propagate across the Santa Monica Mountains (from Liu & Heaton 1984).

interpretation is that this functional form is easiest to explain by anelastic attenuation. They further suggest that this occurs as the waves propagate through the uppermost several kilometers of the crust. This interpretation is reinforced by direct observations of body waves in deep boreholes (Malin & Walker 1985, Hauksson et al 1987). Although we presently favor the attenuation model for f_{max}, there are other models in which f_{max} is explained as a source effect resulting from a characteristic distribution of fault slip (e.g. Aki 1987).

The attenuation model for f_{max} may explain the surprising observation of very high-frequency, vertical-component *P*-waves that are sometimes recorded simultaneously with low-frequency, horizontal-component *S*-

CUCAPAH   85°
June 9, 1980   03:28 GMT

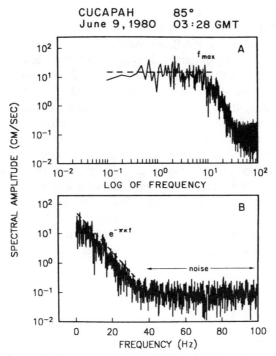

Figure 6 Fourier amplitude spectrum of the N85°E component of ground acceleration recorded at Cucapah from the Mexicali Valley, Mexico, earthquake of 9 June 1980 (M_L = 6.2). (*A*) Log-log axes. (*B*) Linear-log axes. Plot shows that spectral amplitudes decay at very high frequencies in an exponential manner typical of anelastic attenuation (from Anderson & Hough 1984; copyright Seismological Society of America).

waves. Liu & Helmberger (1985) demonstrated that very strong shear-wave attenuation could be used to model high-frequency *P*-waves and low-frequency *S*-waves that were observed in the strong motions of the 15 October 1979 Imperial Valley earthquake and its aftershocks (Brady et al 1982, Anderson & Heaton 1982).

Eastern US vs. Southwestern US vs. Subduction Zone

The propagation of seismic waves through the crust is an important factor that influences the nature of ground motions at any site. Although the detailed description of the physics necessary to fully solve this problem may be very complex, simple distance attenuation laws have been developed to approximate the effect of this complex wave propagation. We compare some representative attenuation laws for response spectral velocity at 1 s and for different regions in Figure 7. Because of the relatively large data set for earthquakes up to magnitude $6\frac{3}{4}$ from the southwestern United States (Joyner & Boore 1982), an average attenuation law is fairly well determined for this region, and values for both "rock" and "soil" sites are given. (The scatter of the data about that law is very large.) In contrast, there are very few strong-motion data available from the eastern US, and consequently there is considerable uncertainty regarding eastern US ground motions. However, it seems clear that earthquakes in the eastern US have been accompanied by relatively large areas of moderate shaking intensity (Nuttli 1973). It has been proposed that regional differences in the characteristics of the phase L_g is the primary reason that eastern US ground motions may decay with distance less rapidly than in the southwestern US (Shin & Hermann 1987). The eastern US attenuation curve in Figure 7 is from Boore & Atkinson (1987), who assumed that eastern US earthquakes are similar to those in the southwestern US and hence that the ground motions are very similar at close distances.

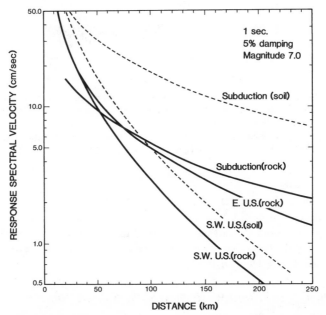

Figure 7 Comparison of distance attenuation laws derived from different regional data sets. The southwestern US curves are derived chiefly from recordings of earthquakes smaller than $M = 7$ (Joyner & Boore 1982); the eastern US curve is derived assuming the southwestern US ground motions in the epicentral region, together with a distance attenuation law derived from small-earthquake data (Boore & Atkinson 1987); and the subduction-zone curves are derived chiefly from large (magnitude greater than 7) Japanese earthquakes (Kawashima et al 1984).

The distance attenuation law derived using ground motions from large Japanese subduction earthquakes (Kawashima et al 1984) is surprisingly different from the attenuation law for the southwestern US. The relatively gentle distance attenuation law is a reflection of the fact that large ground motions have been observed at surprisingly large distances from large subduction earthquakes (Heaton & Hartzell 1987). Curiously though, ground motions for Japanese earthquakes of magnitude less than 7.0 seem roughly comparable to those that have been observed in the southwestern US (Heaton et al 1986). At this point, we do not know whether these observations indicate that the distance attenuation law is fundamentally different for subduction zones, or whether the distance attenuation law changes for very large subduction earthquakes as the source dimensions become large.

Fault Zones as Waveguides

There is evidence that major fault zones, such as the San Andreas, are associated with a zone of relatively low seismic velocities that is as much as several kilometers wide (Feng & McEvilly 1983). Cormier & Spudich (1984) have modeled the propagation of high-frequency body waves from earthquakes occurring in low-velocity fault zones. They find that the low-velocity region serves as a waveguide, and amplification of as much as a factor of 10 can occur for body waves observed at sites within the fault zone. Cormier & Beroza (1987) suggest that the high accelerations observed on the Calaveras fault during the 24 April 1984 Morgan Hill, California, earthquake may have been caused by focusing of energy along the fault zone.

EFFECTS DUE TO THE SEISMIC SOURCE

We use the term *source effects* to signify those phenomena that result from the distribution of slip (in time and space) that occurs during earthquakes. Although this distribution may be quite complex, it is often convenient to assign a single number to an earthquake to indicate the earthquake's size. This number is usually a magnitude, a moment, or sometimes an energy. It is beyond the scope of this paper to discuss the various methods of earthquake quantification, but the subject is discussed in a general way by Kanamori (1978), Båth (1981), and Heaton et al (1986). Although the quantification of an earthquake source with a single number may be convenient, it cannot hope to provide sufficient information to allow a specific prediction of the seismic waves that will be radiated from the source region. As an example of the importance of the seismic source on strong ground motions, we show ground motions in Figure 8 for two

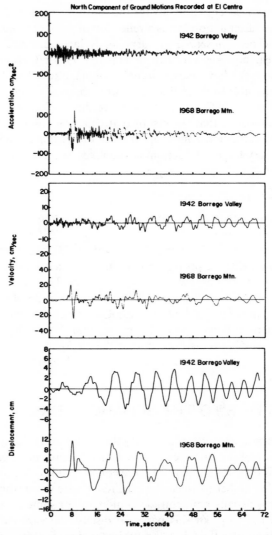

Figure 8   North component of ground motion for two $M = 6.5$ Borrego, California, earthquakes as recorded at El Centro. Note that the 1942 records are plotted on an amplitude
scale half as large as that used to plot the 1968 records. Although these motions were recorded
at the same station and were from comparable earthquakes, the 1968 velocity records are
very different from those in 1942. However, similar long-period surface waves were recorded
for both events (from Heaton et al 1986).

different earthquakes that were recorded at the same site in El Centro, California. These earthquakes have both been assigned surface-wave magnitudes (M_s) of 6.5, and they occurred very close to each other, perhaps on different segments of the same fault system. A close inspection of the records reveals striking similarities and differences; the long-period surface-wave parts of the record are very similar, and yet the higher frequency velocity and acceleration records are very different. Although these records are not yet fully understood, it seems likely that the large pulse seen at the beginning of the record from the 1968 Borrego Mountain earthquake is a direct effect of the rupture process.

We now discuss some of the ways that the seismic source influences ground motions.

Directivity

If an earthquake rupture propagates along a fault at some velocity, then observers that lie in the direction of propagation will record shorter duration (and hence higher amplitude) ground motions than observers located opposite the direction of rupture propagation. This phenomenon is called directivity, and it is similar to the Doppler effect. One of the most convincing examples of this phenomenon is from the 15 October 1979 Imperial Valley, California, earthquake. Low-pass-filtered ground velocities are displayed on a map showing the location of the recording stations and the causative fault in Figure 9 (from Archuleta 1984). Rupture propagated to the northwest from the epicenter, located just south of the Mexico–United States border. Although the stations BCR at the southeastern end of the rupture and E06 at the northwestern end of the rupture are located approximately symmetrical with respect to the fault plane, E06 (and adjacent sites) recorded a large pulse of ground motion that does not appear at BCR. Hartzell & Helmberger (1982), Olsen & Absel (1982), Hartzell & Heaton (1983), and Archuleta (1984) have all successfully modeled these records using directivity. Strong motion data from the 24 January 1980 Livermore Valley, California, earthquake provide another dramatic example of directivity (Boatwright & Boore 1982). Heaton & Helmberger (1979) have also concluded that the large velocity pulse observed on the famous Pacoima Dam recording of the 9 February 1971 San Fernando earthquake is the result of directivity. Although these are some of the most dramatic examples of directivity, this phenomenon is certainly fairly common, since rupture propagation along finite faults is endemic to earthquake sources.

Asperities and Barriers

The 15 October 1979 Imperial Valley earthquake occurred in a relatively simple geologic structure (the Salton trough), and it was well recorded in

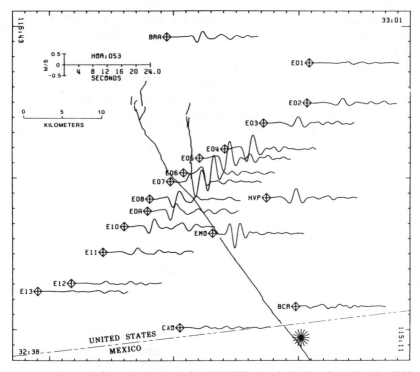

Figure 9   Low-pass-filtered ground velocities (N37°W) recorded during the 15 October 1979 Imperial Valley, California, earthquake. The rupture propagated from its southern epicenter (asterisk) northwestward along the Imperial fault. Stations located along the fault toward the northern end of the rupture recorded much larger ground velocities than did more southern stations (from Archuleta 1984; published by the American Geophysical Union).

the near-source region (Figure 9). Because of these factors, there has been considerable success in the modeling of the rupture process of this earthquake (Hartzell & Heaton 1983, Archuleta 1984). In Figure 10a, we show the final slip distribution for the preferred rupture model of Hartzell & Heaton (1983). Although the total rupture length is nearly 40 km, a large part of the slip is localized to a patch approximately 6 km long. Such patches of higher slip are referred to as asperities, and the asperity in the models of the Imperial Valley earthquake plays a central role in the simulation of the observed ground motions. Furthermore, Spudich & Cranswick (1984) conclude that this asperity is the probable source of much of the very high-frequency *P*-wave motions observed for this earthquake.

Although the geometry of the fault, geologic structure, and recording

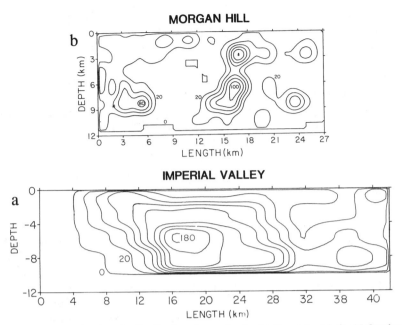

Figure 10 Map showing slip distributions (in centimeters) inferred for (*a*) the 15 October 1979 (*M* = 6.5) Imperial Valley earthquake (Hartzell & Heaton, 1983) and (*b*) the 24 April 1984 (*M* = 6.2) Morgan Hill, California, earthquake (Hartzell & Heaton 1986). Asperities (localized regions of large slip) are thought to play an important role in the radiation of seismic waves from these events. The asterisk in (*b*) shows the hypocentral location.

sites was not quite as advantageous for study of the source of the 24 April 1984 Morgan Hill earthquake, there are clear indications that there were at least two asperities that strongly affect the nature of the ground motions for this earthquake. We show the preferred slip-distribution model of Hartzell & Heaton (1986) in Figure 10*b*. They report that many features of the observed ground motion are explained by the failure of a small asperity near the hypocenter followed by a much larger asperity failure about 12 km to the south.

Asperities have also been introduced into models for large subduction earthquakes. Although the asperities are on a much larger scale (tens of kilometers), there is evidence that the magnitude 8.1 Michoacán, Mexico, earthquake of 19 September 1985 may be thought of as the rupture of two separate large asperities. These two events can be seen in both the strong ground motions (Anderson et al 1987) and in the teleseismic *P*-waves (Houston & Kanamori 1986). The correspondence between asperities inferred from teleseismic *P*-waveforms and ground motions recorded in the near-source region is discussed for a number of large subduction earthquakes by Heaton & Hartzell (1987). Kanamori (1986) and Lay et al (1982) review evidence for asperity distributions in large subduction earthquakes.

Aki et al (1977) and Das & Aki (1977) propose that the rupture along faults may be interrupted by patches that do not rupture in any given earthquake. The patches are called barriers, and if the rupture terminates abruptly, then high-frequency "stopping phases" will be generated. Papageorgiou & Aki (1983a,b) present barrier models for a number of historic earthquakes. One feature of these models is that they constitute an alternative explanation for the f_{max} phenomena discussed previously in this paper.

Roughness

Ground acceleration time histories in the near-source region have many similarities to random time series (Jennings et al 1968, Boore 1983). However, models that specify uniform rupture properties along a fault tend to produce motions that have high-frequency arrivals that result only from the initiation and termination of rupture. Haskell (1966) and Aki (1967) introduced source models in which the rupture process is considered to include statistical irregularities with prescribed correlation distances. Andrews (1981) demonstrated a rupture model in which the stress drop on a fault plane is assumed to be both random (not necessarily uniformly random) and self-affine in space (changes in length scale do not change the statistical nature of the distribution). The spatial distribution of fault slip for a model of this type is shown in Figure 11 (from Andrews 1981). Representative records of ground velocity and displacement that result when such a rupture model is embedded in a homogeneous whole-space are also shown. Boatwright & Quin (1986) discuss the behavior of similar models, but which have the added feature of spontaneous rupture initiation for any point on the fault. Both Andrews (1981) and Boatwright & Quin (1986) demonstrate stress-drop distributions that produce radiated ground motions having constant Fourier amplitude spectra of ground accelerations at frequencies short when compared with the overall rupture duration. We presently do not know how the statistical properties of faults change regionally, or how such variations may affect near-source ground motions.

Stress Drop

The term *stress drop* is both very common in seismology and often quite confusing. Intuitively, it is easy to imagine that the level of stress driving a dislocation should have an important effect on the strength of the radiated

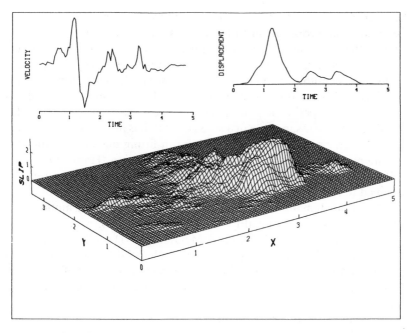

Figure 11 Theoretical model of fault slip assuming that stress drop varies randomly on the fault plane (defined by the X and Y coordinates). Representative body-wave records (both displacement and velocity) are shown for such a model (from Andrews 1981; published by the American Geophysical Union).

seismic waves. However, direct measurements of changes in stress during an earthquake are virtually impossible, and hence this parameter must be deduced through the use of models and indirect measurements. Consequently, there are many different definitions for "stress drop," which would all converge to the same physical parameter if all of the models were appropriate to simulate the Earth. Although it is beyond the scope of this review to comment on individual ways that "stress drop" is measured, we briefly comment on the most common application of this term in strong-motion seismology.

A widely used model for the general quantification of earthquake sources is the ω-squared model of Aki (1967) and Brune (1970, 1971). In this model, ground motions from earthquakes have a Fourier amplitude spectrum given by $S(\omega) = A_0\omega_c^2/(\omega_c^2+\omega^2)$, where ω_c is defined as the spectral corner frequency, and the low-frequency ($\omega < \omega_c$) level A_0 is proportional to the seismic moment M_0. The corner frequency for S-waves is given by $\omega_c = 0.49\beta2\pi(\Delta\sigma/M_0)^{1/3}$, where $\Delta\sigma$ is the stress drop and β is the S-wave velocity at the source. In this model, the corner frequency is assumed to be inversely proportional to the duration of the S-wave, which is, in turn, assumed to be proportional to the dimension of the rupture. The stress drop is then assumed to be proportional to the moment divided by the cube of the source dimension. In practice, this stress drop is measured by fitting asymptotes to log-log spectral amplitude plots, and thus this definition of stress drop gives a measure of the relative amplitude of long- and short-period seismic radiation. The ω-squared spectral model, together with the assumption of a stress drop that is independent of seismic moment, has been remarkably successful in fitting observed earthquake ground motions over a wide range of earthquake sizes (Hanks 1977).

If the ω-squared model is assumed, and if high-frequency motions are assumed to have the same statistical characteristics as random vibrations, then the time-domain amplitude of high-frequency ground motions can be shown to be proportional to $M_0\omega_c^{5/2}$ (Hanks 1979), which in turn is proportional to $M_0^{1/6}\Delta\sigma^{5/6}$. Hanks & McGuire (1981) studied acceleration data from 16 California earthquakes and concluded that these earthquakes had stress drops (sometimes referred to as rms stress drop) of about 100 bars with a standard deviation of a factor of two. In other words, the long-period motions averaged over many records could be used to predict (within a factor of two and a confidence of 85%) the average over many records of the high-frequency motions. As it is used in this sense, stress drop (rms) is really a spectral scaling parameter that gives a measure of the ratio of the high- and low-frequency spectral levels.

As we previously mentioned, stress drop is measured in many ways, none of which are direct measurements of the actual stress drop on the rupture surface. However, the inference that large ground motions often result from localized regions of high slip, and hence high stress drop (asperities), strongly implies that the actual stress drop during a rupture influences the strong ground motions. One question of considerable concern is whether plate-boundary earthquakes [interplate (e.g. those in Califonia)] have systematically lower stress drops than earthquakes away from plate boundaries [intraplate [e.g. those in the eastern United States)]. Kanamori & Anderson (1975) studied seismic moment as a function of fault dimension (defined by aftershocks), and they concluded that intraplate earthquakes have average stress drops (sometimes referred to as static stress drop) that are about twice as large as interplate earthquakes. Kanamori & Allen (1986) show that faults with long recurrence intervals are likely to have shorter rupture lengths for similar-sized (M_s) earthquakes than faults with short recurrence intervals (Figure 12). Both of these observations would imply stronger shaking at the source for eastern US earthquakes. However, Somerville et al (1987) compared teleseismic P-waveforms from moderate-sized earthquakes from the eastern and south-

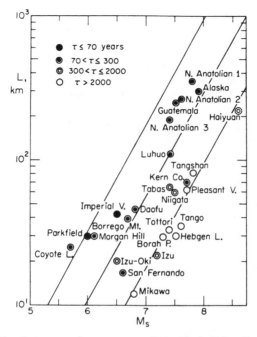

Figure 12 Relation between surface-wave magnitude M_s, fault length, and earthquake recurrence time. The solid lines indicate the trends for constant stress drop. This is evidence that faults with long recurrence intervals may have earthquakes with higher stress drops and perhaps larger high-frequency motions (from Kanamori & Allen 1986; copyright by the American Geophysical Union).

western US and concluded that systematic differences do not exist. The eastern US ground-motion estimates of Boore & Atkinson (1987) shown in Figure 7 were calculated assuming an ω-squared model with stress drops of 100 and 50 bars for the eastern and southwestern US, respectively. However, they also assume that site amplification for rock sites in the eastern US and southwestern US are 1.0 and 2.0, respectively. Thus, epicentral ground motions in the eastern and southwestern US nearly coincide in Boore & Atkinson's (1987) model. At present, there still seems to be some uncertainty about the relationship between static stress drop (a function of seismic moment and overall rupture length) and rms stress drop (a function of the shape of the Fourier amplitude spectrum of body waves).

Radiation Pattern

Haskell (1964) showed that a point shear dislocation in an elastic medium is indistinguishable from a point double-couple force system. Furthermore, the amplitude and polarity of waves radiated from a point double couple varies with the relative orientation of the source and the receiver, and the relationships that describe this variation are referred to as the *radiation pattern*. Liu & Helmberger (1985) discuss the strong ground motions produced by a relatively simple aftershock of the 15 October 1979 Imperial Valley earthquake, and they show convincing evidence of the effect of radiation pattern on the amplitude of motions at periods greater than 1 s. Although there is a clear theoretical basis for the importance of radiation pattern, there are also several examples of instances where relatively large ground motions are observed at locations where a simple application of seismic radiation pattern would predict very small amplitudes. In particular, large high-frequency *P*-waves have been observed at stations located directly along the fault strike for vertical strike-slip earthquakes, an orientation for which the radiation pattern would predict zero amplitude. Cormier & Spudich (1984) suggest that these observations may be due to the lateral refraction of waves within the low velocities that exist along fault zones (see the earlier section on fault zones as waveguides).

Aspect Ratio

The aspect ratio of a rupture is the ratio of its length to its width. Although the rupture lengths and widths are probably roughly comparable for small- and moderate-sized earthquakes, there may be a very large variation in the rupture aspect ratio for very large earthquakes. For instance, the great shallow strike-slip California earthquakes of 1906 and 1857 (about $M = 8$) had rupture lengths of hundreds of kilometers, whereas their rupture widths were probably less than twenty kilometers. However, $M = 8$ earthquakes from subduction zones may have roughly comparable rupture lengths and widths of about 100 km. It seems likely that strong ground motions resulting from a long, narrow rupture are different from those that result from a very wide rupture. Perhaps the aspect ratio of a rupture is one of the reasons that relatively large ground motions are observed at large distances from large subduction earthquakes.

Near-Field

"Near-field" has two separate and very different meanings in seismology. Although it is often used to describe the geographic region within tens or even hundreds of kilometers of an earthquake, it also is used to signify a particular type of ground motion that is best observed very close to the source. We prefer to use the term near-field only in the context of the second meaning and to refer to the geographic region as near-source. In the simple problem of a point dislocation embedded in a homogeneous medium, the complete solution can be expressed in a simple way as the sum of several terms that decay with distance with different integer powers of distance (Harkrider 1976, Heaton 1978). Terms that decay inversely with distance produce motions that are the time derivative of the dislocation, and these parts of the solution are called the far-field waves. Terms that decay inversely as distance squared (or higher powers) produce motions that are proportional to (or time integrals of) the dislocation, and these parts of the solution are called the near-field waves. Many of the generalizations that are made for far-field waves do not apply to near-field waves, and special care must be taken in their interpretation. Heaton (1978) and Heaton & Helmberger (1979) discuss near-field waves in some detail.

Strong ground motions recorded above the rupture surface of the 19 September 1985 Michoacán, Mexico, earthquake provide one of the most dramatic examples of near-field waves yet observed. Vertical ground-displacement records (obtained from doubly integrated acceleration records) are shown in Figure 13 (from Anderson et al 1987). The large ramp-like displacement that results in a net vertical displacement of the ground is almost certainly the result of near-field waves, and the total offset observed in these records is compatible with changes in sea level observed

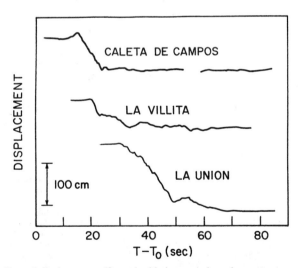

Figure 13   Ground displacements (from doubly integrated accelerometer records) observed at three sites directly above the 19 September 1985 Michoacán ($M_w = 8.1$) earthquake. The large static offset is real and was also observed in changes of sea level adjacent to the recording sites. This is one of the most dramatic examples of near-field waves recorded to date (from Anderson et al 1987; copyright by the American Association for the Advancement of Science).

near these stations. Even larger near-field waves can be inferred for other great earthquakes. For example, dislocations of 7 m have been inferred along some sections of the San Andreas fault during the 1906 California earthquake (Thatcher 1975), and we can thus conclude that near-field waves adjacent to the fault were at least 3.5 m. Even larger horizontal surface displacements of at least 20 m were observed for some regions during the 1964 Alaskan earthquake (Plafker 1965) ($M_w = 9.2$). Estimating the time scale over which such large static displacements occur is still somewhat problematic, but a simple model of spontaneous slip resulting from a stress of 100 bars implies dislocation velocities of about 1 m s^{-1} (Brune 1970).

DISCUSSION

In this review we have shown examples of many different phenomena that affect the nature of ground motions in the near-source region. It is not surprising that such a diversity of ground motions has been observed at comparable distances from earthquakes of comparable size. Although the problem of predicting ground motion is very complex, many of these phenomena can be anticipated—particularly path effects. Great progress has been made toward a better understanding of the physics of these phenomena, and we anticipate that future research will allow us to produce more accurate estimates of the nature of the ground motions that can be expected at particular sites. We are particularly optimistic that the effects of wave propagation can be anticipated through the study of ground motions from relatively numerous small earthquakes and through the modeling of waveforms within complex geologic structures.

Literature Cited

Aki, K. 1967. Scaling law of seismic spectrum. *J. Geophys. Res.* 72: 1217–31
Aki, K. 1987. Magnitude frequency relation for small earthquakes: a clue to the origin of f_{max} of large earthquakes. *J. Geophys. Res.* 92: 1349–55
Aki, K., Bouchon, M., Chouet, B., Das, S. 1977. Quantitative prediction of strong motion for a potential earthquake fault. *Ann. Geofis.* 30: 341–68
Anderson, J. G., Heaton, T. H. 1982. Aftershock accelerograms recorded on a temporary array. *US Geol. Surv. Prof. Paper 1254*, pp. 443–51
Anderson, J. G., Hough, S. E. 1984. A model for the shape of the Fourier amplitude spectrum of acceleration at high frequencies. *Bull. Seismol. Soc. Am.* 74: 1969–94
Anderson, J. G., Brune, J. N., Prince, J., Mena, E., Bodín, P., et al. 1987. Aspects of strong motion from the Michoacán, Mexico, earthquake of September 19, 1985. *Rep.*, Earthquake Eng. Res. Inst. In press
Andrews, D. J. 1981. A stochastic fault model. 2. Time-dependent case. *J. Geophys. Res.* 86: 10,821–34
Archuleta, R. J. 1984. A faulting model for the 1979 Imperial Valley earthquake. *J. Geophys. Res.* 89: 4559–85
Båth, M. 1981. Earthquake magnitude—

recent research and current trends. *Earth-Sci. Rev.* 17: 315–98

Boatwright, J., Boore, D. M. 1982. Analysis of ground accelerations radiated by the 1980 Livermore Valley earthquakes for directivity and dynamic source characteristics. *Bull. Seismol. Soc. Am.* 72: 1843–66

Boatwright, J., Quin, H. 1986. The seismic radiation from a 3-D dynamic model of a complex rupture process. Part I: confined ruptures. In *Earthquake Source Mechanics, Am. Geophys. Union Geophys. Monogr.*, ed. S. Das, J. Boatwright, C. H. Scholz, 37: 97–109. Washington, DC: Am. Geophys. Union

Boore, D. M. 1983. Stochastic simulation of high-frequency ground motions based on seismological models of the radiated spectra. *Bull. Seismol. Soc. Am.* 73: 1865–94

Boore, D. M., Atkinson, G. M. 1987. Prediction of ground motion and spectral response parameters in eastern North America. *Bull. Seismol. Soc. Am.* 77: 440–67

Borcherdt, R. D. 1970. Effects of local geology on ground motion near San Francisco Bay. *Bull. Seismol. Soc. Am.* 60: 29–61

Brady, A. G., Perez, V., Mork, P. N. 1982. Digitization and processing of mainshock ground-motion data from the U.S. Geological Surv. accelerograph network. *US Geol. Surv. Prof. Paper 1254*, pp. 385–406

Brune, J. N. 1970. Tectonic stress and the spectra of seismic shear waves from earthquakes. *J. Geophys. Res.* 75: 4997–5009

Brune, J. N. 1971. Correction. *J. Geophys. Res.* 76: 5002

Cormier, V. F., Beroza, G. C. 1987. Calculation of strong ground motion due to an extended source in a laterally varying structure. *Bull. Seismol. Soc. Am.* 77: 1–13

Cormier, V. F., Spudich, P. 1984. Amplification of ground motion and waveform complexity in fault zones: examples from the San Andreas and Calaveras faults. *Geophys. J. R. Astron. Soc.* 79: 135–52

Das, S., Aki, K. 1977. Fault plane with barriers: a versatile earthquake model. *J. Geophys. Res.* 82: 5658–70

Feng, R., McEvilly, T. V. 1983. Interpretation of seismic reflection profiling data for the structure of the San Andreas fault zone. *Bull. Seismol. Soc. Am.* 73: 1701–20

Hanks, T. C. 1975. Strong ground motion of the San Fernando, California, earthquake: ground displacements. *Bull. Seismol. Soc. Am.* 65: 193–225

Hanks, T. C. 1977. Earthquake stress drops, ambient tectonic stresses and the stresses that drive plate motions. *Pure Appl. Geophys.* 115: 441–58

Hanks, T. C. 1979. *b*-values and $\omega^{-\gamma}$ seismic source models: implications for tectonic stress variations along active crustal fault zones and the estimation of high-frequency strong ground motions. *J. Geophys. Res.* 84: 2235–42

Hanks, T. C. 1982. f_{max}. *Bull. Seismol. Soc. Am.* 72: 1867–80

Hanks, T. C., McGuire, R. K. 1981. The character of high-frequency strong ground motion. *Bull. Seismol. Soc. Am.* 71: 2071–95

Harkrider, D. G. 1976. Potentials and displacements for two theoretical seismic sources. *Geophys. J. R. Astron. Soc.* 47: 97–133

Hartzell, S. H., Heaton, T. H. 1983. Inversion of strong ground motion and teleseismic waveform data for the fault rupture history of the 1979 Imperial Valley, California, earthquake. *Bull. Seismol. Soc. Am.* 73: 1553–83

Hartzell, S. H., Heaton, T. H. 1986. Rupture history of the 1984 Morgan Hill, California, earthquake from the inversion of strong motion records. *Bull. Seismol. Soc. Am.* 76: 649–74

Hartzell, S. H., Helmberger, D. V. 1982. Strong-motion modeling of the Imperial Valley earthquake of 1979. *Bull. Seismol. Soc. Am.* 72: 571–96

Haskell, N. A. 1964. Total energy and energy spectral density of elastic wave radiation from propagatng faults. *Bull. Seismol. Soc. Am.* 54: 1811–41

Haskell, N. A. 1966. Total energy and energy spectral density of elastic wave radiation from propagating faults. II. A statistical source model. *Bull. Seismol. Soc. Am.* 56: 125–40

Hauksson, E., Teng, T., Henyey, T. L. 1987. Results from a 1500 m deep three-level downhole seismometer array: site response, low Q-values and f_{max}. Submitted for publication

Heaton, T. H. 1978. *Generalized ray models of strong ground motion.* PhD thesis. Calif. Inst. Technol., Pasadena. 290 pp

Heaton, T. H., Hartzell, S. H. 1987. Estimation of strong ground motions from hypothetical earthquakes on the Cascadia subduction zone, Pacific Northwest. *Pure Appl. Geophys.* In press

Heaton, T. H., Helmberger, D. V. 1979. Generalized ray models of the San Fernando earthquake. *Bull. Seismol. Soc. Am.* 69: 1311–41

Heaton, T. H., Tajima, F., Mori, A. W. 1986. Estimating ground motions using recorded accelerograms. *Surv. Geophys.* 8: 25–83

Houston, H., Kanamori, H. 1986. Source characteristics of the Michoacán, Mexico, earthquake at periods of 1 to 30 seconds. *Geophys. Res. Lett.* 13: 597–600

Ihnen, S. M., Hadley, D. M. 1986. Prediction of strong ground motion in the Puget Sound region: the 1965 Puget Sound earthquake. *Bull. Seismol. Soc. Am.* 76: 905–22

Ishirara, K. 1985. Stability of deposits during earthquakes. *Proc. Int. Conf. Soil Mech. Found. Eng., 11th.* Rotterdam, Neth: Balkema

Ishihara, K., Shimizu, K., Yamada, Y. 1981. Pore water pressures measured in sand deposits during an earthquake. *Soils Found.* 21(4): 85–100

Jennings, P. C., Housner, G. W., Tsai, N. C. 1968. Simulated earthquake motions. *Rep.*, Earthquake Eng. Res. Lab., Calif. Inst. Technol., Pasadena

Joyner, W. B., Boore, D. M. 1982. Prediction of earthquake response spectra. *US Geol. Surv. Open-File Rep. 82-977*

Joyner, W. B., Warrick, R. E., Oliver, A. A. III. 1976. Analysis of seismograms from a downhole array in sediments near San Francisco Bay. *Bull. Seismol. Soc. Am.* 66: 937–58

Kanamori, H. 1978. Quantification of great earthquakes. *Tectonophysics* 49: 207–18

Kanamori, H. 1986. Rupture process of subduction-zone earthquakes. *Ann. Rev. Earth Planet. Sci.* 14: 293–322

Kanamori, H., Allen, C. R. 1986. Earthquake repeat times and average stress drop. In *Earthquake Source Mechanics, Am. Geophys. Union Geophys. Monogr.*, ed. S. Das, J. Boatwright, C. H. Scholz, 37: 227–35. Washington, DC: Am. Geophys. Union

Kanamori, H., Anderson, D. L. 1975. Theoretical basis of some empirical relations in seismology. *Bull. Seismol. Soc. Am.* 65: 1073–95

Kawashima, K., Aizawa, K., Takahashi, K. 1984. Attenuation of peak ground motion and absolute acceleration response spectra. *Proc. World Conf. Earthquake Eng., 8th*, San Francisco, pp. 257–64

Lay, T., Kanamori, H., Ruff, L. 1982. The asperity model and the nature of large subduction zone earthquakes. *Earthquake Predict. Res.* 1: 3–71

Liu, H. L., Heaton, T. H. 1984. Array analysis of the ground velocities and accelerations from the 1971 San Fernando, California, earthquake. *Bull. Seismol. Soc. Am.* 74: 1951–68

Liu, H. L., Helmberger, D. V. 1985. The 23:19 aftershock of the 15 October 1979 Imperial Valley earthquake: more evidence for an asperity. *Bull. Seismol. Soc. Am.* 75: 689–708.

Malin, P. E., Walker, J. A. 1985. Preliminary results from vertical seismic profiling of Oroville microearthquake *S*-waves. *Geophys. Res. Lett.* 12: 693–722

National Academy of Sciences. 1985. Liquefaction of soils during earthquakes. *Rep.*, Comm. Earthquake Eng. (G. Housner, Chair.), Natl. Res. Counc. Washington, DC: Natl. Acad. Press. 240 pp.

Nuttli, O. W. 1973. Seismic wave attenuation and magnitude relations for eastern North America. *J. Geophys. Res.* 78: 876–85

Olson, A. H., Absel, R. J. 1982. Finite faults and inverse theory with application to the 1979 Imperial Valley earthquake. *Bull. Seismol. Soc. Am.* 72: 1969–2001

Papageorgiou, A. S., Aki, K. 1983a. A specific barrier model for the quantitative description of inhomogeneous faulting and the prediction of strong ground motion. I. Description of the model. *Bull. Seismol. Soc. Am.* 73: 693–722

Papageorgiou, A. S., Aki, K. 1983b. A specific barrier model for the quantitative description of inhomogeneous faulting and the prediction of strong ground motion. II. Applications of the model. *Bull. Seismol. Soc. Am.* 73: 953–78

Plafker, G. 1965. Tectonic deformation associated with the 1964 Alaskan earthquake. *Science* 148: 1675–87

Rial, J. A. 1984. Caustics and focusing produced by sedimentary basins, applications of catastrophe theory to earthquake seismology. *Geophys. J. R. Astron. Soc.* 79: 923–38

Shin, T. C., Herrmann, R. B. 1987. L_g attenuation and source studies using 1982 Miramichi data. *Bull. Seismol. Soc. Am.* 77. 384–97

Singh, S. K., Mena, E., Castro, R. 1987. Some aspects of source characteristics of the 19 September 1985 Michoacán earthquake and ground motion amplification in and near Mexico City from strong motion data. *Bull. Seismol. Soc. Am.* In press

Somerville, P. G., McLaren, J. P., Lefevre, L. V., Burger, R. W., Helmberger, D. V. 1987. Comparison of source scaling relations of eastern and western North American earthquakes. *Bull. Seismol. Soc. Am.* 77: 420–39

Spudich, P., Cranswick, E. 1984. Direct observation of rupture propagation during the 1979 Imperial Valley earthquake using a short baseline accelerometer array. *Bull. Seismol. Soc. Am.* 74: 2083–2114

Thatcher, W. 1975. Strain accumulation and release mechanism of the 1906 San Francisco earthquake. *J. Geophys. Res.* 80: 4862–72

Trifunac, M. D. 1976. Preliminary analysis of the peaks of strong ground motion: dependence of peaks on earthquake magnitude, epicentral distance, and recording site conditions. *Bull. Seismol. Soc. Am.* 66. 189–219

Tsai, N. C. 1969. *Influence of local geology on earthquake ground motion.* PhD thesis. Calif. Inst. Technol., Pasadena.

Vidale, J. 1987. *Application of two-dimensional finite-difference wave simulation to earthquakes, earth structure, and seismic hazard.* PhD thesis. Calif. Inst. Technol., Pasadena. 150 pp.

Zeevaert, L. 1964. Strong ground motions recorded during earthquakes of May the 11th and 19th, 1962, in Mexico City. *Bull. Seismol. Soc. Am.* 54: 209–31

Ann. Rev. Earth Planet. Sci. 1988. 16: 147–71

ORE DEPOSITS AS GUIDES TO GEOLOGIC HISTORY OF THE EARTH

C. Meyer†

Department of Geology and Geophysics, University of California, Berkeley, California 94720

1. INTRODUCTION

The tectonic evolution of the Earth depends largely on changes in the rates and mechanisms for generation and dissipation of its internal heat. By contrast, the chemical evolution at the Earth's surface and in its fluid envelopes is driven chiefly by solar energy, in part regulated by the biomass. Magma-generating processes and magmatic crystallization are both commonly selective of ore-forming ingredients, as are also systems of hydrothermal circulation driven by magmatic heat or regional metamorphism.

Solar energy exerts major control on ores accumulated as detrital concentrations and as chemical sediments. The colossal energy storage by the biomass through photosynthesis has provided most of the diversity of redox potential that characterizes the Earth's epizonal ore-forming environments. The chemical action of free oxygen is evident in many sediments such as red beds and banded iron formations, but, of course, photosynthesis must also sequester chemically equivalent amounts of organic carbon in other sedimentary rocks. At the time of sediment accumulation or diagenesis, organisms or residual carbon may buffer oxidation potential low enough to stabilize sulfide ores. Furthermore, if either strongly reduced or strongly oxidized rocks become recycled through partial melting, the resulting magmas may also selectively incorporate special metal assemblages. These may later be further concentrated into ores by igneous or hydrothermal processes. So tectonic activity and vol-

† Professor Charles Meyer died on November 15, 1987.

147

canism, processes dependent on heat flow from the interior, *interact* near the Earth's surface with weathering, sedimentation, and biogenic processes that are powered chiefly by the Sun (see Wetherill & Drake 1980).

Despite the fact that many ores must have formed at or near the Earth's surface, there are few places where direct observation can now be made of the processes taking place. In the oceans, the Red Sea brine pools and the spectacular "black smokers" at mid-oceanic ridges (Rona et al 1983) provide examples of submarine hydrothermal deposition of metal sulfides. However, these have only inconclusive tectonic analogues in the geologic record of fossil deposits (Skinner 1983). Similarly, continental hot springs may be the outlets for deeper hydrothermal ore-forming fluids, but except for mercury, a few precious metal ores, and the Salton Sea brines, there is little precise information as to what kind of ore deposits (if any) are currently forming at depth. Yet many types of igneous-related ore deposits are localized in Phanerozoic orogenic belts, and inferences of genetic connections with magma systems generated over currently subducting plates are very reasonable (summaries and references in Mitchell & Garson 1981, Sawkins 1984).

It is clear from Figure 1 that the various types of ores show different *temporal* patterns of distribution. For example, while the porphyry coppers are heavily concentrated in the late Phanerozoic, most of the anorogenic igneous ore-forming systems and most of the largest stratiform ores are Proterozoic. Gold veins and large volcanic-hosted base-metal massive sulfides are concentrated in both Archean greenstones and Phanerozoic orogenic belts and near the mid-Proterozoic. This uneven distribution of ore metals through geologic time has been documented repeatedly (references in Laznicka 1973, Meyer 1981). But opinions still differ on how reliable the patterns are as indicators of changing tectonic and geochemical environments. Selective removal by erosion of continental epizonal ore types such as epithermal veins or porphyry coppers undoubtedly con-tributes to the scarcity of these types in Precambrian times, as emphasized by Sawkins (1984, p. 279). Yet for types of ores that are closely linked to specific petrologic assemblages, their "preservation potential" (Mitchell & Garson 1981) would be the same as that for the rest of the rock assemblage.

The potential contributions of ore-forming processes to the interpreta-tion of geologic history depend substantially on the definition and char-acterization of various ore types and on current ideas about the conditions under which they formed (Section 2). The pattern of each type through geologic time may then be used to evaluate how conditions may have changed as the Earth evolved. In Section 3, I discuss how the composite of all the patterns in Figure 1 may help define or constrain the sequence of events for each major division of geologic time.

2. SOME DESCRIPTIVE TYPES OF ORE DEPOSITS

2.1 Stratiform Chromite in Layered Intrusions

Layered intrusions range in age from among the oldest terrestrial rocks, about 3800 Ma in the Akilia association of southern West Greenland, to the mid-Tertiary Skaergaard Complex of East Greenland. All of these contain cumulate layers of chrome-bearing spinels, but the largest and richest chromite deposits are in complexes having early Proterozoic ages— the Great Dyke of Zimbabwe (ca. 2500 Ma) and the Bushveld Complex of the Transvaal (ca. 2000 Ma).

2.2 Podiform or "Alpine" Chromite Deposits in Ultramafic Massifs

Irregularly lenticular or spindle shapes are characteristic of the podiform chromite ores. The crystal assay of the chrome spinels ranges up to about 62% Cr_2O_3, higher by 10% or more than the chromite from layered complexes. Distorted remnant cumulate structures are common in the chromitite, yet the chromite grains show postcumulate growth into an interlocked mosaic texture.

With only a few exceptions, the podiform chromite deposits are in Phanerozoic dunite-harzburgite suites in ophiolites. These ores are most abundant in the Tertiary but are still clearly defined back through the Mesozoic and Paleozoic. There are a few isolated earlier fragmentary examples in the age range of 1000 to 650 Ma along possible suture zones in Egypt and Arabia, but from there back to the Archean there are no podiform chromite ores. In the Archean there are a few deformed and recrystallized small chromitites in the 2700-Ma greenstone belts of Canada, southern Africa, and Australia, plus some of the largest and richest chromite ores in the world in 3300-Ma serpentinites and talc schists near Selukwe, Zimbabwe.

Individual chromite grains at Selukwe do not show overgrowth after accumulation; a simple cumulate texture is preserved as in stratiform examples. However, the high Cr_2O_3 content ($>60\%$) of the chromite grains is more like the podiform grade. So "the Selukwe deposits show a combination of primary stratiform and secondary tectonic alpine characteristics" (Cotterill 1969, p. 186). Sawkins (1984, p. 156) has suggested "that the Selukwe ultramafic sheet may in fact represent part of an Archean ophiolite complex." The sliced-up ultramafic sheet that contains the chromite is part of a succession of basalts and "poorly stratified arkose and graywacke, rare conglomerates," minor quartzites, and banded ironstones (Cotterill 1969). These rocks might suggest what Groves & Batt (1984)

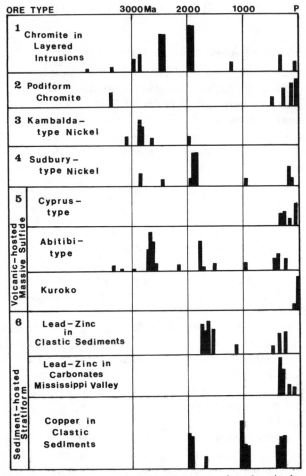

Figure 1 Distribution through geologic time of the groups of ore deposits described in the text. The ore-type numbers in the figure correspond to those used in the Section 2 headings. The width of each vertical bar represents an interval of approximately 50 Myr. The length of each bar is a geological estimate of the quantity of ore formed during that 50 Myr as compared with the total estimated tonnage—the sum of all the bars *for that type of deposit*— for all of geologic time.

have called a platform-phase greenstone assemblage. Genetic ambiguity persists.

2.3 *Nickel Sulfide Ores in Ultramafic Assemblages*

The discovery in 1966 (Woodall & Travis 1969, Ross & Hopkins 1975) of high-grade nickel sulfide ore near Kambalda, Western Australia, set off a

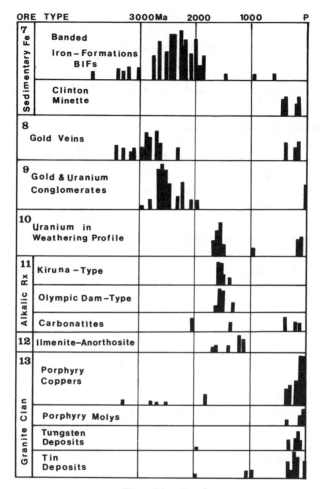

Figure 1 (continued)

worldwide reexamination of ultramafic rocks. The new ore bodies are irregularly conformable lenses of massive pentlandite-pyrrhotite at the bottoms of serpentinized komatiite lava flows and sills. Many of the flows show "spinifex" quench texture, a network of intergrown olivine plates. Disseminated sulfides are present in some of the serpentinites above the massive sulfide.

2.4   Nickel Sulfide Deposits in Gabbroic Intrusions

In amount of sulfide ore, Sudbury, Canada (1850 Ma), easily dominates the zoned intrusions of the Precambrian. Large Phanerozoic deposits of similar metal content are at Norilsk-Talnakh in Siberia, Jin Xiang in China, and Insizwa in Natal. These ores are in large differentiated sills, more closely related to rift-fed flood basalts than to a thick, layered lopolith like Sudbury.

2.5   Volcanic-Associated Massive Base-Metal Sulfide Deposits

These are ores of copper, zinc, and lead in massive iron sulfide lenses that are conformable with enveloping lava flows and volcaniclastic sediments. They were deposited on and below the seafloor by hydrothermal solutions quenched by seawater. Stockworks of ore veinlets in rocks showing extensive hydrothermal alteration mark the path of mineralizing solutions below the seawater interface.

The volcanic rocks may be almost entirely basalt, in which case copper is the main ore metal; this is the *Cyprus type* (see Figure 1). However, if rhyolite domes and felsic aquatuffs are present, copper and zinc are generally both abundant. Lead is commonly present with the zinc, but in Precambrian ore bodies it rarely achieves levels greater than one twentieth of the combined copper and zinc. This is the *Abitibi-type* ore body, typical of the greenstone environments of the Canadian Shield. The rocks enclosing Abitibi-type massive sulfides have generally been metamorphosed to greenschist grade or, in a few cases, to amphibolites.

Tertiary volcanic-associated massive sulfides generally have abundant rhyolitic and dacitic volcaniclastic sediments in a succession of rhyolite flows and basalts, along with abundant bedded anhydrite and gypsum. Lead is commonly the chief ore metal, though proportions of lead, zinc, and copper can be extremely variable. This is the Japanese *kuroko* (black ore) *type* (see Franklin et al 1981).

Wall-rock alteration in the footwall shows a wide chemical range of metasomatism. In Abitibi-type deposits, Mg-rich chlorite is a dominant phase in the core of the footwall stockwork. Chloritic alteration may also be extensive beneath Cyprus-type ore bodies, but it commonly grades upward into sericitic and kaolinitic assemblages above the sulfide mass. Sericite is prevalent in the chalcopyrite-rich stockworks (keiko) of the Japanese massive sulfides, but kaolinite and even pyrophyllite are abundant locally. In sulfur-dominated hydrothermal systems, such extreme acid leaching is charcteristic of oxygenated waters.

2.6   Sediment-Hosted Stratiform Deposits of the Base Metals

These stratiform sulfide deposits of lead, zinc, and copper are not parts of a volcanic edifice. This distinguishes them from the previously cited volcanic-hosted massive sulfides (see Gustafson & Williams 1981, Meyer 1981). There is also a greater diversity in grade of metamorphism and deformation and of metal content of the sulfide ores. The copper ores in this group are nearly all in clastic sediments: The lead and zinc may be in either clastic or carbonate rocks. The clasts may or may not include fine-grained volcanic detritus.

The cluster in mid-Proterozoic time of *lead-zinc ores in clastic sediments* is one of the striking features of Figure 1. This cluster includes the oldest major ore bodies in the world mined primarily for lead. There are at least three distinctive types in the cluster, and all are in clastic rocks. The ore body at Sullivan, British Columbia, is the only one with a well-defined alteration root. The Broken Hill deposits in New South Wales, the Namaqualand group (at Aggeneys, Gamsberg, and other locations in southern Africa), and the Grenville deposits at Balmat, New York, are all highly deformed and metamorphosed to upper amphibolite or granulite grade. Deposits at Mt. Isa and McArthur River in Queensland are folded but not highly metamorphosed. Like the Sullivan deposits, they show spectacular laminations of the sulfides in fine-grained siltstones.

Mississippi Valley lead-zinc deposits in the midcontinent United States are stratabound and concentrated in carbonate rocks at the edges of paleobasins filled with Paleozoic sediments (Ohle 1980). Similar deposits in Mesozoic and younger basins are few in number and much smaller. Variations on the idea of mineralization by basinal brines have received much attention since this hypothesis was first proposed by White (1968), with recent studies reemphasizing the mechanisms of fluid propulsion. Bethke (1986) quantitatively established the feasibility of gravity drive for the Wisconsin-Illinois lead-zinc district. However, other Paleozoic carbonate-lead environments have other characteristics, and Russell (1983) has made a strong case for igneous-related thermal drive for the Irish deposits.

Giant stratiform ores of *copper sulfides in clastic sediments* bracket the Proterozoic lead group in geologic time. The giant Udokan deposit in Siberia is said to be about 2000 Ma (Gustafson & Williams 1981), and the African Copper Belt is between 1200 and 800 Ma. Udokan shows copper sulfides in a complex succession of argillite, siltstone, sandstone, sandy limestone, and conglomerate breccias. There are also clasts of feldspars and "acid eruptives" (Samonov & Pozharisky 1977). The sediments are gray to black, except for the beds above the top ore zone, which are pinkish gray. Fold style and metamorphism, as well as zoning among the sulfide assemblages, resemble that of the African Copper Belt deposits, which are about 1 Gyr younger. Udokan appears near the time of the earliest important red beds and bedded sulfates.

The African Copper Belt (Mendelsohn 1961) is more than 500 km in length. In the southern half, in Zambia, copper sulfides are disseminated in sandstones and shales not far above a basement of lower and mid-Proterozoic schists and granites. Here the sedimentary succession is folded into a pattern of drag folds on the flanks of the large Kafue anticline, and metamorphism reached the stage in which arkosic sediments (and some ores) were remobilized and emplaced as sulfide-bearing quartz-biotite-feldspar gash veins.

Northward, across the boundary into Zaire, the metamorphic grade diminishes and the fold style changes. With the northward plunge of the Kafue anticline, higher parts of the section are preserved, and northward into the westward-curving Lufilian arc, huge nappes and flat "thrusts" imbricate and overturn the beds (François 1974). There is little metamorphism, though cross-cutting ore pipes (Derriks & Vaes 1956) may have resulted from remobilization. Toward Kolwezi, at the northwest end of the Lufilian arc, outliers of Series des Mines rocks ride out over stratigraphically higher Kundelungu rocks.

François (1974) has shown that the detachments and folds of the Kolwezi sector result from gravity sliding. The likely direction of sliding seems to be radially northward and eastward, since the nappes overturn in those directions. Their axes, though quite irregular in detail, generally parallel both the curving long dimension of the Copperbelt basin and the trends of the penetrative folds around the Kafue anticline in Zambia. The range of metamorphic grade and deformation within the limits of this single basin illustrates the wide diversity of these features encountered in Proterozoic basins in general. This is in sharp contrast to the comparatively uniform greenschist metamorphism and mild penetrative deformation prevalent in the greenstone basins of the Archean.

2.7   Sedimentary Iron Ores

Precambrian iron formations have iron-rich layers alternating with silica-rich layers on a scale of millimeters to centimeters. This produces a banded appearance in outcrop (banded iron formation, or BIF; James & Sims 1973, Trendall 1968). Most Phanerozoic iron-rich sediments show less banding and are silica poor.

Most Archean BIFs are parts of greenstone successions. They are commonly individually smaller in both areal extent and total thickness than their Proterozoic counterparts. They are also not as regularly nor as conspicuously banded. These features led to their designation as *Algoma-type* BIFs, as distinguished from the gigantic, uniformly bedded *Superior-type* BIFs that are found more commonly in the Proterozoic (Gross 1965).

But Algoma-type BIFs also distally accompany mid-Proterozoic massive sulfides, such as those at Jerome, Arizona, and even some Phanerozoic massive sulfides.

The striking proliferation of Superior-type BIFs between about 2600 and 1800 Ma is one of the spectacular geochemical anomalies of geologic history. These deposits have no clear association with volcanic activity except perhaps regionally. Their common association with limestones suggests lagoonal, shelf, or epeiric basinal deposition, with shallow-water deposition prevailing at least for the oxide facies. Accumulated ferrous iron in the deep ocean is the most reasonable source for the iron as well as for the silica. These accumulations were transported to shallow-water coastal environments by upwelling currents (Goodwin 1973). But the source for the oxygen is even more difficult to ascertain. Oxygen would be available from photodissociation and photosynthesis, but in the latter process vast amounts of carbon must be sequestered to make the oxygen available for iron precipitation (van Valen 1971). There is little evidence for this in the vicinity of the iron formations, so an atmospheric *reservoir* has instead been advocated as the source.

Oolitic ironstone ores are major examples of iron oxide–rich chemical sediments formed in Phanerozoic seas. The Silurian Clinton ores of eastern United States were apparently deposited on shallow marine tidal flats along with fine-grained clastic rocks and carbonate cement. Solution and transport mechanisms for the iron may have been facilitated by organisms or organic debris. At the site of deposition, iron oxides replace calcareous fossils and calcite oolites.

The low-grade but plentiful Jurassic Minette ores of Europe are generally similar to the Clinton deposits, but they consist largely of chamosite oolites, with sideritic cement. Such ores demonstrate reducing conditions at the site of deposition and/or diagenesis.

2.8   *Gold-Bearing Veins*

Most gold-bearing veins are in Archean greenstone belts, and nearly all major greenstone belts have important gold mines. The veins tend to concentrate in the basaltic or BIF portions of greenstone successions, although the best ore commonly clusters around felsic intrusions (Anhaeusser 1976). In some districts much of the ore is in interflow chemical sediments, especially carbonate-sulfide facies of BIFs (Ridler 1976). There are very few gold veins in the middle and upper Proterozoic, but several major districts are found in Phanerozoic island-arc, volcanic-graywacke, and shale assemblages in California, in Victoria, Australia, and in the world's currently largest vein-gold producer, Muruntau in central Asia.

Gold-quartz veins have enormous vertical range. Some ores have been followed to depths of 3 km without an appreciable change in value. This uniformity of mineralization has led to the important hypothesis that the gold and quartz are derived from associated rocks undergoing metamorphism, either from the walls of the veins (Boyle 1979) or hydrothermally from deeper zones in the metamorphic complex (Phillips et al 1984).

2.9   *Gold and Uranium in Quartz-Pebble Conglomerates*

Gold and uranium are the only two elements that concentrate to ore grade in ancient conglomerates, but these conglomerates are among the world's largest repositories for those elements.

About 13 ancient conglomerates have been mined extensively for gold and/or uranium. All of these are within the age range extending from the Dominion Reef (South Africa) at 2800 Ma to Tarkwa (Ghana) at about 1950 Ma. The age range of the supergiant Witwatersrand deposit is about 2800 to 2500 Ma [see Pretorius (1981) for a summary].

Rounded pyrite is common among the clasts and as sand-sized grains in the matrix. Uraninite, gold, and other heavy clasts are dispersed unevenly in the matrix, but the gold is all recrystallized. Uranium is also concentrated, along with gold, in carbon-rich beds as thucolite. Pretorius (1981) notes that gold is not in the conglomerates after the first red beds, and that detrital uraninite is seldom seen as clasts after the decline of the BIFs. Nearly all placer deposits, fossil and modern, of the Phanerozoic are immediately above, or just downstream from, a lode gold source, as in the Tertiary conglomerates of California.

2.10   *Uranium in the Weathering Profile: Veins and Roll-Fronts*

Numerous large, high-grade veins of pitchblende and uraninite are grouped under the term *unconformity-type uranium* ores because of their distinctive concentration in veins, pipes, and other permeable zones below middle Proterozoic unconformities (Nash et al 1981). The first discovered (and still the largest) accumulations of this ore type are in the Alligator River area of Australia and the Athabaska basin of Saskatchewan.

Though there are differences of opinion about the temperature and origin of the solutions that transported the uranium, the cluster of ages around 1700–1600 Ma is certainly significant with respect to the changing redox potential at weathering surfaces over the mid-Proterozoic. Subsequently there have been many local changes of redox potential in natural habitats, which have kept uranium periodically on the move, right down to the Mesozoic and younger "roll-fronts" and calcrete deposits.

2.11   Ores in Alkalic Granites, Porphyries, and Carbonatites

Compared with granodiorites and quartz monzonites, alkalic granites and equivalent porphyries have more lithophillic ore suites, including the rare-earth elements and uranium along with iron ores. Copper is the only major base metal in this group, and fluorine, phosphates, and barite are present locally. Siderite, ankerite, or dolomite range widely in abundance, up to dominance in carbonatites.

The Kiruna iron ore district in Sweden has the largest known massive magnetite ore body. The main mass is crudely lenticular in shape and broadly conformable with enclosing volcanic rocks. The footwall is syenitic keratophyre, and there is a quartz phenocryst-bearing keratophyre over the ore. Magnetite ore penetrates both hanging wall and footwall as stubby gash-type veins and as matrix to fragments of enclosing rocks (Wright 1986). Ambiguous contact relations and a range of geochemical variations have led to spirited discussions between geologists who have worked in this area in recent years (Parák 1975, Frietsch 1978). Arguments continue as to whether the ores are intrusive or extrusive fluidized ore magmas, hydrothermal additions, or recrystallized exhalites.

There are similar but much smaller deposits of iron ore in mid-Proterozoic iron-oxide-rich red and purple porphyries and epizonal red granites in Missouri, which have long been compared to the Kiruna deposits.

A major stimulus for restudy of ores associated with alkalic rocks came from the discovery in 1975 of the gigantic Olympic Dam ore deposit near Roxby Downs Station in South Australia (Roberts & Hudson 1983). Olympic Dam is essentially a hematite-cemented series of breccias involving several rock types, including red granite and alkalic porphyries. These fill a graben-like trench, the fault boundaries of which relate to major lineaments (O'Driscoll 1982).

In the Olympic Dam ores, uranium, phosphorus, fluorine, rare-earth elements, and gold are present with copper and iron. Among the copper minerals, chalcocite and bornite grade downward into chalcopyrite and pyrite. The hematite-rich matrix of the breccias is so abundant as to be almost ore grade for iron. A magnetic anomaly under the ore may indicate transition to magnetite at depth. Copper, uranium, and gold may all have been enriched during the pre-Adelaidean (late Proterozoic) exposure to weathering. Youles (1984) pointed out resemblances between the Olympic Dam ores and the Mt. Painter uranium deposit on the east side of the Adelaide geosyncline and proposed an anorogenic peralkaline origin for granitic magmas of the same probable age as those at Olympic Dam.

2.12   The Titanium-Anorthosite Connection

Another striking temporal pattern is displayed by the titanium association with anorthosite massifs. Most anorthosite massifs worldwide were emplaced between 1700 and 1100 Ma, with the best ilmenite ores in the later part of that range.

The spatial and temporal association of anorthosites with rapakivi granites and other alkalic rocks is well documented (Emslie 1978). Even though they were probably not differentiated from a common source magma, they do appear to be mutually dependent on a specific tectonic environment that was best developed in the mid-Proterozoic.

2.13   Intermediate Igneous Rocks and Ore Deposits

The association of copper with intermediate igneous rocks is a dominant pattern in world distribution since the Laramide orogeny and is especially pronounced in the circum-Pacific orogenic belt. Most of the world's main concentrations of molybdenum, tungsten, and tin ores are also in Phanerozoic orogenic belts.

Ore-bearing intrusive rocks range from batholiths to small stocks. The stocks, generally having porphyritic texture, account for a disproportionately large percentage of the ores. Copper is the most abundant ore metal in the group, associated with greater or lesser amounts of zinc and lead, which are commonly zoned around the copper. Additionally, many porphyry coppers contain molybdenum, up to about 0.05 or 0.06% Mo at mining grade. In some deposits that are generally more siliceous and potassic, molybdenum may have a greater value than copper in grades of up to about 0.3% Mo. Tungsten may also be present in recoverable amounts. Tungsten as wolframite and tin as cassiterite are major ores associated with intermediate rocks that range toward granitic. In a few instances tin is present partly as sulfide (for example, in Bolivia) in epizonal porphyry similar to the porphyry of the porphyry coppers.

Typical porphyries tend to intrude as clusters of individual plutons 1 or 2 km in diameter. Commonly four or more successive porphyries are in a cluster, but generally only one or two guide mineralization (Gustafson & Hunt 1975).

The ore and wall-rock alteration are generally zoned in inverted concentric cups around a core of biotite-orthoclase-sericite containing low-sulfur sulfide assemblages (Guilbert & Lowell 1974). Sericite-pyrite and pyrophyllite-pyrite alteration assemblages were superimposed on the earlier zoning, presumably during decline of the hydrothermal system with influx of groundwater (Gustafson & Hunt 1975). The full column of mineralization and alteration in a moderate-sized porphyry copper may reach 5 km (possibly 8 km) in total height.

Probably 90% of known porphyry ore has been formed in the last 75 m.y. in locations around the Pacific basin in North and South America, the Philippines, and western Pacific islands. In this circum-Pacific belt there are a few deposits as old as the Triassic, but deposits are more numerous in the Jurassic and Cretaceous. The timing is episodic (Titley & Beane 1981). Generally times of widespread continental uplift seem to have been optimum (Lowell 1974). The time of the Laramide orogeny, at the turn of the Tertiary, was favored in the United States, with another peak at around 30 Ma. Porphyry ore emplacement continued in South America until about 5 Ma. In the western Pacific island arcs, deposits date as young as 1 Ma and are doubtlessly forming there today under active volcanos; in addition, deposits may also be forming now under some active continental hot springs.

Ore-bearing porphyry magmas are generated above subduction zones in island-arc systems. Most of these are porphyry coppers, some with good gold values, but molybdenum is not excluded. Even more porphyry coppers are formed over Andean-type subduction zones under continents, where the magmas must rise through continental crust.

Paleozoic porphyry coppers are small and economically unimportant except in the Soviet Union. But these too are in orogenic belts, with the largest examples in Kazakhstan and the Uzbek region. There are also small, low-grade Paleozoic porphyry coppers in the Appalachians and in the Tasman geosyncline.

Haib, Namibia, is the best known example of porphyry coppers in Proterozoic rocks. This deposit appears to date at about 1800 Ma, about the same age as several similarly mineralized zones in the granitoid complex of central Finland (Gáal & Isohanni 1979). Similar deposits seem to be somewhat more abundant in the Archean than in the Proterozoic, especially in the Superior province of Canada (Nunes & Ayres 1982) and in the Pilbara block of Western Australia. None of these deposits is commercial.

One of the Canadian examples is particularly instructive from the point of view of judging the tectonic significance of the Precambrian porphyry coppers. This is the Pearl Lake porphyry in the McIntyre mine at Timmins, Ontario (Davies & Luhta 1978). The porphyry is a quartzeye albite schist, a stratigraphic unit among Archean basaltic lavas, some of which are cut by gold-quartz veins. Davies & Luhta identified the schists as metatuffs. The schists carry chalcopyrite and bluish molybdenum-bearing quartz veinlets, with coarse anhydrite. The feldspars are dusted with hematite and are sericitized near the veinlets. By hand specimen, the rock *is* a porphyry copper—but in a totally different geologic setting compared with the Phanerozoic habitat of porphyry coppers. Davies & Luhta have convincingly demonstrated how precarious it is to use only mineralogical assemblages to deduce tectonic history.

Porphyry molybdenum deposits (in which Mo is economically dominant over Cu) are present in porphyries ranging from granodiorite to alkalic granite in composition, as long as the redox potential is high enough (in I-type granite) to collect the Mo. Granodioritic molybdenum porphyries are mostly at continental margins. They contain only about 0.1% Mo and minor copper (White et al 1981). Comparatively, molybdenum porphyries such as the classic example at Climax, Colorado, are much more alkalic. They grade up to about 0.5% Mo, and nearly all are younger than most porphyry coppers, dating from about 30 Ma to 5.6 Ma in the Colorado Mineral Belt on projection of the Rio Grande Rift. A-type granites in other settings also host molybdenum.

Granite-clan rocks also host the world's major deposits of tungsten and tin. Generally these ores are in quartz veins fed by the siliceous residues of granitic rocks. Tin and tungsten granites tend to be more equigranular than porphyritic, and the plutons are larger.

Tin veins are not always quartz rich; much cassiterite may be in greisen alteration zones of white mica, tourmaline, topaz, beryl, and fluorite with or without quartz. Tin is generally not abundant as a sulfide, except in the Bolivian tin belt whose habitat in the Andean porphyries led Sillitoe et al (1975) to generate the term "porphyry-tin deposits."

Hercynian granites in Europe make up one of the great tin belts of the world, and the dominantly Mesozoic granites of Indonesia, Malaysia, Burma, Thailand, and southern China comprise another. Smaller tin concentrations are in the Kibaran mobile belt in southern Africa, in Rondonia, Brazil, and in granophyres such as those of the Bushveld Complex.

The largest molybdenum deposits, on the other hand, are less than 50 Myr old. Tungsten peaks in age between the molybdenum and tin ages, largely owing to the great Yanshanian wolframite granites in southeastern China.

Burnham & Ohmoto (1980) have pointed out that the fugacity of O_2 in a magma depends largely on the composition of the rocks from which the magma is partially melted. If these rocks are granodiorites or strongly oxidized sediments, the fO_2 would likely be greater than the fugacity of oxygen in equilibrium with an assemblage of quartz, fayalite, and magnetite (QFM), and the magma will be I type (Chappell & White 1974) or magnetite (Ishihara 1974). But felsic magmas generated by partial melting of carbonaceous sediments—S type or ilmenite type—are commonly below the QFM buffer. The siliceous residues of these may yield tin or tungsten mineralization. The *collector* mechanism by which metals may be selectively incorporated into a partial melt may thus be affected by the carbon content of the rocks undergoing melting. A substrate of I-type granite heated to partial melting would tend to yield magmas providing molybdenum deposits, whereas a substrate of carbonaceous shales would tend to yield magmas providing tin or tungsten deposits in their siliceous residues.

3. TECTONIC AND CHEMICAL EVOLUTION OF THE EARTH

3.1 Demarcations

There are good reasons for both tectono-stratigraphic chronology and isotope chronology. The main subdivisions of geologic time—Archean, Proterozoic, and Phanerozoic—are based on long-standing field criteria that recognize their transitional character and lack of synchroneity. Subdivisions into lower, middle, and upper are admittedly provincial, applicable only for a specific area or topic.

Isotopic dates are synchronous (assuming equivalence of "reset" history) and definitive. They even establish the nonsynchronous nature of, for example, the beginning of the Proterozoic, if that is defined as continent stabilization using the criterion of preserved, little-deformed basins. The two time scales are compatible only if they are not combined. There is little value in sharply defining the Archean-Proterozoic transition at 2500 Ma when some of the best-known Proterozoic basins, like the Witwatersrand, are older than 2500 Ma.

3.2 Archean

The proportion of ore deposits that are dependent on volcanic activity is greater in the Archean than in later segments of geologic time. The only Archean sedimentary ore types are the banded iron formations, and most of those are Algoma type, commonly associated with volcanic-hosted massive sulfide deposits.

Yet, unexpectedly, the oldest known ore deposit is a BIF in the Isua series of Greenland, in a granulite-gneiss belt that has been dated at about 3750 Ma. The Isua iron formation is part of a quartzite-shale-carbonate succession more typical of Superior-type BIFs than of the greenstone or Algoma type (Windley 1986). Yet it is hard to equate to modern settings, because copper-bearing sulfide and carbonate facies BIFs are present. Additionally, associated amphibolites, talc schist, and carbonate-bearing siliceous schist, including a horizon of rhyolite(?) boulders, may have been part of an associated volcanic assemblage (Allaart 1976).

There are also chromitite horizons in layered intrusions within gneiss belts, as at Fiskenaesset, Greenland. Chemically, it appears that these assemblages survive anything short of actual "meltdown" of the gneiss to make granitic magma, so it seems unlikely that ores of any of these types were much more abundant in the gneisses before metamorphism than they now appear.

By contrast, Archean *greenstone* belts, ranging from about 3600–2500 Ma in age, are host to some of the richest ores of Cr, Ni, Cu, Zn, Fe, and Au of any tectonic units. The period from about 2900–2600 Ma was the most productive for all but chromium ores. Algoma-type BIFs, Abitibi-type massive sulfides, Kambalda-type nickel deposits, and gold veins dominate the ore assemblages. But there are also suggestions of other igneous-related types, including showings of copper and molybdenum porphyries, tin greisens, and antimony veins. There are even Ta, Nb, and rare earth element concentrations in granophyres. None of these, however, are large enough to make important mines. However, they do establish that locally, in abundantly preserved epizonal igneous assemblages, the chemical environments were appropriate for these elements to congregate. So most changes in ore types over Earth history were in relative abundances, but these changes were major.

Kambalda-type nickel ores in komatiite flows are almost uniquely Archean. None has been found younger than about 1800 Ma. The type locality is by far the largest district, though there are numerous smaller examples in Canada and Zimbabwe at about the same age as the Yilgarn block in Western Australia. But their absence from the older Barberton, South Africa, type-locality for komatiites is perplexing. The nickel is there, located in the silicates, but the sulfur is not. This has led to proposals that the sulfur is acquired from sulfidic sediments when the komatiite flows come out on the seafloor (Groves & Batt 1984), rather than from the mantle source of the ultramafic rocks (Naldrett 1981). For ultramafic rocks in general, sulfides are singularly scarce throughout geologic history, both before and after the 2900–2700 Ma komatiite flows that carry Kambalda-type ores. At any rate, it seems likely that the komatiite *magmas* at Kambalda reached equilibrium with an immiscible sulfide melt before quenching, a difficult feat at 1600°C on the seafloor.

Gold ores, Kambalda-type nickel, and Abitibi-type base-metal massive sulfides (even Algoma-type BIFs) may all be present in the same greenstone belt, as in the Abitibi of Canada, or alternatively one or more may be absent. Komatiites require large-percentage partial melts, probably with deep mantle penetration on major rifts. Abitibi-type massive sulfides require felsic rocks, which are presumably small-percentage partial melts, subjected to a long period of seawater hydrothermal circulation in a permeable zone over a large heat source. Nickel does not appear to move from silicates to sulfides during such hydrothermal action, even from the basalts in Cyprus-type copper ores.

The volcanic rocks with Abitibi-type Cu-Zn massive sulfides include cyclical tholeiitic basalts, some calc-alkaline suites, and even large rhyolite domes or thick surges of rhyolitic fragmentals. Some of these ore bodies are among the largest sulfide masses in the world, reaching as much as 300×10^6 tonnes of ore. The Archean examples average larger than similar deposits later in geologic time. The fumaroles must have run for a long time, so the subsurface magma pool had to be large. Possibly this required a thicker crust than was available in most places, especially in older belts such as the Barberton and Pilbara.

Like other types of basins, greenstone belts seem to have been deepening as filling was going on. Nearly all of the filling was a mixture of cyclical volcanic rocks and high-energy sediments, with surges of turbidites and soft-sediment deformation. Evidence for recumbent folding and detachment sliding is increasing as mapping progresses, but so far the patterns do not reveal much systematic asymmetry in the stratigraphic and fold patterns, nor in the distribution of types of mineral deposits. Distribution patterns of the belts and of the volcanic centers within the belts seem more linear and parallel than radial. There are few indications of incipient "triple junctions." Ore shoots in the sulfide ore bodies, as demonstrated by stope models at Jerome and Quemont, tend to be stretched vertically, which could be the result of "sagging" between deep linear fractures as a result of separation [see Windley (1986) for discussion and references].

Gold distribution in the greenstone belts is very different from the distribution of massive sulfides and komatiite nickel. Gold veins are present in nearly all Archean greenstone belts, but they are generally stratabound in selected stratigraphic units. Gold's widespread distribution emphasizes the worldwide uniformity of metamorphic and structural evolution among Archean greenstone belts.

3.3 Proterozoic: 2500 Ma (3000 Ma in South Africa)– 600 Ma

The Archean-Proterozoic transition is a tectonic transition. In the first part of the Proterozoic, up to about 1800 ± 200 Ma, gradual stabilization of the continents involved preservation of large epicontinental basins. These have hosted some of the world's largest ore deposits: gold and uranium in quartz-pebble conglomerates such as the Witwatersrand, and the world's principal sources of iron ore, the banded iron formations. These two contrasting types of sediments appear to have been contemporaneously deposited in different basins over about a 600 Myr period, until the tenure of the gold terminated with the Tarkwa (Ghana) deposit around 1950 Ma. This was also near the beginning of the rapid decline of the BIFs and of detrital uraninite and pyrite (Pretorius 1981).

In Laurasia, at least, there was a major surge of volcanic activity centering at about 1800 Ma. The oldest Yavapai rocks in Arizona show pillow basalts at the base of the exposed part of the succession, grading upward through hyaloclastic basalt-andesites to thick, stubby rhyolite flows and domes. These are capped by about 300 m of rhyolitic aquatuff that thins abruptly in at least one direction. At the unconformity above is the great Verde massive sulfide at Jerome (Lindberg 1986). Except for the absence of ultramafic rocks, this is an Archean-style volcanic assemblage emplaced at about 1800 Ma. A classic chloritic alteration pipe and pyritic Cu-Zn ore also identify the deposit as Abitibi type. Other deposits of the same age, in Wisconsin, Canada, and Scandinavia, contain less copper and chlorite but more lead and sericitic alteration. They show transitions to some of the smaller sulfide lenses of the 1500- and 1000-Ma groups in North America and the Near East.

The andesitic and felsic fragmentals *above* the Jerome ore zone show turbidity bedding and soft-sediment deformation, filling in around small rhyolitic "resurgent" domes. These features, plus the abrupt lateral thinning of the aquatuff *below* the ore at Jerome, suggest that deposition of the Verde ores might have taken place in a submarine caldera.

Except for Udokan, in Siberia, the earliest sediment-hosted base-metal sulfide deposits are dominated by lead. Prior to about 1750 Ma, there have been only a few small deposits of any kind mined primarily for lead, and lead is a very minor component in the Abitibi-type massive sulfides. But starting with Broken Hill, New South Wales, and continuing with Mt. Isa and McArthur River in Queensland and the Namaqualand ores near Aggeneys, southern Africa, a substantial percentage of the world's lead was concentrated in sedimentary basins over a period of about 250 Myr. Of this group only the Sullivan, British Columbia, ore body shows a well-defined alteration root, though Broken Hill and the deposits at Gamsberg in Namaqualand are associated with BIFs and other chemical sediments. All these lead ores, even the thinly bedded ones, are high grade compared with the disseminated-type ores of the sedimentary copper deposits. Gustafson & Williams (1981) suggest that this may indicate a major difference in metal supply; the lead was furnished by hydrothermal "exhalative" sources, whereas the more disseminated copper deposition was dominated by sedimentary-diagenetic processes.

This point is supported by the fact that the two giant Proterozoic sediment-hosted disseminated copper deposits bracket the lead surge in geologic time, indicating that the chemistry of sulfide sedimentation did not simply evolve from lead to copper over the mid-Proterozoic interval.

There are also striking contrasts between the volcanic-hosted massive sulfides and the sediment-hosted base-metal sulfide groups. They are not contemporaneous, nor are they even in the same region. Among the major deposits, neither the sediment-hosted lead ores nor the copper ores appear to be distal to volcano-hosted base-metal deposits on the same stratigraphic surfaces. Though some may be fed hydrothermally, the sediment-hosted deposits are apparently not part of large volcanic provinces.

There are also contrasts in structural history and grade of metamorphism. The Abitibi-type copper deposits are nearly all cross-folded into steep plunges but are only rarely metamorphosed at higher than greenschist grade. Broken Hill and Gamsberg are mostly at granulite grade, but while the rocks are tightly folded, the fold axes (F_1) are broadly subhorizontal and undulating. Kinematically, there appears to have been more uniform sheet flow, parallel to primary fold limbs, at a higher temperature. These features are characteristic of Proterozoic mobile belts, which are different from igneous-dominated orogenic provinces. The sedimentary copper deposits and the deposits at Mt. Isa and McArthur River are flexurally folded but not highly metamorphosed. The full contrast of fold style and metamorphism is demonstrated in a single basin in the African Copper Belt.

The reason for the major surge of lead starting at 1750 Ma is still unresolved (Hutchinson 1981, Windley 1986). Primary lead concentration would be likely in felsic rocks, particularly those with abundant potassium. These had been accumulating in continental masses since stabilization of the continents and would be accessible to rifts that penetrated continental crust. The small lead deposits in rocks of the Transvaal basin at about 2300 Ma indicate that the process of lead concentration was probably at work in this older craton. Though the spectacular surge beginning at 1750 Ma correlates well with increasing oxidation levels in the atmosphere, it is difficult to see why, among the base metals, lead alone would be affected to such an extreme degree. But if it is not a chemical effect at the site of deposition, it must be an effect at the site of supply or of access to the supply.

In this regard, preservation of the Proterozoic continental crust has locked-in several innovative petrologic features that seem indicative of tectonic evolution. Both anorthosite massifs and large volumes of K-rich anorogenic granitic rocks are distinctive for the period 1700–1100 Ma, and they are commonly mutually associated. Large basaltic magmas would be required to produce anorthosite massifs, such as those containing ilmenite-hematite ores, by crystal separation. Emplacement of basaltic magmas into continental crust could generate small-percentage partial melts from which alkalic granites might appear as associates of the anorthosites in space and time. Rhyolitic molybdenum-rich magmas, rapakivi granites, tin-granites, and even carbonatites could be generated this way depending on starting materials. Of course, *any* slow crustal heating to partial melt conditions could do the same thing, with or without accompanying anorthosites. Deep continental keels depressed into the mantle by imbrication or other crustal thickening on a large scale might also generate large volumes of K-rich granitic magmas. Still another mechanism for slow heating may result from insulation by large continental masses over long periods.

Some authorities (e.g. Piper 1983) conclude on paleomagnetic evidence that a large continent must have existed over much of the Proterozoic. Alkalic magmas thus generated could collect and concentrate molybdenum, tungsten, or tin, depending on the redox level imposed by the rocks that were subjected to melting. Phosphorus, fluorine, berryllium, tantalum, and the rare earth elements are also commonly concentrated.

Under these conditions, and assuming BIFs were incorporated into the partial melt, the ultimate magmatic concentration of Kiruna-type iron ores might be possible. E. B. Kisvarsanyi (1980) described alkaline granite and porphyry assemblages of this type in Missouri with local enrichment in magnetite, phosphorus, fluorine, tungsten, uranium, niobium, and the rare earth elements. The midcontinent anorogenic basement assemblage apparently extends from the Grenville belt, under the Illinois basin to Missouri (where it is exposed at the surface), and then southward to the Llano region of Texas (Anderson 1983). Local members of the same suite continue almost to the west coast of North America. They are the substrate to the Colorado Mineral Belt. The age range of these rocks is mostly between 1500 and 1350 Ma. It is also an interesting point that the world's largest lead concentration, the Mississippi Valley ores of the midcontinent United States, are concentrated in Paleozoic basins over this substrate of red, potassium-rich granites and porphyries (G. Kisvarsanyi 1977).

Rocks of similar character and age also have wide distribution in other parts of the world. The iron, copper, gold, uranium, and rare earth element assemblage at Olympic Dam in South Australia may be a member of the group. Other mineralized districts in south-central Australia have somewhat similar assemblages and ages, some emphasizing gold (Tennant Creek) and others uranium (Mt. Painter). There is also an iron mine with large rare earth element credits in Inner Mongolia, China. Tectonically, these are all in stable cratonic areas; many are still covered by platform and basin rocks.

Expectably, these assemblages can be generated at any time in geologic history after continent stabilization. Carbonatites, for example, range from the Palabora deposit, Transvaal, South Africa, at about 2100 Ma, to Recent. But the large alkalic suites generated worldwide between about

1600 and 1100 Ma must signify particularly advantageous circumstances during that mid-Proterozoic interval. The supercontinent hypothesis (Piper 1983) thus has good petrologic support. The anorogenic systems may have been major mechanisms for heat dissipation before the breakup of proto-Pangea.

The atmosphere and oceans had also been evolving over much of the same interval. Starting at about 2000 Ma, both the decline of the BIFs, of gold, and of detrital uranium and pyrite in quartz-pebble conglomerates and the inauguration of sediment-hosted base-metal sulfide deposits (with ocean-water $SO_4^=$ as a source for sulfide) suggest an increased rate of free-oxygen accumulation. Additional support for this hypothesis comes also from the first appearance and optimum development of vein-type pitch-blende and uraninite ores beneath unconformities in the 1700–1600 Ma range.

3.4 Phanerozoic: ca. 600 Ma to Present

After a period of minimum ore-forming activity between about 1000 and 500 Ma, most of the major types of ore in both the Archean and Proterozoic were rejuvenated. (Exceptions are komatiitic nickel, anorthositic ilmenite, and BIFs.) And some "new" types were added or greatly increased in abundance: Cyprus-type and Kuroko massive sulfides, and the Paleozoic Besshi massive sulfide ores of Honshu, Japan. Other new types were the abundant podiform chromite and, most importantly, those gigantic providers of much of the world's copper, molybdenum, tungsten, and tin—the porphyry coppers and allied types.

Phanerozoic orogenic belts can practically be outlined by igneous-related ores—obviously by those related to consuming plate boundaries, but also by those that may have formed at midocean ridges, such as podiform chromite and perhaps some Cyprus-type massive sulfides. After "rafting" over to the subduction zone, undoubtedly most of these were destroyed, but a few were obducted onto the continents.

Petrologic interpretations become complicated at consuming plate boundaries as magmas rise from a subducting plate through an overriding oceanic plate on their way to surface volcanism. In oceanic regimes, andes-itic volcanism in island arcs generates some (gold-rich) porphyry coppers. In backarc situations near a continent, crustal extension provides the habitat for deep-water calderas containing Kuroko massive sulfides. Kuroko lead must come from reworked (bimodal?) felsic volcanic rocks or their continental ancestors.

But the situation becomes even more complicated if *continental* crust is in the path of magmas rising from the subducting slab. This may supply some of the ingredients, as well as play host, to epizonal porphyry intrusions, and a diverse group of porphyry copper, molybdenum, tungsten, and tin deposits may evolve, as well as retinues of skarn ores, breccia pipes, and epithermal veins.

Contamination or partial melting of continental crust must provide much of the diversity of redox potential for selection of copper, molybdenum, tungsten, or tin at the site of magma generation: copper and molybdenum if the rocks melt to a moderately oxidized I-type granitic magma, or tungsten and tin if carbon-rich sediments contribute extensively to the melt. Locally, organic carbon was available in Precambrian sediments for such reduction to produce S-type tin granites. But apparently most tin and tungsten had to wait for abundant mid-Phanerozoic carbon to generate major tungsten and tin deposits in Hercynian and later "granites."

The magma generated for granitoid systems is, of course, not all from descending slabs. All that is needed is heat, and this can be applied by submergence of continental crust into the mantle through fault imbrication in tectonically thickened continental crust. Such a regime may have produced the Yanshanian granites of southern China, which are hosts for many of the world's largest tungsten concentrations.

Finally, Climax-type molybdenum deposits in the Colorado Mineral Belt are on projections of the Rio Grande Rift zone, whose bimodal volcanism may generate the anorogenic silicic porphyries that contain the ore.

4. CONCLUSIONS

The retinue of igneous-related ore deposits in the Archean greenstones is more nearly similar to those of the Phanerozoic than to those of the middle and upper Proterozoic (after 1800 Ma). Volcanic-hosted massive sulfides and gold veins, which were rejuvenated in the record of the Phanerozoic after about a 1200-Myr period of minimum incidence, show many similar structural features, including rate, style, and intensity of deformation and metamorphism. The basic processes of deposition of the ores and the deformation mechanics succeeding deposition were broadly similar.

The mid-Proterozoic, however, initiates a new suite of igneous-related deposits in addition to new types of sulfide ores in supracontinental basin sediments. These are the ores associated with anorthosites, particularly those associated with alkalic igneous suites of A-type granites and porphyries. These alkalic suites make up a large percentage of the igneous activity in the platform areas of at least midcontinent North America, Australia, Mongolia, and Scandinavia, mostly rather inconspicuously because they are still covered with later platform rocks. The alkalic suites suggest small degrees of partial melting, such as might be expected from the keel of large continental masses.

Wide ranges of metamorphism and deformation characterize basin-type sulfide ores (and mobile belts) of the mid-Proterozoic. These suggest local releases of energy through the continental crust, perhaps by means of aborted deep rifts (the onset of continental separation). Partial separation may actually have evolved for some of these, and they became the loci of high heat flow while rewelding as mobile belts. Others remained simply as troughs or basins, slowly sinking owing to stretching of continental crust over zones of heat accumulation. The thermal blanket of large, thick, potassic, uranium-rich continental crust may even have been a sufficiently destabilizing heat source. For the mid-Proterozoic, the distribution and extent of metamorphism of sediment-hosted lead deposits of all sizes may help delineate tectonic patterns.

The lithophillic ore suites of mid-Proterozoic alkalic rocks present two contrasting assemblages, one oxidized and the other reduced as compared with the level of oxidation in oceanic basalts and andesites. This diversity of oxidation potential among igneous-related ores suggests that the chemical contributions of free O_2, and equivalent sequestered carbon in continental rocks, are instrumental in determining the redox potential of the magmas. Where the continental rocks that were subjected to partial melting included banded iron formations, the great Kiruna-type ores were segregated from K-rich magmas in many parts of the world, especially in the period about 1500–1300 Ma. But where the magmas were generated by partial melts from carbon-rich sediments, tin or tungsten ore suites emerged. Banded iron formations existed from earliest Archean, peaking in the lower Proterozoic, so they were integral parts of the large continental mass of the mid- and upper Proterozoic, perhaps providing the iron and oxygen for Kiruna-type ores.

The metamorphic activation of gold in ultramafic and mafic suites of Archean greenstones provides much of the world's gold. (If Witwatersrand "placer" gold is credited to the greenstones, this becomes *most* of the world's gold.) Backarc turbidites associated with mafic and ultramafic rocks provide much Phanerozoic gold. But between about 2200 and 500 Ma, there were few ultramafic rocks, and those that are present were not subjected to greenstone-type metamorphism. So the only gold veins in that 1700-Myr period are in iron-rich oxidized assemblages like Tennant Creek and Olympic Dam, products of reworked continental crust possibly containing banded iron formations.

ACKNOWLEDGMENTS

I am grateful to Virginia B. Meyer, who helped with many aspects of this study, including patient preparation of the manuscript.

170   MEYER

Literature Cited

Allaart, J. H. 1976. The pre-3750 m.y. old supracrustal rocks of the Isua area, central West Greenland, and associated occurrence of quartz-banded ironstone. In *The Early History of the Earth*, ed. B. F. Windley, pp. 177–89. London: Wiley

Anderson, J. L. 1983. Proterozoic anorogenic granite plutonism of North America. See Medaris et al 1983, pp. 133–54

Anhaeusser, C. R. 1976. The nature and distribution of Archean gold mineralization in southern Africa. *Miner. Sci. Eng.* 8: 46–84

Bethke, C. M. 1986. Hydrologic constraints on the genesis of the Upper Mississippi Valley mineral district from Illinois Basin brines. *Econ. Geol.* 81: 233–49

Boyle, R. W. 1979. The geochemistry of gold and its deposits. *Geol. Surv. Can. Bull.* 280. 584 pp.

Burnham, C. W., Ohmoto, H. 1980. Late-stage processes of felsic magmatism. *Min. Geol. (Jpn.) Spec. Iss.* 8: 1–11

Chappell, B. W., White, A. J. R. 1974. Two contrasting granite types. *Pac. Geol.* 8: 173–74

Cotterill, P. 1969. The chromite ores of Selukwe, Rhodesia. *Econ. Geol. Monogr.* 4, pp. 154–86

Davies, J. F., Luhta, L. E. 1978. An Archean "porphyry-type" disseminated copper deposit, Timmins, Ontario. *Econ. Geol.* 73: 383–96

Derriks, J. J., Vaes, J. F. 1956. Le gîte d'uranium de Shinkolobwe. *UN Int. Conf. Peaceful Uses of Nucl. Energy, Geneva*, 6: 94 pp.

Emslie, R. F. 1978. Anorthosite massifs, rapakivi granites and late Proterozoic rifting in North America. *Precambrian Res.* 7: 61–98

François, A. 1974. Stratigraphie, tectonique et mineralisation dans l'arc cuprifère du Shaba. In *Gisements Stratiforms et Provinces Cuprifères*, ed. P. Bartholomé, pp. 79–101. Liège: Soc. Geol. Belg.

Franklin, J. M., Lydon, J. W., Sangster, D. F. 1981. Volcanic-associated massive sulfide deposits. See Skinner 1981, pp. 485–627

Frietsch, R. 1978. On the magmatic origin of the iron ores of the Kiruna type. *Econ. Geol.* 73: 478–85

Gáal, G., Isohanni, M. 1979. Characteristics of igneous intrusions and various wall rocks in some Precambrian porphyry copper-molybdenum deposits in Pohjanmaa, Finland. *Econ. Geol.* 74: 1198–1210

Goodwin, A. M. 1973. Archean iron formations and tectonic basins of the Canadian Shield. *Econ. Geol.* 68: 915–33

Gross, G. A. 1965. Geology of the iron deposits of Canada. *Can. Geol. Surv. Econ. Geol. Rep. 22.* 181 pp.

Groves, D. I., Batt, W. D. 1984. Spatial and temporal variations of Archean metallogenic associations in terms of evolution of granitoid-greenstone terrains with particular emphasis on the Western Australian Shield. In *Archean Geochemistry*, ed. A. Kröner, G. N. Hanson, A. M. Goodwin, pp. 74–98. Berlin: Springer-Verlag

Guilbert, J. M., Lowell, J. D. 1974. Variations in zoning patterns in porphyry copper deposits. *Can. Inst. Min. Metall. Trans.* 77: 105–15

Gustafson, L. B., Hunt, J. P. 1975. The porphyry copper deposit at El Salvador, Chile. *Econ. Geol.* 70: 857–912

Gustafson, L. B., Williams, N. 1981. Sediment-hosted stratiform deposits of copper, lead and zinc. See Skinner 1981, pp. 139–78

Hutchinson, R. W. 1981. Metallogenic evolution and Precambrian tectonics. See Kröner 1981, pp. 733–59

Ishihara, S., ed. 1974. *Geology of Kuroko Deposits. Soc. Min. Geol. Jpn. Spec. Iss.* 6. 435 pp.

James, H. L., Sims, P. K. 1973. Precambrian iron formations of the world: introduction. *Econ. Geol.* 68: 913–14

Kisvarsanyi, E. B. 1980. Granitic ring complexes and Precambrian hot spot activity in the St. Francois terrane, mid-continent region United States. *Geology* 8: 43–47

Kisvarsanyi, G. 1977. The role of Precambrian igneous basement in the formation of the stratabound lead-zinc-copper deposits in southeast Missouri. *Econ. Geol.* 72: 435–42

Kröner, A., ed. 1981. *Precambrian Plate Tectonics.* New York: Elsevier. 781 pp.

Laznicka, P. 1973. Development of non-ferrous mineral deposits in geologic time. *Can. J. Earth Sci.* 10: 18–25

Lindberg, P. A. 1986. A brief geologic history and field guide to the Jerome district, Arizona. *Geol. Soc. Am. Rocky Mt. Sect. Guideb.*, ed. J. D. Nations et al, pp. 127–39. Boulder, Colo: Geol. Soc. Am.

Lowell, J. D. 1974. Regional characteristics of porphyry copper deposits of the Southwest. *Econ. Geol.* 69: 601–17

Medaris, L. G., Byers, C. W., Mickelson, D. M., Shanks, W. C., eds. 1983. *Proterozoic Geology: Selected Papers From an International Symposium. Geol. Soc. Am. Mem.* 161. 315 pp.

Mendelsohn, F., ed. 1961. *Geology of the Northern Rhodesian Copper Belt.* London: McDonald. 523 pp.

Meyer, C. 1981. Ore-forming processes in geologic history. See Skinner 1981, pp. 6–41

Mitchell, A. H. F., Garson, M. S. 1981. *Mineral Deposits and Global Tectonic Settings.* New York: Academic. 405 pp.

Naldrett, A. J. 1981. Nickel sulfide deposits: classification, composition, and genesis. See Skinner 1981, pp. 628–85

Nash, J. T., Granger, H. C., Adams, S. S. 1981. Geology and concepts of genesis of important types of uranium deposits. See Skinner 1981, pp. 63–116

Nunes, P. D., Ayres, L. D. 1982. U-Pb zircon ages of the Archean setting Net Lake porphyry molybdenum occurrence, northwest Ontario, Canada. *Econ. Geol.* 77: 1236–39

O'Driscoll, E. S. 1982. Patterns of discovery—the challenge for innovative thinking. *Pet. Explor. Soc. Aust.* 1: 11–31

Ohle, E. L. 1980. Some considerations in determining the origin of ore deposits of the Mississippi Valley type—Part II. *Econ. Geol.* 75: 161–72

Parák, T. 1975. Kiruna iron ores are not "intrusive-magmatic ores of the Kiruna type." *Econ. Geol.* 70: 1242–58

Phillips, G. N., Groves, D. L., Martyn, J. E. 1984. An epigenetic origin for Archean banded-iron-formation-hosted gold deposits. *Econ. Geol.* 79: 162–71

Piper, J. D. A. 1983. Dynamics of the continental crust in Proterozoic times. See Medaris et al 1983, pp. 11–34

Pretorius, D. A. 1981. Gold and uranium in quartz-pebble conglomerates. See Skinner 1981, pp. 117–38

Ridler, R. H. 1976. Stratigraphic keys to the gold metallogeny of the Abitibi belt. *Can. Min. J.* 97: 81–88

Roberts, D. E., Hudson, G. R. T. 1983. The Olympic Dam copper-uranium-gold deposit, Roxby Downs, South Australia. *Econ. Geol.* 78: 799–822

Rona, P. A., Bostrom, K., Laubier, L., Smith, K. L. Jr., eds. 1983. *Hydrothermal Processes at Seafloor Spreading Centers.* New York: Plenum. 798 pp.

Ross, J. R., Hopkins, G. M. 1975. The nickel deposits of Kambalda, Western Australia. In *Economic Geology of Australia and Papua New Guinea, Vol. 1, Metals*, ed. D. G. Knight, pp. 100–21. Parkville, Victoria: Aust. Inst. Min. Metall.

Russell, M. J. 1983. Major sediment-hosted exhalative zinc and lead deposits: formation from hydrothermal convection cells that deepen during crustal extension. In *Sediment-Hosted Stratiform Lead-Zinc Deposits. Mineral. Assoc. Can., Short Course Handb.*, ed. D. G. Sangster, 8: 251–82

Samonov, I. Z., Pozharisky, I. F. 1977. Deposits of copper. In *Ore Deposits of the USSR*, ed. V. I. Smirnov, 2: 170–81. London: Pitman. 424 pp. (From Russian)

Sawkins, F. J. 1984. *Metal Deposits in Relation to Plate Tectonics.* New York: Springer-Verlag. 325 pp.

Sillitoe, R. H., Halls, C., Grant, J. N. 1975. Porphyry tin deposits in Bolivia. *Econ. Geol.* 70: 913–27

Skinner, B. J., ed. 1981. *Economic Geology: 75th Anniversary Volume.* New Haven, Conn: Economic Geology Publ. Co. 964 pp.

Skinner, B. J. 1983. Submarine volcanic exhalations that form mineral deposits: an old idea now proven correct. See Rona et al 1983, pp. 76–81

Titley, S. R., Beane, R. E. 1981. Porphyry copper deposits. See Skinner 1981, pp. 214–69

Trendall, A. F. 1968. Three great basins of Precambrian banded iron formation deposition: a systematic comparison. *Geol. Soc. Am. Bull.* 79: 1527–44

van Valen, L. 1971. The history and stability of atmospheric oxygen. *Science* 171: 439–43

Wetherill, G. W., Drake, C. L. 1980. The Earth and planetary sciences. *Science* 209: 96–104

White, D. E. 1968. Environments of generation of some base-metal ore deposits. *Econ. Geol.* 63: 301–35

White, W. H., Bookstrom, A. A., Kamilli, R. J., Ganster, M. W., Smith, R. P., et al. 1981. Character and origin of Climax-type molybdenum deposits. See Skinner 1981, pp. 270–316

Windley, B. F. 1986. *The Evolving Continents.* London: Wiley. 399 pp. 2nd ed.

Woodall, R., Travis, G. A. 1969. The Kambalda nickel deposits, Western Australia. *Proc. Commonwealth Min. Metall. Cong., 9th, London*, 2: 517–33

Wright, S. F. 1986. On the magmatic origin of the iron ores of the Kiruna type—an additional discussion. *Econ. Geol.* 81: 192–94

Youles, I. P. 1984. The Olympic Dam copper-uranium-gold deposit, Roxby Downs, South Australia—a discussion, with reply by Roberts & Hudson immediately following. *Econ. Geol.* 79: 1941–55

Ann. Rev. Earth Planet. Sci. 1988. 16: 173–200

GEOLOGY OF HIGH-LEVEL NUCLEAR WASTE DISPOSAL

Konrad B. Krauskopf

Department of Geology, Stanford University, Stanford, California 94305

INTRODUCTION

The disposal of radioactive waste, volumewise, is only a small part of the general problem of managing industrial waste, but it is a peculiarly difficult part in both its technical and political aspects. Like much other industrial waste, material containing radioactive elements must be isolated from surface environments for very long times, and any quantity that escapes must be kept very small. The guarantee of effective isolation must be especially firm because of the damage that even small amounts of radioactivity can cause, and because of widespread fears in the popular mind of a mysterious influence somehow associated with mushroom clouds, genetic defects, and delayed incidence of cancer. The necessary extreme degree of isolation can almost certainly be achieved, but convincing the general public that the achievement is possible remains a stubborn political problem.

Geology plays a role in the management of radioactive waste simply because the favored disposal method is subsurface burial. Much low-level waste—material containing only traces of radioactivity (for example, discarded radiation sources and equipment used in radiation therapy)—has already been buried, and is currently being buried, in trenches only a few meters deep. High-level waste requires deeper burial, and no waste of this sort has yet been placed permanently underground. Much effort has gone into planning for deep disposal in all countries with a well-developed nuclear power program, but the plans remain only on paper. Because high-level waste is the most difficult kind to manage, and because a satisfactory method for its disposal has yet to be demonstrated, it is the focus of attention in this paper. Geologic aspects of the disposal problem are

173

0084–6597/88/0515–0173$02.00

emphasized, and engineering aspects are touched on more lightly. Political aspects are barely mentioned, although they may well prove ultimately the most difficult of all to manage.

FORMS OF HIGH-LEVEL WASTE

The material that constitutes high-level radioactive waste comes in two principal varieties: the spent fuel rods that are discarded periodically from operating nuclear reactors, and the liquid that results from dissolving fuel rods in order to recover their contained plutonium and/or uranium. According to one definition, only the second kind is properly called high-level waste, on the grounds that spent fuel rods are still a potential source of nuclear energy and hence not "waste" in the usual sense. Because present plans call for permanent disposal of both kinds by underground burial, they are both considered as high-level waste (HLW) for purposes of this article.

Spent Fuel Rods

The fuel rods in a reactor are slender metal tubes, 4 to 5 m long and a few centimeters in diameter, made of a zirconium alloy, which contain the nuclear fuel in the form of uranium dioxide (Benedict et al 1981, pp. 84–106; OTA 1985, pp. 24–26). After a reactor has operated for several months or a year, the fuel becomes so contaminated with products of the fission reaction—atoms of elements with roughly half the mass of the fissile isotope ^{235}U—that production of energy is no longer efficient. The fuel rods must then be removed from the reactor and replaced by new ones. The spent rods cannot be simply discarded with ordinary garbage, because their contained fission products are highly radioactive; they must be lifted from the reactor by remote control, and for safekeeping they are placed in basins of cool circulating water. Such basins, with a steadily increasing load of discarded fuel rods, are present at all active reactor sites. The basin materials and the water serve to keep the rods cool and to absorb their radiation, so that no hazard results to the surroundings. These accumulations of spent fuel rods at reactor sites are the first form of HLW.

Reprocessing Waste

The second form has a more complicated origin. Spent fuel contains, in addition to the fission products, a substantial amount of heavy elements formed during reactor operation by the addition of neutrons to atoms of the more abundant uranium isotope, ^{238}U. The principal "transuranic" elements are neptunium, plutonium, americium, and curium; they are all radioactive and contribute to the intense radioactivity of spent fuel. Of these elements, the most interesting is plutonium because of its usefulness in nuclear weapons: The isotope ^{239}Pu has the same property of fissility as ^{235}U and is even better suited for military purposes. To obtain plutonium for weapon manufacture, spent fuel is generated in reactors designed to maximize the amount of this isotope; the fuel rods are broken up, their contents are dissolved in nitric acid, and the plutonium is separated by adding an organic reagent (Benedict et al 1981, pp. 457–88). This leaves all the other radioactive elements, both fission products and the other transuranic elements, in the acid solution. To keep this hot and radioactive liquid from contaminating the environment, it is pumped into large, double-walled steel tanks set just under the ground surface. The acidity is neutralized, precipitating a radioactive slurry; then partial evaporation causes some of the remaining solutes to crystallize. At the two places in the United States where plutonium has been produced in large quantity— Hanford in south-central Washington and the Savannah River plant in South Carolina—there are large areas of such tanks filled with mixtures of liquid, slurry, and crystals (NRC/NAS 1978b, 1981, Catlin 1980). This is the second kind of HLW, often called "reprocessing waste" because its generation involves the reprocessing of spent fuel.

Plutonium would serve as well for generating useful energy as for making bombs, and a good case can be made for separating plutonium from *all* spent fuel for this purpose. Along with plutonium, the remaining uranium in the fuel could be recovered also, and new fuel rods could be fabricated using the two fissile isotopes. In this way a larger fraction of the energy contained in the original uranium would be put to use. Early in the nuclear age it was expected that this would be normal procedure: No spent fuel would be stored in water basins for more than a short time, and the only kind of HLW would be the reprocessing waste from the separation of uranium and plutonium. But objections were raised on the grounds that the plutonium generated by reprocessing in many plants around the country would be difficult to guard against pilfering by would-be terrorists. For a time in the United States the reprocessing of spent fuel was forbidden on this account; the ban is now lifted, but under present conditions commercial reprocessing appears to be only marginally economic and has not been started (OTA 1985, pp. 67–73). In some European countries spent fuel is being reprocessed on a small scale.

Thus, spent fuel in the United States at present has an ambiguous status: As long as no reprocessing is done, it is simply a form of high-level waste that must be disposed of, but if reprocessing should seem more attractive in the future, spent fuel would become a valuable potential source of energy. For the moment it is commonly regarded as waste, and it is so treated here.

Why is Disposal Needed?

High-level radioactive waste, then, consists of spent fuel rods in water basins at reactor sites and reprocessing waste in steel tanks such as those at Hanford and Savannah River. Both kinds at present are effectively isolated from contact with the biosphere. Occasional leaks have developed in the steel tanks and water basins, but the amount of released radioactivity is small and no detectable harm to the environment has resulted (Catlin 1980, NRC/NAS 1978b, pp. 36–40). As long as leaks can be quickly repaired, or as long as new containers are available into which the waste can be transferred, the present methods of storage will provide an adequate margin of safety. It is reasonable, then, to inquire why further efforts at disposal are necessary. Why not leave the waste where it is, rather than go to the effort and expense of putting it deep underground?

The query is reasonable except for the problem of time. The waste could indeed be kept safely in its present containers for an indefinite future, *provided that someone is on hand to watch over it.* The proviso is essential: Possible leaks must be monitored, water must be kept circulating, new tanks and water basins must be built as the old ones corrode. Such surveillance must continue as long as the waste remains hazardous, and this means for a very long time. The radionuclides in the waste have a variety of half-lives (Table 1); many decay rapidly, becoming harmless within a few years or decades, but a few have half-lives so long that the waste remains potentially dangerous for hundreds of thousands or even a few millions of years (NRC/NAS 1983, pp. 26–38). It hardly seems likely that adequate surveillance of surface facilities can be counted on to continue for such periods. Hence a more permanent method of disposal is needed, some place to stow the waste where it can be depended on to remain immobile, without further attention, for times on the order of a million years.

This is the essential problem of waste disposal: how to guarantee its immobility for times that are ordinarily of interest only to geologists. It is a new kind of problem for engineers, who normally plan for structures with an expected useful life of no more than several decades or a few centuries at most. There is certainly reason for caution in approaching a solution, and the long delay in getting an underground repository constructed is understandable.

Amounts and Composition of High-Level Waste

If the waste is to be put into a cavity excavated in rocks deep below the surface, how big an engineering operation will this be?

The amount of HLW that will have accumulated in the United States by the year 2000 is estimated to be roughly 42,000 tonnes of spent fuel and 300,000 m³ of reprocessing waste, with a total radioactive content of about 37×10^9 Ci (USDOE 1985). For disposal underground the reprocessing waste will be converted to some kind of solid, most probably a borosilicate glass, and both this solid and the spent fuel rods will be enclosed in metal canisters. The volume changes in these operations are hard to estimate, but the total quantity of material to be placed under the surface will probably be on the order of 2×10^5 m³. To visualize this volume, suppose it could be dumped into a college football stadium: The bottom of the stadium would be covered to a depth of 8 to 10 m, or roughly 30 feet. To get this amount of material underground, with all handling done by remote control, is certainly a major operation but by no means an impossible one. The engineering procedures of mining out a cavity of the necessary size, transporting the waste to the site, moving it into the cavity, and then backfilling cavity and shaft are well within present capabilities. The difficulty of HLW disposal is not a problem of inadequate technical know-how.

The composition of radioactive material in the waste is variable because it depends on the kind of reactor from which the waste comes, on the length of time the fuel was in the reactor, and on the amount of radioactive decay since removal of the fuel from the reactor. To give an idea of relative amounts of the principal isotopes, compositions of specific samples of the two kinds of HLW are shown in Table 1. The compositions are similar, except that most of the uranium and plutonium have been removed from the reprocessing waste, and iodine is low in the reprocessing waste because it was partly volatilized during reprocessing. The table shows clearly the loss with time of radionuclides with short half-lives, especially ^{90}Sr and [137]Cs. Isotopes that appear to increase with time are those that are formed in part by decay of other actinide elements in the waste. The complications in trying to predict the behavior of waste in a geologic environment are manifest from the table, since each element shown has its own chemical peculiarities that determine how it will react with constituents of rock and groundwater.

MINED GEOLOGIC DISPOSAL

Many alternative ways to dispose of HLW have been suggested: send it in rockets into outer space, set it down on the ice cap of Greenland or Antarctica, bury it in a very deep hole where it will melt its surroundings and become part of a new-formed igneous rock, sink it into the red-clay sediment of the deep ocean floor (Altomare et al 1979, Milnes 1985, NASA 1986, Hollister et al 1981, Sandia National Laboratories 1983). Attractive

Table 1   Important radionuclides in two kinds of high-level waste[a]

Isotope	Half-life (yr)	Amount in one spent-fuel assembly from a pressurized water reactor (g)		Amount in one canister of borosilicate glass made from reprocessing waste (g)	
		Time after discharge from reactor		Time after reprocessing	
		10 yr	1000 yr	10 yr	1000 yr
^{90}Sr	28.8	200	0	880	0
^{99}Tc	2.14×10^5	360	360	1600	1600
^{129}I	1.6×10^7	83	83	0.38	0.38
^{135}Cs	3×10^6	140	140	2400	2400
^{137}Cs	30.2	440	0	2000	0
^{234}U	2.45×10^5	88	150	2.9	16
^{238}U	4.47×10^9	440,000	440,000	9900	9900
^{237}Np	2.14×10^6	210	660	930	1000
^{238}Pu	87.7	60	260	13	0.014
^{239}Pu	2.4×10^4	2300	2300	53	68
^{240}Pu	6570	1100	1000	36	56
^{242}Pu	3.76×10^5	210	210	4.8	5.0
^{241}Am	433	230	120	120	26
^{243}Am	7370	40	36	180	160

[a] Source: NRC/NAS (1983, pp. 31, 35).

as these possibilities sound, they all have the disadvantage of prohibitive expense or an inadequate data base for predicting their long-term consequences. In all countries where the problem of managing high-level waste has arisen, careful study has shown that burial in a mined cavity several hundred meters under the surface is the most feasible alternative. In contrast to other options, geologic disposal involves only the uppermost levels of the Earth's solid crust, a part of the planet with which we have long been familiar and about which we can readily obtain detailed information and make reliable predictions.

The construction of a mined geologic repository is conceptually simple (NRC/NAS 1983, pp. 108–22; OTA 1985, pp. 39–50). A shaft will be sunk to a depth of several hundred meters, deep enough that a cavity excavated at its base will be well protected from surface disturbances, either natural or man-made. The cavity will consist of an array of tunnels, much like the tunnels of a mine, and in the floors or walls of the tunnels holes will be drilled for emplacement of the waste containers (Figure 1). These containers will be metal cylinders, or *canisters*, filled with solid waste in the form of either spent fuel rods or glass fabricated from reprocessing waste;

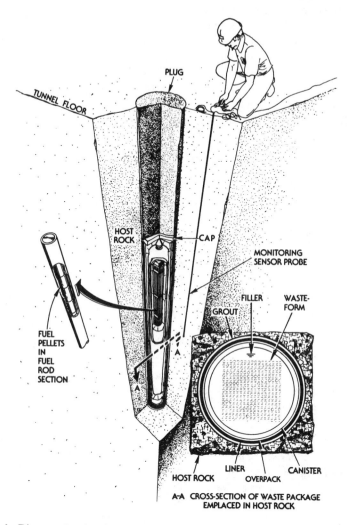

Figure 1 Diagram showing the planned disposal of spent fuel rods in a mined geologic repository. The fuel rods are in a metal canister placed in a hole drilled into the floor of a repository tunnel. The small diagram on the left is an enlargement of a section of a single fuel rod. The square diagram on the lower right is a cross section of the emplacement hole, showing a cluster of fuel rods (the "waste form") enclosed in the canister and overpack (US Department of Energy).

in some repository designs the canisters are surrounded by an additional metal shield, or *overpack*. When all canisters are in place in one section of the repository, the tunnels and holes in that section will be filled with crushed rock or clay (the *backfill*), and after the entire repository is filled the shaft also will be filled and sealed. In effect the waste will become part of the surrounding rock, and seemingly its radioactivity will be adequately isolated from surface environments for the necessary long times. Monitoring devices will provide checks on the effectiveness of the isolation for at least several years after repository closure, so that corrective measures can be taken if necessary.

Criteria for Disposal Sites

Obviously this procedure can ensure permanent disposal only if the site for the repository has been carefully chosen. Because isolation times up to a million years are required, the possibility of geologic disturbance must be considered: the repository should not be located where severe earthquakes or volcanic eruptions are likely, or where deep erosion by running water or ice is expectable. Areas of current mining activity or hydrocarbon production should be avoided, as well as areas where such activity is likely in the future. The rock at repository depth should be strong enough to maintain an opening at least during the few decades needed for repository filling, and its structure should be simple enough to facilitate mining. The site should have topography suitable for a large engineering operation, and it should preferably be at a distance from population centers. These criteria for choosing a repository site seem self-evident, and they can be satisfied at many places in the United States (NRC/NAS 1978a).

One additional criterion is needed, and this is by far the hardest to satisfy: The site must be in an area where groundwater cannot jeopardize the integrity of the repository. At a depth of several hundred meters the rock nearly everywhere is saturated with water; the water almost certainly will be moving, albeit very slowly in most places; and the great danger to a repository over long periods is the probability that canisters in contact with water will eventually corrode, part of the waste will dissolve, and the moving water will carry its radioactive burden to the point where it emerges in springs or seepages or where it is tapped by wells. All other kinds of geologic disturbance can be avoided by proper choice of site, except for extremely improbable events, but the eventual transport of some radioactive material by groundwater is a threat almost anywhere. The geologic problem of radioactive waste disposal is chiefly a matter of efforts to prevent or minimize the movement of dissolved waste constituents.

After a repository has been filled and sealed, the sequence of events in the reestablishment of groundwater flow can be imagined (NRC/NAS

1983, pp. 46–51). The flow would have been interrupted, of course, by construction of the repository, and the tunnels would be kept dry by ventilation as long as waste was being emplaced. When backfilling is completed and ventilation is shut off, water will gradually fill all vacant spaces. Depending on the nature of the backfill and overpack, contact of water with the canisters may be long delayed, but eventually all materials in the emplacement holes will be saturated; the times required will depend on the nature of the rock, but in general they are estimated to be on the order of several decades. Eventually the water will be incorporated into the regional groundwater flow, so that water in contact with canister surfaces will be slowly moving. After decades or centuries the canisters will almost certainly corrode, and water will make contact with the waste itself. The waste will be in a nearly insoluble form and dissolution will be very slow, but over long periods appreciable amounts will dissolve and be carried into the surrounding rock. Some of the dissolved radionuclides will precipitate or be trapped by sorption on mineral surfaces as they move through the rock, but ultimately a small quantity may travel with the water to the ground surface. Details and timing of this predicted sequence will vary from one repository site to another, but at almost any site the eventual escape of some waste constituents with groundwater must be regarded as a strong possibility.

Engineered and Geologic Barriers

To minimize or delay the movement of radioactive substances from a repository to the surface, two strategies are possible: The waste may be protected from contact with groundwater, or the repository site can be so chosen that the nature of the groundwater and the rock through which it must travel will keep the movement of dissolved material very slow (Eisenbud 1982). Protecting the waste from groundwater contact is a question of engineering: Metal for the canisters must be as resistant to corrosion as possible, the backfill must have low permeability, the waste must be in a nearly insoluble form. The choice of site depends on geology—finding a place where groundwater at depth has a noncorrosive composition, where its movement is slow, where its path to the surface is long, and where the rock through which it travels has a large capacity for sorbing dissolved material. In effect, plans call for a series of barriers, partly engineered and partly geological, to groundwater movement. This "multibarrier approach" to controlling the amount of radionuclide release is a part of disposal planning wherever the mined geologic alternative is being considered.

The ideal engineered barrier clearly would be a noncorrodible canister. A canister made of gold or platinum, for example, would certainly endure far beyond the time needed for HLW to decay to harmless levels. In the real world a less exotic metal or alloy must be chosen, and much effort has been devoted to finding an appropriate one. Several steel and titanium alloys have been suggested, which in experiments under simulated repository conditions have been shown to corrode so slowly that canisters should remain intact for at least several hundred years. Such times would permit substantial decay of ^{90}Sr and ^{137}Cs, two of the fission products that would be most harmful to living things if they escape prematurely. Thus there seems no question that the radionuclides in waste can be kept from moving at all (or can be *contained*, in waste-disposal parlance) for several centuries into the future. A containment period of at least 300 yr, and probable containment up to 1000 yr, is one of the requirements in regulations set up by the Nuclear Regulatory Commission for planned repository construction in the United States (USNRC 1983).

After the containment period, immobilization of the radionuclides cannot be guaranteed. In some repositories containment will certainly last more than the predicted 300 to 1000 yr, but in general by the end of this period one must count on corrosion having breached at least some of the canisters, so that waste is directly exposed to groundwater. Some radionuclides will dissolve, and the goal then becomes keeping the amount that reaches the ground surface very small. In the argot of waste disposal, the goal is to keep the waste *isolated*—to keep it so well separated from the environment that the amount escaping is acceptably low. Limits of acceptability are set by regulatory agencies: In the United States, isolation must be effective enough that no more than one part in 10^5 of any radionuclide present in the waste after 1000 yr can escape per annum (USNRC 1983). This requirement and additional limits on concentrations in groundwater in the immediate vicinity of a repository are designed to ensure that additions to radiation in the external environment are no larger than the observed variations in natural background levels.

Isolation is achieved in part by engineered barriers—the low solubility of the waste form, and the slowing of radionuclide movement by backfill— but chiefly by the geologic barrier imposed by the rock through which groundwater must move to reach the surface. The rock barrier operates in several ways: by keeping the movement of groundwater slow, by causing precipitation of some of the radionuclides, by sorbing some on mineral surfaces, and by dispersing the groundwater flow. The effectiveness of the barrier depends on the nature of the rock, and different kinds of rock have widely different abilities to provide isolation. Choosing suitable rocks for repository siting is a major problem for geologists concerned with HLW disposal. The qualities of some promising rock types are described in the next section.

ROCKS SUITABLE AS HOSTS FOR REPOSITORIES
Salt

In the earliest detailed study of disposal possibilities, a committee of the National Research Council in the mid-1950s recommended rock salt as a likely medium for repository construction (NRC/NAS 1957), and salt has remained a prime favorite ever since (e.g. NRC/NAS 1970, APS 1978, Isherwood 1981, Vol. 1). It has several obvious virtues. The mere existence of a body of salt beneath the surface means that groundwater cannot have moved through it in any quantity—if it had, the salt would not be present. Thus excavating a repository in a thick salt bed or in the interior of a salt dome would seemingly avoid the groundwater problem entirely. Furthermore, salt has desirable mechanical properties: It is easily mined, and at depths of several hundred meters, it will maintain an opening for years or a few decades; but ultimately, under the pressure of overlying rock, it flows plastically to fill any underground openings. Thus waste in a salt repository would eventually be encased in a solid mass of crystalline salt, with no openings into which water from an external source could penetrate. An additional merit of salt is a thermal conductivity higher than that of other common rocks, meaning that temperatures in a salt repository (for a given loading and spacing of canisters) will be lower than in repositories elsewhere.

A geological skeptic can raise a few questions about salt. The property of plastic flow, although desirable in the long term, may be a nuisance while the repository is being filled, especially since the flow is accelerated by heat; during the decade or more required for waste emplacement, the repository tunnels will require periodic reaming to counteract contraction. Salt always contains some water, up to a percent or so, both as films between the crystals and as tiny brine inclusions within the crystals; the inclusions are known to move toward a source of heat (Jenks 1979), and fears have been expressed that brine will collect in large amounts around waste canisters and will accelerate corrosion. Experiments with non-radioactive heat sources have shown, however, that the amounts to be expected are too small to have much importance, especially since the brine would not be moving but would simply collect in stagnant pools enclosed by solid salt (Shefelbine 1982). Effects of radiation on salt adjacent to canisters has also been a worry, since strong radiation fields are known to decompose NaCl with the production of free chlorine; but recent work indicates that this effect is probably not serious at the expected radiation levels. Thus, most objections to salt have been answered by experimental studies, and it remains an attractive repository medium.

The United States is fortunate in having abundant salt at appropriate depths, both as bedded salt (for example, in the Permian Basin extending from New Mexico into Kansas, and in the Salina Basin from Michigan to New York) and as the famous salt domes of the Gulf Coast. Both kinds of salt formation have been considered for repository development, and part of a thick salt bed underlying northern Texas has been chosen by the US Department of Energy as one of the three currently most favored sites for detailed study (USDOE 1986a). In several European countries also, salt is regarded as a likely choice for the host rock of an HLW repository.

Crystalline Rock

Crystalline rock, meaning any coarse-grained igneous or metamorphic rock with quartz and/or feldspar as prominent minerals, is another variety with much to recommend it as a repository host. The term is strictly a misnomer, since other rocks, such as limestone, salt, and most shales, also consist largely of crystals, but it has come into common use for granite and all its relatives, gneiss, high-grade schist, and mixtures of all of these. Such rocks are less easily mined than salt, but once a tunnel has been driven it maintains its shape indefinitely. Because the rocks have been formed at high temperatures, they should be little affected by the modest temperature rise expectable from nuclear waste. Groundwater in crystalline rock generally has a composition (dilute, slightly alkaline, weakly reducing) that will ensure a slow rate of canister corrosion and low solubility of many radionuclides. The rock itself commonly has very low permeability, but the joints and shear zones that are almost universally present provide possible paths for groundwater movement. Locating a suitable repository site in crystalline rock is chiefly a matter of finding a place where fractures at repository depths are few, small, and/or sufficiently clogged with gouge and secondary minerals to keep the overall permeability low.

The possibilities of crystalline rock for disposal purposes have been studied intensively in Sweden, where such rocks are almost the only kind available (SKBF 1983, NRC/NAS 1984). Similar rocks of the Canadian Shield are the object of current attention in Canada (Chapman & Sargent 1984), and crystalline rock is the favored medium also in France and Switzerland. In the United States the exploration of crystalline rock has lagged, but large areas in the north-central states, in New England, and in the southeast have been noted as likely targets for future study.

The foreign work has demonstrated that sites in crystalline rock suitably free from major fractures and shear zones can probably be found, but the search is not easy (e.g. Ahlbom et al 1983, Ubbes & Duguid 1985). Mapping of joints on surface exposures, supplemented by study of closely spaced drillholes in areas that show promise, gives some indication of conditions at depth. Geophysical measurements, both from the surface and using instruments lowered into the drillholes, help to locate major fractures. The inferences from such exploration must ultimately be checked by the sinking of a shaft and driving tunnels from its base in order to permit hands-on study of rock structures. Crystalline-rock investigations have not yet progressed to shaft-sinking at likely repository sites, but the detailed surface-plus-drillhole exploration has shown with reasonable certainty that adequate volumes of sound rock exist in which the flow of groundwater through fractures is small enough to be either a minor problem or one controllable by grouting.

Shale and Clay

Argillaceous rocks have some appeal as media for waste disposal because they are relatively impermeable and have a high capacity for sorbing ions out of solutions that move through them. They are notoriously variable in composition and structure, and only a few types would have the necessary strength to maintain mined openings, the necessary freedom from joints and interbeds to ensure low permeability, and the necessary resistance to alteration by a modest temperature increase. The possible deleterious effect of heating is the objection most commonly urged against the use of shale. The clay minerals, it is argued, will lose water and the rock will thereby become more permeable, or the dominant mineral might change from a highly sorptive smectite to a less sorptive illite. Recent experiments indicate that expectable changes would not be seriously damaging if repository temperatures do not greatly exceed 100°C, but further work is needed on the kinetics of the alteration processes.

Field studies of argillaceous rocks are under way in Italy and Belgium. In the United States, the Pierre Shale under parts of the Great Plains and the Chattanooga Shale of the southeastern states are often mentioned as possibilities worth exploring. Reconnaissance surveys have been made of these and other shale units (Isherwood 1981, Vol. 2), but detailed work on clays and shale has been limited so far chiefly to laboratory studies.

Tuff

Like shale, consolidated volcanic ash has the virtues of low permeability and a marked capacity for sorbing ions from solution. If the ash has been partly welded, it loses much of its sorptive ability but becomes better able to maintain an opening underground; if it is partly converted to zeolites, as tuff beds often are, its sorptive capacity may be markedly increased. Ideally, a repository might be located in the welded portion of an ash-flow unit, surrounded on all sides by a buffer of unwelded and zeolitized ash to trap any radionuclides that might escape.

Tuff beds of this ideal character and sufficiently thick to accommodate a repository are not common, but a few possibilities have been found in the western United States (Bedinger et al 1985). One site in particular, in southern Nevada, has an unusual thickness of tuff that accumulated from several volcanic centers during the Tertiary; this site is also attractive because of an extremely low water table, some 600 m below the surface. This opens the possibility of excavating a repository in the unsaturated zone well above the water table (Winograd 1981, Roseboom 1983). In such a location the repository would never fill permanently with flowing groundwater, so that the major uncertainty about geologic disposal would be circumvented. So promising is this site that it has been designated as another of the three locations considered by the Department of Energy as best qualified for intensive exploration (USDOE 1986b).

Objections can be raised to a site in southern Nevada on the grounds that not enough is known about water movement in the unsaturated zone, that the unsaturated zone is at least in part an oxidizing environment where the solubility of some radionuclides might be a problem, and that the danger of severe earthquakes and renewed volcanism, although small, is certainly greater here than in more tectonically stable parts of the continent. The objections are not trivial, but they seem outweighed by the manifest advantages of this area.

Basalt

Still another possible location for a repository is in the interior of a very thick lava flow. The first reaction of most geologists to this suggestion is negative, because lava flows are known to be cut by networks of cracks, cracks so abundant that they often permit water to move through a flow and emerge as conspicuous springs at its base. Nevertheless, among the huge pile of basalt flows that underlie the Columbia River Plateau in southern Washington there are a few with thicknesses up to 80 m and interiors that drillcores show to have only minor fracturing. The rock, of course, would maintain a repository opening admirably; except for the fractures it would be highly impermeable, and any water that penetrates it would be well buffered at a fairly high pH and low Eh by the abundant ferrous minerals in the rock. For these reasons the Department of Energy has chosen south-central Washington as the third of its most favored regions for further study (USDOE 1986c).

Possible reasons for questioning this choice, in addition to the fracturing of the basalt, are the presence of copious aquifers in beds of sediment and volcanic rubble between the lava flows, the demonstrated existence of stress in the rock that causes it to shatter when openings are made, and the high measured temperatures in drillholes at repository depths (up to

57°C) that would require massive cooling to make underground work possible. Almost certainly a successful repository could be constructed here, but the cost would doubtless be greater than at the sites in Texas and Nevada.

Elsewhere in the world, lava has not been suggested as a repository medium. The thick, relatively unfractured flows at appropriate depths in southern Washington may well be unique, just as are the massive tuff beds above the water table in southern Nevada. The United States is singularly fortunate in having these two attractive possibilities for disposal sites, in addition to the many sites that could be developed in salt beds or salt domes.

Summary

For each kind of rock suggested as a suitable medium for repository construction, a lively argument can be joined between those who are impressed with its favorable characteristics and those who see its possible defects (NRC/NAS 1983, pp. 145–94). Some rock types are generally preferable to others, but no one variety can be picked out as the best in all circumstances. Rocks are too variable for broad generalizations: A given type may be adequate in one location but questionable because of impurities or fracturing or complex structure in another. Examples for any of the rocks described can be cited where repositories could be safely located, and other examples where disposal of HLW would be risky. Choosing a good place for a repository will always require the exercise of geologic judgment regarding the local characteristics of a probably suitable rock variety at a particular site.

MODELING RADIOACTIVE RELEASE

Models and Scenarios

Current plans for HLW disposal, as described above, call for setting up engineered barriers that will guarantee containment of all radionuclides for several hundred years, followed by isolation effective enough to ensure that releases of radioactivity will not exceed regulatory limits. The isolation is to be achieved principally by the geologic barrier represented by the rock through which groundwater must travel from the repository to the ground surface. The required period of isolation is hard to specify: Ideally it should be several million years to ensure the decay to harmless levels of all radionuclides in the waste, but geologic predictions for such times are so uncertain that realistic regulatory standards must be set for shorter periods. Current regulations in the United States specify reasonable assurance of isolation for 10,000 yr, with the provision that engineered and geologic barriers must be durable enough that the period of isolation will in all probability be much longer (USEPA 1985).

But predictions about groundwater behavior, even for 10,000 yr, are fraught with difficulty. Regulatory standards look fine on paper, but how can we really be sure that releases will not exceed the standards ten millennia from now? There is no way to provide absolute assurance, of course, but fairly realistic calculations of the amounts of release to be expected at various times in the future can be made by using *models*. These are based on plausible *scenarios*—imagined sequences of events that may occur in and near a repository after it is backfilled and sealed (e.g. Burkholder 1980, NRC/NAS 1983, pp. 246–96; Milnes 1985). One could suppose, for example, that no canister is breached by corrosion for a thousand years, that most canisters remain intact for another thousand years, and that thereafter the movement of groundwater through the repository is no faster than its present measured rate. This would be an *optimistic* scenario, one based on the assumption that all parts of the repository and its surroundings behave entirely as expected. Alternatively, one could imagine that something goes dreadfully wrong—for example, that most canisters are defective and are breached shortly after the repository is closed, so that waste is in contact with groundwater early in the containment period. This is an extremely pessimistic or *conservative* scenario, and models based on it would predict radioactive releases much larger and earlier than those for the optimistic scenario.

Obviously a wide spectrum of scenarios is possible, ranging from highly conservative to rosily optimistic. The most plausible ones are somewhat on the conservative side, in recognition of the fact that human enterprises seldom work entirely according to plan but generally involve mistakes, errors of judgment, or unexpected natural events that cause less than ideal performance. If calculations based on conservative scenarios show that radioactive releases nevertheless remain low, the effectiveness of the repository in isolating waste is confirmed—that is, the barriers to release have enough redundancy to compensate for errors and accidents. In the models used to calculate releases, it is customary to use scenarios on the conservative side, just to make sure that the calculated results are not overly sanguine. In other words, actual releases from a well-sited and well-constructed repository will be in all probability less than the numbers calculated from the usual models.

Scenarios and models all assume movement of groundwater through a repository, since this is the only plausible way for radionuclides to get to the surface in appreciable amounts. For a repository constructed below the water table in crystalline rock, basalt, or shale, this assumption is realistic: Surely groundwater would ultimately invade such a repository, and dissolved radionuclides would be carried into the adjacent rock. For a salt repository, however, or one constructed in tuff above the water table, the assumption seems dubious. Brine might indeed collect in a salt repository, coming out of the adjacent salt or its interbeds, but the brine would not be moving; it would simply remain as a stagnant pocket or pockets trapped in massive salt. A tuff repository in the vadose zone might have a transitory filling of water during rare severe storms but no permanent flow that could transport dissolved material from the waste. One can nevertheless imagine extreme scenarios—say, tectonic movement that opens a path for fluids from the outside into a salt repository or that raises the water table above a repository in tuff—so that even for these favored repository sites an attempt to model possible radioactive releases by groundwater movement is justified.

To model releases at different times after repository closure, numbers are assigned to successive steps in the movement of each radionuclide from the waste to the ground surface, taking account of its steadily decreasing concentration due to radiactive decay. Typically the steps would include breaching of the canisters by corrosion, dissolving of individual radionuclides from the waste, transport of the nuclides through the debris from corrosion and through the surrounding backfill, and then their movement into the adjacent rock. Precipitation of some radionuclides as the groundwater traverses the rock and moves into new chemical environments, sorption of nuclides on mineral surfaces, dispersion of groundwater flow, diffusion of nuclides into the rock along fractures, and dilution by groundwater from other sources would all be part of the model calculations. Complete models would include also the details in movement of radionuclides after they reach the ground surface—their movement into drinking water and water used for irrigation, then their consumption by plants, by animals, and ultimately by humans—all leading to estimates of specific radiation doses to individuals or of general increases in environmental radiation levels (USEPA 1979, Jenne 1981). Needless to say, the calculations for complete models can become very intricate.

Data for the calculations come from many sources. Some depend on field measurements—for example, the rate and direction of groundwater movement, the permeability of rock through which the radionuclides move, and the chemical composition of the groundwater. Such data, of course, are specific for each repository site. More general kinds of data include the rate of canister corrosion, the rate of dissolution of the waste form (glass or spent fuel), the effect of backfill in modifying groundwater composition, and the sorptive capacity of different kinds of rock. This information is best obtained from laboratory experiments, conducted with care to maintain conditions similar to those expected in nature. Some of the data on solubility come from standard tables, modified according to the conditions of Eh, pH, and composition of particular groundwaters. Because solubility and sorption are among the most important geologic variables pertaining to radionuclide movement, they will serve to illustrate some of the complications of quantitative model building.

Solubility

If a radionuclide forms a compound of sufficiently low solubility under repository conditions, it will obviously not travel far with groundwater. For some of the elements in HLW, this property alone is sufficient to ensure that their release to surface environments will be within regulatory limits (Allard 1983). For example, plutonium under the Eh-pH conditions expected in most repositories would form the very insoluble oxide PuO_2. The solubility of this compound is so low that measurement is difficult, and a range of values has been reported; numbers in the high part of the range do not exceed 10^{-5} mg liter^{-1} as the total amount of Pu in solution. The maximum permissible amount of this element in water for ordinary use, as set by regulatory agencies (e.g. USNRC 1976), is about 10 times this figure, so releases of plutonium seem adequately controlled. For strontium, on the other hand, the most insoluble compound to be expected is $SrSO_4$, which would permit a total Sr concentration of about 0.6 mg liter^{-1}; by ordinary standards this is a low solubility, but the maximum permissible concentration of ^{90}Sr is so small (2×10^{-9} mg liter^{-1}) that the amount remaining in solution could evidently pose a serious hazard to living things.

Thus some of the radionuclides in HLW would have concentrations in groundwater that are controlled to safe levels by the insolubility of their compounds, whereas others would be grossly uncontrolled. Table 2 is a summary of pertinent solubility data and includes for comparison the maximum permissible concentrations in water for ordinary use. The number given for each element is the total concentration at equilibrium with its most insoluble compound. Solubilities are shown for four sets of Eh-pH conditions in groundwater; the set in the first column (Eh = -0.2 V, pH = 9) is representative of conditions to be expected in most repository environments. The numbers indicate that low solubility is an effective control only for the heavy actinide elements and technetium, and for most of these the effectiveness is limited to a fairly narrow range of acidity and oxidation potential. All of these elements under moderately reducing conditions form insoluble oxides (NpO_2, PuO_2, Am_2O_3, Tc_3O_4) and all except Am are readily oxidized to more soluble forms if the redox potential increases (for example, NpO_2^+, $PuO_2(OH_3^-)$, TcO_4^-).

Solubility can serve to control concentration both at the immediate surface of the dissolving waste and then later as groundwater traverses

Table 2   Estimates of solubility for important radionuclides, given in concentration (mg liter^{-1}) of each radionuclide in equilibrium with its most insoluble compound under the specified Eh and pH at 25°C and 1 atm[a]

| | Reducing conditions Eh = −0.2 V | | Oxidizing conditions Eh = +0.2 V | | |
	pH = 9	pH = 6	pH = 9	pH = 6	MPC[b] (mg liter^{-1})
Sr	0.6	High[c]	0.6	High[c]	2×10^{-9}
Cs	High	High	High	High	2×10^{-7}
I	High	High	High	High	4×10^{-4}
Tc	10^{-10}	High	High	High	1×10^{-2}
U	10^{-3}	10^{-6}	High	High	5×10^{-3} (for ^{234}U)
Np	10^{-4}	10^{-4}	10^{-2}	10^{-1}	4×10^{-3}
Pu	10^{-5}	10^{-4}	10^{-5}	10^{-3}	8×10^{-5} (for ^{239}Pu)
Am	10^{-8}	10^{-5}	10^{-8}	10^{-5}	1×10^{-6} (for ^{241}Am)

[a] The table is compiled from many sources. The solubility values are in part derived from experimental data on compounds assumed to be the most insoluble under repository conditions, and in part are calculated from thermochemical data.

[b] MPC = maximum permissible concentration in water for ordinary use (USNRC 1976).

[c] The designation "high" means greater than 1 mg liter^{-1}, a solubility so large that it would not be an effective control of concentration.

the interstices of the surrounding rock. Plutonium, for example, might precipitate as PuO_2 just as it is liberated from the waste surface and never move out into the groundwater at all; on the other hand conditions near the waste might be different from those at a distance (higher temperature, lower pH, higher Eh), so that some Pu would go into solution and precipitate only later as groundwater carries it into the rock. The requirement for control is simply that appropriate conditions for precipitation exist somewhere along its path to the surface.

Unhappily, the use of numbers to demonstrate control of concentration by solubility is not as straightforward as this discussion would suggest (Allard 1983). Solubilities calculated from thermochemical data are often different from those determined by direct experiment, especially for very low solubilities like those under consideration here. The numbers in Table 2 are a mixture of values from both sources, in general chosen to represent maximum reported values so as to keep estimates of release rates on the conservative side. Differences in the values have a variety of causes: In experimental work, equilibrium may not always be assured; solubilities may be increased by traces of complex-forming ions or dissolved organic material; apparent high solubilities may result from colloid formation; and thermochemical data for actinide compounds are not as complete or reliable as could be wished. The differences between calculated and experimental values for some elements may be as much as a few orders of magnitude.

Thus the numbers in a compilation like Table 2 can be used in model calculations only with some reservation, but because they are chosen in the high part of reported ranges they should give suitably conservative release estimates. The table shows clearly that concentrations of some elements are effectively limited by solubilities of their compounds, but that the limitation may break down unless the groundwater in a repository remains weakly reducing and slightly alkaline.

Sorption

From Table 2 it is apparent that adequate release control for all radio-nuclides cannot be achieved by low solubility alone. Another potent mechanism for limiting concentrations is the tendency of materials in solution to be retained by mineral surfaces in the rock through which they travel. The retention has a variety of causes—attraction of ions for residual charges on the mineral surface, substitution of an ion for another ion already attached, and replacement of ions that are part of the crystal structure. These processes in general are not easily separable, and they are commonly lumped under the name *sorption*.

To quantify sorption requires experiments in which a solution is allowed to stand in contact with chunks of rock, or is caused to flow through a column of rock, until a steady state is reached between the amount of solute remaining in solution and the amount sorbed on the rock (Anderson et al 1983, Isherwood 1981, Vol. 2, pp. 220–48; Moody 1982). The ratio of concentrations (g g^{-1} of solute in the rock, g mliter^{-1} of solute in the liquid) thereafter remains constant and is designated the *distribution coefficient* K_d with units of mliter g^{-1}. When a solution is flowing through rock, each sorbed particle remains attached only temporarily, going back into solution as another particle comes to take its place; in effect, the movement of solute lags behind the movement of water. The amount of lag can be expressed by a retardation factor R_f, which is related to K_d by the equation

$$R_f = 1 + K_d \times \frac{\text{density of rock}}{\text{porosity of rock}} = \frac{\text{rate of groundwater movement}}{\text{rate of radionuclide movement}}.$$

The values of K_d and R_f depend, of course, on the nature of the rock and the nature of the solute. Because sorption can greatly delay the transit of radionuclides from a repository to the ground surface, measurements of these quantities have been very numerous. Some representative values for the retardation factor are given in Table 3.

Table 3   Retardation factors for important radionuclides in rock materials being considered for repository siting[a]

$$R_f = 1 + K_d \times \frac{\text{density of rock}}{\text{porosity of rock}} = \frac{\text{rate of groundwater movement}}{\text{rate of radionuclide movement}}$$

Element	Granite	Basalt	Tuff (volcanic ash)	Shale or clay	Salt
Sr	20–4,000	50–3,000	100–100,000	100–100,000	10–50
Cs	200–100,000	200–100,000	500–100,000	200–100,000	40–100
I	1	1	1	1	1
Tc	1–40	1–100	1–100	1–40	1–10
U	20–500	50–500	10–400	50–2,000	20–100
Np	10–500	10–200	10–200	40–1,000	10–200
Pu	20–2,000	20–10,000	50–5,000	50–100,000	40–4,000
Am	500–10,000	100–1,000	100–1,000	500–100,000	200–2,000

[a] The table is compiled from reported experimental values in many sources. The high figure in each range is an estimate for assumed most common repository conditions (Eh = −0.2 to −0.3 V, pH = 7 to 9); the low figure is an estimate of the minimum value under less favorable conditions of Eh, pH, or complexing. The numbers for salt do not refer to sorption on salt itself, but rather on ordinary rock material in the vicinity of a salt repository, where the groundwater may be fairly concentrated brine.

Numbers in the table show ranges of values commonly reported in the literature for sorption from dilute solutions (except the last column, which refers to saline waters in rocks near salt deposits) but do not include extremes of reported values. The wide ranges in the numbers are not surprising and are not a reflection on the validity of the experimental work: Sorption depends on so many variables that much scatter in K_d values is to be expected, even in results from the same laboratory. Despite the scatter, the numbers in Table 3 conform in a general way to behavior expected in sorption processes. Cations are sorbed more strongly than anions because residual charges on most mineral surfaces in ordinary rocks are negative; thus the numbers for iodine and technetium, which exist in groundwater chiefly as I^- and TcO_4^-, are much smaller than those for the other elements. Clay minerals, zeolites, and glass shards are especially efficient sorbents, as suggested by the high numbers in the columns for tuff and shale. Sorption is less effective for ions in salt solutions than for those in dilute groundwater, as shown by the low numbers in the last column. It is evident from the table that sorption of strontium and cesium in most rocks is probably sufficient to delay them until decay has made them harmless, even if some waste should escape by accident during the containment period, and likewise that sorption is an important factor in helping to control escape of the heavy actinide elements.

Models using numbers like those in Tables 2 and 3 show that most of the major nuclides in HLW will be sufficiently immobilized by low solubility or sorption or both to keep them from escaping in hazardous amounts (Bengtsson et al 1983). A conspicuous exception is iodine, which does not form insoluble compounds with the common constituents of rocks or groundwater and is not appreciably sorbed by the suggested repository media. Very likely much of the iodine will be removed from HLW before it is buried, or material to trap it will be added to the backfill (NRC/NAS 1983, pp. 39–40). For technetium also, the demonstration of immobility is less than satisfactory because its low solubility depends critically on maintenance of a narrow range of Eh-pH conditions, and sorption contributes only slightly to its retardation.

Other Variables

Precipitation of insoluble compounds and sorption on mineral surfaces are the most important mechanisms that retard the motion of radionuclides, but a complete model includes other factors that influence concentrations and amounts released at any particular time and place. One such factor is *dispersion*—the nonuniformity of groundwater flow, or the tendency of water and solutes to spread out both along and to the sides of a straight-line path. *Diffusion* of ions into rock adjacent to a principal path of groundwater movement aids dispersion in spreading the plume of dissolved radionuclides. Numerical values for the effects of dispersion and diffusion can be estimated for flow in a homogeneous medium, but reliable values are less easy to find for movement along a network of fractures—which is the type of flow that probably predominates in the rock around most planned repository sites (Rasmusson & Neretnieks 1981). Another expectable influence on radionuclide concentrations is *dilution* by mixing of the groundwater that has traversed a repository with water from other sources, an influence that is usually hard to express quantitatively because of uncertainty about details of the groundwater regime.

Summary

The complications of model construction are only hinted at in this cursory discussion. A more complete treatment is given in many recent documents; good examples are the technical background studies for the Environmental Protection Agency's release standards (USEPA 1979), the Swedish calculations of release rates in crystalline rock (Bengtsson et al 1983), the comprehensive National Research Council study of the disposal system (NRC/NAS 1983), and the sections on performance assessment in the Department of Energy's environmental assessments of the three preferred repository sites (USDOE 1986a,b,c).

At best, of course, the calculation of radionuclide releases is beset with uncertainties at every step, from the initial choice of a plausible scenario through the estimates of rates of canister corrosion and waste-form dissolution, then through the calculations of solubility and retardation by sorption, and on to the assignment of plausible values to effects of dispersion and dilution. The amount of uncertainty can be estimated and allowed for, using conservative assumptions throughout, but even so the numerical values obtained by model calculations are always somewhat questionable. They would be a poor basis for prediction except for the fact that even with extreme scenarios and conservative values for all important variables, the model results indicate that releases can be kept very low for the necessary long times. The semiquantitative support that models give to this conclusion is impressive enough to convince most in the technical community that a well-constructed repository in any one of several possible geologic environments can indeed accomplish the safe disposal of high-level waste.

NATURAL ANALOGS

Convincing though the model calculations may be, they still involve gross extrapolation from laboratory experiments and geologic observations during a period of a few years to times on the order of tens and hundreds of millennia. Any way of checking the extrapolations would make the case stronger, and a possible check is provided by geologic and archaeologic observations of materials and situations similar to those that might be expected in a repository many thousands of years after its closure. No exact replica of an ancient repository can be expected in nature, of course, but some parts of the expected repository environment may well be duplicated in places accessible to observation (Chapman et al 1984, NRC/NAS 1983, pp. 300–5).

Glass and metal objects from the ancient world, for example, have survived with only moderate alteration despite burial in caves or exposure to seawater for 2000–3000 yr. Their compositions are different from modern glasses and steel alloys, but they would almost certainly not be more resistant; their survival bodes well for the durability of borosilicate glass and stainless-steel canisters in a repository. Bentonite clay, a smectite-rich material often suggested for use as backfill, is found in beds that in some places can be shown to have existed for millions of years at slightly elevated temperatures without undergoing appreciable alteration of the clay minerals. The limited migration of metals away from low-temperature hydrothermal ore deposits supports the conclusion that radionuclides will be similarly retarded in their movement through rocks near a repository.

Natural analogs of radionuclide movement are most convincing when the analog elements are the same as, or chemically similar to, the radioactive elements. Ordinary strontium and cesium, for example, should be identical to ^{90}Sr and ^{137}Cs in their behavior in groundwater, and the observed minor amounts of these elements in rocks near accumulations of their minerals helps to confirm their predicted sluggish movement away from a repository. The actinide elements have no nonradioactive equivalents, and satisfactory analogs are hard to find. In their trivalent state (the only expectable oxidation state for americium) they are similar to the light rare earth elements, and in their quadrivalent state (an expectable state for neptunium and plutonium) to thorium. The general immobility of thorium and the rare earths in geologic environments is an indication that the actinides will not move far from a repository, but the evidence is incomplete because neptunium and plutonium are known to be mobile in their higher oxidation states. For these higher states uranium is a possible analog, but not a very satisfactory one because its chemistry is somewhat different. The often-reported very minor migration of uranium near ore deposits where conditions have remained reducing, coupled with the improbability of oxidizing conditions in a repository, is good evidence (but not wholly convincing) for the immobility of neptunium and plutonium.

The natural occurrence that comes closest to duplicating an ancient repository is a deposit of uranium ore at Oklo in Gabon (Cowan 1976, Jakubick & Church 1986). A low ^{235}U/^{238}U ratio in the ore suggested to the French operators of the Oklo mine that a natural fission reactor had existed here some 1800 m.y. ago, and the surmise was confirmed by analyses of the ore and nearby rocks that showed the presence of elements formed by the radioactive decay of fission products and heavy actinides. Somehow at that ancient time, the right distribution of UO_2 ore and the moist sediment enclosing it had come about to permit a fission reaction that persisted off and on for 0.5 m.y. Seemingly this would be an ideal place to get definitive evidence about the migration capabilities of the fission products and actinides. In general, the analytical results show the expected relative amounts of migration—greatest for Cs, Sr, Pb, and Ra, least for the actinides—but the analogy is not quantitative because conditions for the fission reaction were unlike those in a present-day reactor and the variations in temperature and exposure to air and water during 1800 m.y. are not known.

Natural analogs cannot prove the quantitative validity of extrapolations from short-term experiments to the times needed for waste disposal, but they do give a strong indication of their general reasonableness. The models of repository behavior have a firm basis, both in the experiments and in a variety of geologic observations.

THE CURRENT STATUS OF HLW DISPOSAL

Since there is widespread agreement among scientists and engineers that a mined geologic repository can be constructed, and since carefully worked-out models show that hazards from such a repository will remain negligible for many millennia, the question may be fairly asked, Why the delay? Why isn't a repository for high-level waste already in operation, or at least under construction? The answer requires a brief look at the sociopolitical side of the disposal problem, which in contrast to the technical side appears to be in a state of chronic disarray.

According to United States law, the disposal of HLW is the responsibility of the federal government and is to be paid for by a levy on the electric utilities that generate nuclear power. The procedure for managing disposal is spelled out in the Nuclear Waste Policy Act, passed by Congress in 1982 (NWPA 1983). The finding of sites and actual construction of repositories is delegated to the Department of Energy (DOE); the construction is to be licensed by the Nuclear Regulatory Commission, which has drawn up a set of rules to which any site and any repository must conform; and the rules are devised to ensure that radioactive releases will not exceed limits established by the Environmental Protection Agency for at least 10,000 yr and probably much longer. The Act includes a schedule of deadlines that the DOE is expected to meet for the various steps leading to repository construction, a schedule that mandates opening of the first repository for receiving waste by 1998. Because of difficulties in meeting some of the early deadlines, the Department has recently requested postponement of repository opening for five years beyond this date (USDOE 1987).

The difficulties arise chiefly from objections by local citizenry whenever DOE announces that a given site appears favorable for repository construction and will be studied intensively toward this end. The objections take the form of appeals to state and national legislators and the filing of lawsuits that can delay further work almost indefinitely (Carter 1987). Fear of radioactivity is so ingrained and so widespread that this response is practically automatic anywhere in the country. No amount of assurance on geologic or engineering grounds that waste can be buried safely suffices to overcome the dread of having radioactive material anywhere in the vicinity of one's habitation.

It does not help, of course, that technical experts are not entirely unanimous in supporting mined geologic disposal as the best disposal method. The great majority of those competent to judge feel that a good case has been made for the practicality and safety of this procedure, but a few with impeccable scientific backgrounds remain skeptical about some details and especially about the validity of geological and technical projections so far into the future. A few others doubt the wisdom of making the waste so inaccessible, even if safety can be assured, on the grounds that future generations may find uses for it. Divided opinion among those with technical knowledge is confusing to citizen groups and helps to support the objections to repository siting, especially since the skeptics and the doubters generally have louder voices than their more conventional colleagues.

The political dilemma is a serious one, and no good solution is in sight. Basically it is a question of weighing different kinds and degrees of risk. Putting high-level waste underground can never be made completely safe, in an absolute sense. There will always be some risk, both the risk of mistakes in engineering details and the risk of faulty judgments about the rates and intensity of geologic processes. These risks can be made smaller by further research, of the sort that is going on today in many laboratories. But research is expensive; how much more is really needed? Risks involved in mined geologic disposal are already lower, in the minds of most in the technical community, than other risks we accept without question in everyday life. Is it worthwhile trying to make the risks still smaller? And even if the risk is made demonstrably smaller, will this be convincing to those who fear radioactivity in all its forms? Such questions can only be answered in the political arena.

From a purely geological point of view, the problem of managing high-level radioactive waste seems well in hand. There are many places where waste could be buried, and for at least brief geologic times there is reasonable assurance that no geologic processes would make it a hazard to the environment. But how much assurance is needed, and how the assurance can be made meaningful to voters who must ultimately decide the issues, are questions a geologist is not equipped to answer.

Literature Cited

Ahlbom, K., Carlsson, L., Olsson, O. 1983. Final disposal of spent nuclear fuel—geological, hydrological, and geophysical methods for site characterization. *KBS TR 83-43*, Swed. Nucl. Fuel Supply Co./Div. KBS, Stockholm

Allard, B. 1983. Actinide solution equilibria and solubilities in geologic systems. *KBS TR 83-35*, Swed. Nucl. Fuel Supply Co./Div. KBS, Stockholm

Altomare, P., Bernardi, R., Gabriel, D., Nainan, D., Parker, W., Pfundstein, R. 1979. Alternative disposal concepts for high-level and transuranic radioactive waste disposal. *USEPA ORP/CSD 79-1*, US

Environ. Prot. Agency, Washington, DC. 270 pp.

American Physical Society (APS). 1978. Report to the American Physical Society by the Study Group on Nuclear Fuel Cycles and Waste Management. *Rev. Mod. Phys.* 50(1): 43 pp.

Anderson, K., Torstenfelt, B., Allard, B. 1983. Sorption of radionuclides in geologic systems. *KBS TR 83-63*, Swed. Nucl. Fuel Supply Co./Div. KBS, Stockholm

Bedinger, M. S., Sargent, K. A., Brady, B. T. 1985. Geologic and hydrologic characterization and evaluation of the Basin and Range Province relative to the disposal of high-level radioactive waste. *US Geol. Surv. Circ. 904-C.* 27 pp.

Benedict, M., Pigford, T. H., Levi, H. W. 1981. *Nuclear Chemical Engineering.* New York: McGraw-Hill. 1008 pp. 2nd ed.

Bengtsson, A., Magnusson, M., Neretnieks, I., Rasmusson, A. 1983. Model calculations of the migration of radionuclides from a repository for spent nuclear fuel. *KBS TR 83-48,* Swed. Nucl. Fuel Supply Co./Div. KBS, Stockholm

Burkholder, H. C. 1980. Waste isolation performance assessment—a status report. In *Scientific Basis for Nuclear Waste Management,* ed. C. F. M. Northrup, 2: 689–702. New York: Plenum

Carter, L. 1987. Nuclear imperatives and public trust: radioactive waste. *Issues Sci. Technol.* 3(2): 46–61

Catlin, R. J. 1980. *Assessment of the Surveillance Program of the High-Level Waste Storage Tanks at Hanford: Report to Assistant Secretary for Environment.* Washington, DC: US Environ. Prot. Agency. 231 pp.

Chapman, N. A., Sargent, F. P., eds. 1984. The geochemistry of high-level waste disposal in granitic rocks. *AECL-8361,* Whiteshell Nucl. Res. Estab., Pinawa, Manitoba

Chapman, N. A., McKinley, I. G., Smellie, J. 1984. The potential of natural analogues in assessing systems for deep disposal of high-level radioactive waste. *NAGRA 84-41,* Natl. Genoss. Lagerung radioakt. Abfälle, Baden, Switz. 103 pp.

Cowan, G. A. 1976. A natural fission reactor. *Sci. Am.* 235: 36–47

Eisenbud, M. 1982. Radioactive waste management. In *Outlook for Science and Technology: The Next Five Years,* pp. 287–320. San Francisco: Freeman. 788 pp.

Hollister, C. D., Anderson, D. R., Heath, G. R. 1981. Subseabed disposal of nuclear wastes. *Science* 213: 1321–26

Isherwood, D. 1981. *Geoscience Data Base Handbook for Modeling a Nuclear Waste Repository,* Vols. 1, 2. *NUREG/CR-0912.* Livermore, Calif: Lawrence Livermore Lab. 315 pp., 331 pp.

Jakubick, A. T., Church, W. 1986. Oklo natural reactors: geological and geochemical conditions—a review: *Res. Rep. INFO-0179,* At. Energy Control Board, Ottawa, Can. 53 pp.

Jenks, G. H. 1979. Effects of temperature, temperature gradients, stress, and irradiation on migration of brine inclusions in a salt repository. *ORNL-5526,* Oak Ridge Natl. Lab., Oak Ridge, Tenn.

Jenne, E. A. 1981. Geochemical modeling: a review. *PNL-3574,* Battelle Pac. Northwest Lab., Richland, Wash.

Milnes, A. G. 1985. *Geology and Radwaste.* London: Academic. 328 pp.

Moody, J. B. 1982. *Radionuclide Migration/Retardation: Research and Development Technology Status Report.* Columbus, Ohio: Battelle Mem. Inst. 61 pp.

National Aeronautics and Space Administration (NASA). 1986. *Technical Report on Preliminary Risk Assessment for Nuclear Waste Disposal in Space,* Vol. 1. *NASA CR 162028.* Columbus, Ohio: Battelle Mem. Inst. 30 pp.

National Research Council (NRC/NAS). 1957. The disposal of radioactive waste on land. *Report of the Committee on Waste Disposal of the Division of Earth Sciences.* Washington, DC: Natl. Acad. Press. 126 pp.

National Research Council (NRC/NAS). 1970. *Disposal of Solid Radioactive Wastes in Bedded Salt Deposits.* Washington, DC: Natl. Acad. Press. 28 pp.

National Research Council (NRC/NAS). 1978a. *Geological Criteria for Repositories for High-Level Radioactive Wastes.* Washington, DC: Natl. Acad. Press. 19 pp.

National Research Council (NRS/NRC). 1978b. *Radioactive Wastes at the Hanford Reservation: A Technical Review.* Washington, DC: Natl. Acad. Press. 269 pp.

National Research Council (NRC/NAS). 1981. *Radioactive Waste Management at the Savannah River Plant: A Technical Review.* Washington, DC: Natl. Acad. Press. 68 pp.

National Research Council (NRC/NAS). 1983. *A Study of the Isolation System for Geologic Disposal of Radioactive Wastes.* Washington, DC: Natl. Acad. Press. 345 pp.

National Research Council (NRC/NAS). 1984. *A Review of the Swedish KBS-3 Plan for Final Storage of Spent Nuclear Fuel.* Washington, DC: Natl. Acad. Press. 77 pp.

Nuclear Waste Policy Act of 1982 (NWPA). 1983. *Public Law 97-425,* Jan. 7, 1983. Washington, DC: US Govt. Print. Off.

Office of Technology Assessment (OTA), US Congress. 1985. *Managing the Nation's Commercial High-Level Radioactive Waste. OTA-0-171.* Washington: Govt. Print. Off. 348 pp.

Rasmusson, A., Neretnieks, I. 1981. Migration of radionuclides in fissured rock: the influence of micropore diffusion and longitudinal dispersion. *J. Geophys. Res.* 86: 3749–58

Roseboom, E. H. Jr. 1983. Disposal of high-level nuclear waste above the water table in arid regions. *US Geol. Surv. Circ. 903.* 21 pp.

Sandia National Laboratories. 1983. *The*

200   KRAUSKOPF

Subseabed Disposal Program: 1983 Status Report. SAND83-1387. Albuquerque, N.Mex: Sandia Natl. Labs. 179 pp.
Shefelbine, H. C. 1982. Brine migration: a summary report. Sandia Rep. SAND82-0152. Albuquerque, N.Mex: Sandia Natl. Labs. 51 pp.
Swedish Nuclear Fuel Supply Co (SKBF). 1983. Final Storage of Spent Nuclear Fuel, Vol. 2, Geology KBS-3. Stockholm: Swed. Nucl. Fuel Supply Co./Div. KBS. 107 pp.
Ubbes, W. F., Duguid, J. O. 1985. Geotechnical assessment and instrumentation needs for isolation of nuclear waste in crystalline rocks. BMI/OCRD-24, Battelle Mem. Inst. Columbus, Ohio. 185 pp.
US Department of Energy (USDOE). 1985. Spent Fuel and Radioactive Waste Inventories, Projections, and Characteristics. DOE/RW-0006. Washington, DC: Govt. Print. Off.
US Department of Energy (USDOE). 1986a. Environmental Assessment, Deaf Smith County Site, Texas. DOE/RW-0069. Washington, DC: US Dep. Energy
US Department of Energy (USDOE). 1986b. Environmental Assessment, Yucca Mountain Site, Nevada Research and Development Area. DOE/RW-0073. Washington, DC: US Dep. Energy
US Department of Energy (USDOE). 1986c. Environmental Assessment, Reference

Repository Location, Hanford Site, Washington. DOE/RW-0070. Washington, DC: US Dep. Energy
US Department of Energy (USDOE). 1987. Draft Mission Plan Amendment, January 1987. Washington, DC: US Dep. Energy
US Environmental Protection Agency (USEPA). 1979. Technical Support of Standards for High-Level Radioactive Waste Management, Vol. C, Migration Pathways: EPA 520/4-79-007C: Washington, DC: Govt. Print. Off. 177 pp.
US Environmental Protection Agency (USEPA). 1985. Environmental Standards for the Management and Disposal of Spent Nuclear Fuel, High-Level and Transuranic Radioactive Wastes. 40 CFR 191. Washington, DC: Govt. Print. Off.
US Nuclear Regulatory Commission (USNRC). 1976. Maximum Concentrations in Effluents in Unrestricted Areas. 10 CFR 20, Append. B, Table II. Washington, DC: Govt. Print. Off.
US Nuclear Regulatory Commission (USNRC). 1983. Disposal of High-Level Radioactive Wastes in Geologic Repositories: Technical Criteria. 10 CFR 60. Washington, DC: Govt. Print. Off.
Winograd, I. J. 1981. Radioactive waste disposal in thick unsaturated zones. Science 212: 1457–64

Ann. Rev. Earth Planet. Sci. 1988. 16: 201–30

TECTONIC EVOLUTION OF THE CARIBBEAN

Kevin Burke

Lunar and Planetary Institute, 3303 NASA Road One, Houston, Texas 77058 and Department of Geosciences, University of Houston, Houston, Texas 77004

INTRODUCTION

The last decade has seen many syntheses of tectonic history based on existing geological information interpreted in the light of the understanding that has followed since the recognition of the plate structure of the lithosphere. Interpretations of Caribbean tectonic history, nearly all of them broadly similar to that of a pioneer synthesis by Pindell & Dewey (1982), have lately proliferated, perhaps in part because of the availability of a superb tectonic map at a scale of 1 : 2,500,000 (Case & Holcombe 1980). I draw attention here to salient results of recent syntheses (such as those in Bonini et al 1984, Mascle 1985, Pindell & Barrett 1988, and Bouysse 1987) and emphasize outstanding problems of Caribbean tectonics. These problems illustrate a contrast between the Caribbean and regions like the Tethysides (Şengör 1987), California (Saleeby 1983), and Western North America (Schermer et al 1984), all of which have been the subject of recent articles in this series. The Caribbean represents neither a mountain belt formed during the gradual closure of a major ocean nor a belt in which numerous blocks and fragments have been attached to an existing continent. It has more in common with Indonesia and the Philippines (Hamilton 1979), where presently evolving island-arc systems dominate. It is in these embryonic orogenic collages that the complexity of the mountain-building process is best revealed. Students of mature mountain systems may have something to learn about the limitations of their syntheses from the study of these evolving arcs. On a more positive note, a consensus is emerging that ancient continents represented in, for example, the Superior Province of Canada and the Kaapvaal of South Africa resulted from numerous

201

0084–6597/88/0515–0201$02.00

collisions of island arcs (Burke & Dewey 1972, deWit & Ashwal 1986). An understanding of the evolution of the Caribbean arcs, which are well exposed and formed on ocean floor over the last 150 m.y., may aid in unraveling the complex processes involved in the evolution of such ancient arc systems.

My review is organized chronologically, starting with the geology of the site on which the Caribbean has formed.

THE SITE ON WHICH THE CARIBBEAN FORMED

The Caribbean and the Gulf of Mexico began to develop less than 200 Ma in the area where western Gondwanaland and Laurasia had collided about 100 Myr earlier to assemble the short-lived great continent of Pangea. There has been some variety of opinion about the exact way in which Africa, North America, and South America fitted together at the time of the assembly of Pangea, but various factors are combining to yield uniformity about this important fit. Revised interpretations of ocean-floor magnetic and fracture-zone data (especially SEASAT-derived data in the latter case) have confirmed how both North and South America fitted against Africa (Pindell et al 1987), and increasing recognition of the unreliability of nearly all published late Paleozoic paleomagnetic data has removed a source of much confusion. Outstanding problems of western Pangean reconstruction relate to four issues: (*a*) the tightness of the fit of Africa to the Americas, (*b*) the position of Yucatán, (*c*) the shape and structure of Mexico during the Permo-Triassic, and (*d*) the shape and structure of northwestern South America during the same interval.

Tightness of Fit of Africa to the Americas

The resolution of this problem depends upon the availability of good geophysical data from the continental margins. Where such data are available, the amount of extension, both before and early during the opening of the ocean, at the continental margins and on intracontinental rifts (such as the Newark rifts; Olsen et al 1982) can be roughly estimated from seismic and gravity data, including data on sediment thicknesses and subsidence rates (e.g. Pindell 1985). A second factor in assessing the tightness to fit is an estimate of the amount of sediment deposited at the continental margin after rupture. Pindell & Dewey (1982, Figure 8) ascribed a particularly large value to this quantity for the Amazon shelf, where erosion from the Andes and deposition by the great river have contributed massive amounts of sediment to the margin. Because of this, their Gulf of Guinea fit (Pindell & Dewey 1982, Figure 8) is much tighter than that of earlier writers.

As more continental margin geophysical data are published, better estimates of the continental margin fit will become possible, but a special
problem remains in the Bahamas and on the western side of Florida, where
the extent and amount of attenuation of Floridian continental basement
are unknown [see Pindell (1985), and Pindell & Barrett (1988) for two
different guesses]. The very existence of a discrete Florida Straits block,
required in some recent syntheses, remains to be clearly established. It
appears possible to do without it in a reconstruction that starts with a
longer Yucatán shortened to its present length at the end of the Cretaceous
by collision and strike-slip motion (Figure 1).

Yucatán

Some reconstructions of Pangea have left a hole at the site of the Gulf of
Mexico or have put Yucatán in the Pacific, but there is increasing support
for Choubert's (1935) suggestion that Yucatán lay at the site of the Gulf
of Mexico before rotating to its present position. Pindell (1985; also Pindell
& Dewey 1982), on the basis of regional geology, has interpreted the
western shore of the Gulf of Mexico as lying close to the site of a major
Jurassic right-lateral strike-slip fault and has suggested that Yucatán
rotated along this boundary to its present location by counterclockwise
movement from a position against the northern shore of the Gulf of

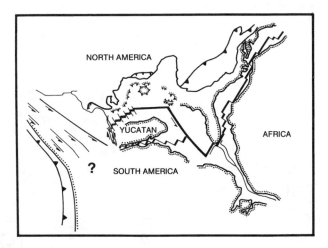

Figure 1   The site on which the Caribbean formed (modified from Pindell 1985). The Gulf
of Mexico formed from this assembly when Yucatán rotated away from the coast of the
Gulf of Mexico, together with South America and Africa, over about 8 Myr at the end of
the Jurassic. Immediately afterward, proto-Caribbean ocean floor began to form between
Yucatán and South America. The creation of a separate Florida Straits block (Pindell 1985)
has been avoided by making Yucatán longer.

Mexico about a pole near the east coast of Florida. Shepherd et al (1987) reached a similar conclusion by interpreting magnetic anomalies in the gulf. Another possibility is that the motion of Yucatán with respect to North America might have been concentric with the contemporaneous motion of Africa (Klitgord & Schouten 1980). This motion would restore Yucatán to a more northwesterly site within the Gulf of Mexico, but it does not seem so satisfactory in explaining either the deep gulf magnetic anomalies or the prerotation fit of the gulf evaporites (White 1980). The fit of Yucatán in the assembly of Pangea depicted in Figures 1 and 2 appears most consistent with presently available geological and geophysical data.

Structure of Mexico During the Permo-Triassic

Mexico has always provided a challenge in Pangean reconstructions because of its overlap with South America. Recognizing this, many authors (e.g. Choubert 1935, Bullard et al 1965) have simply not shown Mexico on their representations. A problem arises because so much of Mexico is covered by Cenozoic volcanic rocks, which makes interpretation of the underlying pre-Cenozoic structure very difficult. Silver & Anderson (1974) suggested a solution by invoking hundreds of kilometers of sinistral offset on the Mohave-Sonora megashear. In a similar way, Pindell & Dewey (1982) and Pindell (1985) suggested that the discontinuity of the Huastecan belt of eastern Mexico (Figure 2) might record 700 km of left-lateral motion on at least three major faults. The Coahuila and Maya blocks of Campa & Coney (1983) would appear to embody parts of the Huastecan

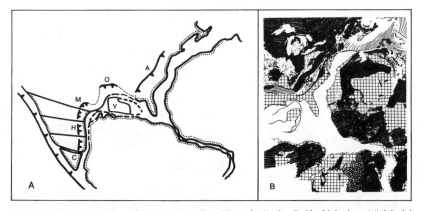

Figure 2   Two assemblies of Pangea that place Yucatán in the Gulf of Mexico. (*A*) Model simplified after Pindell & Dewey (1982). A, Appalachians; O, Ouachitas; M, Marathon; H, Huastecan; C, Chortis; Y, Yucatán. (*B*) A much earlier version simplified from Choubert (1935).

Paleozoic collision zone, but as the work of these writers clearly shows, we appear to be a long way from being able to separate the effects of late Paleozoic collision from those of Mesozoic strike-slip motion in eastern Mexico. All that can be said at present is that most of the blocks making up western Mexico had not joined North America by Permian times, and that much of eastern Mexico appears to have lain farther to the northwest than it does now. Fragments of eastern Mexico moved in fault slices to their present, more southeasterly positions during the Triassic and Early Jurassic before the Gulf of Mexico finished forming.

Shape of Northwestern South America

It has proved difficult to reconstruct the outline of the northwestern corner of South America to the shape it had when Pangea was assembled because so much has subsequently happened to the margin. It appears from reconstructions (Figures 1, 2) that the west-facing coast of northwestern South America rifted from the then eastern margin of Mexico, and that the north-facing coast rifted from Yucatán. Since that time, the west coast has developed as an Andean margin that has experienced two arc collisions, while the north coast has experienced intense deformation. The collisions have been accompanied and followed by substantial deformation of South America inboard of the collisional suture zones, and it is the effects of this deformation, especially thrust and strike-slip motion, that are proving particularly difficult to quantify.

The extent of deformation associated with the late Neogene collision of Panama with South America is relatively well known (e.g. Mann & Burke 1984). About 100 km of strike-slip movement on each of the Santa Marta and Bocono faults has carried the Maracaibo block northward for 70 km in the last 7 Myr as it attempts tectonic escape from the collision [see discussion between Salvador (1986) and Dewey & Pindell (1986)]. The possibility of similar motion on a larger scale in association with the Late Cretaceous arc collision recorded in the Romeral suture in Colombia is illustrated in Figure 3 (modified after Maze 1984). Maze plotted the extent of Jurassic rift facies, here interpreted as marking the crest of a Jurassic Andean arc, from the Rocas Verdes suture zone in southen Chile to the north coast of South America. As is typical of Andean arc crests, the volcanic and rift deposits occupy an extremely narrow zone [see Şengör et al (1987) for Asian examples]. The splitting of the crestal zone into two in the Perija and Mérida Andes is so unusual that I suggest that it should be attributed to the tectonic escape (Burke & Şengör 1986) of northwestern South America by strike-slip faulting in response to the Romeral arc collision. The geometry is similar to that of the current phase of escape, but the required offset of at least 500 km is about seven times bigger. In

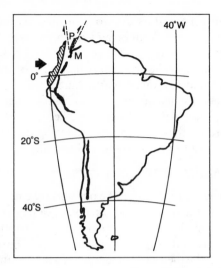

Figure 3 Jurassic red beds and volcanic rocks (indicative of the location of the crest of the Andean arc at the time) extend in a single narrow zone (shown in black) from 40°S to the far north of South America, where they occur in both the Mérida (M) Andes and Perija (P) Andes. This duplication is attributed to tectonic escape along strike-slip faults (dashed) resulting from arc collision (hatched region and black arrow) along the Romeral suture. Modified from Maze (1984).

this interpretation, the late Paleozoic collisional intrusives and volcanic rocks of northernmost South America (e.g. Pindell & Dewey 1982, Figure 2) moved northward hundreds of kilometers by strike-slip motion in the Late Cretaceous and Cenozoic before being caught up in eastward strike-slip motion. This two-stage rotation accounts particularly well for paleomagnetic observations of Jurassic rocks from the area (Maze & Hargraves 1984, MacDonald & Opdyke 1984). Because of the large amounts of deformation, the representation of northwestern South America and Mexico remains the major problem in reconstructing the western Pangean site of formation of the Caribbean.

In spite of these difficulties, the broad structure represented in Figures 1 and 2, first discerned by Choubert (1935) and clearly set down by Pindell & Dewey (1982), represents a useful model of the structure of the site on which the Caribbean was to be established during the Jurassic and Cretaceous.

RIFTED MARGINS OF THE CARIBBEAN

Following protracted intracontinental rifting in later Triassic and Early to Middle Jurassic times, the breakup of western Pangea began when seafloor spreading in the Central Atlantic started to carry Africa and South America away from North America. The oldest seafloor in the Central Atlantic is of Oxfordian age (about 160 Ma; see papers in Sheridan et al 1983), and for 20 Myr the opening of the Central Atlantic was accompanied by extension and seafloor spreading in the Gulf of Mexico while Yucatán remained attached to South America (Figure 1). By the Early Cretaceous, seafloor spreading had ceased within the gulf, which had reached its present size (Shepherd et al 1987), and Yucatán began to separate by seafloor spreading from South America (Pindell 1985). Rifted margins of significance for Caribbean history are thus (*a*) around the Gulf of Mexico [discussed at length in Pindell (1985) and not considered in detail here], (*b*) along the present east coast of Yucatán, (*c*) along the eastern coast of Florida and in northwesternmost Cuba, and (*d*) along the northern coast of South America as far east as the Demerara Rise.

Rocks that might reveal the Jurassic to Early Cretaceous rifting history of the east coast of Yucatán are buried beneath Neogene sediments, and because the Yucatán coast was tectonized as the Cuban arc swept by en route to its collision with Florida and the Bahamas, it is unlikely that a good record of rift history is preserved. Seismic-reflection lines with deep well control will be needed before any clear idea can be obtained about what happened as the eastern Yucatán margin rifted away from South America [see Mascle et al (1985, profile 8) for present knowledge].

The development of the eastern Florida rifted margin of North America is represented by the 5.5-km-deep Great Isaac No. 1 well in the westernmost Bahamas (Burke et al 1984, Figure 4). The well passes out of Cretaceous carbonates into Late Jurassic carbonates overlying volcanogenic red clastic rocks (Jacobs 1977) and appears to indicate that there has been no unusual extension of continental crust beneath the Bahamas. Farther south, evidence of rifting comes from the rocks of the Cayetano formation, which consist of continentally derived marine and nonmarine sandstones of Jurassic age outcropping in the Pinar del Río province of westernmost Cuba. Pindell (1985, Table 1) suggested that this material was formerly part of the Yucatán margin, but an alternative possibility is suggested to me by the Middle Jurassic marine ages of some Cayetano rocks, which appear too old to be associated with Yucatán rifting. I propose that this material represents sediments deposited south of Florida during Jurassic rifting that have been moved westward toward a free-face in the Gulf of Mexico by strike-slip motion associated with the collision between Cuba and southern Florida during the Eocene.

Along the north coast of South America, conjugate with the eastern coast of Yucatán, there are massive evaporite occurrences that I interpret as indicators of marine incursion during rifting. Anhydrite occurs in sporadic outcrops both in the Araya-Paria peninsula within the Early Cretaceous Coriaquito formation and in Trinidad. In these localities, thicknesses of about 1 km have been encountered in three wells (Couva offshore 1, Avocado 1, and Couva marine 2) drilled over an area of about 100 km² within the Gulf of Paria. Bray & Eva (1983) have described material from these wells as being of Early Cretaceous age.

Along the northern coast of South America, marine sediments associated with the formation of the Atlantic-type margin are no older than Kimmeridgian in age, which is consistent with the idea that Yucatán rifted from South America at about 140 Ma. Atlantic-type margin sediments are distributed sporadically through northern Venezuela in tectonized fragments. Eastward from Trinidad as far as the Demerara Rise, corresponding sediments are not tectonized but are mainly confined to offshore areas (for an exception, see Wong & Van Lissa 1978). In Trinidad, the Northern Range consists of highly tectonized and metamorphosed Kimmeridgian and Cretaceous sediments, largely of over-the-shelf-break facies, which have been juxtaposed by late Neogene strike-slip faulting against units of an unmetamorphosed Cretaceous shale-limestone shelf succession (Robertson et al 1985). A major problem of the Cretaceous Atlantic-type margin of northern South America is the difficulty of identifying in tectonically isolated fragments the various shelf and continental slope sedimentary environments and restoring them to their original relative positions. It seems possible from my initial attempts at restoration that the reconstructed continental edge will prove to be rather irregular, perhaps indicating that the conjugate straight coast of Yucatán is not an original shape but rather a product of shearing as Cuba swept by. If so, the Isle of Pines off the south coast of Cuba (Somin & Millan 1972) is a strong candidate to represent a continental fragment torn from Yucatán (Figure 4).

CONSTRAINTS ON CARIBBEAN EVOLUTION FROM THE RELATIVE MOTIONS OF NORTH AND SOUTH AMERICA

Ladd (1976) was first to construct a plate-evolutionary history for the Caribbean. He used magnetic and fracture-zone data from the Central and the South Atlantic to estimate, by the finite difference method, how North and South America have moved with respect to each other since the rifting episodes that led to formation of the Gulf of Mexico and the Caribbean. This method has also been used by Sclater et al (1977), Pindell & Dewey (1982), Pindell (1985), and Pindell et al (1987). The versions by Pindell and his various collaborators differ from those of others because in all of them

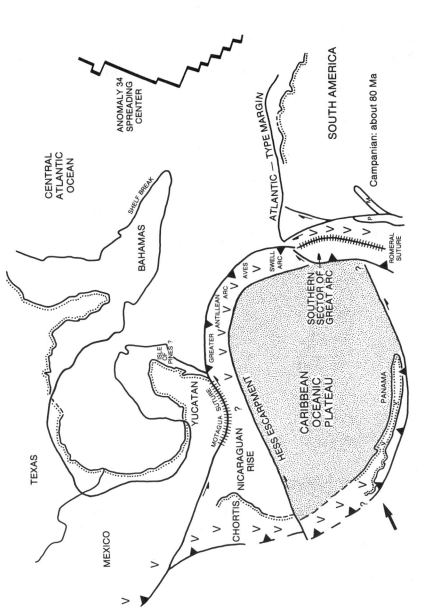

Figure 4 Caribbean oceanic plateau entering the Atlantic at the end of the Cretaceous, carrying the three segments of the Great Arc of the Caribbean on its prow and trailing the Panama–Costa Rican arc on its stern. Collision with North America is taking place at the Motagua suture and that with South America at the Romeral suture. The latter event induced an episode of tectonic escape that sheared off the Perija Andes (P) from the Mérida Andes (M). Modified after Pindell & Barrett (1982).

Africa is considered, following a tentative suggestion by Burke & Dewey (1974), to have behaved as if it lay during the Early Cretaceous on two separate plates, divided by a divergent plate boundary along the Benue Valley. In this interpretation, South America and northwestern Africa are envisaged as having rotated away together from North America in a south-southeasterly direction through about 20° of arc between 180 and 80 Ma. From that time on, as Africa began to behave as a single continent and South America separated from northwest Africa, South American motion with respect to North America has been small. For example, the site of the city of Caracas (on rigid South America) is within about 100 km of where it was 80 Ma. During the first half of the interval since 80 Ma the site moved about 300 km to the east, and since then it has moved back.

The other interpretations, which envisage South America separating from a coherent Africa at about 130 Ma, start with about 15° of rotation of Africa and South America to the east-southeast with respect to North America and then move South America through about 10° in an easterly direction till late in the Cretaceous. After Cretaceous times, these older (and less well-constrained) interpretations show the north coast of South America describing a roughly circular counterclockwise rotation to return it to the position it occupied at about 80 Ma.

It is unclear to what extent Africa was split in two during the Early Cretaceous. Field observations in the Benue Valley continue to reveal evidence of intense Late Cretaceous strike-slip and compressional deformation (e.g. Benkhelil 1982), but it is not possible to assess from local evidence how much extension took place along the Benue trend before compression began. The place to look for evidence of Early Cretaceous extension in Africa is offshore in the South Atlantic. On the two-plate model, fracture zones at the continental margin south of the Niger River delta and on the conjugate shore south of the shoulder of South America should describe small circles about an early pole that relates to the opening of only the South Atlantic. Fracture zones in the Gulf of Guinea and adjacent to the north coast of Brazil should describe small circles of much greater radius. My attempts to confirm this distinction have been frustrated by the sparse distribution of ship-tracks and by the effects of sediment burial close to the shore. Deciding whether the single-plate model or the two-plate model of Africa is better is of continuing importance in Caribbean geology, and it seems likely that careful interpretation of SEASAT data could go far toward resolving this problem.

THE CARIBBEAN OCEANIC PLATEAU

Recognizing a resemblance between "ice-finger rafting" that he had observed while flying over the Arctic Ocean and the shape of the Caribbean, Wilson (1966) suggested that the Caribbean entered the Atlantic Ocean from the Pacific and was bounded against both North and South America by major transform fault systems. Both of these suggestions have proved extremely successful in the interpretation of Caribbean tectonics.

Deep-sea drilling (Donnelly et al 1973, summarized in Table 3 of Burke et al 1984) showed that in places where the Caribbean ocean floor is accessible to the drill, it consists of basaltic rocks, mainly tuffs and fine-grained intrusives, interbedded with roughly 80-Ma pelagic sediment. Unloaded Caribbean ocean floor generally lies at a depth of 4 km and is thus about 1 km too shallow for its known age, a property that is consistent with its anomalous thickness of 15–20 km as measured by refraction [see Burke et al (1978) for a review]. Locally, as in part of the Colombian basin, the ocean floor lies at 5 km and the crust is of more normal thickness (Houtz & Ludwig 1977).

Magnetic anomalies over the Caribbean are of low amplitude, and thus it has not proved easy to identify seafloor-spreading patterns. The most successful interpretations are for the Yucatán (Hall & Yeung 1982, confirmed by Rosencrantz et al 1984), Cayman (Rosencrantz & Sclater 1987, Rosencrantz et al 1987), and Grenada (Pindell & Barrett 1988, Bouysse 1987) basins, all of which are younger than the Caribbean itself. For the Caribbean ocean floor proper there are, as yet, no wholly persuasive identifications of magnetic anomalies, although the work of Ghosh et al (1984) in identifying mid- to Upper Jurassic ocean floor in the Venezuelan basin is perhaps the most successful attempt.

All these unusual properties can be accounted for if the Caribbean represents an oceanic plateau (in the sense of Winterer 1976). In this hypothesis, the Caribbean ocean floor consists of material made in the Pacific during the later Jurassic (perhaps at the Farallon-Phoenix spreading center) that experienced a major episode of offridge volcanism while still in the Pacific at about 80 Ma. This igneous activity, known as the B″ event (e.g. Burke et al 1978), extended over an area in excess of 1×10^6 km^2 and involved emplacement of enough basaltic material to thicken the crust from a typical oceanic value of about 6 km to 15 to 20 km. Locally, as in the Venezuelan basin (Ghosh et al 1984), the original magnetic striping may be discerned at low amplitudes in spite of the massive later igneous event.

Oceanic plateaus are most numerous in the western Pacific (Winterer 1976), and relatively little is known about their geology. Some of them, such as the Shatski and Hess Rises, appear to have been formed at spreading centers and may have had "twins" that later disappeared as a result of subduction (Livacarri et al 1981). Plateaus that formed off the ridge axis, such as the mid-Pacific mountains and the Caribbean, would not have had twins. Because oceanic plateaus represent bodies of basalt 20 or more km thick, they stand isostatically elevated above the ocean floor. It was recognized by Vogt et al (1976) that this buoyant character made them difficult to subduct. Attempts at subducting oceanic plateaus have led to changes in arc geometry or occasionally to reversals of subduction direction. Perhaps the best documented example of this process is in Southern Malaita in the Solomon Islands, where fragments of oceanic plateau are exposed in an accretionary prism choked by the attempted subduction of the Ontong-Java plateau (Hughes & Turner 1977).

Slices in the Malaita accretionary wedge contain no ultramafic material, and this is as expected where accretionary thrusting is picking up slices of a plateau consisting of a thickness of tens of kilometers of basalt. The absence of ultramafic material from an accretionary prism has been used as an indicator of an attempt to subduct an oceanic plateau, both around the Caribbean (Burke et al 1984, Table 4; Pindell & Barrett 1988) and elsewhere (e.g. Pallister 1985).

Recognition that an oceanic plateau occupies the Caribbean is helpful in interpreting much of the later tectonic evolution. For example, the extent of Neogene deformation and of current intraplate earthquake activity in the Caribbean is much greater than is normal for old ocean floor. A possible explanation is that whereas old oceanic lithosphere is very strong because all but the top 6 km of its 70 km thickness is mechanically dominated by the behavior of olivine, the top basaltic 20 km of 70-km-thick Caribbean lithosphere is in contrast dominated by the behavior of plagioclase. This is why the Caribbean lithosphere is presently deforming in a style intermediate between that of strong old ocean floor and weak continent.

The possibility that oceanic plateaus play an important role in the evolution of mountain belts is intriguing. Nur & Ben-Avraham (1982) suggested that oceanic plateaus might contribute significantly to continental evolution, but their definition of "oceanic plateau" was so broad as to include both the microcontinents and extinct island arcs, which have long been recognized as making up the blocks and fragments of the continents (Wilson 1968, Jacobs et al 1974, p. 478). A more focused question asks "What role do oceanic plateaus (in Winterer's sense) play in the evolution of the continents?" In spite of an extensive search (Burke & Şengör 1986), no coherent blocks hundreds or thousands of square kilometers in area have been identified as ancient oceanic plateaus lying within continents. Because slivers of material reminiscent of oceanic plateau rocks (i.e. containing no ultramafic rocks, or only cumulate ultramafic rocks and abundant interbedded pelagic sediments) have been identified in suture zones, it seems probable that ancient oceanic plateaus did exist, but all have been destroyed or removed from the Earths' surface in the course of continental assembly. Caribbean tectonic evolution has been so strongly influenced by its contained oceanic plateau that it seems worth considering the role that ancient plateaus might have played in the development of mountain belts. Moores (1986) has made the bold suggestion, largely on the basis of his observations in Brazil, that older Proterozoic ocean floor everywhere had the thickness of oceanic plateaus, a suggestion that, if confirmed, is likely to be important for tectonic studies of ancient environments. In that case, the Caribbean may provide helpful analogues.

THE GREAT ISLAND-ARC SYSTEM OF THE CARIBBEAN

One of the less-expected results of the recognition that the geology of the Caribbean preserves the record of a coherent sequence of plate-evolutionary events has been that what had been conceived of as a variety of independent island-arc systems, originating in different places at different times (i.e. the Greater Antillean arc, the Lesser Antillean arc, the Aves Swell arc, the Netherlands Antillean–Venezuelan arc, and the Villa de Cura arc; Figure 4)—are beginning to emerge as elements in a single Great Caribbean island-arc system (or Great Arc of the Caribbean) that was constructed in the Pacific and migrated into the Atlantic (see, for example, Beets et al 1984, Duncan & Hargraves 1984, Leclere-Vanhoeve & Stephan 1985, Bouysse 1987, Pindell et al 1987, Pindell & Barrett 1988). On a less grand scale, Mann & Burke (1984) drew attention to continuity between the Greater Antillean and Aves–Lesser Antillean arc systems, and Speed (1985) emphasized continuity between the Lesser Antillean and Netherlands Antillean island-arc systems.

Initiation of the Arc

The history of the Great Arc of the Caribbean appears to have begun, perhaps by nucleation on a fracture zone of the proto-Caribbean spreading system (Pindell & Dewey 1982, Figure 16) or of the Farallon-Phoenix system (Duncan & Hargraves 1984, Figure 2), either in the Late Jurassic or the earliest Cretaceous. There are reported Early Cretaceous metamorphic ages for arc-complex rocks from both Cuba and Hispaniola, and fossiliferous rocks preserved within the Greater Antilles show that accretion was in progress by Albian times. Most interpretations depict the subduction direction of the early arc as being toward the Atlantic, with the Pacific Ocean (Farallon plate?) floor being consumed. This makes the plate geometry readily compatible with that of the contemporary Andean margins of both North and South America. Only the Bermeja complex of southwesternmost Puerto Rico seems to clearly represent accretionary wedge material formed during this first episode in Caribbean arc history, because it contains fragments of normal ocean floor (not of oceanic plateau) as old as Tithonian (end-Jurassic) in age and is unconformably overlain by Turonian (90 Ma) rocks (Mattson & Pessagno 1979). It also appears to record an appropriate polarity of subduction. Other fragments of ocean floor, such as that at Desirade (Bouysse et al 1983) and Siquisique in Venezuela (Bartok et al 1985) may date from this epoch, as may ultramafic rocks in accretionary prisms of the Greater Antillean arc in the Blue Mountains of Jamaica, in Cuba, and in the Cordillera Central of Hispaniola. Ages of formation and accretion are not well known in these areas, and subduction polarity has been much obscured by later tectonism.

Attempts at Subducting the Oceanic Plateau

Before the end of the Cretaceous, and apparently only very shortly after the Caribbean oceanic plateau had been constructed, the Great Arc and the plateau collided while both were still in the Pacific.

Evidence of this encounter is preserved in thrust slices in accretionary wedges that contain abundant basaltic and pelagic material associated with little or no ultramafic rock. Accretionary complexes representing fragmentary records of the collision are distributed from the Romeral suture zone of Colombia in the south (Spadea et al 1987) through the Greater Antilles to the Nicoya complex of Costa Rica. Data summarized in Burke et al (1984, Table 4) and Pindell & Barrett (1988, Table 1) indicate that these attempts to subduct the Caribbean oceanic plateau took place during the Campanian and Maestrichtian and possibly extended into the Paleocene, i.e. over about 20 Myr between 80 and 60 Ma.

Reversal of Polarity

The location of the Great Arc at the time of its collision with the Caribbean oceanic plateau must have been in the Pacific, because material recording the collision outcrops in both western Colombia and western Costa Rica. The continued migration of the arc into the Atlantic requires that after the time of this collision, subduction at the Great Arc was generally to the southwest, a reversal of the previous geometry that resulted from suturing of the arc to the plateau (Figure 5). The ocean floor consumed during this latest (and still continuing) episode has mainly been Atlantic Ocean floor, but it has also included the last vestiges of the ocean floor made on the site of the Caribbean during the separation of Yucatán from South America—the "proto-Caribbean ocean floor" of various authors.

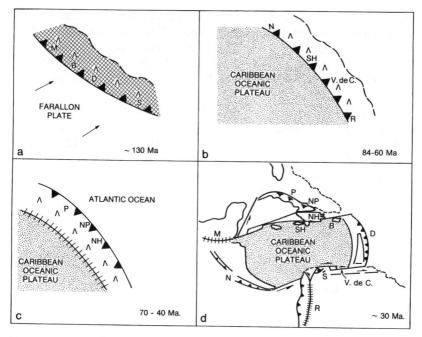

Figure 5 Evolution of the Great Arc of the Caribbean. (*a*) The arc formed at the end of the Jurassic and subducted Pacific Ocean floor. Slivers incorporated in accretionary wedges at this time may include Motagua (M), Bermeja (B), Desirade (D), and Siquisique (S). (*b*) The arc attempted to subduct the Caribbean oceanic plateau late in Cretaceous time, and slivers of this material are represented at Nicoya (N), southern Haiti (SH), and Romeral (R). The Villa de Cura (V. de C.) arc rocks may have formed at this time. (*c*) After a collision between the arc and the plateau, the polarity of subduction reversed, so that slivers of Atlantic Ocean floor began to be incorporated into an accretionary wedge (P, Purial; NP, Nipe; NH, north Hispaniola). (*d*) After the arc collided with Yucatán and Cuba in the north and Colombia in the south, strike-slip motion on both the north and south sides of the Caribbean distributed the fragments of ocean floor in the pattern shown here.

Segmentation of the Great Arc

The Great Arc of the Caribbean broke into three as it entered the Atlantic. The western end of the northern segment collided with southern Yucatán at the end of the Cretaceous, while the rest of that segment, after sweeping past the east coast of Yucatán, collided progressively (from west to east) with southernmost Florida and the Bahamas between the end of the Paleocene and the early Oligocene. A central segment of the arc formed the Aves Swell during the Late Cretaceous and cleared the easternmost Bahamas to enter the Atlantic. After splitting at its southern end to make the Grenada basin, it survives today as the active Lesser Antillean arc.

The southern segment of the Great Arc failed to enter the Atlantic and collided with the northwest corner of South America. Strike-slip motion has tectonized this part of the arc into fragments that are strewn along and offshore of the northern coast of South America in Colombia, Venezuela, the Netherlands Antilles, and Tobago (Rowe & Snoke 1986).

THE FATE OF THE NORTHERN SEGMENT   The collision of the western end of the arc with North America to form the Motagua suture zone in Guatemala at the end of the Cretaceous (Wadge et al 1984) has been described in an unpublished thesis (Rosenfeld 1981) and is apparently represented by the finest of Caribbean ophiolites. Unfortunately, it is difficult to reconstruct exactly what happened in this collision because, although the collisional foreland basin is well preserved (Pindell & Barrett 1988), the part of the arc that was involved has been transported hundreds of kilometers away to the east by later strike-slip motion. Its place has been taken by a fragment from the Mesozoic west coast of North America, the Chortis continental block (Wadge & Burke 1983).

The rest of the northern segment, now preserved in the greater Antilles, swept past Yucatán continuing on into the Atlantic beyond the Motagua suture for another 700 km, subducting Atlantic Ocean floor as it went. There are at least three distinct kinds of ocean floor represented in the accretionary wedges of the greater Antilles: (a) material from the Pacific Ocean (Farallon plate?) of normal oceanic character subducted (to the northeast?) before the collision of the arc with the Caribbean oceanic plateau and perhaps best represented by the Bermeja complex; (b) the oceanic plateau itself, represented by basaltic and pelagic material in the southern peninsula of Haiti and along strike at Baoruco (Bellon et al 1985); and (c) proto-Caribbean and Atlantic Ocean floor best represented by the great ophiolitic masses of Cuba, such as Nipe and Purial. Wadge et al (1984) identified a fourth element in Greater Antillean accretionary wedges representing material made during the Late Cretaceous in the Yucatán basin. This material was caught up during the Paleogene and is exposed in the Oriente province of Cuba, Jamaica, and Hispaniola.

The Yucatán basin formed by extension within the northern segment of the arc as it moved past the Yucatán peninsula (Hall & Yeung 1982), perhaps in response to stresses imposed on the Caribbean plate margin by that passage. Except in Oriente province, which is separated from the rest of Cuba by thin crust (Wadge et al 1984, Figure 2), all the rocks outcropping in Cuba may lie on thrusts. In forming the Yucatán basin, the Greater Antillean arc perhaps split close to the crest of its then active volcanoes, leaving a remnant arc represented within the Cayman ridge.

It has long been clear from an anomalous thickness of sediments in the southeastern Gulf of Mexico (Rainwater 1967) that the collision of the Cuban part of the northern sector of the arc with Florida and the western Bahamas had begun by the end of the Paleocene. This event can be regarded as a typical island-arc–continent collision (e.g. Gealey 1980), although the Bahama Banks appear not to overlie continental crust, but rather to represent an enormous thickness of carbonate bank that has grown on a platform of volcanic rocks, possibly in a compressed fracture zone (see Burke et al 1984, p. 45). The structure of the banks is beginning to be revealed by seismic reflection (Eberli & Ginsberg 1986), and the extent of deformation within the Bahamas associated with the Cuban collision is becoming clear. The structural development of the Cuban collision with ophiolite nappes, back-thrusting, and the development of postcollisional basins (first discerned by Wassal 1956) is also beginning to be well described (e.g. Mossakovskiy & Albear 1978, Pszczolkowski 1978) and is typical of arc-continent collisions. Oil wells in Cuba have encountered Bahamian carbonate platform stratigraphy beneath thrusts carrying ocean-floor rocks, and it seems possible that the entire 200 km width of the island has been thrust over the rocks of the Bahama Banks.

The more easterly parts of the northern segment of the Great Arc record a collision with the more easterly parts of the Bahamas and are preserved in Hispaniola, Puerto Rico, and the Virgin Islands. The strike-slip motion that took fragments of the more westerly parts of the arc away from the Motagua collision zone has carried these fragments more than 1000 km to the east. The southern peninsula of Haiti, Jamaica, and much of the submerged Nicaraguan Rise represent this material, juxtaposed by strike-slip movement against the arc-material involved in the later collision with the Bahamas (e.g. Pindell & Dewey 1982).

Strike-slip motion associated with the opening of the Cayman trough spreading center has dominated the tectonic evolution of the Greater Antilles since the beginning of the Oligocene and possibly earlier (Rosencrantz & Sclater 1987). This motion, which has been distributed across a 250-km-wide strike-slip plate boundary zone, or PBZ (Burke et al 1980), has had a variety of consequences. Although the tectonics of the PBZ have made it difficult to reconstruct the geology of the Bahamian collision, they have exposed Cretaceous rocks in anticlines and overthrust zones, especially at restraining bends (e.g. Mann et al 1984, 1986). Granodiorites and tonalites, representing the roots of Cretaceous andesitic volcanoes, occur in all the larger islands, and other familiar elements of island-arc geology, such as accretionary wedges, forearcs, backarcs, and island-shelf carbonate banks (locally with spectacular rudist faunas; Chubb 1971), are all exposed.

These elements are juxtaposed in very complex ways. In an attempt to resolve this complexity, Paul Mann (of the Institute of Geophysics, University of Texas at Austin) and I have been trying to unravel the effects of PBZ strike-slip motion on the Cretaceous and Paleogene arc rocks of the Greater Antilles. In Figure 6 we show the islands as made up of fragments juxtaposed by strike-slip, with two fragments in Cuba, five in Hispaniola, and one each in Puerto Rico and Jamaica. This is certainly a simplification, but it serves to illustrate the style of the disruption. We have found that the sedimentary basins that formed within the arc system during the collision and in the later strike-slip episodes are extremely useful in guiding our interpretation (e.g. Mann & Burke 1984a, Figure 5; Mann et al 1986).

THE AVES SWELL AND LESSER ANTILLEAN ARC SEGMENT   The central segment of the Great Arc entered the Atlantic through the gap between North and South America at about the beginning of Cenozoic time and cleared the southeastern end of the Bahamas by the beginning of the Oligocene. 20 Myr later (Pindell & Barrett 1988).

The pre-Oligocene history of this part of the arc is not well known because of the lack of outcrops. Rocks from the small island of Desirade in the Lesser Antilles include material of Late Jurassic age that may represent ocean floor as well as possible Cretaceous arc rocks. However, structural relations are obscure, and although this material could be representative of the early stages of development of the Great Arc, other interpretations abound (see, for example, Bouysse et al 1983, Pindell & Barrett 1988). Cretaceous granodiorite and volcanic fragments have been dredged from the Aves Swell (Bouysse et al 1985), and the shape of the swell and its geophysical signatures are typical of those of extinct arcs. The Saba Bank well at the northern end of the Aves Swell (see Figure 6 for location) encountered Eocene siliceous volcanic rocks that indicated a persistence of arc activity into the Cenozoic in that area.

The Grenada basin separates the southern two thirds of the Lesser Antilles from the Aves Swell, although the two are not separated by deep water farther north. Magnetic anomalies on the floor of the Grenada basin (Speed & Westbrook 1984) seem to indicate that it is a marginal basin formed by arc splitting. Bouysse (1987) interprets the magnetics as indicating east-west extension, although Pindell & Barrett (1988) see evidence of north-south extension. Westercamp et al (1985) reported an occurrence of Campanian volcanic rocks from the island of Union in the Grenadines, which indicates that the arc had not split by that time. Very Late Cretaceous or earliest Cenozoic splitting of the central segment of the Great Arc to isolate the Lesser Antilles and leave the southern Aves Swell as a remnant arc is perhaps an indication that the Grenada basin formed somewhat like

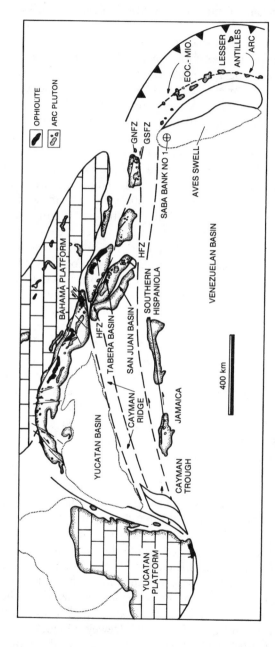

Figure 6 Sketch showing fragmentation of the northern segment of the Great Arc of the Caribbean as the Cayman trough began to open. Nine arc fragments are depicted that are now embodied in the islands of Cuba, Hispaniola, Jamaica, and Puerto Rico (HFZ, Hispaniola fault zone; GNFZ, Great Northern fault zone; GSFZ, Great Southern fault zone). Based on unpublished work by Paul Mann and the author.

the Yucatán basin, in response to stresses applied to the edge of the Caribbean plate as it began to move past South America (Pindell & Barrett 1988).

There is evidence of Cenozoic volcanic activity in the Lesser Antilles from Eocene time on (for example, in the Tufton Hall graywackes of Grenada; Speed & Larue 1985), but only sporadic outcrops record the earlier phases. Thirty-four Neogene and Quaternary volcanic centers have been distinguished (Maury & Westercamp 1985, Table 1), and there is evidence that the current calc-alkaline magmatic activity involves some subducted sedimentary material (e.g. Dupré et al 1985). Volcanism is sporadic in the Lesser Antillean arc, but violent eruptions do occur with a frequency of decades, and one at Mont Pelée in 1902 was reponsible for more fatalities than any other volcanic event so far this century.

At its northern end the Lesser Antilles arc splits into an eastern "limestone Caribbees" and a western "volcanic Caribbees" (Figure 6). The two arcs trend subparallel, and the separation has been attributed to a late Neogene change in inclination of the underlying subducted slab, possibly in response to arrival at the Lesser Antillean trench of buoyant elements of the Atlantic Ocean floor, represented by the Barracuda Rise and the Tiburon Rise (Moretti & Ngokwey 1985, McCann & Sykes 1984, Stein et al 1986).

The forearc and accretionary wedge of the northern Lesser Antilles are of normal width, but farther south they become very wide. Exposures of the accretionary wedge in the Scotland district of Barbados show that this expansion in width can be related to the occurrence of huge volumes of coarse, quartz-rich clastic rocks. The structure of this exceptional accretionary wedge is being worked out in detail at outcrop (e.g. Speed & Larue 1982, Speed 1983, Larue & Speed 1983, Biju-Duval et al 1985), by multichannel seismic studies (e.g. Torrini & Speed 1987, Valery et al 1985, Speed & Westbrook 1984), and by deep ocean drilling (Leg 110 Scientific Drilling Party 1987).

It has long been recognized that the source of the huge volume of quartz-rich sediments must be South America (e.g. Senn 1940), and Dickey (1980) took a bold step when he pointed out that because the outcropping accretionary wedge rocks of the Scotland district contained Eocene fossils, it was necessary in seeking their source to consider where the Lesser Antilles were with respect to South America in the Eocene. Dickey concluded that the Lesser Antillean trench was at that time more than 1000 km farther to the west and that the source of the Barbados sediments was the deltaic fan of a large river, possibly a "proto-Magdalena." Dickey's idea has proved popular (e.g. Pindell & Dewey 1982, Dewey & Pindell 1986, Pindell & Barrett 1988, Pindell et al 1987), but the unexpected discovery from fission-track studies (Baldwin et al 1986) that the rocks in Barbados containing Eocene faunas appear to have been resedimented during the Oligocene has led to the alternative suggestion that the Barbados accretionary prism represents material resedimented from a "proto-Orinoco" delta, which was incorporated into the Lesser Antillean accretionary wedge only in the early Miocene (Figure 7).

THE FATE OF THE SOUTHERN SEGMENT OF THE GREAT ARC   What happened and when to the southern segment of the Great Arc is the subject of lively controversy. Two end-member positions exist: Either the Arc collided only with the west coast of South America, and all the fragments of it that are now distributed along 1000 km of the coast of northern South America as far as Tobago have been transported by strike-slip motion; or the arc

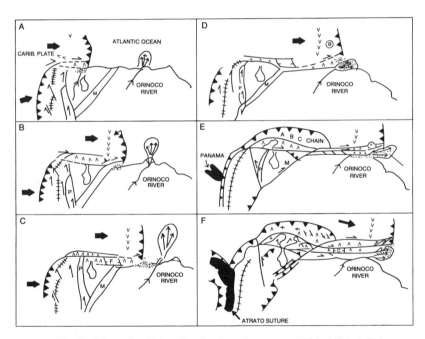

Figure 7   The Caribbean–South America plate boundary zone (PBZ) in (A) early Paleocene, (B) middle Eocene, (C) early Oligocene, (D) early Miocene, (E) late Miocene, and (F) now. Solid triangles: front of accretionary prism and thrusts; railroad tracks: sutures; V: volcanic arc; Λ: arc material carried north by tectonic escape from the Romeral suturing event that was then diverted eastward by strike-slip. P, Perija Andes; M, Mérida Andes; F, Falcon pull-apart basin; C, Cariaco pull-apart basin; B, Scotland district of Barbados. Dots indicate a foreland basin type of environment opposite the migrating Lesser Antillean hinge zone. Modified after Dewey & Pindell (1986).

has been colliding obliquely in Cenozoic times with the South American continent, so that there has been little strike-slip motion. Speed (1985) has emerged as the leading protagonist of a version of the latter hypothesis, and I am disposed to support the former. An intermediate position has been adopted by Dewey & Pindell (1986) and Pindell & Barrett (1988), who interpret the data as indicating collision of the southern end of this segment of the arc with the then west coast of Colombia along a line marked by the Romeral suture at about the end of the Cretaceous, followed by progressive arc collision with the north coast of South America from west to east. These authors define the timing of arc collision with the South American continent in a particular area by the beginning of local foreland basin deposition on the continent. The idea is that loading of the continent by the overthrust arc flexes the lithosphere downward, so that deposition starts in the foreland basin. Using this criterion, they trace the propagating arc collision eastward from the vicinity of Lago de Maracaibo during the Paleocene to eastern Venezuela in the Miocene, since which time there has been only strike-slip motion between the Caribbean plate and South America. Speed's interpretation dates the collision of the Lesser Antillean arc with South America as starting at about 65°W in the early Oligocene and migrating eastward very slowly since that time. His model requires that the volcanic arc represented in the islands offshore of Venezuela was active late in the Cenozoic, and in support he cites whole-rock K/Ar ages (e.g. a 44-Ma age from the islet of Los Testigos; Santamaria & Schubert 1974). Beets et al (1984, p. 107), in reviewing the age of igneous rocks in the same area, concluded that there was no arc activity during the Cenozoic and that later ages were all reset [see Loubet et al (1985) for a similar conclusion]. There is clearly a strong case for redetermining the age of the rocks from Los Testigos by methods (such as $^{40}Ar/^{39}Ar$) that may reveal the complexities of the thermal history.

Paul Robertson and I have been addressing the alternatives of oblique collision and strike-slip by investigating the nature of the southern Caribbean PBZ at its eastern end in Trinidad. Here we find little support for Speed's (1985) interpretation, which requires the El Pilar fault to be an active thrust. Our studies of outcrop and seismic-reflection lines confirm the work of Vierbuchen (1984) in showing that the El Pilar fault has moved as a strike-slip fault since late Quaternary times and by indicating 30 to 40 km of right-lateral offset along the El Pilar fault since the Pliocene (Robertson & Burke 1986). East of Trinidad, there is a substantial flexural depression of the South American continent at the Lesser Antillean subduction zone hinge, and it seems possible that the foreland basin sediments that have been used to locate the migration of Cenozoic arc collision through time (e.g. Pindell & Barrett 1988) could be confused with deposits opposite the migrating hinge. We also confirm evidence from earthquake distribution (e.g. Kafka & Weidner 1981) and mapping in Venezuela (e.g. Munro & Smith 1984) indicating that the strike-slip PBZ extends a long distance south into South America (as far as 9°30′N), so that all Cenozoic thrusts mapped in eastern Venezuela are within the PBZ and may well be secondary to the overall east-west strike-slip motion, rather than recording arc collision.

An interpretation of the history of the southern segment of the Great Arc similar to those of Duncan & Hargraves (1984) and Beets et al (1984) is sketched in Figure 7. In this scenario we assume that the arc collided only with the west coast of South America and that the Aves Swell–Lesser Antillean segment passed by South America to the north.

An essential element of this version of the arc's fate is that after Late Cretaceous collision along the Romeral suture in Colombia (e.g. Pindell & Barrett 1988), there was substantial strike-slip motion of Andean arc material to the north in an episode of tectonic escape. The escape was like that which is presently carrying the Maracaibo block to the north on the Bocono and Santa Marta faults and which is regarded as a consequence of the collision of Panama (Mann & Burke 1984b), but it was on a grander scale. It has, for example, been suggested earlier (p. 206) that a measure of the extent of this strike-slip motion is given by the 500-km displacement of the Perija Andes from the main Andean chain. This displacement was already well underway by Paleocene time (Figure 7A), and the material that had been carried north is shown as being diverted eastward in the southern Caribbean PBZ, depicted here as being initiated early in the Cenozoic. The thick Paleocene sediments of Maracaibo are localized by flexure opposite the passing hinge zone. As time passed, more and more material is envisaged as being carried north to escape the Romeral suture zone, only to be displaced again by up to hundreds of kilometers of strike-slip motion in the PBZ (Figure 7B; cf. Priem et al 1986). By early Oligocene time the Falcon pull-apart basin (Muessig 1984) (F in Figure 7C) developed within the zone, and soon afterward the Orinoco deep-sea fan was picked up to broaden the Lesser Antillean accretionary prism in the Barbados area (B in Figure 7D). Propagation of the PBZ to the east carried arc rocks along the coast, and the combination of elevation of the coastal mountains with migration of the hinge zone has diverted the lower reaches of the Orinoco River from a northeasterly to an easterly flow. From late Miocene times on (Figures 7E,F), the collision of Panama has started a new episode of tectonic escape, and the east-west strike-slip PBZ has continued to operate only in the east. A different phase in PBZ development may be about to begin because the coastline of northern South America farther east has a southeasterly rather than an east-west trend.

CENTRAL AMERICA IN CARIBBEAN HISTORY

Central America, which bounds the Caribbean to the southwest, consists geologically of two parts: the Chortis block extending southward from the southern border of Yucatán to 11°N, and the narrow Costa Rica–Panama isthmus in the shape of an "S" lying on its back.

Chortis

This block, much of which occupies the Republic of Honduras, has Precambrian basement (Case et al 1984) extending offshore to the east to underlie at least part of the Nicaraguan Rise. The rise itself is separated to the southeast from the Colombian basin by the Hess escarpment and is flanked to the north by the Cayman trough (Figures 4 and 6). Wadge & Burke (1983) estimated that 1100 km of young ocean floor underlay the Cayman trough and that Chortis could be rotated to the position it occupied before the Cayman trough opened. Unlike MacDonald (1976), who left an oceanic gore between Mexico and Chortis, they repositioned Chortis flush against the coast of Mexico. Azema et al (1985) attained a similar 1100-km offset by matching Mesozoic platform deposits in Oaxaca with those of Honduras.

These observations indicate that Chortis, restored to its late Mesozoic position, may have been the southwesternmost part of North America while the Caribbean was opening. The occurrence of what has been interpreted to be a large fragment of the Caribbean oceanic plateau in the Nicoya complex at the southernmost tip of Chortis (Schmidt-Effing 1979) seems consistent with this idea. The southwest coast of North America was active as an Andean margin during the Late Cretaceous, and the Great Arc of the Caribbean appears to have interacted with it along a trench-to-trench transform system, perhaps now represented by the Hess escarpment, as it carried the Caribbean oceanic plateau into the Atlantic (Figure 4).

MacDonald (1976) produced a geologically improbable oceanic gore between Mexico and Chortis by rigid body rotation, as plate-tectonic theory demands. Wadge & Burke (1983) decided that there was so much deformation within Chortis during the Cenozoic that plate-tectonic constraints could be relaxed. Mann & Burke (1984b) pursued this idea by showing that the deformation within Chortis and the Nicaraguan Rise consisted mainly of rifts that were progressively younger from east to west. They suggested that parts of Chortis suffered local intense strain during overall strike-slip motion as they rounded the southern tip of Yucatán, and that this strain had generated rifts ranging in age from Paleocene (the Wagwater trough of Jamaica) to recent (the Guatemala City rift).

Part of the Mesozoic Andean margin of North America has been carried south to form the Middle America volcanic province with the rotation of Chortis, and this active and mature volcanic arc has been much studied (e.g. Stoiber & Carr 1974). The broad forearc and trench have also been studied in detail (e.g. Moore et al 1986, von Huene et al 1982, Aubouin et al 1982). Uplift and limited accretion are reported, especially from the vicinity of the site of subduction of the buoyant Cocos ridge (Mann et al 1986).

An outstanding problem of the tectonics of Chortis is exactly how far the block underlain by Precambrian crust extends beneath the Nicaraguan Rise and, in a related question, exactly how the northern sector of the Great Arc of the Caribbean, which is represented in Jamaica and beneath the eastern Nicaraguan Rise, is joined to Chortis [see Pindell & Dewey (1982, Figure 19) for one possible solution].

Costa Rica and Panama

In contrast to Chortis, the southern part of Central America appears to contain no material older than an island-arc complex constructed on the opposite rim of the Caribbean oceanic plateau from that occupied by the Great Arc of the Caribbean (Figure 4). This Costa Rican–Panamanian arc appears to have been established in Late Cretaceous time and to have become inactive in its southern part early in the Cenozoic. Wadge & Burke (1983) attributed this inactivity to the establishment of a transform boundary along which the arc was carried toward South America. Collision of the arc with Colombia to form the Atrato suture at about 7 Ma has been one of the most significant Neogene tectonic events in the evolution of the Caribbean. It was not only responsible for constructing a land bridge between the continents and shutting off the Caribbean from the Pacific, but it also led to the bending of Panama into its peculiar recumbent-"S" shape and to the initiation of strike-slip motion on the Bocono and Santa Marta faults that is allowing the Maracaibo block to escape northward into the Caribbean.

CONCLUSION

There is so much research in the field of neotectonics in the Caribbean (e.g. reviews by Mann & Burke 1984b, Mann et al 1988) that I have decided to omit neotectonics from what has of necessity proved to be a selective and incomplete review.

If I have strongly emphasized the results of modern tectonic syntheses, it is because I believe these results have already proved extremely useful in helping to pose critical questions about the geological evolution of the

Caribbean and that they will continue to help in guiding research toward outstanding problems.

ACKNOWLEDGMENT

My Caribbean studies over the last 25 years have been supported by numerous government agencies, by universities, and by industry in several countries. This review was written while I was the Director of the Lunar and Planetary Institute, which is operated by the Universities Space Research Association under Contract No. NASW-4066 with the National Aeronautics and Space Administration. This paper is Lunar and Planetary Institute Contribution No. 643.

Literature Cited

Aubouin, J., von Huene, R., et al. 1982. Subduction sans accrétion: La marge pacifique du Guatemala, premiers résultats du Leg 84 du DSDP. *C. R. Acad. Sci. Ser. II* 294(13): 803–12

Azema, J., Biju-Duval, B., Bizon, J. J., Carfantan, J. C., Mascle, A., et al. 1985. Le Honduras (Amérique centrale nucléaire) et Le block d'Oaxaca (Sud du Mexique): deux ensembles comparables du continent nord-américain séparés par le jeu décrochant sénestre des failles du systeme Polochic-Motagua. See Mascle 1985, pp. 427–38

Baldwin, S. L., Harrison, T. M., Burke, K. 1986. Fission track evidence for the source of accreted sandstones, Barbados. *Tectonics* 5: 457–68

Bartok, P. E., Renz, O., Westermann, G. E. G. 1985. The Siquisique Ophiolites, northern Lara State, Venezuela: a discussion on their Middle Jurassic ammonites and tectonic implications. *Geol. Soc. Am. Bull.* 96: 1050–55

Beets, D. J., Maresch, W. V., Klaver, G. Th., Mottana, A., Bocchio, R., et al. 1984. Magmatic rock series and high-pressure metamorphism as constraints on the tectonic history of the southern Caribbean. See Bonini et al 1984, pp. 95–130

Bellon, H., Vila, J. M., Mercier de Lepinay, B. 1985. Chronologie ^{40}K-^{40}Ar et affinités géochimiques des manifestations magmatiques au Crétacé et au Paléogène dans l'île d'Hispaniola. See Mascle 1985, pp. 329–40

Benkhelil, J. 1982. Benue Trough and Benue Chain. *Geol. Mag.* 119: 155–68

Biju-Duval, B., Caulet, J. P., Defaure, Ph., Mascle, A., Muller, C., et al. 1985. The terrigenous and pelagic series of Barbados Island: Paleocene to middle Miocene slope deposits accreted to the Lesser Antilles margin. See Mascle 1985, pp. 187–98

Bonini, W. E., Hargraves, R. B., Shagam, R., eds. 1984. *The Caribbean South American Plate Boundary and Regional Tectonics. Geol. Soc. Am. Mem. 162.* 421 pp.

Bouysse, P. 1987. Occurrence of Upper Cretaceous rocks in the Grenadines (southern Lesser Antilles): implications for the opening of Grenada back-arc basin and for Caribbean geodynamics. *Tectonophysics.* In press

Bouysse, P., Schmidt-Effing, R., Westercamp, D. 1983. La Desirade Island (Lesser Antilles) revisited: Lower Cretaceous radiolarian cherts and arguments against an ophiolitic origin for the basal complex. *Geology* 11: 244–47

Bouysse, P., Andreieff, P., Richard, M., Baubron, J. C., Mascle, A., et al. 1985. Aves swell and northern Lesser Antilles Ridge: rock-dredging results from ARCANTE 3 Cruise. See Mascle 1985, pp. 65–76

Bray, R., Eva, A. N. 1983. Age, depositional environment and tectonic significance of the Couva Marine evaporite, offshore Trinidad. *Caribb. Geol. Conf., 10th, Cartagena, Colomb.*, p. 29 (Abstr.)

Bullard, E. C., Everett, J. E., Smith, A. G. 1965. The fit of the continents around the Atlantic: a symposium on continental drift. *Philos. Trans. R. Soc. London Ser. A* 258: 41–51

Burke, K. C., Dewey, J. F. 1972. Orogeny in Africa. *Proc. Conf. Afr. Geol.*, ed. T. F. J. Dessauvagie, A. J., Whiteman, pp. 583–608. Ibadan, Nigeria: Ibadan Univ.

Burke, K., Dewey, J. F. 1974. Two plates in Africa during the Cretaceous? *Nature* 249: 313–16

Burke, K., Şengör, A. M. C. 1986. Tectonic escape in the evolution of the continental crust. In *Reflection Seismology: The Continental Crust. Geodyn. Ser.*, 14: 41–53. Washington, DC: Am. Geophys. Union

Burke, K., Fox, P. J., Şengör, A. M. C. 1978. Buoyant ocean floor and the evolution of the Caribbean. *J. Geophys. Res.* 83: 3949–54

Burke, K., Grippi, J., Şengör, A. M. C. 1980. Neogene structures in Jamaica and the tectonic style of the northern Caribbean plate boundary zone. *J. Geol.* 88: 375–86

Burke, K., Cooper, C., Dewey, J. F., Mann, J. P., Pindell, J. 1984. Caribbean tectonics and relative plate motions. See Bonini et al 1984, pp. 31–64

Campa, M. F., Coney, P. J. 1983. Tectonostratigraphic terranes and mineral resource distributions in Mexico. *Can. J. Earth Sci.* 20: 1040–51

Case, J. E., Holcombe, T. L. 1980. Geologic-tectonic map of the Caribbean region. *US Geol. Surv. Misc. Invest. Ser. Map I-1100*, scale 1 : 2,500,000

Case, J. E., Holcombe, T. L., Martin, R. G. 1984. Map of geologic provinces in the Caribbean region. See Bonini et al 1984, pp. 1–30

Choubert, B. 1935. Recherchés sur la genèse des chaines paleozoiques et anté-cambriennes. *Rev. Geogr. Phys. Geol. Dyn.* 8: 1–50

Chubb, L. J. 1971. Rudists of Jamaica. *Paleontogr. Am.* 7(45): 257

Dewey, J. F., Pindell, J. L. 1986. Reply to Amos Salvador. *Tectonics* 5(4): 703–5

deWit, M. J., Ashwal, L. D., eds. 1986. In *Workshop on Tectonic Evolution of Greenstone Belts. LPI Tech. Rep. 86-10*. Houston, Tex: Lunar Planet. Inst. 227 pp.

Dickey, P. A. 1980. Barbados as a fragment of South America ripped off by continental drift. *Trans. Caribb. Geol. Conf., 9th*, pp. 51–52

Donnelly, T. W., Melson, W., Kay, R., Rogers, J. J. W. 1973. Basalts and dolerites of Late Cretaceous age from the Central Caribbean. In *Initial Reports of the Deep Sea Drilling Program*, 15: 989–1011. Washington, DC: US Govt. Print. Off.

Duncan, R. A., Hargraves, R. B. 1984. Plate tectonic evolution of the Caribbean region in the mantle reference frame. See Bonini et al 1984, pp. 81–93

Dupré, B., White, W. M., Vidal, Ph., Maury, R. C. 1985. Utilisation des traceurs couplés (Pb-Sr-Nd) pour déterminer le rôle des sédiments dans La genèse des basaltes des L'arc des Antilles. See Mascle 1985, pp. 91–98

Eberli, G. P., Ginsberg, R. N. 1986. Segmentation and coalescence of Cenozoic carbonate platforms, northwestern Great Bahama Bank. *Geology* 15: 75–79

Gealey, W. K. 1980. Ophiolite obduction mechanism. *Proc. Int. Ophiolite Symp.*, ed. A. Parayiotou, pp. 228–43. Nicosia: Cyprus Geol. Surv. Dep.

Ghosh, N., Hall, S. A., Casey, J. F. 1984. Seafloor spreading magnetic anomalies in the Venezuelan Basin. See Bonini et al 1984, pp. 65–80

Hall, S. A., Yeung, T. 1982. A study of magnetic anomalies in the Yucatán Basin. *Proc. Caribb. Geol. Conf., 9th, Santo Domingo, Domin. Rep.*, 2: 519–26

Hamilton, W. 1979. Tectonics of the Indonesian region. *US Geol. Surv. Prof. Pap. 1078*. 345 pp.

Houtz, R. E., Ludwig, W. J. 1977. Structure of Colombia basin, Caribbean Sea, from profiler-sonobuoy measurements. *J. Geophys. Res.* 82(30): 4861–67

Hughes, G. W., Turner, C. C. 1977. Upraised Pacific floor, Southern Malaita, Solomon Islands. *Geol. Soc. Am. Bull.* 88: 412–14

Jacobs, C. 1977. Jurassic lithology in Great Isaac 1 well, Bahamas: discussion. *Am. Assoc. Pet. Geol. Bull.* 61: 443

Jacobs, J. A., Russell, R. D., Wilson, J. T., eds. 1974. *Physics and Geology*. New York: McGraw-Hill. 622 pp. 2nd ed.

Kafka, A. L., Weidner, D. J. 1981. Earthquake focal mechanisms and tectonic processes along the southern boundary of the Caribbean plate. *J. Geophys. Res.* 86: 2877–88

Klitgord, K. D., Schouten, H. 1980. Mesozoic evolution of the Atlantic Caribbean and Gulf of Mexico. *Symp. Origin of the Gulf of Mex. and the Early Opening of the Cent. North Atl.*, ed. R. H. Pilger, pp. 100–1. Baton Rouge: La. State Univ.

Ladd, J. W. 1976. Relative motion of North America and Caribbean tectonics. *Geol. Soc. Am. Bull.* 87: 969–76

Larue, D. K., Speed, R. C. 1983. Quartzose turbidites of the accretionary complex of Barbados, 1, Chalky Mount succession. *J. Sediment. Petrol.* 53: 1337–52

Leclere-Vanhoeve, A., Stephan, J. F. 1985. Evolution géodynamique des Caraibes dans le système points chauds. See Mascle 1985, pp. 21–34

Leg 110 Scientific Drilling Party. 1987. Accretionary complex penetrated, defined. *Geotimes* 32: 13–16

Livacarri, R. F., Burke, K., Şengör, A. M. C. 1981. Was the Laramide Orogeny related to subduction of an oceanic plateau? *Nature* 189: 276–78

Loubet, M., Montigny, R., Chahati, B., Duarte, N., Lambret, B., Martin, C., Thuizat, R. 1985. Geochemical and geochronological constraints on the geodynamical development of the Caribbean Chain of Venezuela. See Mascle 1985, pp. 553–66

MacDonald, W. D. 1976. Cretaceous-Tertiary evolution of the Caribbean. *Trans. Caribb. Geol. Conf., 7th, Guadeloupe*, pp. 69–78

MacDonald, W. D., Opdyke, N. D. 1984. Preliminary paleomagnetic results from Jurassic rocks of the Santa Marta massif, Colombia. See Bonini et al 1984, pp. 295–98

Mann, P., Burke, K. 1984a. Cenozoic rift formation in the northern Caribbean. *Geology* 12: 732–36

Mann, P., Burke, K. 1984b. Neotectonics of the Caribbean. *Rev. Geophys. Space Phys.* 22: 309–62

Mann, P., Burke, K., Matumoto, T. 1984. Neotectonics of Hispaniola. *Earth Planet. Sci. Lett.* 70: 311–24

Mann, P., Dolan, J. F., Keller, G., Shiroma, J. 1986. Paleogene basin formation in the Greater Antilles Arc–Bahama, platform collision zone: preliminary results. *Geol. Soc. Am., Ann. Meet. Expo., 99th, San Antonio, Tex.*, 18: 681 (Abstr.)

Mann, P., Schubert, C., Burke, K. 1988. Review of Caribbean neotectonics. In *Decade of North American Geology, Caribbean Region*. Boulder, Colo: Geol. Soc. Am. In press

Mascle, A., ed. 1985. *Géodynamique des Caraïbes*. Paris: Ed. Technip. 566 pp.

Mascle, A., Cazes, M., LeQuellec, P. 1985. Structure des marges et bassins Caraïbes: une revue. See Mascle 1985, pp. 1–20

Mattson, P. H., Pessagno, E. A. Jr. 1979. Jurassic and early Cretaceous radiolarians in Puerto Rican ophiolite—tectonic implications. *Geology* 7: 440–44

Maury, R. C., Westercamp, D. 1985. Variations chronologiques et spatiales des basaltes neogenes des Petites Antilles; implications sur l'evolution de L'arc. See Mascle 1985, pp. 77–89

Maze, W. B. 1984. Jurassic La Quinta formation in the Sierra de Perija, northwest Venezuela: geology and tectonic movement of red beds and volcanic rocks. See Bonini et al 1984, pp. 263–82

Maze, W. B., Hargraves, R. B. 1984. Paleomagnetic results from the Jurassic La Quinta formation in the Perija Range, Venezuela and their tectonic significance. See Bonini et al 1984, pp. 287–93

McCann, W. R., Sykes, L. R. 1984. Subduction of aseismic ridges beneath the Caribbean plate: implications for the tectonics and seismic potential of the northeastern Caribbean. *J. Geophys. Res.* 89: 4493–4519

Moore, G. F., Shipley, T. H., Lonsdale, P. F. 1986. Subduction erosion versus sediment offscraping at the toe of the Middle America Trench off Guatemala. *Tectonics* 5: 513–23

Moores, E. M. 1986. The Proterozoic ophiolite problem, continental emergence, and the Venus connection. *Science* 234: 65–68

Moretti, I., Ngokwey, K. 1985. Aseismic ridge subduction and vertical motion of overriding plate. See Mascle 1985, pp. 245–54

Mossakovskiy, A. A., Albear, J. F. 1978. Nappe structure of western and northern Cuba and history of its emplacement in the light of a study of olistrostromes and molasse. *Geotectonics* 12: 225–36

Muessig, K. W. 1984. Structure and Cenozoic tectonics of the Falcon Basin, Venezuela and adjacent areas. See Bonini et al 1984, pp. 217–30

Munro, S. E., Smith, F. D. Jr. 1984. The Urica fault zone, northeastern Venezuela. See Bonini et al 1984, pp. 213–16

Nur, A., Ben-Avraham, Z. 1982. Oceanic plateaus, the fragmentation of continents and mountain building. *J. Geophys. Res.* 87: 3644–61

Olsen, P. E., McCune, A. R., Thomson, K. S. 1982. Correlation of the early Mesozoic Newark Super Group by vertebrates, principally fishes. *Am. J. Sci.* 282: 1–44

Pallister, J. S. 1985. Pillow basalts from the Angayucham Range, Alaska: chemistry and tectonic implications. *Eos, Trans. Am. Geophys. Union* 66: 1102 (Abstr.)

Pindell, J. L. 1985. Alleghenian reconstruction and the subsequent evolution of the Gulf of Mexico, Bahamas and proto-Caribbean. *Tectonics* 4: 1–39

Pindell, J. L., Barrett, S. F. 1988. Geological evolution of the Caribbean region: a plate tectonic perspective. In *Decade of North American Geology, Caribbean Region*. In press

Pindell, J. L., Dewey, J. F. 1982. Permo-Triassic reconstruction of western Pangea and the evolution of the Gulf of Mexico/Caribbean region. *Tectonics* 1: 179–212

Pindell, J. L., Dewey, J. F., Cande, S. C., Pitman, W. C. III, Rowley, D. B., La Brecque, J. 1987. Plate-kinematic framework for models of Caribbean evolution. *Tectonophysics*. In press

Priem, H. N. A., Beets, D. J., Verdurmen, E. A. Th. 1986. Precambrian rocks in an early Tertiary conglomerate on Bonaire, Netherlands Antilles (South Caribbean borderland): evidence for a 300 km eastward displacement relative to the South American mainland? *Geol. Mijnbouw* 65: 35–40

Pszczolkowski, A. 1978. Geosynclinal

sequences of the Cordillera de Guaniguanico in western Cuba; their lithostratigraphy, facies development, and paleogeography. *Acta Geol. Pol.* 28: 1–96

Rainwater, E. H. 1967. Resumé of Jurassic to Recent sedimentation history of the Gulf of Mexico Basin. *Trans. Gulf Coast Geol. Soc.* 17: 179–210

Robertson, P., Burke, K. 1986. Evolution of a plate boundary hinge zone during the last 20 million years in the southeastern Caribbean. *Eos, Trans. Am. Geophys. Union* 67: 1210 (Abstr.)

Robertson, P., Burke, K., Wadge, G. 1985. Structure of the Melajo clay near Arima, Trinidad and strike-slip motion in the El Pilar fault zone. *Trans. Geol. Conf. Geol. Soc. Trinidad and Tobago, 1st,* pp. 21–33

Rosencrantz, E., Sclater, J. G. 1987. Depth magnetic anomalies and spreading in the Cayman Trough. *Eos, Trans. Am. Geophys. Union* 68: 405–6 (Abstr.)

Rosencrantz, E., Sclater, J. G., Mann, P. 1984. The age of the Yucatán basin. *Eos, Trans. Am. Geophys. Union* 65: 1101 (Abstr.)

Rosencrantz, E., Sclater, J. G., Borner, S. 1987. Basement depths and heat flow in Yucatán and Cayman Trough. In *Marine Heat Flow,* ed. J. A. Wright, K. E. Louden. Boca Raton, Fla: CRC Press. In press

Rosenfeld, J. H. 1981. *Geology of the western Sierra de Santa Cruz, Guatemala, Central America, an ophiolite sequence.* PhD thesis. State Univ. NY, Binghamton. 313 pp.

Rowe, D. W., Snoke, A. W. 1986. North coast schist group, Tobago, West Indies: important constraints on tectonic models for the southeastern Caribbean region. *Geol. Soc. Am. Abstr. With Programs* 18: 734 (Abstr.)

Saleeby, J. B. 1983. Accretionary tectonics of the North American Cordillera. *Ann. Rev. Earth Planet. Sci.* 11: 45–73

Salvador, A. 1986. Comments on "Neogene block tectonics of eastern Turkey and northern South America: continental applications of the finite difference method" by J. F. Dewey and J. L. Pindell. *Tectonics* 5: 697–701

Santamaria, F., Schubert, C. 1974. Chemistry and geochronology of the southern Caribbean–Northern Venezuelan plate boundary. *Geol. Soc. Am. Bull.* 85: 1085–98

Schermer, E., Howell, D. G., Jones, D. L. 1984. The origin of allochthonous terranes: perspectives on the growth and shaping of continents. *Ann. Rev. Earth Planet. Sci.* 12: 107–31

Schmidt-Effing, R. 1979. Geodynamic history of the oceanic crust in southern Central America. *Geol. Rundsch.* 68: 457–92

Sclater, J. G., Hellinger, S., Tapscott, C. 1977. The paleobathymetry of the Atlantic Ocean from the Jurassic to the present. *J. Geol.* 85: 509–22

Şengör, A. M. C. 1987. Tectonics of the Tethysides: orogenic collage development in a collisional setting. *Ann. Rev. Earth Planet. Sci.* 15: 213–44

Şengör, A. M. C., Altiner, D., Cin, A., Ustaomer, T., Hsu, K. J. 1987. Origin and assembly of the Tethyside orogenic collage at the expense of Gondwana-Land. *Proc. First Lyell Symp. Tethyside Gondwana-Land. Geol. Soc. London Spec. Publ.* In press

Senn, A. 1940. Paleogene of Barbados and its bearing on the history and structure of the Antillean-Caribbean region. *Am. Assoc. Pet. Geol. Bull.* 24: 1548–1610

Shepherd, A., Hall, S., Burke, K. 1987. Magnetic anomalies indicate Gulf of Mexico originated by counterclockwise rotation of Yucatán from Gulf Coast. *Am. Assoc. Pet. Geol. Conv., Los Angeles, Calif.,* p. 614 (Abstr.)

Sheridan, R. E., Gradstein, F. M., et al. 1983. *Initial Reports of the Deep Sea Drilling Project,* Vol. 76. Washington, DC: US Govt. Print. Off.

Silver, L. T., Anderson, T. L. 1974. Possible left-lateral early to middle Mesozoic disruption of the southwestern North American craton margin. *Geol. Soc. Am. Abstr. With Programs* 6: 955 (Abstr.)

Somin, M. L., Millan, G. 1972. The metamorphic complexes of the Isle of Pines, Escambray, and Oriente of Cuba and their age. *Izv. Akad. Nauk SSSR Ser. Geol.* 5: 48–57 (In Russian)

Spadea, P., Delaloye, M., Espinosa, A., Orrego, A., Wagner, J. J. 1987. Ophiolite complex from La Tetilla, southwestern Colombia. *J. Geol.* 95: 377–95

Speed, R. C. 1983. Structure of the accretionary complex of Barbados, 1: Chalky Mount. *Geol. Soc. Am. Bull.* 94: 92–116

Speed, R. C. 1985. Cenozoic collision of the Lesser Antilles arc and continental South America and origin of the El Pilar fault. *Tectonics* 4: 41–69

Speed, R. C., Larue, D. K. 1982. Barbados: architecture and implications for accretion. *J. Geophys. Res.* 87: 3633–43

Speed, R. C., Larue, D. K. 1985. Tectonic evolution of Eocene turbidites of Grenada. See Mascle 1985, pp. 101–8

Speed, R. C., Westbrook, G. K. 1984. Lesser Antilles Arc and adjacent terranes. *Ocean Margin Drilling Program, Regional Atlas Ser., Atlas 10.* Woods Hole, Mass: Mar. Sci. Int. 27 pp.

Stein, S., Wiens, D. A., Engeln, J. F., Fujita, K. 1986. Comment on "Subduction of a-

230   BURKE

seismic ridges beneath the Caribbean plate: implications for the tectonics and seismic potential of the northeastern Caribbean" by W. R. McCann and L. R. Sykes. *J. Geophys. Res.* 91: 784–86

Stoiber, R. E., Carr, M. J. 1974. Quaternary volcanic and tectonic segmentation of Central America. *Bull. Volcanol.* 37: 304–25

Torrini, R. Jr., Speed, R. C. 1987. Structure and tectonics of the forearc basin/accretionary prism transition, Lesser Antilles forearc. *Tectonics.* In press

Valery, P., Nely, G., Mascle, A., Biju-Duval, B., LeQuellec, P., et al. 1985. Structure et croissance d'un prisme, d'accrétion tectonique proche d'un continent: La Ride de la Barbade au sud de l'arc antillais. See Mascle 1985, pp. 172–86

Vierbuchen, R. C. 1984. The geology of the El Pilar fault zone and adjacent areas in northeastern Venezuela. See Bonini et al 1984, pp. 189–212

Vogt, P. R., Lowrie, A., Bracey, D. R. 1976. Subduction of aseismic oceanic ridges: effects on shape, seismicity and other characteristics of consuming plate boundaries. *Geol. Soc. Am. Spec. Pap. 172.* 59 pp.

von Huene, R., Aubouin, J., Arnott, R. J., Baltuck, M., Bourgois, J., et al. 1982. *Initial Reports of the Deep Sea Drilling Project*, Vol. 84. Washington, DC: US Govt. Print. Off.

Wadge, G., Burke, K. 1983. Neogene Caribbean plate rotation and associated Central American tectonic evolution. *Tectonics* 2: 633–43

Wadge, G., Draper, G., Lewis, J. F. 1984. Ophiolites of the northern Caribbean: a reappraisal of their roles in the evolution of the Caribbean plate boundary. In *Ophiolites and Oceanic Lithosphere*, ed. I. G. Gass, S. J. Lippard, A. W. Shelton, pp. 367–80. London: Geol. Soc. London

Wassal, H. 1956. The relationship of oil and serpentine in Cuba. *Rep. Int. Geol. Congr., 20th, Mexico,* 3: 65–77

Westercamp, D., Andreieff, P., Bouysse, P., Mascle, A., Baubron, J. C. 1985. The Grenadines, southern Lesser Antilles. Part 1: stratigraphy and volcano-structural evolution. See Mascle 1985, pp. 109–18

White, G. W. 1980. Permian-Triassic continental reconstruction of the Gulf of Mexico–Caribbean area. *Nature* 283: 823–26

Wilson, J. T. 1966. Are the structures of the Caribbean and Scotia arc regions analogous to ice rafting? *Earth Planet. Sci. Lett.* 1: 335–38

Wilson, J. T. 1968. Static or mobile earth: the current scientific revolution. *Proc. Am. Philos. Soc.* 112: 309–20

Winterer, E. L. 1976. Marine geology and tectonics, anomalies in the tectonic evolution of the Pacific. In *The Geophysics of the Pacific Ocean Basin and its Margins*, ed. G. H. Sutton, M. H. Manghnani, R. Moberly, pp. 269–78. Washington, DC: Am. Geophys. Union

Wong, Th. E., Van Lissa, R. V. 1978. Preliminary report on the occurrence of Tertiary gold-bearing gravels in Surinam. *Geol. Mijnbouw* 57: 365–68

Ann. Rev. Earth Planet. Sci. 1988. 16: 231–49

THE EARTH'S ROTATION

John M. Wahr

Department of Physics and Cooperative Institute for Research in Environmental Sciences, Campus Box 390, University of Colorado, Boulder, Colorado 80309

INTRODUCTION

The Earth's rotation is not constant. Instead, both the rate of rotation and the position of the rotation axis vary with time. Changes in the rotation rate are directly proportional to changes in the length of a day (LOD). In addition, the time integral of the LOD variability is proportional to fluctuations in Universal Time, the measure of time as determined by the overhead transits of celestial objects.

Variations in the position of the rotation axis are usually classified either as "polar motion" or as "nutation," where "polar motion" describes motion of the axis with respect to the Earth's surface, and "nutation" denotes motion of the axis with respect to inertial space. The distinction between polar motion and nutation is somewhat artificial, since, in general, nutation cannot occur without some accompanying polar motion, and vice versa. In practice, though, axis motion caused by an individual excitation process is mostly either one or the other, depending on the time scale. Excitation at periods much longer than one day as seen by an observer on the Earth causes mostly polar motion: The rotation axis does not move much with respect to inertial space compared with its motion with respect to the Earth. Thus, since processes originating within the Earth capable of affecting rotation generally have long time scales, they cause polar motion. Conversely, excitation with a nearly diurnal (retrograde) period as seen from the Earth causes axis motion that is mostly nutation. For example, the gravitational attraction of the Sun and Moon causes nutational motion, since the Sun and Moon have nearly diurnal periods as seen from the Earth.

This article is a survey of rotation observations and, especially, of the geophysical implications of those observations for all three types of

231

0084–6597/88/0515–0231$02.00

variations in rotation: LOD, polar motion, and nutation. The subject of rotation touches on diverse fields in the geophysical sciences, including solid Earth geophysics, meteorology, and oceanography, and only a brief summary is presented here. More detailed descriptions can be found in Munk & MacDonald (1975), Lambeck (1980), and Rochester (1984).

OBSERVATIONAL TECHNIQUES AND RESULTS

Variations in rotation are detected by observing the apparent motion of objects in space from fixed points on the Earth. Until recently, all such observations involved using optical telescopes to monitor the apparent angular positions of stars. Detailed observations of this sort were made over the last century or more. Furthermore, long-period terms in the LOD over the last few hundred years and a linear trend over the last two to three thousand years have been resolved from historical records of eclipses and planetary occultations. For example, recorded solar times of Babylonian eclipses differ by up to several hours from the solar times predicted using the present positions of the Sun and Moon and the assumption that the mean rotation rate has remained constant over the last few thousand years (see, for example, Brosche & Sundermann 1978).

Within the last decade or two, however, several new techniques have been implemented that have significantly greater accuracies. These include lunar laser ranging, satellite laser ranging, and very-long-baseline interferometry (VLBI). Lunar laser ranging (LLR) involves the measurement of the distance between powerful Earth-based lasers and the Moon by recording the round-trip travel time of laser pulses reflected from mirrors on the lunar surface. In satellite laser ranging a satellite is tracked by measuring the round-trip travel time of laser pulses originating from the Earth and reflected from mirrors attached to the outside of the satellite. Currently, the most accurate satellite results come from tracking the satellite LAGEOS. The VLBI technique uses widely separated radio antennas to detect signals from distant astronomical radio sources. By comparing the recordings of the same signals detected at two antennas, the length and orientation of the baseline vector between the antennas can be determined.

Although all three of these techniques provide excellent rotation results, VLBI is probably the most versatile. Satellite laser ranging, for example, is not presently as accurate at long periods due to uncertainties in the satellite orbit, although short-period variability is very well determined. LLR has not yet had enough ground-based lasers in simultaneous operation to give routine, reliable polar motion data, although the LOD results have been excellent. Furthermore, both these techniques are clearly inferior to VLBI for determining nutational motion because VLBI is tied directly to an inertial coordinate system defined by the distant radio sources.

The accuracies of the results from these new techniques are improving rapidly. Currently, the LOD can probably be determined from all three techniques to better than 0.1 ms, and polar motion from VLBI and satellite laser ranging to better than 0.002 arcsec, where both these numbers refer to values averaged over 3–5 days or less. The results improve substantially when longer averaging times are used (for an assessment of the accuracies, see Robertson et al 1985). VLBI nutation results at specific nutation frequencies are accurate to better than 0.2 milliarcseconds (mas).

Each of the new techniques involves observations of the propagation times of electromagnetic waves passing through the Earth's atmosphere. The results are, thus, sensitive to uncertainties in the atmosphere's index of refraction. This, in fact, is currently the limiting source of error for the VLBI results. (The error is mostly related to the uncertainty in the atmospheric water vapor content.) However, these errors are nowhere near as large as the refraction errors in the traditional stellar optical results. Uncertainties in the index of refraction have a much greater relevant effect on apparent angular positions than on propagation times. Still, the old results from the stellar optical technique are invaluable when investigating variability at decade and longer time scales.

The observational results for the LOD fall roughly into three categories. First, there is a linear increase in the LOD of about 2 ms per century, as determined from the ancient astronomical record. Second, there are irregular decade fluctuations of about 4 to 5 ms over 20 to 30 yr. Figure 1 shows the sum of the linear increase and the irregular fluctuations, as determined from about 150 yr of astronomical data. Note that even over this long time period it is impossible to cleanly separate the linear increase from the decade fluctuations.

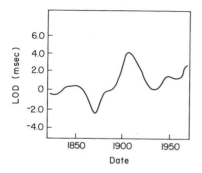

Figure 1   The length of day (LOD) is variable over a wide range of time scales. At long periods are decade fluctuations (caused by the transfer of angular momentum between the Earth's fluid core and solid mantle) and a linear increase in the LOD (caused by a combination of tidal friction in the oceans and the effects of the last ice age). Here we show astronomical results for the long-period variability during 1820–1975, using data from Morrison (1979). Note that even 150 yr of data is inadequate to cleanly separate the linear increase in the LOD from the decade fluctuations.

The third category includes those variations in the LOD with periods shorter than about 5 yr. The solid line in Figure 2 shows a typical example of the short-period LOD variability, as determined from a combination of LLR, VLBI, and LAGEOS results (the effects of tides have been removed, as described below) during 1982–86. Although the linear increase and decade fluctuations have not been removed from the results shown in Figure 2, they are only marginally evident in the data, since the latter cover such a short time span.

The observed variability of polar motion is much less complex. As an example, VLBI results for the position of the pole from November 1983 to April 1987 are shown in Figure 3. Note that the pole follows a roughly circular, counterclockwise path about its mean position. This short-period variability can be separated into an annual oscillation (the "annual wobble") and a 14-month oscillation (called the "Chandler wobble" after the American astronomer S. C. Chandler, who first reported the motion in 1891), both with amplitudes of about 0.1 arcsec. At long periods there is also evidence from a century or so of optical data for a linear drift of the rotation axis and for perhaps a 30-yr periodic variation (Dickman 1981). No other significant variability has been observed.

Nutational motion occurs at discrete, nearly diurnal frequencies determined by the orbital periods of the Earth and Moon. The amplitudes at these frequencies are sensitive to details of the Earth's internal structure, as we discuss in more detail below.

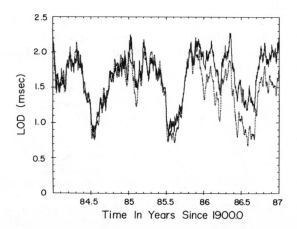

Figure 2 The atmosphere is responsible for much of the short-period variability in the LOD. Here, results from atmospheric wind and pressure data are compared with 1982–86 LOD data obtained with a simultaneous solution using VLBI, LLR, and LAGEOS observations. (Results provided by Marshall Eubanks.) The effects of tides have been removed from the LOD results.

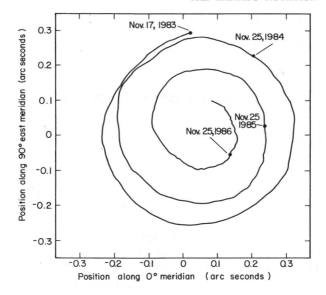

Figure 3   The Earth's rotation axis does not remain fixed with respect to the Earth's surface. The motion of the pole from 17 November 1983 through 19 April 1987 is shown above, using VLBI results that have been reduced by the US National Geodetic Survey. The *x*- and *y*-coordinates represent the amount the pole is tipped along the Greenwich and 90°E meridians, respectively. The pole moves in a counterclockwise direction along a roughly circular path. This motion is a superimposition of annual and 14-month oscillations.

ROTATION THEORY

In the following sections, we discuss what can be learned about the Earth from the observational results. The initial step in understanding an individual variation is to identify the excitation source. Then, the observed fluctuation in rotation can be used either to learn more about the excitation process or, if that process is already well enough understood, to learn about the Earth's deformational response to the excitation source (the amount of deformation sometimes affects the rotational response, as is described below) and thereby to help constrain our knowledge of material properties and structure within the Earth. In this section, we discuss how to compute the effects of a given excitation process on polar motion and the LOD. The theory of nutation is more complicated and is briefly described in a later section.

Suppose we are in a coordinate system rotating with respect to inertial space with angular velocity vector $\omega(t)$. Let $\mathbf{H}(t)$ be the angular momentum vector of the Earth as seen in our coordinate system, and let $\mathbf{L}(t)$ be the external torque on the Earth. Then

$$\partial_t H + \omega \times H = L, \tag{1}$$

and H has the form

$$H = I \cdot \omega + h, \tag{2}$$

where I is the Earth's time-dependent inertia tensor, and

$$h = \int_{earth} \rho r \times v \, d^3 r \tag{3}$$

is the relative angular momentum, representing the net contribution to H of all motion relative to the coordinate system. [The variables ρ, r, and v in (3) are the internal density, particle coordinate, and particle velocity, respectively.]

Let us now remove the arbitrariness of ω and define it as the time-dependent, mean rotation vector of the mantle, which is the quantity detected by the observations. For this choice of ω, there is no net contribution to h from motion in the mantle. There may, however, be contributions to h from the core, the atmosphere, or the oceans.

Suppose that in its equilibrium state, the entire Earth is rotating about the $\hat{z}$ axis with uniform angular velocity $\Omega = \Omega \hat{z}$. In this state, the Earth is elliptical and has diagonal inertia tensor

$$I_0 = \begin{pmatrix} A & 0 & 0 \\ 0 & A & 0 \\ 0 & 0 & C \end{pmatrix}, \tag{4}$$

where A and C are the Earth's principal moments of inertia and differ by about one part in 300. The Earth's equilibrium angular momentum is then

$$H_0 = I_0 \cdot \Omega. \tag{5}$$

Now we do something to perturb the Earth's angular momentum. We can accomplish this by exerting an external torque L on the Earth, by changing the Earth's inertia tensor to

$$I = I_0 + \begin{pmatrix} c_{11} & c_{12} & c_{13} \\ c_{21} & c_{22} & c_{23} \\ c_{31} & c_{32} & c_{33} \end{pmatrix}, \tag{6}$$

or by inducing motion in the fluid portions of the Earth so as to give rise to a relative angular momentum h. The result of any of these perturbations is that ω must change so that (1) and (2) remain satisfied.

Let the perturbed rotation vector for the mantle be

$$\omega = \Omega + \Omega\mathbf{m} = \Omega + \Omega\begin{pmatrix} m_1 \\ m_2 \\ m_3 \end{pmatrix}. \tag{7}$$

Here, m_1 and m_2 represent polar motion (two parameters are needed to describe polar motion, since it takes two parameters to define the angular position of an axis passing through the origin), and m_3 represents a change in the LOD. (The change in the LOD is $-m_3 2\pi/\Omega$.)

Using (2), (4), (6), and (7) in (1) and ignoring second-order terms in the perturbation gives

$$\partial_t m_3 = -\partial_t\left(\frac{h_3}{\Omega C} + \frac{c_{33}}{C}\right) + \frac{L_3}{\Omega C}, \tag{8}$$

$$\Omega A \partial_t m_+ - i\Omega^2(C-A)m_+ + (\partial_t + i\Omega)(h_+ + \Omega c_+) = L_+, \tag{9}$$

where $m_+ = m_1 + im_2$, $h_+ = h_1 + h_2$, $c_+ = c_{13} + ic_{23}$, and $L_+ = L_1 + iL_2$. Here (8) and (9) represent equations for the variation in the LOD and for polar motion, respectively. Note that (9) is a complex equation for m_+ and so represents two real equations for m_1 and m_2.

These two equations allow us to find m_3 and m_+ once we know L_i, h_i, and c_{i3} for a given excitation process. L_i can be found directly from sufficient knowledge of the excitation. However, finding h_i and c_{i3} is more involved, because these quantities depend not only on the mass redistribution and relative motion associated with the excitation process, but also on $\mathbf{m}$. This is because the incremental centrifugal force caused by a change in rotation deforms the Earth.

It is convenient to include this rotational deformation separately by writing c_{i3} and h_i as sums of terms dependent directly on the excitation process and of terms dependent on the components of $\mathbf{m}$. The former terms will be denoted here by $\overline{c_{i3}}$ and $\overline{h_i}$, and the latter terms are linear in $\mathbf{m}$, to first order, and vanish for a nondeformable Earth.

The effects of the $\mathbf{m}$-dependent terms turn out to be negligible on the m_3 equation (8). Thus (8) is approximately valid as written, with h_3 and c_{33} replaced by $\overline{h_3}$ and $\overline{c_{33}}$. The resulting equation is easy to integrate for m_3, once L_3, $\overline{h_3}$, and $\overline{c_{33}}$ are known. The problem of modeling LOD variability, then, reduces to finding the latter quantities from knowledge of the excitation process.

The $\mathbf{m}$-dependent terms *are* important, however, in the polar motion equation (9). After modeling and separating out these terms, transforming to the frequency domain where all time dependence is assumed to be of the form $\exp(i\lambda t)$ (where λ is the angular frequency), and assuming that the forcing period is much longer than one day so that $\lambda \ll \Omega$, Equation (9) reduces to

$$m_+ = -\frac{1}{\Omega A_m}\left[\frac{iL_+ + \Omega^2\overline{c_+} + \Omega\overline{h_+}}{\lambda - \lambda_{CW}}\right],\tag{10}$$

where λ_{CW} is the Chandler wobble eigenfrequency (see the discussion in a later section) given by

$$\lambda_{CW} = \Omega\left[\frac{C - A - \kappa\Omega^2 a^5/3G}{A_m}\right].\tag{11}$$

Here A_m is a principal moment of inertia of the mantle, G is Newton's gravitational constant, and κ is a dimensionless parameter describing the deformation induced by the incremental centrifugal force and depending on the elastic and anelastic parameters within the Earth.

Equation (10) implies that the frequency spectrum for polar motion should be resonant at $\lambda = \lambda_{CW}$, about one cycle per 14 months. By using observations to estimate the resonance frequency, κ can be determined. In fact, the resonant frequency can be estimated without detailed knowledge of L_+, $\overline{h_+}$, or $\overline{c_+}$. If these quantities are reasonably independent of frequency near $\lambda = \lambda_{CW}$, then λ_{CW} can be determined directly from the observed frequency dependence of polar motion near λ_{CW}.

INTERPRETATION

Tidal Friction

Most of the observed linear trend in the LOD is due to gravitational tides in the Earth and oceans caused by the Moon and Sun. The Moon, for example, deforms the Earth and oceans into the ellipsoidal shape shown greatly exaggerated in Figure 4. The orientation of the ellipsoidal bulge is fixed with respect to the Moon, while the Earth rotates at 1 cycle day^{-1} relative to that bulge. The resulting lunar tides are time dependent, with frequencies equal to integral multiples of 1 cycle day^{-1}, modulated by the frequencies of the lunar orbit, such as 1 cycle per 27.7 days and 1 cycle per 13.7 days.

If there were no energy dissipation in the Earth and oceans, the ellipsoidal tidal bulge shown in Figure 4 would be oriented exactly toward the Moon. However, since there is some dissipation, the Earth and oceans take a short time to fully respond to the Moon's gravitational force. The maximum tidal uplift occurs shortly after the Moon is overhead, and the bulge leads the Earth-Moon vector by a small angle δ, shown greatly

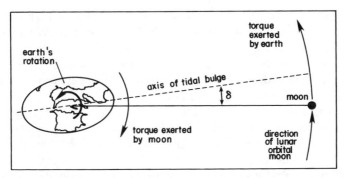

Figure 4   The gravitational force from the Moon deforms the Earth as shown here (greatly exaggerated), producing tides in the solid Earth and oceans. Because of energy dissipation in the oceans, the tidal bulge leads the Earth-Moon vector by the small angle δ. The Moon's gravitational force acts on the bulge to produce a clockwise torque on the Earth, and so to increase the LOD. The bulge causes a counterclockwise torque on the Moon, leading to an increase in the Moon's orbital period.

exaggerated in Figure 4. The Moon's gravitational force acts on the tidal bulge to produce a clockwise torque on the Earth [a time-independent L_3 in Equation (8)], opposite to its rotation. The result is a steady decrease in the rotation rate, and thus an increase in the LOD. There is a similar, although somewhat smaller, effect from the Sun.

Most of the tidal energy dissipation is believed to occur in the oceans. Frictional effects are much more important there than in the solid Earth. It is still not entirely clear, though, whether most of the dissipation occurs in shallow seas or in the deep ocean, or what the dominant frictional mechanisms are.

The lag angle δ can be determined independently of observations of the LOD by ranging to satellites such as LAGEOS. The tidal bulge perturbs the orbit of a satellite, and so δ can be found by solving for the orbit. When the satellite results for δ are used to predict the lunar torque on the Earth, the expected increase in the LOD is about 25% larger than that implied by the historical eclipse record (see, for example, Goad & Douglas 1978, Cazenave & Daillet 1981).

There is other evidence tending to confirm the satellite results. The Earth's tidal bulge acts gravitationally on the Moon, causing a counter-clockwise torque on the Moon in its orbit about the Earth (see Figure 4). The torque is in the direction of the Moon's motion, and so it tends to increase the angular momentum of the Moon. The rate of increase of lunar orbital angular momentum must equal the rate of decrease of the Earth's rotational angular momentum.

The increase in lunar angular momentum causes the Moon to move farther away from the Earth and to increase its orbital period. This increase in period has been determined accurately from LLR data (Williams et al 1978), and the results predict a decrease in the Earth's rotational angular momentum consistent with the satellite estimates of δ as described above but inconsistent with the astronomical results. The likely explanation for the discrepancy is discussed in the next section.

The observed effects on the lunar orbit are surprisingly large. When the dissipation rate inferred from the LLR and satellite ranging results are used in models to extrapolate the present lunar orbit backward in time, the Moon is predicted to have been so close to the Earth 1.5 Gyr ago that it would have been torn apart by gravitational forces from the Earth. The Moon, though, is known to be over 4 Gyr old.

The implication is that tidal friction in the oceans is larger now than it has been over most of the Earth's history. The dissipation is sensitive to the shape of the ocean basins and to the rotation rate itself. Ocean basins, for example, have changed drastically over geological time as a result of continental drift. Whether these effects are large enough to sufficiently affect the oceanic dissipation is currently receiving attention (see, for example, Brosche & Sundermann 1982).

Postglacial Rebound

LOD   The 25% discrepancy between the historical astronomical evidence for the increase in the LOD and the LLR and satellite ranging results for the effects of tidal dissipation implies that some other mechanism is tending to decrease the LOD, thus partially offsetting the effects of tidal friction. This acceleration of the Earth is probably caused by the effects of the last ice age. When the ice over northern Canada and Scandinavia melted several thousand years ago, it left deep depressions now filled by Hudson's Bay and the Baltic Sea. The Earth behaves as a viscous fluid over long time periods, and the depressed areas are slowly uplifting as material deep within the Earth flows horizontally. There is thus a net transfer of material within the Earth toward higher northern latitudes. This decreases the Earth's polar moment of inertia [c_{33} in (8)] and so increases the rotation rate.

This interpretation has recently been independently confirmed using LAGEOS ranging data. The changing internal mass distribution leads to a change in the Earth's gravitational field, which affects the LAGEOS orbit. By solving for the orbit, the change in the moment of inertia can be determined (Yoder et al 1983, Rubincam 1984). The results are consistent with the additional linear decrease in the LOD inferred from the ancient historical record.

The linear decrease in the Earth's moment of inertia depends on the rate at which material is flowing inside the Earth, which in turn depends on the viscosity of the Earth. In fact, the observed linear change in the moment of inertia has been used to place tight bounds on the viscosity of the Earth's lower mantle (see, for example, Peltier 1983).

POLAR MOTION   Postglacial rebound may also be responsible for the linear drift of the pole suggested by the astronomical polar motion data taken over the last century. The horizontal readjustment of material within the Earth causes a steady drift of the Earth's figure axis [represented by $\overline{c_+}$ in (10)] relative to the Earth's surface. To conserve angular momentum, the mean position of the Earth's rotation axis (represented by m_+) remains coincident with the figure axis, and so the pole also drifts. The rate of drift implied by the astronomical data has been used as an additional constraint on mantle viscosity (Yuen et al 1983).

Decade Fluctuations

LOD   The decade fluctuations in the LOD are believed to be due to the transfer of angular momentum between the fluid core and the solid mantle. When the mantle gains angular momentum its rotation rate increases, and so the observed LOD decreases. This variability of m_3 can be computed from (8) either by estimating h_3 from assumptions about core flow or by estimating the torques L_3 responsible for the exchange of angular momentum.

At least two viable mechanisms have been proposed to explain the required torques. One is electromagnetic forcing. The Earth's magnetic field is caused by electric currents in the core. If these currents change with time, there will be changes in the magnetic field and so, by Faraday's Law, an electric field will be produced everywhere, including in the lower mantle. Since the lower mantle is an electrical conductor, the induced electric field gives rise to electric currents in the mantle that interact with the large, time-independent components of the magnetic field through the Lorentz force. The result is, in general, a net torque on the mantle and a resulting change in rotation. Stix & Roberts (1984) found that the electromagnetic torque is probably the right order to explain the observed LOD variability.

An alternative mechanism is topographic coupling, caused by fluid pressure acting against topography at the core-mantle boundary. This idea was first proposed by Hide (1969), but at that time it was not possible to meaningfully estimate the strength of the coupling. Recent developments, however, have now made such estimates feasible. Speith et al (1986) used models for fluid velocities at the top of the core from Voorhies (1986), together with the assumption of geostrophy, to estimate lateral variations in pressure at the core-mantle boundary. They combined their estimates with seismic-related results for the shape of the boundary to estimate the zonal topographic torque on the mantle from the core. Their results are the right order to explain the observed decade fluctuations. (In fact, they are several times too large.)

Whatever the nature of the torque, the assumption that the decade fluctuations are due to core flow suggests that there ought to be some correlation between the observed LOD variability and the observed time dependence in the Earth's magnetic field. It has proven difficult to find such a correlation because the magnetic field variations are attenuated as they travel upward through the conducting mantle. In fact, when correlations have been identified, the time lag between the changes in the LOD and in the observed magnetic field has been used to help constrain the mantle's electrical conductivity (see, for example, Backus 1983).

POLAR MOTION   The one conceivable decade-scale variation in polar motion is the 30-yr oscillation suggested by the astronomical polar motion record. It is not clear what could cause such a variation. It is even possible that this motion might simply reflect poorly modeled local deformation at the telescopes, rather than an actual variation in the position of the rotation pole. The problem, at the moment, is far from resolved.

Short-Period Fluctuations

LOD   The observed short-period variability in the LOD includes fluctuations at monthly and fortnightly periods caused by the lunar tides. Figure 5 is another view of the tidal bulge in the Earth, again greatly exaggerated. As the Moon orbits the Earth, the bulge remains continually

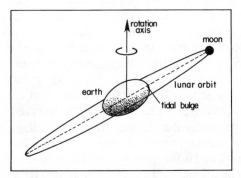

Figure 5   The tidal bulge as seen from the equator. The bulge follows the Moon in its orbit about the Earth, causing fluctuations in the Earth's polar moment of inertia at the lunar orbital periods of 27.7 and 13.7 days. Because of conservation of angular momentum, there are corresponding oscillations in the LOD.

oriented along the Earth-Moon vector. When the Moon is at high declinations, as it is in Figure 5, the Earth's polar moment of inertia [c_{33} in (8)] is decreased slightly as a result of the shift of the bulge away from the equator, and the Earth rotates more quickly. When the Moon lies in the Earth's equatorial plane, the tidal bulge is also in that plane, the polar moment of inertia increases slightly, and the rotation rate decreases. There are also variations in the size of the bulge, and so in the rotation rate, as the distance to the Moon changes during the lunar orbit.

The result is variability in the LOD at the various periods that characterize the Moon's orbit. The largest of these LOD fluctuations occur at 27.7 and 13.7 days, since these are the principal orbital periods. The amplitudes of the fluctuations depend on the Earth's material properties. For example, if there was large energy dissipation within the solid mantle at these periods, the mantle would behave more like a fluid than it would for small dissipation rates, and so the amplitudes of c_{33} and of the LOD variability would be larger: A fluid is more easily deformed than a solid. Dissipation would also introduce a phase lag between the LOD fluctuations and the motion of the Moon. These effects, particularly the increase in amplitude, have been used to help learn about the anelastic behavior of the mantle at these periods (Merriam 1984).

The remaining short-period fluctuations in the LOD consist of large 6- and 12-month periodic terms and smaller, more irregular variations at other periods. Tidal deformation from the Sun is responsible for about 10% of the annual and about 50% of the semiannual variability through the same mechanisms as for the fortnightly and monthly oscillations. There are also small seasonal contributions (less than 5%) from the effects of ocean currents, mostly from the circumpolar current around Antarctica, and probably near-negligible contributions from seasonal changes in ground-water storage.

Instead, most of the annual and semiannual variability is caused by seasonal forcing from the atmosphere, particularly by the seasonal exchange of angular momentum between the solid Earth and atmospheric winds (Lambeck & Hopgood 1981). For exmaple, when the winds increase in strength from west to east, the Earth slows down. This exchange of angular momentum is accomplished by a combination of surface friction torques (due to viscous drag as the winds blow over the surface) and mountain torques (caused by higher pressure on one side of a topographic feature than on the other). These two torques contribute about equally to the coupling (Wahr & Oort 1984).

Seasonal variations in global atmospheric pressure are less important than winds, but they do contribute about 10% of the observed semiannual and annual LOD variability. A change in the atmospheric pressure at a given point implies a change in the amount of atmospheric mass vertically above that point. Pressure data thus reflect the mass distribution in the atmosphere and can be used to determine the atmosphere's polar moment of inertia. Because of conservation of angular momentum, seasonal variations in the moment of inertia $[c_{33}(t)]$ are accompanied by fluctuations in the LOD.

The total effects of the atmosphere can be accurately estimated using global wind and pressure data to find $h_3(t)$ and $c_{33}(t)$, respectively, and then applying these values in (8). These effects are compared in Figure 2 with LOD results for 1982–86. (The effects of tides have been removed from the LOD results.) The agreement is remarkable (see also Barnes et al 1983, Rosen & Salstein 1983, Dickey et al 1986). In fact, the atmosphere appears to be responsible for most of the nonseasonal variations as well. For example, the large maximum during the winter of 1982–83 is probably associated with the extreme El Niño event in the southern Pacific, which has often been blamed for the unusual weather patterns occurring around the globe during that period.

There is also evidence in both data sets of a 50-day oscillation (Langley et al 1981). It is not certain why such an oscillation should exist, but the good agreement between the two data sets suggests that the term is probably real. Meteorologists are presently trying to understand its origin.

POLAR MOTION   The annual wobble evident in polar motion data is mostly due to the effects of annual redistribution of mass within the atmosphere (see, for example, Merriam 1982, Wahr 1983). This causes a perturbation in the inertia tensor of the atmosphere $[\overline{c_+}$ in (10)], which leads to a shift in the position of the rotation pole (m_+). In fact, most of the observed annual wobble is related to the large seasonal atmospheric pressure variation over central Asia: high pressure in winter and low pressure in summer.

Ground-water storage is also important in exciting the annual wobble. The Earth's inertia tensor, and so the position of the rotation pole, are affected by seasonal variations in the amount of water in snow and ice, in the water table, in rivers and lakes, and in the oceans. The effects of water storage are roughly 25% of the effects of atmospheric pressure (Van Hylckama 1970, Hinnov & Wilson 1987). The effects of wind and ocean currents are believed to be negligible (Wahr 1983).

The 14-month Chandler wobble is a free motion of the Earth. An analytical expression for its frequency is given above in Equation (11). The motion is analogous to the free nutation of a top. If the figure axis of a rapidly spinning top is initially displaced slightly from the rotation axis, the figure axis will proceed to move along a conical path about the rotation axis. The frequency of the motion depends on how nonspherical the top is, an effect evident in (11) through the $C-A$ term in the numerator.

The Earth's nonrigidity is responsible for the κ term in (11), which lengthens the period of the Chandler wobble by about 4 months, from the 10-month period expected for a rigid Earth to the observed 14 months. In fact, observations of the Chandler wobble period and decay rate (after initial excitation, the motion damps out in several decades) have been used to solve for κ and thus to constrain the value of mantle anelasticity pertinent to a 14-month period (Smith & Dahlen 1981).

Thus, the period and decay rate of the Chandler wobble are now well understood on theoretical grounds. The primary excitation source, however, has not yet been identified. The problem is to find a mechanism that can produce a large enough offset between the figure and rotation axes to excite the Chandler wobble to observed levels. Fluctuations in atmospheric pressure probably provide only about 25% of the necessary power (Wilson & Haubrich 1976, Wahr 1983). Other effects, including perturbations in the inertia tensor due to earthquakes, appear to be even less important (Dahlen 1973). A recent, intriguing hypothesis is that the excitation may be due to fluid pressure at the top of the core, acting against the elliptical core-mantle boundary to produce a torque on the mantle (Gire & Le Mouel 1986). Still, not much more is known now about the excitation source than was known in Chandler's time, nearly 100 years ago.

Nutations

As described in the introduction, the Earth's nutational motion is caused by the gravitational attraction of the Sun and Moon. The motion can be separated into a discrete sum of periodic terms with frequencies, as seen from the diurnally rotating Earth, of 1 cycle day^{-1} modulated by the lunar and solar orbital frequencies.

The Earth is believed to have a rotational normal mode, called the free core nutation (FCN), with an eigenfrequency within the diurnal band of nutation frequencies. To understand the dynamics of the FCN, suppose the fluid core and solid mantle were tipped about an equatorial axis in opposite directions and then released. If the core's internal density distribution and the core-mantle boundary were spherical, there would be no restoring torque, and the core and mantle would remain tipped relative to each other. But, because the real core/mantle boundary is elliptical and the Earth is elliptically stratified, there are, instead, restoring pressure and gravitational torques between the core and mantle. As a result, the core and mantle execute periodic twisting motion with respect to each other.

This motion is the FCN, and its frequency is 1 cycle day^{-1} plus a factor dependent on the strength of the restoring torque.

Sasao et al (1980) showed that for a hydrostatically prestressed Earth, the FCN eigenfrequency has the form

$$\lambda_{FCN} = \omega \left[1 + \frac{A}{A_m} (e_f - \beta) \right],\tag{12}$$

where $e_f \cong (C_f - A_f)/A_f$ is the dynamical ellipticity of the core, with C_f and A_f the principal moments of the fluid core about the polar and equatorial axes, respectively; and β is a numerical factor that represents the effects of deformation and is effectively independent of any aspherical stratification within the Earth. For a hydrostatically prestressed Earth, β is about 25% of e_f (Sasao et al 1980), and $\lambda_{FCN} \cong (1 + 1/460)$ cycles day^{-1} (Wahr 1981).

Wahr (1987) showed that (12) is valid even if the mantle is not hydrostatically prestressed. In this case, the ellipticity of the core-mantle boundary need not equal the hydrostatic value, and in fact the boundary need not be an ellipse at all. Instead, its radius could depart from its mean spherical value with any arbitrary latitudinal and longitudinal dependence, so long as this departure is small.

As it happens, though, the quantity e_f in (12) depends only on the Y_2^0 spherical harmonic term of the core internal density field and on the Y_2^0 component of the shape of the core-mantle boundary. It does not depend on any other spherical harmonic component of the structure. Thus, observational results for λ_{FCN} could yield information on coefficients of this one spherical harmonic.

The free nutational motion at the period λ_{FCN} has not yet been clearly observed. Evidently, it is not excited sufficiently by any internal process. However, because λ_{FCN} is so close to the frequencies of the lunar-solar nutations, the amplitudes of these forced nutations are affected by the presence of the mode by up to 20 or 30 mas, a perturbation that can be readily detected even by conventional optical observations of nutation. These effects of the FCN are included in the standard forced nutation model adopted by the International Astronomical Union (IAU), which is based on the rigid Earth values of Kinoshita (1977) and corrections for nonrigidity from Wahr (1981), the latter assuming a hydrostatically prestressed Earth.

Herring et al (1986) found that VLBI results for the forced nutations disagree with the IAU adopted theory by almost 2 mas, particularly at the annual frequency of $\lambda = \Omega(1 + 1/365.25)$ cycles day^{-1}. This is the forcing frequency closest to λ_{FCN}, and it suggests that the FCN frequency may be somewhat larger than expected, close to $\lambda_{FCN} = \Omega(1 + 1/435)$ cycles day^{-1} (Gwinn et al 1986).

Equation (12) suggests that λ_{FCN} can be modified by changing either β or e_f. Wahr & Bergen (1986) (see also Dehant 1987) considered the effects of mantle anelasticity on β and concluded that the corresponding perturbation of λ_{FCN} is too small and, more importantly, has the wrong sign. Although the effects of anelasticity are, in principle, observable using the current VLBI observations, the larger discrepancy between observation and theory must first be resolved before the VLBI results can be used to study diurnal anelasticity.

Probably the most likely explanation for the discrepancy is, instead, uncertainty in e_f, as postulated by Gwinn et al (1986). Recent results based on seismic tomography suggest that the shape of the core-mantle boundary may diverge appreciably from hydrostatic equilibrium. It is not straightforward to compute the effects on e_f given only the shape of the core-mantle boundary. A perturbed boundary and an aspherical density distribution in the mantle will cause perturbations in the internl density surfaces in the core, and these will affect e_f. However, current models of the boundary based on seismic data suggest values for the nonhydrostatic portion of e_f that are as large or larger than the result inferred from the nutation observations. In fact, the VLBI nutation observations are proving to be a valuable independent constraint on the seismic models.

SUMMARY

There is now at least some general understanding of what causes most of the various observed fluctuations in rotation. Some aspects of the subject are understood very well. For others, there are many missing details. The hope is that in the process of filling in these details, we will learn more about the Earth and its environment.

In this review we have given some representative examples of ongoing research. One geophysical goal mentioned several times above is to better constrain the values of mantle anelasticity. There are, of course, other ways to study anelasticity. Particularly useful are observations of the attenuation of seismic waves traveling through the Earth following earthquakes. These observations, however, only tell us about dissipation at very short periods, from seconds to minutes. The frictional mechanisms responsible for dissipation are probably different at different time scales and are not clearly understood in any case. Rotational observations offer a unique opportunity to see the effects of anelasticity at much longer time periods.

There are, in fact, many opportunities, only some of which are described above, to use rotation data to learn about the solid Earth. But what about the fluid portions of the Earth? The oceans and, especially, the atmosphere have large effects on the rotation. Can the data be used to learn about them as well?

The answer is yes, but probably only in a limited sense. As an example, an unexplained variation in rotation could be a clue that there are some unknown processes in the atmosphere or oceans that should be studied further. Or, the rotation data could help confirm the reliability of certain meteorological or oceanographic results either as inferred directly from atmospheric or oceanic data or as deduced from numerical or analytical models. For instance, the agreement between the short-period LOD data and meteorological data has convinced meteorologists that the 50-day oscillation apparent in the atmospheric data is probably real and deserves attention. It is unlikely that the rotation data can help constrain many details of an atmospheric or oceanic disturbance. What they can do is to suggest or confirm that a disturbance exists.

Still, it is not clear just what the future holds. Rotation results have improved dramatically over the last few years and have already provided much new valuable information. As more high-quality data become available and as the techniques further improve, we should be able to resolve many of the long-standing problems in the field and, if we are lucky, discover new ones.

Literature Cited

Backus, G. 1983. Application of mantle filter theory to the magnetic jerk of 1969. *Geophys. J. R. Astron. Soc.* 74: 713–46

Barnes, R. T. H., Hide, R., White, A. A., Wilson, C. A. 1983. Atmospheric angular momentum fluctuations correlated with length of day changes and polar motion. *Proc. R. Soc. London Ser. A* 387: 31–73

Brosche, P., Sundermann, J. 1978. *Tidal Friction and the Earth's Rotation.* Berlin: Springer-Verlag. 241 pp.

Brosche, P., Sundermann, J. 1982. *Tidal Friction and the Earth's Rotation II.* Berlin: Springer-Verlag. 345 pp.

Cazenave, A., Daillet, S. 1981. Lunar tidal acceleration from earth satellite orbit analysis. *J. Geohys. Res.* 86: 1659–63

Dahlen, F. A. 1973. A correction to the excitation of the Chandler wobble by earthquakes. *Geophys. J. R. Astron. Soc.* 32: 203–17

Dehant, V. 1987. Nutations and inelasticity of the earth. *Proc. IAU-IAG Symp. No. 128, Earth Rotation and Reference Frames.* In press

Dickey, J. O., Eubanks, T. M., Steppe, J. A. 1986. High accuracy earth rotation and atmospheric angular momentum. In *Earth Rotation: Solved and Unsolved Problems,* ed. A. Cazenave, pp. 137–62. Dordrecht: Reidel

Dickman, S. R. 1981. Investigation of controversial polar motion features using homogeneous International Latitude Service data. *J. Geophys. Res.* 86: 4904–12

Gire, C., Le Mouel, J. L. 1986. Flow in the fluid core and earth's rotation. In *Earth Rotation: Solved and Unsolved Problems,* ed. A. Cazenave, pp. 241–58. Dordrecht: Reidel

Goad, C., Douglas, B. 1978. Lunar tidal acceleration obtained from satellite-derived ocean tide parameters. *J. Geophys. Res.* 83: 2306–10

Gwinn, C. R., Herring, T. A., Shapiro, I. I. 1986. Geodesy by radio interferometry: studies of the forced nutations of the earth, 2. Interpretation. *J. Geophys. Res.* 91: 4755–66

Herring, T. A., Gwinn, C. R., Shapiro, I. I. 1986. Geodesy by radio interferometry: studies of the forced nutations of the earth, 1. Data. *J. Geophys. Res.* 91: 4745–54

Hide, R. 1969. Interaction between the earth's liquid core and solid mantle. *Nature* 222: 1055–56

THE EARTH'S ROTATION 249

Hinnov, L. M., Wilson, C. R. 1987. An estimate of the water storage contribution to the excitation of polar motion. *Geophys. J. R. Astron. Soc.* 88: 437–60

Kinoshita, H. 1977. Theory of the rotation of the rigid earth. *Celestial Mech.* 15: 277–326

Lambeck, K. 1980. *The Earth's Variable Rotation: Geophysical Causes and Consequences.* Cambridge: Cambridge Univ. Press. 449 pp.

Lambeck, K., Hopgood, P. 1981. The earth's rotation and atmospheric circulation from 1963 to 1973. *Geophys. J. R. Astron. Soc.* 64: 67–89

Langley, R. B., King, R. W., Shapiro, I. I., Rosen, R. D., Salstein, D. A. 1981. Atmospheric angular momentum and the length of the day: a common fluctuation with a period near 50 days. *Nature* 294: 730–33

Merriam, J. B. 1982. Meteorological excitation of the annual polar motion. *Geophys. J. R. Astron. Soc.* 70: 41–56

Merriam, J. B. 1984. Tidal terms in universal time: effects of zonal winds and mantle Q. *J. Geophys. Res.* 89: 10,109–14

Morrison, L. V. 1979. Redetermination of the decade fluctuations in the rotation of the earth in the period 1861–1978. *Geophys. J. R. Astron. Soc.* 58: 349–60

Munk, W. H., MacDonald, G. J. F. 1975. *The Rotation of the Earth.* Cambridge: Cambridge Univ. Press. 323 pp.

Peltier, W. R. 1983. Constraint on deep mantle viscosity from LAGEOS acceleration data. *Nature* 304: 434–36

Robertson, D. S., Carter, W. E., Campbell, J. A., Schuh, H. 1985. Daily earth rotation determinations from IRIS very long baseline interferometry. *Nature* 316: 424–27

Rochester, M. G. 1984. Causes of fluctuations in the rotation of the earth. *Philos. Trans. R. Soc. London Ser. A* 313: 95–105

Rosen, R. D., Salstein, D. A. 1983. Variations in atmospheric angular momentum on global and regional scales and the length of day. *J. Geophys. Res.* 88: 5451–70

Rubincam, D. P. 1984. Postglacial rebound observed by Lageos and the effective viscosity of the lower mantle. *J. Geophys. Res.* 89: 1077–88

Sasao, T., Okubo, S., Saito, M. 1980. A simple theory on dynamical effects of stratified fluid core upon nutational motion of the earth. *Proc. IAU Symp. No. 78, Nutation and the Earth's Rotation,* ed. E. Fedorov, M. Smith, P. Bender, pp. 165–83. Dordrecht: Reidel

Smith, M. L., Dahlen, F. A. 1981. The period and Q of the Chandler wobble. *Geophys. J. R. Astron. Soc.* 64: 223–82

Speith, M. A., Hide, R., Clayton, W., Hager, B. H., Voorhies, C. V. 1986. Topographic coupling of core and mantle and changes in the length of day. *Eos, Trans. Am. Geophys. Union* 67: 908 (Abstr.)

Stix, M., Roberts, P. H. 1984. Time-dependent electromagnetic core-mantle coupling. *Phys. Earth Planet. Inter.* 36: 49–60

Van Hylckama, T. E. A. 1970. Water balance and earth unbalance. *Int. Assoc. Sci. Hydrol., Proc. Reading Symp. World Water Balance Publ.* 92, pp. 434–553. Paris: AIHS-UNESCO

Voorhies, C. V. 1986. Steady flows at the tow of the earth's core derived from geomagnetic field models. *J. Geophys. Res.* 91: 12,444–66

Wahr, J. M. 1981. The forced nutations of an elliptical, rotating, elastic and oceanless earth. *Geophys. J. R. Astron. Soc.* 64: 705–28

Wahr, J. M. 1983. The effects of the atmosphere and oceans on the earth's wobble and on the seasonal variations in the length of day—2. Results. *Geophys. J. R. Astron. Soc.* 74: 451–87

Wahr, J. M. 1987. The theory of the earth's orientation, with some new results for nutation. *Proc. IAU Symp. No. 129, The Impact of VLBI on Astrophysics and Geophysics,* ed. J. Moran. In press

Wahr, J., Bergen, Z. 1986. The effects of mantle anelasticity on nutations, earth tides, and tidal variations in rotation rate. *Geophys. J. R. Astron. Soc.* 87: 633–68

Wahr, J. M., Oort, A. H. 1984. Friction- and mountain-torque estimates from global atmospheric data. *J. Atmos. Sci.* 41: 190–204

Williams, J. G., Sinclair, W. S., Yoder, C. F. 1978. Tidal acceleration of the Moon. *Geophys. Res. Lett.* 5: 943–46

Wilson, C. R., Haubrich, R. 1976. Meteorological excitation of the earth's wobble. *Geophys. J. R. Astron. Soc.* 46: 707–43

Yoder, C. F., Williams, J. G., Dickey, J. O., Schutz, B. E., Eanes, R. J., et al. 1983. Secular variation of earth's gravitational harmonic J_2 coefficient from LAGEOS and the nontidal acceleration of earth rotation. *Nature* 303: 757–62

Yuen, D. A., Sabadini, R., Boschi, E. 1983. The dynamical equations of polar wander and the global characteristics of the lithosphere as extracted from rotational data. *Phys. Earth Planet. Inter.* 33: 226–42

Ann. Rev. Earth Planet. Sci. 1988. 16: 251–71

THE GEOPHYSICS OF A RESTLESS CALDERA—LONG VALLEY, CALIFORNIA[1]

John B. Rundle

Division 1541, Sandia National Laboratories, Albuquerque,
New Mexico 87185

David P. Hill

US Geological Survey, Menlo Park, California 94025

INTRODUCTION

At 9:33 PDT on the morning of May 25, 1980, a strong earthquake of
local magnitude $M_L = 6.1$ heralded the onset of an earthquake sequence
that focused worldwide attention on the major geologic and tectonic feature nearest its epicenter, the Long Valley caldera of eastern California.
This sequence included two additional $M_L = 6$ events on May 25 and a
fourth $M_L = 6$ event on May 27 accompanied by more than 300 $M_L > 3$
aftershocks over the next several weeks, all of which were located either
within or to the south of the caldera (Uhrhammer & Ferguson 1980,
Cramer & Toppozada 1980). These were not the first major events to occur
regionally within the last decade. The October 4, 1978, Bishop event
($M_L = 5.7$), located some 20 km to the southeast of Long Valley, initiated
a sequence of swarm-like activity that migrated generally toward the
caldera over the next two years (Ryall & Ryall 1980, Van Wormer & Ryall
1980). In fact, Van Wormer & Ryall (1980), in a manuscript written prior
to the May 1980 earthquakes, identified this sequence of events as a

251

possible precursor to larger events. Moreover, they noted that variations in seismicity during the period 1977–80 occurred over the entire Sierra Nevada–Great Basin, "indicating a regional rather than a local cause for the observed changes."

This persistent seismic activity is only one of the more obvious clues to the geophysical and geological importance of Long Valley in terms of currently active tectonic and volcanic processes. Recognition of this importance within the scientific community has resulted in numerous geophysical and geological investigations focused on Long Valley caldera (see, for example, Muffler & Williams 1976, Hill et al 1985a). Other articles have reviewed selected results of these investigations (Hermance 1983, Hill et al 1985b, Rundle et al 1986, Goldstein 1987). As a well-studied example of restless calderas, Long Valley offers valuable clues to understanding the causes for unrest in many of the other major volcanic calderas around the world [see Newhall et al (1984) for a compilation]. Examples of these include Campi Flegrei in Italy (Lirer et al 1987), Rabaul in Papua New Guinea (Mori & McKee 1987), Yellowstone in the continental United States (Christiansen 1984), and Katmai in Alaska (Hildreth 1983).

BACKGROUND: TECTONIC AND GEOLOGIC SETTING

Long Valley is a 17-km-wide (north-south) by 32-km-long (east-west) elongate structure that sits astride the boundary between the high-heat-flow Basin and Range extensional environment to the east and the uplifted, cold, Sierra Nevada block to the west (Bailey et al 1976, Bailey 1987; see Figure 1). Formed as a result of the great Bishop Tuff eruption at 0.7 Ma, in which some 600 km^3 of rhyolitic magma was erupted in a time scale of not longer than days to weeks as deduced from the welded nature of the Bishop Tuff (Hildreth 1979), the caldera lies in an embayment of the Sierra Nevada, just to the west of the White Mountains and the northern end of the Owens Valley of eastern California. Within the west-central part of the elliptical ring-fracture system lies the resurgent dome, a structural uplift that formed as the result of the inflow of magma into the shallow crustal reservoir partially depleted by the eruption of the Bishop Tuff. Surrounding the dome is a relatively low-lying moat, filled with post-collapse rhyolites, rhyodacites, basalts, and Pleistocene lake deposits (Bailey et al 1976). Volcanism following caldera collapse included episodes of aphyric rhyolite extrusion at 0.68–0.64 Ma; porphyritic hornblende-biotite rhyolite that erupted at 0.5, 0.3, and 0.1 Ma; and porphyritic hornblende-biotite rhyodacite that "leaked up" along the caldera ring fracture system at 0.2 Ma to 50,000 yr ago (Bailey et al 1976).

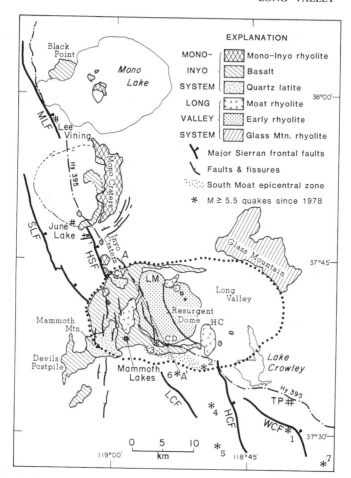

Figure 1 Regional geologic and tectonic setting of Long Valley caldera, and location key (after Hill et al 1985a). Dots indicate the boundary of the caldera. Cross section *A-A'* refers to that shown in Figure 2. HCF = Hilton Creek fault, HC = Hot Creek, LCF = Laurel Canyon fault, MLF = Mono Lake fault (also known as Lee Vining fault), WCF = Wheeler Crest fault, HSF = Hartly Springs fault, SLF = Silver Lake fault, CD = Casa Diablo, LM = Lookout Mountain, TP = Tom's Place, CV = Crestview. Numbers refer to following dates and magnitudes of earthquakes: (1) 4 October 1978, $M = 5.7$; (2) 25 May 1980, $M = 6.1$; (3) 25 May 1980, $M = 6.0$; (4) 25 May 1980, $M = 6.1$; (5) 27 May 1980, $M = 6.2$; (6) 30 September 1981, $M = 5.7$; and (7) 23 November 1984, $M = 5.8$.

From a more regional perspective, the caldera coincides with a left-stepping reentrant, the Mammoth embayment, along the north-south trend of the Sierra Nevada chain (Figure 1). To the north lies the Pliocene-Pleistocene Mono basin, which includes Mono Lake. The Mono Craters chain of Holocene flows, domes, and vents extends south of Mono Lake toward the caldera (Figure 1). The most recent documented eruptions from the Mono Craters occurred between AD 1325 and AD 1365 along the northern third of the chain and perhaps from within the southern end of Mono Lake (Wood 1977, Sieh & Barsik 1986). Still farther south along the same linear trend are the Inyo Craters (Figure 1; Bailey et al 1976, Wood, 1977, Miller 1985, Fink 1985, Eichelberger et al 1985, 1986, Sieh & Bursik 1986). This chain of domes and craters formed during the emplacement of a dike complex roughly contemporaneous with the north Mono Craters eruption (Wood 1977, Sieh & Bursik 1986, Fink 1985), and they are essentially coincident with the Hartly Springs fault, which marks the eastern boundary of the Sierra Nevada and enters Long Valley at its northwest corner. The Hilton Creek fault, a large, east-dipping normal fault with 25 m of offset in the last 11,000 yr, extends from the southern margin of the caldera into the Sierra Nevada block.

Long Valley lies roughly 50 km northwest of the northern terminus of the Owens Valley fault, site of the great 1872 $M \approx 8$ Owens Valley earthquake (see, for example, Cramer & Toppozada 1980). Other major events known to have occurred in or to the south of Long Valley include an $M_L > 5.5$ event in 1889, a $M_L \approx 6$ event in 1927, and three $M_L > 5.5$ events in 1941. To the north, along the central Nevada seismic belt (Richter 1958), major earthquakes have occurred in 1915 (Pleasant Valley; $M_L = 7.6$), 1932 (Cedar Mountain; $M_L = 7.3$), and 1954 (Dixie Valley–Fairview Peak; three events with magnitudes of 6.6, 6.4, and 6.8). Thus, in addition to substantial historic and prehistoric volcanic activity, the Long Valley caldera sits within one of the most active seismic belts in the country.

CALDERA STRUCTURE

The subsurface structure of the caldera, the size and distribution of magma chambers, and the structural environment associated with the adjacent Sierra Nevada block and western Basin and Range province provide important clues to the processes governing current seismic and magmatic activity beneath Long Valley caldera and its vicinity. To date, most of the geophysical investigations in the region have focused on the structure of the upper 5 to 10 km of the crust; our knowledge of mid- to lower-crustal structure remains sketchy at best.

Results From Seismic Investigations

Beginning in the early 1970s, investigators from several institutions carried out a series of studies on the structure of Long Valley using a variety of passive and active seismic techniques. Figure 2 summarizes some of their results in a schematic cross section of the caldera. Detailed color-coded syntheses of these results can be found in Rundle et al (1985, 1986).

PASSIVE SEISMIC STUDIES   Among the first investigations of the structure of Long Valley using passive recording of naturally occurring earthquakes was Steeples & Iyer's (1976) teleseismic P-wave delay study of the low-velocity zone beneath the caldera. Using some 39 teleseisms from a wide variety of azimuths recorded on a portable analog network temporarily installed within the caldera, they found persistent relative P-wave delays of 0.3 s in the west-central part of the caldera. Their processing of the data included an approximate correction for near-surface delays due to the shallow caldera structure. Using a simple graphical ray-tracing technique, Steeples & Iyer (1976) found that the pattern and magnitude of the delays could best be fit by P-wave velocity decreases of 10–15% at depths of 5 to 20 km beneath the caldera. They suggested that this low-velocity volume may be related to the magma chamber underlying the caldera. More recent work (Dawson et al 1986) confirms the existence of significant, azimuthally dependent, teleseismic travel-time delays of as much as 0.5 s. After removal of the effects due to the surficial caldera fill, the remaining delays provide evidence for a low-velocity volume consistent with a magma chamber underlying the western margin of the resurgent dome.

Ryall & Ryall (1981a), Sanders & Ryall (1983), and Sanders (1984) studied the pattern of S-wave attenuation for seismic body waves propagating beneath the caldera to infer the geometry, location, and fine-scale detail of the magma chamber beneath Long Valley. Their technique builds upon a method due to Kubota & Berg (1967) and Matumoto (1971), who observed azimuthally dependent extinction of S-waves, apparently due to screening by molten magma chambers, for rays passing beneath the Katmai, Alaska, chain of volcanic vents. In an analogous study, Ryall & Ryall (1981a) analyzed seismic signals from events to the south of Long Valley recorded at stations of the University of Nevada regional seismic network, which is located generally to the north and east of the caldera. The Ryalls noticed that the seismic signals showed a systematic variation in character that depended on the apparent ray path from the source to the receiver but not on the focal mechanism of the event. For many of the seismograms, the P- and S-wave arrivals are clear and impulsive, and the records contain significant energy at frequencies of 5–10 Hz. For other recordings, the onset of the P-waves is much more emergent, and the frequency content is distinctly lower, more in the range of 2–3 Hz at most. But it is in the S-wave train where the greatest differences lie. S-waves are absent on the most anomalous records and are only weakly present on

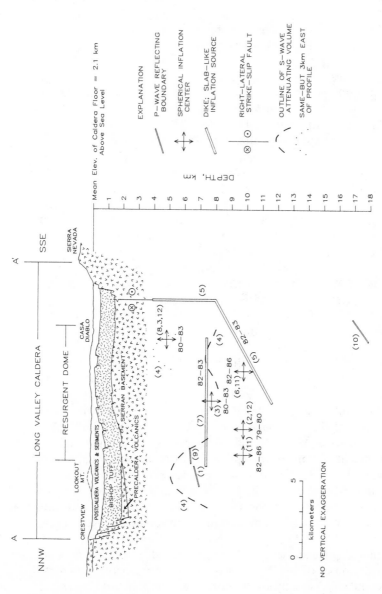

Figure 2 Vertical section through Long Valley caldera along profile *A-A'* in Figure 1 (adapted from Hill et al 1985a, Rundle et al 1986). Shallow caldera structure based on seismic-refraction results of Hill et al (1985c). Deeper structure beneath caldera shows the location and geometry of inflation (dislocation) sources used to model deformation observed during the years indicated (e.g. 79–80), seismic *P*-wave reflecting boundaries, and volumes that produce anomalous *S*-wave attenuation. Parenthetic numbers give references: (1) Hill 1976, (2) Savage & Clark 1982, (3) Rundle & Whitcomb 1984, (4) Sanders 1984, (5) Savage & Cockerham 1984, (6) Castle et al 1984, (7) Denlinger & Riley 1984, (8) Denlinger et al 1985, (9) Hill et al 1985c, (10) Luetgert & Mooney 1985, (11) Savage et al 1987, (12) Langbein et al 1987.

others. Noting that all of the anomalous seismograms were associated with rays that passed beneath the south-central part of Long Valley [near Casa Diablo (CD) on Figure 1], Ryall & Ryall (1981a) concluded that a zone of molten material at depths in excess of 7–8 km in this region could account for the observations. Further systematic studies by Sanders & Ryall (1983) and Sanders (1984) supported and extended these observations, leading to a detailed picture of the attenuating volume beneath the caldera at depths as shallow as 4–5 km beneath the surface.

Other investigations using quite different techniques further support this basic picture of a substantial, low-velocity volume beneath the caldera. Luetgert & Mooney (1985) took advantage of the opportunity afforded by the intense January 1983 earthquake swarm in the south moat (discussion following) to deploy 120 portable seismographs along an essentially north-south line 25–45 km to the north of the caldera. At offset distances between kilometers 32 to 38, they recorded a clear P-wave arrival (which they call P_r) about 2.8–3.0 s after the first crustal P-wave arrival (P_g) but well ahead of the crustal S-wave arrival (S_g). In addition, the dominant frequency of the P_r arrival is significantly less than that of the P_g phase. Combining these observations, Luetgert & Mooney (1985) interpreted the reflection as a wave that propagated downward from the earthquake through normal basement rock into a highly attenuating volume, followed by reflection from a northward-dipping lower boundary of the anomalous volume at a depth of 18–20 km.

To avoid the scattering and attenuating effects of the shallow, poorly consolidated caldera fill on seismic signals, Rundle et al (1985) and Elbring & Rundle (1986) installed a three-component seismometer in a 900-m-deep well located near the northwestern edge of the resurgent dome [near Lookout Mountain (LM) on Figure 1] and recorded earthquakes to the south-southeast, at offsets of 10 to 35 km. A small companion network of surface stations was installed within 5 km of the wellhead, with one three-component instrument located at the wellhead. The observations clearly showed major differences between the waveforms recorded on the uphole and downhole instruments, with significantly higher frequency content downhole. Using all three components of the particle motion, Elbring & Rundle (1986) observed delays of both P- and S-waves that appeared anomalous, although no consistent pattern of S-wave extinction was observed. They explained the delays as a consequence of propagation through, or possibly diffraction around, a small low-velocity body lying between 4 to 6.5 km depth underneath the southern region of the central resurgent dome. Elbring & Rundle (1986) also reported evidence for a major reflecting horizon at about 5–6 km depth beneath the northern extent of the resurgent dome. These interpretations are consistent with those advanced by other investigators, although Hauksson (1987) offers an alternate interpretation of these data that does not require a small low-velocity body at shallow depth.

ACTIVE SEISMIC STUDIES: WIDE ANGLE   Active seismic investigations in Long Valley using wide-angle refraction and reflection techniques have produced a reasonably clear image of the upper 5 km of the caldera structure and its relation to the adjacent crust. The first systematic reflection data were obtained by Hill (1976) on two lines running essentially north-south and east-west, respectively, through the caldera. A more detailed investigation in 1982–83 (Hill et al 1985c) involved 18 shot points recorded on a dense network of profiles within the caldera, with typical station spacings of 0.5 to 1.0 km along the profiles. This investigation also included east- and north-trending profiles running through the Mono Craters system to the north. The resulting data indicate that P-wave velocities in crystalline Mesozoic granitic basement rocks are 5.6 km^{-1} s at 2 km beneath the surface, gradually increasing with depth at a gradient of 0.1 s^{-1}. Offsets in P-wave arrival times across the caldera boundary of up to 0.5 s indicate that the crystalline basement within the caldera was down-dropped along the ring-fracture faults by an average of 2 km during caldera formation. The improved resolution afforded by the dense station spacing along the 1982–83 profiles indicates that earlier estimates of the displacement of the crystalline basement across the northern ring-fracture system of 3 to 4 km were too large and that the estimated 1-km displacement across the eastern ring-fracture system was too small (see Hill 1976, Abers 1985). These seismic-refraction data also indicate that the welded tuff (the Bishop Tuff), which fell back into the newly formed caldera at the time of the eruption, averages about 1 km in thickness, with P-wave velocities of 3.9–4.4 km^{-1} s. The postcaldera volcanic rocks erupted on top of the Bishop Tuff have P-wave velocities of 2.8 to 3.1 km^{-1} s and thicknesses of 200 to 400 m.

An interesting aspect of the 1973 wide–angle data involves a secondary arrival recorded on the east-west profile over a region in the northwest quadrant of the caldera (Hill 1976). This arrival followed the initial P-waves by about 2 s and was similar in amplitude and frequency content, but appeared to be 180° out of phase with respect to the first arrival. Hill (1976) noted that an arrival with these characteristics would result from the reflection of P-waves from a low-velocity horizon, and he suggested that one explanation for the secondary arrival involved a reflection from the roof of a magma chamber at a depth of 7–8 km beneath the western margin of the resurgent dome (near LM in Figure 1). Hill et al (1985c) reported a similar secondary arrival on their Minaret-Bald profile, which also crossed the northwest sector of the caldera, supporting the earlier observation.

ACTIVE SEISMIC STUDIES: REFLECTION   During the summer of 1984, an integrated seismic-imaging experiment was conducted in the northwest region of the caldera by a team of scientists from seven institutions (Rundle et al 1985). Centered around the same well used by Elbring & Rundle (1986) on the northwest corner of the resurgent dome, the investigation was a coordinated attempt to image the reflector first seen by Hill (1976) near LM on Figure 1. As described by Rundle et al (1985) and Deemer (1985), there were three active seismic experiments: (a) an expanding spread (or walk-in) profile about a fixed common midpoint (located about 1 km south of the well), (b) a conventional common depth point (CDP) seismic-reflection profiling experiment, and (c) a vertical seismic profile (VSP) in the well. The vertical seismic-profiling data were inconclusive. However, a small-scale, normal incidence reflection carried out in the immediate vicinity of the wellhead does show clear reflectors at 3.0 and 4.0 s two-way travel time, corresponding to depths of approximately 5 and 8 km, respectively. The lower reflecting horizon is presumably coincident with the reflector identified by Hill et al (1985c).

The principal problem in attempting to image subsurface reflectors in Long Valley is that strong, near-surface velocity variations ("statics") obscure the underlying deeper structure by destroying the coherence of the wave trains [see Telford et al (1976) for a discussion]. Variations of 100 ms due to static shifts are common, as are severe reverberation and attenuation problems (Johnson et al 1986). While strong reflectors such as the Bishop Tuff can be seen clearly in the data (Deemer 1985), weaker reflectors with only small velocity contrast are obscure. With these problems in mind, a second P-wave reflection-profiling experiment was carried out during the summer of 1985 (Murphy et al 1985, Tono & Malin 1986, Johnson et al 1986) in order to explore the anomaly beneath the southern end of the resurgent dome (near CD on Figure 1) by using intensive CDP and wide-angle coverage. The wide-angle data were plagued by problems relating to shadowing and reverse moveouts associated with strongly three-dimensional structures (see Telford et al 1976, Tono & Malin 1986). The wide-angle experiment, however, did record clear reflections at about 2.5 and 4.8 s, confirming the existence of a reflector at 5–10 km beneath the southern resurgent dome. Results from the CDP experiment were less conclusive (Johnson et al 1986).

Results From Other Investigations

STRUCTURE FROM GRAVITY DATA   Interpretations of the gravity field over the caldera (Kane et al 1976, Abers 1985, Jachens & Roberts 1985, Hermance 1987) indicate that the thickness of the caldera fill is somewhat greater beneath the north-northeastern region of the caldera, where it may approach 3 km, and shallowest in the west-southwest quadrant, where it is only 1 to 1.5 km thick. When constrained by density from borehole samples, the density models based on these gravity data show general agreement with the P-wave velocity structure based on the seismic-refraction data (Abers 1985). A detailed gravity survey by Jachens & Roberts (1985) defined two small gravity highs over the south moat of the caldera, which these authors believe may be related to magmatic intrusions near the base of the caldera fill.

HYDROTHERMAL SYSTEM AND ELECTROMAGNETIC STRUCTURE   Heat-flow data are complicated by the transitional position of Long Valley between the low-heat-flow Sierra Nevada block characterized by 0.5 to 1.0 Heat Flow Units (HFU) and the high-heat-flow Basin and Range province characterized by 1.5 to 2.0 HFU (Lachenbruch et al 1976b). Heat flow within the caldera is influenced principally by the flow of hot hydrothermal brines along the predominant faults and fractures within the caldera fill and the underlying basement (Lachenbruch et al 1976a,b, Sorey & Lewis 1976, Sorey 1985, Blackwell 1985). The pattern of heat discharge is governed by a net west-to-east flow. Cold water from the Sierra Nevada enters the caldera at its western margin and is heated by hot water upwelling along intracaldera faults somewhere west of the resurgent dome. These thermal waters are then discharged into Hot Creek gorge and Lake Crowley along the southeastern margin of the caldera, from where it flows into the Owens River drainage system. A portion of the discharge may be recirculated back to the western quadrant of the caldera along deep (>3 km) aquifers (Blackwell 1985). Both Sorey (1985) and Blackwell (1985) argue that the relatively shallow ($\simeq 5$–7 km) magma chamber underlying the resurgent dome, suggested by the deformation and S-wave attenuation data, is not reflected in the superficial heat-flow data; hence, any substantial intrusion at these depths must be younger than 40,000 yr. Blackwell (1985) argues that heat in the western quadrant of the caldera probably originates from the 500-yr-old Inyo Craters chain of vents (Figure 1). However, direct observation of the present-day temperature of the Inyo Craters dike indicates that it is cold, with a maximum temperature of only about 80°C (Eichelberger et al 1985). This observation seems to cast doubt on a model involving the Inyo Craters as the heat source and suggests that perhaps the heat is originating instead from hydrothermal waters percolating up along the western ring-fracture system into the shallow aquifers.

From the resistivity (Stanley et al 1976), audiomagnetototelluric (Hoover et al 1976), and magnetotelluric (Hermance et al 1984, Hermance

1987) data, it is clear that the zones of high electrical conductivity within the western half of the caldera are shallow and associated with the hot hydrothermal aquifers. Visual inspection of the isothermal contours given by Blackwell (1985) and of the contours of relative conductance given in Hermance (1987) clearly demonstrates this association. In the eastern half of the caldera, however, zones of high conductivity are dominated by shallow, clay deposits. In both cases, these shallow zones of high conductivity tend to mask the deeper underlying conductivity anomalies that may be associated with magma.

LITHOLOGIC LOGS FROM DRILLHOLES   Most of the wells in Long Valley have been drilled by companies representing the geothermal industry, and thus until recently most of the data from the wells have been held as proprietary information. Moreover, few of the wells penetrate to depths in excess of 1 km, thus limiting the information from them to the caldera fill. Data are available, however, from the three deepest wells in the caldera: (a) Mammoth 1 drilled on the southwest margin of the resurgent dome (near CD on Figure 1) to a depth of about 1600 m (Hermance 1983); (b) Clay Pit 1 drilled on the northeast margin of the resurgent dome to 1800 m (Hermance 1983); and (c) the Republic well drilled in the southeast region of the caldera to a depth of 2109 m (Smith & Rex 1977). Two of these wells penetrated entirely through the Bishop Tuff. Mammoth 1 penetrated metasediments and roof pendant rocks, and Clay Pit 1 penetrated granophyric-granite porphyry. By contrast, the Republic well penetrated only the lowermost welded unit of the Bishop Tuff. These results confirm surface geophysical measurements indicating that the thickest fill is in the eastern section of the caldera.

SEISMICITY

The region of eastern California between Long Valley caldera and the head of the Owens Valley has sustained one of the highest rates of earthquake activity in California during historic time (Van Wormer & Ryall 1980, Hill et al 1985b). A 37-yr period of relative quiescence that followed a sequence of $M = 5.5$ to 6.0 earthquakes in 1941 near Tom's Place south of the caldera (Figure 1) was interrupted on October 4, 1978, by an $M_L = 5.7$ earthquake located midway between the caldera and the town of Bishop (Figure 3). As pointed out by Ryall & Ryall (1981b), the renewed activity represented by this event was preceded by a year and a half of complete seismic quiescence for $M > 3$ events along much of the eastern front of the Sierra Nevada. The regional and teleseismic P-wave fault plane solution for this event indicated predominantly strike-slip displacement, with the T-axis trending east-northeast (Ryall & Ryall 1981b). Following this event, aftershock activity (up to $M_L \approx 4$) gradually spread to the northwest toward the caldera over the next year and a half. Then, on May 25–27, 1980, a sequence of four $M_L = 6$ events occurred, initiating activity within Long Valley caldera, which (based on limited data) had been regarded as seismically quiescent (Steeples & Pitt 1976, Van Wormer & Ryall 1980). The epicenters of the first and third events on May 25 and the event on May 27 (Nos. 2, 4, and 5, respectively, in Figure 1) define a linear trend extending 15 km south of the caldera. The second $M = 6$ event on May 25 was located within the south moat of the caldera. None of these events and their numerous aftershocks are associated with the surface expression of mapped faults in the region. Indeed, they lie midway between the Hilton Creek fault to the east and the Laurel Canyon fault to the west (Figures 1, 3). As Hill et al (1985b) note, this sequence of events was

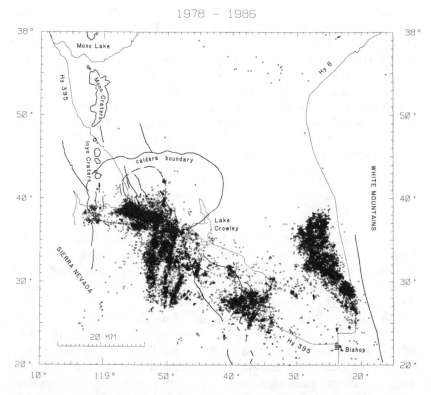

Figure 3 Map of the regional seismicity from 1978 through 1986 (plot courtesy of R. S. Cockerham).

remarkable in its swarm-like character; it was preceded by a substantial increase in local activity and followed by an intense, long-lasting series of aftershocks whose depths ranged to as much as 18 km.

The focal mechanisms of the events of May 25–27 have been a subject of considerable controversy. Initial studies on mechanisms, based upon regional, short-period data alone, showed strike-slip displacements, with T-axes trending northeast (Cramer & Toppozada 1980). Subsequent analysis by Julian (1983) and Julian & Sipkin (1985), using a variety of P- and SH-motions, spectral amplitudes, initial phases, and long-period waveforms, showed that the first and last of the May 25 $M = 6$ events had mechanisms that were not simple double-couples. Ekstrom & Dziewonski (1983), using long-period body waves and mantle waves, found a non-double-couple mechanism for the October 4, 1978, event as well, a result in disagreement with that of Ryall & Ryall (1981b). Wallace et al (1982), Given et al (1982), Barker & Langston (1983), and Wallace (1985) pointed out that fitting both short- and long-period data with one common double-couple mechanism is evidently not possible for these earthquakes. They suggested that the radiation pattern is the result both of simultaneous slip on two planes with different orientations and of regional complexity in structure. Priestly et al (1985) argued that the spectra of the 1980 earthquakes cannot be distinguished from the spectra of "tectonic" earthquakes, and that the former showed few of the characteristics one might expect from the sudden extension of a fluid-filled crack. By contrast, Julian (1983), Julian & Sipkin (1985), and Chouet & Julian (1985) argued that the non-double-couple solutions were a consequence of the expansion of a fluid-filled dike due to magma injection. In the absence of further data, it seems unlikely that this particular issue can be unambiguously decided.

In the two years following the May 1980 events, eight moderate swarms occurred within the south moat of the caldera (Ryall & Ryall 1983). Five of these contained one or more events of $M_L > 4$, and each swarm typically lasted several days. A ninth swarm began on January 6, 1983, in the south moat of the caldera with a sequence of small earthquakes, but it quickly grew to major importance. Two events had magnitudes in excess of 5, and depths of the swarm events were in the range of 2 to 10 km below the surface (Savage & Cockerham 1984). In horizontal projection, the swarm roughly defined an elongate (10 km by 2 km) structure trending west-northwest within the south moat. Fault plane solutions showed simple right-lateral strike-slip on the eastern extent of the swarm area (T-axes east-northeast) but considerably more complexity on the western part (T-axes generally northeast; Savage & Cockerham 1984). Some of the western events showed normal faulting, and one even had a reverse mechanism.

Over the succeeding months, regional seismicity diminished, with only a few tens of $M_L > 1$ events occurring per month. This gradual decline continued until November 23, 1984, when an $M_L = 5.8$ event occurred in Round Valley just a few kilometers to the southeast of the 1978 Bishop event, initiating another cycle of activity. The main shock, which had a focal depth of 13 km, was preceded by 4 s by a single $M_L = 2.2$ foreshock, and the aftershock sequence included an $M_L = 5.2$ event, as well as several $M_L \geqslant 4$ events (Barker & Wallace 1985, Corbett et al 1985, Smith et al 1985). Over the next several days, the aftershocks migrated in a north-northeasterly direction from the epicenter, defining several vertical planes oriented along this azimuth. Three days after the main event, the after-shocks also began migrating in a north-northwesterly direction toward the caldera (Corbett et al 1985). Using teleseismic and regional body wave data, Corbett et al (1985) found the fault plane solution to be vertical strike slip on a N30°E fault plane (T-axis N70°E).

Following the Round Valley event, seismic activity throughout the region returned to the relatively low-level characteristic of the preceding year. But once again, on July 20, 1986, the relative quiescence was broken by a sequence of magnitude 6 events (the Chalfant sequence; Cockerham 1986, Uhrhammer 1986, Corbett et al 1986). This time the sequence occurred on the White Mountains fault zone, 30 km southeast of the caldera. The sequence began on July 3 with a series of small foreshocks and culminated in a $M_L = 6$ event on July 20, a $M_L = 6.4$ event on July 21, and an $M_L = 6$ event on July 31 accompanied by thousands of after-shocks. Focal depths of the swarm were generally in the range of 4 to 12 km. Fault plane solutions again showed principally strike-slip displacements, although in this case the displacements involved right-lateral slip accompanied by a small component of normal slip along a plane striking N28°W and dipping 72° southwest. The associated T-axes showed a general east-west orientation (Uhrhammer 1986, Corbett et al 1986), in contrast to the more northeasterly T-axis orientations associated with the earlier $M > 5.5$ sequences. While it seems clear that the spatially diffuse, quasi-periodic behavior of the activity that began in October 1978 must involve a perturbation in crustal stresses of regional extent, the processes behind this stress perturbation are not clear (Savage & Cockerham 1987).

DEFORMATION

Because Long Valley caldera lies on a major transportation route (Highway 395), we have a valuable history of vertical deformation data across the caldera based on the routinely repeated leveling surveys along this highway. The earliest surveys, which date from 1905 and are somewhat less precise than modern surveys (Castle et al 1984), combined with modern geodolite data (Savage 8 Clark 1982), reveal that the region was essentially stable until at least 1979, and perhaps until the 1980 earthquakes. Leveling surveys along Highway 395 in 1905, 1914, 1932, 1957, and 1975 show little change beyond that attributable to random error. However, the leveling surveys of 1980 and 1982–85 show considerable differences relative to the survey of 1975 (Figure 4). In all of these later surveys, the data are presented as elevation changes relative to Tom's Place, which is located south of Lake Crowley (Figure 1). The evidence available from historic and recent surveys (Castle et al 1984) indicates that Tom's Place is relatively stable, and thus the uplift shown in Figure 4 is probably close to the total uplift. The later surveys, beginning with the leveling of 1982, were carried out over an extensive network of lines within the caldera (Castle et al 1984, Rundle & Whitcomb 1984, Savage et al 1987). These leveling data were

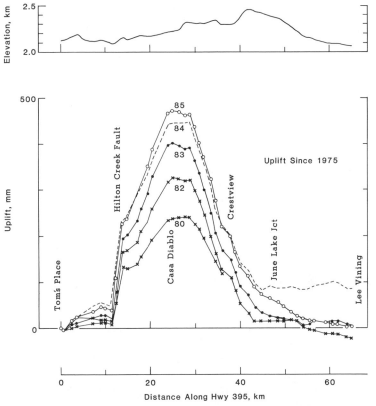

Figure 4   Leveling data along Highway 395, from Tom's Place south of Lake Crowley to Lee Vining on the west shore of Mono Lake. Refer to Figure 1 for location of Highway 395 (after Savage et al 1987).

supplemented by observations of surface gravity change carried out, in some cases, from 1980 until the present (Jachens & Roberts 1985, Whitcomb & Rundle 1985, Rundle & Whitcomb 1986). While the 1982 and 1984 surveys showed evidence of contamination by systematic survey error, it appears that these problems can be at least partially corrected (Savage et al 1987).

In addition to the leveling observations, precise local and regional single-frequency geodolite surveys have been carried out near or within the caldera during the summers of 1979, 1980, and 1982–86. The purpose of these observations is to measure lengths of lines connecting networks of monuments with an accuracy of several tenths of a part per million. When combined with uplift observations from leveling or gravity change measurements, these data provide a three-dimensional picture of the state of deformation of the crust. In addition to the regular summer measurements, a partial geodolite survey concentrating on the existing intracaldera lines was carried out immediately following the January 1983 earthquake swarm (Savage & Clark 1982, Savage & Cockerham 1984, Savage et al 1987). The single-frequency geodolite data indicated that significant deformation did not commence within the caldera until sometime between mid-1979 and mid-1980 (Savage & Clark 1982). Monitoring efforts within the caldera were intensified following the January 1983 earthquake swarm (Hill 1984). These efforts included establishing both a dense network of trilateration lines across the southwestern section of the caldera, measured nightly using a two-color laser geodimeter (Linker et al 1986), and a telemetered array of seven borehole tiltmeters and a borehole dilatometer.

The deformation data collected to date clearly indicate that the physical processes responsible for the observed deformation are spatially complex and vary with time (Savage & Clark 1982, Rundle & Whitcomb 1984, 1986, Savage & Cockerham 1984, Denlinger et al 1985, Linker et al 1986, Savage et al 1987). Just as clearly, the underlying mechanism responsible for deformation in the caldera involves continued inflation of one or more magma chambers underlying various regions of the resurgent dome. In turn, the stresses generated by the emplacement of magma within the host rock have induced slip on the active faults associated with the 1983 earthquake swarm (the South Moat fault zone), and possibly on the Hilton Creek fault as well. While the details of the magmatic emplacement mechanisms differ, in some cases substantially (Rundle & Whitcomb 1984, Savage & Cockerham 1984, Denlinger et al 1985), all investigations conclude that as much as 200×10^6 m^3 of magma must have been emplaced since 1980 beneath the caldera. These data, though relatively extensive and detailed by most standards, are still too sparse to ensure a unique physical model of the entire process. They are adequate, however, to resolve total volumes of injected magma. In particular, they indicate that about half of the 200×10^6 m^3 cumulative total was emplaced during the events of 1980, and that the remainder was apparently injected rather uniformly over the succeeding years.

CONCLUSIONS

Long Valley is one of several restless calderas dominated by recurring earthquake swarms and uplift of the caldera floor (see Newhall et al 1984). Two of the larger of these restless calderas, Yellowstone (Pelton & Smith 1982, Smith & Braile 1984) and Long Valley, have sustained large earthquakes ($M \gtrsim 6$) within adjacent sections of the crust and have produced cumulative uplifts of 0.8 and 0.5 m, respectively. In contrast, two somewhat smaller calderas, Rabaul in Papua New Guinea (Mori & McGee 1987) and the Campi Flegrei in Italy (Lirer et al 1987), have produced much larger cumulative uplifts of 2.5 to 3 m, respectively, but with seismic activity limited to repeated swarms of moderate ($M \lesssim 5$) earthquakes within the confines of the caldera boundaries. To be sure, repeated swarms of moderate ($M \lesssim 5$) intracaldera earthquakes are common to all four calderas. In principle, the unrest in any of these calderas may eventually culminate in an explosive volcanic eruption; the challenge is to understand the processes driving the unrest and to be able to recognize clearly the signs precursory to an eruption.

In the case of Long Valley, the areal extent of the seismicity from 1978 to the present and the relative regularity of the major seismic episodes suggest that the current unrest is associated with a regional crustal strain event occurring at a relatively uniform rate (Savage & Cockerham 1987). Although we have made significant advances in understanding the structural setting and the geologic history of Long Valley caldera in recent years, our understanding of the underlying processes is still inadequate to reliably project the course the current unrest may take over the next few years to decades. Given the history of youthful volcanism and continuing tectonism in this region, however, there can be little doubt that volcanism will eventually return to Long Valley caldera or the adjacent Inyo-Mono Craters.

ACKNOWLEDGMENTS

We are grateful to our many colleagues who have given us their help, time, and encouragement over the past decade in our studies of the physical processes operating within Long Valley caldera. This research was supported by the US Department of Energy under contract DE-AC04-

76DP00789, and by the US Geological Survey under the Volcano Hazards and Geothermal Research Programs.

Literature Cited

Abers, G. 1985. The subsurface structure of Long Valley caldera, Mono County, California. *J. Geophys. Res.* 90: 3627–36

Bailey, R. A. 1987. Geologic map of Long Valley Caldera, Mono-Inyo Craters volcanic chain and vicinity, Mono County, California. *US Geol. Surv. Misc. Invest. Map I-1933.* In press

Bailey, R. A., Dalrymple, G. B., Lanphere, M. A. 1976. Volcanism, structure, and geochronology of Long Valley caldera, Mono County, California. *J. Geophys. Res.* 81: 725–44

Barker, J. S., Langston, C. A. 1983. A teleseismic body-wave analysis of the May 1980 Mammoth Lakes, California, earthquakes. *Bull. Seismol. Soc. Am.* 73: 419–34

Barker, J. S., Wallace, T. C. 1985. Inversion of the teleseismic body waves for the moment tensor of the November 23, 1984, Round Valley, California, earthquake. *Eos, Trans. Am. Geophys. Union* 66: 952 (Abstr.)

Blackwell, D. D. 1985. A transient model of the geothermal system of the Long Valley caldera, California. *J. Geophys. Res.* 90: 11,229–42

Castle, R. O., Estrem, J. E., Savage, J. C. 1984. Uplift across the Long Valley caldera, California. *J. Geophys. Res.* 89: 11,507–16

Chouet, B., Julian, B. R., 1985. Dynamics of an expanding fluid-filled crack. *J. Geophys. Res.* 90: 11,187–98

Christansen, R. L. 1984. Yellowstone magmatic evolution: its bearing on understanding large-volume explosive volcanism. In *Explosive Volcanism: Inception, Evolution, and Hazards,* pp. 84–95. Washington, DC: Natl. Acad. Press. 176 pp.

Cockerham, R. S. 1986. The Chalfant earthquake sequence, eastern California, of July and August 1986: preliminary results. *Eos, Trans. Am. Geophys. Union* 67: 1106 (Abstr.)

Corbett, E. J., Martinelli, D. M., Smith, K. D. 1985. Aftershock locations of the November 23, 1984 Round Valley, California earthquakes. *Eos, Trans. Am. Geophys. Union* 66: 952 (Abstr.)

Corbett, E. J., Depolo, D. M., Delaplain, T. W. 1986. The July 1986 Chalfant Valley, California earthquake sequence: major events and temporal development of the

aftershock zone. *Eos, Trans. Am. Geophys. Union* 67: 1106 (Abstr.)

Cramer, C. H., Toppozada, T. R. 1980. A seismological study of the May, 1980, and earlier earthquake activity near Mammoth Lakes, California. In *Special Report 150: Mammoth Lakes, California Earthquakes of May 1980,* ed. R. W. Sherburne, pp. 91–130. Sacramento. Calif. Div. Mines Geol. 141 pp.

Dawson, P. B., Evans, J. R., Iyer, H. M. 1986. Preliminary results from teleseismic traveltime residuals in the Long Valley, California, region. *Eos, Trans. Am. Geophys. Union* 67: 1101 (Abstr.)

Deemer, S. J. 1985. *Seismic reflection profiling in the Long Valley caldera, California: data acquisition, processing, and interpretation.* PhD thesis. Univ. Wyo., Laramie. 195 pp.

Denlinger, R. P., Riley, F. 1984. Deformation of Long Valley caldera, Mono County, California, from 1975 to 1982. *J. Geophys. Res.* 89: 8303–14

Denlinger, R. P., Riley, F. S., Boling, J. K., Carpenter, M. C. 1985. Deformation of Long Valley caldera between August 1982 and August 1985. *J. Geophys. Res.* 90: 11,199–11,209

Eichelberger, J. C., Lysne, P. C., Miller, C. D., Younker, L. W. 1985. Research drilling at Inyo Domes, California: 1984 results. *Eos, Trans. Am. Geophys. Union* 65: 186–87

Eichelberger, J. C., Carrigan, C. R., Westrich, H. R., Price, R. H. 1986. Nonexplosive silicic volcanism. *Nature* 323: 598–602

Ekstrom, G., Dziewonski, A. M. 1983. Moment tensor solutions of Mammoth Lakes earthquakes. *Eos, Trans. Am. Geophys. Union* 64: 262 (Abstr.)

Elbring, G. J., Rundle, J. B. 1986. Analysis of borehole seismograms from Long Valley, California: implications for caldera structure. *J. Geophys. Res.* 91: 12,651–60

Fink, J. H. 1985. Geometry of silicic dikes beneath the Inyo Domes, California. *J. Geophys. Res.* 90: 11,127–34

Given, J., Wallace, T. C., Kanamori, H. 1982. Teleseismic analysis of the 1980 Mammoth Lakes earthquake sequence. *Bull. Seismol. Soc. Am.* 72: 1093–1109

Goldstein, N. E. 1987. Pre-drilling data review and synthesis for the Long Valley caldera, California. *Eos, Trans. Am. Geophys. Union.* In press

Hauksson, E. 1987. Absence of evidence for a magma chamber in downhole and surface seismograms from Long Valley caldera, eastern California. *J. Geophys. Res.* In press

Hermance, J. F. 1983. The Long Valley–Mono Basin volcanic complex in eastern California: status of present knowledge and future research needs. *Rev. Geophys. Space Phys.* 21: 1545–65

Hermance, J. F. 1987. Delineating the subsurface mega-structure of Long Valley caldera: regional gravity and magnetotelluric constraints. *Geol. Soc. Am. Bull.* In press

Hermance, J. F., Slocum, W. M., Neumann, G. A. 1984. The Long Valley–Mono Basin volcanic complex: a preliminary magnetotelluric and magnetic variation interpretation. *J. Geophys. Res.* 89: 8325–37

Hildreth, W. 1979. The Bishop Tuff: evidence for the origin of compositionally zoned magma chambers. *Geol. Soc. Am. Spec. Pap. No. 180*, pp. 43–75

Hildreth, W. 1983. The compositionally zoned eruption of 1912 in the Valley of Ten Thousand Smokes, Katmai National Park, Alaska. *J. Volcanol. Geotherm. Res.* 18: 1–56

Hill, D. P. 1976. Structure of Long Valley caldera, California, from a seismic refraction experiment. *J. Geophys. Res.* 81: 745–53

Hill, D. P. 1984. Monitoring unrest in a large silicic caldera, the Long Valley–Inyo Craters volcanic complex in east-central California. *Bull. Volcanol.* 47–2: 371–95

Hill, D. P., Bailey, R. A., Ryall, A. S. 1985a. Active tectonic and magmatic processes beneath Long Valley caldera, eastern California: an overview. *J. Geophys. Res.* 90: 11,111–20

Hill, D. P., Wallace, R. E., Cockerham, R. S. 1985b. Review of evidence on the potential for major earthquakes and volcanism in the Long Valley–Mono Craters–White Mountains regions of eastern California. *Earthquake Predict. Res.* 3: 571–94

Hill, D. P., Kissling, E., Luetgert, J. H., Kradolfer, U. 1985c. Constraints on the upper crustal structure of the Long Valley–Mono Craters volcanic complex, eastern California, from seismic refraction measurements. *J. Geophys. Res.* 90: 11,135–50

Hoover, D. B., Frischknecht, F. C., Tippens, C. L. 1976. Audiomagnetotelluric sounding as a reconnaissance exploration technique in Long Valley, California. *J. Geophys. Res.* 81: 801–9

Jachens, R. C., Roberts, C. W. 1985. Temporal and areal gravity investigations at

Long Valley caldera, California. *J. Geophys. Res.* 90: 11,210–18

Johnson, R. A., Deemer, S., Berg, R., Burke, M., Smithson, S. B., Rundle, J. B. 1986. Seismic reflection profiling in the Long Valley caldera. *Eos, Trans. Am. Geophys. Union* 67: 313 (Abstr.)

Julian, B. R. 1983. Evidence for dike intrusion earthquake mechanisms near Long Valley caldera, California. *Nature* 303: 323–25

Julian, B. R., Sipkin, S. A. 1985. Earthquake processes in the Long Valley caldera area, California. *J. Geophys. Res.* 90: 11,115–70

Kane, M. F., Mabey, D. R., Brace, R.-L. 1976. A gravity and magnetic investigation of the Long Valley caldera, Mono County, California. *J. Geophys. Res.* 81: 754–62

Kubota, S., Berg, E. 1967. Evidence for magma in the Katmai volcanic range. *Bull. Volcanol.* 31: 175–214

Lachenbruch, A. H., Sorey, M. L., Lewis, R. E., Sass, J. H. 1976a. The near-surface hydrothermal regime of Long Valley caldera. *J. Geophys. Res.* 81: 763–68

Lachenbruch, A. H., Sass, J. H., Munroe, R. J., Moses, T. H. 1976b. Geothermal setting and simple heat conduction models for the Long Valley caldera. *J. Geophys. Res.* 81: 769–84

Langbein, J., Linker, M., Tupper, D. 1987. Analysis of two-color Geodimeter measurements of deformation within the Long Valley caldera: June 1983 to October 1985. *J. Geophys. Res.* 92: 9423–42

Linker, M. F., Langbein, J. O., McGarr, A. 1986. Decrease in deformation rate observed by two-color laser ranging in Long Valley caldera, eastern California, 1983–1984. *Science* 232: 213–16

Lirer, L., Luongo, G., Scandone, R. 1987. On the volcanological evolution of Campi Flegrei. *Eos, Trans. Am. Geophys. Union* 68: 226–34

Luetgert, J. H., Mooney, W. D. 1985. Crustal refraction profile of the Long Valley caldera, California, from the January 1983 Mammoth Lakes earthquake swarm. *Bull. Seismol. Soc. Am.* 75: 211–21

Matumoto, T. 1971. Seismic body waves observed in the vicinity of Mount Katmai, Alaska, and evidence for the existence of molten chambers. *Geol. Soc. Am. Bull.* 82: 2905–20

Miller, C. D. 1985. Holocene eruptions at the Inyo volcanic chain, California—implications for possible eruptions in Long Valley caldera. *Geology* 13: 14–17

Mori, J., McKee, C. 1987. Outward-dipping ring-fault structure at Rabaul caldera as shown by earthquake locations. *Science* 235: 193–95

Muffler, L. J. P., Williams, D. L. 1976. Geothermal investigations of the US Geological Survey in Long Valley, California, 1972–1973, 1976. *J. Geophys. Res.* 81: 721–24

Murphy, W. J., Renaker, E., Robertson, M., Martin, A., Malin, P. E. 1985. The 1985 Mammoth wide-angle reflection survey. *Eos, Trans. Am. Geophys. Union* 66: 960 (Abstr.)

Newhall, C. G., Dzurisin, D., Mullineaux, L. S. 1984. Historical unrest at large Quaternary calderas of the world. *US Geol. Surv. Open-File Rep. 84–939*, 2: 714–42

Pelton, J. R., Smith, R. B. 1982. Contemporary vertical surface displacements in Yellowstone National Park. *J. Geophys. Res.* 87: 2745–51

Priestly, K. F., Brune, J. N., Anderson, J. G. 1985. Surface wave excitation and source mechanisms of the Mammoth Lakes earthquake sequence. *J. Geophys. Res.* 90: 11,177–86

Richter, C. F. 1958. *Elementary Seismology.* San Francisco: Freeman. 768 pp.

Rundle, J. B., Whitcomb, J. H. 1984. A model for deformation in Long Valley, California, 1980–1983. *J. Geophys. Res.* 89: 9371–80

Rundle, J. B., Whitcomb, J. H. 1986. Modeling gravity and trilateration data in Long Valley, California, 1983–1984. *J. Geophys. Res.* 91: 12,675–82

Rundle, J. B., Elbring, G. J., Striker, R. P., Finger, J. T., Carson, C. C. et al. 1985. Seismic imaging in Long Valley, California, by surface and borehole techniques: an investigation of active tectonics. *Eos, Trans. Am. Geophys. Union* 66: 194–200

Rundle, J. B., Carrigan, C. R., Hardee, H. C., Luth, W. C: 1986. Deep drilling to the magmatic environment in Long Valley caldera. *Eos, Trans. Am. Geophys. Union* 67: 490–91

Ryall, A., Ryall, F. 1980. Spatial-temporal variations in the seismicity preceding the May, 1980, Mammoth Lakes, California, earthquakes. In *Special Report 150: Mammoth Lakes, California Earthquakes of May 1980*, ed. R. W. Sherburne, pp. 27–40. Sacramento: Calif. Div. Mines Geol. 141 pp.

Ryall, A., Ryall, F. 1981b. Spatial-temporal variations in seismicity preceding the May 1980 Mammoth Lakes earthquakes, California. *Bull. Seismol. Soc. Am.* 71: 747–60

Ryall, A., Ryall, F. 1983. Spasmodic tremor and possible magma injection in Long Valley caldera, eastern California. *Science* 219: 1432–33

Ryall, F., Ryall, A. 1981a. Attenuation of *P* and *S* waves in a magma chamber in Long Valley caldera, California. *Geophys. Res.*

Lett. 8: 557–60

Sanders, C. O. 1984. Location and configuration of magma bodies beneath Long Valley, California, determined from anomalous earthquake signals. *J. Geophys. Res.* 89: 8287–8302

Sanders, C. O., Ryall, F. 1983. Geometry of magma bodies beneath Long Valley, California determined from anomalous earthquake signals. *Geophys. Res. Lett.* 10: 690–92

Savage, J. C., Clark, M. M. 1982. Magmatic resurgence in Long Valley caldera, California: possible cause of the 1980 Mammoth Lakes earthquakes. *Science* 217: 531–33.

Savage, J. C., Cockerham, R. S. 1984. Earthquake swarm in Long Valley, California, January 1983: evidence for dike injection. *J. Geophys. Res.* 89: 8315–24

Savage, J. C., Cockerham, R. S. 1987. Quasiperiodic occurrence of earthquakes in the 1978–1986 Bishop–Mammoth Lakes sequence, eastern California. *Bull. Seismol. Soc. Am.* 77: 1347–58

Savage, J. C., Cockerham, R. S., Estrem, J. E., Moore, L. R. 1987. Deformation near the Long Valley caldera, eastern California, 1982–1986. *J. Geophys. Res.* 92: 2721–46

Sieh, K. E., Bursik, M. 1986. Most recent eruption of the Mono Craters, eastern central California. *J. Geophys. Res.* 91: 12,539–71

Smith, J. L., Rex, R. W. 1977. Drilling results from the eastern Long Valley caldera. *Am. Nucl. Soc. Conf. Energy and Mineral Resour. Recovery, Golden, Colo.*, pp. 529–40

Smith, K. D., Martinelli, D. M., Corbett, E. J. 1985. Focal mechanisms of the November 23, 1984, Round Valley, California earthquakes. *Eos, Trans. Am. Geophys. Union* 66: 952 (Abstr.)

Smith, R. B., Braile, L. W. 1984. Crustal structure and evolution of an explosive silicic volcanic system at Yellowstone National Park. In *Explosive Volcanism: Inception, Evolution, and Hazards*, pp. 96–109. Washington, DC: Natl. Acad. Press. 176 pp.

Sorey, M. L. 1985. Evolution and present state of the hydrothermal system in Long Valley caldera. *J. Geophys. Res.* 90: 11,219–29

Sorey, M. L., Lewis, R. E. 1976. Convective heat flow from hot springs in the Long Valley caldera, Mono County, California. *J. Geophys. Res.* 8: 785–91

Stanley, W. D., Jackson, D. B., Zohdy, A. A. R. 1976. Deep electrical investigations in the Long Valley geothermal area, California. *J. Geophys. Res.* 81: 810–20

Steeples, D. W., Iyer, H. M. 1976. Low-velocity zone under Long Valley as determined from teleseismic events. *J. Geophys. Res.* 81: 849–60

Steeples, D. W., Pitt, A. M. 1976. Microearthquakes in and near Long Valley, California. *J. Geophys. Res.* 81: 841–47

Telford, W. M., Geldart, L. P., Sheriff, R. E., Keys, D. A. 1976. *Applied Geophysics.* Cambridge: Cambridge Univ. Press. 860 pp.

Tono, H., Malin, P. E. 1986. Shallow crustal modeling of Mammoth wide angle reflection data. *Eos, Trans. Am. Geophys. Union* 67: 1101 (Abstr.)

Uhrhammer, R. A. 1986. The 1986 Chalfant Valley earthquake sequence. *Eos, Trans. Am. Geophys. Union* 67: 1106 (Abstr.)

Uhrhammer, R. A., Ferguson, R. W. 1980. The 1980 Mammoth Lakes earthquake sequence. In *Special Report 150: Mammoth Lakes, California Earthquakes of May 1980*, ed. R. W. Sherburne, pp. 131–36. Sacramento: Calif. Div. Mines Geol. 141 pp.

Van Wormer, J. D., Ryall, A. S. 1980. Sierra Nevada–Great Basin boundary zone: earthquake hazard related structure, active tectonic processes, and anomalous patterns of earthquake occurrence. *Bull. Seismol. Soc. Am.* 70: 1557–72

Wallace, T. C. 1985. A reexamination of the moment tensor solutions of the 1980 Mammoth Lakes earthquakes. *J. Geophys. Res.* 90: 11,171–76

Wallace, T. C., Given, J., Kanamori, H. 1982. A discrepancy between long- and short-period mechanisms of earthquakes near the Long Valley caldera. *Geophys. Res. Lett.* 9: 1131–34

Whitcomb, J. H., Rundle, J. B. 1985. Vertical distortion in Long Valley, California, associated with the January 1983 earthquake swarm. *Geophys. Res. Lett.* 12: 522–25

Wood, S. H. 1977. Distribution, correlation, and radiocarbon dating of late Holocene tephra, Mono and Inyo Craters, eastern California. *Geol. Soc. Am. Bull.* 88: 89–95

Ann. Rev. Earth Planet. Sci. 1988. 16: 273–93

OBSERVATIONS OF COMETARY NUCLEI

Michael F. A'Hearn

Astronomy Program, University of Maryland, College Park, Maryland 20742

INTRODUCTION

Despite the wealth of information obtained on the nucleus of comet P/Halley by the various spacecraft that visited the comet, the study of cometary nuclei is still in its infancy; our direct knowledge of their properties is almost negligible. In fact, there are probably more theories describing details of cometary nuclei than there are well-determined properties. There are two reasons for this dearth of directly observed data. On the one hand, when a comet is near the Sun, the nucleus is almost always lost in the central condensation of the coma. On the other hand, when a comet is far from the Sun and the nucleus is presumably unobscured, the nucleus is too faint to be observed. In fact a crude estimate is that only about half the periodic comets have nuclei that would be detectable at aphelion even with the Hubble Space Telescope. Nevertheless, the study of cometary nuclei has important implications for our understanding of the solar system.

One of the obvious reasons for studying cometary nuclei is to understand many of the phenomena in cometary comae. Thus far, research has gone in the direction of using observed phenomena in comae to infer properties of nuclei. It is quite likely that studies of cometary nuclei will indeed lead to a new understanding of how comae and their phenomena are produced, although it is likely that most future work on comets will continue in the older vein of inferring nuclear properties from observations of comae. This is merely because it is easier to observe phenomena in the coma than it is to observe nuclei. A more important reason for studying cometary nuclei is to use them as clues to more far-reaching cosmogonical questions tied to the origin and evolution of the solar system. If the usual paradigm, in which comets are coeval with the planetary system and have since remained

273

0084–6597/88/0515–0273$02.00

forever cold, is indeed correct, then the molecular composition and the micro- and macroscopic physical characteristics of cometary nuclei will provide a wealth of information about the conditions under which the solar system formed. The birth of comets is thus tied directly to the birth of the solar system. The death of comets is also intimately tied to our understanding of the subsequent evolution of the solar system. Until quite recently it was thought that most of the Earth-approaching Amor-Apollo asteroids were necessarily extinct cometary nuclei rather than asteroids perturbed from the asteroid belt (e.g. Wetherill 1979). Recent work by Wisdom (1985) has shown that this need not be the case, but it is still likely that some fraction of these asteroids are composed of extinct comets. Furthermore, observational results in the last few years, particularly the spacecraft images of Halley's nucleus, argue strongly for the formation of inert mantles on comets, implying that cometary nuclei must leave behind inert, asteroidal bodies when they exhaust their volatiles. The ultimate goal, then, in studying cometary nuclei is to understand the birth and death of comets and thereby relate them to the formation and evolution of our planetary system.

It is in this context of the birth and death of comets that one even calls into question the distinction between comets and asteroids. With the discovery of asteroidal bodies in orbits associated with meteor streams, the observational distinction becomes somewhat fuzzy; in this review we reserve the term *cometary nuclei* to bodies that, at least at some time, have been observed as fuzzy (i.e. that have exhibited detectable outgassing). The term *asteroids* then encompasses all bodies that have not exhibited outgassing, whether the body is an extinct cometary nucleus or, like Chiron, one that has never come close enough to the Sun for us to determine whether or not it is a cometary nucleus.

In this review we concentrate on the attempts to observe cometary nuclei, particularly observations that are used to determine fundamental physical parameters of these bodies and that are relevant to the relationship between comets and asteroids. We do not discuss models of nuclei in great detail. Although one might expect the spacecraft encounters with P/Halley to have solved many outstanding questions, it turns out that the thrust of this review is to emphasize the uncertainties and even inconsistencies that have arisen in the last few years, particularly with the numerous results on P/Halley. Two other recent reviews of comets in general include sections on cometary nuclei. Mendis et al (1985) devote nearly 100 pages of their massive treatise on comets to the subject of nuclei, while Spinrad (1987) also devotes a section of his review on cometary composition to the question of nuclei. The models of cometary nuclei are discussed extensively in the articles by A. H. Delsemme, J. M. Greenberg, and B. Donn & J.

Rahe in *Comets* (Wilkening 1982). The exciting recent results on comet P/Halley are discussed in many different papers in the proceedings of the 20th ESLAB Symposium on the Exploration of Halley's Comet (Battrick et al 1986).

PHYSICAL PROPERTIES

We begin with a discussion of the basic physical properties of cometary nuclei—their size, shape, reflectivity, and rotation. These basic properties are all rather poorly known except for a very few comets where they have been more or less directly measured. Inferences about other comets are based almost exclusively on assumptions about one or another of these properties.

Size and Albedo

Although these two properties are physically unrelated, they are tied so closely together observationally that it is most convenient to discuss them together. Since cometary nuclei have never been spatially resolved except by the spacecraft that visited P/Halley, nearly all early estimates of the sizes of cometary nuclei were based on measured optical brightnesses combined with assumptions about the albedo. Unfortunately, the albedo was totally unknown until recently, so that one was forced to cite a rather wide range of albedos. As an example, Roemer (1966) followed earlier analyses to derive nuclear radii of many comets from nuclear magnitudes measured at the Flagstaff station of the US Naval Observatory. She was forced to assume albedos and chose to use what were thought at the time to be the extreme possible values. This led to radii differing by a factor of 6, and even at that she deduced a maximum radius for comet P/Arend-Rigaux only 60% of the now well-established value. On the other hand, her maximum radius for comet P/Encke neatly bracketed the most probable, although still not yet well-determined, value in use today. These discrepancies may have been partly due to the choice of albedo or partly due to inadequate nuclear magnitudes.

Delsemme & Rud (1973) were the first to attempt to determine independently the radius and albedo of cometary nuclei. Their remarkable insight was to use the brightness at large distances from the Sun to estimate the product of cross section and albedo and to use estimates of the total vaporization rate near the Sun to estimate the product of cross section and energy absorbed (i.e. $1 - A$, where A is the albedo). They used nuclear magnitudes by Roemer to determine AS, where S is the cross section. To determine $(1 - A)S$ they assumed that the gas production when the comet was near the Sun was dominated by equilibrium vaporization of water that used nearly all the absorbed energy. They then used ultraviolet measurements of hydrogen and hydroxyl to estimate the production rate of gas. They found rather high albedos (65% for the Bond albedo) for the nuclei of comets Tago-Sato-Kosaka 1969 IX and Bennett 1970 II. The corresponding radii were 2.2 and 3.8 km, respectively. Application of the same method to comet P/Encke led to inconsistencies, which caused Delsemme & Rud to conclude that the vaporization must occur from only a relatively small fraction of the surface, a conclusion supported by all subsequent work. O'Dell (1976) applied the Delsemme-Rud method to comet Kohoutek. Using postperihelion magnitudes at large distances from the Sun, he deduced a radius of 2.0 km and an albedo of 0.73. He also noted that the preperihelion magnitudes contradicted this and were unrealistic, implying that the preperihelion magnitudes did not refer to the nucleus.

With our current understanding of cometary nuclei, vaporization probably occurs from only a small fraction of the surface and thus invalidates the method for nearly all periodic comets; not only is the effective area different for the reflected light and for the vaporization equilibrium, but also the inactive areas of the nucleus become very hot and therefore reradiate a nonnegligible fraction of the total absorbed energy even for an active comet like P/Halley. There are not sufficient data on new and long-period comets to decide whether or not the method is valid for them, but for dynamically new comets it is almost certain that all preperihelion magnitudes, no matter what the heliocentric distance, will be contaminated by a halo of grains.

Another approach to determining the radius of cometary nuclei is to use back-scattered radar measurements. If one simply measures the returned power, a radar albedo must be assumed, leaving us no further ahead than in assuming optical albedos. However, if the spin vector (both rate and orientation) is known, then the bandwidth of the returned signal can be used to determine the equatorial radius of the nucleus. Radar signals have been successfully bounced off comets Encke and IRAS-Araki-Alcock. In the case of comet Encke, the spin vector had been inferred by Whipple & Sekanina (1979), and this result was used by Kamoun et al (1982) with their radar measurements to deduce a radius of 1.5(+2.3, −1.0) km. If the spin axis is correctly known, the main uncertainty here is in fitting the spectral profile of the returned radar signal, since one must measure the full width of the points where it merges with the noise. A strong radar return from comet IRAS-Araki-Alcock exhibited a spectral profile implying that much of the signal was actually returned from boulders orbiting the nucleus (Goldstein et al 1984). Goldstein et al inferred a radius of 3 to 4 km, but without knowing the orientation of the axis, they were not able to determine a unique radius.

More recently, several investigators have attempted to use infrared observations of the nuclei of a variety of comets in order to determine the size and albedo of the nucleus directly. Again the reflected light of the nucleus gives the quantity AS directly, and if the vaporization by the nucleus does not take up a significant fraction of the absorbed energy, then the thermal emission (typically peaked in the 10–20 μm range) is directly proportional to $(1 - A)S$. This method has been widely applied to determining the sizes of asteroids where all the absorbed energy is reradiated as thermal energy [see Lebofsky et al (1986, and references therein) for a discussion of the methods]. The most significant problem in this approach is the separation of the nuclear brightness from that of the coma. Even at large distances from the Sun, one cannot rely on the absence of a coma, as pointed out by Roemer (1966) and many others. In the few years immediately preceding Halley's current apparition, a number of comets were observed specifically for the purpose of determining nuclear properties. These were comets selected specifically because their nuclei were thought to dominate the light of negligible or nonexistent comae.

Cruikshank & Brown (1983) measured the thermal flux from Schwassmann-Wachmann 1, a comet in a nearly circular orbit just outside the orbit of Jupiter that is normally stellar in appearance but undergoes large, unpredictable outbursts in brightness with an associated coma. Although they did not attempt to separate the coma from the nucleus in the infrared, nor did they simultaneously measure the reflected light, they argued that the comet was quiescent at this time and that the last outburst had taken place months earlier. They therefore were probably measuring the nucleus, and their interpolation of other data for the optical brightness was reasonable. They found a radius of 40 km and a geometric albedo of 0.13, making this nucleus one of the largest known.

Our group (H. Campins, L.-A. McFadden, R. L. Millis, and myself) has concentrated on the study of relatively nearby comets with faint comae, and we have attempted to explicitly separate the contributions by the coma and the nucleus. Furthermore, we have emphasized simultaneity of the measurements in the thermal infrared and the reflected optical. Thus far we have studied in detail the nuclei of comets P/Neujmin 1 (Campins et al 1987) and P/Arend-Rigaux (Millis et al 1987) (see also the preliminary summary in A'Hearn 1986). In both cases it is clear that there was significant variability over a short time scale. The better-studied comet was P/Arend-Rigaux, and Figure 1 shows the simultaneous reflected and thermal light curves. We have argued that these light curves can only be interpreted as resulting from the varying cross section of the nucleus. For both comets, we also measured the variation in brightness with aperture size to remove the contribution by the coma; in the case of Neujmin 1, we

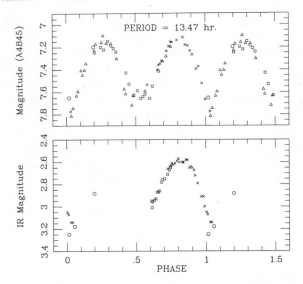

Figure 1   Light curve of comet P/Arend-Rigaux (after Millis et al 1987). The lower curve is thermal emission at 10 μm, while the upper curve is simultaneously measured reflected sunlight at 4845 Å. Data from three nights are combined with a period of 13.47 hr. The lack of a phase shift between the two curves shows that the variation is due to variable cross section rather than variable albedo. Millis et al give a variety of arguments to show that the cross section of the nucleus is varying, rather than the cross section of dust in the coma.

measured only an upper limit to the brightness of the coma in the continuum. After correcting for the coma, we derived effective radii at maximum light of 10.6 and 5.2 km, lower limits to the axial ratios of 1.45 and 1.6, and geometric V-band albedos of 0.02–0.03 and 0.028 for Neujmin 1 and Arend-Rigaux, respectively. Bond albedos are smaller than the geometric albedos by a factor that depends on the phase function. These comets are therefore much blacker, and thus larger, than inferred in previous studies. The radius and albedo of Arend-Rigaux are similar to those derived in less-complete studies by Tokunaga & Hanner (1985), Brooke & Knacke (1986), Veeder et al (1987), and Birkett et al (1987). The variability was also in agreement with that found by other groups (see below).

The most dramatic results for the size of a cometary nucleus have come, of course, from the spacecraft that flew through the coma of P/Halley. With the images from those spacecraft one can directly measure the projected dimensions, and by using the various aspects seen by the three spacecraft with imaging systems (Vega 1 and 2 and Giotto) one can also reconstruct that actual three-dimensional shape. Figure 2 shows one of the best images from Giotto, which has been processed specifically to show the limb of the

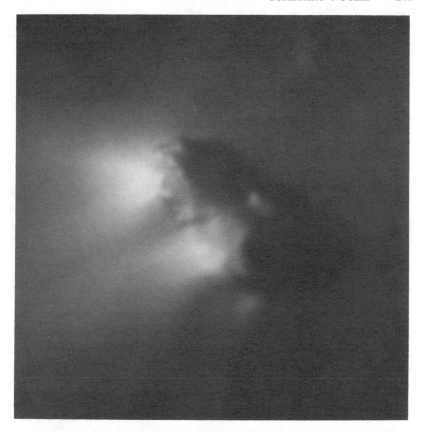

Figure 2   Composite image of the nucleus of Comet P/Halley, synthesized from 60 different images taken with the Halley Multicolour Camera (HMC) on Giotto. The Sun is to the left, and north is toward the top. The dark side of the nucleus is silhouetted against the light scattered by dust in the coma. The terminator is clearly defined, as is the complex surface structure in the daylit regions. The bright region on the dark side of the terminator is caused by a "hill" that extends into morning sunlight. The bright jets are seen by scattering of sunlight from the dust in the jets. Projected dimensions are 15 × 8 km. Copyright 1986 by the Max-Planck-Institüt. Image provided by Harold Reitsema on behalf of the Giotto HMC team headed by Uwe Keller.

nucleus. It is clear that the shape is irregular, characterized by various people as "peanut-" or "potato-shaped." Most investigators have given numerical values for a triaxial ellipsoid, although it is clear that one "end" of the nucleus is larger than the other. Using the best images from Giotto, Reitsema et al (1986) found projected dimensions of 14.9 by 8.2 km for an effective projected radius of 5.5 km. When images from more than one spacecraft are used, typical reconstructed dimensions are 16 × 8 × 6 km

(Möhlmann et al 1986), $16 \times 10 \times 9$ km (Wilhelm et al 1986), and $16.0 \times 8.2 \times 7.5$ km (Sagdeev et al 1986a, Figure 9) for maximum projected effective radii near 6 km. Sagdeev et al (1986b) have derived the surface reflectivity using Vega images over phase angles of 20–100° and inferred a geometric albedo of 0.04. Similarly, Delamere et al (1986) have used images from Giotto to derive an average geometric albedo of 0.035, although the extrapolation in phase was much larger than with the Vega images because Giotto approached from the dark side of the nucleus.

Thus the images from the spacecraft dramatically confirm what had begun to be deduced from ground-based observations in the few years immediately preceding the encounters: Cometary nuclei, at least those of periodic comets, are remarkably dark and elongated. Since most estimates of their sizes made more than a few years ago assumed or inferred rather high albedos, we infer that nearly all previous estimates of nuclear sizes, other than those listed by Roemer (1966) as assuming an albedo of 0.02, are too small. It remains to be seen whether new comets are also dark, although such a hypothesis is quite plausible because only a relatively small amount of dark material mixed with ice makes the mixture extremely dark (see, for example, Clark 1982).

Shape and Rotation

There have been two relatively recent reviews of the rotational properties of cometary nuclei, one each by Whipple (1982) and Sekanina (1981), the two individuals who have obtained most of the results. Whipple gives an extensive table of the 47 rotational periods known at that time from his method. Recent results, however, have called into question many of the earlier results and may require a complete rethinking of the whole approach to determining rotational properties. We therefore discuss the older results briefly.

Whipple's original method involves looking for recurrent features in the coma, particularly halos, and has led to the largest number of rotational periods (specifically the list in his review). The limitations of this method include the fact that frequently one must assume an expansion speed for the halos using an empirical relationship with heliocentric distance that was established using multiple images of halos in some of the comets. Furthermore, it implicitly assumes, in the absence of self-contradictory results, that the halos originate from a single active area on the nucleus. As pointed out by Mendis et al (1985), a discrepancy with Larson & Minton (1972) on the period for comet Bennett is probably due solely to the assumed expansion velocity. Whipple's period for P/Halley of 10.3 hr is more difficult to reconcile, but there are a variety of other problems in understanding the rotation of this comet, which we address below.

Sekanina's models of the nongravitational forces, originally developed in collaboration with Whipple, usually involve the precession of an oblate spheroid to explain the variations from one apparition to the next of the nongravitational force. The most recent evidence, however, suggests that a prolate shape may be more common among nuclei; this is certainly the case for P/Halley and is probably the case for those comets that show significant variation in cross section, such as P/Neujmin 1 and P/Arend-Rigaux.

Sekanina (1985) has applied the technique, for example, to comet P/Giacobini-Zinner and found that the nucleus must be remarkably flat (dimensions 1 × 8 × 8 km) and rotating nearly at breakup velocity, with a period of only 1.66 hr, the shortest period on record. Subsequent measurements during the 1985 apparition of this comet showed that the peak in production of OH occurred a month earlier than predicted by this model (McFadden et al 1987) and that structures seen in infrared images implied an equatorial plane deviating from that predicted (C. M. Telesco, private communication). It is likely, however, that these discrepant observations can be reconciled with only a small change of the orientation of the pole, and that the basic model is consistent with all subsequent data.

A case of particular interest is comet P/Encke, for which Whipple & Sekanina (1979) derived not only the rotational period but also the precessional history of the pole over the entire known history of the comet. Note that Whipple (1950) originally developed the dirty snowball model of cometary nuclei specifically to explain the nongravitational motions of this comet. Subsequently A'Hearn et al (1985) showed that one of their critical assumptions, a major asymmetry about perihelion in the production of gas, was invalid. Sekanina (1986a) then found evidence that the lag of maximum vaporization from the subsolar meridian was quite asymmetric about perihelion and recalculated the model. Interestingly, the orientation of the pole and its precession turned out very similar to that found previously with a quite different assumption about physical properties. This suggested that determination of the polar orientation was insensitive to the model used, so that one might learn the polar orientation without learning any other properties of the nucleus. More recent photometric results, discussed below, question even the orientation of the pole.

We infer from these results that even if there is significant variation with time in the nongravitational force, there may be a whole suite of models that will reproduce the nongravitational forces.

Recent developments in determining rotational periods have centered on photometric variability and on extensions of the method originally applied by Larson & Minton (1972) to comet Bennett. The discussion of

size and albedo above has shown unambiguously that many cometary
nuclei are far from spherical. Furthermore, it is clear that they do not
generally exhibit symmetry about their rotation axes. Asteroids, of course,
commonly exhibit brightness fluctuations due to their varying cross section
as they rotate, and observations of this variation as the viewing aspect
changes allow one to determine the spin vector. Because of the difficulty
in making the observations, it is not known whether or not rotational
modulation is also common for cometary nuclei, but the latest results
suggest that this is the case.

Prior to the reviews cited above, the only photometric determination of
a rotational period was that of P/d'Arrest by Fay & Wisniewski (1978),
who found a period of 5.2 hr. The signal-to-noise ratio of this result was
not high, and Whipple (1981) found a rather different period of 8.9 hr
using his halo method. More recently, Whipple (1982) has adopted a
period of 5.2 hr, but the initial discrepancy is a cause for concern that
either method might have significant uncertainties. Wisniewski et al (1986)
have derived photometric periods for comets P/Neujmin 1 and P/Arend-
Rigaux of 1.053 and 1.138 days, respectively. For both of these cases, they
show clearly convincing, repetitive light curves, although both curves
exhibit 4 peaks per period. In the case of P/Arend-Rigaux, Millis et al
(1987) have used the light curve of Figure 1 to deduce a period of 13.47
hr with 2 peaks per period, while Jewitt & Meech (1985) have used optical
data to deduce, as one of two possibilities, a period of 6.78 hr (or some
multiple thereof). Note that the three deduced periods for P/Arend-Rigaux
are all nearly commensurate; our period is almost exactly twice that of
Jewitt & Meech, while that of Wisniewski et al is roughly twice ours.
Furthermore, the light curves exhibit 2, 1, and 4 peaks per period, respectively. We believe that the arguments presented by Millis et al clearly
justify the choice of a double-peaked curve for this comet and thus a period
of 13.47 hr. For both these comets, the arguments by Millis et al and by
Campins et al (1987) convincingly show that most of the observed variation
in brightness is due to variations in the projected cross section of the
nucleus, suggesting that perhaps the true rotational period of P/Neujmin
1 may be near 12.6 hr, i.e. about half the period found by Wisniewski et
al. The variations in the shape of the light curve from one cycle to the next
could be due to residual activity of the nucleus. For these two comets, the
amplitudes of the variation imply minimum axial ratios near 1.5.

Jewitt & Meech (1987) have studied the variability of P/Encke near
aphelion and found variations with amplitude 0.8 mag and a period
(assuming a double-peaked light curve) of 22.43 hr. The amplitude, if it is
totally due to variation in the projected cross section, implies a minimum
for the axial ratio near 2. These results are totally inconsistent with the

precessional model (discussed above) of Whipple & Sekanina (1979) and Sekanina (1986a), which predicts an oblate spheroid of axial ratio near unity and utilizes a period of 6.25 hr derived by Whipple's halo method. Furthermore, the rotational pole of that model was nearly parallel (within 15°) to the line of sight during one of the observing runs when Jewitt & Meech detected the variability. This result reinforces the conclusion above that different physical models can equally well reproduce the variations in the nongravitational forces. Furthermore, it suggests that even the orientation of the pole may not be correctly determined. This is a very troubling conclusion, since it requires rethinking a large body of results. Perhaps all oblate nuclear models would lead to similar positions for the pole, whereas all prolate models would lead to a set of polar orientations quite similar to each other but quite different from the set found for oblate models.

By far the most dramatic results regarding rotation of cometary nuclei have come from the studies of P/Halley. Prior to mid-1986, Sekanina & Larson (1986, and references therein) had extensively developed the method originated by Larson & Minton (1972) to determine the spin vector of P/Halley and map the distribution of active areas on the nucleus. This method utilizes the times of occurrence, the positions, and the amount of curvature in narrow jets of dust observed in the inner coma. Their final deduced period for P/Halley was 52.1 hr, and the obliquity of the pole to the orbital plane was 30°. They even mapped the distribution of active areas on the surface of the nucleus. The shape of the nucleus is immaterial in this approach and was therefore assumed to be spherical. Their period was quite different from the 10.3 hr found by Whipple (1982) using the halo method.

Several attempts to obtain a period photometrically while the comet was still far from the Sun were unsuccessful because any periodic variability apparently was masked by sporadic activity and insufficient signal-to-noise ratio in the photometric data. Closer to perihelion, however, numerous types of data (ground-based photometry of many types and covering different periods, halos on photographs, ultraviolet brightness, etc) were used to infer a periodicity near 52 hr and a variety of other data (sunward spike and tailward jets) were used to infer the orientation of the pole. Many of these results, however, were not convincing, since they were really based on investigations to determine whether or not a 52-hr periodicity could be found in the various sets of data. As noted above, the spacecraft provided images showing an extremely prolate (axial ratio roughly 2) nucleus with its long axis close to both the ecliptic and the orbital plane. Comparison of the orientations of the nucleus in the images from the three spacecraft yielded a period near the 52 hr of Sekanina & Larson (1986).

The results to that point have been summarized in more detail by Sekanina (1986b).

Using narrow-band photometric observations of the coma, Millis & Schleicher (1986) found a very clear periodicity in the outgassing with a period of 7.37 days. When announced at the Heidelberg symposium on comet Halley, this led to two immediate effects. The observers rushed off to search their data for the longer period, and the theoreticians instantly produced models with combined rotation and precession in order to explain the two periods. The situation is still not understood. Festou et al (1987) showed that a 7.3-day (or 14.6-day) periodicity is present in much older photometric data, and we demonstrated (Samarasinha et al 1986, Hoban et al 1987) that the morphology of the jets of free radicals, which is quite different from the morphology of the jets of dust, is dominated by the 7.37-day period. Schloerb et al (1986) found that the production of HCN observed with radio telescopes was also correlated with fluctuations in morphology of jets and production of optically observed gases. Smith et al (1987) have reexamined the images from the spacecraft and concluded that the orientations of the nucleus observed at the times of the three encounters require that the principal motion have a 2.2-day period around an axis more or less orthogonal to the long axis of the nucleus. They suggest that the 7.4-day period might be associated with a small-amplitude precession.

One characteristic of the images of P/Halley was the existence of discrete jets emanating from a small fraction of the surface. Since these jets are distributed far from uniformly about the surface, they should produce torques. Samarasinha et al (1986) pointed out that the torques should lead to significant changes in the rotational period. Wilhelm (1987) has since calculated a model for the nuclear motion that exhibits forced precession of about 30° over the apparition. If the nucleus is deformable and dissipative, it could then also have a free-precessional period of 14.8 days, just twice the period found by Millis & Schleicher (1986). This model is very attractive in explaining many of the observed phenomena. It is difficult, however, to reconcile the forced precession with the fact that the dust jets imply a polar orientation in 1986 similar to that for 1910, as found by Sekanina & Larson (1986). It is also difficult to reconcile large torques, which produce both changes in period and precession, with the fact that P/Halley's nongravitational forces appear to be remarkably constant in time (Yeomans & Kiang 1981).

The principal conclusion to be drawn from these recent studies is that cometary nuclei are often prolate rather than oblate. Furthermore, the motions are apparently much more complex than assumed in the models, leading to significant questions about the physical source of periodicities found in observed data. One must therefore suspend judgment on most of the various determinations of spin vectors until the proper model has been found.

Composition

The chemical composition of cometary nuclei is critical to understanding the origin of comets. Unfortunately this review is being completed about one year too early to understand the wealth of data acquired from P/ Halley. Most of the relevant data have not been fully reduced, let alone interpreted in terms of the origin of nuclei. Prior to P/Halley's current apparition, the reasonably direct knowledge of nuclear composition was limited to knowing that H_2O, HCN, S_2, and perhaps CO or CO_2 and CH_3CN were present in the volatile fraction, with silicates in the refractory fraction. Atomic abundances showed that carbon appeared to be depleted by a factor of several from the volatile fraction.

We now know a number of additional facts. It is clear that CO is present in the nucleus at a level of 5–10% and that a comparable amount of CO is produced in the inner coma by an as yet unknown process (Woods et al 1986, Eberhardt et al 1986a). This argues for very low temperatures of cometary formation, as does the presence of S_2. CO_2 is also present in the coma as about 3% of the volatiles (Combes et al 1986, Krankowsky et al 1986), but its abundance in the nucleus has not yet been reported. Isotope ratios of easily measured gaseous species ($^{18}O/^{16}O$, $^{15}N/^{14}N$, $^{13}C/^{12}C$, and $^{34}S/^{32}S$) seem to be terrestrial, but the significance of these results has not been determined. It has been suggested (Solć et al 1986) that ^{12}C is enriched in the grains. The D/H ratio is clearly enhanced compared with interstellar values but by much less than would be expected either from equilibrium at tens of degrees Kelvin or from models of interstellar chemistry on the surfaces of grains (Eberhardt et al 1986b, Schleicher & A'Hearn 1986, Thielens 1983). Clearly further modeling is required to understand the D/H ratio.

The more dramatic discoveries have come in our understanding of the refractory composition. Initial discoveries were reported by Kissel et al (1986a,b). The silicates turn out to be Mg- and Fe-silicates, as expected. More importantly, the spacecraft found both a large number of particles with compositions similar to that of carbonaceous chondrites and a large number of "CHON" particles composed purely of the elements Carbon, Hydrogen, Oxygen, and Nitrogen. These particles have been interpreted as organic polymers, and they have been invoked as the source of CO in the coma (Eberhardt et al 1986a) and of jets of free radicals (A'Hearn et al 1986a,b). There is much carbon in these particles, but it is very difficult to determine the absolute amount and therefore it is hard to determine whether or not the carbon missing from the volatiles is present in the CHON particles.

Because of the volatile state of our understanding, a proper review of the composition of cometary nuclei must be postponed for a year or more.

THE CONNECTION WITH ASTEROIDS

As noted in the introduction, it was thought until quite recently that Amor-Apollo (A-A) asteroids were too short-lived to be replenished with asteroids from the belt. Öpik (1963) was probably the first to suggest that therefore most of the A-A asteroids were extinct cometary nuclei. Wetherill (1979) summarized the dynamical arguments for this scenario. It was only in 1982 that Wisdom (1983) showed that the motion of asteroids near the 3/1 commensurability with Jupiter eventually (i.e. after millions of years) becomes chaotic and can lead to large eccentricities. He subsequently showed (Wisdom 1985; see also Wetherill 1984) that this chaotic behavior could replenish the supply of A-A asteroids, thus removing the "requirement" that many of the A-A asteroids be cometary nuclei. Nevertheless, some of the A-A asteroids might still be extinct cometary nuclei, as might some of the non-A-A asteroids, particularly those with $T < 3$, where T is the Tisserand invariant [the integral of the motion that remains constant during close encounters with Jupiter (Kresak 1979)]. Furthermore, the discovery of Chiron, an apparently asteroidal object orbiting between 8 and 19 AU from the Sun, has led to speculation about the possibility of observing cometary nuclei that have not yet begun to vaporize. It is therefore very important to determine whether cometary nuclei are physically similar to any asteroids.

Mantle Buildup

Over the last few years, numerous models have been calculated in which an inert mantle is allowed to build up on the surface of a cometary nucleus, thereby choking off vaporization. Models of this type have been calculated by Mendis & Brin (1977), Brin & Mendis (1979), Horanyi et al (1984), and Fanale & Salvail (1984). The implication of these models is that in some comets the mantles will be periodically blown off, but that in other comets the mantle will monotonically increase in thickness until vaporization is completely shut off. One of the most dramatic aspects of Figure 2 is the fact that all of the outgassing from the nucleus of P/Halley is confined to a few prominent jets that probably cover no more than 10% of the surface of the nucleus. A straightforward interpretation of this phenomenon is that a mantle has built up over most of the surface and that the jets occur where the mantle has not yet built up. Millis et al (1987) pointed out that the average vaporization per unit area of P/Halley was roughly 6×10^{15} $cm^{-2} s^{-1}$ at $r = 1.5$ AU. If this were concentrated in 10% of the surface, this would still be in good agreement with equilibrium vaporization (see, for example, Mendis et al 1985, Figure A19). Millis et al also pointed out that the average rates were more than an order of magnitude lower for comets P/Arend-Rigaux and P/Neujmin 1, thus arguing for virtually complete mantle coverage on these comets. The existence of mantles as predicted by the models is apparently confirmed. If this is correct, then many comets must necessarily leave behind inert bodies that should appear asteroidal.

Color and Surface Properties

Over the last several years, W. K. Hartmann, D. P. Cruikshank, and various collaborators have pioneered in measuring the infrared and visible colors of a number of comets in an attempt to relate comets to asteroids and satellites in the outer solar system. Their initial work was aimed at showing a general trend in *VJHK* colors and albedo with heliocentric distance and showing that comets were an extension of the sequence (e.g. Hartmann et al 1982, Hartmann & Cruikshank 1984). Most of their cometary data clearly referred to dust in the comae, but they argued that these data for the comae favored a nucleus composed of icy grains colored by carbonaceous dirt like that found in D-type (then known as RD) asteroids. More recently they have also looked at spectrophotometric data and expanded their data set (Hartmann et al 1987). In particular, they have compared cometary colors and albedos with those of a variety of asteroids selected for two criteria: The asteroid must have been identified as a likely extinct comet, preferably on dynamical grounds; and there must exist sufficient data to classify the asteroid in the standard taxonomy. Their conclusion is that asteroids that are likely on dynamical grounds to be cometary are all dark and red (classes D, C, and P in the current notation). Unfortunately, their selection criteria eliminated several A-A asteroids that are thought by some to be extinct cometary nuclei—specifically 2201 Oljato, which they reject on dynamical grounds despite evidence for an atmosphere; and 3200 Phaeton = 1983TB, which they also reject on dynamical grounds despite its association with the Geminid meteor stream. This leaves only two A-A asteroids ($q < 1.3$ AU) in their table (although the text of their paper says four), and these are the asteroids into which one might expect comets to normally evolve. On the other hand, none of the three "bare" cometary nuclei in their sample has $q < 1.3$ either, owing to observational selection. The ten asteroids in their sample all turn out to be of types D, C (or its subclass B), or P or at least "like" these classes.

They also argue that the three bare cometary nuclei are either D or D-like, with reddish colors and low albedos.

While Hartmann et al are clearly correct in arguing that cometary nuclei are dark and reddish, as are asteroids of types C and D, the physically interesting question is whether or not there is a physical connection between the C- or D-type asteroids and the cometary nuclei. Tholen (1984) has argued for the importance of using the same type of data for all objects when doing classification work and has pioneered in this effort using the eight-color photometry of asteroids. While this is clearly the correct approach when actually doing classifications, it is not a sufficient approach when testing the hypothesis that cometary nuclei bear a physical relationship to certain types of asteroids. For that, one must also use all other existing data to test for agreement or disagreement. Hartmann et al have also recognized this and therefore use, for example, the albedo, which is not a part of the classification system. Likewise, they correctly note that overemphasis on the color we reported for P/Neujmin 1 led to the erroneous impression that this nucleus was like an S-type asteroid, which would have a far higher albedo than does Neujmin 1. In our own work (Campins et al 1987, Millis et al 1987), which provided part of the data for Hartmann et al, we have sought a "complete match" and therefore emphasized the differences rather than the similarities between cometary nuclei and asteroids. We conclude that there are both significant differences among the cometary nuclei and significant differences between the cometary nuclei and each of the asteroidal types. This leads us to question the association between cometary nuclei and C- or D-type asteroids, despite the fact that we agree with Hartmann et al on all the basic similarities.

Some of the key differences between the three cometary nuclei and asteroids are as follows. Comet P/Schwassmann-Wachmann 1 has an albedo of 0.13, which is brighter than the albedos of any C- or D-type asteroids for which albedos have been measured. Furthermore, in plots of $J-H$ vs $H-K$ and of $V-J$ vs $J-K$ the values for Schwassmann-Wachmann 1 fall outside the regions defined by the C- and D-type asteroids (Hartmann et al 1982, Figures 7, 8). Using our data, on the $V-J$ vs $J-K$ plot, Arend-Rigaux would fall at the edge of the D region, while Neujmin 1 would fall in the C region; in the $J-H$ vs $H-K$ plot they both fall in the center of the spectrally degenerate C-D region. On the other hand the optical data show that Neujmin 1 has a turndown in the near-ultraviolet, which makes it at least as red in that spectral region as the reddest D-type asteroid, while Arend-Rigaux is more nearly gray in this region, more like C- than D-type asteroids. This inconsistency suggests to us that there may be significant physical differences between the two nuclei themselves and between them and any known asteroidal type. The only other nucleus studied in any detail is that of P/Halley, and the colors deduced from the images taken with Giotto are sufficiently uncertain (H. J. Reitsema, private communication) that no useful conclusions can yet be drawn about similarity to asteroids or even other cometary nuclei. We would stress that the observations are sufficiently difficult that substantial uncertainties remain in the data. Obviously much more work is needed to obtain observations of cometary nuclei in order to show how they are related to asteroids of various types.

Rotational Distribution

As discussed above, rotational vectors have been determined for a number of comets, albeit with some uncertainty in the validity of some of the results. Whipple (1982) inferred that the distribution of rotational periods (P) for comets was significantly different from that of small, A-A asteroids, in that the cometary distribution of log P was flatter and had a longer average period than the asteroidal one. Whipple therefore concluded that the comets were formed by a different process, i.e. that they were not collisional fragments. Farinella et al (1985) have recently argued that there are selection effects in the use of the A-A asteroids because they are typically more prolate than belt asteroids. They show that the distribution of cometary rotational periods does not differ significantly from that of small main-belt asteroids, which typically have light curves with rather small amplitudes. On the other hand, the recent evidence discussed above shows that comets are frequently prolate, so Whipple's original comparison with A-A asteroids is the appropriate one.

If the distribution of rotational periods is to be of use in understanding the relationship between comets and asteroids, we must also understand whether or not the distribution of rotational vectors is primordial. Whipple (1982) argued that there was some evidence, although not clear-cut, for a gradual increase of period due to sublimation and the resultant decrease in the moment of inertia. Furthermore, the torques due to asymmetric outgassing will change the distribution of rotational vectors. The torques normally considered in the precessional models lead only to precession, i.e. only to changes in the polar orientations. Samarasinha et al (1986) have pointed out, however, that a prolate nucleus spinning about its minor axis is likely to exhibit substantial torques acting to change the magnitude of the rotational vector. In the case of P/Halley, they estimated that the characteristic time to change the angular momentum $L/\dot{L}$ would be 3 yr. This implies substantial changes in the rotational period over a single apparition. Wilhelm (1987) has done a much more extensive calculation and predicted changes of the total angular momentum by several percent on a time scale of tens of days. These factors lead us to conclude that the distribution of rotational periods is unlikely to be determined by the process of cometary formation and must be used carefully for inferences about the origin of comets. It may be that the only sensible comparison is between the distribution for cometary nuclei and the distribution for asteroids that might plausibly be extinct comets, e.g. the C- and D-type A-A asteroids.

SUMMARY

1. Cometary nuclei, at least of periodic comets, are bigger and blacker than generally thought as recently as five years ago. Geometric albedos may be typically 3%, almost unbelievably dark, and typical radii are probably of order 5 km. For only a very few comets are there independent determinations of albedo and size.
2. Nuclei of periodic comets are probably highly prolate unless they are both oblate and rotating about one of the major axes, which is unlikely.
3. There are significant discrepancies among the various methods for determining the rotational vectors of nuclei. The motions are probably much more complex for most comets than is assumed in models.
4. The images of P/Halley provide convincing evidence for the existence of mantles discussed in many models. Data for nearly "extinct" comets suggest that these comets are almost totally covered with refractory mantles, as predicted by models.
5. There are numerous pieces of evidence suggesting a connection between cometary nuclei and A-A asteroids of types D and C, but there are still a number of differences that must be understood before the connection is definite.
6. Future attempts to study cometary nuclei must be encouraged, since it is likely to be many years before another spacecraft views a cometary nucleus at short range.

ACKNOWLEDGMENTS

I would like to particularly thank my collaborators on the programs to study cometary nuclei: Lucy McFadden, who has tried to make me understand asteroids and geology; Humberto Campins, whose enthusiasm has pushed much of this work through; and Bob Millis, who made sure we were doing things carefully and correctly by finding my omissions and mistakes. Other astronomers who have helped with discussions are too numerous to mention, with the exceptions of Bill Hartmann and Dale Cruikshank, who have been invaluable both because of their own work on cometary nuclei, which has been particularly stimulating, and because Dale has assisted in arranging our observations at Mauna Kea. My own work on comets and thus on this paper has been possible only because of continuing funding through grant NSG-7322 from the NASA Planetary Astronomy Program.

Literature Cited

A'Hearn, M. F. 1986. Are cometary nuclei like asteroids? See Lagerkvist et al 1986, pp. 263–67

A'Hearn, M. F., Birch, P. V., Feldman, P. D., Millis, R. L. 1985. Comet Encke: gas production and light curve. *Icarus* 64: 1–10

A'Hearn, M. F., Hoban, S., Birch, P. V., Bowers, C., Martin, R., et al. 1986a. Cyanogen jets in comet Halley. *Nature* 324: 649–51

A'Hearn, M. F., Hoban, S., Birch, P. V., Bowers, C., Martin, R., et al. 1986b. Gaseous jets in comet P/Halley. See Battrick et al 1986, 1: 483–86

Battrick, B., Rolfe, E. J., Reinhard, R., eds. 1986. *ESLAB Symposium on the Exploration of Halley's Comet, 20th*, Vols. 1, 2, *ESA SP-250*. Noordwijk, Neth: ESA Publ. Div. 618 pp., 466 pp.

Birkett, C. M., Green, S. F., Zarnecki, J. C., Russell, K. S. 1987. Infrared and optical observations of low activity comets, P/Arend-Rigaux (1984k) and P/Neujmin 1 (1984c). *Mon. Not. R. Astron. Soc.* 225: 285–96

Brin, G. D., Mendis, D. A. 1979. Dust release and mantle development in comets. *Astrophys. J.* 229: 402–8

Brooke, T. Y., Knacke, R. F. 1986. The nucleus of comet P/Arend-Rigaux. *Icarus* 67: 80–87

Campins, H., A'Hearn, M. F., McFadden, L.-A. 1987. The bare nucleus of comet Neujmin 1. *Astrophys. J.* 316: 847–57

Clark, R. N. 1982. Implications of using broadband photometry for composition remote sensing of icy objects. *Icarus* 49: 244–57

Combes, M., Moroz, V., Crifo, J. F., Bibring, J. P., Coron, N., et al. 1986. Detection of parent molecules in comet Halley from the IKS-Vega experiment. See Battrick et al 1986, 1: 353–58

Cruikshank, D. P., Brown, R. H. 1983. The nucleus of comet P/Schwassmann-Wachmann 1. *Icarus* 56: 377–80

Delamere, W. A., Reitsema, H. J., Huebner, W. F., Schmidt, H. U., Keller, H. U., et al. 1986. Radiometric observations of the nucleus of comet Halley. See Battrick et al 1986, 2: 355–57

Delsemme, A. H., Rud, D. A. 1973. Albedos and cross-sections for the nuclei of comets 1969 IX, 1970 II and 1971 I. *Astron. Astrophys.* 28: 1–6

Eberhardt, P., Krankowsky, D., Schulte, W., Dolder, U., Lämmerzahl, P., et al. 1986a. On the CO and N₂ abundance in comet Halley. See Battrick et al 1986, 2: 383–86

Eberhardt, P., Dolder, U., Schulte, W., Krankowsky, D., Lämmerzahl, P., et al. 1986b. The D/H ratio in water from Halley. See Battrick et al 1986, 1: 539–41

Fanale, F. P., Salvail, J. R. 1984. An idealized short period comet model: surface insolation, H₂O flux, dust flux and mantle development. *Icarus* 60: 476–511

Farinella, P., Paolicchi, P., Zappala, V. 1985. On the rotation of cometary nuclei and small asteroids. In *Dynamics of Comets: Their Origin and Evolution*, ed. A. Carusi, G. B. Valsecchi, pp. 173–78. Dordrecht: Reidel. 439 pp

Fay, T. D., Jr., Wisniewski, W. 1978. The light curve of the nucleus of comet d'Arrest. *Icarus* 34: 1–9

Festou, M. C., Drossart, P., Lecacheux, J., Encrenaz, T., Puel, F., et al. 1987. Periodicities in the light curve of P/Halley and the rotation of its nucleus. *Astron. Astrophys.* In press

Goldstein, R. M., Jurgens, R. F., Sekanina, Z. 1984. A radar study of comet IRAS-Araki-Alcock 1983d. *Astron. J.* 89: 1745–54

Hartmann, W. K., Cruikshank, D. P. 1984. Comet color changes with solar distance. *Icarus* 57: 55–62

Hartmann, W. K., Cruikshank, D. P., Degewij, J. 1982. Remote comets and related bodies: *VJHK* colorimetry and surface materials. *Icarus* 52: 377–408

Hartmann, W. K., Tholen, D. J., Cruikshank, D. P. 1987. The relationship of active comets, "extinct" comets, and dark asteroids. *Icarus* 69: 33–50

Hoban, S., Samarasinha, N. H., A'Hearn, M. F., Klinglesmith, D. A. 1987. An investigation into periodicities in the morphology of CN jets in comet P/Halley. *Astron. Astrophys.* In press

Horanyi, M., Gombosi, T. I., Cravens, T. E., Korosmezey, A., Kecskemety, K., et al. 1984. The friable sponge model of a cometary nucleus. *Astrophys. J.* 278: 449–55

Jewitt, D., Meech, K. 1985. Rotation of the nucleus of comet P/Arend-Rigaux. *Icarus* 64: 329–35

Jewitt, D., Meech, K. 1987. CCD photometry of comet P/Encke. *Astron. J.* 93: 1542–48

Kamoun, P. D., Campbell, D. B., Ostro, S. J., Pettengill, G. H., Shapiro, I. I. 1982. Comet Encke: radar detection of nucleus. *Science* 216: 293–96

Kissel, J., Sagdeev, R. Z., Bertaux, J. L., Angarov, V. N., Audouze, J., et al. 1986a. Composition of comet Halley dust particles from Vega observations. *Nature* 321: 280–82

Kissel, J., Brownlee, D. E., Büchler, K., Clark, B. C., Vechtig, H., et al. 1986b. Composition of comet Halley dust particles from Giotto observations. *Nature* 321: 336–37

Krankowsky, D., Lämmerzahl, P., Herrwerth, I., Woweries, J., Eberhardt, P., et al. 1986. In situ gas and ion measurements at comet Halley. *Nature* 321: 326–29

Kresak, L. 1979. Dynamical interrelations among comets and asteroids. In *Asteroids*, ed. T. Gehrels, pp. 289–309. Tucson: Univ. Ariz. Press. 1181 pp.

Lagerkvist, C-I., Lindblad, B. A., Lundstedt, H., Rickman, H., eds. 1986. *Asteroids, Comets, Meteors II*. Uppsala, Swed: Uppsala Univ. 620 pp.

Larson, S. M., Minton, R. B. 1972. Photographic observations of comet Bennett 1970 II. In *Comets: Scientific Data and Missions*, ed. G. P. Kuiper, E. Roemer, pp. 183–208. Tucson: Univ. Ariz. Press. 222 pp.

Lebofsky, L. A., Sykes, M. V., Tedesco, E. F., Veeder, G. J., Matson, D. L., et al. 1986. A refined "standard" thermal model for asteroids based on observations of 1 Ceres and 2 Pallas. *Icarus* 68: 239–51

McFadden, L.-A., A'Hearn, M. F., Feldman, P. D., Böhnhardt, H., Rahe, J., et al. 1987. Ultraviolet spectrophotometry of comet Giacobini-Zinner during the ICE encounter. *Icarus* 69: 329–37

Mendis, D. A., Brin, G. D. 1977. The monochromatic brightness variations of comets—II. The core-mantle model. *The Moon and the Planets* 17: 359–72

Mendis, D. A., Houpis, H. L. F., Marconi, M. L. 1985. The physics of comets. *Fundam. Cosmic Phys.* 10: 1–380

Millis, R. L., Schleicher, D. G. 1986.

Rotational period of comet Halley. *Nature* 324: 646–49

Millis, R. L., A'Hearn, M. F., Campins, H. 1987. An investigation of the nucleus and coma of comet P/Arend-Rigaux. *Astrophys. J.* In press

Möhlmann, D., Borner, H., Danz, M., Elter, G., Mangoldt, T., et al. 1986. Physical properties of P/Halley derived from Vega-images. See Battrick et al 1986, 2: 339–40

O'Dell, C. R. 1976. Physical processes in comet Kohoutek. *Publ. Astron. Soc. Pac.* 88: 342–48

Öpik, E. J. 1963. Survival of comet nuclei and the asteroids. *Adv. Astron. Astrophys.* 2: 219–62

Reitsema, H. J., Delamere, W. A., Huebner, W. F., Keller, H. U., Schmidt, W. K. H., et al. 1986. Nucleus morphology of comet Halley. See Battrick et al 1986, 2: 351–54

Roemer, E. 1966. The dimensions of cometary nuclei. *Mem. Soc. R. Sci. Liège 15th Ser.* 15: 23–28

Sagdeev, R. Z., Avanesov, G. A., Shamis, V. A., Ziman, Ya. L., Krasikov, V. A., et al. 1986a. TV experiment in Vega mission: image processing technique and some results. See Battrick et al 1986, 2: 295–305

Sagdeev, R. Z., Avanesov, G. A., Ziman, Ya. L., Moroz, V. I., Tarnopolsky, V. I., et al. 1986b. TV experiment of the Vega mission: photometry of the nucleus and inner coma. See Battrick et al 1986, 2: 317–

Samarasinha, N. H., A'Hearn, M. F., Hoban, S., Klinglesmith, D. A. 1986. CN jets of comet Halley—rotational properties. See Battrick et al 1986, 1: 487–91

Schleicher, D. G., A'Hearn, M. F. 1986. Comets P/Giacobini-Zinner and P/Halley at high dispersion. In *New Insights in Astrophysics. ESA SP-263*, pp. 31–33. Noordwijk, Neth: ESA Publ. Div.

Schloerb, F. P., Kinzel, W. M., Swade, D. A., Irvine, W. M. 1986. HCN production from comet Halley. See Battrick et al 1986, 1: 577–81

Sekanina, Z. 1981. Rotation and precession of cometary nuclei. *Ann. Rev. Earth Planet. Sci.* 9: 113–45

Sekanina, Z. 1985. Precession model for the nucleus of periodic comet Giacobini-Zinner. *Astron. J.* 90: 827–45

Sekanina, Z. 1986a. Effects of the law for nongravitational forces on the precession model of comet Encke. *Astron. J.* 91: 422–31

Sekanina, Z. 1986b. Nucleus studies of comet Halley. *Adv. Space Res.* 5: 307–16

Sekanina, Z., Larson, S. M. 1986. Coma morphology and dust-emission pattern of periodic comet Halley. IV. Spin vector refinement and map of discrete dust sources for 1910. *Astron. J.* 92: 462–82

Smith, B. A., Larson, S. M., Szego, K., Sagdeev, R. Z. 1987. Rejection of a proposed 7.4-day rotation period of the comet Halley nucleus. *Nature* 326: 573–74

Solć, M., Vanýsek, V., Kissel, J. 1986. Carbon stable isotopes in comets after encounters with P/Halley. See Battrick et al 1986, 2: 373–76

Spinrad, H. 1987. Comets and their composition. *Ann. Rev. Astron. Astrophys.* 25: 231–69

Thielens, A. G. G. M. 1983. Surface chemistry of deuterated molecules. *Astron. Astrophys.* 229: 177–84

Tholen, D. J. 1984. *Asteroid taxonomy from cluster analysis of photometry.* PhD. dissertation. Univ. Ariz., Tucson

Tokanuga, A. T., Hanner, M. S. 1985. Does comet P/Arend-Rigaux have a large dark nucleus? *Astrophys. J.* 296: L13–16

Veeder, G., Hanner, M. S., Tholen, D. J. 1987. The nucleus of comet P/Arend-Rigaux. *Astron. J.* 94: 169–73

Wetherill, G. W. 1979. Steady-state populations of Apollo-Amor objects. *Icarus* 37: 96–112

Wetherill, G. W. 1984. The asteroidal source region of ordinary chondrites. *Meteoritics* 19: 335

Whipple, F. L. 1950. A comet model. I. The acceleration of comet Encke. *Astrophys. J.* 111: 375–94

Whipple, F. L. 1981. On observing comets for nuclear rotation. In *Modern Observational Techniques for Comets*, ed. J. C. Brandt, B. Donn, J. M. Greenberg, J. Rahe, pp. 191–201. Pasadena, Calif: Jet Propul. Lab. 319 pp.

Whipple, F. L. 1982. Rotation of comet nuclei. See Wilkening 1982, pp. 227–50

Whipple, F.L., Sekanina, Z. 1979. Comet Encke: precession of the spin axis, nongravitational motion and sublimation. *Astron. J.* 84: 1894–1909

Wilhelm, K. 1987. Rotation and precession of comet Halley. *Nature* 328: 27–30

Wilhelm, K., Cosmovici, C. B., Delamere, W. A., Huebner, W. F., Keller, H. U., et al. 1986. A three-dimensional model of the nucleus of comet Halley. See Battrick et al 1986, 2: 367–69

Wilkening, L. L., ed. 1982. *Comets.* Tucson: Univ. Ariz. Press. 766 pp.

Wisdom, J. 1983. Chaotic behavior and the origin of the 3/1 Kirkwood gap. *Icarus* 56: 51–74

Wisdom, J. 1985. Meteorites may follow a chaotic route to Earth. *Nature* 315: 731–33

Wisniewski, W. Z., Fay, T., Gehrels, T. 1986. Light variations of comets. See Lagerkvist et al 1986, pp. 337–39

Woods, T. N., Feldman, P. D., Dymond, K. F. 1986. The atomic carbon distribution in the coma of comet Halley. See Battrick et al 1986, 2: 431–35

Yeomans, D. K., Kiang, T. 1981. The long-term motion of comet Halley. *Mon. Not. R. Astron. Soc.* 197: 633–46

Ann. Rev. Earth Planet. Sci. 1988. 16: 295–317

THE GEOLOGY OF VENUS

Alexander T. Basilevsky

V. I. Vernadsky Institute of Geochemistry and Analytical Chemistry, Academy of Sciences of the USSR, Moscow, USSR

James W. Head, III

Department of Geological Sciences, Brown University, Providence, Rhode Island 02912

1. Introduction

The nature of the surface of Venus is one of the keys to answering fundamental questions about the origin and evolution of the terrestrial planets and is of critical significance to comparative planetology. The last 25 years of solar system exploration have provided unprecedented views of the Earth, Moon, Mars, and Mercury. These views have shown that the smaller terrestrial planetary bodies (those one half the radius of the Earth or less: the Moon, Mercury, and Mars) are characterized by globally continuous, unsegmented lithospheres that stabilized very early in the history of the solar system and by ancient surfaces that preserve the several-billion-year-old record of early heavy bombardment and early heating and volcanism (Head & Solomon 1981). In contrast, the Earth is characterized by a globally segmented, laterally moving lithosphere, which is created at divergent plate boundaries and destroyed at convergent plate boundaries. Movement is measured in centimeters per year, and the average age of the surface of the planet is less than 2 b.y. Plate tectonics and plate recycling are fundamental mechanisms of heat loss for the Earth, in contrast to conduction, which is the dominant mechanism for the smaller, one-plate planetary bodies (Solomon & Head 1982). What are the reasons for these differences? Is size, or perhaps position in the solar system and initial conditions, the key? Venus, which is approximately the same size and density as Earth and is the closest planet to Earth, offers an opportunity to test these ideas.

0084–6597/88/0515–0295$02.00

The dense cloud cover of Venus has kept global and regional visible-wavelength panoramas of its surface from view, however, and until the development of high-resolution Earth-based radar telescopes and orbital radar altimetry and imaging systems, the geological nature of the surface of Venus had literally been a mystery. Over the last few years, data have been accumulating and a picture of the nature of the surface of Venus is very slowly emerging (McGill et al 1983, Florensky et al 1983a, Surkov 1983, Phillips & Malin 1983). Recent missions have provided global low-resolution data on the general surface characteristics (Pettengill et al 1980), and local high-resolution images have begun to provide a sense of the geologic structure of parts of the planet (Campbell et al 1983, 1984), as well as detailed panoramas and chemical composition measurements of the actual surface materials (Florensky et al 1983b, Surkov et al 1984). The most exciting developments in the last several years have been (a) the Venera 15/16 imaging radar missions (Kotelnikov et al 1984), which have provided high-resolution imaging coverage of the northern mid- to high latitudes (approximately 20–25% of the planet); (b) the continuing analysis of the Earth-based radar data; (c) the new discoveries from the Pioneer-Venus data; and (d) the new chemical data from the Vega 1 and 2 landers. This review summarizes the emerging picture of the characteristics of the surface of Venus from these data and provides a progress report on the nature and significance of geological processes operating there. We conclude with an assessment of the type of information necessary to complete this emerging picture, so that the themes of terrestrial planet formation and evolution can be understood.

2. Global Characteristics, Composition, and the Nature of Local Surfaces

The Pioneer-Venus mission radar experiment obtained near-global data for the surface of Venus, from which altimetry, roughness, and reflectivity values have been determined at approximately 100-km average horizontal resolution (Figure 1). On the basis of these data, the following features were established: global hypsometry (Venus is distinctly unimodal in contrast to the bimodality of the Earth), the major physiographic provinces [Venus can be subdivided into lowlands (about 27% of the surface), rolling uplands (about 65%), and highlands (about 8%)], and the distribution of areas of anomalous roughness and reflectivity (Pettengill et al 1980, Masursky et al 1980, Garvin et al 1985).

Using these three data sets, map units were compiled and their characteristics were interpreted in terms of geological processes and the nature of the surface (Head et al 1985). It appears that the vast majority of the surface of Venus is made up of regionally contiguous block-covered and

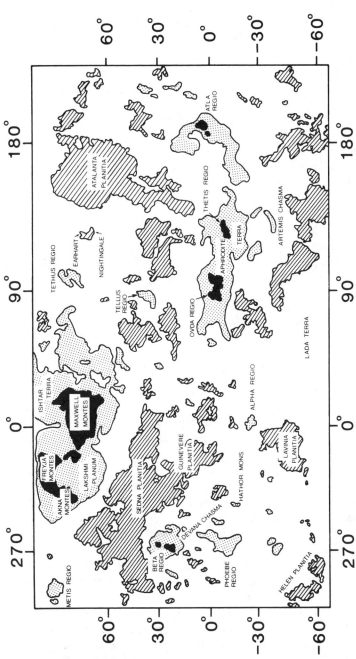

Figure 1 Reference map of Venus, showing the geographic location of various features mentioned in the text. The lowlands are indicated by oblique ruling, the rolling plains by white areas, and the highlands by stippled and black patterns (Masursky et al 1980). The black pattern indicates the location of the highest mountainous regions in the highlands (those regions in excess of 4 km above the mean planetary radius of 6051.0 km).

bedrock surfaces, while less than one fourth of the surface contains porous and unconsolidated soillike material. The distribution of soillike deposits does not support the presence of large areas of ancient, impact-produced regolith or regional pyroclastic deposits. A small percentage of the surface is characterized by very high dielectric materials. These areas occur predominantly at high altitudes in regions interpreted to be of both tectonic (Maxwell Montes) and volcanic (Rhea and Theia Mons in Beta Regio) origin, but they also occur at lower elevations in patterns indicating a possible volcanic flow origin (Head et al 1985).

The nature and distribution of regional topographic slopes have been analyzed and compared with those of the Earth (Sharpton & Head 1985, 1986). Although regional slopes on the Earth and Venus span the same range (0–2.4°), the slope frequency distributions are distinctly different, with Earth characterized by an excess of extremely low slopes due to abundant regions of planation and deposition. Venus has a distinct peak in slope frequency at about 0.09°, probably related to the lack of atmospheric/hydrospheric erosional processes and associated planation and deposition as seen on Earth. Approximately twice as much of the Earth's surface is characterized by slopes in excess of 0.24° as Venus', a difference primarily attributable to the presence of continental margins on Earth. The regional distribution of slopes shows that highland areas of Venus are different, with Ishtar Terra characterized by steep bounding slopes, and Aphrodite Terra by more symmetrical, generally shallower slopes.

The series of Soviet Venera lander missions provided insight into the nature of the surface of Venus at a local scale from the point of view of imaging and surface composition. Chemical analyses of surface materials at several landing sites have been interpreted in terms of terrestrial mafic rocks with normal alkalinity, such as tholeiitic basalts or gabbros (Surkov et al 1987, Barsukov et al 1982, 1986a). At two sites the composition of the surface suggests the presence of more differentiated material close in composition to terrestrial subalkaline basaltoids (Venera 13) and perhaps similar to syenites (Venera 8). The work of C. P. Florensky and his colleagues characterized the geology of the surface of Venus and showed that it was dominated by blocks and a layered or laminated pavement (interpreted to be of erosional, duricrust, or pyroclastic origin), with small amounts of surface soil cover (Florensky et al 1977, 1983a,b). Additional studies interpreted the pavement to be of volcanic lava-flow origin (Garvin et al 1984). In a recent joint study between Soviet and US investigators (Pieters et al 1986), multispectral images of the basaltic surface of Venus obtained by Venera 13 were processed to remove the effects of the orange-colored incident radiation resulting from interactions with the thick atmosphere. At visible wavelengths, the surface of Venus is shown to be dark and without significant color. High-temperature laboratory reflectance spectra of basaltic materials indicate that these results are consistent with either ferric or ferrous mineral assemblages. A high reflectance in the near-infrared observed for neighboring Venera 9 and 10 sites, however, suggests that the basaltic surface material contains ferric minerals and thus may be relatively oxidized.

3. Regional Distribution of Units: Venera 15/16 Results

Acquisition of high-resolution radar images by the Soviet Venera 15/16 spacecraft (Kotelnikov et al 1984) permitted the geologic characterization of the topographic features revealed by the Pioneer-Venus data (Figure 2) and an understanding of their regional distribution for the northern mid- to high latitudes (Barsukov et al 1986b, Basilevsky et al 1986). The Venera 15/16 data, in conjunction with the high-resolution Earth-based data (e.g. Campbell et al 1983, 1984, Stofan et al 1987a), revealed the presence of abundant volcanism, extremely complex tectonic deformation, unusual large ovoidal features of apparent volcano-tectonic origin, and an impact crater density providing an estimated age for the northern part of Venus of 0.5 to 1.0 b.y. (Ivanov et al 1986). Correlation of the Venera 15/16 geologic map and the Pioneer-Venus data permitted the derivation of roughness and reflectivity characteristics for the geologic units (Bind-schadler & Head 1986a,b). Over 70% of the surface imaged by Venera 15/16 consists of plains units interpreted to be of volcanic origin, while about 25% of the surface is characterized by highly deformed units of tectonic origin (Barsukov et al 1986b, Basilevsky et al 1986, Bindschadler & Head 1987). In the next section we review the characteristics of geologic processes interpreted from these data.

4. Evidence for Geological Processes

On the basis of available data, several types of geological processes can be identified as presently acting or having previously acted on Venus: volcanism, tectonism, impact cratering, gravity-induced downslope move-ment of surface material, eolian erosion/sedimentation, and chemical weathering.

Volcanism is evidently responsible for the formation of the venusian plains, i.e. for over 70% of Venus' surface. Venera 15/16 and Arecibo Observatory radar images of venusian plains display many radar-bright and radar-dark flowlike features up to 100–200 km long (Figure 3a,b). Their morphology and association with some volcanic centers (calderalike depressions, domes, constructs) and fault zones leave little doubt that they are solidified lava flows of relatively low viscosity (Barsukov et al 1984, 1985a,b, 1986b, Pronin 1986, Pronin et al 1986) typical of plains-forming

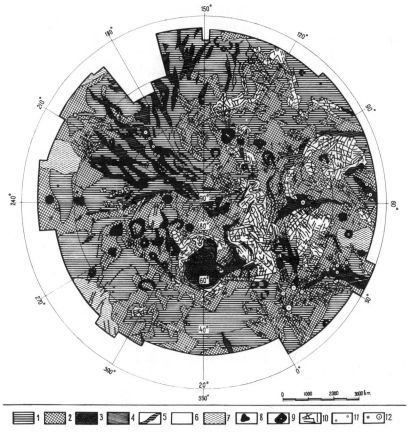

Figure 2   Geological-morphological map of the northern latitudes of Venus compiled from radar images obtained by the Venera 15/16 spacecraft by a mapping team of associates from the Vernadsky Institute of Geochemistry and Analytical Chemistry and the Geological Institute of the USSR Academy of Sciences (see Barsukov & Basilevsky 1986). The key is as follows: 1 (lowland smooth plains); 2 (lowland rolling plains); 3 (highland plains; Lakshmi Planum); 4 (mountain belts around Lakshmi Planum); 5 (ridge belts); 6 (tesserae, or "parquet" terrain); 7 (large dome uplifts); 8 (volcanoes); 9 (ovoids, or coronae); 10 (faults); 11 (impact craters); 12 ("arachnoids").

basaltic volcanism. This morphologic evidence is in good accordance with the previously discussed basaltic composition of the surface determined at the landing sites of the Venera 10, 13, and 14, and Vega 1 and 2 space-probes. The plains are evidently composed of a sequence of basaltic flows with some admixture of basaltic debris having eolian and/or pyroclastic origin.

In many places on the plains, numerous domes are observed with diameters of several to 15–20 km (Figure 3b), and a summit crater can be seen on the top of some of them. Spatial distribution of the domes is irregular (Slyuta et al 1987). Clusters of domes alternate with dome-free areas, indicating that their presence is not a required aspect of plains-forming volcanism.

Within the plains, Venera 15/16 images also show several tens of generally circular, gently sloping rises, 50 to 300 km in diameter and usually less than 1 km in height (Figures 2, 4). These structures have summit craters and calderas and also sometimes display radial systems of flowlike features. These characteristics suggest an origin as shield volcanoes. Theoretical considerations of volcanic processes in the Venus environment lead to the predictions that there will be less cooling of magma in the final stages of ascent and that once the magma reaches the surface, convective heat losses will be much more important than in the subaerial terrestrial environment because of the high atmospheric gas density. There appear to be no reasons, however, to expect large systematic differences between lava-flow morphologies on Venus and on Earth. On the other hand, conditions on Venus will tend to inhibit the subsurface exsolution of volatiles, and pyroclastic eruptions involving continuous magma disruption by gas bubble growth may not occur at all unless the exsolved magma volatile content exceeds several weight percent (Head & Wilson 1986).

In addition to the domes and larger constructs mentioned above, the venusian plains are often complicated by narrow (up to 10–25 km), linear features (low ridges, shallow grooves, radar-bright bands of unclear nature), as well as impact craters and buttes of parquetlike terrain. Combinations of these landforms are responsible for the morphological variability of venusian plains. Barsukov et al (1986b) classified the plains into several types: (a) ridge-and-band plains, (b) band-and-ring plains, (c) patchy rolling plains, (d) dome-and-butte plains, and (e) smooth plains. The age relations of plains belonging to the various types are not yet clear, and more analysis is required. The average crater retention age of venusian plains within the Venera 15/16 survey area is about 1 b.y. (Ivanov et al 1986).

Tectonic processes on Venus are interpreted from consideration of elevation and morphologic characteristics available from Pioneer-Venus and Venera 15/16 radar surveys and from Arecibo and Goldstone Station Earth-based radar observations. On the basis of these data, four partly overlapping groups of terrains whose origin is evidently due to tectonism (or a combination of tectonism and volcanism) can be tentatively distinguished within the region surveyed by Venera 15/16.

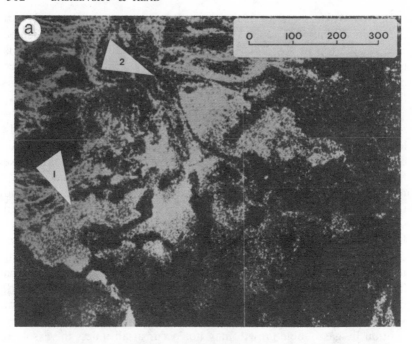

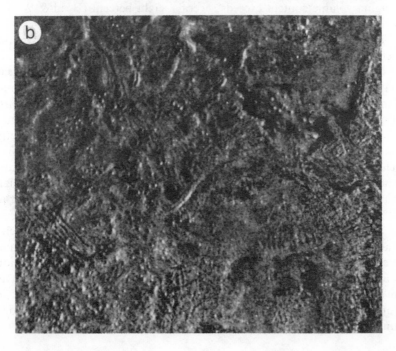

The first group is represented by *tesserae* (tiles in Greek), also often informally called "parquet" terrain. Tesserae are uplands whose morphology is dominated by densely packed systems of ridges and grooves transecting each other in diagonal, chevronlike, orthogonal, and/or chaotic manner (Figure 5). Typical ridge crest-to-crest spacing is about 5 to 20 km. The height of individual ridges over their base is no more than several hundred meters. Within the Venera 15/16 coverage, terrain of this type is mostly concentrated within the Ishtar Terra highlands [Tessera Fortuna (about 4000 by 1500 km), Tessera Laima (about 2000 by 1500 km), and several smaller features] and in some other uplands nearby (Tellus Regio, Tethus Regio). Taken together, the Venera 15/16 and Pioneer Venus data suggest a wide distribution of tesserae within the upland area outside Venera 15/16 coverage (Kreslavsky et al 1987). The morphology of this terrain seems to be related to deformation acting over broad areas, with stress and/or strain having predominantly horizontal components. The origin of the deformation is still controversial. Gravitational spreading of a surface layer of upraised areas (Sukhanov 1986), dragging of the base of the lithosphere by asthenospheric currents (Basilevsky 1986, Pronin, 1986), and general gravitational relaxation processes (Bindschadler et al 1987) have been proposed as mechanisms.

The second group (Figure 6) is represented by terrain whose morphology is dominated by systems of *subparallel ridges and grooves*. Their typical spacing is 5 to 20 km, and heights of individual ridges are typically no more than several hundred meters, similar to the tesserae mentioned above. This group is tentatively subdivided into two subgroups. Subgroup 1 (Figure 6a) is a system of subparallel ridges and grooves surrounding the upland of Lakshmi Planum and forming the *highland mountain belts* of Maxwell, Freyja, and Akna that stand above the adjacent plateau by several kilometers (Campbell et al 1983, Pronin et al 1986). Detailed analysis of Akna and Freyja Montes revealed the presence of anticlines and synclines, thrust faults, and strike-slip faults, and these characteristics were interpreted by Crumpler et al (1986) to indicate the presence of orogenic belts on Venus. The more equidimensional shape of Maxwell Montes has been interpreted to be due to several stages of deformation in

---

Figure 3 Volcanic plains. (*a*) Arecibo Observatory radar image of volcanic plains in Guinevere Planitia, southeast of Ishtar Terra. Arrow 1 points to a radar-bright oval feature about 200 km in length. A series of radar-dark flowlike features emerge from the center and extend into and merge with the surrounding plains. Arrow 2 indicates another area that appears to be a center of radar-dark, flowlike features. (*b*) Venera 15/16 image of western Atalanta Planitia (part of quadrangle 15-11), showing plains and abundant domes and cones. Image is 500 km in width. Scale in kilometers.

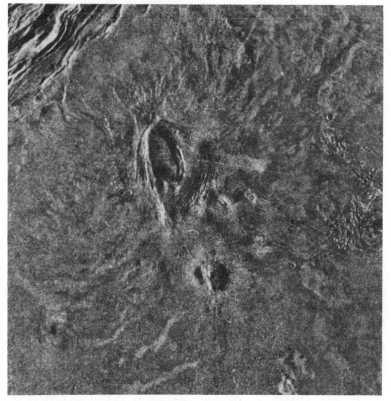

Figure 4 Venera 15/16 image (part of quadrangle 4-32) of Collette, a volcanic construct and caldera on Lakshmi Planum in Ishtar Terra. The oval calderalike structure is about 100 by 200 km, is about 1–3 km deep, and has a rim that rises less than 2 km above the surrounding plains. Numerous radar-bright and dark flowlike features extend away from Collette and flow for several hundreds of kilometers toward the surrounding plains.

which the banded terrain is offset by right-lateral movement along linear strike-slip faults, with offset over distances measured in the tens of kilometers (Vorder Bruegge et al 1985, 1986). Retrodeformation suggests that Maxwell had an original shape that was more linear, like that of Akna and Freyja, but that north-south compressional stress and strike-slip movement deformed it into its present form. The combined data in the Ishtar Terra region suggest that there is large-scale tectonic convergence and crustal thickening occurring there (Head 1986). An alternate view holds that the deformation is due to the local lateral displacement related to hotspot plumes centered in the Lakshmi Planum region (Pronin et al 1986).

Subgroup 2 (Figure 6*b*) is represented by *ridge-and-groove belts* on the

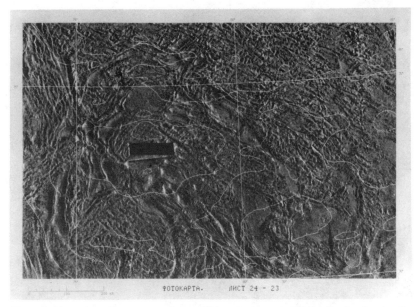

Figure 5   A section of tesserae, or "parquet" terrain, in central Tellus Regio, showing the complex patterns of deformation. The width of the Venera 15/16 subquadrangle (24-23) is 1000 km. Topographic contours are shown at 500-m contour interval.

plains. They are mostly abundant within the longitude range 150–230°, forming an approximately rhomboidal network (Figure 2) with about 300 to 500 km belt-to-belt spacing. Ridges and grooves within these belts are very similar to those in the highland mountain belts in their morphology, vertical amplitude, and spacing. The elevation of subgroup 2 ridge-and-groove belts above the surrounding plains is usually not more than several hundred meters, and in some cases they are even located within shallow troughs. In earlier publications, ridge-and-groove belts of both subgroups were considered as of compressional origin, resembling in some degree terrestrial folded belts (Barsukov et al 1984, 1985a,b, 1986b, Crumpler et al 1986). Subsequently, some workers have proposed that the belts of subgroup 2 may be extensional features formed by stretching and linear diapirism (Sukhanov & Pronin 1987).

The third group is represented by circular features, mainly so-called *coronae* for which the term *ovoids* is also used (Figure 7). These are ring-like systems of essentially concentric subparallel ridges and grooves generally higher than the surrounding plains. Sometimes radial features are also present in addition to the concentric pattern. The diameter of these rings ranges from 150 to 600 km. The morphology, vertical amplitude, and

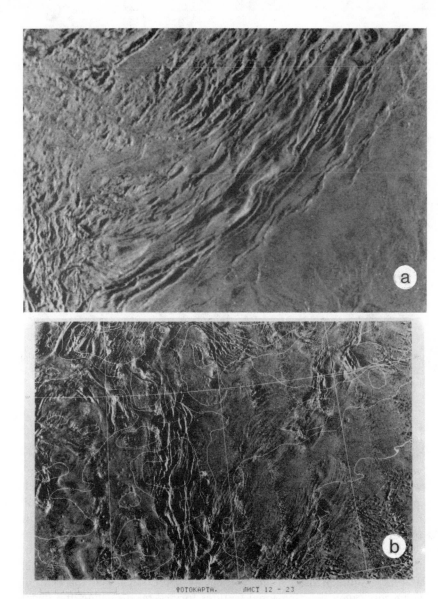

ФОТОКАРТА.   ЛИСТ 12 - 23

Figure 6 Terrain showing subparallel ridges and grooves. Subgroup 1 (highland mountain belts) is exemplified by Akna Montes, shown in (*a*). Here, parallel ridges are concentrated in the high topography of the mountain belt rising several kilometers above the surrounding plain and have been interpreted to be of compressional origin, comprising a component of orogenic belts (Crumpler et al 1986) (portion of quadrangle 4-32; width of image is 650 km). Examples of subgroup 2 (ridges and grooves) are shown in (*b*) and are characterized by parallel ridges and grooves that form in belts up to several hundred kilometers wide, extending across the surface for many hundreds of kilometers (quadrangle 12-23; width is 1000 km). These features have been interpreted to be of extensional or compressional origin by different workers. Topographic contours are shown at 500-m contour interval.

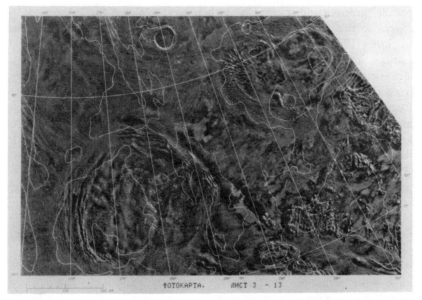

Figure 7   Two different types of coronae or ovoids. In the lower left is Anahit Corona, 400–500 km in diameter and showing an annulus of concentric ridges and a variety of surrounding flowlike features. Pomona Corona, in the upper right-hand area, has a distinctive radial pattern of grooves and ridges, as well as a more subdued concentric annulus. The width of the Venera 15/16 subquadrangle 3-13 is 1000 km.

spacing of ridges within these ringlike systems are mostly similar to those in the previously discussed ridge belts. The area inside the rings is usually lower in elevation than the surrounding ring and distinguished by somewhat chaotic morphology. As the coronae decrease in diameter they merge more or less gradually into another species of ringlike features, the so-called arachnoids, which are 50 to 200 km in diameter and are made up of concentric and concentric-radial systems of narrow ridges. At the larger end of the diameter range, the structure of Lakshmi Planum together with the surrounding ridge-and-groove mountain belts may be considered as a megacorona. The origin of coronae may be related to updoming over upwelling plumes or mantle diapirs, surficial deformation on the flank of the uprising, and subsequent collapse of the core. Stofan & Head (1986) have outlined a range of possible origins of coronae and have investigated gravitational relaxation and diapiric models for their origin and evolution. The origin of arachnoids is not clear.

The fourth group of terrains is represented by large uplands that display predominantly plainslike surface morphology and have systems of subparallel grooves and scarps along their crests. Within the Venera 15/16 coverage, this type of terrain is exemplified by Beta, Bell, and Ulfrum Regiones (Figures 1, 2). The elevation, morphology, and structure of these features leave little doubt that they result from rifting associated with tectonic updoming, or construction and loading. High-resolution images from the Arecibo Observatory revealed the detailed structure of Beta Regio and showed that the central linear depression is a rift zone several hundred kilometers wide containing multiple linear faults spaced 10–20 km apart within it (Campbell et al 1984). Volcanism is also associated with the rifting, and Theia Mons in Beta Regio is seen to be a large shield volcano that is superposed on the western bounding fault of the rift, partly flooding the rift valley. Further analysis of these data shows the relationships between the Arecibo images and the Venera 15/16 coverage in the northern part of Beta (Figure 8). Based on the full pattern of faults revealed by these two data sets, it seems that uplift has been a dominant process in the formation of Beta Regio (Stofan et al 1987b).

Aphrodite Terra (Figure 1), which is not covered by the Venera 15/16 data, is the largest highlands region on the planet and is characterized by large linear troughs interpreted to be of extensional origin (Schaber 1982). Detailed mapping of the Pioneer-Venus topography and imaging, as well as high-resolution Arecibo altimetry, has revealed the presence of major linear discontinuities striking across the topographic trend of Aphrodite (Figure 9). These features are several thousand kilometers long, strike N20°W, are parallel to one another, are separated by distances of 200–800 km, and are the location of sharp topographic discontinuities (Crumpler et al 1987). The characteristics of these features are similar to fracture zones and transform faults found in the terrestrial oceanic crustal environment. In addition, topographic profiles taken between but parallel to the discontinuities are highly symmetrical in broad form around a central axial high and also contain mirror-image shorter wavelength topographic elements across the high (Figure 9; Crumpler & Head 1987). On the basis of these observations and the comparison of many of the features of Aphrodite and terrestrial oceanic divergent plate boundaries, Aphrodite Terra appears to mark the location of extensional deformation and possible crustal spreading on Venus (Head & Crumpler 1987). According to Kreslavsky et al (1987), parts of western Aphrodite may be composed of tesseralike terrain.

The detailed mapping of the nature and spacing of many of the tectonic features described above and the recognition of two scales of deformation (about 100–300 km and 10–20 km) suggest that these length scales may be controlled by dominant wavelengths resulting from unstable compression or extension of the lithosphere. Modeling suggests that these patterns

Figure 8 Northern Beta Regio, showing the northernmost portion of Devana Chasma, the linear trough occurring along the rise crest of Beta Regio. Readily visible are the fault scarps defining the flanks of the rift zone, and the splaying pattern to the north (Stofan et al 1987b). The width of Arecibo Observatory image is 800 km.

could result from a lithosphere that at the time of deformation consisted of a crust that was relatively strong near the surface and weak at its base, and an upper mantle that was stronger than or of nearly comparable strength to the upper crust (Zuber 1987).

In summary, the data obtained thus far indicate that both horizontal and vertical tectonic movements are typical for the surface of Venus. The abundance of horizontal deformation is more similar to that observed on Earth than that commonly observed on the smaller terrestrial bodies, i.e. the Moon, Mercury, and Mars. However, for the area mapped by Venera 15/16, the general tectonic style is evidently different than that of the Earth because there appear to be few recognizable analogs of the key elements of global plate tectonics on Earth (e.g. interrelated planet-wide trenches

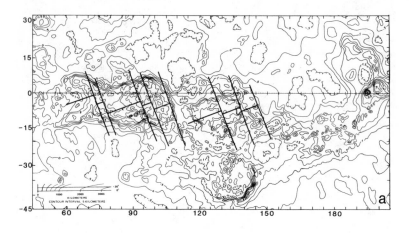

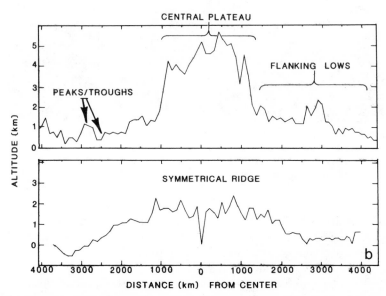

Figure 9   Topographic map of Aphrodite Terra (*a*) showing the location of the linear cross-strike discontinuities (bold lines), the bilateral symmetry of topography parallel to these features (*b*), and the lines connecting the centers of symmetry (rise crests) within domains between discontinuities (from Crumpler et al 1987, Crumpler & Head 1987).

and island arcs). The recent discoveries in Aphrodite Terra, however, suggest that divergence and crustal spreading may be occurring there (Crumpler & Head 1987, Head & Crumpler 1987). Thus, major remaining questions are linked to the nature and global distribution of tectonic styles on Venus, the manner in which these are linked to heat-transfer mechanisms, the way in which the Venus environment influences tectonic activity, and the similarity or lack of similarity in styles of tectonism on Venus and Earth.

Impact cratering on the venusian surface has been identified based on the morphological similarity of a number of craters observed on Venera 15/16 images with impact craters on other planetary bodies. Within the territory surveyed by Venera 15/16, about 150 craters (8 to 140 km in diameter) with impactlike morphology have been identified (Basilevsky et al 1985, 1987, Ivanov et al 1986, Kryuchkov 1987). They are superimposed upon terrains of all types, and their spatial distribution appears to be relatively even. The size-frequency distribution is unimodal with a mode lying in the 16–22.6 km diameter interval. The left branch of the distribution curve reflects the influence of the atmosphere in destroying projectiles. The right branch corresponds predominantly to the production function. On the basis of Hartmann's (1987) calibration curve, the average age of accumulation of the observed population is estimated as 1.0 ± 0.5 b.y.

The smallest of the craters observed have a bowl-shaped morphology. As the diameter increases, the craters display morphological transitions to knobby bottoms, then to central-peaked, and finally to ringed basins, similar to impact craters on other terrestrial and icy bodies (Florensky et al 1976, Basilevsky 1981, Basilevsky & Ivanov 1982, Basilevsky et al 1983, Pike 1977). Impact craters were subdivided into three classes according to the degree of morphological freshness (Figure 10): Class 1 are the freshest ones, with a surrounding radar-bright halo interpreted to be ejecta; class 2 have no halo, but the primary crater morphology is practically undisturbed; and class 3 have visible traces of modification, which, however, are not sufficient to cast into doubt the impact nature of these features. A large number of craterlike features that have been highly reworked by volcanic and tectonic processes can be seen on the Venera 15/16 images. Part of them may be a population of destroyed impact craters. On the basis of Hartmann's (1987) calibration curve, this population is estimated to have formed approximately between 3 and 1 b.y. ago (Nikolaeva et al 1986).

Gravitational downslope movement is thought to act on venusian slopes as a universal process of relaxation of steep slopes produced by other processes such as tectonics, volcanism, and impact cratering. The absence

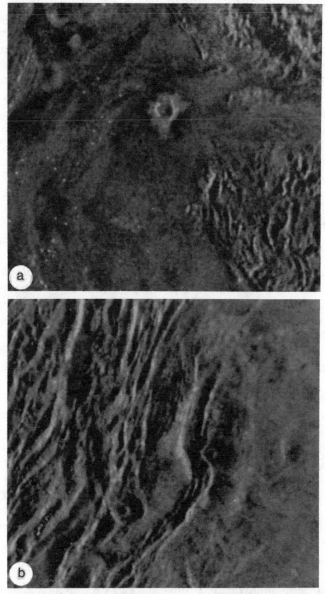

Figure 10   Impact craters on Venus. (*a*) Class 1, fresh crater (Ivka, about 18 km diameter; subquadrangle 4-31) with bright ejecta; (*b*) class 2, morphologically unmodified crater (*left*, Osiponko, about 30 km diameter; subquadrangle 4-22) with no bright ejecta, and class 3, morphologically modified crater (*right*, Vanda, about 30 km diameter; subquadrangle 4-22).

of significant systematic widening of slopes of impact craters belonging to different morphological (age) classes (Ivanov et al 1986) gives evidence of the lack of effectiveness of the process in this size range for at least the time of accumulation of the observed crater population (0.5–1.0 b.y.). Rock-fragment talus seen on the panorama at the Venera 9 landing site suggests that downslope movement of fragmental material occurs (Florensky et al 1983a).

The presence of eolian erosion/sedimentation processes on Venus is deduced from the observation of loose soil material, which is theoretically capable of being involved in eolian mobilization and transportation, and of winds, which are theoretically capable of mobilizing and transporting this material (Greeley et al 1984). The reality of such transport was effectively demonstrated by TV observations at the Venera 13 landing site. Several sequential TV pictures showed clearly that a soil clod of several centimeters in size that was thrown upon the lander's supporting ring was gradually removed during the approximately one-hour observation time (Selivanov et al 1983). Bare soil-free surfaces of local topographic prominences and the presence of soil in the wind-shadow lows observed on TV panoramas of the Venera 10, 13, and 14 landing sites are evidently the result of small-scale eolian processes on Venus (Florensky et al 1983a,b, Basilevsky et al 1986).

Among the features visible on Venera 15/16 images, only one type was suspected as having an eolian origin: radar-dark spots and bands localized at the local topographic lows that may imply wind-shadow conditions and eolian accumulations of loose material (dust?, sand?) (Barsukov et al 1986b). However, these accumulations are often observed in association with typical volcanic features such as clusters of domes and flowlike features, indicating that the origin of this radar-dark material may be volcanic but its distribution over the surface may be at least partly controlled by eolian processes.

Possible indirect evidence of eolian processes seen on Venera 15/16 images is the disappearance of radar-bright halos around impact craters undergoing the process of morphological maturation. The halos are zones of impact-induced surface roughness of decimeter-decameter scale and are present only around the youngest craters, totaling about one fourth of the crater population (whose total accumulation time is about 1 b.y.). These observations suggest that the decimeter-decameter roughness is smoothed out during a time period as long as 100–200 m.y. This leads to an estimation of an average rate of eolian (?) resurfacing of less than a few centimeters per million years.

Chemical weathering is another geological process whose presence on Venus is deduced from fundamental principles rather than from direct observations. Thermodynamic analysis shows that igneous basaltic mineral assemblages should be unstable in the venusian surface environment (Khodakovsky et al 1978, 1979, Barsukov et al 1982, 1986c, Volkov et al 1986). The estimates indicate that stable mineral associations should involve anhydrite and magnetite. Sulfatization and oxidation may be the main chemical processes on the venusian surface:

$$CaSiO_3 + 1.5SO_2 = CaSO_4 + SiO_2 + 0.25S_2, \qquad (1)$$

$$3FeSiO_3 + H_2O = Fe_3O_4 + 3SiO_2 + H_2. \qquad (2)$$

The high content of sulfur measured in the surface material by the Venera 13 and 14 and Vega 2 landers seems to be in accordance with these thermodynamic predictions and may be a result of incorporation of sulfur from atmospheric gases into the weathering products. The kinetics of this weathering is not well known, and this also makes unclear the scale and intensity of these processes on the surface of Venus.

5. *Summary*

Studies show Venus to be a dynamic planet whose surface is relatively young. (The average age of the observed part is more like that of the Earth than the ages of the smaller terrestrial planetary bodies.) The surface does not appear to be significantly modified by terrestrial-style erosion and deposition, but it is highly modified by volcanic resurfacing and a wide range of tectonic activity. The wide range of tectonic styles and the distribution of tectonic activity indicate that both regional extension/compression and vertical/horizontal tectonic deformation (including possible crustal spreading) operate there. The global distribution and integration of these features, and thus the global tectonic style and mechanisms of heat transfer, are, however, not yet clear.

The high-resolution data from Venera 15/16 have provided a major advance in our understanding of the nature of the surface of Venus, and continued analysis of these data will bring further important advances. The true picture of the global characteristics and distribution of geologic structures and features must await further data, however. The Magellan mission of the National Aeronautics and Space Administration is designed to obtain radar images over 90% of the surface of Venus at a typical resolution of several hundred meters, and this mission is scheduled to fly in the late 1980s. The data from this mission, combined with our emerging view of the planet, will provide global information on the surface of Venus that will be better than that presently available for the Earth. We eagerly anticipate the acquisition of these data, which should help us to better understand Venus, its relationship to the Earth, and its relevance to the basic themes of comparative planetology.

Literature Cited

Barsukov, V. L., Basilevsky, A. T. 1986. The geology of Venus. *Priroda*, pp. 24–35 (In Russian)

Barsukov, V. L., Volkov, V. P., Khodakovsky, I. L. 1982. The crust of Venus: theoretical models of chemical and mineral composition. *Proc. Lunar Planet. Sci. Conf., 13th, J. Geophys. Res.* 87: A3–9

Barsukov, V. L., Basilevsky, A. T., Kuzmin, R. O., Pronin, A. A., Kryuchkov, V. P., et al. 1984. Geology of Venus as revealed by analysis of radar images obtained by Venera 15 and Venera 16 (preliminary data). *Geokhimiya* 12: 1811–20 (In Russian)

Barsukov, V. L., Basilevsky, A. T., Kuzmin, R. O., other authors. 1985a. The major types of structures in the northern Venus hemisphere. *Astron. Vestn.* 19: 3–14 (In Russian)

Barsukov, V. L., Basilevsky, A. T., Kryuchkov, V. P., other authors. 1985b. Types of structures on the Venus surface. *Astron. Vestn.*, pp. 16–31 (In Russian)

Barsukov, V. L., Surkov, Yu. A., Dimitriev, L., Khodakovsky, I. L. 1986a. Geochemical study of Venus by landers of Vega-1 and Vega-2 probes. *Geokhimiya*, pp. 275–89 (In Russian)

Barsukov, V. L., Basilevsky, A. T., Burba, G. A., Bobina, N. N., Kryuchkov, V. P., et al. 1986b. The geology and geomorphology of the Venus surface as revealed by the radar images obtained by Veneras 15 and 16. *Proc. Lunar Planet. Sci. Conf., 17th, Part 2, J. Geophys. Res.* 91: D378–98

Barsukov, V. L., Borunov, S. P., Volkov, V. P., Dorofeyeva, V. A., Zolotov, M. Yu., et al. 1986c. Estimation of mineral composition of soil at the landing sites of Venera-13, 14 and Vega-2 landers according to thermodynamical calculations. *Dokl. Akad. Nauk SSSR* 287: 415–17 (In Russian)

Basilevsky, A. T. 1981. On some peculiarities on the structure of impact craters on planets and satellites of the solar system. *Dokl. Akad. Nauk SSSR* 258: 323–25 (In Russian)

Basilevsky, A. T. 1986. Structure of central and eastern areas of Ishtar Terra and some problems of venusian tectonics. *Geotektonika* 20: 282–88 (In Russian)

Basilevsky, A. T., Ivanov, B. A. 1982. Impact cratering on stony and icy bodies: different mechanisms of central peak formation? *Lunar Planet. Sci. XIII*, pp. 27–28 (Abstr.)

Basilevsky, A. T., Ivanov, B. A., Florensky, C. P., Yakovlev, O. I., Fel'dman, V. I., Granovsky, L. B. 1983. *Impact Craters on the Moon and Planets.* Moscow: Nauka. 200 pp. (In Russian)

Basilevsky, A. T., Ivanov, B. A., Kryuchkov, V. P., Kuzmin, R. O., Pronin, A. A., Chernaya, I. M. 1985. Impact craters on Venus based on radar images from Venera 15/16. *Dokl. Akad. Nauk SSSR* 282: 671–74 (In Russian)

Basilevsky, A. T., Pronin, A. A., Ronca, L. B., Kryuchkov, V. P., Sukhanov, A. L. 1986. Styles of tectonic deformations on Venus: analysis of Venera 15 and 16 data. *J. Geophys. Res.* 91: D399–411

Basilevsky, A. T., Ivanov, B. A., Burba, G. A., Chernaya, I. M., Kryuchkov, V. P., et al. 1987. Impact craters of Venus: a continuation of the analysis of data from the Venera 15 and 16 spacecraft. *J. Geophys. Res.* In press

Bindschadler, D. L., Head, J. W. 1986a. Characterization of Venera 15/16 units using Pioneer Venus reflectivity and RMS slope. *Lunar Planet. Sci. XVII*, pp. 50–51 (Abstr.)

Bindschadler, D. L., Head, J. W. 1986b. Pioneer Venus radar characteristics of the parquet terrain. *Lunar Planet. Sci. XVII* (*Suppl.*), pp. 1025–26 (Abstr.)

Bindschadler, D. L., Head, J. W. 1987. Characterization of Venera 15/16 geologic units from Pioneer Venus reflectivity and roughness data. Submitted for publication

Bindschadler, D. L., Head, J. W., Parmentier, E. M. 1987. Preliminary results of mapping and modeling of the parquet terrain, Venus. Submitted for publication (Abstr.)

Campbell, D. B., Head, J. W., Harmon, J. K., Hine, A. A. 1983. Venus: identification of banded terrain in the mountains of Ishtar Terra. *Science* 221: 644–47

Campbell, D. B., Head, J. W., Harmon, J. K., Hine, A. A. 1984. Venus: volcanism and rift formation in Beta Regio. *Science* 226: 167–70

Crumpler, L. S., Head, J. W. 1987. Bilateral topographic symmetry across Aphrodite Terra, Venus. *J. Geophys. Res.* In press

Crumpler, L. S., Head, J. W., Campbell, D.

316   BASILEVSKY & HEAD

B. 1986. Orogenic belts on Venus. *Geology* 14: 1031–34

Crumpler, L. S., Head, J. W., Harmon, J. K. 1987. Regional linear cross-strike discontinuities in western Aphrodite Terra, Venus. *Geophys. Res. Lett.* 14: 607–10

Florensky, C. P., Basilevsky, A. T., Grebennik, N. N. 1976. The relationship between lunar crater morphology and crater size. *The Moon* 16: 59–70

Florensky, C. P., Ronca, L. B., Basilevsky, A. T., Burba, G. A., Nikolaeva, O. V., et al. 1977. The surface of Venus as revealed by Soviet Venera 9 and 10. *Geol. Soc. Am. Bull.* 88: 1537–45

Florensky, C. P., Basilevsky, A. T., Burba, G. A., Nikolaeva, O. V., Pronin, A. A., et al. 1983a. Panorama of Venera 9 and 10 landing sites. See Hunten et al 1983, pp. 137–53

Florensky, C. P., Basilevsky, A. T., Kryuchkov, V. P., Kuzmin, R. O., Nikolaeva, O. V., et al. 1983b. Venera 13 and 14: sedimentary rocks on Venus? *Science* 221: 57–59

Garvin, J. B., Head, J. W., Zuber, M. T., Helfenstein, P. 1984. Venus: the nature of the surface from Venera panoramas. *J. Geophys. Res.* 89: 3381–99

Garvin, J. B., Head, J. W., Pettengill, G. H., Zisk, S. H. 1985. Venus global radar reflectivity and correlations with elevation. *J. Geophys. Res.* 90: 6859–71

Greeley, R., Iversen, J., Leach, R., Marshall, J., White, B., et al. 1984. Wind-blown sand on Venus: preliminary results of laboratory simulations. *Icarus* 57: 112–24

Hartmann, W. K. 1987. Determination of crater retention ages for Venus: a methodology. *Izv. Akad. Nauk SSSR Ser. Geol.*, pp. 67–74 (In Russian)

Head, J. W. 1986. Ishtar Terra, Venus: a simple model of large-scale tectonic convergence, crustal thickening and possible delamination. *Lunar Planet. Sci. XVII*, pp. 323–24 (Abstr.)

Head, J. W., Crumpler, L. S. 1987. Evidence for divergent plate boundary characteristics and crustal spreading on Venus. *Science* 238: 1380–85

Head, J. W., Solomon, S. C. 1981. Tectonic evolution of the terrestrial planets. *Science* 213: 62–76

Head, J. W., Wilson, L. 1986. Volcanic processes and landforms on Venus: theory, predictions, and observations. *J. Geophys. Res.* 91: 9407–46

Head, J. W., Peterfreund, A. R., Garvin, J. B., Zisk, S. H. 1985. Surface characteristics of Venus derived from Pioneer-Venus altimetry, roughness and reflectivity measurements. *J. Geophys. Res.* 90: 6873–85

Hunten, D. M., Colin, L., Donahue, T. M., Moroz, V. I., eds. 1983. *Venus.* Tucson: Univ. Ariz. Press. 1143 pp.

Ivanov, B. A., Basilevsky, A. T., Kryuchkov, V. P., Chernaya, I. M. 1986. Impact craters of Venus: analysis of Venera 15 and 16 data. *Proc. Lunar Planet. Sci. Conf., 17th, J. Geophys. Res.* 91: D413–30

Khodakovsky, I. L., Volkov, V. P., Sidorov, Yu. I., Borisov, M. V. 1978. The preliminary prediction of the mineral composition of surface rocks and the hydration and oxidation processes of the Venus outer shells. *Geokhimiya* 12: 1821–35 (In Russian)

Khodakovsky, I. L., Volkov, V. P., Siderov, Yu. I., Dorofeeva, V. A., Borisov, M. V., Barsukov, V. L. 1979. Geochemical model of the Venus troposphere and crust according to new data. *Geokhimiya* 12: 1747–58 (In Russian)

Kotelnikov, V. A., Akim, E. L., Aleksandrov, Yu. N., Armand, N. A., Basilevsky, A. T., et al. 1984. Study of the Maxwell Montes region on Venus by Venera 15 and Venera 16 spacecraft. *Sov. Astron. Lett.* 10: 883–89

Kreslavsky, M. A., Basilevsky, A. T., Shkuratov, Yu. G. 1987. Prognosis of the distribution of the tessera terrain on Venus using Pioneer Venus and Venera 15/16 data. *Astron. Vestn.* In press (In Russian)

Kryuchkov, V. P. 1987. Analysis of impact crater dimensions on the surface of Venus. *Izv. Akad. Nauk SSSR Ser. Geol.* 6: 75–83 (In Russian)

Masursky, H., Eliason, E., Ford, P. G., McGill, G. E., Pettengill, G. H., et al. 1980. Pioneer-Venus radar results: geology from images and altimetry. *J. Geophys. Res.* 85: 8232–60

McGill, G. E., Warner, J. L., Malin, M. C., Arvidson, R. E., Eliason, E., et al. 1983. Topography, surface properties, and tectonic evolution. See Hunten et al 1983, pp. 69–130

Nikolaeva, O. V., Ronca, L. B., Basilevsky, A. T. 1986. Circular features on the plains of Venus as evidence of its geologic history. *Geokhimiya*, pp. 579–89 (In Russian)

Pettengill, G. H., Eliason, E., Ford, P. G., Loriot, G. B., Masursky, H., et al. 1980. Pioneer-Venus radar results: altimetry and surface properties. *J. Geophys. Res.* 85: 8261–70

Phillips, R. J., Malin, M. C. 1983. The interior of Venus and tectonic implications. See Hunten et al 1983, pp. 159–214

Pieters, C. M., Head, J. W., Patterson, W., Pratt, S., Garvin, J. B., et al. 1986. The color of the surface of Venus. *Science* 234: 1379–83

Pike, R. J. 1977. Size-dependence in the shape of fresh impact craters on the Moon. In *Impact and Explosion Cratering*, ed. R. J. Roddy, R. O. Pepin, R. B. Merrill, pp. 489–509. New York: Pergamon

Pronin, A. A. 1986. The structure of Lakshmi Planum, an indication of horizontal asthenospheric flows on Venus. *Geotektonika* 20: 271–81 (In Russian)

Pronin, A. A., Sukhanov, A. L., Tyuflin, Yu. S., Kadnichanskij, S. A., Kotelnikov, V. A., et al. 1986. Geological-morphological description of the Lakshmi Planum. *Astron. Vestn.* 20: 82 (In Russian)

Schaber, G. G. 1982. Venus: limited extension and volcanism along zones of lithospheric weakness. *Geophys. Res. Lett.* 9: 499–502

Selivanov, A. S., Getkin, Yu. M., Naraeva, M. K., Panfilov, A. S., Fokin, A. B. 1983. Dynamic phenomena detected in panoramas of the surface of Venus transmitted by the Venera 13–14 spacecraft. *Kosm. Issled.* 21: 200–4 (In Russian)

Sharpton, V. L., Head, J. W. 1985. Analysis of regional slope characteristics on Venus and Earth. *J. Geophys. Res.* 90: 3733–40

Sharpton, V. L., Head, J. W. 1986. A comparison of regional slope characteristics of Venus and Earth: implications for geologic processes on Venus. *J. Geophys. Res.* 91: 7545–54

Slyuta, E. W., Nikolaeva, O. V., Kreslavsky, M. A. 1987. Distribution of volcanic domes on the venusian plains: Venera 15/16 results. In preparation

Solomon, S. C., Head, J. W. 1982. Mechanisms for lithospheric heat transport on Venus: implications for tectonic style and volcanism. *J. Geophys. Res.* 87: 9236–46

Stofan, E. R., Head, J. W. 1986. Pioneer-Venus characteristics of ovoids on Venus. *Lunar Planet. Sci. XVII (Suppl.)*, pp. 1033–34 (Abstr.)

Stofan, E. R., Head, J. W., Campbell, D. B. 1987a. Geology of the southern Ishtar Terra/Guinevere and Sedna Planitae region on Venus. *Earth, Moon and Planets* 38: 183–207

Stofan, E. R., Head, J. W., Campbell, D. B., Zisk, S. H., Bogomolov, A. F., et al. 1987b. Venus rift zones: geology of Beta Regio and Devana Chasma. Submitted for publication

Sukhanov, A. L. 1986. Parquet: regions of areal plastic dislocations. *Geotektonika* 20: 294–305 (In Russian)

Sukhanov, A., Pronin, A. A. 1987. Signs of spreading on Venus. *Dokl. Akad. Nauk SSSR*. In press (In Russian)

Surkov, Yu. A. 1983. Studies of Venus rocks by Veneras 8, 9, and 10. See Hunten et al 1983, pp. 154–58

Surkov, Yu. A., Barsukov, V. L., Moskalyova, L. P., Kharyukova, V. P., Kemurdzhian, A. L. 1984. New data on the composition, structure, and properties of Venus rock obtained by Venera 13 and 14. *Proc. Lunar Planet. Sci. Conf., 14th, J. Geophys. Res.* 89: B393–402

Surkov, Yu. A., Kirnozov, F. F., Glazov, V. N., Dunchenko, A. G., Tatsy, L. P., et al. 1987. Uranium, thorium, and potassium in the venusian rocks at the landing sites of Vega 1 and 2. *Proc. Lunar Planet. Sci. Conf., 17th, Part 2, J. Geophys. Res.* 92: E537–40

Volkov, V. P., Zolotov, M. Yu., Khodakovsky, I. L. 1986. Lithospheric-atmospheric interaction on Venus. In *Chemistry and Physics of Terrestrial Planets*, ed. S. U. Saxena, 6: 136–90. New York: Springer-Verlag

Vorder Bruegge, R. W., Head, J. W., Campbell, D. B. 1985. Evidence for major cross-strike structural discontinuities: Maxwell Montes, Venus. *NASA Tech. Memo. 88383*, pp. 369–70

Vorder Bruegge, R. W., Head, J. W., Campbell, D. B. 1986. Cross-strike discontinuities on Maxwell Montes, Venus: evidence for large-scale strike-slip faulting. *Lunar Planet. Sci. XVII*, pp. 917–18 (Abstr.)

Zuber, M. T. 1987. Constraints on the lithospheric structure of Venus from mechanical models and tectonic surface features. *Proc. Lunar Planet. Sci. Conf., 17th, Part 2, J. Geophys. Res.* 92: E541–51

Ann. Rev. Earth Planet. Sci. 1988. 16: 319–54

SEISMIC STRATIGRAPHY

Timothy A. Cross

Department of Geology and Geological Engineering,
Colorado School of Mines, Golden, Colorado 80401

Margaret A. Lessenger

Department of Geophysics, Colorado School of Mines, Golden,
Colorado 80401

INTRODUCTION

Seismic stratigraphy, the science of interpreting or modeling stratigraphy, sedimentary facies, and geologic history from seismic reflection data, has been practiced for at least three decades. However, the term seismic stratigraphy became commonplace in the vocabulary of geologists and seismic intepreters only after publication in 1977 of the authoritative AAPG Memoir 26, *Seismic Stratigraphy—Applications to Hydrocarbon Exploration* (Payton 1977). Subsequently, seismic stratigraphic studies have increased significantly in number, establishing the importance of the discipline to geological interpretation of sedimentary rocks and demonstrating its potential for substantially influencing the nature and methodology of stratigraphic analysis in general. Seismic stratigraphy has initiated a revolution in stratigraphic analysis as profound as that caused by plate tectonics. As a result, almost all disciplines dealing with sedimentary rocks are being reexamined and refocused, and significant new questions are being raised.

Much of the development, emphasis, and structure of seismic stratigraphy was driven by the petroleum industry's increased reliance on seismic data as exploration moved into frontier regions and into deeper parts of sedimentary basins where well control is limited or lacking. There was an accompanying change in industrial focus toward use of seismic data in obtaining more detailed and accurate subsurface images in mature

319

0084–6597/88/0515–0319$02.00

and unexplored basins. Ancillary to these new directions was an expanded use of seismic data for identifying subtle stratigraphic traps, characterizing and predicting occurrences of source and reservoir rocks, and determining fluid distributions in the subsurface. Concomitant advances in seismic data acquisition, processing, and display techniques have provided the resolution necessary for more detailed and sophisticated geological interpretation of seismic data.

Although many of the major advances occurred within the industrial sector, interest in and use of seismic stratigraphy have spread to academic and governmental domains, and applications have been extended from sedimentary basins to modern marine environments. In addition, seismic stratigraphic concepts are being incorporated with more traditional approaches to stratigraphic analysis, with a resulting amalgamation of concepts that is likely to permanently change the science and the manner by which sedimentary rocks are interpreted and understood. Exxon's global sea-level chart (Haq et al 1987) and the attendant presumption of worldwide correlatable events has kindled controversies and invigorated research in numerous disciplines. Scientists are engaged in multidisciplinary efforts to calibrate the sea-level curve, to evaluate the degree of synchroneity of stratigraphic events related to sea-level changes, and to determine the rates, magnitudes and frequencies of eustatic change, tectonic movement, and sediment supply or production. Seismic stratigraphy caused the rediscovery of and generated renewed interest in physical and geochronological correlation techniques. These techniques promise a temporal resolution of the stratigraphic record beyond that previously attainable, and they provide a method of correlating Proterozoic strata for the first time. The infusion of seismic stratigraphy into traditional approaches of stratigraphic analysis has provided the conceptual framework for developing deterministic numerical models of stratigraphic architecture, sedimentary facies distributions, and basin evolution. This contribution of seismic stratigraphy, perhaps more than any other, is most important to future directions in stratigraphic analysis because it is the impetus causing the science to pass from the qualitative into the quantitative domain.

This review presents seismic stratigraphy as consisting of four major approaches. Each comprises a different set of objectives, questions, scales of observation, and, to some extent, methodologies. Although this choice is somewhat arbitrary, it corresponds generally to divisions recognized by others.

The first approach, termed *seismic sequence analysis*, is to define the regional stratigraphic architecture and establish the time-stratigraphic framework of a sedimentary basin through recognition of unconformity-bounded depositional sequences. In seismic sequence analysis, a seismic section is divided into packages of relatively conformable or concordant seismic reflections. These packages are separated from each other by surfaces of unconformity, which are recognized in seismic sections by systematic discordances or terminations of reflections against these surfaces. The philosophical underpinning of seismic sequence analysis is that the strata of a sedimentary basin were deposited in geographically and temporally confined episodes, each separated by a period of erosion or nondeposition. During each episode there was continuity in depositional topography, across which contiguous environments or depositional systems advanced and retreated forming an essentially continuous and conformable sedimentary record. But between episodes, substantive changes in sea level, tectonic movement, or sediment supply caused commensurate changes in the depositional profile, water depths, types of sedimentary environments, and sites of sediment accumulation. Thus, unconformity-bounded depositional sequences represent natural temporal divisions of the stratigraphic record, and correlation of sequences provides an essential, first-order time-stratigraphic framework of a sedimentary basin.

After seismic sequences are defined, the configuration and character of reflections within sequences are studied in an approach termed *seismic facies analysis*. The objective of this approach is to interpret the depositional environments, lithologies, and geological history represented by strata in different positions within a sequence. In exploration, the goal of seismic facies analysis is to determine or predict the occurrence and distribution of specific lithologies, such as potential source or reservoir rocks. A seismic facies consists of a group of seismic reflections whose character differs from adjacent groups within a sequence (Mitchum et al 1977a). Reflection characteristics that are used most commonly to distinguish one seismic facies from another include the geometry of reflections or reflection terminations with respect to the two unconformity surfaces bounding the sequence, the external geometry of the facies, and the internal configuration and character of the reflections. In seismic facies analysis, it is assumed that there are direct, if not unique, relations between the position of a seismic facies within a sequence, the external form and internal configuration of its reflections, and the sedimentary environments represented by that facies. Using this assumption, one may plot seismic facies on a seismic section or transfer them to a map view to be interpreted in the same manner as sedimentary facies information.

The third approach, *seismic lithology analysis*, is the oldest and by far the most commonly applied in seismic stratigraphic studies. It involves detailed analysis and modeling of an individual seismic reflection or a small number of adjacent reflections. The objective of seismic lithology analysis is to relate the response of the acoustic waveform to the properties, geometries, and distributions of particular lithologies and fluids. Technological advances during the past decade such as new data acquisition methods, processing and modeling techniques, utilization of shear-wave and borehole information, and advances in color displays have phenomenally increased the capability of extracting detailed stratigraphic information from seismic reflection character. Today, two different approaches dominate attempts to identify reflector characteristics and causes of changes in those characteristics along reflectors. One uses a set of geological and geophysical assumptions to create a synthetic seismic model that can then be compared with observed reflection character. In this case the question asked is, "given known stratigraphic relations, what seismic response will be recorded at the surface?" The other approach is to model geologically reasonable combinations of lithologies and fluids that would produce the observed signal. In this case the question asked is, "given an observed seismic record, what is the most likely stratigraphic relation that would produce it?"

The final approach, *integrated seismic stratigraphic analysis*, was spawned by the recognized need to integrate geological and geophysical observations and analytical techniques at all scales from seismic sequences to seismic lithologies. Rapidly advancing computer technologies are providing the means by which such integration is implemented. Fundamental to this approach is the geophysical workstation consisting of a minicomputer, specialized software packages, and an interactive color graphics monitor. Although in routine use only a few years, the workstation already has made possible more efficient and detailed structural and stratigraphic analysis of seismic data. The workstation makes available to the interpreter a variety of graphical displays of seismic, well log, cross-section, and map data along with the software to interactively manipulate the data and create new presentations. Color display is a singularly important enhancement for interpretation because it adds a dynamic range and visual dimension to which the human mind is keenly sensitive. It has proven effective in enhancing visual recognition of changes in waveform character along individual reflections. The interactive capability of the workstation allows the interpreter to change processing schemes, redefine geological parameters or assumptions, or present data in a different form, and to have the results of these changes rapidly redisplayed for immediate evaluation. Because the response time is rapid, the workstation allows the interpreter to experiment with multiple ideas and working hypotheses, to test these hypotheses against observed data and synthetic models, and thus to gain greater insight into the geology portrayed by the seismic image. This increased intellectual freedom derived through increased efficiency is perhaps the most important element afforded by contemporary workstations. Future workstations, driven by expert systems and incorporating advanced empirical, visual, and analytical methods, have even greater promise for effecting more accurate and sophisticated seismic stratigraphic analysis. They will facilitate increased integration of geology and geophysics and will be an important element by which substantive new advances and conceptual understanding are accomplished.

SEISMIC SEQUENCE ANALYSIS

The concept of seismic sequences is an offspring of stratigraphic sequence concepts enunciated by L. L. Sloss (Sloss et al 1949, Krumbein & Sloss 1951, Sloss 1963) and H. E. Wheeler (1958, 1959a,b, 1963) and a grandchild of earlier stratigraphic concepts, especially those of Grabau (1924, pp. 723–45) and Caster (1934). Sloss and Wheeler independently recognized that Phanerozoic strata of the North American craton are divisible into major depositional intervals (represented by sequences) that are separated by periods of continental-scale erosion and/or nondeposition (represented by unconformities). These unconformity-bounded sequences span one or more geological periods and comprise essentially conformable strata. Because each sequence represents deposition during similar tectonic and relative sea-level conditions, whereas their bounding unconformities represent substantial changes in one or both, sequences were considered the fundamental, natural divisions of the North American stratigraphic record. Krumbein & Sloss (1951) argued that a sequence is the rock record of a major tectonic cycle.

Among the eight authors of the ground-breaking Exxon contributions in AAPG Memoir 26 (Payton 1977), four were schooled in sequence concepts at Northwestern University, where Sloss taught. It is not surprising that, when confronted with interpreting scores of regional seismic sections, these scientists applied sequence concepts in their analyses. The sequence concepts of Sloss and Wheeler had been painstakingly extracted from geographically discontinuous stratigraphic data across an entire craton. By contrast, because seismic sections are continuous images of the subsurface, the geometric relationships of strata are more readily recognizable. During the gestation of seismic sequence concepts at Exxon (then Esso) in the 1960s and early 1970s (Vail & Wilbur 1966, Vail & Sangree 1971, Vail 1975, Mitchum et al 1976), seismic interpreters recognized unconformity-bounded stratal packages similar to those described by Sloss and Wheeler, but of much smaller temporal and geographical scales. Because these smaller scale sequences pervaded the sedimentary basins they were analyzing and seemed to constitute the basic building blocks of basin stratigraphy, they termed them *depositional sequences* and renamed the sequences of Sloss and Wheeler *supersequences* (Mitchum et al 1977b). A depositional sequence is the lithostratigraphic equivalent of a seismic sequence.

Seismic sequence analysis is based upon the identification of depositional sequences, their three-dimensional form, and their positions with respect to one another and to the sedimentary basin. A depositional sequence, as explained by Mitchum et al (1977b), is a stratigraphic unit composed of a relatively conformable succession of strata that are genetically related by approximate contemporaneity of deposition. The top and bottom boundaries of depositional sequences are surfaces of unconformity, including their downdip continuations into correlative surfaces of conformity. An unconformity is an expression of a relative sea-level fall and concomitant seaward shift in sites of sediment accumulation. Depositional sequences are identified on seismic sections by systematic terminations or discordances of reflections (Figure 1). One particular geometric pattern, termed onlap, is recognized by systematic oblique updip terminations of reflectors against an underlying seaward dipping reflection. Onlap represents a seaward shift in sites of sediment accumulation during a relative sea-level fall and the subsequent landward encroachment of deposition upon the unconformity surface as relative sea level rises. It is argued (Mitchum et al 1977b, Vail et al 1977b) that primary seismic reflections correspond to unconformity or stratal surfaces and, therefore, are temporally significant in the respect that all strata above such a surface or reflection are younger than those below it. All strata within a sequence were deposited during an interval of time that differs from the time intervals represented by strata of overlying and underlying sequences. The temporal limits of a sequence are defined by the ages of the bounding unconformities where they become conformable. In summary, the identification of seismic sequences is based upon the objective criterion of geometrical relations of reflections, and it is

Figure 1 Identification of seismic sequences by systematic terminations or discordances of reflections. From Bally (1983a, p. 2.2.3-55).

independent of lithologies, fossils, or other constituents of the strata within sequences.

There are many elements to this abbreviated definition of depositional sequences, methodology of recognition, and significance for geological interpretations. Most of these elements are generally accepted by seismic stratigraphers and are comprehensively discussed in AAPG Memoir 26 (Payton 1977) and in other texts and course notes (e.g. Brown & Fisher 1980, Sheriff 1980, Macurda 1985, 1986, Hardage 1987) to which the reader is referred. This review concentrates on some of these elements that are more controversial or that, from our experience, are less well understood.

Time-Stratigraphic Significance of Seismic Reflections

Seismic reflections are generated along surfaces of acoustic impedance contrast. Acoustic impedance, a bulk physical property of a rock and its contained fluids, is obtained by multiplying the density of a rock by the velocity of acoustic waves passing through it. The strength of a reflection increases as the difference in impedance between two superposed rock layers increases. A fundamental precept of seismic sequence analysis is that primary seismic reflections originate from unconformities and stratal surfaces, and therefore that seismic reflections are chronostratigraphically significant. Specifically excluded from the category of primary reflections are reflections that are not stratigraphic in origin. The validity of this precept, a subject of recent controversy, is examined by considering first which surfaces generate reflections and then the time-stratigraphic significance of those reflections.

Reflections are generated along surfaces of unconformity because they represent hiatuses in deposition and geographic shifts in sites of sediment accumulation, and consequently they commonly separate strata of different lithologies or structural attitudes. In some instances an unconformity surface may generate a reflection that is not resolved on a seismic section (e.g. Bally 1983b, p. 1.2.2-4). Even when a reflection originates from an unconformity, the amplitude and phase of the reflection may vary laterally as different lithologies with different impedances are superimposed across the surface.

The chronostratigraphic significance of a reflection generated by an unconformity, as advocated by Mitchum et al (1977b) and Vail et al (1977b, 1984), is that it separates a younger from an older depositional sequence, and no other time line or stratal surface crosses it. A surface of unconformity represents the duration of erosion or nondeposition that occurred between two depositional events. Unconformities occur over many geographic and temporal scales, from local to continent-wide and from a few hours to millennia. But a property common to all unconformities is that the duration of their formation changes laterally along the surface. For example, an unconformity in a landward part of a basin, where sediments were exposed to subaerial erosion, may be traced seaward to a position where deposition was continuous and the unconformity merges into a surface of conformity. Along that path, the duration of the unconformity continuously decreases. Although the time interval represented by the unconformity changes from place to place, it is assumed that at all places strata above the unconformity are younger than strata below it. By analogy, a reflection generated at an unconformity is time-significant in that it everywhere separates a younger from an older depositional sequence.

The assumption that all unconformities everywhere separate older from younger strata was challenged by Christie-Blick et al (1987), who showed two examples from deep marine and alluvial-fan environments in which a stratum above an unconformity at one locality would be older than a stratum below it at another locality. They concluded that in most instances the possibility of diachronous unconformities should not raise a significant difficulty for seismic sequence interpretation, but cautioned against indiscriminate application of the assumption. We offer a new definition of the temporal relations of strata relative to unconformities and the chronostratigraphic significance of unconformities: A surface of unconformity everywhere separates overlying strata that are younger than or equal to the age of strata beneath the unconformity. This new definition removes the ambiguities and inconsistencies shown in the examples of Christie-Blick et al and conforms to our own work on ravinement processes (Swift 1968) and transgressive surfaces of unconformity, another instance where rocks above an unconformity at one locality may be older than rocks below the unconformity at another locality.

Unlike unconformities, it is not intuitively apparent that stratal surfaces should generate reflections, as advocated by Vail et al (1977b, pp. 100–6). Moreover, the validity of this precept has been questioned using the following reasoning:

Premise 1 Seismic reflections are generated by surfaces of acoustic impedance contrast. Because the impedances of different layers within the Earth are controlled to a great extent by lithology, surfaces of impedance contrast should occur at lithologic, or sedimentary facies, boundaries.

Premise 2 In many cases sedimentary facies transgress time. That is, if a particular sedimentary facies is followed in a landward or seaward direction, its age or time of deposition will change.

Conclusion Because surfaces of impedance contrast occur at sedimentary facies boundaries, in those instances that these boundaries cross time lines, seismic reflections must also cross time lines.

These two opposing views, that stratal surfaces either do or do not generate reflections, may be resolved by deriving answers to two questions: What are the requirements for generating seismic reflections? Under what geological conditions are these requirements satisfied? The normal answer to the first question is that reflections are generated from physical surfaces along which there is a sharp impedance contrast relative to the wavelength of the acoustic signal; impedance contrasts over diffuse boundaries generate either a weak or no reflection (Neidell 1979, p. 106). However, there is another, equally fundamental requirement: The surface of sharp impedance contrast must be laterally continuous to be resolved by the acoustic wave and the interpreter. If the surface is continuous over a distance of less than one Fresnel zone, it will not be resolved by the acoustic wave (Neidell 1979, p. 83). The surface also must be continuous over a distance of many Fresnel zones to create a coherent pattern of reflections on a seismic section. In the absence of lateral continuity, reflections generated from a surface will appear on only one or a few seismic traces and the eye will not detect a pattern of reflection continuity.

Which stratigraphic surfaces fulfill both requirements of lateral continuity and sharp impedance contrast? An essential reference to understanding is Campbell (1967), in which the genesis of beds and bedding surfaces and the distribution of sedimentary facies along beds are discussed. In many depositional settings there is little or no topographic relief on the depositional surface, and sediments aggrade layer upon layer through time. In these instances, beds are essentially horizontal, parallel, and continuous; sedimentary facies boundaries correspond with bedding surfaces; and bed boundaries are concordant with time lines. In this common situation, impedance contrasts along bedding surfaces will generate reflections parallel with both bedding and time, and the two requirements are satisfied.

However, there are other cases in which sedimentary facies transgress time and cross bedding surfaces. These occur where the depositional surface has topographic relief, such as along a shoreline, reef margin, or seaward of the shelf-slope break, or where there is a major geographic shift of contiguous depositional environments, either gradual or punctuated, as in the case of a regional transgression. As an illustration, consider a depositional surface traced from the coastal plain, across the shoreline and shoreface, and onto the marine shelf. Along this path depositional environments change gradationally, as do the sedimentary characteristics of each. Thus, a depositional surface is synchronous and sediments deposited along it change facies gradationally. This relation is directly translated to rocks in the form of bedding surfaces, which are relict depositional surfaces produced during periods of nondeposition or abrupt change in depositional conditions (Figure 2). A stratum may contain multiple intergradational facies as it is traced laterally (for example, from delta top, through the delta front, to basin environments), but in all cases it is bounded by continuous, synchronous bedding surfaces (e.g. Grabau 1924, Van Siclen 1958, Busch 1959, 1974, Jackson 1964, Campbell 1967, 1979, Asquith 1970, Galloway & Brown 1973, Frazier 1974, Brown & Fisher 1977, Vail et al 1977b).

In contrast to bedding surfaces, sedimentary facies boundaries are gradational, diachronous, and discontinuous (Figure 2). Time-equivalent facies —that is, those contained within a single bed—gradually change from one lithology to another over distances generally ranging from tens to hundreds of meters. A facies boundary within a single bed is not a laterally continuous physical surface of contrasting impedance, as is a bedding surface. Even if facies transitions at the bed scale could be resolved by the seismic method, the transitions involve such gradual variations in acoustic impedance that they are not a significant source of reflections. Similar facies contained within adjacent beds gradually rise or fall with respect to synchronous bedding surfaces and, therefore, are diachronous. Surfaces of impedance contrast also rise or fall, leading to incongruent convolution of wavelets and, more importantly, lack of laterally continuous reflections. Such facies boundaries among beds, presumably resolvable with high-

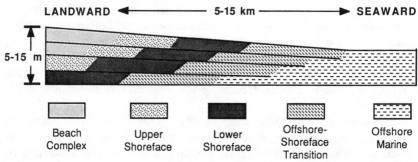

LANDWARD ◄────── 5-15 km ──────► SEAWARD

5-15 m

| Beach Complex | Upper Shoreface | Lower Shoreface | Offshore-Shoreface Transition | Offshore Marine |

Figure 2 Cross-sectional geometry of beds, bedding surfaces, and sedimentary facies along a beach-to-offshore marine transect. In this depositional system, which has topographic relief, bed boundaries (heavy lines) follow depositional topography and are laterally continuous, essentially isochronous physical surfaces. Facies changes within a bed are gradational. Similar facies traced from bed to bed are diachronous and laterally offset across bedding surfaces, and no discrete, laterally continuous physical surface separates one facies from another. Figure modified from Campbell (1979).

resolution seismic methods under suitable conditions, do not form a laterally continuous physical surface of contrasting impedance, as do bedding surfaces or unconformities of any scale. This distinction is the essence of the argument that reflections are generated along time-significant surfaces.

Numerous beds form a larger scale stratigraphic entity, the progradational unit (Figure 3). For example, most sediments in coastal plain

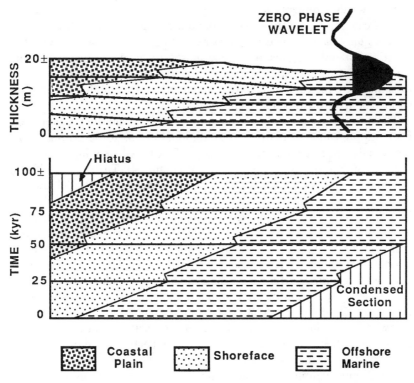

Figure 3 Cross-sectional geometry of a progradational unit, consisting of sets of beds such as those shown in Figure 2. The top lithostratigraphic diagram shows geometry in space, whereas the bottom chronostratigraphic diagram shows geometry in time and space. In the top diagram, sigmoidal geometry time lines, essentially horizontal in coastal plain and offshore marine environments and inclined in beach-to-shoreface environments where there is depositional topography, are approximated by straight-line segments. Facies zigzags indicate episodicity of sedimentation due to autogenic processes, such as shifts of delta lobes. The schematic zero-phase wavelet (symmetrical and centered about the reflection interface) shows the approximate scale of seismic resolution with respect to progradational units. A basal condensed section (representing a period of nondeposition toward the seaward margin of a progradational unit) and a hiatus at the top (representing a period of sediment bypassing) are surfaces in the upper diagram but time-space areas in the lower diagram.

to marine shelf environments are deposited during progradational events, or seaward movement of the shoreline and all contiguous environments. Within a progradational unit, time lines through coastal plain to marine shelf facies approximate the sigmoidal depositional profile of the coastal plain-beach-shoreface-shelf transition. The termination of a progradational event is punctuated by a deepening episode and concomitant landward shift of facies tracts, such that at a particular geographic location the shallowest water sediments of one event are abruptly overlain by the deepest water sediments of a subsequent event. Thus, dissimilar facies often are juxtaposed at these event boundaries, and although the bounding surfaces may have formed over a period of several thousand years, they constitute a time-significant surface at the scale of seismic resolution. This situation is acoustically identical to that of unconformities in that dissimilar lithologies are juxtaposed, creating a sharp, well-defined physical surface of acoustic impedance contrast.

The interactions of sea-level fluctuations, tectonic subsidence, and sediment supply cause successive progradational events to be arranged systematically in a geometric hierarchy of seaward-stepping, vertically stacked, and landward-stepping patterns (Figure 4). The time-significant bounding surfaces of stacked progradational events are parallel and laterally continuous, and they form sharp surfaces of laterally continuous impedance contrasts. Depending upon the amount of displacement of sedimentary facies across event boundaries, impedance contrasts along these surfaces will vary but generally will be greater, and always more laterally continuous, than those along facies boundaries. Moreover, in contrast to bedding surfaces, the thickness of a progradational event and the physical separation of event boundaries often are at or near seismic resolution. Gradual facies changes within a progradational unit are often below the resolving capability of the seismic method and are exhibited primarily in the form of lateral variations in reflection waveform.

Perhaps more than any other geological feature, large-scale progradational units exhibiting distinct clinoform and downlapping geometries might be expected to generate reflections from facies boundaries. Because of their size they are easily resolvable by the seismic method, and such progradational packages have been observed on seismic records worldwide from strata of many geologic ages. The sedimentary facies composing these units should be approximately parallel and subhorizontal. Moreover, lithologies on opposite sides of bedding surfaces are likely to be similar along much of their distance because the magnitude of facies offset across stratal surfaces during short-term progradational episodes is minimal compared with the scale of the clinoforms. This is an ideal geometry for anticipating a reflection from facies boundaries, even though

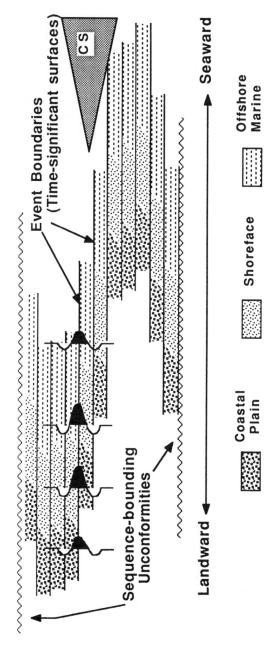

Figure 4 Depiction of hierarchical stacking patterns of progradational units within the context of seismic sequences. Above a basal sequence-bounding unconformity, progradational units step progressively seaward, then become vertically stacked and then step landward. The condensed section (labeled CS) corresponds in time to the landward-stepping events, and its duration expands in a seaward direction. Subsequent progradational units are vertically stacked and begin a seaward-stepping phase prior to the development of the upper sequence-bounding unconformity. Durations of unconformities expand in a landward direction. Schematic zero-phase wavelets at the scale shown in Figure 3 show that reflections are generated at the time-significant surfaces bounding the progradational units. The amplitude or phase of the wavelet may vary along those surfaces as a result of differences in impedance contrasts generated by superposition of different sedimentary facies across event boundaries.

the boundaries are diachronous and gradational. Why are the reflections in these cases parallel to time-significant, clinoform stratal surfaces, and why are reflections from facies boundaries generally not displayed on seismic sections? The answer is similar to that previously given. Stratal surfaces are sharp and laterally continuous and provide laterally continuous surfaces of impedance contrast, even though the impedance contrasts may vary or change sign along the surfaces. By contrast, transitions between facies are gradual, and boundaries between adjacent facies are not discrete physical surfaces. The three-dimensional seismic reflection wavefield responds most noticeably to the sharp acoustic contrasts of different depositional episodes, rather than to the gradual lateral change in sedimentary facies.

In summary, seismic reflections are generated from time-significant surfaces at many different scales. Reflections are apparent along sequence-bounding unconformities. At a smaller scale, reflections are generated along the time-significant, laterally continuous, progradational event boundaries. If seismic resolution is sufficiently high, reflections generated from stratal surfaces are recognizable. Both time-significant surfaces of all scales and facies boundaries occur in rocks. Acoustic responses from both types of boundaries are contained within seismic information, although only one response (a reflected waveform) is exhibited on conventional seismic sections. Conventional seismic "wiggle trace" displays emphasize laterally continuous surfaces of sharp impedance contrasts, and consequently time-significant surfaces are emphasized. Instead, facies changes and boundaries are detected by observing subtle waveform changes along or between reflectors. Facies boundaries and time-significant surfaces can be distinguished by displaying amplitude-derived velocity sections. One need only carefully examine such a display to see that responses to both types of boundaries are contained in seismic information (e.g. Figure 1 of Neidell et al 1985).

New Perspectives on Unconformities

Before seismic sequence concepts, there were four types of unconformities. Each was defined by its geometrical and lithological relations with adjacent rock units, assigned criteria for recognition, and ascribed geologic significance. Although the physical aspects of surfaces of unconformity have not changed, seismic sequence analysis has altered our perspective on their genesis and geologic significance.

In the original formulation of seismic sequence concepts (Mitchum et al 1977b), unconformity designations were relegated to existing compartments of the four unconformity types. In retrospect, two fundamentally different types of unconformities had been illustrated by different geometries of discordant reflections, but at that time the distinctions between the two were not recognized. The first type, labeled surfaces showing evidence of erosion and nondeposition, was defined by onlap geometries. The second type, labeled surfaces showing evidence of only nondeposition, was defined by oblique downdip reflection terminations against an underlying horizontal or seaward-dipping reflection, a geometry termed downlap. It was tacitly assumed that these two surfaces were identical and continuous from basin center to basin margin.

By 1980, after discussions with M. T. Jervey and D. E. Frazier (P. R. Vail, personal communication, 1987), a conceptual distinction between the two types was recognized and shortly thereafter formalized. Vail & Todd (1981) and Vail et al (1984) restricted the term "unconformity" to those surfaces representing a significant hiatus with evidence of erosional truncation (subaerial or subaqueous) and/or subaerial exposure. The other type of unconformity, marine surfaces with significant hiatuses but without evidence of erosion, was termed a *condensed section*. In this revised usage, unconformities are sequence boundaries formed during periods of relative sea-level fall and are defined by onlap, toplap, or erosional truncation reflection geometries in seismic sections. Condensed sections are defined by reflections downlapping onto horizontal or seaward-dipping concordant reflections in seismic sections, a result of progradation across a depositional surface previously starved of sediment. A condensed section is developed when the rate of relative sea-level rise is greater than the rate of sediment supply, resulting in landward shift in sites of sediment accumulation and starved conditions further basinward. Condensed sections occur within a sequence, generally toward its stratigraphic base, and between the bounding unconformities.

Embry (1983) confirmed this new view of unconformities by showing from outcrop relations that the two types of unconformities were distinct in both genesis and relative time of formation, and that they normally were restricted to different parts of a basin. Those showing physical evidence of subaerial exposure, erosion, truncation, and/or sedimentary bypass occurred toward basin margins. By contrast, those displaying evidence of continuous marine submersion and marked only by hiatus in deposition normally occurred in more basinal localities. He further noted that where the two types were present in the same stratigraphic section, they were not contiguous but instead were separated by a "succession of transgressive strata."

It is interesting to note that in many cases, depending on whether reflections are images of deep or shallow marine deposits, condensed sections are more apparent than unconformities on seismic sections.

Downlapping reflections often terminate against a condensed section reflection at greater angles than do onlap reflections against an unconformity reflection. Why were stratal units bounded by unconformities, as opposed to those bounded by condensed sections, selected as the building block of seismic stratigraphic architecture? Frazier (1974) had previously recognized stratal sequences bounded by hiatuses (equivalent to condensed sections in genesis), which he termed *depositional episodes*, in Quaternary strata of the Gulf Coast and had demonstrated their utility in stratigraphic correlation. There was a debate among Exxon seismic interpreters about whether to use unconformities (revised usage) or hiatuses to divide strata into time-bounded packages. Ultimately, unconformities were chosen because (a) they could be traced more easily across multiple sedimentary facies of shallow marine and nonmarine deposits, where most seismic stratigraphic studies of the time were focused; (b) they appeared less sensitive to changes in sediment supply rates than condensed sections, and therefore were more useful for defining baselevel changes and erecting a relative sea-level curve; and (c) the priority and influence of earlier sequence concepts of Sloss and Wheeler largely predisposed their selection (P. R. Vail, personal communication, 1987).

One aspect of seismic sequences and their bounding unconformities that has attracted much attention and criticism is the contention that they are globally synchronous (Vail et al 1977a, 1984, Haq et al 1987). This contention originates from the view that although the genesis of seismic sequences is a product of the interactions among sea-level fluctuations, tectonic movement, and sediment supply, the eustatic component overprints and dominates the other two controls. Christie-Blick et al (1987) have critically evaluated the evidence for this contention. They concluded (a) that many of the major sequence-bounding unconformities persist in numerous basins and their synchroneity is corroborated by conventional stratigraphic evidence; (b) that the ages of lesser, or third-order, sequence boundaries are at or beyond the scale of biostratigraphic resolution, and therefore their synchroneity cannot be independently documented by conventional methods; and (c) that tectonic mechanisms proposed for generating sequence-bounding unconformities result in pronounced diachrony, contrary to the best stratigraphic evidence indicating synchrony of major unconformities.

There is additional controversy about the magnitudes of sea-level change required to produce their bounding unconformities. Originally Vail et al (1977a, 1984) estimated amplitudes of global sea-level change of 100 m or more, occurring over periods of a few million years, by measuring the vertical thickness represented by shifts in coastal onlap. Amplitudes of this magnitude, particularly during nonglacial periods, have been questioned because the implied amplitudes and rates of sea-level change are as much as an order of magnitude greater than those of known nonglacial mechanisms. Pitman (1978), Pitman & Golovchenko (1983), and Christie-Blick et al (1987) showed analytically that a downward shift in coastal onlap reflects only the rate of sea-level fall relative to the rate of basin subsidence, rather than a response to a sea-level fall alone. Downward shifts may be produced by some combination of an increased rate of sea-level fall, a decreased rate of subsidence, or a decreased rate of sediment supply. Christie-Blick et al concluded that global onlap charts and their derivative eustatic curves do not adequately separate eustatic from tectonic controls and almost certainly are incorrect in the inferred magnitudes of sea-level variation. Recent paleogeographic models suggest that the originally estimated amplitudes of eustatic variations may be four to eight times too high (Shaw & Hay 1987).

Extensions of Seismic Sequence Concepts to Stratigraphic Analysis

Seismic sequence concepts and methodologies are stimulating research and adding new dimensions to conventional stratigraphic and sedimentologic studies. There is renewed interest in testing ideas of eustatic control of depositional sequences, cyclic versus episodic deposition, and synchrony of major unconformities in the stratigraphic record. Studies are being designed to distinguish between the fundamental eustatic, tectonic, and sediment supply controls on deposition. The creation and calibration of global sea-level curves is the focus of much research involving a variety of biostratigraphic, sedimentologic, and geochronologic methods. There are new attempts to discriminate between autogenic and allogenic deposition and, consequently, to determine at which scales stratigraphic architecture is ordered and predictable and at which scales it is characterized by a haphazard arrangement of sedimentary facies mosaics. Because an ordered arrangement of strata at several hierarchical scales is inferred from seismic sequence concepts, a conceptual framework has been advanced that will allow the development of deterministic geological models that more accurately simulate stratigraphic architecture, sedimentary facies distributions, and basin evolution. This provides renewed hope that the discipline of stratigraphy will become more predictive and quantitative.

Two aspects of seismic sequence concepts are likely to permanently modify traditional approaches to stratigraphic analysis. The first is a resurrection of ideas from the 1930s that strata containing diverse lithologies and representing a variety of depositional environments may be correlated at a temporal resolution beyond the best biostratigraphic or geochronologic methods. Because both sequence-bounding unconformities and hiatuses separating progradational events represent natural breaks in the stratigraphic record and because they are regional in extent, it is possible to use these surfaces as the basis for correlation in both outcrop and geophysical well logs (e.g. Wanless & Weller 1932, Busch 1959, Campbell 1967, Frazier 1974, Cant 1984, Goodwin & Anderson 1985). This method will profoundly influence stratigraphic correlation of Proterozoic strata, where conventional correlation tools are lacking or limited in resolution (e.g. Grotzinger 1986, Christie-Blick et al 1988). The second aspect constitutes a departure from traditional lithostratigraphic analysis, in which strata are considered and interpreted primarily in the context of sedimentary facies. In the future, most surface and subsurface stratigraphic studies will focus on recognizing depositional sequences, their bounding unconformities, and their internal progradational events as the first-order elements of stratal architecture. Distributions and geometries of sedimentary facies tracts will be placed within the temporal and architectural context of sequences. The term *sequence stratigraphy* has been coined to represent this approach, and examples are presented by Abbott (1985), Muir et al (1985), and Sarg & Lehmann (1986).

SEISMIC FACIES ANALYSIS

After seismic sequences are defined and the stratigraphic architecture and time-stratigraphic framework of a sedimentary basin established, the geometry and character of reflections within sequences are studied by the methods of seismic facies analysis. In this phase of analysis, the lithologies, depositional environments, and geological history of strata contained within depositional sequences are interpreted. In addition, paleotopographic elements, directions of progradation or fluvial channel flow, and relative changes of sea level can be detected. Associations and distributions of seismic facies obtained from a grid of seismic sections may be transferred to map views for further analysis and interpretation.

Sangree & Widmier (1977, 1979) and Mitchum et al (1977a) defined a seismic facies as an areally restricted group of seismic reflections whose appearance and characteristics are distinguishable from those of adjacent groups. They suggested that some combination of the following attributes might be used to distinguish one seismic facies from another: reflection configuration and continuity, amplitude and frequency spectra of reflections, interval velocity, geometrical relations of reflections to the unconformities bounding the sequences, and three-dimensional form of the seismic facies. Except for interval velocity, these parameters can be evaluated visually on a seismic section without special computer enhancements of the seismic data, although such enhancements are becoming routine. Because reflections follow stratal surfaces, reflection geometry parallels depositional geometry. Therefore the areal distributions and associations of seismic facies allow inferences about paleoenvironments by analogy with geologic depositional models. It is important to note that the original discussions of seismic facies analysis explicitly distinguished the description of seismic facies by objective criteria of reflection characteristics from the geological interpretation of the defined seismic facies. Although this distinction has not always been followed, an interpretation procedure should follow these guidelines of first mapping objective seismic parameters and then inferring their geological significance.

From experience, three criteria have been selected and used successfully for identifying, classifying, and mapping seismic facies, although other, equally useful combinations of attributes are possible. One is the geometry of reflections and reflection terminations with respect to the two unconformity surfaces that bound the sequences. This geometry places the seismic facies within the context of seismic sequences and helps limit the range of geologically possible depositional settings and lithologies that might produce similar internal reflection characteristics.

Another criterion, reflection configuration, is the pattern formed by groups of reflections within the seismic facies. Reflection configurations reveal stratification patterns from which depositional processes and depositional topography can be inferred. Four commonly occurring reflection configurations have been recognized, each of which is associated with, but not limited to, a set of depositional environments. For example, parallel and basinward diverging patterns are associated with alluvial plains, shallow marine shelf deposits, delta platforms, and marine basin plains. Progradational or clinoform patterns, in which reflections are inclined with respect to underlying and overlying reflections, are characteristic of alluvial fans, prograding shelf platforms, prodelta environments, and the transition from outer shelf-slope-rise environments. Mounded and draped patterns are characteristic of reefs, banks, or deltas in shallow marine shelf and platform environments, and of slumps, submarine canyon fans or aprons, and turbidites or hemipelagic deposits in deeper marine settings. Onlap and fill patterns are produced by onlapping strata of coastal-plain and paralic environments, by deposits that fill submarine canyons or drowned alluvial valleys, and by marine strata onlapping the continental rise or slope. This is but a partial listing of the types of reflection configurations and their inferred geological significance. Interested readers are encouraged to consult more comprehensive works for further explanation and examples (e.g. Mitchum et al 1977a, Sangree & Widmier 1977, 1979, Bubb & Hatlelid 1977, Brown & Fisher 1977, 1980, Roksandic 1978, Macurda 1985, 1986, Hardage 1987).

A third criterion found useful for characterizing a seismic facies is its three-dimensional form. Some of the common forms are sheet, wedge, lens, bank, mound, and fill. Like the other two criteria, the external form of a seismic facies is thought to be representative of one or more depositional geometries and environments.

The areal variation of reflection character and seismic facies can be presented and studied on seismic facies maps. These maps may show some property of reflections (e.g. amplitude or phase), the distribution of seismic facies units, the direction of progradation, the paleotopography or paleo-slope, or some combination of these or other attributes. Any recognizable feature that appears variable and geologically significant may be incorporated in a seismic facies designation. In order that seismic facies be limited to a manageable and interpretable number and spatial variability, the types and distributions of seismic facies shown on a seismic facies map are typically restricted to those occurring within a seismic sequence.

Most seismic facies maps published have followed the convention proposed by Sangree & Widmier (1977, 1979) and Mitchum et al (1977a), in which three attributes are combined into a single mappable variable. These are the geometric relations of the reflections with respect to the upper and lower sequence boundaries (for example, toplap and downlap, respectively) and the internal reflection configuration (for example, progradational oblique). Other modifiers may be added to these basic attributes, such as reflector continuity, amplitude strength, or frequency content (e.g. McGovney & Radovich 1985). By assuming relations between seismic and sedimentary facies, the resulting seismic facies maps may be interpreted in much the same manner as geologic maps. Often the three-dimensional shapes of the facies, their map view, and their distribution within the sequence and with respect to each other mimic known stratigraphic patterns. Thus, even if specific relations between seismic and sedimentary facies are not known or assumed, a reasonable geological interpretation may be advanced. Examples of seismic facies analysis and the construction and interpretation of seismic facies maps, in addition to those already cited, are contained in papers by Ramsayer (1979), Berg (1982), Conticini (1985), Hubbard et al (1985a,b), Kirk (1985), and Mitchum (1985).

Perhaps the most significant limitation of seismic facies analysis is the relatively weak documentation of relations between seismic and sedimentary facies. The conventional manner of validating seismic facies interpretations is through prediction and subsequent drilling. This type of blind testing, although decisive, too often fails to produce new insights or bases for alternative interpretations when predictions are not confirmed. More efficient approaches of establishing relations between seismic and sedimentary facies must be attempted through combinations of outcrop or well-log analyses tied to seismic sections, and through analysis of forward and inverse geological models. In the former, depositional sequences can be identified and the geometry of stratification patterns and distribution of sedimentary facies mapped. From established surface or subsurface geologic relations, synthetic seismic sections can be created for comparison with observed data. In the latter, synthetic geological models derived from first principles can be treated in the same manner as outcrop or well-log data and used to create synthetic seismic sections. Significant advances in seismic facies interpretations will be achieved through studies that establish more definitive relations between seismic and sedimentary facies.

SEISMIC LITHOLOGY ANALYSIS

Seismic lithology analysis, the standard approach of the petroleum exploration and production geophysicist, constitutes a set of methods designed to extract information about the properties, geometries, and distributions of rocks and fluids from the character of individual or small groups of reflections. Geophysicists directly measure seismic traveltimes and waveforms and infer the configuration of surfaces of acoustic impedance contrast in the subsurface. From acoustic images, they interpret the combinations of lithology, porosity, and fluid content that can produce the observed signal. Seismic lithology analysis is conducted primarily at the scale of the progradational event. To achieve this degree of stratigraphic definition and to use the seismic response as a lithology and fluid-content indicator, a variety of data acquisition, processing, and modeling techniques have been developed that improve the spatial resolution of the seismic method.

Since the earliest days of reflection seismology, a primary goal of seismic data processors and interpreters has been to extract stratigraphic information from seismic data and to distinguish that information from noise. In the 1940s and 1950s, geophysicists applied geologic concepts to seismic data by calculating reflection geometries and constructing "seismic cross sections" by hand. Although the primary focus during that period was on structural interpretation, reflection divergences and discontinuities were used to map stratigraphic pinchouts and reefs (e.g. Jasinski 1957). In addition, seismic records were tied to subsurface geology by utilizing velocity information from well check shot surveys and geophysical well logs. Generation of the continuous seismic record section by computer was a milestone in interpretational geophysics in the late 1950s. Previously, a continuous seismic profile was tediously constructed by first calculating the temporal position of each trace and then pasting the individual traces on long sheets of paper. The advent of computer-generated continuous record sections circumvented this laborious task. It became evident to interpreters that reflection geometries displayed on a wiggle-trace seismic section resembled a geologic cross section and facilitated the visualization of stratigraphic information in seismic data.

The primary objectives of technological advances since the late 1950s have been to increase the fidelity of the seismic record section, thereby improving the stratigraphic image, and to incorporate petrophysical information in calculating lithologies and fluid content. Formulation of the Common Depth Point (CDP) method (Mayne 1962) and the digital revolution were milestones in the development of modern seismic analysis. In the early 1960s, the Geophysical Analysis Group at the Massachusetts Institute of Technology provided essential research into geophysical information theory. During this time, the potential of extracting detailed stratigraphic information from seismic data was not aggressively pursued, although a few interpreters had shown empirically that variations in reflection character were related to variations in lithology or fluid content. Investigators in the Soviet Union had begun essential research into the effect of petrophysical properties on reflection character (Churlin & Sergeyev 1963), but their work was largely unknown elsewhere. In the 1970s, geophysicists began to deterministically analyze seismic responses in terms of lithology, porosity, and fluid type. Development of "bright spot," or "hydrocarbon indicator," analysis established the importance of fluids and lithology in controlling waveform character. It also emphasized the need to develop more sophisticated acquisition and processing methods that would allow causal inference from subtle changes in reflection character.

Significant technological advances have occurred in the past decade by using the CDP method, information theory, and rock physics to improve the fidelity of the seismic record section and to estimate detailed stratigraphic properties. Despite these advances, the presence of noise and the nonunique nature of seismic data continue to limit seismic stratigraphic definition. Since the solution is often ambiguous, geophysical and geologic models are essential for a complete interpretation.

Seismic Fidelity

Geophysicists have strived to improve the fidelity of the seismic record section so that it more closely resembles a stratigraphic cross section. Fidelity is improved by increasing the signal-to-noise content and resolving power of seismic data. Resolution, the ability to distinguish between two narrowly separated features, has two aspects in geological analysis of seismic data. Vertical resolution is the distance between two reflecting surfaces, generally bedding or event surfaces, that are discerned on seismic records. Lateral resolution is the subhorizontal distance between two features distinguished on seismic records. Since the spatial resolution of seismic data determines the limits to which stratigraphic geometries can be delineated, geophysicists have developed acquisition, processing, and interpretation techniques that improve both lateral and vertical resolution. For enhanced vertical resolution, a temporally compressed symmetrical waveform is desired, and therefore broad-band, zero-phase data are preferred. Lateral resolution depends on the size of the Fresnel zone at the subsurface reflector and the extent to which reflections have been migrated to their correct spatial position. The Fresnel zone is the region of constructive interference formed when a radiating wave encounters a flat reflector. To minimize the size of the Fresnel zone and increase lateral resolution, high-frequency signals are desired. Lateral discontinuities in reflectors diffract or defocus the acoustic wave, but data may be migrated to refocus the signal and increase lateral resolution.

New designs in data acquisition programs for stratigraphic analysis have occurred concomitantly with advances in electronics technology (Laster 1985) and have increased signal-to-noise ratios, spatial sampling, and frequency content. Some important advances in data acquisition resulted from a variety of improved seismic sources, including horizontally polarized shear wave sources, land air guns, and Vibroseis™. Dramatic advances are attributable to the increased volume of data that can be recorded economically. With larger recording capability, receiver-group spacings can be reduced, increasing spatial sampling and resolution of subsurface images. Placing receivers beneath the surface reduces cultural noise and ameliorates signal degradation and attenuation in unconsolidated layers. Three-component geophones are used to determine the polarization of incoming waves and thus improve the ability to distinguish and correlate incoming events across receiver arrays.

Advances in data processing have resulted from new insights into the seismic method and from the explosion in computational power that permitted new applications of older ideas (Schultz 1985). Processing programs that increase resolution retrieve the true amplitude of the data, increase the signal-to-noise ratio, and shape the waveform. Amplitudes of the recorded data are corrected for transmission losses, spherical spreading, and attenuation with propagation distance. Despite progress in estimating attenuation parameters, mechanisms causing attenuation are still poorly understood (White 1983). Severe velocity and thickness variations within the surficial weathering layer commonly degrade the recorded signal by attenuating high frequencies and delaying signals from deeper reflectors.

Refracted arrivals are used to analyze the extent of the weathering layer, and refraction statics software is applied to correct for temporal discontinuities.

Shaping the waveform by wavelet processing and deconvolution to achieve better resolution is one of the most significant advances in seismic stratigraphic processing. In wavelet processing, the propagating wavelet is converted to a more desired shape, usually zero-phase. A noncausal, zero-phase (symmetrical) wavelet, when convolved with subsurface reflection coefficients, allows the human eye to distinguish between reflections much more easily than those produced with the minimum phase wavelet, even though the frequency content is equivalent. With the exception of Vibroseis™ data, seismic data are assumed to be minimum phase. To perform the conversion from minimum to zero-phase, the shape and phase of the original propagating wavelet must be determined. The wavelet is influenced by many factors, including the source type, recording instrumentation, ground coupling of the source and geophone, and absorption during propagation. The shape and phase of the original wavelet presently are derived by either statistical or deterministic methods or by directly measuring the wavelet, a method presently applicable only to deep marine environments (Neidell 1979).

Many sedimentary units of interest to interpreters are at the scale of the progradational event and have thicknesses less than the dominant wavelength of typical seismic data. The ability to detect or measure thin beds with seismic data is crucial for describing the geometries of progradational events and sedimentary facies distributions within them. Resolution of thin beds is aided by analyzing the true amplitudes of reflections. By seismic forward modeling, Widess (1973) and others (e.g. Neidell & Poggiagliolmi 1977, Meckel & Nath 1977, Sheriff 1985) showed that the critical thickness for which the geometry of a thin bed can be distinguished is approximately 1/8 to 1/4 of the dominant wavelength with zero-phase, noise-free data. The top and bottom of a bed thicker than the critical thickness can be distinguished on the seismic record. For beds thinner than 1/4 wavelength, the thicknesses can be determined by measuring the peak-to-trough amplitude from the top to the bottom of the reflection. When a bed is approximately 1/4 wavelength in thickness, the amplitude increases or "tunes" by constructive interference. Below this tuning thickness, the amplitude will decrease owing to destructive interference. Forward models are used to calibrate the amplitude response of a thin bed of interest. A similar resolution problem exists for laterally discontinuous stratigraphic properties. Forward models indicate that the critical width of a laterally discontinuous feature is approximately one Fresnel zone, below which reflection amplitudes decrease (Neidell 1979). As with vertical resolution, the amplitude response to lateral variations can be calibrated with forward models.

Development of three-dimensional seismic surveys and resulting interpretational methodologies during the past decade have greatly improved spatial definition of subsurface stratigraphy. Two-dimensional seismic surveys acquire data with receivers oriented along a line and in-line with the source. In three-dimensional surveys, data are acquired within a two-dimensional surface grid of receiver and/or source positions. The data contain information from a three-dimensional subsurface grid, the third dimension resulting from the recorded reflection traveltimes. The ability to display variations in reflection character in map and cross-section views, especially in color, has enhanced stratigraphic interpretations. Reflection continuity is sometimes best developed horizontally in subtle stratigraphic features. For example, Brown (1985) illustrated fluvial channels and reservoir geometry with three-dimensional data attributes displayed in color on maps and cross sections.

Methods of measuring seismic waveforms and traveltimes in boreholes to calibrate the seismic response, determine acoustic properties of known lithologies, and tie stratal depths to seismic traveltimes were introduced as early as 1931. The most common method, Vertical Seismic Profiles (VSPs), provides good estimates of attenuation, wavelet form, and interbed multiples, information that may be added to standard processing routines to improve resolution and geologic accuracy of the seismic image (Balch & Lee 1984). Because of their expense, most VSPs use only one source offset from a well, thus providing only a one-dimensional view of the subsurface. There is increasing use of VSPs conducted at many different offsets (walk-away VSPs) that provide information lateral to the borehole. The development of a feasible downhole source, such as a downhole air gun, with sufficient energy to return information to a two-dimensional array of surface receivers will be a significant development, since it will improve three-dimensional subsurface definition (Lee et al 1984). Tomographic inversion of seismic traveltimes for the velocity field between boreholes is a method that is becoming more common because it can increase the resolution of stratal configurations.

Analysis of Lithology and Fluid Content

Lateral variations of different attributes of the seismic signal in temporal, spectral, and spatial domains often may indicate changes in discontinuous stratigraphic properties. Deterministic evaluation of lateral variations or anomalies along reflections began with "bright spot" technologies in the 1970s and has since blossomed in scope and popularity. Reflection amplitudes are primarily dependent upon contrasts in acoustic impedance. A significant impedance contrast developed along a reflector can cause the amplitude to increase dramatically and thus become "bright," or more noticeable, on a seismic section. Significant impedance contrasts commonly occur within a rock unit along a gas/water contact. Depending on local acoustic conditions, other subtle variations in reflection character, such as lateral phase changes and amplitude decreases, also indicate lateral and vertical boundaries of gas and liquids. This technology has been extended to detecting other stratigraphically induced variations in acoustic impedance, such as those caused by porosity development and lithology changes.

Empirical and deterministic models of petrophysical properties developed by Gassmann (1951), Biot (1956), and Gardner et al (1974) are used to model subsurface anomalies. Combined with either measured or empirical velocity-density relations, the compressional wavefield measures only the plane-wave modulus. Since two elastic moduli are required to define a homogeneous isotropic solid, moduli are obtained experimentally, empirically, or measured directly. Analysis of horizontally polarized shear-wave data has facilitated measurement of variations in elastic moduli for these models and detection of subsurface anisotropy.

Seismic data collected with different source-receiver offsets record travel paths of acoustic waves that have different angles of incidence with respect to a single reflector (Figure 5). Associated with variations in angles of incidence are changes in reflection character expressed as changes in amplitude and/or phase along the reflector. These measured variations are related to contrasts in Poisson's ratio between the strata at the reflecting interface that in turn may be used to infer lithology and fluid content. A new technology, termed "Amplitude Variations with Offset," or "AVO," is developing that promises to increase the resolution of lithology and fluid definition. As an example, Ostrander (1984) analyzed "bright spot" anomalies along reflections using the AVO method. One "bright spot" in the Sacramento Valley, California, was identified as a gas-bearing sand, but another, similar "bright spot" in Nevada was demonstrated to be a basalt. To successfully model AVO anomalies, Poisson's ratio for the targets of interest must be known, estimated, or calculated.

Expanded use of the Hilbert transform of the seismic trace has also increased the ability of the seismic reflection method to discern lateral stratigraphic variations (Taner et al 1979). The recorded seismic trace is treated as an analytical time series that is the real part of a complex trace. In contrast to the Fourier transform used to analyze the frequency and phase characteristics over large portions of a trace, the Hilbert transform allows analysis of local variations in attributes of a trace. Each attribute, such as reflection strength, phase, and instantaneous frequency, empha-

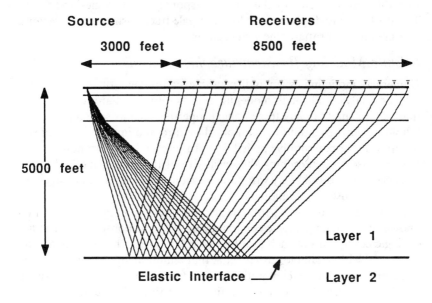

Source
3000 feet

Receivers
8500 feet

5000 feet

Layer 1

Elastic Interface ⟋   Layer 2

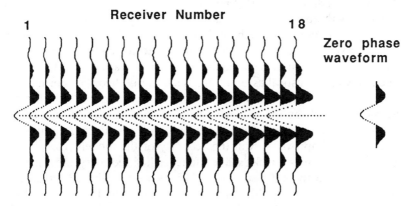

Receiver Number
1   18

Zero phase
waveform

Figure 5 Recording geometry and synthetic shot record illustrating amplitude changes with different source-receiver offsets. Associated with each source-receiver offset is an angle of incidence at the elastic interface. Variations in elastic parameters between the two layers at the interface cause variations in reflection amplitude and/or phase with different angles of incidence. This example shows a shale/gas sand interface with an increase in amplitude associated with increased source-receiver offset. The shale (Layer 1) has a compressional velocity of 9000 ft s^{-1}, density of 2.4 g cm^{-3}, and a Poisson's ratio of 0.4. The gas sand (Layer 2) has a compressional velocity of 7500 ft s^{-1}, density of 2.14 g cm^{-3}, and a Poisson's ratio of 0.1.

sizes different elements of the seismic response to lithologies and fluids. The attributes are displayed in color on wiggle-trace seismic record sections, thus enhancing stratigraphic visualization.

Seismic Modeling for Stratrigraphic Analysis

Numerical modeling is an indispensable component of seismic stratigraphic analysis. Through modeling, seismic responses are related quantitatively or qualitatively to geologic properties such as stratal geometry, lithology, porosity, and fluid content. Forward and inverse modeling are the approaches commonly applied. In both cases, it is important to qualify the response by determining the sensitivity of the seismic method to the presence of noise and likely variations in geologic configurations, since the seismic response is not unique.

In forward modeling a synthetic seismic response is calculated from a known or assumed geologic configuration. If the response to a particular geologic configuration such as a known reservoir can be determined and delineated with available seismic data, that temporal and spatial signature can then be used as an exploration tool. Attempts are made to determine which seismic attribute or combination of attributes will uniquely detect the geologic configuration of interest. More commonly, forward models of several possible geologic configurations are run and compared for similarity with observed seismic data.

There are a number of methods for calculating seismic forward models, each with varying degrees of approximation to the complete calculation of the wavefield. Some algorithms are based on the wave equation, whereas others rely on acoustical assumptions, such as ray theory or normal incidence reflection. Each relies upon geologic assumptions for configuring the models, such as the values and geometries of acoustic impedance contrasts from either well logs or petrophysical estimates. As geological specifications improve, so will the accuracy of the models. Geophysical assumptions, such as wavelet shape and recording geometry, are also used. One-, two-, and three-dimensional solutions are modeled (Peterson et al 1955, Wuenschel 1960, Hilterman 1970, Taner et al 1970, Torey 1970, Shah 1973, Kelly et al 1976), depending upon the perceived complexity of the geologic configuration.

In the second approach, inverse models are used to deduce the actual geologic configuration from seismic data. Velocity information contained in seismic amplitudes and traveltimes is also used to interpret lithology and fluids. Reflection amplitudes are inverted to derive impedance contrasts and then combined with empirically derived velocity-density relations to obtain acoustic velocities. Changes in lithology, porosity, and fluid content may produce changes in acoustic impedance that are detectable seismically. The derived velocity sections may be displayed in color, providing increased resolution from conventional wiggle-trace sections (Savit & Wu 1982, Neidell & Beard 1984). Often the presence of noise and lack of low-frequency information inhibits inversion of amplitude data. Seismic data are typically recorded with different source-receiver offsets, resulting in different traveltimes for the same reflector. This geometric relationship is the basis for conventional velocity analysis (Taner & Koehler 1969) in which derived velocities are the root-mean-square or "stacking" velocities for nondipping reflectors. Neidell & Beard (1984) derived velocity from traveltimes and amplitudes and analyzed stacking velocities at a fine scale to detect anomalous zones of porosity development.

Stratigraphic Sequences and Seismic Lithology Analysis

Geophysical research for seismic stratigraphic analysis has been devoted almost exclusively to technological developments that improve the stratigraphic image and invert the seismic data for lithologic parameters. Often the aspect providing the most uncertainty in lithology analysis is the assumed geologic configuration. Since a geologic configuration inferred from seismic data is not unique, a narrow range of independently derived probable configurations is imperative to an accurate solution. Geologic models are essential as independent constraints for seismic data acquisition design, processing and inversion algorithms, and as an input to seismic forward models. Stratigraphic sequence analysis and quantitative simulation of stratigraphic architecture promise to provide accurate geologic models at scales ranging from the depositional sequence to the progradational event.

Seismic stratigraphy has come full circle. The development of high-fidelity continuous record sections provided the basis for extending sequence concepts of Sloss and Wheeler to analysis of seismic data. This approach to seismic data analysis produced seismic sequence and seismic facies concepts and methodologies. As stratigraphic images were progressively more refined by advances in geophysical acquisition and processing techniques, interpretations of stratal architecture, lithology, and fluid content at the scale of the progradational event became viable. However, stratigraphic interpretations from seismic data alone have remained ambiguous, owing to the multiple geologic configurations that may produce an observed seismic record. To close the circle, the increased accuracy in defining and predicting geologic configurations through stratigraphic sequence analysis and deterministic stratigraphic models promises to provide geologic constraints for seismic data acquisition, processing, modeling, and the estimation of seismic lithology.

INTEGRATED SEISMIC STRATIGRAPHIC ANALYSIS

Integrated seismic stratigraphic analysis—the qualitative and quantitative incorporation of diverse geological and geophysical concepts, data, and methods—is rapidly advancing and has the potential to significantly increase the accuracy of stratigraphic interpretations and predictions. Stratigraphic interpretations of seismic data are elusive and no single technique provides a unique solution. By integrating multiple methods and maximizing their individual strengths, a solution may be derived that is more accurate and consistent with all types of data than is obtainable from any one method alone. In the future, seismic stratigraphic analysis will incorporate and utilize geological information in ways similar to current treatments of geophysical data. Whereas quantitative geophysical stratigraphic analysis is well established, deterministic geological models are in their infancy. The development of such geological models will facilitate the symbolic and numerical integration of geophysical and geological data and consequently increase the accuracy of predicting stratal architecture and lithology distribution.

Deterministic geological models will complement and constrain seismic interpretations at all scales. Geological and geophysical inverse models may be used to extract the fundamental components of sea-level fluctuations, tectonics, and sediment supply in areas where the stratal architecture is known. Using the extracted components, forward geological models will predict stratal architecture and lithologies in areas beyond geologic control. A synthetic seismic response to the predicted geologic configuration can then be calculated and compared with seismic data. Observed differences can be used to revise the interpetation. Successive iterations should provide multiple possible stratigraphic configurations along with associated uncertainty limits. The final integrated solutions should be far more accurate than those obtained using present methodologies.

Computer workstations, using both numerical and symbolic processing techniques, will provide the means to accomplish qualitative and quantitative integration. Presently, interactive geophysical workstations play an integral role in seismic stratigraphic analysis by providing rapid manipulation of seismic and well-log data as well as display capabilities not commonly available before the 1980s. Rapid access to data displayed in cross-section and map views enables interpreters to investigate multiple ideas quickly and efficiently. Data may be displayed at different scales to visually enhance reflection geometries and waveform character. Seismic sequence boundaries not resolved in conventional displays often become apparent through compression of the horizontal scale. Subtle stratigraphic variations may be discerned at larger scales by color displays of waveform attributes and velocity sections. For example, time-slice maps that show variations in waveform attributes along a single reflector have significantly enhanced detection of subtle lateral variations of lithology and fluids (Brown 1985).

Future geoscientific workstations will incorporate artificial intelligence techniques and forward and inverse models to integrate site-specific geophysical and geological data, general empirical knowledge, analytical models, and fundamental concepts far beyond present capabilities. This will provide increased confidence over interpretations based on methods that singly have fundamental limitations. Perhaps the most significant advances in integrated seismic stratigraphic analysis will result from the incorporation of artificial intelligence drivers in geoscientific workstations (Figure 6). In the area of artificial intelligence, both pattern recognition and expert systems will be used. Pattern recognition will aid data correlation and searches for unique configurations. Expert systems will provide the geological and geophysical knowledge necessary to facilitate analysis. Expert-system drivers will suggest possible interpretations or solution paths as well as their associated degrees of uncertainty or reliability. Interpreters will access and manipulate large volumes of empirical or "rule-of-thumb" knowledge and fundamental concepts as well as geophysical data and geological observations. In the future, all scales of seismic stratigraphic analysis will be integrated and questions will be asked that are not now capable of resolution.

ACKNOWLEDGMENTS

We have benefited from discussions with several individuals concerning various aspects of seismic stratigraphic methodologies and concepts. We especially thank D. Bradford Macurda for his suggestions on essential material to include in this review; Frank Hadsell, Ray Sengbush, Dale Turner, and Peter Vail for their historical perspectives on milestones and developments in seismic stratigraphy; Barbara Radovich for comments on the time-stratigraphic significance of reflections and examples of reflections transecting sedimentary facies boundaries, and for sharing her perspectives on the impact that seismic stratigraphy has had in petroleum exploration and geology; and Nick Christie-Blick for providing preprints of manuscripts. TAC benefited from exposure to seismic sequence and seismic facies concepts and their development while employed by Exxon Production Research Co. and thanks the many individuals in that organization who shared their experiences and insights with him. The draft manuscript

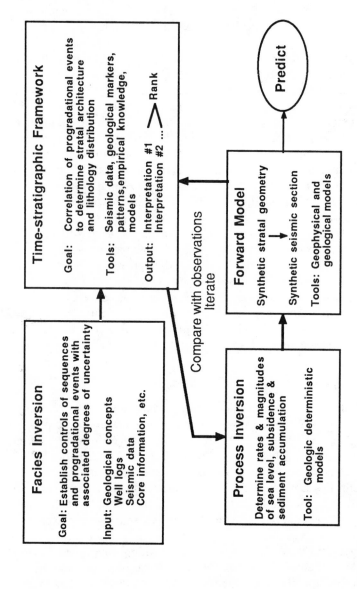

Figure 6 Schematic configuration of a future geoscientific workstation. Two key elements distinguish this from contemporary workstations. One is the full integration of manipulable visual displays of geologic and geophysical data with analytical programs to model and modify the data and test ideas and assumptions. The other, the expert-system driver, will retain and utilize information derived from each step to modify solutions and guide the user in selecting among possible interpretations.

was reviewed by Nick Christie-Blick, Frank Hadsell, Ian Lerche, Don Loomis, D. Bradford Macurda, Barbara Radovich and Peter Vail. Their comments and criticisms were exceptionally insightful and helpful and are much appreciated.

Literature Cited

Abbott, W. O. 1985. The recognition and mapping of a basal transgressive sand from outcrop, subsurface, and seismic data. In *Seismic Stratigraphy II—An Integrated Approach*, Am. Assoc. Pet. Geol. Mem., ed. O. R. Berg, D. G. Woolverton, 39: 157–67

Asquith, D. O. 1970. Depositional topography and major marine environments, Late Cretaceous, Wyoming. *Am. Assoc. Pet. Geol. Bull.* 54: 1184–1224

Balch, A. H., Lee, M. W. eds. 1984. *Vertical Seismic Profiling Technique, Applications, and Case Histories.* Boston: Int. Human Resour. Dev. Corp. 488 pp.

Bally, A. W. 1983a. Seismic expression of structural styles. *Am. Assoc. Pet. Geol. Studies in Geol. 15*, Vol. 2

Bally, A. W. 1983b. Seismic expression of structural styles. *Am. Assoc. Pet. Geol. Studies in Geol. 15*, Vol. 1

Berg, O.R. 1982. Seismic detection and evaluation of delta and turbidite sequences: their application to exploration for the subtle trap. *Am. Assoc. Pet. Geol. Bull.* 66: 1271–88

Biot, M. A. 1956. Theory of propagation of elastic waves in fluid-saturated porous solid, I: Low-frequency range. *J. Acoust. Soc. Am.* 28: 168–78

Brown, A. R. 1985. The role of horizontal seismic sections in stratigraphic interpretation. In *Seismic Stratigraphy II—An Integrated Approach*, Am. Assoc. Pet. Geol. Mem., ed. O. R. Berg, D. G. Woolverton, 39: 37–47

Brown, L. F. Jr., Fisher, W. L. 1977. Seismic-stratigraphic interpretation of depositional systems: examples from Brazilian rift and pull-apart basins. See Payton 1977, pp. 213–48

Brown, L. F. Jr., Fisher, W. L. 1980. Seismic stratigraphic interpretation and petroleum exploration. *Am. Assoc. Pet. Geol. Contin. Ed. Course Note Ser. 16.* 125 pp.

Bubb, J. N., Hatlelid, W. G. 1977. Seismic stratigraphy and global changes of sea level, Part 10: Seismic recognition of carbonate buildups. See Payton 1977, pp. 185–204

Busch, D. A. 1959. Prospecting for stratigraphic traps. *Am. Assoc. Pet. Geol. Bull.* 43: 2829–43

Busch, D. A. 1974. Stratigraphic traps in sandstones—exploration techniques. *Am. Assoc. Pet. Geol. Mem. 21.* 174 pp.

Campbell, C. V. 1967. Lamina, laminaset, bed and bedset. *Sedimentology* 8: 7–26

Campbell, C. V. 1979. Model for beach shoreline in Gallup Sandstone (Upper Cretaceous) of northwestern New Mexico. *N. Mex. Bur. Mines Miner. Resour. Circ. 164.* 32 pp.

Cant, D. J. 1984. Development of shoreline-shelf sand bodies in a Cretaceous epeiric sea deposit. *J. Sediment. Petrol.* 54: 541–56

Caster, K. E. 1934. The stratigraphy and paleontology of northwestern Pennsylvania. *Bull. Am. Paleontol.* 21(71): 19–31

Churlin, V. V., Sergeyev, L. A. 1963. Application of seismic surveying to recognition of productive part of gas-oil strata. *Geol. Nefti Gaza* 7(11)

Christie-Blick, N., Mountain, G. S., Miller, K. G. 1987. Seismic stratigraphic record of sea-level change. In *Sea-Level Change, Studies in Geophysics.* Washington, DC: Natl. Acad. Press. In press

Christie-Blick, N., Grotzinger, J. P., von der Borch, C. C. 1988. Sequence stratigraphy in Proterozoic successions. *Geology.* In press

Conticini, F. 1985. Seismic facies quantitative analysis—new tool in stratigraphic interpretation. *Oil Gas Dig.* 7(2): 22–24

Embry, A. F. 1983. Depositional sequences —impractical lithostratigraphic units. *Geol. Soc. Am. Abstr. With Programs* 15(6): 567 (Abstr.)

Frazier, D. E. 1974. Depositional-episodes: their relationship to the Quaternary stratigraphic framework in the northwestern portion of the Gulf basin. *Tex. Bur. Econ. Geol. Circ. 74-1.* 27 pp.

Galloway, W. E., Brown, L. F. Jr. 1973. Depositional systems and shelf-slope relations on cratonic basin margin, uppermost Pennsylvanian of north-central Texas. *Am. Assoc. Pet. Geol. Bull.* 57: 1185–1218

Gardner, G. H. F., Gardner, L. W., Gregory, A. R. 1974. Formation velocity and density—the diagnostic basis for stratigraphic traps. *Geophysics* 39(6): 770–80

Gassmann, F. 1951. Elastic waves through a packing of spheres. *Geophysics* 16: 673–85

(see also *Geophysics* 18: 269 for corrections)

Goodwin, P. W., Anderson, E. J. 1985. Punctuated aggradational cycles: a general hypothesis of episodic stratigraphic accumulation. *J. Geol.* 93: 515–33

Grabau, A. 1924. *Principles of Stratigraphy.* New York: Seiler. 1185 pp. 2nd ed.

Grotzinger, J. P. 1986. Cyclicity and paleoenvironmental dynamics, Rocknest platform, northwest Canada. *Geol. Soc. Am. Bull.* 97: 1208–31

Hardage, B. A., ed. 1987. *Seismic Stratigraphy.* London: Geophys. Press. 432 pp.

Haq, B. U., Hardenbol, J., Vail, P. R. 1987. Chronology of fluctuating sea levels since the Triassic. *Science* 235: 1156–66

Hilterman, F. J. 1970. Three-dimensional seismic modeling. *Geophysics* 35(6): 1020–37

Hubbard, R. J., Pape, J., Roberts, D. G. 1985a. Depositional sequence mapping as a technique to establish tectonic and stratigraphic framework and evaluate hydrocarbon potential on a passive continental margin. In *Seismic Stratigraphy II—An Integrated Approach, Am. Assoc. Pet. Geol. Mem.,* ed. O. R. Berg, D. G. Woolverton, 39: 79–91

Hubbard, R. J., Pape, J., Roberts, D. G. 1985b. Depositional sequence mapping to illustrate the evolution of a passive continental margin. In *Seismic Stratigraphy II—An Integrated Approach, Am. Assoc. Pet. Geol. Mem.,* ed. O. R. Berg, D. G. Woolverton, 39: 93–115

Jackson, W. E. 1964. Depositional topography and cyclic deposition in west-central Texas. *Am. Assoc. Pet. Geol. Bull.* 48: 317–28

Jasinski, E. J. 1957. Delineation of reef reservoirs by seismic control. *Ann. Spring Conf. Exploration Division Guidepost to the Future—A Symposium, 11th,* pp. 20–38. Dallas: Atlantic Refining Co.

Kelly, K. R., Ward, R. W., Treitel, S., Alford, R. M. 1976. Synthetic seismograms: a finite difference approach. *Geophysics* 41(1): 2–27

Kirk, R. B. 1985. A seismic stratigraphic case history in the eastern Barrow subbasin, north west shelf, Australia. In *Seismic Stratigraphy II—An Integrated Approach, Am. Assoc. Pet. Geol. Mem.,* ed. O. R. Berg, D. G. Woolverton, 39: 183–207

Krumbein, W. C., Sloss, L. L. 1951. *Stratigraphy and Sedimentation.* San Francisco: Freeman. 497 pp.

Laster, S. J. 1985. The present state of seismic data acquisition: one view. *Geophysics* 50(12): 2443–51

Lee, M. W., Balch, A. H., Parrott, K. R. 1984. Radiation from a downhole air gun source. *Geophysics* 49(1): 27–36

Macurda, D. B. Jr. 1985. *A Workshop in Seismic Stratigraphy.* Houston: GeoQuest Int., Inc.

Macurda, D. B. Jr. 1986. *Seismic Facies Analysis.* Houston: GeoQuest Int., Inc.

Mayne, H. W. 1962. Common reflection point horizontal data stacking techniques. *Geophysics* 27(6): 927–38

McGovney, J. E., Radovich, B. J. 1985. Seismic stratigraphy and facies of the Frigg fan complex. In *Seismic Stratigraphy II—An Integrated Approach, Am. Assoc. Pet. Geol. Mem.,* ed. O. R. Berg, D. G. Woolverton, 39: 139–54

Meckel, L. D. Jr., Nath, A. K. 1977. Geologic considerations: stratigraphic modeling, interpretation. See Payton 1977, pp. 417–38

Mitchum, R. M. Jr. 1985. Seismic stratigraphic expression of submarine fans. In *Seismic Stratigraphy II—An Integrated Approach, Am. Assoc. Pet. Geol. Mem.,* ed. O. R. Berg, D. G. Woolverton, 39: 117–36

Mitchum, R. M. Jr., Vail, P. R., Todd, R. G., Sangree, J. B. 1976. Regional seismic interpretation using sequences and eustatic cycles. *Am. Assoc. Pet. Geol. Bull.* 60: 669 (Abstr.)

Mitchum, R. M. Jr., Vail, P. R., Sangree, J. B. 1977a. Seismic stratigraphy and global changes of sea level, Part 6: Stratigraphic interpretation of seismic reflection patterns in depositional sequences. See Payton 1977, pp. 117–33

Mitchum, R. M. Jr., Vail, P. R., Thompson, S. III 1977b. Seismic stratigraphy and global changes of sea level, Part 2: The depositional sequence as a basic unit for stratigraphic analysis. See Payton 1977, pp. 53–62

Muir, I., Wong, P., Wendte, J. 1985. Devonian Hare Indian-Ramparts (Kee Scarp) evolution, MacKenzie Mountains and subsurface Norman Wells, N. W. T.: basin-fill and platform reef development. *Soc. Econ. Paleontol. Mineral. Core Workshop* 7, pp. 311–41

Neidell, N. S. 1979. Stratigraphic modeling and interpretation: geophysical principles and techniques. *Am. Assoc. Pet. Geol. Contin. Ed. Course Note Ser. 13.* 145 pp.

Neidell, N. S., Beard, J. H. 1984. Progress in stratigraphic seismic exploration and the definition of reservoirs. *J. Pet. Technol.* 36(5): 709–26

Neidell, N. S., Poggiagliolmi, E. 1977. Stratigraphic modeling and interpretation. See Payton 1977, pp. 389–416

Neidell, N. S., Beard, J. H., Cook, E. E. 1985. Use of seismic-derived velocities for stratigraphic exploration on land: seismic porosity and direct gas detection. In *Seismic Stratigraphy II—An Integrated Approach*, Am. Assoc. Pet. Geol. Mem., ed. O. R. Berg, D. G. Woolverton, 39: 49–77

Ostrander, W. J. 1984. Plane-wave reflection coefficients for gas sands at non-normal angles of incidence. *Geophysics* 49(10): 1637–48

Payton, C. E., ed. 1977. *Seismic Stratigraphy—Applications to Hydrocarbon Exploration*. Am. Assoc. Pet. Geol. Mem. 26. 516 pp.

Peterson, R. A., Fillippone, W. R., Coker, F. B. 1955. The synthesis of seismograms from well log data. *Geophysics* 20(3): 516–38

Pitman, W. C. III. 1978. Relationship between eustacy and stratigraphic sequences of passive margins. *Geol. Soc. Am. Bull.* 89: 1389–1403

Pitman, W. C. III., Golovchenko, X. 1983. The effect of sealevel change on the shelf-edge and slope of passive margins. In *The Shelfbreak: Critical Interface on Continental Margins. Soc. Econ. Paleontol. Mineral. Spec. Publ.*, ed. D. J. Stanley, G. T. Moore, 33: 41–58

Ramsayer, G. R. 1979. Seismic stratigraphy, a fundamental exploration tool. *Offshore Tech. Conf.* 11: 1859–62

Roksandic, M. M. 1978. Seismic facies analysis concepts. *Geophys. Prospect.* 26: 383–98

Sangree, J. B., Widmier, J. M. 1977. Seismic stratigraphy and global changes of sea level, Part 9: Seismic interpretation of clastic depositional facies. See Payton 1977, pp. 165–84

Sangree, J. B., Widmier, J. M. 1979. Interpetation of depositional facies from seismic data. *Geophysics* 44: 131–60

Sarg, J. F., Lehmann, P. J. 1986. Lower-middle Guadalupian facies and stratigraphy, San Andres/Grayburg formations, Permian basin, Guadalupe Mountains, New Mexico. *Permian Basin Sect., Soc. Econ. Paleontol. Mineral. Publ. 86-25*, pp. 1–8

Savit, C. H., Wu, C. 1982. Geophysical characterization of lithology—application to subtle traps. In *The Deliberate Search for the Subtle Trap, Am. Assoc. Pet. Geol. Mem.*, ed. M. Halbouty, 32: 11–30

Schultz, P. S. 1985. Seismic data processing: current industry practice and new directions. *Geophysics* 50(12): 2452–57

Shah, P. M. 1973. Ray tracing in three dimensions. *Geophysics* 38(3): 600–4

Shaw, C. A., Hay, W. W. 1987. Refining eustatic sea level curves with balanced paleogeographic models. *Geol. Soc. Am. Abstr. With Programs* 19(7): 840 (Abstr.)

Sheriff, R. E. 1980. *Seismic Stratigraphy*. Boston: Int. Human Resour. Dev. Corp. 227 pp.

Sheriff, R. E. 1985. Aspects of seismic resolution. In *Seismic Stratigraphy II—An Integrated Approach, Am. Assoc. Pet. Geol. Mem.*, ed. O. R. Berg, D. G. Woolverton, 39: 1–10

Sloss, L. L. 1963. Sequences in the cratonic interior of North America. *Geol. Soc. Am. Bull.* 74: 93–114

Sloss, L. L., Krumbein, W. C., Dapples, E. C. 1949. Integrated facies analysis. In *Sedimentary Facies in Geologic History, Geol. Soc. Am. Mem.*, C. R. Longwell, Chairman, 39: 91–124

Swift, D. J. P. 1968. Coastal erosion and transgressive stratigraphy. *J. Geol.* 75: 444–56

Taner, M. T., Koehler, R. 1969. Velocity spectra—digital computer derivation and applications of velocity functions. *Geophysics* 34(6): 859–81

Taner, M. T., Cook, E. E., Neidell, N. S. 1970. Limitations of the reflection seismic method—lessons from computer simulations. *Geophysics* 35: 551–73

Taner, M. T., Koehler, R., Sheriff, R. E. 1979. Complex seismic trace analysis. *Geophysics* 44(6): 1041–63

Torey, A. W. 1970. A simple theory for seismic diffractions. *Geophysics* 35(5): 762–84

Vail, P. R. 1975. Eustatic cycles from seismic data for global stratigraphic analysis. *Am. Assoc. Pet. Geol. Bull.* 59: 2198–99 (Abstr.)

Vail, P. R., Sangree, J. B. 1971. Time stratigraphy from seismic data. *Am. Assoc. Pet. Geol. Bull.* 55: 367–38 (Abstr.)

Vail, P. R., Todd, R. G. 1981. Northern North Sea Jurassic unconformities, chronostratigraphy and sea-level changes from seismic stratigraphy. In *Petroleum Geology of the Continental Shelf of North-West Europe*, ed. L. V. Illing, G. D. Hobson, pp. 216–35. London: Heyden & Son

Vail, P. R., Wilbur, R. O. 1966. Onlap, key to worldwide unconformities and depositional cycles. *Am. Assoc. Pet. Geol. Bull.* 50: 638–39 (Abstr.)

Vail, P. R., Mitchum, R. M., Thompson, S. III. 1977a. Seismic stratigraphy and global changes of sea level, Part 4: Global cycles of relative changes of sea level. See Payton 1977, pp. 83–97

Vail, P. R., Todd, R. G., Sangree, J. B. 1977b. Seismic stratigraphy and global changes of sea level, Part 5: Chronostratigraphic significance of seismic reflections. See Payton 1977, pp. 99–116

Vail, P. R., Hardenbol, J., Todd, R. G. 1984. Jurassic unconformities, chronostratigraphy, and sea-level changes from seismic stratigraphy and biostratigraphy.

354   CROSS & LESSENGER

In *Interregional Unconformities and Hydrocarbon Accumulation, Am. Assoc. Pet. Geol. Mem.*, ed. J. S. Schlee, 36: 129–44

Van Siclen, D. C. 1958. Depositional topography—examples and theory. *Am. Assoc. Pet. Geol. Bull.* 42: 1897–1913

Wanless, H. R., Weller, J. M. 1932. Correlation and extent of Pennsylvanian cyclothems. *Geol. Soc. Am. Bull.* 43: 1003–16

Wheeler, H. E. 1958. Time-stratigraphy. *Am. Assoc. Pet. Geol. Bull.* 42: 1047–63

Wheeler, H. E. 1959a. Note 24—unconformity-bounded units in stratigraphy. *Am. Assoc. Pet. Geol. Bull.* 43: 1975–77

Wheeler, H. E. 1959b. Stratigraphic units in time and space. *Am. J. Sci.* 257: 692–706

Wheeler, H. E. 1963. Post-Sauk and pre-Absaroka Paleozoic stratigraphic patterns in North America. *Am. Assoc. Pet. Geol. Bull.* 47: 1497–1526

White, J. E. 1983. *Underground Sound.* New York: Elsevier. 253 pp.

Widess, M. B. 1973. How thin is a thin bed? *Geophysics* 38(6): 1176–80

Wuenschel, P. C. 1960. Seismogram synthesis including multiples and transmission coefficients. *Geophysics* 25(1): 106–29

Ann. Rev. Earth Planet. Sci. 1988. 16: 355–88

IN SITU–PRODUCED COSMOGENIC ISOTOPES IN TERRESTRIAL ROCKS

D. Lal

Scripps Institution of Oceanography, Geological Research Division, La Jolla, California 92093 and Physical Research Laboratory, Ahmedabad 380009, India

> *There is something fascinating about science. One gets such wholesale returns of conjecture out of such a trifling investment of fact.*
>
> Mark Twain, *Life on the Mississippi*

In the late forties, Libby and his collaborators detected the naturally occurring radioactive isotope ^{14}C, produced by cosmic radiation in the Earth's atmosphere (Libby 1946, Anderson et al 1947). This marked the onset of the search for cosmic-ray-produced (cosmogenic) isotopic changes in terrestrial and extraterrestrial samples and in the cosmic rays themselves. During the four decades since Libby's discovery, some four dozen cosmogenic stable and radioactive isotopes in extraterrestrial samples (Reedy et al 1983) and two dozen in terrestrial samples (Lal & Peters 1967) have been discovered. These isotopes have found applications in cosmic-ray physics (in the study of acceleration and propagation of galactic and solar cosmic rays), in solar system astrophysics (in determining the evolutionary history of meteorites, the lunar surface, and interplanetary dust particles), and in the Earth sciences (in archaeology, meteorology, glaciology and oceanography). For reviews on these topics, reference is made to Simpson (1983); Reedy et al (1983); and Lal & Peters (1967), Oeschger et al (1970), and Faure (1986), respectively. The process of discovery continues, propelled by advances in the techniques used to measure these isotopes. Even now, examples of new isotopic effects are being discovered, and new applications are being made of the various isotopic effects. When one looks at the history of the development of the terrestrial cosmogenic field, one observes that it grew in waves: After an idea is implemented, the field

355

0084–6597/88/0515–0355$02.00

matures soon thereafter, followed by a quiescent period until the next technical advance produces a quantum jump in detection capability.

But why is technology the limiting factor? And wherein lie the potentialities of the fields of cosmic-ray physics and of cosmogenic isotopes in the solar system? Cosmic rays were discovered 75 years ago by Victor F. Hess during a series of 10 manned balloon flights, made during day and night, up to altitudes of 5350 m. The ionization chambers used by him established that the radiation was of extraterrestrial origin. H. C. van de Hulst once remarked that the word cosmic rays, at the time the name was given, was a curious double confession of ignorance—cosmic meaning that we do not know from where they were coming and ray meaning that we did not know what they were! The rapid development of a variety of cosmic-ray detectors during the 1930s and 1940s removed much of the mystery of the "rays," but we are still struggling to discover their origin. New techniques that have increased our knowledge of their composition and energy spectrum have also improved our understanding of galactic astronomy. Carl Anderson once noted that when we measure something that could not be measured before, or learn how to measure it more accurately, we almost always find out something interesting. The continuing discoveries in the field of cosmic rays are apt illustrations of this hypothesis.

The interconnection between cosmic rays and geophysics was established with the discovery of cosmogenic ^{14}C on Earth. Improvements in techniques added new isotopes to the list of useful tracers in Earth sciences, and similar developments occurred in the study of meteorites. However, in the latter case, the improvements came more rapidly because of the generally higher nuclear reaction rates and longer effective exposure times.

Implicit in the above is the single fact that makes cosmic rays a very powerful tool in astronomy, astrophysics, and geophysics. During its lifetime, the Galaxy has been continuously filled with high-energy cosmic rays, up to energies of $\sim 10^{20}$ eV. The long history of cosmic-ray effects, going back to close to the time of formation of the Galaxy, and the significant depths in matter down to which cosmic-ray effects can be investigated give the field of cosmogenic isotopes its versatility. In the case of extraterrestrial objects without an appreciable atmosphere, the nuclear effects of cosmic rays are discernible up to depths of $\sim 10^3$ g cm^{-2}. On Earth, the effects extend deeper, up to 10^6 g cm^{-2}, owing to decay of charged pi-mesons in the Earth's atmosphere, giving rise to the penetrating muons. The penetrating component has been used as cosmic "X rays" in a variety of applications: for example, to determine the geometry of pyramids or, in civil engineering, to determine the overburden in different directions. We discuss in this article the intensity of nuclear disintegrations caused by this component in rocks at depths down to 10^4 g cm^{-2} below sea level.

Before getting on with the topic of the present paper, it seems appropriate to comment briefly on another aspect of the process of discovery. A generally valid statement that can be made here is that no ideas were ever entirely new! They all had been given an expression to earlier. Ideas evolve; most ideas do not gel when they first appear on the scene. Even feasible ones do not catch on immediately. Let us cite some examples relevant to the subject of this review:

1. The possibility of production of ^{14}C in the Earth's atmosphere by cosmic-ray neutrons was first suggested by Montgomery & Montgomery (1939), soon after discovery of cosmic-ray neutrons (Locker 1933, Rumbaugh & Locker 1936). Korff (1940) pointed out that most of the slow cosmic-ray neutrons would lead to production of ^{14}C because of the rather high thermal neutron cross section for the ^{14}N (n, p)^{14}C reaction. The search for radiocarbon in nature was made by Libby and his collaborators in 1946. This soon culminated in the detection of ^{14}C in nature (Anderson et al 1947). Soon thereafter, Arnold & Libby (1949) demonstrated the feasibility of using terrestrial cosmogenic ^{14}C for dating archaeological samples.

2. The production of isotopes by nuclear interactions of cosmic rays was first discussed by Grosse (1934), five years prior to the realization of cosmic-ray neutron production of ^{14}C on Earth. For an excellent historical account of development of the cosmogenic isotope field, see Faure (1986).

3. The accelerator mass spectrometry (AMS) technique as we know it was developed a decade ago; it has lowered the detection limit for several long-lived isotopes (e.g. ^{14}C, ^{36}Cl, ^{10}Be, ^{26}Al) by a factor of a million. The first application of this technique had, however, been made earlier by Alvarez to determine if ^{3}He or ^{3}H was stable (Alvarez & Cornog 1939). His colleague, Richard Muller, elaborated on this idea and made the first proposal for AMS studies of ^{14}C (Muller 1977).

4. Studies of ^{10}Be, ^{26}Al, ^{36}Cl, ^{3}He, and other isotopes produced by cosmic rays in rocks have now opened up the scope of terrestrial cosmogenic isotopes to studies of geophysical processes during the Pleistocene; these developments are discussed later in this paper. Their usefulness had, however, been pointed out long ago, and attempts had been made to detect them (see, for example, Davis & Schaeffer 1955, Tanaka et al 1968, Takagi et al 1974, Srinivasan 1976).

5. Only recently was it realized that cosmic rays are the principal agency for producing Li, Be, and B in our Galaxy, after two decades of studies of cosmic-ray spallations in nature. The "galactogenesis" hypothesis was first

proposed by Reeves et al (1970), following a suggestion by B. Peters (Audouze & Vauclair 1980, Ch. 7). For a recent treatment of the subject, see Walker et al (1985).

An idea that is feasible often gets stalled. There was no great urgency to develop the AMS technique during the 1950s and 1960s because the counting techniques available then appeared quite adequate! The beta activity of small solid samples could easily be measured at levels of disintegrations per hour (Lal & Schink 1960). The gas proportional counting technique had then been considerably improved for ^{14}C counting (Oeschger et al 1972). For samples of nominally 1 cm^3 STP volume, Davis (1968) had demonstrated measurement sensitivity of 1 disintegration per day for ^{37}Ar. A number of ingenious low-level counting systems, employing coincidence-anticoincidence schemes, were available for X-ray and gamma-ray emitters. (For a comprehensive review of the state-of-the-art counting techniques in the 1970s, see Oeschger & Wahlen 1975.)

The principal motivation to develop the AMS technique was technological, possibly to use particle accelerators that had otherwise outlived their usefulness! As mentioned earlier, the idea of using in situ isotopes preceded technological developments. When the AMS method worked for ^{14}C (Nelson et al 1977, Bennett et al 1977), there was then sufficient motivation to develop it for ^{10}Be, ^{26}Al, etc, and successes in their detection and new applications led to rapid flowering of the terrestrial "atmospheric" and "in situ" cosmogenic isotope studies using the AMS technique. The initial developments occurred contemporaneously in Canada, the USA, and France, and the technique spread in the 1980s to Israel, Switzerland, and Italy. For a comprehensive discussion of early AMS studies, reference is made to Hedges (1979), Litherland (1984), Raisbeck & Yiou (1984), Brown (1984), Suter et al (1984), and other articles in the *Proceedings of the Third International Symposium on AMS* (Wölfli et al 1984). For recent studies, see Lal (1987a), Elmore (1987), and Elmore & Phillips (1987).

IN SITU VS THE ATMOSPHERIC COSMOGENIC ISOTOPES

The AMS technique has widened the scope of applications of both "atmospheric" and "in situ" cosmogenic isotopes produced on the Earth. The principal applications of the former are given in Table 1. We shall from here on confine our attention to the latter. The atmospheric isotopes in fact often become a hindrance to the study of the in situ isotopes. The former are transported through the upper layers of the Earth by geophysical/ geochemical processes and in some cases (e.g. ^{10}Be) can be incorpo-

Table 1   Principal "atmospheric" cosmic-ray-produced isotopes and their applications[a]

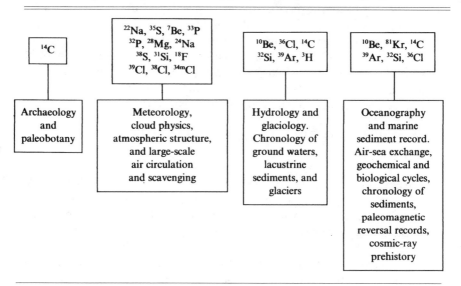

^{14}C	^{22}Na, ^{35}S, ^{7}Be, ^{33}P ^{32}P, ^{28}Mg, ^{24}Na ^{38}S, ^{31}Si, ^{18}F ^{39}Cl, ^{38}Cl, ^{34m}Cl	^{10}Be, ^{36}Cl, ^{14}C ^{32}Si, ^{39}Ar, ^{3}H	^{10}Be, ^{81}Kr, ^{14}C ^{39}Ar, ^{32}Si, ^{36}Cl
Archaeology and paleobotany	Meteorology, cloud physics, atmospheric structure, and large-scale air circulation and scavenging	Hydrology and glaciology. Chronology of ground waters, lacustrine sediments, and glaciers	Oceanography and marine sediment record. Air-sea exchange, geochemical and biological cycles, chronology of sediments, paleomagnetic reversal records, cosmic-ray prehistory

[a] In some cases, an appreciable "artificial" production of radionuclides has significantly changed the global inventories (e.g. ^{3}H, ^{14}C, and ^{36}Cl).

rated into the rocks to overwhelm the amount that is directly produced in situ. Cosmic-ray fluxes in the troposphere are lower than the mean fluxes in the stratosphere by factors of up to 500. Consequently, we confront two problems: (a) low isotope concentrations, and (b) the possibility of an appreciable amount of the isotope produced in the atmosphere being incorporated into the rock by chemical or physical processes. This effect is clearly applicable only for those isotopes that can be scavenged from the atmosphere by wet precipitations. For convenience, the "atmospheric" component is termed the "garden variety" (Nishiizumi et al 1986).

Special precautions have to be taken to insure that an in situ measurement is not affected by the garden variety isotope (Lal & Arnold 1985). Further, one has to work at a high level of sensitivity to insure that sources of contamination are minimized and that there exist no other unrecognized (nuclear) sources of production of the isotope. For example, ^{10}Be can be produced in situ in rocks by radiogenic nuclear reactions (Middleton & Klein 1987); the term radiogenic refers to as due to alpha particles from nuclear transmutations or from their interactions, e.g. neutrons. The carrier beryllium salt obtained commercially has been found to contain an appreciable amount of ^{10}Be (Middleton et al 1984).

Terrestrial in situ cosmogenic studies began two decades after the cosmogenic meteorite (in situ) studies, primarily because the former studies are more difficult. The terrestrial cosmogenic studies require sensitivities 3-4 orders of magnitude higher. But there are fruits to be gathered from such studies because one learns about geophysical events that cannot be studied by other means! And the principal promise of the cosmogenic terrestrial in situ studies is that they will allow us to examine recent Earth history—the geological and geophysical events during the Pleistocene.

IN SITU ISOTOPES DETECTED IN ROCKS SO FAR

We have listed in Table 2 the cosmogenic in situ isotopes detected in rocks to date. The number that can be detected is much greater than the number on this list, and it is also much more than the number of isotopes produced in the atmosphere because of the presence of higher mass number target elements in rocks. A list of often-studied isotopes with half-lives exceeding two weeks is presented in Table 3 (Lal 1987a). Even considering the small terrestrial in situ isotope production rates in rocks exposed at sea level and

Table 2   In situ terrestrial cosmogenic isotopes detected so far

Isotope	Half-life (yr)	Technique used[a]	Reference(s)[b]
Radioactive			
^{10}Be	1.6×10^6	AMS	1, 2, 3,
^{26}Al	7.1×10^5	AMS	1, 2, 3, 4, 5
^{36}Cl	3.0×10^5	C, AMS	6, 7, 8
^{37}Ar	35 days	C	9, 10
^{39}Ar	269	C	9, 10
^{41}Ca	1.0×10^5	AMS	11, 12
^{92}Nb, ^{94}Nb	$3.6 \times 10^7, 2.0 \times 10^4$	C	13, 14
^{129}I	1.6×10^5	C, AMS	15, 16
Stable			
^{3}He	Stable	MS	17, 18
^{21}Ne	Stable	MS	19
^{126}Xe	Stable	MS	20

[a] Abbreviations: C = counting, MS = mass spectrometry, AMS = accelerator mass spectrometry.
[b] References: (1) Jha & Lal (1982), (2) Yiou et al (1984), (3) Nishiizumi et al (1986), (4) Tanaka et al (1968), (5) Hampel et al (1975), (6) Davis & Schaeffer (1955), (7) Kubik et al (1984), (8) Phillips et al (1986), (9) R. Davis (private communication, 1984), (10) Lal et al (1986), (11) Raisbeck & Yiou (1979), (12) Henning et al (1987), (13) Apt et al (1974), (14) Clayton & Morgan (1977), (15) Takagi et al (1974), (16) Fehn et al (1986), (17) Craig & Poreda (1986), (18) Kurz (1986a), (19) Marti & Craig (1987), (20) Srinivasan (1976).

Table 3 Often-studied cosmogenic nuclides with half-lives exceeding two weeks

Isotope(s)	Half-life (yr) (S = stable)	Main targets[a]
^{3}H[b]	12.3	O, Mg, Si, Fe (N, O)
^{3}He, ^{4}He	S	O, Mg, Si, Fe (N, O)
^{7}Be[b]	53 days	O, Mg, Si, Fe (N, O)
^{10}Be[b]	1.6×10^6	O, Mg, Si, Fe (N, O)
^{14}C[b]	5730	O, Mg, Si, Fe (N)
^{20}Ne, ^{21}Ne, ^{22}Ne	S	Mg, Al, Si, Fe
^{22}Na[b]	2.6	Mg, Al, Si, Fe (Ar)
^{26}Al	7.1×10^5	Si, Al, Fe (Ar)
^{32}Si[b]	100–200	(Ar)
^{35}S[b]	87 days	Fe, Ca, K, Cl (Ar)
^{36}Cl[b]	3.0×10^5	Fe, Ca, K, Cl (Ar)
^{36}Ar, ^{38}Ar	S	Fe, Ca, K
^{37}Ar[b]	35 days	Fe, Ca, K (Ar)
^{39}Ar[b]	269	Fe, Ca, K (Ar)
^{40}K	1.3×10^9	Fe
^{39}K, ^{41}K	S	Fe
^{41}Ca	1.0×10^5	Ca, Fe
^{46}Sc	84 days	Fe
^{48}V	16 days	Fe, Ti
^{53}Mn	3.7×10^6	Fe
^{54}Mn	312 days	Fe
^{55}Fe	2.7	Fe
^{56}Co	79 days	Fe
^{59}Ni	7.6×10^4	Ni, Fe
^{60}Fe	1.5×10^6	Ni
^{60}Co	5.27	Co, Ni
^{81}Kr[b]	2.1×10^5	Rb, Sr, Zr (Kr)
^{78}Kr, ^{80}Kr, ^{82}Kr, ^{83}Kr	S	Rb, Sr, Zr
^{129}I[b]	1.6×10^7	Te, Ba, La, Ce (Xe)
$^{124-132}$Xe	S	Te, Ba, La, Ce, I

[a] Elements from which most production occurs; those in parentheses are for the Earth's atmosphere.
[b] Atmospheric cosmogenic isotopes.

at depths underground ($< 10^4$ g cm^{-2}), most of these isotopes can be measured, and it is primarily the promise of the applications that decides whether or not attempts are being made today to look for an isotope.

IN SITU ISOTOPE SOURCE FUNCTIONS

In order to fully exploit the potential of cosmogenic in situ isotopes, it is necessary to understand their production mechanisms and source functions and their expected time variability in relation to terrestrial and solar system influences. The problem of propagation of the cosmic rays through the Earth's atmosphere was dealt with in great detail earlier (Lal & Peters 1967) to estimate the source strengths of cosmic-ray-produced isotopes in the atmosphere. The composition of the secondary cosmic rays also undergoes a continuous change as one goes to deeper depths in the atmosphere. This is illustrated in Table 4. In the troposphere, most nuclear disintegrations are due to neutrons. At sea level, negative muon captures become significant. At depths underground below sea level, exceeding ~ 1-m rock equivalent, negative muon captures, fast muon disintegrations, and the secondary neutrons are responsible for the cosmogenic products. Reference is made to an earlier publication on the subject (Lal & Peters 1967) and a recent update on production rates underground (Lal 1987b). Table 4 provides a qualitative guide to the types of nuclear reactions that must be considered for estimating the production rate of an isotope from a target.

Isotope production, of course, depends on the composition of the target. In the case of the atmosphere, which can be regarded to be composed of only N, O, Ar, and Kr for the present purposes, life is simple. Atmospheric isotopes are produced primarily either by high-energy spallation or by neutron capture. The neutron capture reactions yield ^{14}C and ^{81}Kr; the remainder of the two dozen or so atmospheric isotopes (Table 1) are all produced by high-energy spallation, principally by neutrons. In the case of ^{39}Cl, an appreciable contribution arises from μ^- capture in ^{40}Ar (Winsberg 1956). For all practical purposes, one could ignore μ^- reactions for all other isotopes produced in the atmosphere. However, the situation is more complex when it comes to in situ production in rocks. Several isotopes can be produced effectively in spallation reactions as well as by thermal neutrons and slow muons, as discussed below.

ISOTOPE PRODUCTION BY FAST NUCLEONS, SLOW NEUTRONS, AND MUONS

The production rates of thermal and fast neutrons in the atmosphere have been estimated fairly accurately by Lal & Peters (1967), who based their rates largely on experimental measurements of (a) slow and fast neutrons in the atmosphere, (b) nuclear disintegrations in nuclear emulsions and cloud chambers, (c) energy spectra of protons and the penetrating component in the atmosphere, and (d) energy spectra of various charged particles at production in nuclear disintegrations at high altitudes in the atmosphere. These data allow one to determine the flux and energy spectrum of the nucleons and mu-mesons within the atmosphere at all latitudes

Table 4 Cosmic-ray primary and secondary particles in the atmosphere and at depths underground at 45° latitude (based on Lal & Peters 1967)

	Depth in the Atmosphere (g cm^{-2})				Depth Underground Below Sea-level (kg cm^{-2})		
	0	200	500	1030 (sea level)	1	50	500
Cosmic ray primary and secondary particles[a]	p α CNO Si, Fe	e, γ $\mu^\pm$ n p	e, γ $\mu^\pm$ n p	μ e, γ n p	$\mu^\pm$ n p	$\mu^\pm$ n p	$\mu^\pm$ n p
Nuclear disintegrations[a]	p α CNO Si, Fe	n p γ μ $\mu^\pm$	n μ_c p γ $\mu^\pm$	n μ_c p $\mu^\pm$	μ_c $\mu^\pm$ n p	$\mu^\pm$ μ_c n p	$\mu^\pm$ μ_c n p

Approximate Total Rates of Nuclear Disintegrations

	0	200	500	1030	1	50	500
(1) /g s	10^{-2}	5×10^{-3}	10^{-3}	2×10^{-5}	10^{-6}	10^{-9}	5×10^{-12}
(2) /g yr	3×10^5	1.5×10^5	3×10^4	6×10^2	30	3×10^{-2}	2×10^{-4}

[a] Particles contributing to total cosmic-ray fluxes and nuclear disintegrations are shown in decreasing order of importance. For the present purposes, we define nuclear interactions as those involving nuclear excitation of > 10 MeV. Symbols are p = proton, n = neutron, α = ^{4}He, e = electron, γ = photon, $\mu^\pm$ = positive and negative mu-mesons, μ_c = nuclear capture of negative mu-meson. C, N, O, Si, and Fe refer to these nuclei.

and altitudes. Direct measurements of isotope production rates in targets exposed at different altitudes (cf. Lal et al 1960, Rama & Honda 1961) and the observed fallout of long- and short-lived isotopes are in good agreement with the family of curves obtained for the rate of production of nuclear disintegrations in the atmosphere (Figure 1). The energy spectrum at production, based on observations of nuclear disintegrations in nuclear emulsions exposed at 68,000 feet over England, is shown in Figure 2 (Powell et al 1959). Based on these data, Rossi (1952) calculated the expected energy spectrum of neutrons and protons in the troposphere at 700 g cm^{-2} depth. He deduced the differential energy spectrum to be of the form

$$\frac{dN}{dE} = \text{constant} \times \frac{1}{(50+E)^2},\tag{1}$$

where E is the kinetic energy.

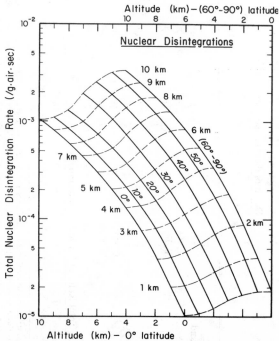

Figure 1 The total rate of nuclear disintegrations in the atmosphere (with energy release >40 MeV) plotted as a function of altitude and geomagnetic latitudes (0–90°, in steps of 10°). Curves for different latitudes have been successively displaced along the abscissa by 1 km (based on Lal & Peters 1967).

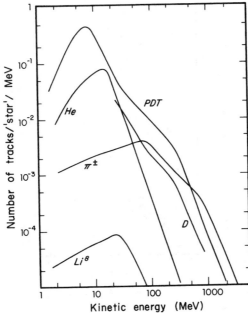

Figure 2   Kinetic energy spectrum of different classes of secondary particles emitted in nuclear disintegrations in "nuclear emulsions" exposed at 68,000 ft over England (based on Powell et al 1959). The curve labeled PDT refers to the total spectrum of protons, deuterons, and tritons emitted.

There exists a more direct way of obtaining the neutron energy spectrum in the troposphere—by combining data on the measured proton energy spectrum with the ratio of neutron- to proton-produced nuclear disintegrations of different sizes (Lal 1958). Using this technique, Lal (1958) has deduced the derived absolute neutron energy spectrum at $\geq 45°$ latitude, 680 g cm^{-2} depth (Figure 3). The spectrum is of the form

$$\frac{dN}{dE} = \text{constant} \times \frac{1}{(60+E)^{2.45}} \tag{2}$$

for $40 < E < 500$ MeV (Figure 3) and is in good agreement with the predictions of Rossi (1952), based on the energy spectrum of nucleons at production (Figure 2). Rossi's predicted proton energy spectrum also agrees well with the measured proton spectrum.

Further, since the relative size distributions of nuclear disintegrations of different sizes as observed in nuclear emulsions and cloud chambers have been measured in the atmosphere at several locations, it is possible to deduce energy spectra of neutrons at different locations in the atmo-

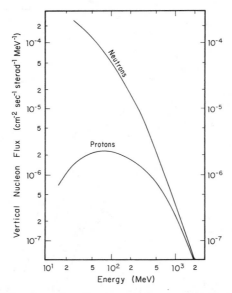

Figure 3   The omni-directional flux of neutrons and protons at latitude $\lambda = 50°$ and depth 680 g cm^{-2} (based on Lal 1958).

sphere from the known spectrum at any point (Lal 1958). Reference is made to a compilation of cosmic-ray data in the atmosphere (Allkofer & Grieder 1984). Thus we have a fair understanding of the behavior of the nucleonic and the penetrating muons in the atmosphere. The isotope production rates within a rock can therefore be calculated fairly reliably from the altitude-latitude isotope production curves given by Lal & Peters (1967), as long as the total sample shielding depth does not exceed ~ 1000 g cm^{-2}. (The total sample shielding depth is given by the sum of atmospheric pressure and the vertical shielding within the rock.) When the total shielding depth exceeds 10^3 g cm^{-2}, one has to carefully estimate relative neutron and muon fluxes. The energy spectrum of fast muons changes appreciably within the troposphere. The stopping muon flux at shallow depths underground (below sea level) does not show the usual exponential absorption, up to depths of about 2 kg cm^{-2}. Within the troposphere, a simple absorption behavior is found, and the stopping flux of negative muons, S_{μ^-}, can be well represented by the relation (Rossi 1948, Winsberg 1956, Lal 1958)

$$S_{\mu^-} \, (\mathrm{g}^{-1}\,\mathrm{s}^{-1}) = 5.1 \times 10^{-6} \exp\left(-X/L\right), \tag{3}$$

where X is the atmospheric depth (g cm^{-2}) and L is the absorption mean free path for slow muons. The value of L is determined (Rossi 1948, Lal 1958) to be 247 g cm^{-2} for $160 < X < 1030$ g cm^{-2}, the depth range of validity of Equation (1). Equation (3) gives the rate of stopping of negative muons in matter; the rate of interaction would depend on the composition of the target. In light nuclei, an appreciable fraction of negative muons can undergo decay in orbit in the K-shell before being captured by the nucleus. The capture lifetimes have been measured in different elements. A great deal of work has been reported in the literature on the muonic capture probabilities in binary and more complex compounds (cf. Eckhause et al 1966). The capture rates in an element are expected to depend on the nature of the compound and thus deviate from the expected probabilities calculated on the basis of geometrical calculations. We note, however, that generally the magnitude of the deviation is small (von Egidy & Hartmann 1982). For the present purposes we therefore assume that μ^- capture in the K-shell of an atom depends on the geometric cross sections of the nucleus and the lifetime for capture of μ^- by the nucleus, τ_c. The mean lifetime of muon for decay, τ_d, is 2.2 μs. The probability that a muon captured by an atom will interact (i.e. be captured by the nucleus) rather than decay in the K-shell of the nucleus is then given by τ_m/τ_c, where τ_m (the mean lifetime of μ^- for decay or capture) is a measured quantity and is related to decay mean lifetime, τ_d, by the equation

$$(1/\tau_m) = (1/\tau_c) + (1/\tau_d). \tag{4}$$

In Figure 4 we show the measured values of τ_m for different elements as a function of the atomic number Z. The data are based on measurements by Eckhause et al (1966), who have also reviewed all earlier published data. The data in Figure 4 do not fall along a smooth curve. Capture probabilities are not direct functions of Z, but rather of the effective charge. Also, hyperfine effects are expected for some nuclei (cf Eckhause et al 1966, and references therein). Again, for our present purposes, we choose the smoothened capture probabilities; at a later date, when refinements are needed, we can reopen this issue. The rock-averaged μ^- capture probabilities are given in Table 5.

PRODUCTION RATES OF ISOTOPES

As discussed, there are three principal mechanisms by which the isotopes are produced in terrestrial rocks: (a) by high-energy spallation by nucleons, (b) by neutron capture reactions, and (c) by muon-induced nuclear disintegrations. The latter include nuclear capture of μ^- and coulombic interactions of fast muons. Within the troposphere, nuclear reactions of protons and alpha particles contribute to less than 10% of the nuclear reactions (Figure 3); photons are an order of magnitude less significant. Here, fast neutrons are responsible for most of the nuclear interactions

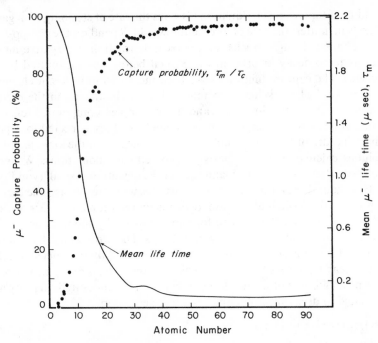

Figure 4   The calculated percentages of stopping negative muons captured by the nucleus are plotted as a function of the atomic number Z. The values are based on the measured mean lifetimes τ_m of negative mu-mesons brought to rest in different elements, shown by the smooth line (data from Eckhause et al 1966). Note that in some cases, τ_m values have been measured in isotopes (e.g. ^{6}Li, ^{7}Li, ^{10}B, and ^{11}B).

(Table 4). In nucleon or fast-muon-induced spallation reactions, the product nuclei are lighter than or equal to the target. The product closest to the target has either a lower A or Z value. In the case of negative muon capture, the heaviest product has the same A value as the target but its Z value is lower by unity. Neutron interactions can lead to a product with an A value higher by unity than the target; this is in fact the most frequent mode for thermal neutron capture reactions.

Thus, whereas generally a product can form from the target (Z, A) by any of the three nuclear actions, in the troposphere (and at depths underground) a product with $A = A + 1$ can form only from neutron capture because proton- and alpha-induced reactions are not important. Negative mu-captures lead to low nuclear excitation (a few tens of MeV), since the major part of the available energy (107 MeV) is liberated by the capture of μ^- with a proton, leading to a nucleus of $(Z-1, A)$ and emission of a neutrino. The neutrino takes away most of the energy. Fast muon inter-

Table 5 Some useful physical parameters for common rock types[a]

Rock type	(ρ) Assumed density (g cm^{-3})	(R_σ) Geometric cross section relative to air nuclei	(f_c) Fraction of stopping negative muons captured	(σ_m) Macroscopic thermal neutron absorption cross section (mole barns g^{-1})	(q_n) Radiogenic neutron production rate (10^{-6} cm^{-3} s^{-1})	$(\bar{\sigma})$ Mean thermal neutron cross section per atom (barns)
Granite	2.6	1.23	0.395	0.0120	1.6	0.61
Basalt	2.9	1.27	0.422	0.0084	0.44	0.76
Ultramafic	3.0	1.22	0.369	0.0073	5.7×10^{-4}	0.44
Shale	2.4	1.12	0.383	0.0163	0.76	0.73
Sandstone	2.3	1.22	0.378	0.0102	8.3×10^{-2}	0.47
Limestone	2.7	1.16	0.352	0.0067	—	0.35

[a] Neutron parameters σ_m and q_n are based on calculations of Andrews & Kay (1982) and J. N. Andrews et al (preprint).

actions lead to much higher nuclear excitation (George 1952), depending on the kinetic energy of the muon.

The above considerations have to be taken into account in deciding which reactions would be principal contributors to a product in rocks. In Table 6 we have listed some isotopes for which thermal neutron and muon reactions are significant as examples of reactions that have been considered.

The isotope production rates P $(g^{-1} s^{-1})$ can be conveniently written as

$$P_{sp} = \sum_i q_i Y_i, \tag{5}$$

$$P_n = F_n \frac{\sigma_i n_i}{\sum_j \sigma_j n_j} = F_n \frac{\sigma_i n_i}{\sigma_m}, \tag{6}$$

$$P_{\mu_c^-} = \sum_i J_i(\mu^-) \cdot Y_i, \tag{7}$$

where the subscripts sp, n, and μ_c^- refer to the three reaction types. The rate of nuclear disintegration by fast nucleons or muons is designated as q $(g^{-1} s^{-1})$; $J(\mu^-)$ is the capture rate of μ^- mesons $(g^{-1} s^{-1})$; Y is the yield of the isotope per disintegration; σ is the neutron capture cross section (barns); and n is the concentration of target atoms (mole g^{-1}). The subscript i refers to the target element or the atomic species leading to the production of the isotope; j refers to different elements in the target. The macroscopic thermal neutron absorption cross-section, σ_m, is thus the summation of $\sigma_j n_j$ (mole barns g^{-1}). F_n is the production rate of neutrons in the nuclear disintegrations (n $g^{-1} s^{-1}$).

Table 6 Selected isotopes with an appreciable production in terrestrial rocks[a] due to neutron capture, mu-meson, and low-energy alpha particle reactions

	Thermal neutrons		Capture of μ^-	Low-energy α-particles	
Isotope(s)	Target(s)	Reaction(s)	Principal target(s)	Target(s)	Reaction(s)
^{3}He	^{6}Li	(n, α)	—	^{11}B	(α, t)
^{10}Be	—	—	^{10}B, C, N, O	^{7}Li	(α, p)
^{14}C	^{14}N; ^{17}O	(n, p); (n, α)	N, O	^{11}B	(α, p)
^{21}Ne, ^{22}Ne	—	—	Na, Mg, Al	^{17}O, ^{18}O, ^{19}F	(α, n)
^{26}Al	—	—	Si, S	^{23}Na; ^{25}Mg	(α, n); (α, t)
^{36}Cl	^{35}Cl; ^{39}K	(n, γ); (n, α)	K, Ca, Sc	^{33}S	(α, p)
^{129}I	^{128}Te	(n, γ)	^{130}Te, Ba	—	—

[a] Contingent, of course, on the presence of the target element in the rock.

The rates of nuclear disintegrations produced by fast nucleons and fast muons have been determined for the atmosphere (Figure 1) and for rocks exposed below sea level (Lal 1987b). For rocks exposed within the troposphere, nucleon spallation rates can be estimated fairly accurately by scaling the rates with essentially the same absorption mean free path as in the atmosphere. This rate is latitude dependent, with the value changing from 200 g cm^{-2} at the equator to 150 g cm^{-2} at 50–90° latitude (Lal & Peters 1967).

Thermal neutron fluxes T_n (cm^{-2} s^{-1}) within a rock can be estimated quite accurately from (6) with the following relation for T_n:

$$F_n = P_{sp}\bar{Y}_n, \tag{8}$$

$$T_n = (F_n/\sigma_m), \tag{9}$$

where $\bar{Y}_n$ is the average yield of neutrons in the spallation of target elements in the rock, and ρ is the mean density of the rock. Estimated values of σ_m for different rock types are given in Table 5. We estimate that for a typical rock composition, 3.2 neutrons are produced; the corresponding yield for air nuclei is 2.9 (Lal 1987b). The value of $J_i(\mu^-)$ is given by

$$J_i(\mu^-) = S(\mu^-) \cdot f_i p_i, \tag{10}$$

where $S(\mu^-)$ is the μ^- stopping rate (g^{-1} s^{-1}), and f_i is the fraction stopped by the target nucleus i; this quantity can be deduced by approximating a geometric capture cross section. The value of p_i, the probability of capture of μ^- by the nucleus i, is given by direct measurements of τ_m (Figure 4). The mean fractions of stopping μ^- mesons that are captured, $f_c = \Sigma f_i p_i$, for different rock types are listed in Table 5.

By far the largest uncertainties in estimating isotope production rates at present are in evaluating the various yield factors in Equations (5), (7), and (8). At first sight it may appear that this should be a fairly easy task, since extensive data are available for most isotopes of interest for different meteorite types. This is not true for several reasons. The composition and energy spectrum of the nuclear interacting particles in the meteorites differ considerably from those in the atmosphere (cf. Lal & Peters 1967). First, most of the pions (π mesons) decay in the atmosphere while they interact in the meteorite; pions are produced efficiently in nucleon interactions (Figure 2). Second, the path length in the atmosphere (1030 g cm^{-2}) is considerably larger than that in the case of meteorites (typically < 100 g cm^{-2}). Finally, the nucleonic cascade in meteorites is not in equilibrium. In the atmosphere, for depths exceeding 200 g cm^{-2}, an equilibrium is reached between fast and slow neutrons. In the atmosphere, the spectrum of fast neutrons that produce most of the nuclear disintegrations is much softer compared with that in meteorites. The energy spectrum of tropospheric neutrons for $40 < E < 500$ MeV was stated to be of the form

$$\frac{\mathrm{d}N}{\mathrm{d}E} = \text{constant} \times (60+E)^{-2.5}, \tag{11}$$

where E is expressed in MeV (Lal 1958). The primary cosmic-ray proton spectrum for energies above 500 MeV is well expressed by the relation

$$\frac{\mathrm{d}N}{\mathrm{d}E} = \text{constant} \times (\alpha+E)^{-2.5}, \tag{12}$$

with a value of 1000 MeV for α. The energy spectrum becomes softer as nucleonic cascade develops within the meteorite. The value of α continuously decreases. For energies exceeding 100 MeV, the nucleonic spectra in meteorites are well represented by Equation (12) with 250 MeV $< \alpha <$ 600 MeV for a 5–100 cm range of meteorite radii (Reedy & Arnold 1972, Bhattacharya et al 1980, Bhandari & Potdar 1982, Reedy 1985). Thus, the atmospheric neutron spectra are much softer than the nucleonic spectra in meteorites. (Nucleon spectral shapes in meteorites lie in between the neutron and proton curves in Figure 3.) A greater proportion of the fast nucleons in the atmosphere are neutrons. Even in meteorites, isotope production rates based on proton excitation functions are often found to be grossly in error. The case of ^{10}Be production in chondrites is a good example (Tuniz et al 1984).

Neutron cross-section data are, however, very limited to date. Considering their importance in in-situ terrestrial cosmogenic studies, these should certainly be determined in the near future.

The yields of several isotopes (^{14}C, ^{10}Be, ^{26}Al, ^{36}Cl, ^{37}Ar, ^{39}Ar, and ^{129}I) are now being determined in irradiations of suitable targets with beam-stop neutrons arising from proton bombardments of several targets and with slow and fast muons. This work is being carried out jointly between scientists of the University of California, San Diego, Los Alamos National Laboratory, San Jose State University, University of Pennsylvania, Brookhaven National Laboratory, and University of Arizona. In some cases, one can use natural production of isotopes in well-documented rock samples to obtain isotope production rates. An example of this is the recent study by Nishiizumi et al (1987) of ^{10}Be and ^{26}Al in glacially polished rocks exposed since the retreat of the Wisconsin glacial era.

It is also feasible to measure isotope production rates directly by exposing suitable targets to cosmic rays. Production rates of several isotopes can be determined in this manner. This method was first adopted by Lal et al (1960) to determine absolute isotope production rates and has since been successfully employed by several investigators (Rama & Honda 1961, Bhandari et al 1969, Yokoyama et al 1977, Reyss et al 1981).

A large number of in situ cosmogenic isotopes promise to find useful applications in geophysical problems. Table 3 lists isotopes with half-lives exceeding two weeks. Several of these isotopes have multiple production mechanisms, and some examples of these are given in Table 6. Thermal neutron cross sections are available in most cases. The yields in μ^- capture can be estimated reasonably well from available data (cf. Charalambus 1971, and references therein). At present, the principal uncertainties in estimating reaction yields are for fast neutron and muon interactions; cross-section data are limited for neutrons and not available for fast muons. Some attempts have been made earlier to obtain the best-guess estimates by combining information on the experimental data on isotope production in the atmosphere, meteorites, and lunar samples; on observations of the nature of nuclear disintegrations in visual detectors; and on reaction systematics (Rudstam 1955, Lal 1958, Arnold et al 1961).

Production rates of ^{3}He in rocks up to underground depths of 10^4 g cm^{-2} have recently been estimated (Lal 1987b). These estimates include production by fast muons and thermal neutrons. The surface production rates of a dozen long lived isotopes in rocks exposed in the troposphere have been estimated by N. Bhandari & D. Lal (in preparation).

In rocks shielded by more than 3 kg cm^{-2} (including the path length in the atmosphere), the radiogenic neutron production in typical rock types becomes an important source of thermal neutrons (Lal 1963, 1987b). The production rates of ^{3}He by neutron capture in ^{6}Li and ^{10}B (Lal 1987b) and of a number of radioactive isotopes (^{3}H, ^{14}C, ^{36}Cl, ^{37}Ar, ^{39}Ar, ^{81}Kr, ^{85}Kr, and ^{129}I) due to radiogenic reactions have been estimated by Zito et al (1980) and Andrews et al (1987). Production of some of these isotopes in spontaneous and neutron-induced fission of uranium have also been considered by these authors.

In Figure 5, we reproduce the calculations of Lal (1987b) for rates of nuclear disintegrations due to nucleons and muons in standard rock (density = 3.09 g cm^{-3}, $Z^2/A = 6.3$) exposed at sea level (latitudes $\geq 45°$) as a function of depth (0–10 kg cm^{-2}). The corresponding estimates (Lal 1987b) for ^{3}He production, including cosmogenic and radiogenic thermal neutrons, are shown in Figure 6.

TEMPORAL VARIATIONS IN ISOTOPE PRODUCTION RATES IN ROCKS

Temporal changes in the terrestrial cosmic-ray flux are expected to arise due to (a) changes in the Earth's dipole field, (b) solar modulation of

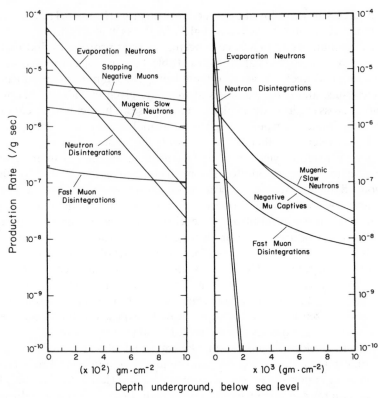

Figure 5 The rates of occurrence of nuclear disintegrations due to nucleons and muons, and production rates of evaporation neutrons in fast-neutron-produced disintegrations, in standard rock (density = 3.09 g cm^{-3}, Z^2/A = 6.3) as a function of depth underground (below sea level) for geomagnetic latitudes $\geq 45°$. The left side shows on an enlarged scale the production rates for 0–1 kg cm^{-2} depth (0–3.3 m rock equivalent) (based on Lal 1987b).

cosmic-ray flux, and (*c*) sporadic acceleration of cosmic rays in solar flares. In in-situ studies, we are concerned primarily with the first two. Solar flare cosmic radiation has a very soft spectrum, i.e. dN/dE = constant $\times\ E^{-\gamma}$ with $3 < \gamma < 6$ (Lal 1972, Reedy et al 1983). This radiation produces significant in situ isotopic changes in cosmic dust grains, outer layers of meteorites, and the lunar regolith. Since most of the solar cosmic-ray flux is confined to particles below a few hundred MeV kinetic energies, the secondary particle flux in the troposphere is not increased generally. A few exceptions are the unusually hard spectrum and/or high fluence events such as the flare events during 23 February 1956, 14 July 1959, and August 1972 (see Reedy 1977). Solar cosmic-ray production rates in the

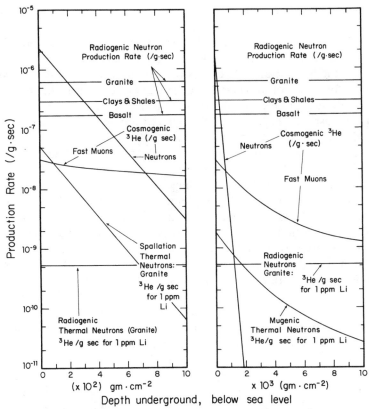

Figure 6 Production rates of ^{3}He in rocks by cosmogenic fast neutrons and muons, and cosmogenic/radiogenic thermal neutrons in a rock containing 1 ppm Li, are given for underground depths (0–10 kg cm^{-2}). The three horizontal lines in the upper part give the radiogenic production rates of neutrons in three rock types (based on Lal 1987b).

atmosphere for ^{3}H, ^{14}C, and ^{10}Be have been estimated (Lingenfelter 1963, Lal & Peters 1967, Lal 1987c).

Changes in the dipole field can appreciably change the depth altitude production rates in the atmosphere. The minimum energy of particles incident at the top of the atmosphere at any given latitude depends on the Earth's dipole field. Since the altitude-latitude production rates are known for the present dipole field, they can be easily calculated for any specified dipole field strength. The procedure has been discussed earlier (Lal & Peters 1967) and applied for the case of ^{14}C and ^{10}Be (Lal & Venkatavaradan 1970, O'Brien 1979, Castagnoli & Lal 1980, Lal 1987c). The minimum cutoff rigidity R_c (momentum to charge ratio) at any latitude λ is given by the relation

$$R_c(\lambda) = 14.9 \, (M/M_0) \cos^4 \lambda \; (\text{GV}), \tag{13}$$

where M_0 is the present dipole field intensity and M is the new dipole field intensity. Since production rates as in Figure 1 are known for all latitudes and M_0, the latitude-altitude curves for any given M value can be constructed for values of $M/M_0 < 1$. For values higher than 1, the R_c values at low latitudes are higher than those at $0°$ for $M/M_0 = 1$. Suitable corrections must then be made for the reduced flux at $0°$, according to the procedure adopted by Castagnoli & Lal (1980).

Changes in cosmic-ray flux incident at the top of the atmosphere due to changes in the level of solar modulation can now be modeled quite satisfactorily. Detailed calculations have been made for ^{14}C, ^{3}H, and ^{10}Be (Lal & Peters 1962, Lingenfelter & Flamm 1964, O'Brien 1979, Castagnoli & Lal 1980, Lal 1987c). As our understanding improves of the character of solar modulation, which is to a first approximation related to solar activity (indexed by a number of indices, such as sunspot number and geomagnetic indices K_p), we can improve on the expected variations in isotope production rates due to changes in solar activity.

In Figure 7 we show the measured changes in the differential primary cosmic-ray proton flux at the top of the atmosphere during different years. For each of the curves we have also specified the value of the modulation parameter ϕ, which appears as the single parameter in the force field approximate solution for the inward transport of cosmic rays through the solar plasma in interplanetary space (Garcia-Munoz et al 1975, Castagnoli & Lal 1980, Lal 1987c). The higher the value of ϕ, the lower the incident cosmic-ray flux; there exists a reasonable correlation between ϕ and sunspot number (see also Lingenfelter & Ramaty 1970). However, according to our present understanding, ϕ is not related in a straightforward manner with sunspot number (Garcia-Munoz et al 1977).

Extensive continuous data exist for counting rates of low-energy neutrons at several neutron monitor stations, mostly since 1955. These and the latitude-altitude ^{14}C production rates calculated by Lingenfelter (1963) allow one to obtain changes in isotope production rates in rocks exposed in the troposphere [see also the discussions by Lal (1987c) for ^{14}C and ^{10}Be solar modulation–dependent production rates in the atmosphere].

Observational data pertinent to temporal variations in cosmic-ray flux are (a) ^{14}C in tree rings, and (b) ^{10}Be in ice and in marine deposits. The tree ring ^{14}C data now go back to 10,100 yr B.P. (Becker & Kromer 1986, Stuiver et al 1986), and the ^{10}Be ice data go back to 140,000 yr B.P. (Yiou et al 1985). The ^{10}Be data in marine sediments go back to ~ 10 m.y. B.P. (Inoue & Tanaka 1979, Raisbeck & Yiou 1984), and in manganese crusts they also go back to ~ 10 m.y. B.P. (Ku et al 1982). The records of

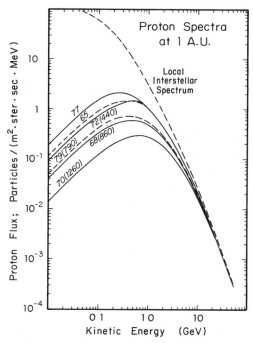

Figure 7 Measured differential kinetic energy spectra of protons near the Earth during different years of observation. The last two digits of the year (A.D.) are marked on the curves; the numbers within parentheses give the effective value of the modulation parameter ϕ. For sources of data, calculated ϕ values, and the predicted local interstellar spectrum, reference is made to Garcia-Munoz et al (1975).

atmospheric ^{14}C and ^{10}Be have to be deconvoluted suitably to obtain variations in the cosmic-ray flux; geophysical and geochemical processes introduce appreciable modulations in the record. Reference is made to Beer et al (1984) and Lal (1985) for discussions of the complexities of the ^{14}C record in tree rings, to Lal (1987d) for a discussion of the ^{10}Be record in ice, and to Somayajulu (1977) for a review of the ^{10}Be record in marine sediments. Our present techniques do not allow determination of changes in the terrestrial cosmic-ray flux due to dipole field from these records. The solar modulation of cosmic-ray flux is, however, seen in the ^{14}C tree ring record, and the long-term record of ^{10}Be in marine deposits does show that, averaged over periods of the order of 10,000 yr, the gross cosmic-ray flux incident in the Earth has not changed by more than 30–50% during the past 10 m.y.

It is to be expected that our techniques for the deconvolution of the record of atmospheric isotopes will improve substantially in the near future to allow estimation of changes in the terrestrial cosmic-ray flux due to the three causes discussed above. However, the studies of in situ isotopes in rocks and trees from different latitudes show greater promise (Lal et al 1985).

APPLICATIONS OF IN SITU COSMOGENIC ISOTOPES

At sea level the production rates of ^{14}C, ^{10}Be, and ^{26}Al in suitable targets are of the order of 10–50 atoms g^{-1} yr^{-1}. The production rates at 2 and 5 km altitude are higher by factors of ~ 4 and ~ 30, respectively. If the effective cosmic-ray irradiation time is of the order of 10^3 yr at sea level, one would expect detectable amounts using AMS ($\sim 10^6$ atoms) in samples of a few tens of gram weight. This rough calculation forms the basis for the wide range of applications of the in situ method. The potential applications of in situ isotopes were pointed out earlier by Jha & Lal (1982) and Lal (1986a) for a variety of geophysical processes, e.g. erosion and tectonic movements.

Closed System

Besides the primary question of detection, there are several other questions that decide the usefulness of an in situ isotope as a geophysical tracer. These questions are the same that have been asked earlier in absolute age-dating techniques and in studies of cosmogenic isotopes in meteorites. The central requirement is that the rock/mineral system being dated should be a "closed system" with respect to its losing an isotope to or gaining an isotope from the environment. In age dating of rocks, the location or disposition of the rock is not important; however, the in situ cosmogenic isotope amount produced depends sensitively on the irradiation site and geometry. One has, therefore, to confine studies to rocks/sediments where sufficient geological information is available to make plausible models of the past geometric configurations.

To avoid possible interferences from atmospheric ^{10}Be adsorbed in rock matrix via seeping rainwater, studies of in situ ^{10}Be and ^{26}Al can be done in quartz mineral grains. The idea (Lal & Arnold 1985) has been tested and has passed stringent tests (Nishiizumi et al 1987). Quartz is a ubiquitous mineral in terrestrial settings; it serves as an ideal target for in situ production of three long-lived isotopes: ^{14}C, ^{10}Be, and ^{26}Al. Other minerals should likewise be explored for in situ applications.

The applications of the in situ cosmogenic production can be conveniently divided under two headings: (*a*) samples whose past evolution geometry is well known a priori, and (*b*) samples whose past evolution geometry is known qualitatively but the time constants are not known. The expected isotope concentrations due to in situ production can be determined if both the cosmic-ray intensity $I(t)$ and the sample geometry evolution $G(t)$ are known. Therefore, if only one of these is known, the other can be modeled; the degree of resolution on the parameter evaluated would depend on the number of isotopes studied. This, in short, is the basis of application of the in situ isotope method. We now discuss the two sample types and illustrate some of their applications.

Well-Characterized Samples

Examples of samples in this class are trees, monuments, and glacial polished-rock surfaces. In the case of trees, the bombardment geometry changes with time in a precisely known manner (i.e. it can be fully characterized from the tree rings). In the other two cases, the geometry remains fixed, since the objects were exposed in that orientation. Such samples clearly would yield information on the changes in cosmic-ray flux, which in turn could be due to any of the following three causes:

1. Changes in the geomagnetic dipole field of the Earth.
2. Movement of the sample due to change in the altitude and/or latitude.
3. Changes in the primary cosmic-ray flux incident at the top of the atmosphere due to solar modulation.

These questions have been discussed, with special reference to (1) and (3), by Lal et al (1985).

Noncharacterized Samples

A variety of geological samples fall in this category. Some of the samples (for example, sands) will have had a very complex "irradiation" past, but a number of samples can be assigned a qualitative irradiation history based on geophysical/geological evidence. Examples of the latter are accreting sediments and eroding rock surfaces. The goal of in situ isotopes in such cases would be to study either of the following: (a) rates of accumulation of sediments, or (b) rates of erosion of rock surfaces.

The cosmic-ray intensity variations cannot be deduced generally from such samples. To illustrate the use of noncharacterized samples, we consider the expected concentration of an isotope C in a rock surface eroding at a constant rate ε at a fixed location. Let us further assume that the rock formed from igneous activity at some time T in the past; in this case, the initial isotope concentration is zero. After T years of irradiation, we would expect the following depth profile for the concentration $C(x)$ of the isotope (Lal & Arnold 1985):

$$C(X) = \frac{P_0 e^{-\mu X}}{\lambda + \varepsilon\rho\mu} (1 - e^{-\lambda T}),$$ (14)

where P_0 is the isotope production rate at the rock surface, λ is the disintegration constant of the isotope, ρ is the mean density of the rock, μ is the inverse of the mean absorption length for cosmic-ray particles in the rock, and the units of X, μ, and ε are g cm^{-2}, cm^2 g^{-1}, and cm s^{-1}, respectively.

There are several interesting aspects of Equation (14), which is a special solution for the expected in situ isotope concentration in a rock as a function of depth (Lal & Arnold 1985, Lal 1986b): It assumes constant isotope production and erosion rates and no preirradiation history of the rock. To bring out certain interesting geophysical aspects of in situ irradiation, we further assume that several meters of rock have been eroded away. The isotope concentrations in the rock are then in a quasi-steady state, and the surface concentration C_s becomes

$$C_s(X = 0) = \frac{P_0}{\lambda + \varepsilon\rho\mu},$$ (15)

with the concentration at depth X decreasing exponentially, as in Equation (14).

The resulting surface isotope concentration depends on the isotope half-life, the mean erosion rate, and of course the isotope production rate. We may define two irradiation time periods: the equivalent time T_{eq} and the effective time T_{eff}. The former is the time it would take to result in a certain isotope production if the rock mass was irradiated in a fixed geometry, and the latter is the effective time corresponding to the ratio of the concentration to production rate. We obtain

$$T_{eq} = \frac{1}{\lambda} \ln \left(1 + \frac{\lambda}{\mu\rho\varepsilon} \right),$$ (16)

$$T_{eff} = \frac{1}{\lambda + \mu\rho\varepsilon}.$$ (17)

The effect of erosion is to set a new equilibrium value, in effect lowering the mean lifetime of the isotope by a factor $1/(1 + \mu\rho\varepsilon)$. This is equivalent to saying that the erosion mean lifetime is $(1/\mu\rho\varepsilon)$. The characteristic mean erosion period T_{ero} is defined by the ray attenuation length $(1/\mu)$, the density ρ, and the erosion rate ε:

$$T_{ero} = \frac{1}{\mu\rho\varepsilon}.$$ (18)

If the mean rock density is taken to be 3.5 g cm^{-3} and $1/\mu = 150$ g cm^{-2}, we obtain

$$T_{ero} \, (yr) = 43/\varepsilon, \qquad (19)$$

where ε is expressed in cm yr^{-1} (Lal 1986b). The mean erosion period T_{ero} becomes comparable to 10^6 yr (the order of half-lives of ^{10}Be and ^{26}Al) when $\varepsilon \sim 5 \times 10^{-5}$ cm yr^{-1}. For smaller erosion rates, the effective irradiation time will be determined by the isotope half-life, and for higher erosion rates, by the erosion rate.

From a study of concentrations of isotopes of different half-lives, an exposed surface can be characterized and the erosion rate determined. The erosion time scales accessible are simple functions of the isotope half-life. If different isotope pairs yield similar erosion rates, one can determine the validity of the steady-state uniform erosion rate model adopted. Deviations between results for different isotope pairs can conversely be used to model time variations in the erosion rate $\varepsilon(t)$.

About two-dozen rock surfaces have been studied for erosion by our group (Nishiizumi et al 1986; K. Nishiizumi, D. Lal, J. R. Arnold, J. Klein & R. Middleton, unpublished results). The in situ ^{10}Be and ^{26}Al erosion rates fall in the range of 10^{-2}–10^{-5} cm yr^{-1}, corresponding to effective irradiation periods of 5×10^3–10^6 yr.

A very interesting application of the in situ irradiation method is to the determination of episodic surface losses. After a discrete loss of surface up to a certain depth, (say X) following a steady-state situation, all isotope concentrations still follow the exponential decrease with depth, but the absolute isotope concentrations depart considerably from the steady-state situation, depending on the isotope half-life and the value of X. From a comparison of surface isotope concentrations, it should be possible to decide whether the horizon has suffered episodic losses.

In principle, from in situ studies one should be able to measure rates of uplift (which change the surface production rate). However, in practice this may be a difficult task, especially for rapidly eroding rocks. It can easily be shown that if the erosion rate is lower than 10^{-4} cm yr^{-1}, it should be possible to determine uplift rates exceeding 1 cm yr^{-1}.

Finally, can one date a surface using the in situ method? Let us reconsider the case of an igneous rock formation T years ago, for which Equation (14) applies. If now two in situ isotopes are studied in which, at least for one, the term within parentheses departs significantly from unity (i.e. $\lambda \sim 1/T$), both the date of eruption and the erosion rate can be determined.

The principal applications of different in situ radioisotopes in recent geophysical problems are shown schematically in Table 7.

Table 7 Principal in situ cosmic-ray-produced isotopes and their geophysical applications

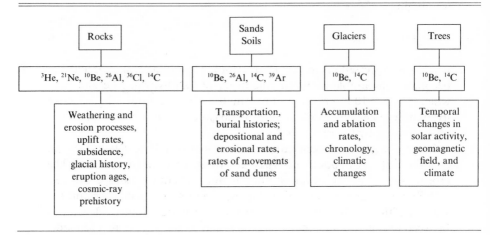

DISCUSSION

We have discussed the potential of the newly emerging field of terrestrial cosmogenic in situ isotopes and isotope production mechanisms and rates. The AMS technique, originally motivated by the necessity to measure the long-lived atmospheric cosmogenic isotopes, has now made this field tractable and attractive. These developments have in turn led to a resurgence in the use of ultra-sensitive counters and mass spectrometers for studying in situ cosmogenic short-lived and stable isotopes, respectively, that cannot be studied using AMS (for example, ^{39}Ar, ^{3}He, and ^{21}Ne).

The experimental field of in situ isotopes using AMS is only a few years old. In the initial stages it becomes necessary to establish the technique from the study of selected samples and also to determine the cosmic-ray production rates experimentally. Work is now in progress in these directions. The study of glacial polished-rock samples from the Sierra Nevada ranges (Nishiizumi et al 1987) seems to solve these problems. These data show that (*a*) quartz is an ideal closed-system mineral, and that (*b*) the amounts of ^{10}Be and ^{26}Al are in good agreement with expectations for cosmic-ray irradiation for 12,000 yr. Accelerator irradiations are now being carried out (Klein et al 1987) to determine the production rates of these isotopes by neutron irradiation and in μ^{-}-capture in SiO_2.

Use of the stable isotopes ^{3}He and ^{21}Ne seems attractive for the study of erosion and other geophysical processes (Table 7). The stable isotopes are useful for determining very slow erosion rates and thus supplement erosion studies based on ^{10}Be and ^{26}Al in quartz. Additionally, they can be used to characterize multiple exposure histories. The effective irradiation period for a stable isotope would of course be expected to be greater than that for a radioactive isotope. However, if $\lambda \ll (1/T_{\text{eff}})$ and T_{eff}(stable isotope) > T_{eff}(radioactive isotope), then clearly the sample has received an earlier irradiation several half-lives ago. A particular advantage of using in situ ^{3}He may lie in the possibility of making a large number of measurements conveniently. AMS studies of ^{10}Be and ^{26}Al are time consuming and expensive; an annual output of 50 samples per year for one project is large. In the case of ^{3}He, the corresponding yield could be an order of magnitude higher, judging from the performance of two groups (Craig & Poreda 1986, Kurz 1986a,b). Attempts have been made to estimate ^{3}He production rates in a radiocarbon-dated Hawaiian lava flow (Kurz 1986b). More work in this direction is needed, including simultaneous studies of ^{3}He and other radioactive isotopes (e.g. ^{10}Be and ^{26}Al).

Studies of in situ ^{129}I, ^{41}Ca, ^{14}C, and ^{39}Ar are just beginning. In the cases of ^{41}Ca and ^{14}C, techniques for their detection have yet to be perfected. It is certainly advantageous to study several in situ isotopes in the same sample, but this may not be practical. However, the potentials of these isotopes in various studies (cf. Table 7) are enormous. Thus, the applications of these isotopes would depend largely on the advances that would be made in the near future in high sensitivity and rapid analyses of several isotopes in a sample.

With an increase in the sensitivity and ease of detection of isotopes, we are sure to see new uses of particular in situ isotopes for the study of specific problems. Raisbeck & Yiou (1979) discussed the possible applications of ^{41}Ca in the dating of $CaCO_3$ sediments and animal remains. Phillips et al (1986) discussed the possibility of age dating young volcanic rocks using in situ ^{36}Cl. Age determinations are possible if two or more isotopes are used (Lal 1987a); unless the magnitude of erosion is known a priori, one has to determine both the erosion rate and the time of exposure of the surface.

An interesting recent application of in situ ^{10}Be was made by Lal et al (1987) in placing a lower limit to the amount of cosmogenic ^{3}He in alluvial diamonds from Zaire. Their results showed that a minimum of 29% of the ^{3}He in the alluvial diamonds was cosmogenic. They surmised that before concluding that the high ^{3}He/^{4}He ratios observed in some diamonds, corresponding to the ratios in solar wind, existed in the mantle at some time (Ozima et al 1983), it was necessary to study diamonds collected by underground mining methods. Such diamonds would be free from cosmogenic in situ ^{3}He. The high ^{3}He/^{4}He ratios reported by Ozima and his collaborators are all for undocumented diamonds, and thus an appreciable amount of cosmogenic ^{3}He cannot be ruled out in these diamonds.

During the next decade we should be able to obtain a wealth of geophysical data and information about changes in the cosmic-ray intensity in the past 10^2–10^6 yr due to changes in the geomagnetic field intensity and in solar activity (cf. Lal et al 1985). Cosmic rays are ideal probes for sensing changes in the dipole field of the Earth. Studies of isotopes deep underground should provide valuable information on the long-term secular variations in the flux of energetic cosmic-ray particles ($> 10^4$–10^5 GeV).

I have discussed the principal applications of in situ isotopes as they appear to me from the present-day perspective. The history of science is replete with examples of rapid and drastic changes in outlook in a field. It may be interesting to recall here the statement ascribed to F. W. Aston, the father of mass spectrometry. In 1935 Aston reportedly said that the job of mass spectrometry was done and that the field of research would die away (Cadogan 1986). This example suggests that I may have too narrowly constrained the potential applications of in situ cosmogenic isotopes.

ACKNOWLEDGMENTS

It is my pleasure to acknowledge continued exciting and stimulating collaboration during this work with J. R. Arnold and K. Nishiizumi at UCSD, with R. C. Reedy at the Los Alamos National Laboratory, with R. Middleton and J. Klein at the University of Pennsylvania, and with W. Wölfli at ETH Zurich. I am especially thankful to J. R. Arnold, R. E. Lingenfelter, and R. C. Reedy for critical comments on the manuscript.

This work was funded in part by grants received from the Los Alamos National Laboratory (LANL) IGPP funds. It is also a pleasure to acknowledge the use of the LANL proton beam-stop and stopped muon channel facilities. I also gratefully acknowledge a travel grant received from the Smithsonian US-India Foreign Currency Program to visit the Scripps Institution of Oceanography.

Literature Cited

Allkofer, O. C., Grieder, P. K. F. 1984. Cosmic rays on Earth. *Physik Daten/Physics Data* 25(1): 1–379

Alvarez, L. W., Cornog, R. 1939. Helium and hydrogen of mass 3. *Phys. Rev.* 56: 513

Anderson, E. C., Libby, W. F., Weinhouse, S., Reid, A. F., Kirshenbaum, A. D., Grosse, A. V. 1947. Natural radiocarbon from cosmic radiation. *Phys. Rev.* 72: 931–36

Andrews, J. N., Kay, R. L. F. 1982. Natural production of tritium in permeable rocks. *Nature* 298: 361–63

Andrews, J. N., Davis, S., Fabryka-Martin, J., Fontes, J.-C., Lehmann, B. E., et al. 1987. The in-situ production of radioisotopes in rock matrices with particular reference to the stripa granite. Submitted for publication

Apt, K. E., Knight, J. D., Camp, D. C., Perkins, R. W. 1974. On the observation of ^{92}Nb and ^{94}Nb in nature. *Geochim. Cosmochim. Acta* 38: 1485–88

Arnold, J. R., Libby, W. F. 1949. Age determinations by radiocarbon content: checks with samples of known age. *Science* 110: 678–80

Arnold, J. R., Honda, M., Lal, D. 1961. Record of cosmic ray intensity in the meteorites. *J. Geophys. Res.* 66: 3519–31

Audouze, J., Vauclair, S. 1980. *An Introduction to Nuclear Astrophysics.* Dordrecht: Reidel

Becker, B., Kromer, B. 1986. Extension of the Holocene dendochronology by the preboreal pine series, 8800 to 10,100 B.P. *Radiocarbon* 28: 961–67

Beer, J., Andree, M., Oeschger, H., Stauffer, B., Balzer, R., et al. 1984. The Camp-Century ^{10}Be record: implications for long-term variations of the geomagnetic dipole moment. See Wölfli et al 1984, pp. 380–84

Bennett, C. L., Beukens, R. P., Clover, M. R., Gove, H. E., Liebert, R. B., et al. 1977. Radiocarbon dating using electrostatic accelerators: negative ions provide the key. *Science* 198: 508–10

Bhandari, N., Potdar, M. B. 1982. Cosmogenic ^{21}Ne and ^{22}Ne depth profiles in chondrites. *Earth Planet. Sci. Lett.* 58: 116–28

Bhandari, N., Fruchter, J., Evans, J. 1969. Rates of production of ^{24}Na and ^{28}Mg in the atmosphere by cosmic radiation. *Earth Planet. Sci. Lett.* 7: 89–92

Bhattacharya, S. K., Imamura, M., Sinha, N., Bhandari, N. 1980. Depth and size dependence of ^{53}Mn activity in chondrites. *Earth Planet. Sci. Lett.* 51: 45–50

Brown, L. 1984. Applications of accelerator mass spectrometry. *Ann. Rev. Earth Planet. Sci.* 12: 39–59

Castagnoli, C., Lal, D. 1980. Solar modulation effects in terrestrial production of carbon-14. *Radiocarbon* 22: 133–58

Charalambus, S. 1971. Nuclear transmutation by negative stopped muons and the activity induced by the cosmic-ray muons. *Nucl. Phys. A* 166: 145–61

Clayton, D. D., Morgan, J. A. 1977. Muon production of ^{92,94}Nb in the Earth's crust. *Nature* 266: 712–13

Cadogan, J. I. G. 1986. Opening addresses. In *Advances in Mass Spectrometry 1985*, ed. J. F. J. Todd, Part A, pp. xxxvii–xxxix. New York: Wiley

Craig, H., Poreda, R. J. 1986. Cosmogenic ^{3}He in terrestrial rocks: the summit of lavas of Maui. *Proc. Natl. Acad. Sci. USA* 83: 1970–74

Davis, R. Jr. 1968. A search for neutrinos from the sun. *Proc. Int. Semin. Neutrino Phys. and Neutrino Astrophys., Moscow,* 2: 99–128

Davis, R. Jr., Schaeffer, O. A. 1955. Chlorine-36 in nature. *Ann. NY Acad Sci.* 62: 105–22

Eckhause, M., Siegel, R. T., Welsh, R. E., Filippas, T. A. 1966. Muon capture rates in complex nuclei. *Nucl. Phys.* 81: 575–84

Elmore, D. 1987. Ultrasensitive radioisotopes, stable isotope, and trace-element analysis in the biological sciences using Tandem Accelerator Mass Spectrometry. *Biol. Trace Elem. Res.* 12: 231–45

Elmore, D., Phillips, F. M. 1987. Accelerator mass spectrometry for measurement of long-lived radioisotopes. *Science* 236: 543–50

Faure, G. 1986. *Principles of Isotope Geology.* New York: Wiley. 589 pp. 2nd ed.

Fehn, U., Holdsen, G. R., Elmore, D., Brunelle, T., Teng, R., Kubik, P. W. 1986. Determination of natural and anthropogenic ^{129}I in marine sediments. *Geophys. Res. Lett.* 13: 137–39

Garcia-Munoz, M., Mason, G. M., Simpson, J. A. 1975. The anomalous ^{4}He component in the cosmic-ray spectrum at ≤ 50 MeV per nucleon during 1972–1974. *Astrophys. J.* 202: 265–75

Garcia-Munoz, M., Mason, G. M., Simpson, J. A. 1977. New aspects of the cosmic ray modulation in 1974–1975 near solar minimum. *Astrophys. J.* 213: 263–68

George, E. P. 1952. Observations of cosmic rays underground and their interpretation. *Prog. Cosmic Ray Phys.* 1: 395–454

Grosse, A. V. 1934. An unknown radioactivity. *J. Am. Chem. Soc.* 56: 1922–23

Hampel, W., Takagi, J., Sakamoto, K., Tanaka, S. 1975. Measurement of muon-induced ^{26}Al in terrestrial silicate rock. *J. Geophys. Res.* 80: 3757–60

Hedges, R. E. M. 1979. Radioisotope clocks in archaeology. *Nature* 281: 19–24

Henning, W., Bell, W. A., Billquist, P. J., Glagola, B. G., Kutschera, W., et al. 1987. Calcium-41 concentration in terrestrial materials: prospects for dating of Pleistocene samples. *Science* 236: 725–27

Inoue, T., Tanaka, S. 1979. ^{10}Be in marine sediments, Earth's environment and cosmic rays. *Nature* 277: 209–10

Jha, R., Lal, D. 1982. On cosmic ray production of isotopes in surface rocks. *Proc. Symp. Nat. Radiat. Environ., 2nd,* ed. K. G. Vohra, K. C. Pillai, M. C. Misra, pp. 629–35. New Delhi: Wiley-Eastern. 691 pp.

Klein, J., Nishiizumi, K., Reedy, R. C., Middleton, R., Lal, D., Arnold, J. R. 1987. ^{10}Be and ^{26}Al production in quartz: laboratory measurements of cross sections. *Int. Symp. Accel. Mass Spectrom. Niagara-on-the-Lake, Ontario, Can., Abstr.,* ed. H. E. Gove, p. 37

386 LAL

Korff, S. A. 1940. On the contribution to the ionization at sea level produced by the neutrons in the cosmic radiation. *J. Geophys. Res.* 45: 133–34

Ku, T. L., Kusakabe, M., Nelson, D. E., Southon, J. R., Korteling, R. G., et al. 1982. Constancy of oceanic deposition of ^{10}Be as recorded in manganese crusts. *Nature* 299: 240–42

Kubik, P. W., Korschinek, G., Nolte, E., Ratzinger, U., Ernst, H., et al. 1984. Accelerator mass spectrometry of ^{36}Cl in limestone and some paleontological samples using completely stripped ions. See Wölfli et al 1984, pp. 326–30

Kurz, M. D. 1986a. Cosmogenic helium in a terrestrial igneous rock. *Nature* 320: 435–39

Kurz, M. D. 1986b. In situ production of terrestrial cosmogenic helium and some applications to geochronology. *Geochim. Cosmochim. Acta* 50: 2855–62

Lal, D. 1958. Investigations of nuclear interactions produced by cosmic rays. PhD thesis. Tata Inst. Fundam. Res., Bombay, India. 90 pp.

Lal, D. 1963. Isotopic changes induced by neutrinos. *Proc. Cosmic Ray Conf., 6th, Jaipur*, pp. 190–96

Lal, D. 1972. Hard rock cosmic ray archaeology. *Space Sci. Rev.* 14: 3–102

Lal, D. 1985. Carbon cycle variations during the past 50,000 years: atmospheric ^{14}C/^{12}C ratio as an isotopic indicator. In *The Carbon Cycle and Atmospheric CO$_2$: Natural Variations, Archean to Present. Geophys. Monogr.* 32: 221–33. Washington DC: Am. Geophys. Union

Lal, D. 1986a. Cosmic ray interactions in the ground: temporal variations in cosmic ray intensities and geophysical studies. *Lunar Planet. Inst. Tech. Rep. 86-06*, pp. 43–45

Lal, D. 1986b. On the study of continental erosion rates and cycles using cosmogenic ^{10}Be and ^{26}Al and other isotopes. *Proc. Workshop Dating Young Sediments, Beijing*, ed. A. J. Hurford, E. Jäger, J. A. M. Ten Cate, pp. 285–98. Bangkok: CCOP Tech. Secretariat

Lal, D. 1987a. Cosmogenic nuclides produced in situ in terrestrial rocks. *Nucl. Instrum. Methods Phys. Res.* B29: 238–45. Amsterdam: North-Holland

Lal, D. 1987b. Production of ^{3}He in terrestrial rocks. *Chem. Geol. (Isot. Geosci. Sect.)* 66: 89–98

Lal, D. 1987c. Theoretically expected variations in the terrestrial cosmic ray production rates of isotopes. *Proc. Fermi Sch. Phys., 95th, Varenna, 1985.* Ital. Acad. Sci. In press

Lal, D. 1987d. ^{10}Be in polar ice: data reflect changes in cosmic ray flux or polar meteorology. *Geophys. Res. Lett.* 14: 785–88

Lal, D., Arnold, J. R. 1985. Tracing quartz through the environment. *Proc. Indian Acad. Sci. (Earth Planet. Sci.)* 94: 1–5

Lal, D., Peters, B. 1962. Cosmic ray produced isotopes and their application to problems in geophysics. *Prog. Elem. Part. Cosmic Ray Phys.* 6: 1–74

Lal, D., Peters, B. 1967. Cosmic-ray produced radioactivity on the earth. In *Handbook of Physics*, 46/2: 551–612. Berlin: Springer-Verlag

Lal, D., Schink, D. R. 1960. Low background thin wall flow counters for measuring beta activity of solids. *Rev. Sci. Instrum.* 31: 395–98

Lal, D., Venkatavaradan, V. S. 1970. Analysis of causes of ^{14}C variations in the atmosphere. In *Radiocarbon Variations and Absolute Chronology*, ed. I. U. Olsson, pp. 549–70. New York: Wiley. 652 pp.

Lal, D., Arnold, J. R., Honda, M. 1960. Cosmic ray production rates of ^{7}Be in oxygen and ^{32}P, ^{33}P, ^{35}S in argon at mountain altitudes. *Phys. Rev.* 118: 1626–32

Lal, D., Arnold, J. R., Nishiizumi, K. 1985. Geophysical records of a tree: new applications for studying geomagnetic field and solar activity changes during the past 10^4 years. *Meteoritics* 20: 403–14

Lal, D., Venkatesan, T., Davis, R. Jr. 1986. Cosmogenic ^{37}Ar, ^{39}Ar in terrestrial rocks. *Proc. ICOG, 11th, Cambridge, Terra Cognita*, 6: 250

Lal, D., Nishiizumi, K., Klein, J., Middleton, R., Craig, H. 1987. Cosmogenic ^{10}Be in Zaire alluvial diamonds: implications for ^{3}He contents of diamonds. *Nature* 328: 139–41

Libby, W. F. 1946. Atmospheric helium-three and radiocarbon from cosmic radiation. *Phys. Rev.* 69: 671–72

Lingenfelter, R. E. 1963. Production of carbon-14 by cosmic-ray neutrons. *Rev. Geophys.* 1: 35–55

Lingenfelter, R. E., Flamm, E. J. 1964. Production of carbon 14 by solar protons. *J. Atmos. Sci.* 21(2): 134–40

Lingenfelter, R. E., Ramaty, R. 1970. Astrophysical and geophysical variations in ^{14}C production. In *Radiocarbon Variations and Absolute Chronology*, ed. I. U. Olsson, pp. 513–17. New York: Wiley. 652 pp.

Litherland, A. E. 1984. Accelerator mass spectrometry. See Wölfli et al 1984, pp. 100–8

Locker, G. L. 1933. Neutrons from cosmic-ray stösse. *Phys. Rev.* 44: 779–81

Marti, K., Craig, H. 1987. Cosmic-ray-produced neon and helium in the summit lavas of Maui. *Nature* 325: 335–37

Middleton, R., Klein, J. 1987. ^{26}Al: measurement and applications. *Philos. Trans. R. Soc. London Ser. A* 323: 121–43

Middleton, R., Klein, J., Brown, L., Tera, F. 1984. ^{10}Be in commercial beryllium. See Wölfli et al 1984, pp. 511–13

Montgomery, C. G., Montgomery, D. D. 1939. The intensity of neutrons of thermal energy in the atmosphere at sea level. *Phys. Rev.* 56: 10–12

Muller, R. A. 1977. Radioisotope dating with a cyclotron. *Science* 196: 489–94

Nelson, D. E., Korteling, R. G., Stott, W. R. 1977. Carbon-14: direct detection at natural concentrations. *Science* 198: 507–8

Nishiizumi, K., Lal, D., Klein, J., Middleton, R., Arnold, J. R. 1986. Production of ^{10}Be and ^{26}Al by cosmic rays in terrestrial quartz in situ and implications for erosion rates. *Nature* 319: 134–36

Nishiizumi, K., Kohl, C. P., Klein, J., Middleton, R., Winterer, E. L., et al. 1987. In situ ^{10}Be and ^{26}Al in quartz: calibration and application. *Int. Symp. Accel. Mass Spectrom., 4th, Niagara-on-the-Lake, Ontario, Can., Abstr.*, ed. H. E. Gove, p. 38

O'Brien, K. 1979. Secular variation in the production of cosmogenic isotopes in the Earth's atmosphere. *J. Geophys. Res.* 84: 423–31

Oeschger, H., Wahlen, M. 1975. Low level counting techniques. *Ann. Rev. Nucl. Sci.* 25: 423–61

Oeschger, H., Houtermans, J., Loosli, H., Wahlen, M. 1970. The constancy of cosmic radiation from isotope studies in meteorites and on the Earth. In *Radiocarbon Variations and Absolute Chronology*, ed. I. U. Olsson, pp. 471–98. New York: Wiley. 652 pp.

Oeschger, H., Stauffer, B., Bucher, P., Frommer, H., Moll, M., et al. 1972. *Proc. Int. Conf. Radiocarbon Dating, 8th, Lower Hutt, N.Z.*, 1: D70–90. Wellington: R. Soc. N.Z.

Ozima, M., Zashu, S., Nitoh, O. 1983. ^{3}He/^{4}He ratio, noble gas abundance and K-Ar dating of diamonds—an attempt to search for the records of early terrestrial history. *Geochim. Cosmochim. Acta* 47: 2217–24

Phillips, F. M., Leavy, B. D., Jannik, N. O., Elmore, D., Kubik, P. W. 1986. The accumulation of cosmogenic chlorine-36 in rocks: a method for surface exposure dating. *Science* 231: 41–43

Powell, C. F., Fowler, P. H., Perkins, D. H., eds. 1959. *The Study of Elementary Particles by the Photographic Method.* New York/London: Pergamon. 669 pp.

Raisbeck, G. M., Yiou, F. 1979. Possible use of ^{41}Ca for radioactive dating. *Nature* 227: 42–43

Raisbeck, G. M., Yiou, F. 1984. Production of long-lived cosmogenic nuclei and their applications. See Wölfli et al 1984, pp. 91–99

Rama, Honda, M. 1961. Cosmic-ray-induced radioactivity in terrestrial materials. *J. Geophys. Res.* 66: 3533–39

Reedy, R. C. 1977. Solar proton fluxes since 1956. *Proc. Lunar Sci. Conf., 8th*, pp. 825–39

Reedy, R. C. 1985. A model for GCR-particle fluxes in stony meteorites and production rates of cosmogenic nuclides. *Proc. Lunar Planet. Sci. Conf., 15th, Part 2, J. Geophys. Res.* 90: C722–28

Reedy, R. C., Arnold, J. R. 1972. Interaction of solar and galactic cosmic-ray particles with the moon. *J. Geophys. Res.* 77: 537–55

Reedy, R. C., Arnold, J. R., Lal, D. 1983. Cosmic ray record in solar system matter. *Ann. Rev. Nucl. Part. Sci.* 35: 505–37

Reeves, H., Fowler, W. A., Hoyle, F. A. 1970. Galactic cosmic ray origin of Li, Be, B in stars. *Nature* 226: 727–29

Reyss, J.-L., Yokoyama, Y., Guichard, F. 1981. Production cross sections of ^{26}Al, ^{22}Na, ^{7}Be from argon and of ^{10}Be, ^{7}Be from nitrogen: implications for production rates of ^{26}Al and ^{10}Be in the atmosphere. *Earth Planet. Sci. Lett.* 53: 203–10

Rossi, B. 1948. Interpretation of cosmic-ray phenomena. *Rev. Mod. Phys.* 20: 537–83

Rossi, B. 1952. *High-Energy Particles.* Englewood Cliffs, NJ: Prentice-Hall. 569 pp.

Rudstam, G. 1955. Spallation of elements in mass range 51–75. *Philos. Mag.* 46: 344–56

Rumbaugh, L. H., Locker, G. L. 1936. The specific ionization of cosmic ray particles as determined by Geiger-Muller counter efficiency. *Phys. Rev.* 49: 854–55

Simpson, J. A. 1983. Elemental and isotopic composition of the galactic cosmic rays. *Ann. Rev. Nucl. Part. Sci.* 33: 323–81

Somayajulu, B. L. K. 1977. Analysis of causes for the ^{10}Be variations in deep sea sediments. *Geochim. Cosmochim. Acta* 41: 909–13

Srinivasan, B. 1976. Barites: anomalous xenon from spallation and neutron-induced reactions. *Earth Planet. Sci. Lett.* 31: 129–41

Stuiver, M., Pearson, G. W., Braziunas, T. 1986. Radiocarbon age calibration of marine samples back to 9000 CAL YR BP. *Radiocarbon* 28: 980–1021

Suter, M., Balzer, R., Bonani, G., Hofmann, H., Morenzoni, E., et al. 1984. Precision measurements of ^{14}C in AMS—some results and prospects. See Wölfli et al 1984, pp. 117–22

388   LAL

Takagi, J., Hampel, W., Kirsten, T. 1974. Cosmic-ray muon-induced [129]I in tellurium ores. *Earth Planet. Sci. Lett.* 24: 141–50

Tanaka, S., Sakamoto, K., Takagi, J., Tsuchimoto, M. 1968. Search for aluminum-26 induced by cosmic-ray muons in terrestrial rock. *J. Geophys. Res.* 73: 3303–9

Tuniz, C., Smith, C. M., Moniot, R. K., Kruse, T. H., Savin, W., et al. 1984. Beryllium-10 contents of core samples from St. Severin meteorite. *Geochim. Cosmochim. Acta* 46: 955–65

von Egidy, T., Hartmann, F. J. 1982. Average muonic Coulomb capture probabilities for 65 elements. *Phys. Rev. A* 26: 2355–60

Walker, T. P., Mathews, G. J., Viola, V. E. 1985. Astrophysical production rates for Li, Be and B isotopes from energetic ^{1}H and [4]He reactions with HeCNO nuclei. *Astrophys. J.* 299: 745–51

Winsberg, L. 1956. The production of chlorine-39 in the lower atmosphere by cosmic radiation. *Geochim. Cosmochim. Acta* 9: 183–89

Wölfli, W., Polach, H. A., Andersen, H. H., eds. 1984. *Accelerator Mass Spectrometry, AMS '84. Proc. Int. Symp., 3rd, Zurich, 1984. Nucl. Instrum. Methods Phys. Res.,* Vol. 233, No. B5. Amsterdam: North-Holland. 448 pp.

Yiou, F., Raisbeck, G. M., Klein, J., Middleton, R. 1984. [26]Al/[10]Be in terrestrial impact glasses. *J. Non-Cryst. Solids* 67: 503–9

Yiou, F., Raisbeck, G. M., Bourles, D., Lorius, C., Barkov, N. I. 1985. [10]Be in ice at Vostok, Antarctica during the last climatic cycle. *Nature* 316: 616–17

Yokoyama, Y., Reyss, J.-L., Guichard, F. 1977. Production of radionuclides by cosmic rays at mountain altitudes. *Earth Planet. Sci. Lett.* 36: 44–50

Zito, R., Donahue, D. J., Davis, S. N., Bentley, H. W., Fritz, P. 1980. Possible sub-surface production of carbon-14. *Geophys. Res. Lett.* 7: 235–38

Ann. Rev. Earth Planet. Sci. 1988. 16:389–476

TIME VARIATIONS OF THE EARTH'S MAGNETIC FIELD: From Daily to Secular

Vincent Courtillot and Jean Louis Le Mouël

Laboratoire de Géomagnetisme et Paléomagnétisme, Institut de
Physique du Globe, Place Jussieu, 75005 Paris, France

1. INTRODUCTION

Variations in time of the Earth's natural magnetic field have a frightfully
wide spectrum, ranging over more than 20 orders of magnitude: They
extend from well over 10^3 Hz, where they merge into the radioelectric noise
spectrum, to probably more than 100 m.y. In rough summary, the higher
frequencies stem from external sources in the ionosphere and above into
the magnetosphere, whereas the longer periods arise from internal sources,
primarily fluid motions in the core at depths in excess of 2900 km. Methods
for separating external and internal signals have been available since the
time of Gauss (1838), but the accuracy, the space distribution, and the
time duration of available data are often not sufficient to unambiguously
solve the problem for all types of variations. Although periods shorter
than 1 yr should clearly be assigned to external sources, the extent to which
the spectra of the external and internal parts overlap is the subject of recent
debate and progress. Currie (1966, 1968) proposed that there was little
overlap and a cutoff of ~ 3.7 yr. A long debated ~ 11-yr contribution due
to the solar cycle was clearly demonstrated by Currie (1973a), Courtillot
& Le Mouël (1976a,b) and Alldredge (1976), and the cutoff period was
thus rejected to at least ~ 20 yr. The situation seemed to be confused again
when Courtillot et al (1978a) provided evidence for sharp impulses of 1-
yr duration in the internal, so-called secular, variation. The debate that
followed over the reality of brief internal events is one of the topics
addressed in this review.

We first provide a short historical account of successive discoveries

389

0084–6597/88/0515–0389$02.00

of time variations in the geomagnetic field, with an emphasis on the observational developments that often preceded theoretical advances. We next examine the instruments that are used to monitor geomagnetic variations and the precision of these instruments. We discuss the status of permanent magnetic observatories and of data acquisition at relocatable stations and, most recently, with satellites. The bulk of the paper then reviews our knowledge of the external and internal contributions to geomagnetic variations in the period range from less than 1 day to several hundred years: Emphasis is placed on recent progress in the period range where these two types of variations are now known to overlap, particularly on the solar-cycle effects and on secular variation impulses (also known as jerks). We review historical measurements made over the last 400 yr in European observatories, examine the recent worldwide compilations of Gubbins & Bloxham (1985) from 1715 onward, and briefly summarize the important observations of paleosecular variation, with which archaeo- and paleomagnetism allow us to extend our investigations from the 10^2 to the 10^6 yr time scale. A discussion of the use and interpretations of geomagnetic variations follows, in which our emphasis is on topics more relevant to internal geophysicists: External variations are used to probe the electrical conductivity of the Earth's crust and mantle, and internal variations allow one to infer motions in the convecting fluid core.

2. HISTORICAL BACKGROUND

Unfortunately, there still appears to be no comprehensive history of the evolution of ideas and discoveries in geomagnetism. For a summary of facts relevant to the study of geomagnetic variation, we rely on textbooks such as Chapman & Bartels (1940, 1962) or, more recently, Merrill & McElhinny (1983). The reader is referred to these books for further references.

The first observation that the geomagnetic field varies in time is due to Henry Gellibrand, Professor of Astronomy at Gresham College: Gellibrand found in June 1634 that the declination at London was over 7° smaller than when first measured in October 1580 by William Borough. Two theoretical papers in 1683 and 1692, followed by two voyages in the Atlantic Ocean between 1698 and 1700, led Edmund Halley to discover the westward drift of some geomagnetic features. Halley suggested that one could identify four magnetic poles, two attached to the solid crust and two to a drifting core below the crust.

In 1722, a London instrument maker by the name of George Graham performed very careful observations of a compass needle: Aided with a microscope with which he could resolve intervals of 2′ (arcminutes), he discovered continuous variations in excess of 30′ in the same day, sometimes in the course of only a few hours. In addition to discovering *nonsecular* magnetic variations, Graham introduced the distinction between quiet and disturbed days. In 1741, Olav Peter Hiorter and Andreas Celsius in Uppsala made the connection between magnetic storms and polar lights, observing changes in declination over 4° within 4 min of time. Coordinated observations were made in London the same year by Graham. Hiorter also observed that the needle did not return to its prestorm position and suggested that this might be the cause of the so-called *monthly* magnetic motions. In 1759, Canton found in London that the mean range of the daily oscillation of the needle on quiet days was larger in summer (over 13′) than in winter (less than 7′), the first mention of an *annual* period. In 1782 in Paris, Dominique Cassini showed that the daily variation could not be due to a daily variation in temperature.

Frequent and detailed observations became common in the early part of the nineteenth century with the work of Alexander von Humbold in Berlin, Carl Friedrich Gauss and Wilhelm Weber in Göttingen, and François Arago in Paris. Von Humbold organized numerous simultaneous observations all over the globe and in 1828 constructed one of the first observatories in Berlin. In 1834, Gauss and Weber joined what was to become the Magnetische Verein (Göttingen Magnetic Union, 1836–41), making observations in Göttingen at 1-hr and then 5-min intervals. In addition to his theoretical work, Gauss paid great attention to the quality of observations, making it possible to detect changes of a few arcseconds over a few minutes of time by use of Poggendorf's method of reading (1832). In his *Allgemeine Theorie des Erdmagnetismus* in 1838, Gauss introduced means of measuring absolute magnetic intensity, allowing one to follow variations in the full vector field. He and Weber subsequently discovered the daily and longer variations of all remaining geomagnetic elements (i.e. those other than D or I). As is well known, Gauss derived the technique to separate internal and external sources. He ascribed secular variation to cooling and thickening of the Earth's solid crust. The Göttingen Magnetic Union introduced the important concepts of baseline and scale values, allowing the reduction of all magnetometer readings to absolute values, and also developed the magnetic inductor.

Edward Sabine (1857, 1868–77) summarized the British effort to establish magnetic observatories in British colonies. He used time averaging to eliminate temporary fluctuations, clearly distinguishing between them and secular variation. [He separated "the history of all that is not permanent in the phenomena, whether it appear in the form of momentary, daily, monthly or annual change and restoration; or in progressive changes not compensated by counterchanges, but going on continually accumulating in one direction, so as in the course of many years to alter the mean amount of the quantities observed" (quoted by Chapman & Bartels 1962, p. 934)]. Sabine also discovered in 1852 that during the years 1841–48, magnetic disturbances varied in parallel with the $\sim$ 10-yr *sunspot* variation. He calculated the *lunar* daily magnetic variation, which had been discovered in 1850 by C. Kreil at Prague. [The lunar semidiurnal range is 44" (arcseconds) in summer months.] Sabine noted that these discoveries were "the result of the system of observations enjoined, and so carefully and patiently maintained."

A key instrumental advance was made in 1846 at Greenwich, under the direction of G. B. Airy, when C. Brooke introduced photographic recording. Much of the progress in the following decades was related to the identification of the external sources of transient variations. In 1882, B. Stewart suggested that the quiet-day solar variation (Sq) must have its origin in electric currents in the upper atmosphere. A. Schuster in 1889 applied Gauss's method of spherical harmonic analysis to Sq and confirmed that its primary source was external; he also found an internal part, which he correctly ascribed to induction in the solid conducting Earth. Schuster's work forms the basis of all subsequent work on global electromagnetic induction and the search for the distribution of mantle conductivity.

Soundings with radio waves led to the discovery of the *ionosphere* in the 1920s. The study of the disturbance daily variation SD and the relation between storms and auroras had led K. Birkeland in 1896 to study streams of electrons in a magnetic field and F. A. Lindemann in 1919 to study streams of neutral but ionized particles. In the early 1930s, S. Chapman and V. C. A. Ferraro studied the impact of such a stream coming from the Sun on the geomagnetic field and discovered that such a stream confined the geomagnetic field to a *magnetosphere*.

In the meantime, A. Delesse in 1849 and M. Melloni in 1853 had discovered that rocks could become magnetized in the direction of the magnetic field, paving the way for the study of paleomagnetism. This was extended in 1899 by G. Folgerhaiter to archaeological material that he thought recorded historical changes in the magnetic field. P. David in 1904 and B. Brunhes in 1906 discovered that natural remanent magnetization could be opposed to that of the present field; P. L. Mercanton in 1926 subsequently found that reversely magnetized rocks could be found all over the world, establishing the global nature of the process. In 1929 M. Matuyama noted that in his collection, reversed lavas were always older than normal ones. This was the beginning of work on geomagnetic reversals and their time scale. In 1925, R. Chevallier attempted to recover secular variation from the volcanic flows of Mount Etna. The study of archaeomagnetism, in particular of intensity variations in the historical past, was pioneered by E. Thellier in the mid-1930s. The first attempts to determine paleosecular variation from sediments was made by E. A. Johnson et al in 1948 and D. H. Griffiths in 1953. These advances provided the tools to close the gap between already mature geomagnetism and nascent paleomagnetism.

3. MAGNETIC MEASUREMENTS

There are approximately 180 permanent magnetic observatories in operation, with a very uneven distribution (Figure 1). Between 50 and 100 of

Figure 1   World distribution of permanent magnetic observatories. Digital observatories are shown as open circles with oblique three-letter codes; observatories without high resolution digital results are shown as closed circles with upright three-letter codes. Names of several European digital observatories have been omitted because of crowding. After World Data Center for Solid Earth Geophysics (1987), where code names can be found.

them are used in most current global geomagnetic analyses. Both the number of observatories and the reliability of the data they provide have increased with time, although the Southern Hemisphere still lacks a proper network of observation points (Figure 2). As a typical example of the nature and accuracy of measurements, we briefly review the operations at the French observatory of Chambon-la-Forêt. Chambon is located in Western Europe, within what is probably the densest array of world observatories. Actually, 10 of them have formed an informal network where any malfunction in one is immediately spotted (actually within a month) by the others. This has in part been responsible for the correct identification of solar-related effects and more prominently for ascertaining that the 1969.5 secular variation impulse was not an artifact (see Section 5.1).

Absolute measurements of the intensity F, declination D, and inclination I are performed approximately once a week. F is measured with a proton precession magnetometer, I and D with a Zeiss theodolite equipped with a flux-gate sensor (Gilbert et al 1987, Bitterly et al 1987). Absolute accuracies are about 0.5 nT for F, 5″ for D and I, and 0.5 nT for derived values of the horizontal H and vertical Z components. Two separate variograph systems are used: a classical optical-magnet LaCour (since 1936) and a three-component flux gate (since 1981). The flux-gate system provides both a magnetic tape and an analog recording with a sampling frequency of one measurement per minute. Baselines for the flux-gate variograph are shown in Figure 3 for a typical year (1986). Only the H baseline shows a significant fluctuation, an annual wave due to remaining temperature effects. Finally, reduced absolute hourly mean values are calculated and made available in magnetic tape and yearbook format through world data centers.

Figure 4 displays typical photographic LaCour recordings on a quiet day, a disturbed day, and a day with a sudden storm commencement. From the hourly mean values one can either derive daily mean values and display them for a typical year (Figure 5) or derive monthly and annual mean values and display them over several decades (Figures 6, 7). More elaborate averaging filters can be constructed (see Sections 4 and 5).

Measurements made at permanent magnetic observatories are often complemented with measurements made on a regular basis at relocatable stations. This is desirable because of the need to better interpolate secular variation between observatories. As an example, the French array consists of ~30 repeat stations where elements are measured every five years. In the latest surveys, instruments included a flux gate mounted on a theodolite for D and I, and two proton precession magnetometers for F. Careful allowance for reduction errors and instrumental accuracy allow one to

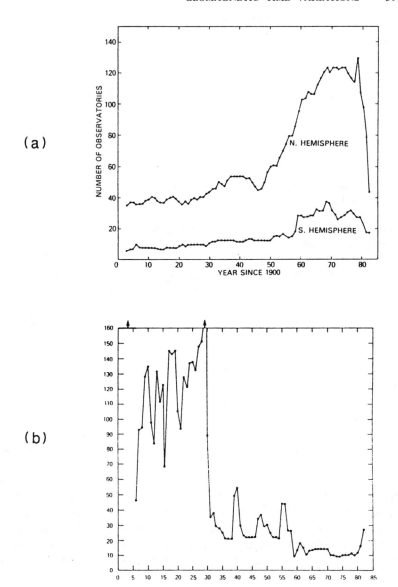

Figure 2   (*a*) Number of observatories contributing to secular variation models. (*b*) Lower bound of condition number, indicating overall quality and number of data used in producing secular variation models (after Langel et al 1986).

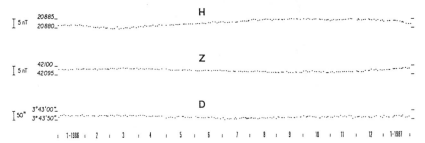

Figure 3   Baselines for the variometers at the Chambon-la-Forêt magnetic observatory in France. Numbers are for months of the year 1986.

estimate the maximum uncertainties of 5-yr secular variation estimates at 1' for D, and about 4 nT for H, F, and Z. Figure 8 illustrates the secular variation of H and Z between 1972.5 and 1977.5 in the French array and neighboring European observatories, compared with the predictions for the same epoch (1975.0) of the International Geomagnetic Reference Field (IGRF) model.

Observatories and repeat stations are not enough to avoid the drawbacks of an insufficiently dense network of observations. One possible way to fill in the gaps is through satellite measurements. The most recent and complete such experiment was MAGSAT, which gathered dense, high-quality data over a 7.5-month period in 1979–80 (Langel et al 1982). F was measured with a cesium vapor magnetometer, and vector components X (north), Y (east), and Z were measured with flux-gate magnetometers. The resulting measurement accuracy is better than 2 nT for F and 6 nT for the vector components. Although MAGSAT provided a main field model at the surface of the Earth with an accuracy better than 20 nT (Langel & Estes 1985a,b), it was not of sufficient duration to provide a secular variation model of similar quality (Cain et al 1983, Langel & Estes 1985a,b, Barraclough 1985). It did provide an estimate of the lowermost order spherical harmonic coefficients of the external field and evidence for their short- and longer-term variations (Section 4.5). A new permanent magnetic satellite observatory will be needed for significant further observational progress on geomagnetic time variations.

4. VARIATIONS WITH AN EXTERNAL ORIGIN

Proving that a given magnetic signal originates from sources external to the Earth requires application of an analysis of the type introduced by Gauss. Often the density of surface measurements, or their accuracy, either

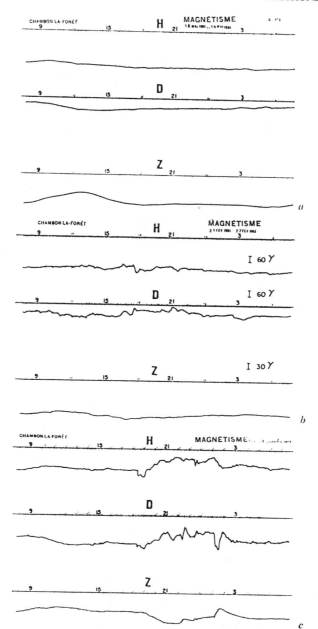

Figure 4 Typical magnetograms (LaCour variometers) recorded at the Chambon-la-Forêt magnetic observatory: top, quiet day; middle, disturbed day; bottom, sudden storm commencement.

CHAMBON 1986

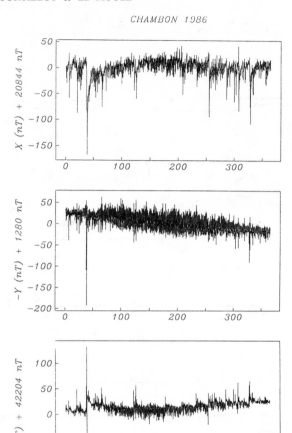

Figure 5 Daily mean values for the three components X, Y, and Z at Chambon-la-Forêt during the year 1986.

does not permit such an analysis to be made, or, if made, renders the results inconclusive. Analysis of variations usually relies on (*a*) an incomplete spherical harmonic analysis; (*b*) a visual analysis of data at selected locations in the time domain; (*c*) a spectral analysis of such data, aided (or sometimes actually impaired) by filtering; or (*d*) a correlation with other phenomena of known external origin, including comparison with some type of forward modeling. In this section on external signals, we first summarize the various current systems in the ionosphere and magnetosphere that are thought to be the source of the variations. We

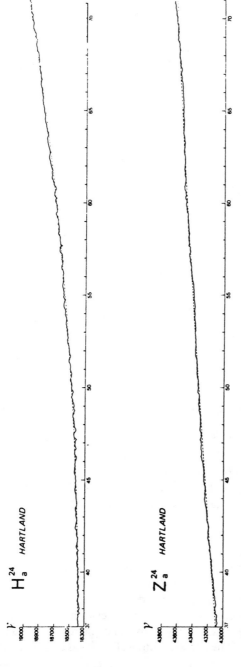

Figure 6 Monthly mean values for all days of the month (a) and all hours of the day (24) for components *H* and *Z* at the Abinger (1937–56) and Hartland (1957–71) observatories in the United Kingdom. A parabolic fit to the data is shown as a dashed curve. Ordinates are in nanoteslas (after Courtillot & Le Mouël 1976a,b).

CHAMBON OBSERVATORY (CLF)

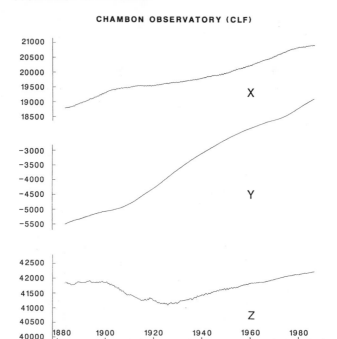

Figure 7   Over a century of monthly mean values for the three components X, Y, and Z at the Chambon-la-Forêt observatory (actually St. Maur 1883–1900, Val Joyeux 1901–35, and Chambon-la-Forêt 1936–86).

then review observational evidence of the various variations, going from the shorter (1 day or less) to the longer (11 yr or more) periods.

4.1   *External Current Systems*

The external field is primarily linked with solar activity through two main kinds of interaction processes: energy transfer between the solar wind and the magnetosphere, and ionization of the E-layer in the ionosphere (see, for example, the summary by Gavoret et al 1986).

The solar wind–magnetosphere interaction is modulated by the B_z component of the interplanetary magnetic field. When B_z is positive (on the order of 2–4 nT) for several hours, the dynamo interaction is in a ground state and the energy of particles within the plasma sheet decays and becomes less than 1 keV. When B_z becomes negative, energy from the solar wind is stored in the magnetotail and convection in the auroral electrojets is enhanced. This energy is released in magnetospheric substorms, resulting in increased auroral jets, heating of the ionosphere, and enhanced currents in the plasma sheet. The corresponding effects make up most of the irregu-

Figure 8 A comparison of observed secular variation of the horizontal vector **H** and vertical component *Z* for the period 1972–77 with values computed from the 1975 International Geomagnetic Reference Field (IGRF) in European observatories and the French array of repeat stations (values in nT/5 yr, indicated by a bar for **H**; after Courtillot et al 1978b). Dashed vectors and values in parentheses are the predictions.

lar geomagnetic activity, which is monitored by a number of indices (Mayaud 1980). Activity in the auroral electrojets is most prominent at high latitudes, where it generates much of the disturbance daily variation SD. It is monitored by the AE index. At midlatitudes, auroral jets, field-aligned currents, and the ring current combine their effects: They are monitored by the 3-hr am and aa indices. At low latitudes, the influence of the ring current, located in the geomagnetic equatorial plane at about three Earth radii, becomes preponderant. The corresponding storm time disturbance variation is monitored by the Dst index.

Ionospheric ionization depends both on precipitation of electrons from the magnetosphere and on the incoming flux of extreme ultraviolet and X rays from the solar wind. Ionization changes related to the electron precipitation during substorms can also be monitored with the aa index. Variations in the ionization of the daytime ionosphere are closely related to those of the Wolf relative sunspot number R, which can be used to monitor the range of the quiet part of the solar daily variation Sq (Reid 1972).

Gavoret et al (1986) have found that, at least over a 3–30 yr range, the various types of indices are remarkably correlated. Figure 9a displays the correlation of aa, Dst, and AE [where the anomalous behavior of Dst between 1973 and 1976 is explained to be due to a contamination by rapid internal secular variation (see Ducruix et al 1980, Gavoret et al 1986, and Section 5.1 of this paper)], Figure 9b the correlation of R with the daily range of Y, and Figure 9c the correlation of aa and R (including the 11- and 22-yr cycles discussed, for instance, by Courtillot et al 1977). These correlations stem from the fact that all these variations are ultimately related to the same energy input from the solar wind (Gavoret et al 1986). This is very convenient, since a single homogeneous and reliable series is available from 1868 onward for the aa index (Mayaud 1980).

4.2 The Daily Variation

The daily variation for quiet days, Sq, is best known for the International Geophysical Year (IGY 1957–59), when a large number of stations operated simultaneously. Malin & Gupta (1977) have analyzed this variation in terms of a spherical harmonic model as a function of position on Earth and Universal Time. They found it possible to use the fundamental 24-hr period, together with the first three harmonics (12 hr, 8 hr and 6 hr) of components X, Y, and Z at 108 stations. Stations perturbed by the localized equatorial electrojet were omitted. Variations were measured from their midnight level, under the reasonable assumption that no ionospheric currents flow through the midnight meridian. Sq varies smoothly both with location and with Universal Time (UT). Fourier components of order p

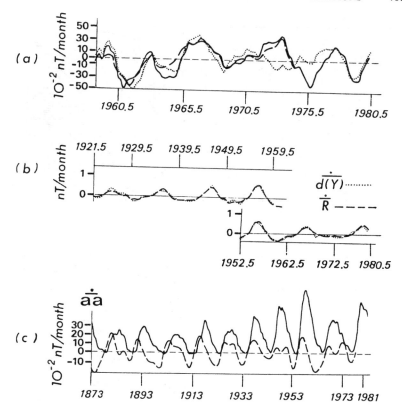

Figure 9 (*a*) First-order time derivative of the geomagnetic indices *aa* (solid line), *Dst* (dotted line), and *AE* (dashed line). *Dst* and *AE* are actually normalized to *aa* amplitudes. (*b*) Same as (*a*) for the Wolf sunspot number *R* (dashed line) and the daily range of *Y* at Chambon-la-Forêt 1920–61 and Niemegk 1951–80 (dotted line). (*c*) Same as (*a*) for the *aa* index (dashed line) and *R* 1873–1981 (solid line) [after Gavoret et al (1986), who describe the filtering used].

(1 to 4) are dominated by spherical harmonic coefficients of degree $p+1$ and order p (e.g. Malin & Gupta 1977, Tables 1 and 2). Malin & Gupta were able to separate internal (i) and external (e) contributions (their table 2; $e/i = 3.0 \pm 0.3$ for $p = 1$ with a phase difference of $-6 \pm 5°$). The external current system for *Sq* at 0 hr and 6 hr UT is illustrated in Figure 10. The internal (induced) pattern is on average 20 min in advance (i.e. to the west) of the external current system and in the opposite sense (as a consequence of Lenz's law). Mayaud (1965, 1980) has extensively described the day-to-day variability of the daily variation field, and more recently Counil et al (1987) have studied its place-to-place variability due to heterogeneities in induced internal currents.

Oh UT

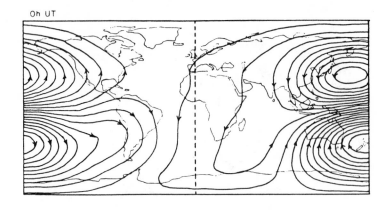

6h UT

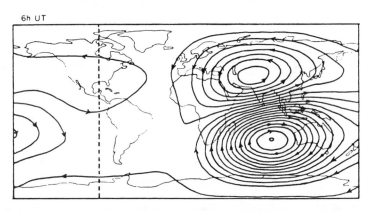

Figure 10 External current system for *Sq* based on magnetic data measured from the smoothed local midnight value. Current flows parallel to the contours in the direction of the arrows; 20 kA between adjacent contours. Currents flowing across the midnight meridian (dashed line) are small (after Malin & Gupta 1977).

4.3   *The 27-Day Variation and its Harmonics — the 2–200 Day Background*

Roberts (1984) analyzed 14-yr-long records of the daily mean values of H, D, and Z recorded by 18 permanent observatories, all but one in the Northern Hemisphere (illustrating the ever-present problem of observatory distribution). Figure 11 (from Roberts 1984) displays the power spectra for various components in the period range from 2 to 200 days at the Fürstenfeldbruck observatory in Europe; because secular variation

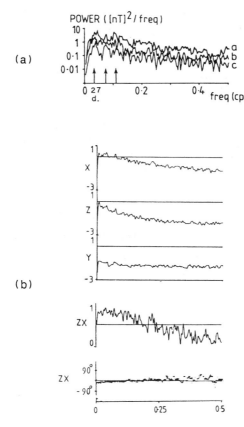

Figure 11   (*a*) Power spectra of the *X*-component data from Fürstenfeldbruck for three 2-yr-long sections of data (a, 1957–58; b, 1968–69; c, 1964–65). The 27-day line and first two harmonics are indicated by arrows. (*b*) Power spectra (for *X*, *Y*, and *Z*) and *ZX* coherence and phase (bottom) for 14 yr of daily mean data (1957–70) at Fürstenfeldbruck (after Roberts 1984).

was removed by application of a high-pass filter with half-point at 200 days, spectral estimates with period longer than 2 months are essentially suppressed. Figure 11*a* shows peaks at periods of 27, 27/2, and 27/3 days. Each spectrum corresponds to a 2-yr section of the data at the times of a strong solar maximum (1957–58), a moderate solar maximum (1968–69), and a solar minimum (1964–65) (see Figure 9*c* for the *R* sunspot index). Power at all frequencies increases with solar activity. The peaks are better identified at the time of the sunspot minimum. The *Y* spectrum (Figure 11*b*) has low energy and is flat, whereas the *X* and *Z* spectra are well correlated (at least for periods larger than 6 days). The 27-day variation is due to the recurrence tendency of magnetic storms, which itself is related to the solar rotation period; the current system involved in the various phases of magnetic storms is the ring current. Because of the Sun's differential rotation, the spectral lines are actually spread over the 26–29 day range (and harmonics). Following Banks (1969) and others, Roberts (1984) shows that the variation together with the intervening continuum are well fit by a $P_1 + P_3$ zonal geometry (i.e. first- and third-order Legendre polynomials of colatitude) with the axis of the uniform external field lying parallel to the main dipole.

Roberts estimated the response functions Z/X (which are simply related to i/e—see Section 6.1) in 11 frequency bands, from 3 to 71 days. Responses at different stations were found to be very different, which Roberts attributed to lateral conductivity inhomogeneities within the Earth and, in particular, to induced electric currents flowing in the oceans.

4.4   Variations With Periods 1 Yr, 6 Months, and Harmonics

In an early analysis, McDonald (1970) studied monthly mean values for selected quiet days at a number of observatories and found that there were increased fluctuations in the 0.2–0.5 yr period range during sunspot maximum years. He also isolated an annual variation, most visible during sunspot minimum years, which he found to be consistent with a ring-current origin. As a preliminary step before analyzing longer periods, Courtillot & Le Mouël (1976a,b) also studied monthly mean values at three observatories (Abinger-Hartland and Chambon-la-Forêt in Europe, Hermanus in South Africa) over a period of ~ 35 yr (1936–71). Rather than applying directly some linear filtering technique, they insisted on the importance of a visual inspection of the data. The smooth internal secular variation masks almost any other signal (Figure 6) and was removed successfully with a quadratic model (see Sections 4.5 and 5.1). The remaining signal is shown in Figure 12a for the Abinger-Hartland observatory. Variations in D are quite small, which indicates a zonal field geometry. Annual oscillations are clearly visible in the H record, superimposed on slower 10–12 yr variations of similar amplitude (20–40 nT). The behavior of the Z-component is similar, with a smaller amplitude for the 1-yr variation. H and Z are strongly correlated at all periods (Courtillot & Le Mouël 1976a,b). Spectra of the residual curves are given in Figure 12b: A 1-yr peak stands out clearly in both H and Z. A 6-month peak is found in H but apparently not in Z. (This is true for all three of the observatories studied.) A study of coherency between stations shows that the annual line in H has opposite phase in Europe and South Africa. This implies that the

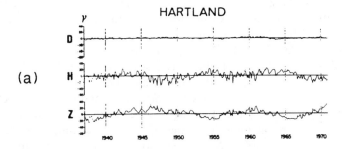

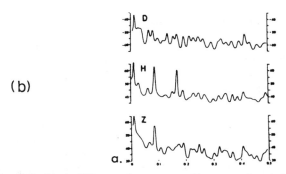

Figure 12 (*a*) Residuals (differences between monthly mean values and a parabolic fit to them) and (*b*) maximum entropy power spectra for the three components *D*, *H*, and *Z* at Abinger-Hartland. For the spectra, abscissae are frequency f in cycles per month, and ordinates are $10 \log P(f)$, where $P(f)$ is the power density in (nanoteslas)^2 per cycle per month. Correspondences are 6 months $= 0.167$, 1 yr $= 0.083$ and 11 yr $= 0.008$ cycle month^{-1} (after Courtillot & Le Mouël 1976a).

annual line on the one hand and the 6-month line and continuum on the other cannot be generated by the same mechanism. Currie (1966) suggested that the semiannual line might, together with most other variations in the range from 2 days to over 2 yr, be due to fluctuations in the ring current because of the varying inclination of the Sun's axis with respect to the Sun-Earth line. He attributed the annual line to an ionospheric dynamo mechanism related to the amplitude modulation of incoming solar radiation (hence ionospheric conductivity) received by the summer and winter hemispheres. Banks (1969) also proposed that the semiannual line could be fit with a P_1 geometry. Courtillot & Le Mouël (1976a) compared the intensities of the 12-, 6-, and 3-month lines using separately monthly mean values for all days but including night hours only (written E_a^n for element E), and for quiet days, disturbed days, or all days including all 24 hours of the day (written, respectively, as E_q^{24}, E_d^{24} and E_a^{24}). To a first approximation, values for E_a^n do not include Sq effects, and differences between E_a^{24} and E_a^n do not include Dst effects (see Courtillot & Le Mouël 1976a, Table 1). Because the intensity of the H_a^n line is three times stronger than that of $H_a^{24} - H_a^n$, Courtillot & Le Mouël concluded that Sq probably played only a small part in both the annual and semiannual lines, which leaves SD as a probable factor. Phase inversion of the H annual line between the Northern and Southern Hemispheres would rule out a Dst origin. Work remains to be done on these spectral peaks: The three main current systems are probably involved, which may account for some misfits when attempting to model observations with either a simple P_1 (6-month line) or P_2 (1-yr line) geometry (Banks 1969).

4.5   The ~11-Yr Solar Cycle and Harmonics

Attempts to isolate a solar-cycle-related variation in geomagnetic elements can be traced back to at least the early part of this century (for a list of early references, see Courtillot & Le Mouël 1976a, 1979, Alldredge 1976, 1982). Techniques have included the removal of a trend, the use of finite differences, and Fourier analysis. The suggestion to use a quadratic trend to remove internal secular variation can be traced back to Schmidt (1916). In one of the first conclusive studies, Fisk (1931) isolated a clear solar-cycle effect with an amplitude of about 10 nT in the H-component from 10 observatories for the period 1903–27. Vestine et al's (1947) monumental work elaborated on Fisk's work; the former included tables to correct internal secular variations from external variations ranging up to the solar-cycle period. Vestine et al noted that better instrumentation and several decades of further observation would probably be required before Z could be included in the analysis.

   Many recent studies have been devoted to analyses of annual mean values of geomagnetic elements over several decades [Yukutake 1965, Currie 1973a, Rivin 1974, 1976, Malin & Clark 1974, Malin & Isikara 1976, Harwood & Malin 1976, 1977, Courtillot & Le Mouël 1976a,b, Alldredge, 1976, 1977a, Yukutake & Cain 1979, Gavoret et al 1986; see Alldredge (1982) for a partial review]. The studies of Currie and Alldredge resort to spectral analysis, the former to maximum entropy and the latter to classical Fourier (periodogram) analysis. In his study of data from 49 observatories, Currie (1973a) identified peaks in maximum-entropy spectra and used them to construct histograms. In these, he finds maxima at the solar cycle and its first four harmonics, and also at the double solar cycle and its first nine harmonics (see also Section 4.6). McLeod (1982) showed that the use of annual means could seriously alias the upper half of the frequency range covered ($T < 4$ yr) and that, as a result, a number of peaks reported by Currie (1973a) could be spurious. Alldredge (1976) synthesized H- and Z-components from 20 observatories for the period 1900–72 in the band from 4 to 13 yr. The solar-cycle effect is readily apparent in his figures and reasonably correlated from one observatory to another. Alldredge also attempted to separate the first-order external and internal spherical harmonic coefficients (see Section 6.1).

Courtillot & Le Mouël (1976a,b) preferred to use the quadratic-fit method. They studied annual mean values of X and Z at 30 observatories for the 35-yr period from 1947 to 1972; Figure 13 shows the residual signals after trend removal. The solar-cycle effect is readily observed in both X and Z, although the latter is more noisy. There is a quite remarkable correlation of residuals from widely separated observatories, although the correlation is most obvious for the neighboring European observatories. Maximum-entropy spectral analysis reveals broad power maxima at ~ 11 yr and its first two harmonics (~ 5.5 yr, ~ 3.7 yr): The sunspot effect is not exactly sinusoidal, and harmonics are expected. McLeod (1982) recalls that broad peaks should be expected, since the solar cycle suffers from low-frequency modulation in both amplitude and duration, which widens the spectral peaks. Actually, there appears to be a random variation of peak amplitude from one cycle to the next (over the period for which data are available), which leads to the presence of subharmonics. No evidence was found for the 6-yr peak suggested by Currie (1973a) to be of internal origin and distinct from solar-related effects. Courtillot & Le Mouël found, however, that the correlation between observatory residuals and external indices was significantly degraded on the younger side of the data then available, from 1967 to 1972 (their Figure 8). They suggested that "the effect could be due to a slight change in the second derivative of the true internal secular variation of one of the two components. Such changes have indeed been observed, for example, in the D component around 1910 at Chambon-la-Forêt." This effect eventually led to the discovery of the 1969.5 secular variation impulse (see Section 5.1) and was understood to be a result of it.

The analysis was resumed by Ducruix et al (1980; see also Ducruix 1980), who studied data from about 80 observatories over a 30-yr period from 1947 to 1977. When a quadratic trend was fitted to the section of the data prior to the impulse (1947–69), Ducruix et al (1980) found a clear solar cycle both in X and Z and an excellent correlation over the entire

410   COURTILLOT & LE MOUEL

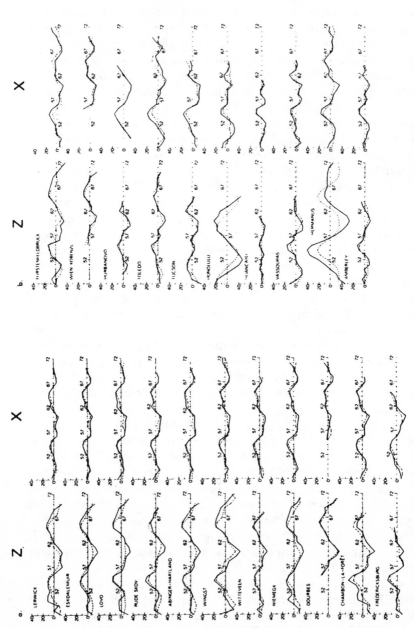

Figure 13 Residuals (as in Figure 12) of X and Z for 21 observatories. The left column is mostly European observatories, and Z data from Northern Hemisphere geomagnetic hemisphere have been reversed. Ordinates in nanoteslas. Dashed curves are least-squares-fit sinusoidal curves containing the 11-yr period and its first two harmonics (after Courtillot & Le Mouël 1976a).

period with the *aa* or *Dst* indices. Figure 14 illustrates this for the averaged residuals of 10 European observatories: The data are very well modeled, including the characteristic double-peak aspect of the cycles, by a fundamental 11-yr wave and its third harmonic (Ducruix et al 1980, Figure 8). At this point, the comment by McLeod (1982) on aliasing due to the use of annual mean values can again be recalled, although McLeod believes that this does not affect the conclusions of Courtillot & Le Mouël (1976a,b). Courtillot & Le Mouël suggested a ring-current P_1 geometry for the 11-yr variation, but failure to isolate the 1969.5 secular variation impulse led them to nonphysical estimates of the Z/X ratio (see Section 6.1). The suggestion of a P_1^0 geometry was based on Z vs X phase relations

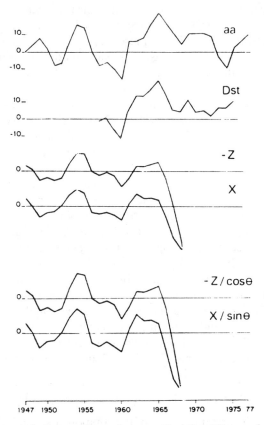

Figure 14   From top to bottom, *aa*, *Dst*, and mean residual X and Z curves for 10 European observatories with and without correction for geomagnetic colatitude (after Ducruix et al 1980).

in both hemispheres and on the amplitude of variations in the H_a^n time series: This amplitude was found to be as large as that in the corresponding H_a^{24} series, leading to the conclusion that Sq does not enter appreciably in either series but rather Dst and possibly Sd.

Following an approach that had become standard (see Section 5.1), Alldredge (1982) analyzed first differences of X annual means from the Rude Skov observatory in Europe from 1900 to 1980. He demonstrated a clear correlation with the sunspot number R and showed that the remaining secular variation is well accounted for by a grossly linear trend (his Figure 1; see Sections 5.1 and 5.2). Alldredge noted that the phase relation between the two curves (X residual and R) is what would be expected if a sudden increase in sunspot number resulted in a sudden increase of the westward-flowing ring current. Alldredge found irregularities in the Z residuals, which he attributed to greater error in Z measurement and problems in baseline control prior to the introduction of proton precession magnetometers; he therefore did not present results for this component.

Gavoret et al (1986) followed a forward modeling approach in their attempt to separate external and internal effects. They reasoned that most of the long-term external variations $\mathbf{B}$ should correlate with either the aa or the R index (Section 4.1). Indeed, the part of $\mathbf{B}$ related to the ring current should be monitored by Dst, whose variations have been shown to be nearly identical to those of aa (with a scale factor of -2; see Figure 9a). Parts related to the field-aligned currents and auroral jets should also be well described by aa. On the other hand, $\mathbf{B}$ variations of ionospheric origin are directly related to the daily variation S and hence to the Wolf number R. Gavoret et al first adjust a set of coefficients b_i such that $B_i - aa \cdot b_i$ ($i = 1$–3 corresponding to the components X, Y, and Z) be as decorrelated as possible from aa. The residual daily variations in monthly means (calculated from the night level) are then scaled to R. For a check of consistency, this technique was applied to monthly mean values of components from four European observatories (Niemegk, Eskdalemuir, Abringer-Hartland, and Chambon-la-Forêt), spanning the period from 1927 to 1982. The data were first filtered with two simple moving-average-centered filters (12 and 24 months wide), which eliminated most periods shorter than 30 months (The impulse response and transfer function are shown by Gavoret et al in their Figures 2 and 3.) They then estimated the external contributions to be

$$X_e = -aa - 0.03\ R, \qquad Y_e = 0.4\ aa, \qquad Z_e = aa + 0.01\ R. \qquad (1)$$

In Figure 15, the first differences of the components $-dX_e/dt$, dY_e/dt, and dZ_e/dt are compared with the corresponding differences of the total field components dX/dt, dY/dt, and dZ/dt. The external signal is most promi-

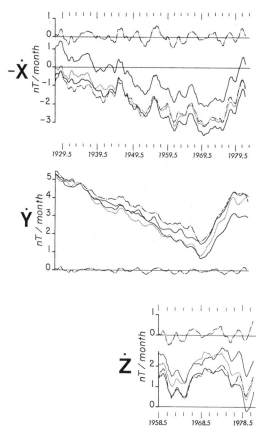

Figure 15 First-order time derivative (filtered monthly mean values) of the $-X$-, Y-, and Z-components at four European observatories (Chambon-la-Forêt, Hartland, Eskdalemuir, Niemegk) compared with the coupled aa and R regressions of Equation (1) (after Gavoret et al 1986).

nent in the X- and Z-components (Z data are taken only from 1957, the time of introduction of proton and optical pumping magnetometers, onward), where it reaches peak-to-trough amplitudes in excess of 1 nT yr^{-1}; it is far smaller, yet quite clear, in Y with a peak-to-trough amplitude of less than 0.5 nT yr^{-1}. By construction, the synthetic signal incorporates the solar-cycle effects seen in external indices in Figure 9. The striking correlation between the synthesized external signal $d\mathbf{B}_e/dt$ and much of the variations in $d\mathbf{B}/dt$ is ample evidence that the external effect is well modeled. It is clear from Figure 15 that this external signal is only a small fraction of the total (mostly internal) secular variation. This figure is discussed further in Section 5.1. Alldredge (1984) shows examples of modeling similar to those of Gavoret et al (1986), and he successfully isolates external effects due to auroral disturbances from internal effects in dY/dt data from Eskdalemuir and Sitka (Alaska) for the period 1962–76 (Figure 19).

Separation of external and internal contributions to the solar-cycle effect by means of spherical harmonic analysis has been attempted by Yukutake & Cain (1979), using annual mean data from 34 observatories for the period 1940–73, and by Alldredge et al (1979) for a larger number of observatories (54), but restricting the analysis to the 12 yr centered on 1958.5, i.e. to the time of the strongest solar cycle in the last century (Figure 9c). Yukutake & Cain's (1979) spherical harmonic analysis was limited to degree 4 for X and Y, and to only degree 3 for Z (always for the same reasons of greater noise level). They present only degree 1 terms (Figure 16). After high-pass filtering with a cutoff near 10 yr, they find no significant variations in the internal dipole g_1^0 (they analyze the total field data; see Section 5), but they do find a very significant (20-nT peak-to-trough)

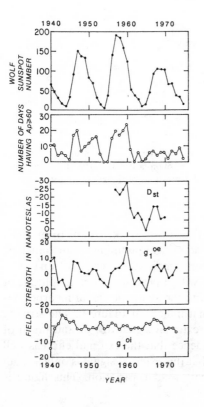

Figure 16 Comparison of variations in the axial dipole field with those of other geomagnetic indices. From top down are sunspot number R, number of days when index Ap exceeded 60, Dst, and the axial dipole components of external and internal origin (high-pass filtered) (after Yukutake & Cain 1979).

external term, which correlates well with external indices and further demonstrates the global nature of the phenomenon. It is also, of course, remarkably well correlated with the mean residual X from European observatories (Figure 14), vindicating the value of localized but well-controlled analyses. Using all degree 1 spherical harmonic coefficients, Yukutake & Cain further demonstrated that the uniform external field was aligned parallel to the geocentric dipole, in agreement with a ring-current origin in which the ring current is located in the geomagnetic equator. Yukutake & Cain also undertook the analysis of a subset of 21 observatories for the much longer period from 1900 to 1973. This analysis confirmed their more detailed results over a shorter interval of time. Using Fourier analysis, they estimated the amplitude of the 11-yr line to be 5.5 nT. It is not clear whether the first harmonic and a peak that they found at 7.1 yr are actually significant; this would require a more careful analysis of the prefiltering that they use, along the lines of McLeod (1985). The analysis of Alldredge et al (1979) on unfiltered data also revealed a very clear global solar-cycle effect in external g_1^0. All of the previous analyses emphasize the double-peaked nature of the solar cycle in geomagnetic elements, a fact that is related to the presence of higher harmonics, particularly the second one, already visible in the external indices (Courtillot et al 1977, Ducruix et al 1980, Yukutake & Cain 1979).

4.6   Longer External Periods—In Particular, 22 and 60 Yr

In a study that we have already discussed, Currie (1973a) further finds evidence for the ~22-yr double solar cycle and its first nine harmonics and for a ~60-yr line (see also Currie 1973b). Although a ~22-yr peak is present in analyses of some external indices (for aa, see, for example, Courtillot et al 1977, Gavoret et al 1986), further work is required to show that ~22-yr lines in geomagnetic components can actually be resolved. Courtillot & Le Mouël (1976a) pointed out that the 7.1- and 4.35-yr peaks interpreted by Currie (1973a) as harmonics of the double solar cycle could be interpreted as parts of the broad 11-yr peak and its harmonics, and that the 22-yr peak was not significantly above noise level.

Rotanova et al (1985) have recently restated the case for 60- and 30-yr lines. The 60-yr variation has particular significance in the framework of Braginsky's (1970) theoretical analysis of oscillations of a hydrodynamic core dynamo (see also Braginsky 1972, 1984). Courtillot & Le Mouël (1976a) note that the average length of time series used by Currie (1973a,b) is 61 yr (with a standard deviation of 18 yr) and that spurious peaks are often found at a period close to the length of the data sample when using maximum entropy. More recently, Gavoret et al (1986) attribute the presence of a 60-yr line to the occurrence of internal secular variation features, which have repeated with an interval of ~ 60 yr in the last 100 yr (see Section 5.2). In an analysis of 400 yr of data collected near London and Paris, Malin (1984) finds no evidence to support a claim for a 60-yr cycle over this long period of time (see Section 5.3).

Although this section is concerned only with external signals, it is appropriate to mention at this point the work of Alldredge (1977a). Alldredge analyzed Z and H at a number of observatories with a classical Fourier (periodogram) technique; he then synthesized curves from Fourier components in the 13–30 yr period range. Alldredge found variations of tens of nanoteslas in amplitude, not very well correlated from one observatory to another, and concluded that these were of internal (core) origin. McLeod (1982) made a careful analysis of the time-series analysis techniques used by Alldredge (1976, 1977a) and also by Currie (1973a) and Courtillot & Le Mouël (1976a,b). He pointed out that the trends removed by Alldredge and Courtillot & Le Mouël amounted to time-varying linear filters, and that the processing used by Alldredge might well be inadequate to prevent very long period core signals from "masquerading" as signals in the period range he investigated. McLeod concluded that part of the synthesized 13–30 yr variations, perhaps the major portion, might stem from longer-period core signals and that the remainder could well be of external origin. Finally, McLeod (1982) observed that the background spectra in the period range from 2 months to 20 yr reported by Courtillot & Le Mouël (1976a,b) were approximately of $1/f^2$ type (i.e. corresponding to the random process known as the "drunkard's walk") and agreed numerically with the spectra obtained by Campbell (1976) for the vastly different period range from 4 hr to 5 min.

5.   VARIATIONS WITH AN INTERNAL ORIGIN (SECULAR VARIATION)

5.1   The 1969.5 Secular Variation Impulse (Jerk)

5.1.1   EARLY RESULTS   Weber & Roberts (1951) and Walker & O'Dea (1952) assumed that secular variation was roughly constant over short intervals of about 5 yr for the practical purpose of establishing a simple short-term prediction of field components. Discontinuities in first derivatives of these components resulted from this modeling, and they were called secular variation impulses. Alldredge (1975) showed that these secular variation impulses were correlated to the sunspot cycle and were an artifact of fitting smooth (quasi-periodic) external variations to a bilinear curve (see also Courtillot et al 1978a, 1979, Ducruix et al 1980, Malin et al 1983;

we define a bilinear curve as a succession of two linear segments with continuous value and discontinuous slope at some point, by analogy to the biquadratic curve defined by Backus 1983). We saw in Section 4.5 that as a part of more recent studies of external variations in long series of annual means from world observatories, it was discovered that an entirely different and very unusual phenomenon had taken place in the late 1960s (Courtillot & Le Mouël 1976a,b, Courtillot et al 1978a). The present review is to a large extent prompted by the desire to assess the present status of this discovery. Courtillot & Le Mouël (1976a,b) had noted that whereas long-term time variations in geomagnetic components could be remarkably well modeled by simple quadratics (parabolas) over significant durations of several decades (their Figures 5 and 6, and Section 4.5), such a fit apparently became a poor one in the late 1960s. Courtillot et al (1978a) found, for example, that for the Y-component in Europe for the period from the late 1940s to 1977, a biquadratic model was necessary, with a discontinuous change in curvature (not slope) in 1970. Alldredge (1979) suggested that this might be an artifact due to the use of a given quadratic for extrapolation beyond its range of definition. In response, Courtillot et al (1979) proposed that a display of secular variation (i.e. the first time derivative) gave the most convincing evidence for the phenomenon. This figure was reproduced by Ducruix et al (1980) and Achache et al (1980) and is shown here as Figure 17. Most researchers have since come to recognize that this biquadratic fit (one in which the first derivative is bilinear or V-shaped) is spectacular; some still discuss the extent to which this representation applies to other components than Y and to the rest of the world outside Europe.

5.1.2 TERMINOLOGY   A note on terminology seems warranted at this point. Secular variation is defined as the rate of change or first time derivative of the internal part of the geomagnetic field, related to core sources and excluding induced effects due to external sources. Secular acceleration is the field's second time derivative. Malin et al (1983) refer to the rate of change of acceleration, i.e. the third time derivative, as the *jerk* (Schott 1978). The secular variation model suggested by Courtillot et al (1978a, 1979) and Malin et al (1983) is one in which the field components are described by two parabolas with continuous value and first derivative, but with discontinuous (Heaviside-distributed or steplike) acceleration and an impulsive (Dirac distribution) jerk in 1969.5 (Figure 18). This same feature was first called a *saut de variation séculaire* (translated as secular variation jump or impulse) by Courtillot et al (1978a). Following Malin et al (1983), it should be called an *impulsive jerk*, not a very appealing name. Malin & Hodder (1982) restricted the use of the term *jerk* to the

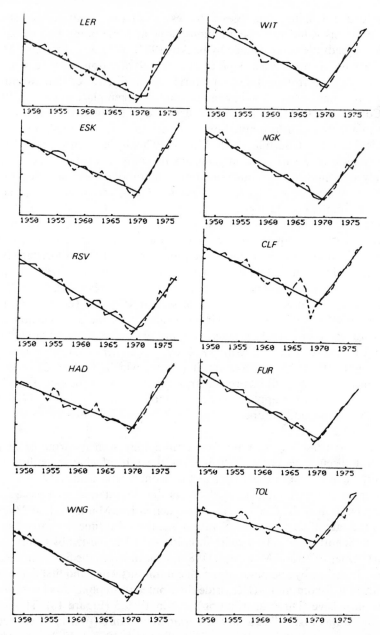

Figure 17   Secular variation (first differences of annual means) of Y at 10 European observatories (intervals of 10 nT yr^{-1}). Solid lines are least-squares bilinear fits to the data, with a jerk in 1969 (after Courtillot et al 1979).

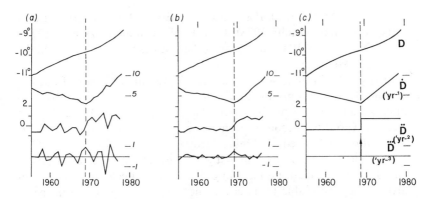

Figure 18 Declination *D* at Eskdalemuir, with its first, second, and third derivatives, obtained by differencing. (*a*) Annual mean values; (*b*) 3-yr running means (after Malin et al 1983); (*c*) theoretical (noiseless) shape of a jerk, or third-order impulse.

step change in acceleration itself. Backus called it an impulse of order 3 (a *n*th order impulse implying a Dirac *n*th derivative), and Courtillot & Le Mouël (1984) termed it an impulsive acceleration in secular variation. The term *jerk* has become the most popular and widely used, and we follow this usage here.

5.1.3 DATE AND WORLDWIDE EXTENT OF THE 1969.5 JERK This subsection updates the review by Courtillot & Le Mouël (1984), highlighting in particular recent contributions of McLeod (1985), Gavoret et al (1986), and Backus et al (1987). We also review the controversy based on Alldredge's (1979, 1984, 1985a,b) opposition to the work of the above authors; Alldredge maintains that the 1969.5 event is partly an artifact and in any case not a feature of internal geomagnetic changes.

The date of the jerk was established as 1969.5 (with an uncertainty of only a few months) by Ducruix et al (1980), Le Mouël et al (1982), and several authors since, including Backus (1983), McLeod (1985), and Gavoret et al (1986). The jerk is particularly spectacular in European *Y* data (Figure 17) because of the small contribution of external currents due to their meridional geometry (Section 4.5). The sunspot effect makes the jerk harder to identify on the *X*- and *Z*-components, all the more so because the jerk amplitudes on these components are small in the Western European observatories. Still, the jerk was identified in at least 38 observatories from the Northern Hemisphere by Courtillot et al (1978a), and careful analysis of data from an increasing number of observatories suggested that it was indeed worldwide and was in particular recorded in the Southern Hemisphere (Chau et al 1981, Le Mouël et al 1982, Malin &

Hodder 1982, Malin et al 1983, Gubbins 1984, McLeod 1985). Local reports from Japan (Fujita 1973), Ulan Bator (Chimiddorzh et al 1977), Finland (Nevanlinna 1985), and many other locales have strengthened the case for the 1969.5 jerk.

Alldredge (1984) and Nevanlinna (1985) present the cases of auroral observatories. When only the quietest days are used and a suitable correction is made for measurable external auroral phenomena, the V-shape of secular variation becomes even more pronounced [Figure 19; this interpretation of Alldredge's Figure 11 by Courtillot & Le Mouël (1985) and McLeod (1985) is not shared by Alldredge]. Gavoret et al (1986) resorted to more elaborate modeling of external effects and time-series analysis. They were able to ascertain the date of the jerk in Europe to within a few months of 1969.5 and to isolate the jerk very clearly in the apparently noisier Z record. They reasoned that secular variation might

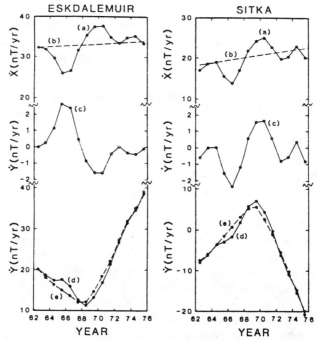

Figure 19 Curves (a) are secular variations (first differences) of X, and (b) their straight-line fits at Eskdalemuir and Sitka. Curves (c) are estimates of the polar disturbances computed from differences between curves (a) and (b) multiplied by appropriate ratios. Curves (d) are observed secular variations of Y and curves (e) are (d) minus (c), which are estimates of dY/dt from the core (after Alldredge 1984).

result from the superposition of a simple worldwide jerk signal and a more regional and more smoothly (time-)varying secular variation, and that in areas close to zero lines of the Z jerk component the spatial gradient of the jerk might be large, possibly larger than the unrelated regional secular variation with smoother time variation. This was confirmed by computing the difference between two observatories (Eskdalemuir and Niemegk; see Figure 20). The sharpness of the jerk and the similarity of the Z difference curves with Y (Figures 15, 17) are quite striking.

Gavoret et al (1986) also calculated the Y acceleration after smoothing. Although some external effects and smoother nonjerk internal signals may remain, the acceleration is consistent with a step function in 1969.5. Smoothing due to various filtering steps is unavoidable and is discussed by McLeod (1985) and Gavoret et al (1986). All European observatories are affected by the impulse simultaneously (actually in less than 6 months, which can to a large degree be due to other remaining errors and in any case implies a phase velocity of the jerk in excess of 4000 km yr^{-1}). Figures 15 and 20 also reveal a more recent jerk in 1979, which is discussed in Section 5.2.1.

Chau et al (1981) and Le Mouël et al (1982) noted that in observations from 80 out of 130 observatories that they studied, the best determined V-shaped curves of secular variation seemed affine (their Figure 11). This observation would imply a natural separation between the space and time parts of secular variation, which could then be written as the product of a function of time only (the jerk) with a purely spatial one (see also Malin et al 1983). Le Mouël et al (1982) and Malin et al (1983), elaborating on the first, rough world charts of Courtillot et al (1978a), demonstrated the worldwide extent of the jerk and the simple features of its three components (Figure 21).

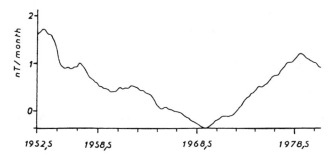

Figure 20 Filtered first-order derivative of the difference between Z-components at Niemegk and Eskdalemuir (after Gavoret et al 1986).

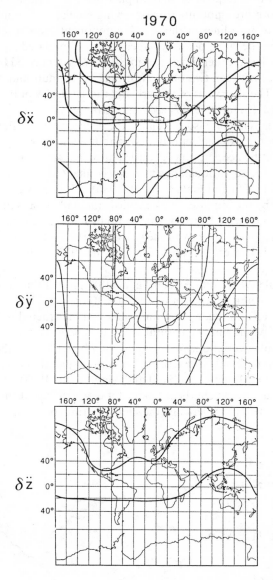

Figure 21   Zero-lines of the *X*-, *Y*-, and *Z*-components of the 1969.5 jerk (after Ducruix et al 1983).

5.1.4 SPHERICAL HARMONIC ANALYSIS   The X- and Z-components appear to be dominated by a zonal geometry, whereas Y is more sectorial (Figure 21). The flat plateau and sharp gradients close to the zero isolines are reminiscent of characteristic functions and might require rather high-degree spherical harmonic expansions for a truly accurate description. Various attempts at separating the external and internal parts of the jerk field have been made. Malin & Hodder (1982) used data from 83 observatories over the period 1961–78 for the first spherical harmonic analysis up to degree 4. They found the mean squared value of the internal part to exceed the external one by a factor of about 3. Gubbins (1984) used data from 160 observatories from 1959–78 (or 1980 when available) for a constrained spherical harmonic analysis up to degree 10, in which the field was forced to obey the frozen-flux approximation (see Section 6.2). McLeod (1985) performed an analysis of Langel et al's (1982) GSFC9/80 field model and obtained jerk spherical harmonic coefficients up to degree 4. McLeod also reviewed previous analyses. The various sets of spherical harmonic coefficients up to degree 4 are shown in Table 1. McLeod argues

Table 1   Spherical harmonic coefficients for the 1969.5 jerk and amplitude for each degree $(S)^{1/2}$, all in nT yr^{-1}, as estimated by Malin & Hodder (1982, MH82), Gubbins (1984, G84), and McLeod (1985, M85)[a]

		MH82			G84			M85		
n	m	g_n^m	h_n^m	$(S_n)^{1/2}$	g_n^m	h_n^m	$(S_n)^{1/2}$	g_n^m	h_n^m	$(S_n)^{1/2}$
1	0	−2.44			−0.9			−0.83		
1	1	*	1.04		−0.0	1.1		−0.43	1.75	
1				3.75			2.0			2.80
2	0	1.03			0.6			0.35		
2	1	0.59	−2.40		0.0	−2.4		0.24	−2.63	
2	2	−0.68	*		0.0	−0.7		0.18	−0.73	
2				4.78			4.45			4.79
3	0	1.04			0.3			1.04		
3	1	0.83	−2.23		0.7	−1.2		1.04	−1.68	
3	2	1.04	−0.43		0.8	−0.2		0.49	−0.15	
3	3	*	*		0.7	−0.3		0.77	−0.63	
3				5.66			3.62			4.99
4	0	*			−0.1			−0.22		
4	1	0.28	*		0.2	−0.6		−0.01	−0.33	
4	2	−0.77	−0.41		−0.4	−0.0		−0.46	0.08	
4	3	0.26	0.95		0.4	0.3		−0.05	1.00	
4	4	−0.9	*		0.1	−0.5		−0.32	−0.17	
4				3.62			2.32			2.74

[a] Asterisks indicate values considered not to be significant and set to zero.

that the external spherical harmonic coefficient of Malin & Hodder can simply be interpreted as noise due to remaining external currents unrelated to the jerk, so that the jerk may not have any significant external components. He finds his own coefficients to be highly correlated with those of Gubbins (1984) but larger, which he attributes tentatively to the fact that Gubbins imposed frozen-flux constraints that may not be justified by the accuracy of the models. In all cases, Table 1 shows that the jerk is dominated by order 1 coefficients, essentially h_2^1 and h_3^1. Calculation of the energy or mean squared value of terms of a given degree $[S_n = (n+1)^2 \Sigma_m (g_n^{m^2} + h_n^{m^2})$; e.g. Lowes 1966] clearly shows that the $n = 2$ and 3 terms dominate $n = 1$ and 4 terms by a factor of about 4. Discrepancies between the various analyses include the large g_1^0 term of Malin & Hodder not found by others and the small g_2^0 term of McLeod.

McLeod finds that the times of the minima for individual S_n terms are between 1968.5 and 1969.4 for the $n = 2$–4 terms, but he obtains a later date of 1973.3 for the dipole term. This may actually be due to a remaining unrelated external contamination. Langel & Estes (1985a) attempted to separate external and internal parts of secular variation using quiet-time MAGSAT data (11/79–4/80); MAGSAT data alone were found to be unable to determine secular variation beyond degree 4, and the data were supplemented by those from 91 magnetic observatories over the period 1978–82 to obtain a degree 10 secular variation model at 1980.0 (termed GSFC12/83). This analysis included an external term of degree 1 that was linearly related to Dst (Langel & Estes 1985b), in the form

$$\mathbf{q}_1(t) = \mathbf{a} + Dst(t) \cdot \mathbf{h}. \qquad (2)$$

As expected for a ring-current origin for most of these external variations, both $\mathbf{a}$ and $\mathbf{b}$ are found to be essentially antiparallel to the internal dipole (the angle with it being less than 5° for $\mathbf{b}$ and 20° for $\mathbf{a}$). A similar result has been obtained more recently by Backus et al (1987). Note that the secular variation model of Langel & Estes (1985a) fits observatory data well for the period 1978–82 for which it is determined but not before that. Some of the secular variation spherical harmonic coefficients are shown by Langel & Estes (Figure 22). Not enough coefficients are listed for a full comparison with Table 1, but the jerk is obvious in h_2^1 and seems to appear in g_2^0. A remaining external signal is likely responsible for the wavelike appearance of g_1^0 variations [see the modeling of Gavoret et al (1986) and Section 4.5].

5.1.5   DISCUSSION   Alldredge (1979, 1981, 1984, 1985a,b) has repeatedly criticized most of the results stated above on the basis that they were artifacts related to the methods used; he considers that the jerk is forced

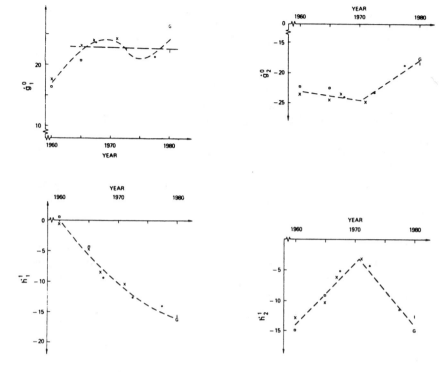

Figure 22   Spherical harmonic coefficients from selected field models for the period 1960–80, in nanoteslas per year [after Langel & Estes (1985b), who list the various models].

on the data. Alldredge contends that the 1970 phenomenon is not as sharp as suggested, that it is contaminated by external signals whenever it seems sharp, and that the 1970 effects are in any case mostly external. Alldredge (1985a) finds no jerk in the GSFC9/80 model of Langel et al (1982) or in the DGRF 65, 70, and 75 models (for the 1985 models, see Quinn et al 1986). He states that "discontinuities are not determined by the analysis, but constitute an a priori assumption" (see, however, Langel & Estes 1985a). Replies and further comments have been offered by Courtillot & Le Mouël (1979, 1984, 1985), Courtillot et al (1977), Malin et al (1983), and McLeod (1985). What seems to us to be a key observation is the order of derivation that is best suited to display the jerk signal (Courtillot et al 1978a, 1979, Malin et al 1983; Figure 18). Because of the strength of secular variation and the long time intervals considered, the original components hide the jerk, which appears as an inflection point. Because of remaining sunspot-related external variations, mostly in X and Z, the second derivative is very noisy: The higher the derivative order and the harmonic number of the basic 11-yr cycle, the higher the corresponding signal that swamps the jerk. [This is particularly true of the 11/3-yr harmonic (see Section 4.5).] This is very clear in Alldredge's (1984) Figure 3 [see McLeod (1985) for a detailed discussion]. It so happens that the jerk is best revealed in first-order derivatives, where the parabolic effect is attenuated and yet external signals, however obvious, do not swamp the internal signal. This is why most useful figures display the first derivative, i.e. secular variation itself (Figures 15, 17–20). McLeod (1985) addresses a very thorough criticism of all earlier work based on time-series analysis. He finds that Alldredge's plots with a duration of ~ 14 yr are often too short to allow one to isolate the jerk from external contributions; 20 of the 49 observatories for which Alldredge plots data do provide clear evidence of the jerk, i.e. dY/dt plots are well approximated by a V-shape. McLeod finds no connection between these plots and a sinusoidal curve of ~ 16-yr period advocated by Alldredge. He concludes that "none of Alldredge's conclusions . . . are required or even supported by his data or arguments." McLeod confirms that the jerk is worldwide, has no resolvable external component, and occurs in 1969.5 with a time scale of 1–2 yr.

Backus & Hough (1985) and Backus et al (1987) have compared the fit of secular variation data with other models more continuous than the biquadratic one but with the same number of adjustable parameters. Backus & Hough (1985) fit the data with a model in which the internal (core) jerk is either biquadratic or quintic and superimposed on an external sunspot signal (based on 2-yr moving averages of monthly sunspot counts R) and additive white noise. As expected, they find a strong sunspot contribution to X and Z but not to Y; at the same time, the biquadratic model is found to fit Y better than a quintic one (but not X and Z). Jerk amplitudes for X and Z are therefore strongly dependent on whether a sunspot correction is made and are less reliable than for Y. Rather than being like white noise, the residuals that they plot (their Figure 5) appear to be still well correlated with external indices, for instance aa (Section 4.1). Gavoret et al (1986) show that both aa and R must be used in a regression model of external contributions. In a broader analysis, Backus et al (1987) have analyzed 3300 satellite values of F and 7000 observatory annual mean values of X, Y, and Z from 1960–78: this is ~ 9.5 yr before and ~ 8.5 yr after the jerk (i.e. less than a full solar cycle in both cases). They find that a biquadratic model is not a statistically better fit to the data than a cubic or a quintic model with the same number of free parameters. In their model, Backus et al use a degree 1 model of the external field, this time related to Dst. Again, we point out the need for a joint R/aa regression: From Equation (1) in Section 4.5 we see that the contributions of aa and R to X and Z can be in the ratio 2/1 (for peak-to-peak amplitudes of 10 nT month^{-1} for daa/dt and 10 month^{-1} for dR/dt). We also note that Dst might lead to serious problems if used in the period 1972–78 (Figure 9a). As expected, the residuals in Backus et al's (1987) Figure 3 are this time more correlated with R. We therefore suggest that there may still be some significant correlated external noise in the analysis. The best fits are obtained for the Y-component. If one attempts to extend them back in time, they diverge very significantly depending on which jerk model is used. Only the biquadratic fit applies to data prior to 1960 (see Figures 20, 23, and 25, and Section 5.2.2), as recognized by Backus et al. In conclusion, the biquadratic model is both the simplest and apparently the

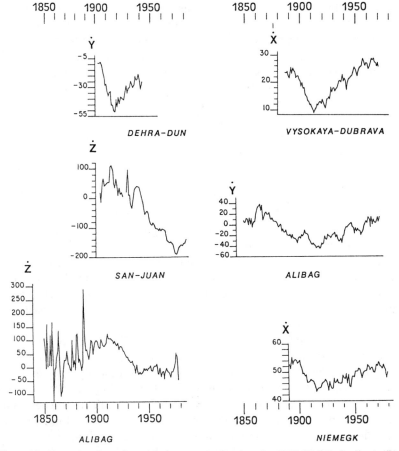

Figure 23 Example of secular variation curves showing the 1912–13 jerk (ordinates in nanoteslas per year; after Ducruix et al 1983).

most successful in describing many features of the jerk field and in separating it from remaining external contributions.

5.2   Secular Variation in This Century

5.2.1   OTHER JERKS   As opposed to the earlier concept of secular variation impulses of Walker & O'Dea (1952), the 1969.5 jerk described in the previous section has many unique features and is certainly not a frequent occurrence in recent decades. Kerridge & Barraclough (1985) searched for other jerks in the period 1931.5–71.5 and concluded that the 1969.5 event was both well resolved and unique in the time span they considered. Courtillot et al (1978a) suggested that a previous jerk occurred around 1910. They sketched its worldwide distribution and found it to be quite similar to that of the 1969.5 event. This was carried further by Ducruix et al (1983), who used first differences of annual mean values of components from 50 observatories, 40 of them in the Northern Hemisphere, during the period 1850–1980. They proposed that a worldwide jerk occurred in 1912–13, with no other such occurrence between 1913 and 1969 (Figure 23). The zero isolines of this jerk (Figure 24) are remarkably similar to those of the 1969.5 event, particularly for Y, the best-determined component (cf. Figures 21 and 23). Gire et al (1984a) confirmed this and established that the sign of the 1912–13 jerk was opposite that in 1969.5. The characteristics of the jerks are clear on the plots of Gavoret et al (1986) for Chambon-la-Forêt (Figure 25). The Y-component there is disturbed by a more regional secular variation event around 1925 (see Section 5.2.2). A possible jerk in 1900 would deserve further analysis, but this is hindered by the lack of early records: Only 19 observatories were in operation in 1900, 12 of them in European countries (Figures 1, 2), where older events might still be uncovered (see Section 5.3.1).

Gavoret et al also identified a more recent jerk in 1979 (Figures 15, 20, 25). The event is quite apparent in the smoothed secular acceleration curve $d^2 Y/dt^2$ (Gavoret et al 1986, their Figure 18). It deserves further analysis, but its worldwide significance is confirmed by other reports, in particular Langel et al (1986).

Langel et al (1986) have analyzed data from ~ 50 to ~ 150 observatories over the period 1899.5–1982.5. Based on the value of the condition number (Figure 2b), they have performed spherical harmonic analyses up to degree $n = 5$ from 1903 to 1930, to $n = 6$ from 1931 to 1958 (and 1981–82), and to $n = 8$ from 1958 to 1980. They chose to fit their calculated coefficients by cubic splines for various reasons. However, cubic splines produce significant smoothing and cannot resolve jerks, unless additional knots are introduced in the analysis (Langel et al 1986). Figure 26 shows their selection of spherical harmonic coefficients, where points are from indi-

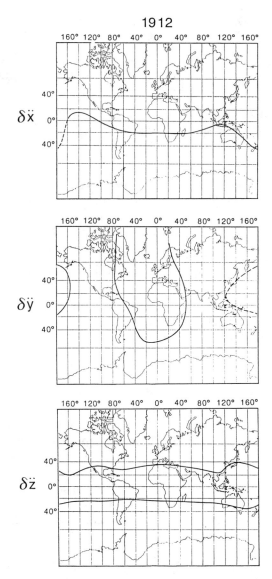

Figure 24   Zero lines of the *X*-, *Y*-, and *Z*-components of the 1912–13 jerk (after Ducruix et al 1983).

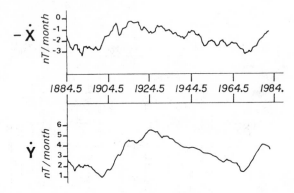

Figure 25 Filtered first-order derivatives of the $-X$- and Y-components [corrected for external effects as described by Equation (1)] at Chambon-la-Forêt 1884–1985 (after Gavoret et al 1986).

vidual data-based spherical harmonic models (one per year); the solid lines are the spline representations. The 1969.5 and 1979 jerks are conspicuous, particularly in the nonsmoothed data. As previously found for the 1969.5 event alone, coefficients h_2^1 and h_3^1 are the most prominent, but the jerks appear possibly also in g_1^0, g_2^0, h_2^2, and g_3^0. The 1912–13 jerk is less obvious (with stronger noise), but it may be seen in g_2^0, g_2^1, h_2^1, g_2^2, g_3^3, and h_3^3.

5.2.2 THE MULTILINEAR SECULAR VARIATION MODEL   A number of the studies that have uncovered jerks or contributed to their description did so as part of more extensive studies of secular variation in this century (Gire et al 1984a, Kerridge & Barraclough 1985, Gavoret et al 1986, Langel et al 1986). Much of the previous discussion and related figures show that secular variation can be modeled to a first approximation by multilinear curves, i.e. a much simpler representation than previously recognized. (This is still, however, the subject of ongoing debate.) Although external signals are obvious, and there may well remain smooth regional internal contributions (see below), one can model many components of secular variation in many observatories as a sequence of continuous linear segments, with slope discontinuities only at ~ 1900 (?), 1912–13, 1969.5, and 1979 (e.g. Figure 3 in Courtillot & Le Mouël 1984). This has been done by Ducruix et al (1983) and Gire et al (1984a) and is described in more detail by Gire (1985, pp. 57–73). Gire represents secular variation

Figure 26 Secular variation of selected spherical harmonic coefficients versus time (1900–84). Points are from models, and solid lines from a spline representation (units are nanoteslas per year; after Langel et al 1986).

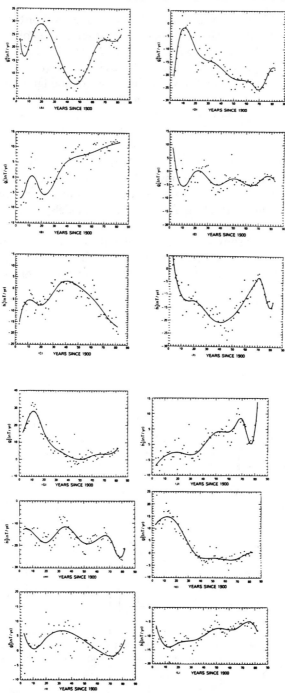

432   COURTILLOT & LE MOUEL

for the period 1900–80 by a series of 3 linear segments. For the 23 observatories in the Northern Hemisphere that she studied, the normalized mean squared residuals are, respectively, 0.24 for X, 0.08 for Y, and 0.31 for Z. Many of the larger values for X and Z are clearly due to external signals, Z being larger because of poorer baseline control and larger instrumental noise. The Y fit is remarkable, and overall the multilinear secular variation model accounts for more than 75% of the field power between 1913 and 1980. The quality of the fit should not be surprising after a simple inspection of Figures 15, 17, 20, and 25.

Gire et al (1984a,b) and Gire (1985) tried to relate this simple global behavior to simple global geophysical parameters such as the energy (or norm) of secular variation S_n (see Section 5.1.4) or the rate of westward drift (see Section 5.2.4). Using Hodder's (1981) models, Gire computed the spherical harmonic spectra of secular variation for six periods from 1913 to 1980. Figure 27 shows the dominance of the $n = 2$ terms, already noted for the spherical harmonic representation of the 1969.5 jerk; the overall decrease in energy (both global and for most individual degrees) from 1913 to 1930; and the broad minimum until 1960 and the notable increase following the 1969.5 jerk. In 1980 secular variation energy is about twice

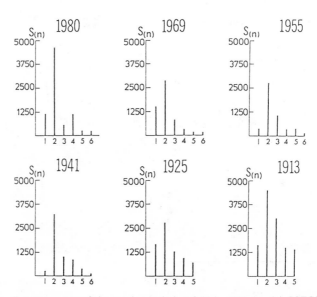

Figure 27   Energy spectra of the secular variation field in 1980 (model GSFC80) and at earlier epochs [Hodder's (1981) models], illustrating the dominance of degree 2 terms and the general time evolution (i.e. a decay to 1969, followed by a sharp increase) (after Gire 1985).

as large as 10 years before. (Note that uncertainties in energy estimates are certainly quite large.) The secular variation maps (Figure 28) confirm and illustrate these observations.

Based on the Chambon-la-Forêt data, which are argued to be reasonably representative of global secular variation (except for the 1925 regional feature in Y), Gavoret et al (1986) emphasize that the behavior of secular variation between 1970 and 1980 is very reminiscent of that between 1900 and 1910. A pair of two opposite jerks would form a "percussion" (1900/1912 and 1969.5/1979), or a period of intense secular variation. The separation of these two intense periods by about 60 yr is believed to be responsible for previous identifications of an external or internal ~ 60-yr spectral line (see Section 4.6).

Gire (1985) studied the geographical and temporal distribution of secular variation residuals to the multilinear fit after correction for some external effects. In an analysis limited to data from 24 observatories in the Northern Hemisphere, she found that the residuals had a regional (as opposed to global) distribution, with amplitudes of some 10 nT yr^{-1} and time constants of ~ 10–20 yr: Gire referred to these residuals as secular variation "bumps." The residuals are again smallest in Y, which might suggest remaining (uncorrected) external effects. The secular variation bumps, such as the 1925 one in Europe, are likely related to motions of local character in the external parts of the core, as opposed to global-scale motions with rather simple geometry and time dependence, which are thought to produce the jerks and linear segments of secular variation (see Section 6.2).

Langel et al (1986) argue that the multilinear secular variation model is an oversimplification and is not adequate (see also Alldredge 1983a). When due account is taken of the smoothing character of their cubic spline representation, we believe that many of their spherical harmonic coefficients can actually be reasonably well fit with a multilinear model. The most prominent coefficients in secular variation are $n = 2$ (then $n = 1$ and 3) and for the 1969.5 jerk $n = 2$ (and 3). In Figure 26, we see that g_2^0, g_2^1, h_2^2, h_3^1, and h_3^3 can be reasonably well fit with a multilinear model. The same is true for only post-1940 values of h_2^1, g_2^2, g_3^0, and g_3^3. The multilinear fit is best for (a) the more recent period starting in the 1940s, when there was a significant increase in global coverage and data quality (Figure 2); (b) the Y-component, which is least affected by still incompletely modeled, mostly meridional external current systems; and (c) the European observatories, which provide the oldest, densest network of reliable data. This leads us to suggest that this simple secular variation model may apply better than generally recognized, even on a global scale, and can be used as a reference to extract other, less regular phenomena.

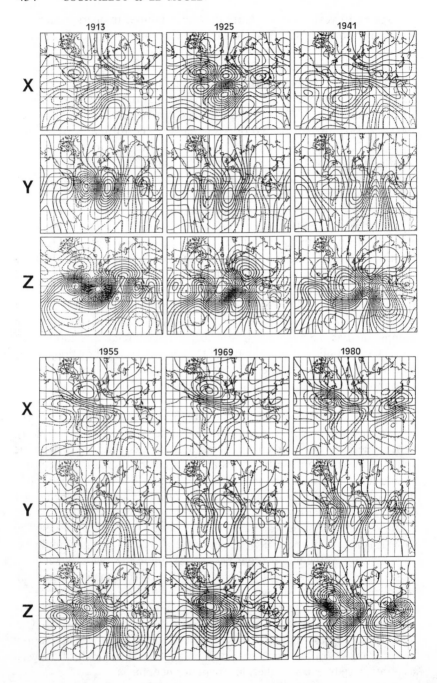

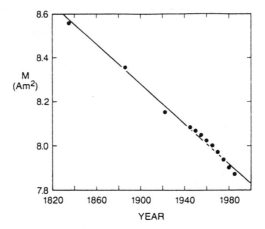

Figure 29   Decay of the dipole moment from 1835 to 1985 (after Fraser-Smith 1987).

5.2.3   SECULAR VARIATION OF THE DIPOLE FIELD   It was found shortly after the first determination by Gauss in 1832 that the dipolar moment was decreasing steadily. This decrease amounts to ∼ 5% per century (e.g. Leaton & Malin 1967, Alldredge 1983b), which has led some to speculate on the dipolar moment vanishing some 2000 yr from now (Figure 29). Departures from this linear trend would require further analysis; some fluctuations are evident on the derivative of g_1^0 (Figure 26). The recent trend of accelerated decrease that started around 1960 appears to be real. In contrast, there has been very little change in the position of the dipole axis (the so-called Gauss poles) since 1850 (Figure 30): Very slight westward drift of under $0.01°$ yr^{-1} may have been preceded by faster motion (see Sections 5.3.2 and 5.4.2). As the dipolar term decays, higher-order terms appear to grow, suggesting some transfer of energy but no significant loss of global energy (e.g. Verosub & Cox 1971).

5.2.4   WESTWARD DRIFT   As noted earlier in Section 2, Halley (1683, 1692) was the first to notice the westward drift of some magnetic features in and around the Atlantic. Bullard et al (1950) may have been the first to undertake a detailed quantitative analysis (for the period 1907–45; see also Vestine & Kahle 1968). They established a distinction between *westward drift of the nondipole field*, which they found to be ∼$0.2°$ yr^{-1}, and *westward drift of the secular variation field*, found to be ∼$0.3°$ yr^{-1}.

Figure 28   Time evolution of secular variation for epochs 1913, 1925, 1941, 1955, 1969 and 1980, and for components X, Y, and Z. Contour intervals are 20 nT yr^{-1} (data from Hodder 1981; after Gire et al 1984a).

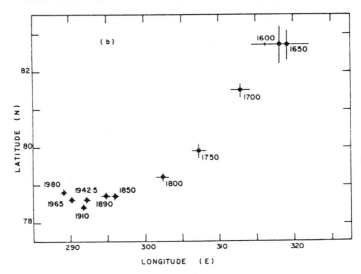

Figure 30   Variations in the dipole axis since 1600 as represented by the change in position of the north magnetic pole (after Merrill & McElhinny 1983; see also Fraser-Smith 1987).

Nagata (1962, 1965), Yukutake (1962, 1967, 1968, 1973), and Yukutake & Tachinaka (1968a,b) proposed an explanation for this discrepancy by separating the field into drifting and nondrifting (standing) components (Figure 31). The lower rate (or even sometimes lack) of westward drift of the nondipole field was attributed to the presence of the standing anomalies, which would be eliminated in secular variation maps. Yukutake & Tachinaka (1986a,b) distinguished between standing features that change in intensity and those that do not. James (1970) relaxed the hypothesis that drift should occur about the geographical axis. More recently, Yukutake (1985) included a third, eastward-drifting component in the analysis. All these analyses confirm that as far as the secular variation field is concerned, drifting features are more important than nondrifting ones, that the drift is dominantly westward, and that it takes place about axes not very far from the present geographic axis.

Actually, it is not a trivial matter to define and measure the rate of westward drift, as it can be based on the total field, on the nondipole field, on the secular variation field, or on some particular features such as the eccentric dipole, specific harmonic coefficients, or particular features averaged, for instance, in latitude bands. Cain & Hendricks (1968) noted that the terms that exhibit the largest drift are, in the following order, $n = 2$, $m = 1$ (terms already found to be dominant both for secular variation and

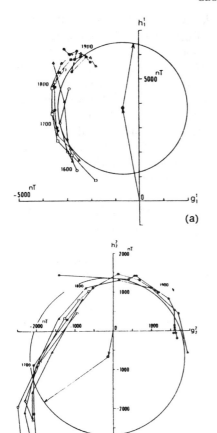

(a)

(b)

Figure 31 Westward drift of spherical harmonic coefficients (g_1^1, h_1^1) and (g_2^2, h_2^2) and the separation between standing (centers of the circles) and drifting (circles) components [after Yukutake (1985), in which the various data sets are listed].

for jerks) and $n = 1$, $m = 1$ (i.e. the equatorial dipole). Cain et al (1985) have calculated the drift of the eccentric dipole, and Fraser-Smith (1987) has recently reviewed variations in the poles of the centered and eccentric dipoles for the period 1600–1985.

Le Mouël et al (1983) and Gire et al (1984b) have recently estimated westward drift by maximizing the contribution of the term $u \cdot d\mathbf{B}/d\Phi$ (where u is the angular velocity of westward drift and Φ is longitude) to the secular variation $d\mathbf{B}/dt$. Westward drift can be evaluated using a single field component or the total intensity, either from existing spherical harmonic models or directly from observatory data (Hodder 1983, Le Mouël et al 1983). The best results were obtained with observatory data: They show a rather clear bilinear slope, with a decrease from 1955 to 1970 and an increase since then, i.e. since the time of the 1969.5 jerk (Figure 32). This would seem to confirm the first suggestion of Courtillot et al (1978a), Ducruix et al (1980), and Chau et al (1981) that the 1969.5 jerk may partly be due to a jerk in the westward drift itself. Gire et al (1984a) found that the ($n = 2, m = 2$) term actually provided the largest contribution to westward drift, followed by ($n = 2, m = 1$). Gire et al (1984b) further noted that estimates of u based on the various components were always in the order $u_Y > u_B > u_Z > u_X$. They suggested that all estimates were smaller than the actual u, with u_Y providing the best estimate (but see Section 6.3.1). In 1980, u was back to its pre-1950 value, twice as large as the value at the time of the minimum in 1970, i.e. at the time of the jerk.

However, Gubbins (1984) and McLeod (1985) conclude from their studies that the jerk spherical harmonic coefficients are not compatible with a simple change in acceleration of the core relative to the mantle about some axis, because the accelerations of g_1^0 and h_1^1 are large and that of g_1^1 is zero. Also, Backus & Hough (1985) point out that if there was a magnetic jerk produced mainly by a jump in the angular acceleration of westward drift in the upper core, then the jerk amplitudes should be proportional to $d\mathbf{B}/d\Phi$ (in Europe, for which their study applies). They find that this works for the amplitudes, but that the signs of the X and Z jumps are wrong. Actually, Le Mouël et al (1985a) find, when computing the flow at the core-mantle boundary (see Section 6.2), that the flow acceleration following the 1969.5 jerk does not solely involve the westward drift term.

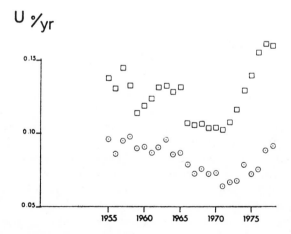

U °/yr

Figure 32 Values of the westward drift rate u computed from raw secular variation data for B (circles) and Y (squares) (after Le Mouël et al 1983).

5.3 *Historical Measurements Since the Sixteenth Century*

5.3.1 EUROPEAN OBSERVATORIES When one turns to a more remote past, the data base becomes extremely scarce. The bulk of early measurements of observatory quality come from Europe. Long series of data from these observatories and from sea voyages have been compiled since the early nineteenth century. We can mention the work of Hansteen (1819), Becquerel (1840), Bauer (1895), Bemmelen (1899), and Fritsche (1900), and, more recently, of Gaibar-Puertas (1953), Yukutake & Tachinaka (1968a,b), Veinberg & Shibaev (1969), and Barraclough (1974) [for further references, see Malin & Bullard (1981) or Bloxham (1986b)]. These compilations are rather unequal in scope, intent, and quality: Some are recompilations of earlier compilations and may be substantially degraded compared with the original data. Recent studies have demonstrated the importance of going back to the original data sources, despite the large amount of work and difficulties involved.

Probably the two longest series come from Paris and London, and secular variation at these observatories is represented in many textbooks [e.g. Le Mouël (1976a, Figure 26.24) or Merrill & McElhinny (1983, Figure 2.7)]. Malin & Bullard (1981) have recently published a very extensive compilation of data from London for the period 1570–1975. A detailed search of original sources has provided 149 observations of declination and 88 of dip (i.e. inclination) dating between the sixteenth century and 1849. Data come from no less than 50 different sites, which can be enclosed in a circle of radius 36 km in the London area. Thus, site corrections must be carefully evaluated. This is done by Malin & Bullard, who present detailed tables together with much delightful and informative historical, technical, and biographical material. The resulting reduced declination data are shown in Figure 33 (lower curve, circles). We see that declination decreased steadily from 11°E in the late sixteenth century to 24°W in the late eighteenth century, was stable from ∼1790 to ∼1830, and then increased with a slope similar (though opposite in sign) to that of the pre-1780 period. It is obvious from Figure 33 that data quality and hence resolution increase with time. Some details of secular variation, such as inflection points or potential jerks, cannot be resolved before ∼1840. Malin & Bullard point out that short-term oscillations near 1840 are probably spurious, and that scatter near 1890 results from including survey observations. There is also an unfortunate data gap of more than 10 yr around 1830.

These problems can to a large extent be circumvented by using data from the Paris observatory (e.g. Rayet, 1876, Malin & Bullard 1981, J. L. Le Mouël et al, in preparation). Actually, observations there are both

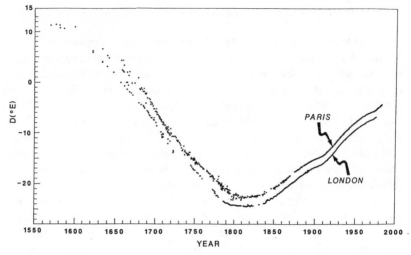

Figure 33 Compilation of declination data from the London (after Malin & Bullard 1981) and Paris (in preparation) areas since the late sixteenth century.

more systematic and more numerous, with far fewer site changes and therefore less need for site corrections. The Observatoire Astronomique Royal was the only site from the late sixteenth century to 1873; after a 10-yr period of measurements in the meteorological observatory at Mont-souris, a magnetic observatory was established in 1883 in St. Maur. It was transferred in 1900 (with appropriate tying) to Val Joyeux, and then to its present location at Chambon-la-Forêt in 1936. The Paris data have been compiled partly independently by E. Bullard & S. R. C. Malin (personal communication, 1980) and our laboratory, but they have not yet been published. The results (Figure 33) are remarkably similar to those of London, with times of gaps or more scattered data that are fortunately different. The major secular variation (slope) changes at about 1795 and 1830 are confirmed. The data are of sufficient quality to allow calculation of first-order differences without being overpowered by noise. This reveals several events of jerklike appearance occurring around 1720, 1735, 1795, and 1830, in addition to the 1912, 1969.5, and 1979 events (J. L. Le Mouël et al, in preparation). Le Mouël et al (1981) and Courtillot & Le Mouël (1984) noted that secular variation seemed to be less intense, i.e. to have fewer sharp changes, prior to 1870 than afterward: A look at better long time series (Figure 33) leads to the conclusion that the quiet period may be limited to 1830–1900. In other terms, there may be no unusually quiet long period in the interval from 1700 to the present, a period during which seven jerks may have occurred. The situation is somewhat reminiscent of geomagnetic reversals in the early 1950s, when only the last few reversals were known.

The inclination data are more scattered and less complete; a maximum around 1720 and a uniform decrease until 1860, followed by rather stationary values until the present, are well-resolved features. Malin & Bullard (1981) have smoothed and interpolated the D and I data using cubic splines and have combined the two in the form of a Bauer (1895) plot (Figure 34). These features can also be found in Abrahamsen's (1973) Bauer plots, together with a review of data from Copenhagen and Denmark from 1500 to 1970. Malin & Bullard find that secular variation is reasonably well fit by a simple model involving a small radial dipole drifting westward at the surface of the core at a rate of $\sim 0.17° \, \text{yr}^{-1}$. Power spectral analysis reveals no evidence for a ~ 60 yr or any other peak, but only a rather uniform falloff with frequency between periods of 15 and 200 yr.

5.3.2 WORLD CHARTS FROM 1715 ONWARDS   Measurements at only a few sites do not allow us to have a global picture of secular variation and to model it. Attempts at such modeling for historical times have been reviewed by Barraclough (1978). More recently, Bloxham (1986b) has compiled original observations from the eighteenth and nineteenth centuries in

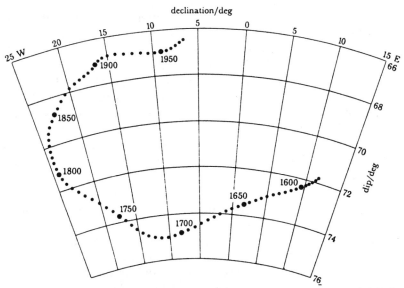

Figure 34 The direction of the geomagnetic field at Greenwich (Bauer plot, zenithal equidistant projection), plotted every fifth year after spline fitting of the original D, I data (after Malin & Bullard 1981).

unprecedented detail, having recourse to original data sources (i.e. voyage reports) and including critical reevaluation of ship positions. The data were assembled in three separate groups to compute geomagnetic field models at three epochs. The 1715.0 model is based on 2575 D-values and 142 I-values measured between 1695 and 1735. Errors are assumed to be of $\sim 1°$. For the 1775.5 model, 4733 D-values and 1382 I-values with estimated 30' errors spanning the period 1765–90 were used. Finally, the 1842.5 model is based on 12,398 measurements of D, I, and F (assumed errors are 15' and 500 nT, respectively) spanning the period 1830–60. For the two earlier epochs, when no intensity measurements were available, Bloxham (1986b) assumed that the absolute values of spherical harmonic coefficients could be determined in proportion to the axial dipole g_1^0, as extrapolated linearly backward by Barraclough (1974; see also Figure 28 and Gubbins 1986). Bloxham believes that these assumptions do not cause severe nonuniqueness problems. Bloxham (1986b) next inverted the data using stochastic inversion and plotted synthesized maps after downward continuation to the core-mantle boundary (see also Gubbins 1983, Bloxham & Gubbins 1985). The resulting picture of secular variation inferred from comparing the different maps at the core-mantle boundary is quite different from the surface studies reviewed above. Prominent static features are identified, both with intense (Siberia, Canada, Antarctica, central Pacific Ocean) and with very low (near both poles, Easter Island, northern Pacific Ocean) magnetic flux. Westward drift appears to be confined to a region extending from Indonesia to the Atlantic Ocean, with strong secular variation in southern Africa and the southern Atlantic Ocean. The core motion models of Le Mouël et al (1985a, and Section 6.2) confirm that in recent times westward drift has mainly affected the Western Hemisphere as a result of geostrophic flow at the core-mantle boundary. To Bloxham & Gubbins, the static core spots and the oscillations under Indonesia appear to be inconsistent with the frozen-flux approximation (e.g. Roberts & Scott 1965, Backus 1968; see Section 6.2). This would imply that the effects of magnetic diffusion are important over the scale of space and time considered here.

Recently, Backus (1987) has shown that the approach of Bloxham & Gubbins (1985) in regularizing the inverse problem involved values of the damping parameter more appropriate to the *existence* than to the *uniqueness* part of the problem. If, for instance, the value of heat flow across the core-mantle boundary is taken as an external bound, Backus (1987) shows that the damping parameter should be taken to be over three orders of magnitude larger. Backus concludes that the averaging disk imposed on the core-mantle boundary has a diameter of $\sim 35°$; comparison with models using the method of harmonic splines (Shure et al 1982, 1985) suggests that many of the small-scale features discussed by Bloxham &

Gubbins are not resolvable. Until new inversions with adequate values of regularizing parameters become available, it may be safer to use the maps that can be computed at the Earth's surface: Figure 35 (J. Bloxham, personal communication, 1987) displays the world charts of the three elements D, X, and Z from 1715 to the present.

5.4   Paleosecular Variation

Significant information on past geomagnetic field behavior can be recovered prior to the time when direct instrumental records are available using the methods of archaeomagnetism and paleomagnetism. Human artifacts, such as bricks and hearths, or natural rocks, such as lava flows or lake sediments, can record past field directions in the form of thermal or depositional remanent magnetization (e.g. Stacey & Banerjee 1974). Only thermal remanent magnetization is readily amenable to paleo-intensity measurements, using methods that are essentially versions of the Thellier & Thellier (1959) method (e.g. Fuller 1987). Only relative measurements are possible in the case of depositional remanent magnetization, a topic of ongoing discussions and great promise (e.g. Banerjee 1983). Gubbins (1986) has noted that it would be sufficient to supplement directional data (essentially D) available between ~ 1600 and 1832 with paleomagnetic measurements of I and F at only ~ 10 sites, evenly spaced in latitude and at ~ 50-yr intervals, in order to recover spherical harmonic models for these early periods. Recent reviews of data pertaining to paleo-secular variation include Merrill & McElhinny (1983), Barton & Merrill (1983), Creer et al (1983), and Creer (1985).

5.4.1 TERMINOLOGY   Some ambiguity exists in the various terms that are commonly used. As noted by Creer et al (1983), the terms *archae-omagnetism* and *paleomagnetism* refer to the study of the remanent magnetization of, respectively, historically dated archaeological relicts and geologically dated rocks, definitions partly in contradiction with their etymology. Also, the term *paleosecular variation* is sometimes restricted to the studies of the latitude dependence of the angular dispersion of the geomagnetic field as measured from paleomagnetic results (e.g. McFadden & Merrill 1984), whereas we use it here in the broader sense of secular variation recorded prior to the availability of direct measurements. Clearly, the major limitation of paleosecular variation studies lies in the increasing nonuniqueness of interpretations in terms of field models as the data base loses resolution both in space and in time.

5.4.2 ARCHAEOMAGNETIC RESULTS   For the last few thousands of years, archaeomagnetism provides complete vector field determinations from selected historical sites. Figure 36 is an example of the data from the

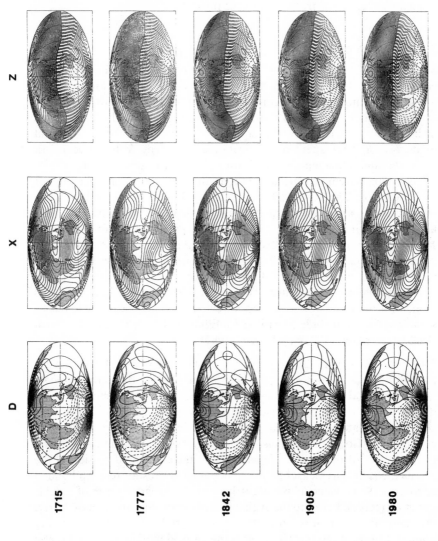

Figure 35 Charts of components *D*, *X*, and *Z* calculated at the Earth surface for five epochs. 1715, 1777, 1842, 1905, and 1980 (J. Bloxham, personal communication, 1987).

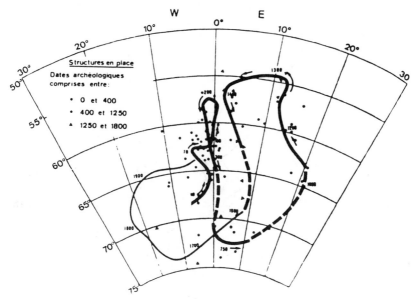

Figure 36 Bauer plot of archaeomagnetic data for the last 2000 yr in the Paris area (after Thellier 1981; see also Tanguy et al 1985).

Western European area, essentially France, reduced to Paris and compiled by Thellier (1981; see also Tanguy et al 1985). This extends the historical records from Paris and London (Figures 33, 34) back to 1800 yr B.P. (before present). The regular loop found for the last centuries is not found in the older data, which speaks against any kind of periodicity. Similar curves have been constructed for southeastern Europe, Egypt and the Near East, the USSR, China, Japan, North America, Peru, and Australia (see Creer et al 1983). Data from ^{14}C-dated lava flows can be used to complement the record in space and extend it into the past (e.g. for the last 30,000 yr in Hawaii; see McWilliams et al 1982). Merrill & McElhinny (1983) have updated a synthesis of Champion (1980) of the motion of the dipole axis over the last 2000 yr. Figure 37 extends Figure 29 into the past, and shows that the mean virtual geomagnetic pole has remained within 10° of (and, as is the case at present, often close to 10° from) the north geographic pole. The overall mean pole for the last 2000 yr lies very close to the North Pole, a fundamental observation for paleomagnetism, although one that cannot be extrapolated back without precaution (Merrill & McElhinny 1983).

McElhinny & Senanayake (1982; see also McElhinny 1983) have compiled a set of 1167 paleointensity measurements from all over the world,

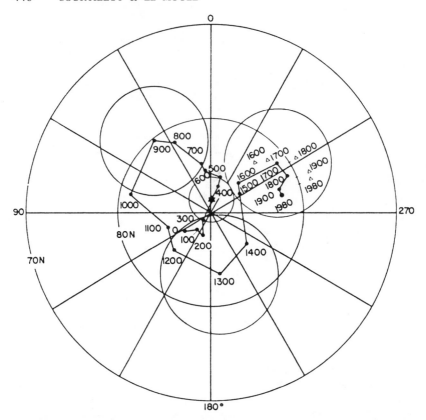

Figure 37   Location of the dipole axis over the last 2000 yr. One-hundred-year means from eight regions are given with the present value (1980). Although successive values are not significantly different from each other, the extreme locations on the path are significantly distinct (see 95% confidence intervals of 1700-, 1300-, and 900-yr means) (after Merrill & McElhinny 1983).

covering the last 12,000 yr, which they averaged in 500–1000-yr intervals. Their results (Figure 38) show that the dipole moment has decreased since 2000 yr B.P., an extension of historical observations (Figure 29) that confirms particularly fast decay over the last 1000 yr. From ∼ 6000 to ∼ 3000 yr B.P. there was a significant increase in dipole strength. The average over the last 10,000 yr is 8.75×10^{22} A m², only slightly larger than the present value. Extension of this analysis back to 50,000 yr B.P. shows much lower dipole strength in the older period. However, McFadden & McElhinny (1982) find a long-term average over the past 5 m.y. of

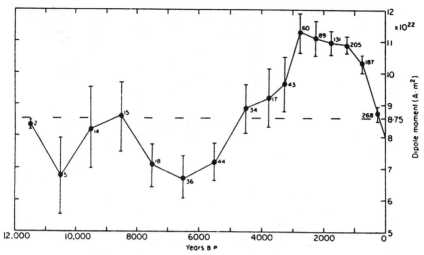

Figure 38 Global mean dipole moments with 95% confidence limits for 500-yr intervals (to 2000 BC) or 1000-yr intervals (prior to 2000 BC) from archaeomagnetic data (after McElhinny & Senanayake 1982).

8.7×10^{22} A m^2, which suggests very long-term fluctuations of dipole intensity with a range of at least 200%. There is no evidence for a period of ~ 8000 yr, which had been previously suggested (see also Barton et al 1979).

5.4.3 LAKE SEDIMENTS   Archaeomagnetic findings can be complemented by paleomagnetic records in lake sediments. Although they do provide interesting results, dry sediments are more prone to alteration than wet ones (e.g. Barton & Merrill 1983, Creer 1985). Some of the earliest wet sediment results come from British lakes, where the sedimentation rate has been on the order of 1 mm yr^{-1}. An average paleosecular variation curve has been obtained for the United Kingdom, spanning the last 10,000 yr (Figure 39, after Turner & Thompson 1982). Separate Bauer plots for the periods 0–5000 and 5000–10,000 yr B.P. (Figure 40) show large swings of typically 30° in amplitude for both *D* and *I*. The more recent 2000 yr are in excellent agreement with archaeomagnetic results (Figure 36), and with other paleosecular variation results from elsewhere in Europe [e.g. Kovacheva (1980) for southeastern Europe]. Apart from a short period with anticlockwise looping (1100–600 yr B.P.), the plots display dominantly clockwise motion, which is often interpreted as reflecting westward drift of nondipole field sources (Runcorn 1969). Skiles (1970) and Dodson (1979) have shown that when certain conditions of relative sign and ampli-

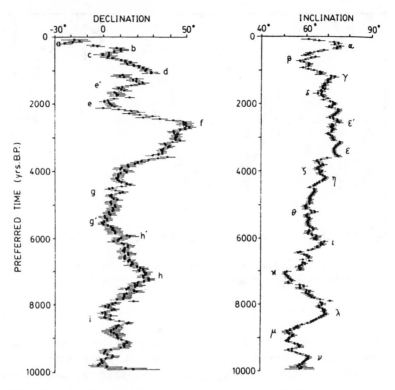

Figure 39 Average curves of secular variation of *D* and *I* from British lake sediments for the last 10,000 yr (after Turner & Thompson 1981; see also Creer 1985).

tude of various nondipole spherical harmonic coefficients are fulfilled, anticlockwise motion can actually result from westward-drifting sources. However, Runcorn's interpretation is still generally regarded as applying, at least in a statistical sense (Merrill & McElhinny 1983).

Creer (1981) and Creer & Tucholka (1983) have also produced a paleosecular variation curve for the last 12,000 yr for North American lake sediments. They find good overall agreement with European results and deduce that westward drift has persisted over the entire period at an average rate of $0.13°$ yr^{-1}, with values as high as $0.5°$ yr^{-1}. However, Creer & Tucholka (1983) believe that details of the curves on a ~ 500-yr time scale cannot be explained simply in terms of sources that are homogeneously drifting in one direction. They further interpret the remarkable similarity of smoothed Western European and North American virtual geomagnetic pole paths in terms of interference between drifting and waxing-and-waning standing sources. Verosub et al (1986) have recently

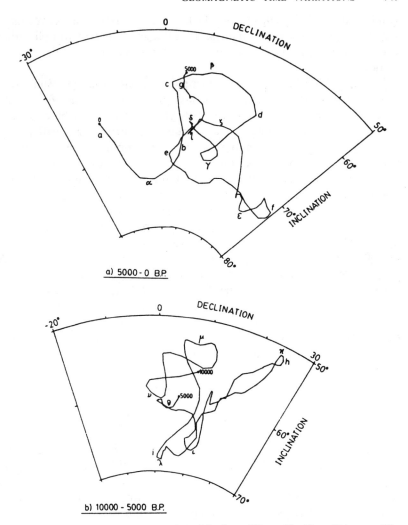

Figure 40   Bauer (1899) plot representation of the data of Figure 39 with an 80-yr smoothing filter for the periods (*a*) 5000–0 B.P., (*b*) 10,000–5000 B.P. (after Turner & Thompson 1981).

published a new North American record for which they find the average field direction over the last 10,000 yr to be identical to that of a geocentric axial dipole. They interpret the clockwise motion of the field direction in terms of slow (0.06° yr⁻¹) westward drift.

Comparison of some European and Japanese data led Yukutake (1967) to document westward drift at 0.3° yr⁻¹ over the last 1400 yr. On the other hand, Turner & Thompson (1981) find little correlation between up-dated versions of the two data sets and suggest that this implies that changes in the nondipole field are more important than dipole drift (or "wobble").

Barton & McElhinny (1982) report paleosecular variation results from southeastern Australia spanning the last 10,000 yr. They also find clock-wise precession with a period of ~2800 yr, which they attribute to west-ward drift of the nondipole field, and anticlockwise precession with period ~5000 yr, which they assign to precession of the dipole. Creer & Tucholka (1983) find a lack of correlation between curves from the Northern and Southern Hemispheres but believe that this may be due to remaining uncertainties in correlation, particularly in absolute dating.

Creer et al (1986) have recently been able to extend the European curve back to ~22,000 yr B.P. using sediments from a volcanic lake in the French Massif Central. They report clockwise rotation from 22,000 to 13,000 yr B.P. and anticlockwise rotation from 13,000 to 8000 yr B.P. A comparison with archaeomagnetic data leads them to estimate that the sedimentary curves involve some 30–50% smoothing due to the recording process, and they conclude that the actual amplitude of paleosecular varia-tion (both in D and I) must have been on the order of 40°. Although Creer (1981) states that the intensity of the nondipole field is abnormally low compared with the dipole intensity in historical time, Creer et al (1986) find that throughout the last 20,000 yr the magnitude of non-dipole foci relative to that of the main dipole has remained similar. In a recent review, Creer (1985) expresses doubt that reliable estimates of paleo-westward drift can be obtained at present from paleosecular variation records.

Creer (1983) and Creer et al (1983) have noted differences in the shapes of maxima and minima of the paleosecular variation D and I curves (Figure 39), with some being pointed and others more rounded. This is interpreted through forward modeling as being due to a drifting pattern of deep-seated sources (radial dipoles or current loops in the upper core).

5.4.4 SPECTRAL ANALYSIS   Many authors have attempted to submit paleosecular variation records to spectral analysis and have proposed the existence of resolvable spectral peaks. An incomplete list includes the following periods: 500, 600, 800, 1000, 1200, 1800, 2400, 3000, 4500, and 9000 yr. Analyses are plagued by the difficult problems of data quality and age resolution. We agree with Turner & Thompson (1981) and Barton & Merrill (1983) that the contradictions between various analyses, the many problems of stability of spectral estimates, and the lack of overall consensus strongly suggest that there are actually no discrete periods in the geo-magnetic record in this spectral range. The process is nonperiodic and possibly nonstationary, with broad energy bands at 9000–3000 and 3000–2000 yr B.P., which might be associated, respectively, with dipole fluc-tuations on the one hand, and drift and nondipole fluctuations on the other. However, there is no evidence for a clear-cut separation of the frequency bands of the two kinds of paleosecular variation.

5.4.5 EXCURSIONS AND REVERSALS Some very large features of paleo-secular variation of unknown geographical extent have become known as "excursions" and have been interpreted either as unusually large secular variations or as aborted dipole field reversals. These are out of the scope of the present review, but we can say that recent discussions of their existence (e.g. Merrill & McElhinny 1983, Creer 1985) do not provide compelling pictures. Reversals themselves are a fundamental feature of the time variations of the geomagnetic field, also out of the scope of this review. Their typical duration (10^3–10^4 yr) is clearly in the range of secular variation we are interested in, and there have been recent tantalizing suggestions that a record of secular variation could be recovered during a reversal (Valet 1985, Valet et al 1986). Confirmation of such observations and combination with the aspects of secular variation reported in this paper hold great promise for an interpretation in a joint geodynamo framework.

5.4.6 STATISTICAL PALEOSECULAR VARIATION On an even longer time scale of millions of years, paleosecular variation can no longer be studied in a deterministic way and only some of its statistical features can be recovered. The main source of data is the record of the angular dispersion (S) of paleomagnetic directions or virtual geomagnetic poles, which is traditionally interpreted to be the superposition of angular dispersion caused by dipole wobble (S_D) and a latitude-dependent contribution due to nondipole secular variation, including drift (S_N). A review of various models and a new paleosecular variation statistical model, in which varia-tions in the nondipole field are assumed to remain in constant proportion to the dipole field, are presented by McFadden & McElhinny (1984; see also Merrill & McElhinny 1983). For the last 5 m.y., S_D and S_N (at the equator, where they are smallest) are respectively, 7° and 11° (for virtual geomagnetic poles). As a result, dispersion of the virtual geomagnetic pole varies between 12 and 20° from the equator to the pole. McFadden & McElhinny conclude that over this time range, the average energy density of the nondipole field has been about 9% of the energy density of the dipole field at the Earth's surface.

6. INTERPRETATION OF GEOMAGNETIC VARIATIONS

The geomagnetic variations of external and internal origin described in the previous sections are not only of interest per se, they also allow one to probe various physical properties of the Earth's deep interior. For instance, the separate inducing and induced fields, and the propagation of secular variation through the mantle, can be used to infer the distribution of electrical conductivity in the mantle. This may provide some of the best information on the thermal state of the mantle and may also help constrain its chemical composition. Secular variation also provides our only image of the pattern of motions in the fluid core, at least close to its surface, and constrains models of field generation by dynamo action. Finally, secular variation appears to be correlated with other global geophysical parameters, such as the rotation velocity of the mantle (or, equivalently, the length of the day), thereby providing information on the couplings between the core and mantle. Each one of these topics would warrant a review in itself; we only present a brief summary of the main problems and recent progress in these fields in order to clarify uses of the phenomena described above and to suggest which new data are desirable to further the solution of some key geophysical questions.

6.1 *The Electrical Conductivity of the Mantle*

Most of the variations described in Sections 4 and 5.1 can be used to constrain the conductivity of the mantle. (The conductivity of the very heterogeneous crust is not discussed here.) The basic theory of electromagnetic induction in a radially symmetric conducting sphere follows the work of Schuster (1889) and Lahiri & Price (1939) and is described in many recent articles (e.g. Banks 1969, Le Mouël 1976b; see also Eckhardt et al 1963, Currie 1967). Variations with a primary external (e) origin induce electric currents within the Earth, which in turn produce magnetic variations of internal (i) origin. The basic data derived from the observations (Section 4) are the ratios i/e, or related functionals called the response, admittance, impedance, and apparent resistivity (e.g. Le Mouël 1976b, Roberts 1986b) at various, hopefully all, frequencies and for various, generally simple harmonic forms (i.e. P_1^0 for *Dst*, 27 days, 11 yr, and their harmonics; P_m^{m-1} for *Sq*). Many new techniques have been proposed for the inversion of the response data (e.g. Banks 1969, 1972, 1981, Bailey 1970, Parker 1970, 1980, 1983, Weidelt 1972, Schmucker 1979). Some of these techniques and applications to new sets of data are discussed, for example, by Ducruix et al (1980), Achache et al (1981), Hobbs (1983), and Jady & Paterson (1983), and have been recently reviewed by Roberts

(1986b). The periods used for the response range from less than a day to 11 yr, with correspondingly increasing penetration depth. Although the results (and, in particular, the depth resolution and uncertainty estimates) depend on the reliability of the data and also on the inversion method, most models proposed in the depth range from 50 to 2000 km are reasonably consistent (within an order of magnitude at a given depth). Ducruix et al (1980) and Achache et al (1981) point out that the 11-yr variation constrains the middle mantle conductivity at rather low values of a few hundred Ω m^{-1} (Figure 41). Hobbs (1983) finds from a Schmucker (1979) inversion of response data from 100 s to 11 yr (Figure 42) that there is good resolution between 50 and 150 km and between 500 and 1100 km. In these depth ranges, conductivity is well described by a simple power law ($\sim 10^{-6}z^2$, with depth z in kilometers and conductivity in Ω^{-1} m^{-1}). Conductivity reaches 1 Ω^{-1} m^{-1} at a depth of 1000 km, and the power law would predict ~ 10 Ω^{-1} m^{-1} at the core-mantle boundary. However, Hobbs (1983) points out that resolution is very low below 1500 km and that the low frequency data do not provide strong constraints. The method is not able to resolve either conductivity in the uppermost mantle or any sharp conductivity variations. Jady & Paterson (1983) apply the Bailey (1970), Parker (1980), and Fischer et al (1981) inversion schemes to *Dst* data for durations up to 10 days following the storm commencement (i.e. periods of 0.5 to 14 days). They find the results from the three inversions

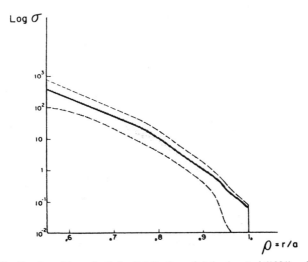

Figure 41 The "preferred" conductivity distribution of Achache et al (1981), with uncertainty indicated by dashed lines. Conductivity given in Ω^{-1} m^{-1}; relative radius reduced to that of the Earth ($a = 6371$ km).

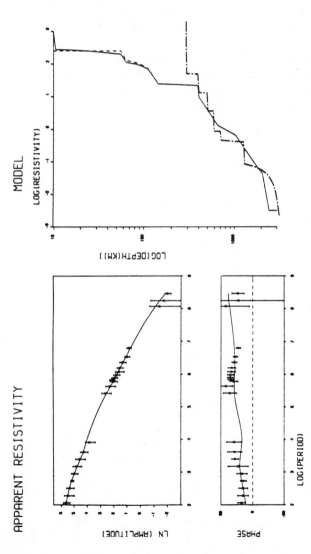

Figure 42   The starting resistivity model (dashed line) and final model (solid line) of Hobbs (1983). For comparison, the chain line is from Banks (1972). The amplitude and phase of the original data, discussed in the text, are shown to the left (period in seconds).

to be consistent. Poor resolution in the upper mantle only allows the suggestion that conductivity there is very low (less than 10^{-2}–10^{-3} Ω^{-1} m^{-1}); however, deeper structures are detected—in particular, two sharp increases in conductivity at ~ 650–750 and ~ 1000-km depths. Jady & Paterson suggest that the former correlates with the well-known 700-km seismic discontinuity, rather than with the 400-km one, as had been previously proposed. The deeper resolvable values at ~ 1200 km are about $10 \, \Omega^{-1} \, m^{-1}$. Parker's (1980) method requires an insulating layer from the surface down to depths of ~ 700 km. It also generates zones of decreasing downward conductivity below major increases, a feature that may not be physical and is not present in other inversions. Roberts (1986a,b) actually finds a more resistive zone below the zone of major conductivity increase at ~ 1000-km depth, but this is found only for data from European observatories and, if real, does not seem to be a global feature. Roberts' (1986b) discussion of the various external variation data is a useful complement to our Section 4. The main emphasis of his review is on the "unquestioned" influence of lateral heterogeneities in conductivity. Roberts believes that this may extend to depths of several hundred kilometers and to rather long periods. However, he states that the assumption of radial symmetry may lead to reasonable first approximations to reality. He concludes from his own studies that resolution estimates quoted in the literature are often overoptimistic. Roberts confirms a rapid increase in conductivity between 400 and 800 km but finds that the structure of this increase cannot be resolved; he does not believe that the correlation with the 700-km seismic discontinuity is warranted. Like Hobbs (1983), Roberts finds the long-period data to be potentially of great significance but comparatively unreliable. Finally, he insists on the importance of lateral conductivity heterogeneities at great depth, such as, for instance, the difference in conductivity profile between the continental United States and elsewhere. Counil et al (1987) also resolve differences in conductivity at depths of some 400 km between Western Europe and Australia. Some authors have also suggested the existence of a systematic difference between deep oceanic (more conductive) and continental (less conductive) structures.

Ducruix et al (1980) and Achache et al (1980, 1981) point out that further constraints can be placed on deep-mantle conductivity by using the propagation of the jerk signals. This follows the pioneering work of Runcorn (1955) and McDonald (1957). Deep-mantle conductivity is found not to exceed a few hundred $\Omega^{-1} \, m^{-1}$ (Figure 41), contrary to much higher values proposed, for instance, by Alldredge (1977b). Backus (1983) recognizes that constraints on deep-mantle conductivity stem from the apparently short duration of the 1969.5 jerk but suggests that this brevity, first observed in European observatories, could actually be a regional phenomenon due to harmonic mode mixing. As we have noted above (Section 5.1.5; see also Courtillot et al 1984), the sharpness of the secular variation V is real and worldwide; the pure internal signal may actually be even sharper, since it is smoothed by remaining superimposed, sunspot-related signals. McLeod (1985) supports this analysis in concluding that the duration (smoothing time in the terms of Backus 1983) of the 1969.5 jerk is shorter than a few years and is globally simultaneous. As a whole, despite suggestions to the contrary (e.g. Alldredge 1977b) and acknowledging poor resolution, it appears to us that the conductivity of the deep mantle may be fairly low, apparently less than $10^3 \ \Omega^{-1} \ m^{-1}$ down to the core-mantle boundary (see also Courtillot & Le Mouël 1979, Courtillot et al 1984).

Direct experiments in pressurized cells have been used to justify the increase in electrical conductivity at either the 400- or the 700-km seismic discontinuities, which are interpreted to be major phase changes. Ongoing experiments are now attaining the temperatures and pressures of the middle and lower mantle. X. Li & R. Jeanloz (personal communication, 1987) have subjected an assemblage of minerals representative of the mantle (pyroxene, olivine, and garnet with Fe/Mg ratios of ~ 0.14) to pressures of 20–90 GPa (~ 700–2000 km depths) and temperatures of 1500–4000 K. Interestingly, they find the assemblage in these conditions to be almost insulating (conductivity less than a few $\times 10^{-3} \ \Omega^{-1} \ m^{-1}$)!

6.2   Motions of the Fluid Core

Joint observations of the main geomagnetic field **B** and its secular variation **Ḃ** can, in principle, be used to derive information on fluid core motions **u**, at least in the uppermost part of the core (see, for example, the early work of Kahle et al 1967). This is seen in the electromagnetic induction equation (e.g. Le Mouël 1976c), whose radial component at the core-mantle boundary [where radial velocity vanishes ($u_r = 0$)] can be written as

$$\dot{B}_r + \text{div}_H(B_r \cdot \mathbf{u}) + \eta \cdot r^{-1} \cdot \text{lap}(rB_r) = 0, \tag{3}$$

where η is the magnetic diffusivity, div_H the horizontal part of the divergence, "lap" the Laplacian, and subscript "r" denotes radial components. With accepted values of core conductivity at $\sim 10^6 \ \Omega^{-1} \ m^{-1}$, η is on the order of 1 $m^2 \ s^{-1}$. From the previous section, we see that the mantle can be considered as insulating to a first approximation, in which case B_r (i.e. the vertical component Z with a change in sign) and $\dot{B}_r$ can be downward-continued to the core-mantle boundary, with a resolution of about $9°$ for harmonic degree 12 (Booker 1969). Using probable values of mantle conductivity (Section 6.1), Benton & Whaler (1983) have shown that this approximation is good to about 2% for B_r and 10% for $\dot{B}_r$. Both quantities are therefore available in the core, below a thin kinematic boundary layer (Roberts & Scott 1965). Roberts & Scott (1965) suggested that the last term of Equation (3), which measures flux diffusion, could be neglected, provided that the vertical scale of **B** in the core is a sizeable fraction of the Earth's radius. This is the so-called frozen-flux approximation, which has now been shown to be a very good one on global-length and decade time scales. It provides a magnetic estimate of the core radius that is within 2% of the seismological determination (Hide 1978, Hide & Malin 1981, Voorhies & Benton 1982; see also Voorhies 1986). Benton & Voorhies (1987) have analyzed 28 main field models for the time span 1945–85. They find that the rate of decline of the Earth's total pole strength (i.e. the absolute flux linking the surface of the Earth $\int\int |B_r| a^2 \sin\theta \, d\theta \, d\Phi$) increased rather suddenly around 1960. This is principally attributed to the dipole and quadrupole terms (see Section 5.2.3). Yet all models but one conserve the absolute magnetic flux linking the core-mantle boundary. There is an overall decrease for truncation levels up to $n = 4$, but this disappears at $n = 5$. A slight increase might even be indicated for $n = 9$ and 10 (Benton & Voorhies 1987). This confirms that the secular decay of the dipole is compensated for by an increase in terms of higher (yet not very high) degree. The analysis further substantiates the frozen-flux approximation over 40 years. More precisely, Backus & Le Mouël (1986) have shown that the frozen-flux approximation holds at the core-mantle boundary outside a reasonably narrow "leaking belt" centered on a "leaky curve" where the second (advection) term of Equation (3) becomes zero. This leaky curve includes points where null flux curves (those on which B_r is zero) can appear and disappear. Recent opposite interpretations of Bloxham & Gubbins (1985, 1986) and Bloxham (1986a) have met with criticism from Voorhies (1986) and Backus (1987; see Section 5.3.2).

Under the frozen-flux approximation, Equation (3) simplifies to

$$\dot{B}_r + \mathrm{div}_H(B_r \cdot \mathbf{u}) = 0 \qquad (4)$$

and leads one to hope that **u** can be determined from B_r and $\dot{B}_r$. However, Backus (1968) pointed out that (4) was a single scalar equation for two unknowns (the horizontal components of velocity), and that no unique solution could be found without additional assumptions. One such assumption is the lack of upwelling (i.e. a purely toroidal fluid motion) at the top of the core (Whaler 1980, 1982, Gubbins 1982). This now appears to be very unlikely (Madden & Le Mouël 1982, Whaler 1984, 1986, Gire et al 1986, Le Mouël et al 1985a, Voorhies 1986), and in any case it does not fully resolve the nonuniqueness. Gire et al (1986) showed that purely toroidal (or purely poloidal) motion models of low degree can be constructed to fit secular variation data, but these have considerably more energy than some combined poloidal-toroidal models and show no decay of higher-degree terms (see their Figure 10). Another assumption is that of steady flow, for which uniqueness has been demonstrated (Voorhies & Backus 1985). The secular variation data cannot be fit with a purely toroidal steady flow, and some amount of upwelling near the top of the core is required (Voorhies 1986). Flow calculations point to a bulk westward drift at about $0.11°$ yr^{-1}, combined with jets and gyres with typical upwelling velocity of some 10^{-4} m s^{-1}. However, Voorhies notes changes in the flow pattern on time scales of 15 yr at most, and the observed secular variation field varies too rapidly for the steady model to be fully appealing. Furthermore, this hypothesis does not resolve the ambiguity in practice.

A third approach involves a comparison of the relative strengths of the terms entering the Navier-Stokes equation at the core-mantle boundary. Hills (1979), Le Mouël (1984), and Le Mouël et al (1985a) argue that the horizontal component of the Coriolis force is balanced by the horizontal gradient of pressure p. The simplified Navier-Stokes equation then yields

$$\mathbf{u} = (2\rho\Omega \cos \theta)^{-1}\mathbf{n} \times \mathbf{grad}\ p, \tag{5}$$

where ρ is density, Ω is the Earth's rotational velocity, θ is colatitude, and $\mathbf{n}$ is the outward radial unit vector. This flow is said to be tangentially geostrophic. The geostrophic approximation certainly fails at the geographical equator, where the horizontal component of the Coriolis force vanishes. Backus & Le Mouël (1986) show that the failure domain is a narrow belt centered on the equator with a width (in degrees) on the order of the dimensionless radial number of $\mathbf{B}$. If the flow $\mathbf{u}$ is indeed geostrophic, Equation (5) shows that it then depends only on the single scalar function p, and Equation (4) can be expected to determine it. In fact, the non-uniqueness is still not entirely resolved. The flow $\mathbf{u}$ is determined uniquely only in a "visible belt" made of all points connected to at least one point of the geographic equator where $B_r = 0$ (i.e. where it intersects a null flux curve) by a level curve of $B_r \cos \theta$ on the core-mantle boundary (Backus & Le Mouël 1986). Backus & Le Mouël display the ambiguous patches, which are outside the visible belt.

With these limitations in mind, it is possible to invert secular variation data using (4) and (5). The usual difficulties arise from the limited amount and accuracy of the secular variation data. The amount of detail resolved by the data is a delicate problem, which we do not discuss here (see Le Mouël et al 1985a). The bulk of the data are found to be compatible with a large-scale (low-degree) flow, and the worldwide occurrence of the 1969.5 jerk (Section 5.1.3) is in favor of the existence of such large-scale flow.

The horizontal vector $\mathbf{u}$ can be written in the form of the sum of poloidal (or scaloidal) and toroidal parts (Roberts & Scott 1965, Backus 1968):

$$\mathbf{u} = \mathbf{grad}_H(s) + \mathbf{n} \times \mathbf{grad}_H(t). \tag{6}$$

The motion derived from the main and secular variation fields given by the IGRF 1980 model is shown on Figure 43, where the expansion of $\mathbf{u}$ has been limited to degree 10 (Le Mouël et al 1985a). Figure 43a represents the isovalue lines of the poloidal scalar s; the poloidal flow is everywhere orthogonal to these lines and is indicated by arrows. Figure 43b represents level lines of the toroidal scalar t, and these are the toroidal flow lines. The overall features of the computed geostrophic motions are unexpectedly simple. The poloidal flow, which results from exchanges of fluid with the deeper core, is characterized by two main jets: an upwelling in the equatorial Indian Ocean (surrounded by lesser downwelling centers), and a downwelling off the coast of Peru (surrounded by lesser upwelling centers), almost antipodal to the previous one. As a result, poloidal motion is dominated by the s_1^1 term. The toroidal flow displays intense westward drift in the Western Hemisphere (from $90°W$ to $90°E$). In the Pacific hemisphere, flow lines are pushed away from the equator to the higher latitudes. This results from superposition of a zonal toroidal westward flow and of a sectorial toroidal flow linked to the poloidal jets that ensures geostrophy. Both the well-known enhancement of westward drift at low latitudes in the Western Hemisphere (Section 5.2.4) and the lack of secular variation in the Pacific hemisphere (e.g. Doell & Cox 1972) therefore appear to be explained. The maps of Bloxham & Gubbins (1985; see Section 5.3.2) indeed display clear evidence of westward drift in the Western Hemisphere but none in the Pacific hemisphere. Overall, the toroidal motion is dominated by the t_1^0 and t_2^1 terms. The t_1^0 term amounts to $0.10°$ yr^{-1}, a value smaller than most listed in Section 5.2.4.

The secular variation field changes in time much faster (in relative value) than the main field itself (Section 5). As a result, the geostrophic motion $\mathbf{u}$ also changes with time [Equation (4)] and so does the pressure field p [Equation (5)]. Le Mouël et al (1985a) compare the geostrophic solutions for 1970 and 1980. They find that the pattern of motion hardly changes but that the total energy of the motion nearly doubles in a decade. The relative variation of p is on the order of 40%. This is thought to reflect in some way the 1969.5 and possibly 1979 jerks.

From Equation (5), one immediately derives

$$\mathrm{div}_H(\mathbf{u} \cos \theta) = 0. \tag{7}$$

A comparison of (7) with the induction equation (4) shows that geostrophic motion cannot generate secular variation of fields with geometry $B_r = \cos \theta$, i.e. an axial dipole field (Le Mouël 1984). Therefore, only the nondipole field is involved in secular variation of both the dipole and

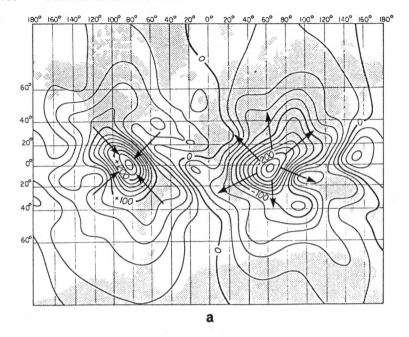

a

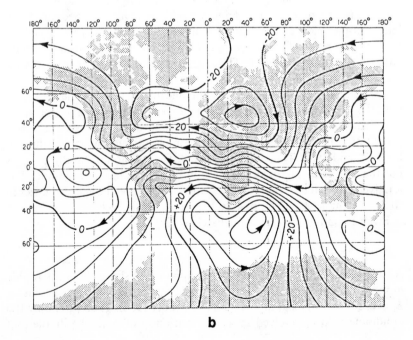

b

nondipole parts of the main field. This may account for the long-standing paleomagnetic observation that dipolar and nondipolar secular variations do not involve the same time constants (Section 5.4). Should the frozen-flux assumption still hold, even approximately, over the timespan of a few thousand years, this would have interesting implications for magnetic reversals [Le Mouël 1984; see also Backus & Le Mouël (1986) and, for an alternate view, Gubbins (1987)].

6.3 Core Motions and Irregularities in Earth Rotation

The Earth's rotation is a far more irregular process than might at first be suspected (Lambeck 1980). Both the rotational speed (or, equivalently, the length of day) and the direction of the rotational axis suffer continuous changes. Most variations with a period shorter than about 2 yr can be attributed to atmospheric motion through exchange of angular momentum between the atmosphere and mantle (Hide et al 1980). Fluctuations with longer time constants, the so-called decade fluctuations, cannot be related to the atmosphere (Hide 1982) and must be due to transfer of angular momentum from the core to the mantle. More precisely, they must in some way be related to the motions of the fluid core discussed in the previous section. This has long been recognized for the length of day, but the role of the fluid core in altering the motion of the pole, and particularly in sustaining the Chandler wobble, is more controversial. This approach, pioneered by Runcorn (e.g. 1970), has recently received renewed attention.

6.3.1 LENGTH OF DAY Decade fluctuations in the length of day based on the work of Morrison (1979), complemented by data from the Bureau International de l'Heure (BIH) from 1956 onward, are shown in Figure 44. Stoyko (e.g. 1951, and references in Le Mouël et al 1981) appears to have been the first (as early as 1937) to point out a correlation between these and variations in magnetic field intensity in Paris. Vestine (1952) elaborated on this and proposed that length-of-day variations preceded magnetic variations by about a decade. More recently, Backus (1983) has suggested a correlation between the 1969.5 jerk and a 1956 slowdown in Earth rotation (for a discussion, see Courtillot et al 1984). Cain et al (1985) have suggested a similar correlation and a lag between length of day and

Figure 43 Motions at the top of the fluid core obtained from secular variation data under the geostrophic hypothesis. (*a*) Poloidal motion—flow lines are everywhere orthogonal to the contours and are shown diverging from the minimum southwest of India and converging to the maximum west of Peru. (*b*) Toroidal motion—flow lines indicate strong westward motion in the Western Hemisphere, and smaller motion in the Pacific hemisphere. Amplitudes are given in kilometers per year (after Le Mouël 1986).

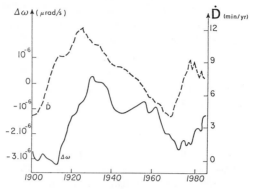

Figure 44 Comparison of the secular variation of *D* in Europe (solid curve and right-hand scale in arcminutes per year) with relative changes in the Earth's rotation rate (dashed curve and left-hand scale) (after Gire & Le Mouël 1985; see Courtillot et al 1983).

westward drift of the eccentric dipole, and Langel et al (1986) have proposed a lag between length of day and dg_1^0/dt. Ambiguities may result from the choice of what is thought to be a global parameter monitoring core motions. The most often-chosen parameter, westward drift rate, can unfortunately not be estimated in any simple, direct, and unique way (Section 5.2.4, and Gire et al 1986): Le Mouël et al (1981, 1983; see corrigendum in Courtillot et al 1983) have selected the time derivative of *Y* in data from European observatories. Although it can be argued that this is not an adequate global parameter (Langel et al 1986), isopors (lines of equal value of components of secular variation) of *Y* in Europe have just the right north-south orientation to provide at least a reasonable estimate of relative fluctuations in westward drift. (Estimates of the total rate and the 1925 local European secular variation feature are discussed in Section 5.2.) The correlation is shown in Figure 44. Many of the shorter period fluctuations in length of day are related to South Pacific oscillation-type events (El Niño). We argue that the two curves present striking similarities, with length of day lagging magnetic secular variation by some 10–12 years. With the data available up to 1978, we predicted that an acceleration of the Earth's rotation rate would take place in the early 1980s (Le Mouël et al 1981). We believe that this has indeed occurred (Figure 44). The 1979 secular variation jerk would now lead us to predict a new change in length-of-day trend around 1990–92. Cain et al, using westward drift of the eccentric dipole, find that Vestine's original (and opposite) correlation is "not confirmed, but the possibility is not discounted." Langel et al (1986) accept that the correlation suggested by Le Mouël et al (1981) is possible, but because they use yet another westward drift estimator, they find it to be not so clear. They calculate correlation coefficients of 0.58 for the 10-yr *lead* of secular variation over length of day, and 0.43 for an 8-yr *lag*.

The interpretation of these and similar other correlations is still an open question. Le Mouël & Courtillot (1981, 1982) first sought an explanation in terms of a sudden transfer of angular momentum between an inner part of the fluid core and an outer layer (a Bullard model) taking place at the times of the jerks, with the former accelerating and the latter slowing down. After some delay, the mantle would in turn be accelerated by electromagnetic coupling, principally with the inner core. The main difficulty with this model is the need for a fast transportation of the toroidal field generated by the differential rotation of the two parts of the core to the core-mantle boundary and the weakly conducting lower mantle. Another possibility is that of topographic coupling across the core-mantle boundary, as first proposed by Hide (1966, 1969, 1970; see also Hide & Stewartson 1977). If the core-mantle boundary is not axisymmetric, the time-varying pressure field $p(t)$ associated with the geostrophic motion at the core-mantle boundary (Section 6.2) exerts a time-varying net torque on the mantle about the polar axis. For example, if the core-mantle boundary is a triaxial ellipsoid with equatorial flattening f' [which seems plausible given the observed gravity potential or the recent seismic studies of the core-mantle boundary (Morelli & Dziewonski 1987)] and if the pressure field contains a term of degree and order 2 (p_2^2), a net torque of the above type proportional to p_2^2 and f' results (Gire & Le Mouël 1985, Le Mouël et al 1985b). Secular variation in fluid motions (Section 6.2) would therefore lead to secular variation of the pressure field, and hence of the length of day. The orders of magnitude inferred from secular variation data appear to be large enough to account for observed fluctuations in length of day (Speith et al 1986, Hinderer 1987, Hinderer et al 1987). It seems appealing and reasonable to associate westward acceleration of the outer layers of the core with eastward acceleration of the mantle. The explanation of the 10–12 yr lag, if confirmed by further data and study, remains to be found: This lag may actually result from the choice of the westward drift estimator itself.

6.3.2   THE CHANDLER WOBBLE   The rotation axis of the Earth describes a complex curve of a few meters in amplitude about its mean position. Superimposed on the annual fluctuation, which is forced by seasonal displacements of air masses in the atmosphere, is the free oscillation with a period of 435 days, known as the Chandler wobble (e.g. Lambeck 1980). Two principal mechanisms by which the Chandler wobble may be sustained against dissipation have been considered. The atmosphere doubtlessly contributes significantly in the range of periods shorter than

2 yr (Hide et al 1980), but it is not clear whether it can be active alone at the longer periods. Earthquakes also do not appear to be an acceptable mechanism: It has been shown that the seismic excitation was too weak, at least for the last 30 yr, by two orders of magnitude to contribute to the wobble (Souriau & Cazenave 1985). The potential importance of the core was first recognized by Runcorn (1970), who suggested some kind of impulsive equatorial electromagnetic torque exerted by the core on the mantle. The same mechanism invoked to account for decade length-of-day changes can contribute to exciting the Chandler wobble by exerting a net torque about an equatorial axis, while at the same time modifying the products of inertia of the elastic mantle. More specifically, the pressure term of degree 2 and order 1 (p_2^1) acts efficiently on the ellipsoidal core-mantle boundary (Gire & Le Mouël 1985, Le Mouël et al 1985b, Hinderer 1987, Hinderer et al 1987). The order of magnitude of the corresponding excitation function, which accounts for both the torque and the modification of the products of inertia, is large enough to significantly contribute to the wobble, provided that the time constants of the pressure field p are favorable. Such would indeed be the case if the large quasi-monotonous variations of secular variation observed between the jerks (Section 5.2) were made up of short, steplike variations (i.e. accomplished in less than 1 yr): Figures 15 and 25 give some indications that such variations may occur at the core-mantle boundary (Gire & Le Mouël 1985, Gavoret et al 1986), and are recorded at the Earth's surface as smoother features as a result of diffusion through the conducting mantle (Section 6.1).

6.3.3   CORRELATION WITH CLIMATIC VARIATIONS   We note briefly in passing that Courtillot et al (1982) have further suggested a correlation between magnetic secular variation and long-term climatic changes, with the latter lagging the former by some 20 yr (see, for example, Chapter 7 in Jacobs 1984).

7.   CONCLUSIONS AND PERSPECTIVES

The many studies of geomagnetic secular variation that we have reviewed here encompass over 10 orders of magnitude in period. We can attempt to incorporate them in a single (semi-quantitative) geomagnetic power density spectrum. Similar attempts over different period ranges have been proposed, for instance, by Banks (1969), Filloux (1980), and Barton (1982). Published results use disparate units and spectral techniques and do not always refer to the same components. Barton (1982) proposes to add the power density spectra of the three components X, Y, and Z. We have summarized some results in Figure 45. McLeod (1982) remarks that both Campbell (1976) and Courtillot & Le Mouël (1976a) find spectra with a $1/f^2$ dependence (where f is frequency), for the very different ranges of 5 min to 4 hr and 2 months to 10 yr (Section 4.6). The composite spectrum of Filloux (1980) has a $1/f^4$ dependence on the high-frequency side, tapering to $1/f$ on the low-frequency side (1–10 yr). Its central portion agrees with a $1/f^2$ behavior. The various external lines at 11 yr, 1 yr, 6 months, 27 days, 24 hr, and their harmonics are superimposed on this background spectrum. The spectrum given by Barton (1982) happens to have a shape similar to that of Filloux but is displaced by six orders of magnitude in period. It begins in the range of observatory records with a $1/f^4$ slope (see also Malin & Bullard 1981, Courtillot & Le Mouël 1976a) and tapers to a $1/f$ slope for the longest paleosecular variation records of lake sediments and marine cores. Again the central portion is more like $1/f^2$. Lowes (in Barton 1982) recalls that the spectrum of Poisson-distributed dipole reversals should be of the form $(2A^2/f_0)/[1 + (\pi f/f_0)^2]$ for an average reversal frequency of $2f_0$. [A is the amplitude of signal for a given component (magnetic field units).] With $f_0 = 2 \times 10^{-6}$ yr^{-1} and values of f larger than f_0 this is essentially a $1/f^2$ spectrum, which is found to be just in the continuation of power values for periods shorter than, say, 100 yr. At periods in excess of 10^3 yr, the paleosecular variation records appear to have powers two orders of magnitude lower than the reversal spectrum.

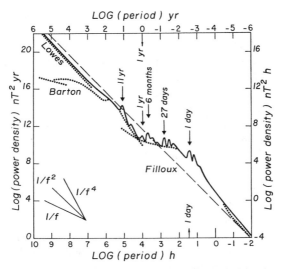

Figure 45   Schematic power spectrum of the geomagnetic field combining data from Filloux (1980; see Roberts 1986b), Barton (1982, 1983) and the suggestion of F. Lowes (in Barton 1982) at the longest periods.

This might in part be due to remaining inaccuracies and smoothing in the recorded data (Section 5.4.3). Also, the tapering of both the Filloux and Barton spectra at what is for each of them long periods may be due to parts of the prefiltering and spectral analysis. It is therefore tempting to suggest that the geomagnetic spectrum might roughly be of $1/f^2$ type over the remarkably large range of periods from 1 min to a million years (12 orders of magnitude), despite major changes from internal to external sources somewhere in the spectrum: However, there seems to be no particular reason why this should be so. This would mean that the rms. amplitude of the background geomagnetic "noise" is roughly proportional to the period. This large increase in power as a function of period is quite different from earlier pictures (e.g. Banks 1969). We recall that there is no evidence for spectral lines, either internal or external, for periods in excess of a few decades and, in particular, no compelling evidence for natural spectral separation between secular variation of the dipole and nondipole fields. The exercise displayed in Figure 45 is only meant to emphasize the extent of the geomagnetic spectrum and to point out both recent progress and directions where future improvement could be sought. Considerable work has confirmed that there is a fairly wide region of the spectrum where the field contributions of internal and external origins overlap. This overlap extends from 1-yr to over 10-yr, and possibly to 100-yr periods. Separation between the two is still largely based on morphology rather than potential field theory. Although confirmation of some inferences and progress in data analysis are desirable, some conclusions have emerged that appear to be fairly robust. A long-period variation related to solar activity and the ~11-yr solar cycle is clearly demonstrated in all components and at all locations. Suggestions for longer periods do not seem to be as well founded, except for the double (magnetic) ~22-yr solar cycle. Much excitement in recent years has been concentrated on the discovery of brief secular variation impulses, or jerks. Such impulses had been suggested 30 years ago but then shown to be artifacts related to a misrepresentation of the solar cycle. Despite heated, but healthy, controversy, the reality of internal impulses is now generally accepted, based on new and diferent data. That such brief (~1 yr) phenomena can be observed at the Earth's surface implies low values of mantle conductivity. It now seems probable that conductivity never exceeds a few hundred $\Omega^{-1} m^{-1}$, and laboratory experiments suggest it could be even less.

Geomagnetic jerks of global significance occurred in 1969 and maybe 1979; earlier ones may also have occurred in 1900 and 1912. Compilation and reanalysis of earlier data may one day allow us to recover earlier jerks, but there are very few places where such work can be attempted (e.g. Paris, London). Jerks appear to be the result of short-term instabilities in motions of the external part of the fluid core. Jerks, and secular variation patterns between jerks, can be used to constrain these motions. In order to make these constraints tractable, a number of hypotheses have been proposed, each of which has been the subject of some controversy. The first hypothesis states that the core is an (almost) perfect conductor, and the mantle an (almost) perfect insulator. It now seems reasonable to assume that the ratio of core to mantle conductivity may reach or exceed 10^4, vindicating this first hypothesis. (Corrections can be made in order to take some finite value of conductivity into account.) High core conductivity leads to the frozen-flux approximation, which assumes that dissipation can be neglected compared with advection of field lines on the decade time scale. The period at which this occurs, which depends essentially on the vertical scale of the field in the core, is still a matter of debate and research. We have seen that some authors think that historical data (for the last few centuries) show the frozen-flux approximation to fail on this decade to century time scale, whereas others find it to hold within the available resolution of the data (considering also the nonuniqueness of the inversion techniques). A second hypothesis pertains to the balance of forces acting on the core fluid. Arguments can be made that the Coriolis force dominates the budget and is the force-balancing the pressure gradient: This is the so-called geostrophic approximation. Under these two approximations, the pattern of core motions obtained by inverting observed secular variation is remarkably simple. It has a strong equatorial "dipolar" component, with an ascending jet to the south of India and a descending limb off the coast of Peru. This remarkable simplicity is an argument in favor of its reality, and it has interesting consequences on the circulation of the deeper core and on core-mantle coupling and the pattern of lower-mantle convection. For instance, Le Mouël et al (1986) have shown a striking correlation between observed long-wavelength geoid anomalies and anomalies computed from density differences induced by the geostrophic motion based on magnetic secular variation.

Ongoing research points to the need for better coverage and data acquisition in space and in time, at scales covering many orders of magnitude. Routine operation of magnetic observatories is an absolute requirement that is often ill understood. Only a large number of high-quality observatories with improved global coverage can provide the long-term data that are the basis for the advances described in this review. International encouragement and cooperation are needed in this respect, so that additional high-quality automated observatories can be installed, and so that existing ones are not discontinued for funding reasons without serious analysis. Also vital to progress in understanding geomagnetic secular variation is the launching of a series of magnetic satellites with long-observation periods. Following the initial snapshot provided by MAGSAT, we now urgently need some kind of permanent magnetic space observatory. The return in terms of understanding the dynamics of the Earth's core would be invaluable, for a comparatively modest investment. An international working group has been established by the International Association of Geomagnetism and Aeronomy (member of IUGG-ICSU), and joint teams (for example, a Franco-American one with representatives of the French and US space agencies) are already at work.

Improvement of data bases back in time is also essential. Important information can be gained by compiling existing historical information. This work, of course, cannot extend the data base back beyond the first magnetic measurements in the sixteenth century. Before this, one must decipher magnetic information stored in natural and artificial rocks, the domain of archaeomagnetism and paleomagnetism. That the gap between geo- and paleomagnetism should be bridged has been advocated for a long time but for several reasons is now more urgent, obvious, and also feasible than ever. Records of paleosecular variation in lakes are more numerous and reliable. A most promising topic is the highly detailed study of geomagnetic reversals in sedimentary or volcanic sequences. An increase in the number of high-quality records of (the same) recent reversals can be expected in coming years. The time scale of reversals ($\sim 10^4$ yr) clearly merges with that of secular variation, and phenomenological models of reversals are now proposed. We have the first indications that secular variation (in its usual sense at the 100-yr scale) may be recovered within a reversal, although this requires confirmation. The same hypotheses used on shorter time scales, such as the frozen-flux approximation, might become testable with such data. It seems safe to predict that major progress will come from such joint work of geo- and paleomagneticians, dynamo specialists, and data gatherers. Progress made in the last 10 years and reported in this review foretells even more exciting progress in the coming decade.

Acknowledgements

This review was written while one of us (VC) enjoyed a sabbatical leave at the University of California, Santa Barbara. The Department of Geology and its chairman, Mike Fuller, are thanked for their kind support. Jeremy Bloxham generously provided the plots shown in Figure 35. George Backus, Robert Langel, and Coerte Voorhies provided preprints. Daniel Gilbert, Camille Gire, Michel Menvielle, and Pascal Tarits kindly helped at various stages of data gathering, and Randy Enkin helped in improving the manuscript. Guy Aveline, J. Distrophe, Gisèle Dupin, and E. Lesur redrafted some of the figures. This article is IPGP Contribution NS 980.

Literature Cited

Abrahamsen, N. 1973. Magnetic secular variation in Denmark, 1500–1970. *J. Geomagn. Geoelectr.* 25: 105–11

Achache, J., Courtillot, V., Ducruix, J., Le Mouël, J. L. 1980. The late 1960's secular variation impulse: further constraints on deep mantle conductivity. *Phys. Earth Planet. Inter.* 23: 72–75

Achache, J., Le Mouël, J. L., Courtillot, V. 1981. Long-period geomagnetic variations and mantle conductivity: an inversion using Bailey's method. *Geophys. J. R. Astron. Soc.* 65: 579–601

Alldredge, L. R. 1975. A hypothesis for the source of impulses in geomagnetic secular variations. *J. Geophys. Res.* 80: 1571–78

Alldredge, J. R. 1976. Effects of solar activity on annual means of geomagnetic components. *J. Geophys. Res.* 81: 2990–96

Alldredge, L. R. 1977a. Geomagnetic variations with periods from 13 to 30 years. *J. Geomagn. Geoelectr.* 29: 123–35

Alldredge, L. R: 1977b. Deep mantle conductivity. *J. Geophys. Res.* 82: 5427–31

Alldredge, L. R. 1979. Commentaire sur: "Sur une accélération recente de la variation séculaire du champ magnétique terrestre" de V. Courtillot, J. Ducruix et J. L. Le Mouël. *C. R. Acad. Sci. Paris Ser. B* 289: 169–71

Alldredge, L. R. 1981. Magnetic signals from the core of the Earth and secular variation. *J. Geophys. Res.* 86: 7957–65

Alldredge, L. R. 1982. Geomagnetic models and the solar cycle effect. *Rev. Geophys. Space Phys.* 20: 965–70

Alldredge, L. R. 1983a. Varying geomagnetic anomalies and secular variation. *J. Geophys. Res.* 88: 9443–51

Alldredge, L. R. 1983b. Long term and recent variation of the geomagnetic dipole moment. *J. Geomagn. Geoelectr.* 35: 215–20

Alldredge, L. R. 1984. A discussion of impulses and jerks in the geomagnetic field. *J. Geophys. Res.* 89: 4403–12

Alldredge, L. R., 1985a. More on the alleged 1970 geomagnetic jerk. *Phys. Earth Planet. Inter.* 39: 255–64

Alldredge, L. R. 1985b. Reply. *J. Geophys. Res.* 90: 6899–6901

Alldredge, L. R., Stearns, C. O., Sugiura, M. 1979. Solar cycle variation in geomagnetic external spherical harmonic coefficients. *J. Geomagn. Geoelectr.* 31: 495–508

Backus, G. 1968. Kinematics of geomagnetic secular variation in a perfectly conducting core. *Philos. Trans. R. Soc. London Ser. A* 263: 239–66

Backus, G. 1983. Application of mantle filter theory to the magnetic jerk of 1969. *Geophys. J. R. Astron. Soc.* 74: 713–46

Backus, G. 1987. Bayesian inference in geomagnetism. *Geophys. J. R. Astron. Soc.* In press

Backus, G., Hough, S. 1985. Some models of the geomagnetic field in western Europe from 1960 to 1980. *Phys. Earth Planet. Inter.* 39: 243–54

Backus, G. E., Le Mouël, J. L. 1986. The region of the core-mantle boundary where a geostrophic velocity field can be determined from frozen-flux magnetic data. *Geophys. J. R. Astron. Soc.* 85: 617–28

Backus, G., Estes, R. H., Chinn, D., Langel, R. A. 1987. Comparing the jerk with other global models of the geomagnetic field from 1960 to 1978. *J. Geophys. Res.* 92: 3615–22

Bailey, R. C. 1970. Inversion of the geomagnetic induction problem. *Proc. R. Soc. London* 315: 185–94

Banerjee, S. K. 1983. The Holocene paleomagnetic record in the U.S.A. In *Late Quaternary of the United States*, ed. S. C. Porter, H. E. Wright, Jr., Minneapolis: Univ. Minn. Press

Banks, R. J. 1969. Geomagnetic variations and the electrical conductivity of the upper mantle. *Geophys. J. R. Astron. Soc.* 17: 457–87

Banks, R. J. 1972. The overall conductivity distribution of the Earth. *J. Geomagn. Geolectr.* 24: 337–51

Banks, R. J. 1981. Strategies for improved global electromagnetic response estimates. *J. Geomagn. Geoelectr.* 33: 569–85

Barraclough, D. R. 1974. Spherical harmonic models of the geomagnetic field for eight epochs between 1610 and 1900. *Geophys. J. R. Astron. Soc.* 36: 497–513

Barraclough, D. R. 1978. Spherical harmonic models of the geomagnetic field. *Geomag. Bull. Inst. Geol. Sci.*, Vol. 8

Barraclough, D. R. 1985. A comparison of satellite and observatory estimates of geomagnetic secular variation. *J. Geophys. Res.* 90: 2523–26

Barton, C. E. 1982. Spectral analysis of paleomagnetic time series and the geomagnetic spectrum. *Philos. Trans. R. Soc. London Ser. A* 306: 203–9

Barton, C. E. 1983. Analysis of paleomagnetic time-series—techniques and applications. *Geophys. Surv.* 5: 335–68

Barton, C. E., McElhinny, M. W. 1982. Time series analysis of the 10,000 yr geomagnetic secular variation record from SE Australia. *Geophys. J. R. Astron. Soc.* 68: 709–24

Barton, C. E., Merrill, R. T. 1983. Archaeo- and paleosecular variation, and long-term asymmetries of the geomagnetic field. *Rev. Geophys. Space Phys.* 21: 603–14

Barton, C. E., Merrill, R. T., Barbetti, M. 1979. Intensity of the Earth's magnetic field over the last 10,000 years. *Phys. Earth Planet. Inter.* 20: 96–110

Bauer, L. A. 1895. On the distribution and the secular variation of terrestrial magnetism, No. III. *Am. J. Sci.* 50: 314–25

Bauer, L. A. 1899. On the secular variation of a free magnetic needle. *Phys. Rev.* 3: 34–48

Becquerel, A. C. 1840. *Traite Expérimental de l'Electricité et du Magnétisme.* Paris: Fermin Didot Frères

Bemmelen, W. van. 1899. Die abweichung der magnetnadel: Beobachtungen, Saecular-variation, wert- und isogonen-systeme bis zur mitte des XVIIIten jahrhunderts. *Magn. Meteorol. Obs. Batavia,* Vol. 21 (Suppl.)

Benton, E. R., Whaler, K. A. 1983. Rapid diffusion of the poloidal geomagnetic field through a weakly conducting mantle: a perturbation solution. *Geophys. J. R. Astron. Soc.* 75. 77–100

Bitterly, J., Cantin, J. M., Burdin, J., Schlich, R., Gilbert, D. 1987. Digital recording of geomagnetic variations in French observatories: results for the 1972–1986 period. *IUGG Gen. Assem., 19th, Vancouver,* 2: 670 (Abstr.)

Bloxham, J. 1986a. The expulsion of magnetic flux from the Earth's core. *Geophys. J. R. Astron. Soc.* 87: 669–78

Bloxham, J. 1986b. Models of the magnetic field at the core-mantle boundary for 1715, 1777 and 1842. *J. Geophys. Res.* 91: 13,954–66

Bloxham, J., Gubbins, D. 1985. The secular variation of Earth's magnetic field. *Nature* 317: 777–81

Bloxham, J., Gubbins, D. 1986. Geomagnetic field analysis—IV. Testing the frozen-flux hypothesis. *Geophys. J. R. Astron. Soc.* 84: 139–52

Booker, J. R. 1969. Geomagnetic data and core motions. *Proc. R. Soc. London Ser. A* 309: 27–40

Braginsky, S. I. 1970. Oscillation spectrum of the hydromagnetic dynamo of the Earth. *Geomagn. Aeron.* 10: 172–81

Braginsky, S. I. 1972 Spherical analyses of the main geomagnetic field of 1550–1800. *Geomagn. Aeron.* 12: 464–68

Braginsky, S. I. 1984. Short-period geomagnetic secular variation. *Geophys. Astrophys. Fluid Dyn.* 30: 1–78.

Bullard, E. C., Freedman, C., Gellman, H., Nixon, J. 1950. The westward drift of the Earth's magnetic field. *Philos. Trans. R. Soc. London Ser. A* 243: 67–92

Cain, J. C., Hendricks, S. 1968. The geomagnetic secular variation 1900–65. *NASA TN D-4527.* 221 pp.

Cain, J. C., Frayser, J., Muth, L., Schmitz, D. 1983. The use of Magsat data to determine secular variation. *J. Geophys. Res.* 88: 5903–10

Cain, J. C., Schmitz, D. R., Kluth, C. 1985. Eccentric geomagnetic dipole drift. *Phys. Earth Planet. Inter.* 39: 237–42

Campbell, W. H. 1976. An analysis of the spectra of geomagnetic variations having periods from five minutes to four hours. *J. Geophys. Res.* 81: 1369–90

Champion, D. E. 1980. Holocene geomagnetic secular variation in the Western United States: implications for the global geomagnetic field. *US Geol. Surv. Open-File Rep. 80–824.* 314 pp.

Chapman, S., Bartels, J. 1940, 1962. *Geomagnetism,* Vols. 1, 2. New York: Oxford Univ. Press. 1049 pp. 2nd ed.

Chau, H. D., Ducruix, J., Le Mouël, J. L. 1981. Sur le caractère planétaire du saut de variation séculaire de 1969–1970. *C. R. Acad. Sci. Paris Ser. B* 293: 157–60

Chimiddorzh, G., Byamba, Ch., Gunchin-Ish, A., Sukhbaator, U., Afanas'yeva, V. I. 1977. Secular variation of the geomagnetic field during 1966–1975 according to observations at Ulan-Bator observatory. *Geomagn. Aeron.* 17: 250–51

Counil, J. L., Menvielle, M., Le Mouël, J. L. 1987. Upper mantle lateral heterogeneities and magnetotelluric daily variation data. *Pure Appl. Geophys.* 125: 319–40

Courtillot, V., Le Mouël, J. L. 1976a. On the long period variations of the Earth's magnetic field from 2 months to 20 years. *J. Geophys. Res.* 81: 2941–50

Courtillot, V., Le Mouël, J. L. 1976b. Time variations of the Earth's magnetic field with a period longer than two months. *Phys. Earth Planet. Inter.* 12: 237–40

Courtillot, V., Le Mouël, J. L. 1979. Comment on "Deep mantle conductivity" by L. R. Alldredge. *J. Geophys. Res.* 84: 4785–90

Courtillot, V., Le Mouël, J. L. 1984. Geomagnetic secular variation impulses: a review of observational evidence and geophysical consequences. *Nature* 311: 709–16

Courtillot, V., Le Mouël, J. L. 1985. Comment on "A discussion of impulses and jerks in the geomagnetic field" by L. R. Alldredge. *J. Geophys. Res.* 90: 6897–98

Courtillot, V., Le Mouël, J. L., Mayaud, P. N. 1977. Maximum entropy spectral analysis of the geomagnetic activity index *aa* over a 107-yr interval. *J. Geophys. Res.* 82: 2641–49

Courtillot, V., Ducruix, J., Le Mouël, J. L. 1978a. Sur une accélération récente de la variation séculaire du champ magnétique terrestre. *C. R. Acad. Sci. Paris. Ser. D* 287: 1095–98

Courtillot, V., Le Mouël, J. L., Leprêtre, B. 1978b. Réseau magnétique de répétition de la France, campagne 1977. In *Observations Magnétiques*, 35: 1–28. Paris: Inst. Phys. Globe

Courtillot, V., Ducruix, J., Le Mouël, J. L. 1979. Réponse aux commentaires de L. R. Alldredge "Sur une accélération récente de la variation séculaire du champ magnétique terrestre." *C. R. Acad. Sci. Paris Ser. B* 289: 173–75

Courtillot, V., Le Mouël, J. L., Ducruix, J., Cazenave, A. 1982. Geomagnetic secular variation as a precursor of climatic change. *Nature* 297: 386–87

Courtillot, V., Le Mouël, J. L., Ducruix, J., Cazenave, A. 1983. Corrigendum. *Nature* 303: 638

Courtillot, V., Le Mouël, J. L., Ducruix, J. 1984. On Backus' mantle filter theory and the 1969 geomagnetic impulse. *Geophys. J. R. Astron. Soc.* 78: 619–25

Creer, K. M. 1981. Long-period geomagnetic secular variations since 12,000 yr B.P. *Nature* 292: 208–12

Creer, K. M. 1983. Computer synthesis of geomagnetic palaeosecular variations. *Nature* 304: 695–99

Creer, K. M. 1985. Review of lake sediment palaeomagnetic data. *Geophys. Surv.* 7: 125–60

Creer, K. M., Tucholka, P. 1983. Epilogue. See Creer et al 1983, pp. 273–306

Creer, K. M., Tucholka, P., Barton, C. E., eds. 1983. *Geomagnetism of Baked Clays and Recent Sediments.* Amsterdam: Elsevier. 324 pp.

Creer, K. M., Smith, G., Tucholka, P., Bonifay, E., Thouveny, N., Truze, E. 1986. A preliminary palaeomagnetic study of the Holocene and late Wurmian sediments of Lac du Bouchet (Haute-Loire, France). *Geophys. J. R. Astron. Soc.* 86: 943–64

Currie, R. G. 1966. The geomagnetic spectrum—40 days to 5.5 years. *J. Geophys. Res.* 71: 4579–98

Currie, R. G. 1967. Magnetic shielding properties of the Earth's mantle. *J. Geophys. Res.* 72: 2623–33

Currie, R. G. 1968. Geomagnetic spectrum of internal origin and lower mantle conductivity. *J. Geophys. Res.* 73: 2779–86

Currie, R. G. 1973a. Geomagnetic line spectra—2 to 70 years. *Astrophys. Space. Sci.* 21: 425–38

Currie, R. G. 1973b. Pacific region anomaly in the geomagnetic spectrum at ~60 years *S. Afr. J. Sci.* 69: 379–83

Dodson, R. E. 1979. Counterclockwise precession of the geomagnetic field vector and westward drift of the non-dipole field. *J. Geophys. Res.* 84: 637–44

Doell, R., Cox, A. V. 1972. The Pacific geomagnetic secular variation anomaly and the question of lateral uniformity in the lower mantle. In *Nature of the Solid Earth*, ed. E. C. Robertson, pp. 245–84. New York: McGraw-Hill

Ducruix, J. 1980. *Description et analyse de quelques variations spatiales et temporelles du champ géomagnétique.* PhD thesis. Univ. Paris VII, Fr. 164 pp.

Ducruix, J., Courtillot, V., Le Mouël, J. L. 1980. The late 1960's secular variation impulse, the eleven year magnetic variation and the electrical conductivity of the deep mantle. *Geophys. J. R. Astron. Soc.* 61: 73–94

Ducruix, J., Gire, C., Le Mouël, J. L. 1983. Existence et caractère planétaire d'une secousse de la variation séculaire du champ géomagnétique en 1912–1913. *C. R. Acad. Sci. Paris Ser. B* 296: 1419–24

Eckhardt, D., Larner, K., Madden, T. 1963. Long period magnetic fluctuations and mantle electrical conductivity estimates. *J. Geophys. Res.* 68: 6279–86

Filloux, J. H. 1980. Observation of very low frequency electromagnetic signals in the ocean. *J. Geomagn. Geoelectr.* 32: I-1–12 (Suppl.)

Fischer, G., Schnegg, P. A., Le Quang, B. V. 1981. An analytic one-dimensional inversion scheme. *Geophys. J. R. Astron. Soc.* 67: 257–78

Fisk, H. W. 1931. Magnetic secular variation and solar activity *Int. Res. Counc.—Comm. Sol. Terrest. Relation. Rep., 3rd,* pp. 52–59

Fraser-Smith, A. C. 1987. Centered and eccentric geomagnetic dipoles and their poles, 1600–1985. *Rev. Geophys.* 25: 1–16

Fritsche, H. 1900. *Die Elemente des Erdmagnetismus und ihre saekularen Aenderungen waehrend des Zeitraums 1550 bis 1900.* St. Petersburg: A. Jacobson

Fujita, N. 1973. Secular change of the geomagnetic total force in Japan for 1970.0. *J. Geomagn. Geoelectr.* 25: 181–94

Fuller, M. 1987. Experimental methods in rock magnetism and paleomagnetism. In *Methods of Experimental Physics*, ed. C.

G. Sammis, T. L. Henyey, 24A: 303–471. New York: Academic

Gaibar-Puertas, C. 1953. Variacion secular del campo geomagnetico. *Obs. Ebro Mem.* No. 11, Tarragona, Spain

Gauss, C. R. 1838. *Allgemeine Theorie des Erdmagnetismus.* In Resultate magn. verein (Reprinted in *Werke,* Vol. 5, p. 121)

Gavoret, J., Gibert, D., Menvielle, M., Le Mouël, J. L. 1986. Long-term variations of the external and internal components of the Earth's magnetic field. *J. Geophys. Res.* 91: 4787–96

Gilbert, D., Cantin, J. M., Bitterly, J., Schlich, R., Folques, J. 1987. Absolute measurements of the Earth's magnetic field in French observatories. *IUGG Gen. Assem., 19th, Vancouver,* 2: 669 (Abstr.)

Gire, C. 1985. *Sur la variation séculaire du champ magnétique terrestre et les mouvements des couches externes du noyau fluide.* PhD thesis. Univ. Paris VII, Fr. 289 pp.

Gire, C., Le Mouël, J. L. 1985. Flow in the fluid core and Earth's rotation. In *Earth Rotation: Solved and Unsolved Problems,* ed. A. Cazenave, pp. 241–58. Dordrecht: D. Reidel. 330 pp.

Gire, C., Le Mouël, J. L., Ducruix, J. 1984a. Evolution of the geomagnetic secular variation field from the beginning of the century. *Nature* 307: 349–52

Gire, C., Le Mouël, J. L., Madden, T. 1984b. The recent westward drift rate of the geomagnetic field and the body drift of external layers of the core. *Ann. Géophys.* 2: 37–46

Gire, C., Le Mouël, J. L., Madden, T. 1986. Motions at the core surface derived from SV data. *Geophys. J. R. Astron. Soc.* 84: 1–29

Gubbins, D. 1982. Finding core motions from magnetic observations. *Philos. Trans. R. Soc. London Ser. A* 306: 247–54

Gubbins, D. 1983. Geomagnetic field analysis—I. Stochastic inversion. *Geophys. J. R. Astron. Soc.* 73: 641–52

Gubbins, D. 1984. Geomagnetic field analysis—II. Secular variation consistent with a perfectly conducting core. *Geophys. J. R. Astron. Soc.* 77: 753–66

Gubbins, D. 1986. Global models of the magnetic field in historical times: augmenting declination observations with archeo- and paleomagnetic data. *J. Geomagn. Geoelectr.* 38: 715–20

Gubbins, D. 1987. Mechanism for geomagnetic polarity reversals. *Nature* 326: 167–69

Gubbins, D., Bloxham, J. 1985. Geomagnetic field analysis—III. Magnetic fields on the core-mantle boundary. *Geophys. J. R. Astron. Soc.* 80: 695–713

Halley, E. 1683. A theory of the variation of the magnetic compass. *Philos. Trans. R. Soc. London* 12: 208–21

Halley, E. 1692. An account of the cause of the change of the variation of the magnetic needle, with an hypothesis of the structure of the earth. *Philos. Trans. R. Soc. London* 17: 563–78

Hansteen, C. 1819. *Untersuchungen ueber den Magnetismus der Erde.* Christiana, Norw: Lehmann and Groendahl

Harwood, J. M., Malin, S. R. C. 1976. Present trends in the Earth's magnetic field. *Nature* 259: 469–71

Harwood, J. M., Malin, S. R. C. 1977. Sunspot cycle influence on the geomagnetic field. *Geophys. J. R. Astron. Soc.* 50: 605–19

Hide, R. 1966. Free hydromagnetic oscillations of the earth's core and the theory of geomagnetic secular variation. *Philos. Trans. R. Soc. London Ser. A* 259: 615–47

Hide, R. 1969. Interaction between the Earth's liquid core and solid mantle. *Nature* 222: 1055–56

Hide, R. 1970. On the Earth's core-mantle interface. *Q. J. R. Meteorol. Soc.* 96: 579–90

Hide, R. 1978. How to locate the electrically conducting fluid core of a planet from external magnetic observations. *Nature* 271: 640–41

Hide, R. 1982. On the role of rotation in the generation of magnetic fields by fluid motion. *Philos. Trans. R. Soc. London Ser. A* 306: 223–34

Hide, R., Malin, S. R. C. 1981. On the determination of the size of earth's core from observations of the geomagnetic secular variation. *Proc. R. Soc. London Ser. A* 374: 15–33

Hide, R., Stewartson, K. 1977. Hydromagnetic oscillations of the Earth's core. *Rev. Geophys. Space Phys.* 10: 579–98

Hide, R., Birch, N. T., Morrison, L. V., Shea, D. J., White, A. A. 1980. Atmospheric angular momentum fluctuations and changes in length of day. *Nature* 286: 114–17

Hills, R. G. 1979. *Convection in the earth's mantle due to viscous shear at the core-mantle interface and due to large scale buoyancy.* PhD thesis. N. Mex. State Univ., Las Cruces

Hinderer, J. 1987. *Sur quelques effets en rotation et déformation d'une planète à noyau liquide, manteau élastique et couche fluide superficielle.* Thèse d'Etat. Univ. Louis Pasteur, Strasbourg. 361 pp.

Hinderer, J., Gire, C., Legros, H., Le Mouël, J. L. 1987. Geomagnetic secular variation, core motions and implications for the Earth's wobbles. *Phys. Earth Planet. Inter.* 49: 121–32

Hobbs, B. A. 1983. Inversion of broad frequency band geomagnetic response data. *J. Geomagn. Geoelectr.* 35: 723–32

Hodder, B. 1981. Geomagnetic secular variation since 1901. *Geophys. J. R. Astron. Soc.* 65: 763–76

Hodder, B. 1983. Geomagnetic westward drift using the correlation coefficient. *Nature* 306: 136–37

Jacobs, J. A. 1984. *Reversals of the Earth's Magnetic Field.* Bristol, Engl: Adam Hilger. 230 pp.

Jady, R. J. Paterson, G. A. 1983. Inversion methods applied to *Dst* data. *J. Geomagn. Geoelectr.* 35: 733–46

James, R. W. 1970. Decomposition of geomagnetic secular variation into drifting and non-drifting components. *J. Geomagn. Geoelectr.* 22: 241–52

Kahle, A. B., Ball, R. H., Vestine, E. H. 1967. Comparison of estimates of surface fluid motions of the Earth's core for various epochs. *J. Geophys. Res.* 72: 4917–25

Kerridge, D. J., Barraclough, D. R. 1985. Evidence for geomagnetic jerks from 1931 to 1971. *Phys. Earth Planet. Inter.* 39: 228–36

Kovacheva, M. 1980. Summarized results of the archaeomagnetic investigations of the geomagnetic field variation for the last 8000 yr in southeastern Europe. *Geophys. J. R. Astron. Soc.* 61: 57–64

Lahiri, B. N., Price, A. T. 1939. Electromagnetic induction in non-uniform conductors, and the determination of the conductivity of the Earth from terrestrial magnetic variations. *Philos. Trans. R. Soc. London Ser. A* 237: 509–40

Lambeck, K. 1980. *The Earth's Variable Rotation: Geophysical Causes and Consequences.* Cambridge: Cambridge Univ. Press. 449 pp.

Langel, R. A., Estes, R. H. 1985a. Large-scale, near-Earth magnetic fields from external sources and the corresponding induced internal field. *J. Geophys. Res.* 90: 2487–94

Langel, R. A., Estes, R. H. 1985b. The near-Earth magnetic field at 1980 determined from MAGSAT data. *J. Geophys. Res.* 90: 2495–2509

Langel, R. A., Estes, R. H., Mead, G. D. 1982. Some new methods in geomagnetic field modelling applied to the 1960–1980 epoch. *J. Geomagn. Geoelectr.* 34: 327–49

Langel, R. A., Kerridge, D. J., Barraclough, D. R., Malin, S. R. C. 1986. Geomagnetic temporal change: 1903–1982, a spline representation. *J. Geomagn. Geoelectr.* 38: 573–97

Leaton, B. R., Malin, S. R. C. 1967. Recent changes in the magnetic dipole moment of the Earth. *Nature* 213: 1110

Le Mouël, J. L. 1976a. Le champ geomagnétique. In *Traité de Géophysique Interne*, ed. J. Coulomb, G. Jobert, 2: 9–67. Paris: Masson

Le Mouël, J. L. 1976b. L'induction dans le globe. In *Traité de Géophysique Interne*, ed. J. Coulomb, G. Jobert, 2: 129–59. Paris: Masson

Le Mouël, J. L. 1976c. L'origine du champ magnétique terrestre. In *Traité de Géophysique Interne*, ed. J. Coulomb, G. Jobert, 2: 161–200. Paris: Masson

Le Mouël, J. L. 1984. Outer-core geostrophic flow and secular variation of earth's geomagnetic field. *Nature* 311: 734–35

Le Mouël, J. L. 1986. La variation séculaire du champ magnétique terrestre et les mouvements au noyau. *La vie des Sciences, C. R. Acad. Sci. Paris* 3(5): 451–68

Le Mouël, J. L., Courtillot, V. 1981. Core motions, electromagnetic core-mantle coupling and variations in the Earth's rotation: new constraints from geomagnetic secular variation impulses. *Phys. Earth Planet. Inter.* 24: 236–41

Le Mouël, J. L., Courtillot, V. 1982. On the outer layers of the core and geomagnetic secular variation. *J. Geophys. Res.* 87: 4103–8

Le Mouël, J. L., Madden, T. R., Ducruix, J., Courtillot, V. 1981. Decade fluctuations in geomagnetic westward drift and Earth rotation. *Nature* 290: 763–65

Le Mouël, J. L., Ducruix, J., Chau, H. D. 1982. The worldwide character of the 1969–1970 impulse of the secular acceleration rate. *Phys. Earth Planet. Inter.* 28: 337–50

Le Mouël, J. L., Ducruix, J., Chau, H. D. 1983. On the recent variations of the apparent westward drift rate. *Geophys. Res. Lett.* 10: 369–72

Le Mouël, J. L., Gire, C., Madden, T. 1985a. Motions at the core surface in the geostrophic approximation. *Phys. Earth Planet. Inter.* 39: 270–87

Le Mouël, J. L., Gire, C., Hinderer, J. 1985b. Sur l'excitation possible de l'oscillation chandlerienne par les mouvements à la surface du noyau. *C. R. Acad. Sci. Paris Ser. B* 301: 27–32

Le Mouël, J. L., Gire, C., Jaupart, C. 1986. Sur la corrélation entre la figure de la terre et les mouvements animant les couches externes de son noyau. *C. R. Acad. Sci. Paris Ser. B* 303: 613–18

Lowes, F. J. 1966. Mean square values on sphere of spherical harmonic vector fields. *J. Geophys. Res.* 71: 2179

Madden, T., Le Mouël, J. L. 1982. The recent secular variation and the motions at the core surface. *Philos. Trans. R. Soc. London Ser. A* 306: 271–80

474   COURTILLOT & LE MOUEL

Malin, S. R. C. 1984. The observational data on GSV over the past 400 years. *Eos, Trans. Am. Geophys. Union,* 65: 203 (Abstr.)

Malin, S. R. C., Bullard, E. 1981. The direction of the Earth's magnetic field at London, 1570–1975. *Philos. Trans. R. Soc. London Ser. A* 299: 357–423

Malin, S. R. C., Clark, A. D. 1974. Geomagnetic secular variation, 1962.5 to 1967.5. *Geophys. J. R. Astron. Soc.* 36: 11–20

Malin, S. R. C., Gupta, J. C. 1977. The Sq current system during the International Geophysical Year. *Geophys. J. R. Astron. Soc.* 49: 515–29

Malin, S. R. C., Hodder, B. M. 1982. Was the 1970 geomagnetic jerk of internal or external origin? *Nature* 296: 726–28

Malin, S. R. C., Isikara, A. M. 1976. Annual variations of the geomagnetic field. *Geophys. J. R. Astron. Soc.* 48: 515–29

Malin, S. R. C., Hodder, B. M., Barraclough, D. R. 1983. Geomagnetic secular variation: a jerk in 1970. In *75th Anniversary Volume of Ebro Observatory,* ed. J. O. Cardus, pp. 239–56. Tarragona, Spain: Ebro Obs.

Mayaud, P. N. 1965. Analyse morphologique de la variabilité jour-à-jour de la variation journalière "régulière" S_R du champ magnétique terrestre. *Ann. Géophys.* 21: 369–401, 514–44

Mayaud, P. N. 1980. *Derivation, Meaning and Use of Geomagnetic Indices. Geophys. Monogr. Ser.,* Vol. 22. Washington DC: Am. Geophys. Union. 154 pp.

McDonald, K. L. 1957. Penetration of the geomagnetic secular field through a mantle with variable conductivity. *J. Geophys. Res.* 62: 117–41

McDonald, K. L. 1970. Monthly means of the geomagnetic field. *J. Geophys. Res.* 75: 5631–33

McElhinny, M. 1983. Analysis of global intensities for the past 50,000 years. See Creer et al 1983, pp. 176–81

McElhinny, M., Senanayake, W. E. 1982. Variations of the geomagnetic dipole 1: the past 50,000 years. *J. Geomagn. Geoelectr.* 34: 39–51

McFadden, P. L., McElhinny, M. W. 1982. Variations in the geomagnetic dipole 2: statistical analysis of VDM's for the past 5 million years. *J. Geomagn. Geoelectr.* 34: 163–89

McFadden, P. L., McElhinny, M. W. 1984. A physical model for paleosecular variation. *Geophys. J. R. Astron. Soc.* 78: 809–30

McFadden, P. L., Merrill, R. T. 1984. Lower mantle convection and geomagnetism. *J. Geophys. Res.* 89: 3354–62

McLeod, M. G. 1982. A note on the secular variation of the geomagnetic field. *Phys. Earth Planet. Inter.* 29: 119–34

McLeod, M. G. 1985. On the geomagnetic jerk of 1969. *J. Geophys. Res.* 90: 4597–4610

McWilliams, M. O., Holcomb, R. T., Champion, D. E. 1982. Geomagnetic secular variation from ^{14}C-dated lava flows in Hawaii and the question of the Pacific non-dipole low. *Philos. Trans. R. Soc. London Ser. A* 306: 211–22

Merrill, R. T., McElhinny, M. W. 1983. *The Earth's Magnetic Field—Its History, Origin and Planetary Perspective.* London: Academic. 401 pp.

Morelli, A., Dziewonski, A. M. 1987. Topography of the core-mantle boundary and lateral homogeneity of the liquid core. *Nature* 325: 678–83

Morrison, L. V. 1979. Re-determination of the decade fluctuations in the rotation of the Earth in the period 1861–1978. *Geophys. J. R. Astron. Soc.* 58: 349–60

Nagata, T. 1962. Two main aspects of geomagnetic secular variation—westward drift and non-drifting components. *Proc. Benedum Earth Magn. Symp., Pittsburgh,* pp. 39–55

Nagata, T. 1965. Main characteristics of recent geomagnetic secular variation. *J. Geomagn. Geoelectr.* 17: 263–76

Nevanlinna, H. 1985. On external and internal parts of the geomagnetic jerk of 1970. *Phys. Earth Planet. Inter.* 39: 265–69

Parker, R. L. 1970. The inverse problem of electrical conductivity in the mantle. *Geophys. J. R. Astron. Soc.* 22: 121–38

Parker, R. L. 1980. The inverse problem of electromagnetic induction: existence and construction of solutions based upon incomplete data. *J. Geophys. Res.* 85: 4421–28

Parker, R. L. 1983. The magnetotelluric inverse problem. *Geophys. Surv.* 6: 5–26

Quinn, J. M., Kerridge, D. J., Barraclough, D. R. 1986. World magnetic charts for 1985—spherical harmonic models of the geomagnetic field and its secular variation. *Geophys. J. R. Astron. Soc.* 87: 1143–57

Rayet, G. 1876. Recherches sur les observations magnétiques faites à l'observatoire de Paris. *Ann. Obs. Paris Mém.* 13: 1–60

Reid, G. C. 1972. Ionospheric effects of solar activity. In *Solar Activity Observations and Predictions. Prog. Astronaut. Aeronaut,* ed. P. S. MacIntosh, M. Druer. 30: 293–312

Rivin, Y. R. 1974. 11-year periodicity in the horizontal component of the geomagnetic field. *Geomagn. Aeron.* 14: 97–100

Rivin, Y. R. 1976. Spherical analysis of the cyclic variation of the horizontal geomagnetic field component. *Geomagn. Aeron.* 16: 531–32

Roberts, R. G. 1984. The long-period electromagnetic response of the Earth. *Geophys. J. R. Astron. Soc.* 78: 547–72

Roberts, R. G. 1986a. The deep electrical structure of the Earth. *Geophys. J. R. Astron. Soc.* 85: 583–600

Roberts, R. G. 1986b. Global electromagnetic induction. *Geophys. Surv.* 8: 339–74

Roberts, P. H., Scott, S. 1965. On analysis of the secular variation. 1. A hydromagnetic constraint: theory. *J. Geomagn. Geoelectr.* 17: 137–51

Rotanova, N. M., Papitashvili, N. E., Pushkov, A. N., Fishman, V. M. 1985. Spectral statistical spatial analysis of 60 and 30 year geomagnetic field variations and conductivity of the lower mantle. *Ann. Géophys.* 3: 225–38

Runcorn, S. K. 1955. The electrical conductivity of the Earth's mantle. *Trans. Am. Geophys. Union* 36: 191–98

Runcorn, S. K. 1969. On the theory of the geomagnetic secular variation. *Ann. Géophys.* 15: 87–92

Runcorn, S. K. 1970. A possible cause of the correlation between earthquakes and polar motion. In *Earthquake Displacement Fields and the Rotation of the Earth*, ed. L. Mansinha et al, pp. 181–87. Dordrecht: D. Reidel

Sabine, E. 1857. On what the colonial magnetic observatories have accomplished. *Proc. R. Soc. London.* 19 pp.

Sabine, E. 1868–77. Contributions to terrestrial magnetism, No. XI to XV. *Philos. Trans. R. Soc. London* (see Bloxham 1986b)

Schmidt, A. 1916. Ergebnisse der magnetischen Beobachtungen in Potsdam und Seddin in der Jahren 1900–1910. *Abh. K. Preuss. Meteorol. Inst.* 5(3): 90 pp.

Schmucker, U. 1979. Erdmagnetische Variationen und die elektrische Leitfaehigkeit in tieferen Schichten der Erde. In *Sitzungsberichte und Mitteilungen der Braunschweigischen Wissenschaftlichen Gesellschaft*, 4: 45–102. Göttingen: Goltze-Verlag

Schott, S. H. 1978. Jerks: the time rate of change of acceleration. *Am. J. Phys.* 46: 1090–94

Schuster, A. 1889. The diurnal variation of terrestrial magnetism. *Philos. Trans. R. Soc. London Ser. A* 180: 467–518

Shure, L., Parker, R. L., Backus, G. E. 1982. Harmonic splines for geomagnetic field modelling. *Phys. Earth Planet. Inter.* 28: 215–29

Shure, L., Parker, R. L., Langel, R. A. 1985. A preliminary harmonic spline model

from MAGSAT data *J. Geophys. Res.* 90: 11,505–12

Skiles, D. D. 1970. A method of inferring the direction of drift of the geomagnetic field from paleomagnetic data. *J. Geomagn. Geoelectr.* 22: 441–62

Souriau, A., Cazenave, A. 1985. Re-evaluation of the Chandler wobble seismic excitation from recent data. *Earth Planet. Sci. Lett.* 75: 410–16

Speith, M. A., Hide, R., Clayton, R. W., Hager, B. H., Voorhies, C. V. 1986. Topographic coupling of core and mantle and changes in length of day. *Eos, Trans. Am. Geophys. Union* 67: 908 (Abstr.)

Stacey, F. D., Banerjee, S. K. 1974. *The Physical Principles of Rock Magnetism.* Amsterdam: Elsevier. 195 pp.

Stoyko, N. 1951. Sur les variations du champ magnétique et de la rotation de la Terre. *C. R. Acad. Sci. Paris Ser. A* 233: 80–82

Tanguy, J. C., Bucur, I., Thompson, J. F. C. 1985. Geomagnetic secular variation in Sicily and revised ages of historic lavas from Mount Etna. *Nature* 318: 453–55

Thellier, E. 1981. Sur la direction du champ magnétique terrestre en France, durant les deux derniers millénaires. *Phys. Earth Planet. Inter.* 24: 89–132

Thellier, E., Thellier, O. 1959. Sur l'intensité du champ magnétique terrestre dans le passé historique et géologique. *Ann. Géophys.* 15: 285–376

Turner, G. M., Thompson, R,. 1981. Lake sediment record of the geomagnetic secular variation in Britain during Holocene times. *Geophys. J. R. Astron. Soc.* 65: 703–25

Turner, G. M., Thompson, R. 1982. Detransformation of the British geomagnetic secular variation record for Holocene times. *Geophys. J. R. Astron. Soc.* 70: 789–92

Valet, J. P. 1985. *Inversions géomagnétiques du Miocène supérieur en Crète—Modalités du renversement et caractéristiques du champ de transition.* PhD thesis. Univ. Paris Sud, Fr. 195 pp.

Valet, J. P., Laj, C., Tucholka, P. 1986. High resolution sedimentary record of a geomagnetic reversal. *Nature* 322: 27–32

Veinberg, B. P., Shibaev, V. P. 1969. *Catalogue—The Results of Magnetic Determinations at Equidistant Points and Epochs. 1500–1940*, ed. A. N. Pushkov. Moscow: IZMIRAN

Verosub, K. L., Cox, A. V. 1971. Changes in the total magnetic energy external to the earth's core. *J. Geomagn. Geoelectr.* 23: 235–42

Verosub, K. L., Mehringer, P. J. Jr., Waterstraat, P. 1986. Holocene secular variation in Western North America: paleomagnetic record from Fish Lake, Harney County, Oregon. *J. Geophys. Res.* 91: 3609–23

Vestine, E. H. 1952. On the variations of the geomagnetic field, fluid motions and the rate of the Earth's rotation. *Proc. Natl. Acad. Sci. USA* 38: 1030–38

Vestine, E. H., Kahle, E. 1968. The westward drift and geomagnetic secular change. *Geophys. J. R. Astron. Soc.* 15: 29–37

Vestine, E. H., Lange, I., Laporte, L., Scott, W. E. 1947. The geomagnetic field, its description and analysis. *Carnegie Inst. Washington Publ. No. 580*

Voorhies, C. V. 1986. Steady flows at the top of Earth's core derived from geomagnetic field models. *J. Geophys. Res.* 91: 12,444–66

Voorhies, C. V., Backus, G. E. 1985. Steady flows at the top of the core from geomagnetic field models: the steady motions theorem. *Geophys. Astrophys. Fluid Dyn.* 32: 163–73

Voorhies, C. V., Benton, E. R. 1982. Pole strength of the earth from MAGSAT and magnetic determination of core radius. *Geophys. Res. Lett.* 9: 258–61

Walker, J. B., O'Dea, P. L. 1952. Geomagnetic secular change impulses. *Trans. Am. Geophys. Union* 33: 797–800

Whaler, K. A. 1980. Does the whole of the Earth's core convect? *Nature* 287: 528–30

Whaler, K. A. 1982. Geomagnetic secular variation and fluid motion at the core surface. *Philos. Trans. R. Soc. London Ser. A* 306: 235–46

Whaler, K. A. 1984. Fluid upwelling at the core-mantle boundary: resolvability from surface geomagnetic data. *Geophys. J. R. Astron. Soc.* 28: 453–73

Whaler, K. A. 1986. Geomagnetic evidence for fluid upwelling at the core-mantle boundary. *Geophys. J. R. Astron. Soc.* 86: 563–88

Weidelt, P. 1972. The inverse problem of geomagnetic induction. *Z. Geophys.* 38: 257–89

Weber, A. M., Roberts, E. B. 1951. The 1950 world isogonic chart. *J. Geophys. Res.* 56: 81–84

World Data Center for Solid Earth Geophysics. 1987. A directory of geomagnetic observatories with high resolution digital results, 1987. *Natl. Geod. Cent. Rep. SE-43.* 66 pp.

Yukutake, T. 1962. The westward drift of magnetic field of the Earth. *Bull. Earthq. Res. Inst. Univ. Tokyo* 40: 1–65

Yukutake, T. 1965. The solar cycle contribution to the secular change in the geomagnetic field. *J. Geomagn. Geoelectr.* 17: 287–309

Yukutake, T. 1967. The westward drift of the Earth's magnetic field in historic times. *J. Geomagn. Geoelectr.* 19: 103–16

Yukutake, T. 1968. The drift velocity of the geomagnetic secular variation. *J. Geomagn. Geoelectr.* 20: 403–14

Yukutake, T. 1973. Fluctuations in the Earth's rate of rotation related to changes in the geomagnetic dipole field. *J. Geomagn. Geoelectr.* 25: 195–212

Yukutake, T. 1985. A preliminary study on variations in the Gauss coefficients of the geomagnetic potential over several hundred years. *Phys. Earth Planet. Inter.* 39: 217–27

Yukutake, T., Cain, J. C. 1979. Solar cycle variations of the first-degree spherical harmonic components of the geomagnetic field. *J. Geomagn. Geoelectr.* 31: 509–44

Yukutake, T., Tachinaka, H. 1968a. The non-dipole part of the Earth's magnetic field. *Bull. Earthq. Res. Inst. Univ. Tokyo* 46: 1027–74

Yukutake, T., Tachinaka, H. 1968b. The westward drift of the geomagnetic secular variation. *Bull. Earthq. Res. Inst. Univ. Tokyo* 46: 1075–1102

Ann. Rev. Earth Planet. Sci. 1988. 16: 477–541

DEEP SLABS, GEOCHEMICAL HETEROGENEITY, AND THE LARGE-SCALE STRUCTURE OF MANTLE CONVECTION: Investigation of an Enduring Paradox

Paul G. Silver and Richard W. Carlson

Department of Terrestrial Magnetism, Carnegie Institution of Washington, Washington, DC 20015

Peter Olson

Department of Earth and Planetary Sciences, Johns Hopkins University, Baltimore, Maryland 21218

INTRODUCTION

The study of the elastic and compositional structure of the mantle has had a long history (e.g. Adams & Williamson 1925, Jeffreys 1939, Birch 1952). But the notion of a mantle everywhere in motion is much more recent. With the emergence of plate-tectonic theory in the late 1960s, the continual motion of the surface plates was seen to imply that the mantle was convecting at depth. This in turn provided an essential connection between the structural properties of the mantle and its internal motions. Emphasis in research gradually shifted from the earlier question of *whether* the mantle was convecting at all (e.g. Urey 1952) to the characterization of convective style.

More than 20 years of post-plate-tectonic Earth sciences has seen the development of two basic models for mantle convection: one in which the

477

0084–6597/88/0515–0477$02.00

scale of convection involves the whole mantle, and the other in which the upper mantle and lower mantle are convecting separately. The intensity of the debate on this subject has been fueled not only by the geodynamic question of convective style, but also by the closely coupled question of whether the Earth's mantle is chemically stratified. This is primarily because the two-layer models almost invariably invoke a compositional density increase between the upper and lower mantle to enforce the two-layer flow, whereas whole-mantle models are dominated by thermal buoyancy and are characterized by mantle-wide overturn with no dynamically significant chemical stratification. This latter issue of stratification has profound implications for many branches of the Earth sciences, especially for the study of the Earth's evolution and initial state; thus it may be viewed as a feature at least as important as the mode of convection itself.

Probably the strongest argument for whole-mantle convection is the seismological evidence that subducted oceanic lithosphere penetrates into the lower mantle (Jordan 1977, Creager & Jordan 1984, 1986a). Since this implies a substantial amount of flux between the upper and lower mantle, it casts considerable doubt on rigorously stratified two-layer convection models. Other observations, primarily geophysical, also argue against two-layer convection—for example, the absence of evidence for either a large thermal boundary layer between the upper and lower mantle or a low-viscosity, partially molten lower mantle, both of which might be expected for the elevated lower-mantle temperatures implied by such a model.

The evidence most often invoked to support two-layer convection, though more indirect, has been the observations of isotopic heterogeneity in the mantle. The existence of this undeniably ancient variability in trace-element abundances has been interpreted as evidence for distinct, mantle-sized geochemical reservoirs that ostensibly have been preserved by major-element stratification and two-layer convection. In the classic two-layer model (Wasserburg & DePaolo 1979), the upper mantle is a reservoir associated with the plate-tectonic-related chemical differentiation produced by seafloor spreading, whereas the lower mantle is a reservoir relatively enriched in incompatible (and radiogenic) elements that is decoupled from plate tectonics and consequently rarely sampled at the surface. With the realization that more than two isotopic components are necessary to describe the continually increasing isotopic data set, this model has been modified so that the plate-tectonic reservoir is now heterogeneous, consisting of the products of plate-tectonic chemical differentiation.

Taking these geophysical and geochemical observations at face value, we are led to the paradox that the mantle appears to have properties of both whole-mantle and two-layer convection. In recent years, researchers have begun to address this paradox by searching for models that are consistent with both the geochemical and geophysical data. Thus far, the most common approach has been to retain the standard models (whole-mantle or two-layer) and reinterpret the seemingly incompatible observations. For example, Anderson (1984) argues that the geophysical evidence against the hot lower mantle implied by two-layer models can actually be satisfied if the lower mantle is *depleted* in the radiogenic, heat-producing elements rather than enriched, as in the two-layer model of Wasserburg & DePaolo (1979). In addition, he argues (Anderson 1987) that the depth of slab penetration has been grossly overestimated by not properly taking into account the effects of anisotropy or phase changes on seismic waves traveling through the slabs.

Whole-mantle models must reinterpret the geochemical observation of long-lived chemical heterogeneity. The most developed of such models is that proposed by Davies (1984; see also Gurnis & Davies 1986), a schematic of which is shown in Figure 1 (taken from Davies 1984). He argues that the geochemical heterogeneity used as evidence for large mantle reservoirs can equally well be explained by *distributed* heterogeneity produced by plate-tectonic-related chemical differentiation and dispersed randomly throughout the whole mantle. The basic idea of distributed heterogeneity is that all heterogeneous bodies are assumed to be both passive and very small compared with the size of the mantle. Some whole-mantle models do allow for compositional stratification (e.g. Davies 1984, Loper 1985); however, the feature that distinguishes these models from those involving two-layer convection is that the stratification does not control the dynamics and is instead the passive, cumulative result of thermally induced plate-tectonic recycling of the mantle over the age of the Earth.

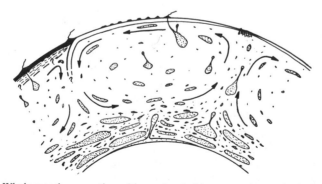

Figure 1   Whole-mantle convection with geochemical heterogeneity randomly distributed throughout the mantle (from Davies 1984). Schematic is meant to illustrate a model of whole-mantle convection that accounts for isotopic heterogeneity without appealing to large reservoirs.

The present contribution, while a review of most of the relevant observations in the study of the mantle, is not a discussion of various arguments for whole-mantle or two-layer convection. Rather, it is concerned primarily with identifying key observations and finding a model or models that satisfy them. Instead of reinterpreting the basic data sets, we show a tendency, at least initially, to accept the "face value" interpretations in spite of their apparent incompatibility and allow for the possibility that the mantle may indeed have properties traditionally associated with both whole-mantle and two-layer convection. If nothing else, this may provide a fresh perspective from which to view the problem.

In this spirit, we focus on the consequences of two lines of evidence: (a) slab penetration into the lower mantle along with its implied return flow, and (b) the geochemical heterogeneity in the Earth's interior. They not only represent the strongest arguments for whole-mantle and two-layer models, respectively, but they also are supported by a relatively firm observational base.

The first sections of the paper are organized according to the major constraints on mantle structure and dynamics: (a) the evidence for lower-mantle slab penetration from seismology; (b) the data on the return flow complementary to subduction from seismic tomography, from the long-wavelength component of the geoid, and from the surficial indications of upwelling (mid-ocean ridges and hotspots); (c) the age of plate-tectonic recycling from Pb isotope systematics; (d) the geochemical evidence for a non-plate-tectonic component; and (e) the constraints on stratification from seismology/mineral physics.

Several mantle models are then evaluated initially by simple mixing calculations in which two constraints on mantle flow are regarded as postulates: (a) that a substantial fraction of subducted lithospheric material (slabs) penetrates well into the lower mantle, and (b) that the plate-tectonic reservoir (i.e. the region of the mantle containing the products of plate-tectonic recycling) has a mean age of 1.7 b.y. The models are then further evaluated in view of the other geophysical and geochemical constraints.

It is found that these two postulates generate a new and rather interesting mantle model that possesses features of both two-layer and whole-mantle convection and is referred to as *penetrative* convection. The basic departure from the standard models is that the subduction-related downwelling from the upper into the lower mantle does not provoke mantle-wide overturn. This is accomplished by the presence of a $\sim 2\%$ intrinsic, compositional density contrast between the upper and lower mantle that serves to minimize entrained flow and hasten the return of the penetrating slab material back into the upper mantle. Although initially motivated by the mixing calculations, this model appears to be highly successful in reconciling many of the available constraints on mantle structure and dynamics. While all the dynamical and geophysical consequences of this model have yet to be established, its potential success argues for careful evaluation.

THE FATE OF SUBDUCTED SLABS

Observations

The fate of subducted slabs is probably the most direct constraint on the mode of mantle convection. Once subduction zones were recognized as the location where lithospheric plates descended into the mantle, their behavior after subduction became a topic of conjecture and active research. Early seismological evidence suggested that slabs remained in the upper mantle. Studies of subduction seismicity found that there were no earthquakes below about 700-km depth (Isacks et al 1968, Isacks & Molnar 1971). Studies of focal mechanisms of deep-focus earthquakes indicated that the slabs just above the seismicity cutoff were almost invariably in down-dip compression (Isacks & Molnar 1971). The most straightforward explanation of these data was that slabs encountered a barrier at this depth and descended no further (e.g. Richter 1979). More recent evidence of a similar sort has been given by Giardini & Woodhouse (1984), who observed severe distortion of the slab beneath Tonga just above 650-km depth that, they suggested, indicated no deeper penetration. However, inferences based on seismicity and focal mechanisms involve two assumptions: (a) that termination of seismicity implies termination of the slab, and (b) that resistance to subduction implies stoppage of the slab altogether.

More direct lines of seismological evidence argue against this interpretation. Slabs have other seismological properties that can and have been used to address the issue of slab penetration into the lower mantle—in particular, their effect on the propagation of seismic waves. The 5–10% increase in seismic velocities within the cold slab represents the largest source of seismic velocity heterogeneity in the mantle. The slab thus has profound effects on the travel times and ray trajectories of seismic waves.

The strongest evidence for lower-mantle slab penetration has come from the residual-sphere analysis of body-wave travel times for deep-focus earthquakes as performed by Jordan (1977) and Creager & Jordan (1984, 1986a). If there is slab material below a deep-focus event, then for a vertically dipping slab, the high seismic velocities within the slab will produce a characteristic pattern of body-wave travel-time residuals corresponding to relatively early arrivals (fast paths) for stations along the strike of the slab and later arrivals (slower paths) for stations at other azimuths. A slightly more complicated but readily predictable pattern is expected for slabs dipping at other angles. The distribution of residuals is projected onto a focal sphere, after corrections for earthquake location and near-receiver heterogeneity, and smoothed in order to enhance near-source heterogeneity. Creager & Jordan (1984, 1986a) have found that the travel-time residuals for deep-focus events are well matched and most easily explained by a lower-mantle, aseismic extension of subducted slabs for most of the subduction zones in the western Pacific and for all of the regions they studied. The residual spheres for some of these events are illustrated in Figure 2 (from Creager & Jordan 1986a). While these studies have generated controversy, this controversy may be due more to their direct consequence of casting doubt on rigorous two-layer convection models than because of serious challenges on seismological grounds. There have been attempts to explain these observations by means other than deep slab penetration, such as other forms of lower-mantle heterogeneity far from the source (Lay 1983), anisotropy (Anderson 1987), or thermal coupling between the upper and lower mantle (no mass flux across the boundary). Without going into the details of these arguments, which may be seen in the above references, it suffices to say that the success of the slab interpretation in explaining the size and character of the travel-time anomaly, the consistency from one subduction zone to the next, and the absence of an equally compelling, quantitative hypothesis argue strongly for this interpretation.

There recently has been independent support for lower-mantle slab penetration from observations of slab-induced waveform complexity. Silver & Chan (1986) have presented a suite of unusually broad and complex shear waves from two deep-focus events ($\sim$ 600-km depth) in the Sea of Okhotsk for stations approximately along the strike of the slab (Figure 3). They find that the simplest explanation for these phases is slab-induced multipathing—that is, the arrival of a body-wave phase by more than one path as a result of strong interaction with the subducted slab. This would appear to require, for deep-focus events, slab material to be present some 300–400 km below these events. Similar complexity has been observed subsequently for other azimuths on the down-dip side of the slab by Beck & Lay (1986), with enhanced complexity along the strike of the slab. It is possible that the pulse complexity may have other causes, such as velocity heterogeneity at other points along the path of the ray. However, recent attempts by several investigators (e.g. Cormier 1987, Vidale 1987, Witte 1987) to theoretically model slab-interaction phases have indeed shown that the slab will produce such pulse complexity. This arises from the presence of an additional arrival diffracted along the bottom surface of the slab for stations on the down-dip side. Furthermore, it has been shown (V. Cormier, personal communication; see Engdahl et

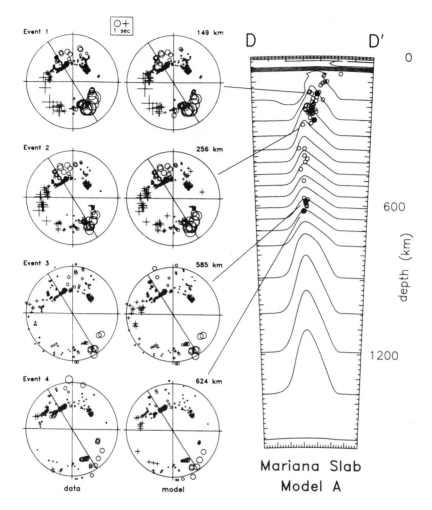

Figure 2   Constraints on lower-mantle slab penetration from travel-time residuals. Residual sphere diagrams for several intermediate- and deep-focus events in the Mariana slab. Travel-time residuals are plotted on a unit sphere around the earthquake as a function of takeoff angle and azimuth (lower-hemisphere projection) and then smoothed to enhance near-source heterogeneity. Pluses indicate slow directions, circles indicate fast directions. Diagonal line gives strike direction of the slab. If there is high-velocity slab material below an earthquake, then for a vertically dipping slab, the high seismic velocities will produce a characteristic pattern of fast arrivals along the strike of the slab and slow velocities elsewhere. This pattern is clearly seen in the data (*left*) and is well predicted by the residual spheres (*center*) produced from a model of the Mariana slab penetrating down to 1300 km depth (from Creager & Jordan 1986a).

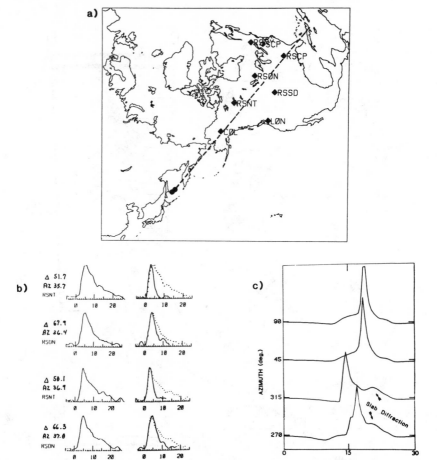

Time (s)   Time (s)

Figure 3 Constraints on lower-mantle slab penetration from waveform complexity (from Silver & Chan 1986). (*a*) Map in azimuthal equidistant projection showing locations of two deep-focus ($\sim$ 600-km depth) events below the Sea of Okhotsk (solid circles) and recording stations (solid diamonds). Dashed line indicates the great circle corresponding to the approximate strike-azimuth of the southern Kuril slab in the vicinity of these two events. (*b*) Displacement seismograms of the body-wave phases S (*left*) and ScS (*right*) for the stations RSNT and RSON (both events). The ScS-phase has the shape and duration expected for this type of earthquake. The S-phases, by comparison, are very broad (up to 20 s in duration) and complex. (*c*) (from Cormier 1987) Waveform modeling of the S-phase that interacts with a slab. For azimuths (measured from slab-strike direction) on the down-dip side of the slab (azimuth $= 315°$ and azimuth $= 270°$), a second phase is clearly visible that is diffracted along the bottom of the slab. Note the qualitative similarity in shape between observed phases and synthetic waveforms.

al 1988) that pulse complexity is expected to be maximized along the strike azimuth of the slab, with an abrupt cutoff of the effect for azimuths even slightly on the up-dip side. This would not only explain the extremely distorted pulse shapes observed by Silver & Chan (1986) along strike (Figure 3), but it would also explain the increased variability in pulse broadening observed by Beck & Lay (1986) for this azimuth.

If it is accepted that there is a substantial slab flux into the lower mantle, the emphasis shifts to the slab's fate once in the lower mantle. The studies mentioned above are most sensitive to slab properties near the earthquake and thus can only "track" the slabs to a depth of perhaps 1000–1200 km, at which point other sources of information must be sought. The recent three-dimensional tomographic images of the lower mantle, obtained from the analysis of the body-wave travel times in the *Bulletin of the International Seismological Centre* (BISC), are the most likely possibility. Dziewonski (1984), representing the lower mantle P-wave velocity as spherical harmonics up to angular order 6 and radial order 4, obtained a model from the least-squares inversion of about 500,000 P-wave travel-time arrivals; this may be viewed as a smoothed model of the true structure. Other modeling approaches have been used, most notably the back-projection method with a block parameterization by Comer & Clayton (1984). While there are advantages and disadvantages to both approaches, they appear to be in agreement at very long wavelengths (Hager et al 1985). Because the Dziewonski model is in the most convenient parameterization to compare with other geophysical observables, and because all the details of that model have been published, we use the Dziewonski model in our discussions.

Slabs do appear to represent an observable signal in the lower mantle. According to Dziewonski (1984), "Certain features show continuity over a large range of depths. For example, a high-velocity zone circumscribing the Pacific seems to stretch from 1000 km to the core-mantle boundary." The more recent waveform modeling study of Woodhouse & Dziewonski (1987) to estimate shear velocities confirms the existence of the same anomaly in the upper 1000 km of the lower mantle, which, they state, appears to be associated with the location of subduction in the present and recent geologic past. In one region, below the Caribbean Sea, this high-velocity feature approximately coincides with a previously reported high-velocity anomaly inferred from S- and P-wave travel times (Jordan & Lynn 1974, Lay 1983). This anomaly recently has been imaged in detail, revealing a vertical, slablike structure extending throughout the top 1000 km of the lower mantle, hypothesized to be associated with past subduction of the Farallon plate (Grand 1987). These observations do not necessarily confirm the existence of slab material in the lower mantle, but they are consistent with it and would represent the simplest explanation for this feature.

Another potential indicator of the presence of slabs is from the recent studies of three-dimensional P-wave velocity variations near the core-mantle boundary by Creager & Jordan (1986b) and Morelli & Dziewonski (1987) on the basis of BISC travel-time anomalies. While the observed travel-time anomalies may have multiple causes, much of the signal appears to arise from topography on the core-mantle boundary (CMB) itself, since the core-reflecting phase (PcP) shows a significant travel-time anomaly opposite in sign to that found for the phase that passes through the outer core (PKP), which is the behavior expected for topography (Morelli & Dziewonski 1987). The presence of boundary topography appears to be corroborated by a study of the same data-set using a very different (spatially localized) parameterization (Gudmundsson et al 1986). Considering the Morelli & Dziewonski (1987) model of topography, the most striking feature is that areas of depressed CMB appear to be associated with regions of present-day subduction. In fact, the correlation of the model of topography with the locations of the trenches (expanded in spherical harmonics up to degree and order 4, the same as the CMB model) is highly significant (98% confidence level). Some caution should be taken, however, in interpreting the travel-time data on CMB topography. The most serious problem is the lack of adequate sampling for the Pacific basin. We note in particular that there is a significant correlation of the topography model with the distribution of oceans (CMB elevated under oceans), although it is not as high as the correlation with trenches. Thus, an unambiguous association with slabs cannot be made until this model has been tested further. If the connection with subduction does indeed stand up, it would, taken literally, suggest that the effects of slabs are felt all the way down to the core-mantle boundary.

Implications

The mass flux for subducted lithosphere in the western Pacific, where the best evidence has been obtained for slab penetration, is about 70 km^3 yr^{-1}, or 300 km^3 yr^{-1} if all slabs are included and assumed to have a plate thickness of 125 km (Creager & Jordan 1986a). This would fill a volume equal to the entire upper mantle in 1 b.y. (300 km^3 yr^{-1}) to 4 b.y. (70 km^3 yr^{-1}). By itself, this amount of flux argues for a mantle with no large-scale stratification associated with the transition zone (the depth range $\sim$ 350–800 km), since any preexisting layering would be destroyed by slab flux well before the present time. However, in attempting to simultaneously satisfy other constraints on mantle convection, in particular the persistence and character of geochemical heterogeneity, it is important to carefully assess the class of models that are and are not excluded by lower-mantle slab penetration. First, strict two-layer convection models, where there effectively is no mass flux between the upper and lower mantle, are the most clearly excluded. Also, models in which the velocity and density jumps at the 650-km discontinuity are predominantly compositional can be ruled out, since the seismically inferred density jump of 6–10% would prevent slabs from penetrating into the lower mantle and the velocity jump would virtually remove the observed lower mantle slab anomaly. However, as pointed out by Christensen & Yuen (1984), a compositionally induced increase in density of less than 3% between the upper and lower mantle would not be expected to prevent slab penetration, and thus some of the density increase could be compositional. However, to maintain such a compositional boundary in the face of this large amount of slab flux, one clearly would require either a stabilizing mechanism or appeal to an evolutionary model, where the mixing between an initially distinct upper and lower mantle has been incomplete.

RETURN FLOW

The subduction-related downwelling discussed in the previous section raises the issue of complementary return flow. We address this question by summarizing recent results examining the relationships between three basic sources of information on mantle flow: (a) the three-dimensional elastic structure of the Earth's interior from seismic tomography, (b) the geoid, and (c) the surface features of plate tectonics. Regions of downwelling and upwelling are expected to reveal themselves seismically as having fast and slow velocities, respectively, due to the expected differences in temperature. The geoid should also reflect such motions, although, as pointed out recently by Richards & Hager (1984) and Hager (1984), the size and even the sign of the geoid anomaly depends on mantle rheology due to the perturbation of compositional boundaries in a non-rigid Earth. What can be said about return flow strictly from plate tectonics? First, at the shallowest depths of the mantle, the ridges clearly are the loci of upwelling material complementary to subduction, one of the basic tenets of plate tectonics. Hotspots must also represent some form of localized upwelling, although the relationship of this upwelling to large-scale mantle convection remains one of the important unsolved problems in geodynamics. The relation of hotspots to ridges is ambiguous. A large number of hotspots are found near ridges, with Iceland being the most obvious example. Some, however (notably Hawaii), are found in ocean basins far from ridges and apparently decoupled from the motions of the plates, while others are found on continents. Thus, the flows associated with ridges and hotspots appear to interact but are not identical and do not have the same origin.

In recent years, several studies have noted intriguing correlations between these three groups of data (seismic tomography, the geoid, and plate-tectonic features). Many of these are listed in Table 1 and summarized pictorially in Figure 4. The least-explored connection is between plate-tectonic features and the results of seismic tomography; these have been calculated by us when necessary. In most cases the correlations have

Table 1 Significant correlations

(a) Seismic tomography vs. tectonic features

Model	Angular order	Depth (km)	Feature	Correlation[a]	Reference[b]
P2	2	400–650	Slabs	+	1
M84A	2–3	550	Slabs	+	1
M84A	1–8	150	Ridges	−	1
L02.56	2–3	LMA[c]	Hotspots	−	2
L02.56	2–3	LMA	Ocean basins	−	1

(b) Seismic tomography (equivalent geoid) vs. geoid

Model	Angular order	Type of geoid	Correlation	Reference
P2	2	Observed	+*	3
M84A	2–3	Observed	+	4
L02.56	2–3	Observed	−	5
L02.56	2–3	Residual[e]	−	2

(c) Tectonic features vs. geoid

Feature	Angular order	Type of geoid	Correlation	Reference
Hotspots	2	Residual	+	6
Hotspots	—	Residual	+[d]	7
Slabs	4–10	Observed	+	8

[a] Significant positive (+) and negative (−) correlations between seismic velocity, the geoid, and plate-tectonic features at given range of spherical-harmonic angular order. All correlations are significant to at least the 95% confidence level except (*), which is at 94%.

[b] (1) This study, (2) Hager et al 1985, (3) Masters et al 1982, (4) Woodhouse & Dziewonski 1984, (5) Dziewonski 1984, (6) Crough & Jurdy 1980, (7) Chase 1979, (8) Hager 1984.

[c] Lower mantle average.

[d] Correlation was not done with spherical harmonics.

[e] Defined as the observed geoid minus the assumed contribution of subducted slabs.

Plate Tectonic
Features

3-D Seismic
Velocity

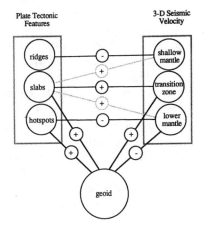

Figure 4 A pictorial summary of the various correlations that have been found between (*a*) seismic tomography, (*b*) the geoid, and (*c*) plate-tectonic features (from Table 1). A line connecting two categories indicates a significant positive (+) or negative (−) correlation. For example, there is a significant negative correlation between seismic velocities in the shallow mantle and the distribution of mid-ocean ridges, meaning that mantle seismic velocities are low where ridges are found. The line between slabs and lower-mantle seismic velocity is dashed because, although there is no formal correlation between trenches and the lower-mantle-averaged model L02.56, there is strong seismological evidence for slab penetration into the lower mantle (see section on the fate of subducted slabs). The line between slabs and shallow-mantle seismic velocity is dashed for the same reason (no formal correlation with model M84A but strong seismic evidence for slab material there).

been calculated using spherical-harmonic coefficients, since this is the most common representation for both the geoid and seismic tomography. Thus, for comparison, the tectonic features—in particular, subduction zones and ridges (both represented as lines), hotspots (represented as points), and the distribution of continents and oceans—are expanded in spherical harmonics as well. The correlations are computed for a specified range of angular order *l* [see Dziewonski et al (1977) for a discussion of this procedure in a geophysical context]. While one must exercise caution in drawing conclusions from these correlations, they are nevertheless useful guides in attempting to characterize the general properties of mantle flow.

   The study of return-flow certainly is more difficult than tracing the path of descent for subducted slabs. Slabs often can be detected through much of the upper mantle by the seismicity of Benioff zones and by the very large velocity contrasts associated with the slabs. As we have seen in the previous section, these anomalies can be traced at least into the upper portion of the lower mantle and possibly to the CMB. There are no deep earthquakes associated with upwelling, so we must rely more heavily on the three-dimensional seismic velocity structure.

   One of the surprises to come out of seismic tomography is that the expected low-velocity expression of upwelling beneath ridges, although dominant at about 150 km depth, is no longer detectable below about

350 km. This observation is based on the shear velocity model M84A of Woodhouse & Dziewonski (1984), derived from the waveform modeling of long-period Rayleigh and Love waves. In the transition zone, the pattern of heterogeneity abruptly changes, bearing little resemblance to the ridge-continent heterogeneity prevalent in the shallow mantle. The lowest order component of this distribution, discovered by Masters et al (1982) using free-oscillation apparent center frequencies (Jordan 1978, Silver & Jordan 1981) and referred to as P2, was later confirmed by Woodhouse & Dziewonski (1984). This transition-zone heterogeneity has been shown to be positively correlated with the observed geoid at the very lowest angular orders l ($l = 2$ for P2, $l = 2$–3 for M84A). We have found that these models also are significantly positively correlated with the global distribution of subduction zones. A close correspondence between P2 and a model for the distribution of subducted slabs also was pointed out by Hager (1984). The presence of a dominant slab component would not be expected from the slabs themselves. Even though slabs have a large (5–10%) velocity contrast, they are highly localized (about 100 km thick), whereas the effective wavelength of degree 2 and 3 heterogeneity is of order 10,000 km. Thus, one would expect less than a 0.1% anomaly, but the observed anomaly is roughly an order of magnitude larger ($\sim$ 1.5–2%). This would suggest that while slabs probably are closely associated with the observed heterogeneity, they are physically only a small part of it. In any event, neither indicator of upwelling—the distribution of hotspots nor ridges— is significantly correlated with transition-zone heterogeneity.

Yet another pattern emerges in the lower mantle. Both the three-dimensional P-wave velocity model of Dziewonski (1984)—for example, L02.56—and the shear velocity results (Woodhouse & Dziewonski 1987) indicate that the dominant slow region of the lower mantle is beneath the Pacific basin with a ring of high velocities at its margins. At angular orders 2 and 3, the equivalent geoid of L02.56 is significantly negatively correlated with the observed geoid (Dziewonski 1984), where the equivalent geoid is obtained from an assumed proportionality between density and velocity. Hager et al (1985) have shown that this correlation is enhanced if the residual geoid is used (the slab contribution to the observed geoid is modeled and removed) and the effects of a nonrigid rheology (including boundary deformation) are considered. In addition, these authors find that the radially averaged heterogeneity is strongly correlated with the degree 2 distribution of hotspots, i.e. the lower mantle is seismically slow where hotspots are more frequent. Taken literally, this would suggest that the distribution of hotspots statistically reflects regions of large-scale upwelling in the lower mantle. One should exercise some caution in evaluating this correlation, however. We have found that if degrees

2 and 3 are considered, there is also a significant correlation with the distribution of ocean basins. This is not surprising, since, as noted and as is clear in Figure 5, the dominant feature in the radially averaged lower-mantle P-wave velocity distribution, at angular orders 2 and 3, consists of low velocities below the entire Pacific basin. However, as one goes from the list of 24 hotspots (Chase 1979) that are most closely associated with ocean basins to the expanded list of 42 hotspots (Crough & Jurdy 1980) containing several continental hotspots, the correlation actually *increases*. In addition, at degree two, where the distribution of hotspots has a peak in power (Crough & Jurdy 1980), the correlation of seismic velocity with hotspots is -0.98, whereas the correlation with the continent-ocean function is not significant. This suggests that there is indeed a connection between slow seismic velocities in the lower mantle and hotspots.

Some investigators have directly examined the relation between the geoid and plate-tectonic features, most notably Chase (1979) and Crough & Jurdy (1980). These authors find that the two largest signals in the geoid are associated with subducted slabs and hotspots. They compare a residual geoid with the global distribution of hotspots and find that the great majority of hotspots are located in geoid highs. Both studies show that the deviation from a random distribution of hotspots is highly significant.

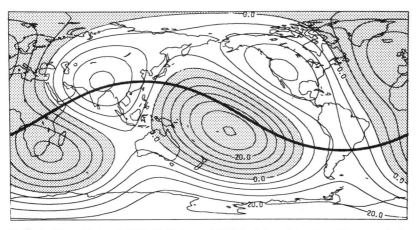

Figure 5   Plot of the model L02.56 (Dziewonski 1984) of three-dimensional P-wave velocity distribution in the lower mantle obtained from BISC travel times. The full model, expanded up to spherical harmonic angular order 6 and radial order 4, has been averaged over radius to enhance radially coherent features, and only the angular orders 2 and 3 are shown. Relatively slow regions are shaded. The two dominant slow regions correspond to the entire Pacific basin and a region centered over southern Africa, and they are interpreted as indicating regions of lower-mantle upwelling. Both the distribution of hotspots and the geoid are strongly correlated with this pattern. Contour interval is 5 m s^{-1}.

What can be concluded from these correlations (see Figure 4)? First, the mantle can be separated into three regimes: the shallow mantle, down to about 350 km depth; the transition zone, between 350 and 800 km; and the lower mantle, down to the core-mantle boundary.

In the lower mantle, the distribution of high seismic velocities indicates descending flow in a nearly great circular ring around the Pacific basin, closely associated with present-day subduction. On the basis of the distribution of low seismic velocities, ascending flow may be more diffuse, extending beneath most of the Pacific basin and also beneath the sector centered on Africa. Of course, this interpretation applies only to the largest scales of motion. The association between seismic velocity, the residual geoid, and hotspots suggests that this represents the dominant large-scale pattern of active upwelling in the mantle.

This style of lower-mantle flow is not entirely unexpected. The three-dimensional planform of global-scale mantle convection is a problem that is just now being approached by numerical modelers. Preliminary results (Baumgardener & Schubert 1987) indicate that the combined effects of sphericity and internal heat production destroy the midplane symmetry found in most laboratory experiments on convection planforms and result in a flow with descending sheets arranged in circular rings, surrounding isolated columns of ascending return flow. The spherical-harmonic spectrum of this flow is dominated by degrees 2–4. While presently limited to the consideration of only modest Rayleigh numbers (below 10^5), the patterns realized in these calculations are nevertheless qualitatively similar to the apparent pattern of convection in the lower mantle, namely the nearly great circular ring of downwelling (around the Pacific) with upwelling on either side. This pattern is illustrated schematically in Figure 6 for a cross section close to the equator with a pole at (60°N, 60°W). This great circle was chosen because it traverses two of the major trench systems, three major ridge systems, and the two major concentrations of hotspots. In addition, it passes near the two seismic-velocity minima in Figure 5 (beneath the central Pacific and southern Africa).

The inferred shallow-mantle upwelling is very different. It is closely associated with mid-ocean ridges and appears to be more spatially localized. The correlation between ridges and the model M84A, while significant for angular orders 1–8, is not significant if only the lowest angular orders are considered. Furthermore, the ridges are not a detectable component in the long-wavelength component of the geoid. This suggests that the active upwelling beneath ridges is restricted to the shallow mantle, and its large-scale geometry is controlled by plate motions.

The transition zone is the most difficult of the three regimes to interpret.

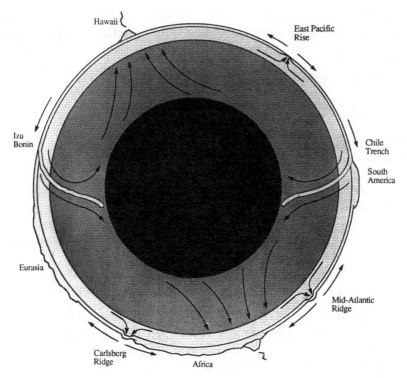

Figure 6 A schematic drawing indicating the apparently different upwelling regimes in the upper and lower mantle (and coherent slab-related downwelling) on the basis of the results of seismic tomography, the geoid, and the distributions of trenches, ocean ridges, and hotspots. Upper-mantle upwelling is associated with the ridges and appears to be passive. The largest scale component of lower-mantle upwelling appears to be associated with the statistical distribution of hotspots. Cross section corresponds to the great circle (pole at 60°N, 60°W) shown in Figure 5. It was chosen because it traverses two of the major trench systems, three major ridge systems, the two major concentrations of hotspots (Stefanick & Jurdy 1984), and passes close to both of the velocity minima of the radially averaged model L02.56 in Figure 5. The radial distances are to scale with the exception of the two volcanoes, indicating the hotspot concentrations, and the continents. This figure shows slabs descending to the core-mantle boundary. While seismic results are suggestive of this, strong evidence from residual sphere analysis (Jordan 1977, Creager & Jordan 1984, 1986a) is available only to a depth of 1200–1300 km.

The pattern that would be expected to be present for unrestrained lower-mantle return flow, namely a continuation of the lower-mantle pattern, does not appear to be present; instead the dominant large-scale feature is associated with subduction, a feature that would not be expected solely from the slabs themselves. Taken literally, the absence of the lower-mantle signal would suggest that the transition zone is a partial barrier to lower-

mantle upwelling. Clearly, it is not a total barrier, since the upwelling must
at least equal the downward slab flux, and the presence of hotspots argues
for some form of communication with the Earth's surface. Nevertheless,
the general pattern of coherent downwelling and partially obstructed
upwelling by the transition zone is suggested by the data.

AGE OF PLATE-TECTONIC RECYCLING

The basalts sampled at mid-ocean ridges (MORB) and ocean islands (OIB)
exhibit heterogeneity in several geochemical and isotopic systems and on
a variety of spatial scales. The variation is by no means random. For
example, in the ^{207}Pb-^{206}Pb system, the data are dominated by a linear
trend, as shown in Figure 7, that may be accounted for by the presence of
two distinct mantle components. One component, dominant in mid-ocean
ridge basalts, is characterized by a depletion in incompatible elements.
This depletion appears to be a long-term feature of its source and not a
product of recent elemental fractionation, as shown by the low ^{87}Sr/^{86}Sr
and high ^{143}Nd/^{144}Nd ratios common to MORB (e.g. Tatsumoto et al 1965,
DePaolo & Wasserburg 1976). We refer to this end-member, incompatible-element-depleted component as MORB-source, or MORB-S.

The second most abundant component, preferentially sampled at ocean
islands, tends towards more radiogenic Sr and Pb and less radiogenic Nd
isotopic compositions. With MORB, this component forms a linear array
in ^{207}Pb-^{206}Pb systematics that if interpreted as an isochron corresponds to
an age of about 1.8 b.y.; also, the radiogenic isotopic end member of this
trend for OIB has Nd, Sr, and Pb isotopic properties most characteristic
of ancient oceanic crust (Chase 1981, Hofmann & White 1982).

If the linear trend is regarded as having true age significance and the
identification with oceanic crust is accepted, then a straightforward
interpretation of the trend is that it reflects the recycling of oceanic
crust by plate-tectonic chemical differentiation, with the 1.8-b.y. age representing the recycling or residence time. The important consequence of
accepting this interpretation is that the residence time may then be used
to constrain the characteristics of the recycling process. Given a value for
the flux of chemically differentiated mantle, it is possible in principle to
determine the portion of the mantle associated with plate-tectonic-related
differentiation—specifically, whether the whole mantle is involved, or
whether plate-tectonic differentiation is only associated with some fraction
of the mantle.

None of the above interpretations is free of controversy. While there is
no doubt about the existence of a dominant linear trend in the isotopic
data, many have argued that it has no real age significance but rather is
an artifact due to the mixing of unrelated components; others object to

the characterization of the end-member component as recycled oceanic crust. The interpretation given above does, however, have the virtue of simplicity, and there are few competing hypotheses at the present time to explain this most prominent feature of oceanic basalt heterogeneity. We therefore accept *as a working hypothesis* the interpretation that the linear trend is primarily the result of the recycling of oceanic crust. The justification for this position, as well as a discussion of the oceanic basalt data set, is given below.

Geochemical Heterogeneity in Oceanic Basalts

The geochemical heterogeneity observed in oceanic basalts may be described as being produced by at least five distinct mantle components (see Table 2). Based on simultaneous consideration of the isotopic variation in a variety of elements (Sr, Nd, Hf, Pb, He, Ar, and Xe) in MORB and OIB, at least four mantle components distinctly more enriched in incompatible elements than MORB-S appear to be present in the mantle (e.g. White 1985, Zindler & Hart 1986, Allègre et al 1987). Three of the components (Table 2), needed to account for the Sr, Nd, and Pb isotopic variation in

Table 2 Geochemical heterogeneity

| | Age (b.y.) | | Recycle | Sampling location[d] |
Source[a]	Recycle[b]	Reservoir[c]	mechanism	(oceanic)
MORB-S	1.7	> 3.8	Subduction	*MOR*, OI
Recycled oceanic crust	1.7	0.1	Subduction	MOR, *OI*
Subcontinental mantle	?	0–3.8	Delamination subduction erosion	Southern Hemisphere (DUPAL)[e]
Continental crust	1.7	0–4.2	Sediment subduction	Southern Hemisphere (DUPAL)
Bulk Silicate Earth	—	4.55	—	Large hotspots

[a] The five primary components thought to be responsible for the isotopic variability of oceanic basalts. The components are ranked in the order of how often they are encountered in the oceanic data set, from the most frequent to the least frequent. See Zindler & Hart (1986) for a more complete description of the components.

[b] Time between injection of component into sublithospheric mantle and its reappearance in lithosphere.

[c] Time between initial differentiation and injection into sublithospheric mantle.

[d] If two sampling locations are given, the italicized location is the predominant. Abbreviations: MOR, mid-ocean ridges; OI, ocean islands.

[e] Associated with the DUPAL anomaly (Dupré & Allègre 1983, Hart 1984).

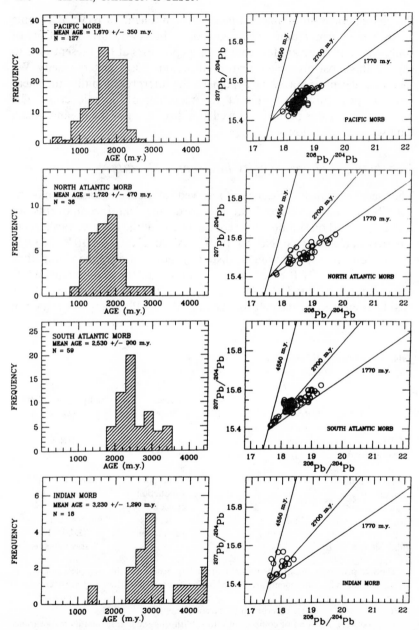

Figure 7 Uranogenic Pb isotopic data for mid-ocean ridge and ocean island basalts from the three major ocean basins. Secondary isochrons of 4550 and 2700 m.y. are shown for reference. The line labeled 1770 m.y. is the Northern Hemisphere oceanic regression line (NHRL) of Hart (1984). Corresponding histograms show the distributions of Pb model ages calculated by connecting each data point with the intersection of the NHRL with the 4550-

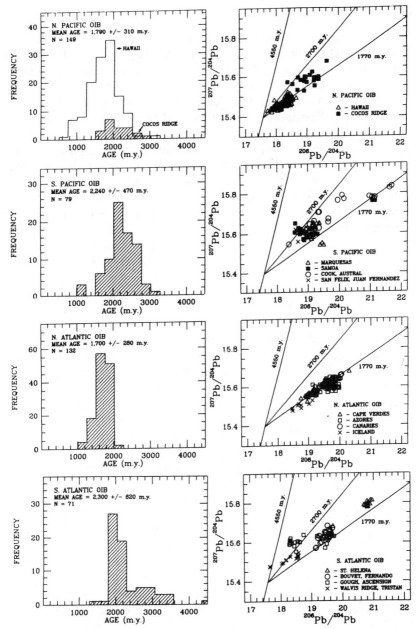

m.y. isochron. Number of samples included in each histogram is given by *N*. Data from Sun & Jahn (1975), Tatsumoto (1978), Sun (1980), Cohen & O'Nions (1982), Richardson et al (1982), Dupré & Allègre (1983), Stille et al (1983), Weis (1983), Hamelin et al (1984, 1985), Staudigel et al (1984), Vidal et al (1984), Chen & Frey (1985), White (1985), Duncan et al (1986), Gerlach et al (1986), Hanan et al (1986), Liotard et al (1986), Shirey et al (1987), and White et al (1987).

OIB, have many of the expected properties of "recycled" near-surface materials produced in the plate-tectonic cycle (e.g. Davies 1984, White 1985, Zindler & Hart 1986, Allègre et al 1987) and have been proposed to represent subducted oceanic crust (Chase 1981, Hofmann & White 1982), subducted pelagic or terrigenous sediments (Armstrong 1981, Hofmann & White 1982, Weaver et al 1986), or recycled pieces of the subcontinental mantle (McKenzie & O'Nions 1983, Hart et al 1986, Hawkesworth et al 1986, Shirey et al 1987). The fourth component has been described as relatively undifferentiated material similar in composition to that expected for the bulk silicate Earth (BSE) (e.g. DePaolo & Wasserburg 1976, Richardson et al 1982, Chen & Frey 1985). In the total oceanic basalt data set, however, this component represents only a very small percentage.

Isotopic model ages for MORB and OIB show a general clustering in the 1–2 b.y. time interval for the Pb-Pb, Rb-Sr, and Sm-Nd systems (Brooks et al 1976b, Carlson et al 1978, Tatsumoto 1978, Sun 1980, Chase 1981, White et al 1987). Of the model ages, those determined from the Pb isotopic composition of MORB and OIB are the most robust because they do not involve measured values for the parent-daughter ratio and hence will not be affected by recent chemical changes caused by melt production and fractionation. In addition, they will not be disturbed by mixing between different components as long as those components were originally derived from a single source, in a single differentiation event. The general agreement among the model ages determined by the various isotopic systems, in spite of the potential problems in the other systems, argues strongly for the idea that the model ages have true age significance.

Table 3 lists the Pb-Pb secondary isochron ages derived from least-squares regression lines fit through Pb isotopic data for various ocean island and ridge basalt groups. [Not all oceanic basalts form linear arrays on Pb-Pb diagrams, and for those that do not (e.g. Indian MORB, Samoa, and Gough), secondary isochron ages cannot be calculated.] The data show linear arrays with slopes suggestive of "ages" between 690 and 2080 m.y. with a mean for all OIB of 1770 m.y. and for MORB (excluding Indian Ocean MORB) of 1550 m.y. The time-averaged ratio $^{238}U/^{204}Pb$ (μ_1) from 4.55 b.y. ago up to the isochron age generally shows a very narrow range (± 0.08) around 8.00, with notable exceptions being the higher μ_1 indicated for Indian Ocean MORB (8.10), South Pacific OIB (8.14 and 8.35), and Gough (8.28) and the Walvis Ridge (8.10) from the South Atlantic.

Linear arrays on $^{207}Pb/^{204}Pb$ vs. $^{206}Pb/^{204}Pb$ diagrams need not have age significance if they result from mixing between two or more *unrelated* components. Mixing lines, however, *will* have time significance if the two mixing components were derived from a common parent source in a single

Table 3 Pb-Pb secondary isochron ages for oceanic basalts

Locale	Number of samples	Isochron Age (m.y.)[a]	μ_1[b]
MORB:			
North Atlantic	57	1200 ± 300	8.03
South Atlantic	57	1650 ± 280	7.97
Pacific	43	2080 ± 350	7.92
Indian	37	—	8.10
OIB:			
Hawaii	129	1230 ± 450	7.99
Galapagos	49	1560 ± 320	8.03
Marquesas-Cook-Austral	41	1280 ± 190	8.14
Samoa	28	—	8.35
Iceland	10	1200 ± 970	7.99
Canaries	37	1650 ± 480	7.99
St. Helena	10	1740 ± 300	7.97
Gough	8	—	8.28
Walvis Ridge	8	700	8.10
All OIB		1770 ± 60	7.99
MORB except Indian		1550 ± 160	8.02

[a] Errors give 95% confidence interval.
[b] μ_1 is defined as the time-averaged $^{238}U/^{204}Pb$ ratio from 4.55 b.y. ago up to the isochron age extrapolated for comparison at the present day.

event. Noting the extremely narrow range in initial μ_1 observed for most oceanic basalt Pb data sets, we can calculate Pb "model ages" for individual oceanic basalts in the manner shown in Figure 7.

In a way, Pb model ages calculated this way provide information similar to Nd model ages (Lugmair et al 1975, DePaolo 1981), except that they depend on the coupled ^{207}Pb-^{206}Pb systems rather than on a comparison of present day isotopic composition with the elemental abundance ratio of parent to daughter (i.e. Sm/Nd). In the simple case of only one differentiation from the presumed MORB source, the mean and standard deviation of a histogram of model ages calculated in this manner will yield equivalent age information to a secondary isochron fit through the data. These model ages, however, offer the advantage over secondary isochron ages in that multiple extraction events from a source that maintained constant U/Pb will produce multiple peaks on the age histogram, each corresponding to the age of an individual extraction event.

Histograms of Pb model ages for MORB and OIB are presented in Figure 7. North Atlantic and Pacific MORB and OIB form tightly peaked distributions with mean ages between 1.7 and 1.9 b.y. Data for southern Pacific OIB, including Samoa, the Societies, Austral, Cook, and Marquesas islands, show a component of the 1.8-b.y. model ages, but they also indicate the presence of older components, in the neighborhood of 2.4–2.7 b.y. age, which are well separated from the 1.8-b.y. trend. These older ages are an expression of the general isotopic feature of Southern Hemisphere ocean island basalts that has been labeled the DUPAL anomaly (Dupré & Allègre 1983, Hart 1984). Similar concentrations of older Pb model ages are seen in South Atlantic and Indian Ocean MORB and OIB (Figure 7), but not in East Pacific Rise MORB from the Southern Hemisphere (Macdougall & Lugmair 1985, 1986, White et al 1987). The limited presence of the DUPAL component in other oceanic islands will result in slightly younger Pb-Pb isochron ages and higher μ, for individual islands (Table 3), whereas the overall data set is much less affected.

That isotopic data for OIB and MORB give model ages near 2 b.y. has been known for some time (Brooks et al 1976b, Carlson et al 1978, Tatsumoto 1978, Sun 1980, Chase 1981). The meaning of these ages, however, has remained somewhat controversial. The "older" (i.e. > 2.2 b.y.) components are represented in their most extreme form by islands like Gough and Samoa. Pb model ages calculated in the above manner for pelagic sediments are in the range 2.6–3.0 b.y., consistent with the idea that a recycled continental component may be responsible for this signature in OIB. Based on the data for Nb/U, Ce/Pb (Hofmann et al 1986), and Lu/Hf (Patchett et al 1984), the "continental" component does not appear to be subducted pelagic or terrigenous sediment, although Nb/U and Ce/Pb data for islands such as Gough and Samoa are limited. An alternative explanation is that this old signature is due to the delamination and recycling of subcontinental lithosphere (McKenzie & O'Nions 1983, Hawkesworth et al 1986, Shirey et al 1987, Hart et al 1986). This latter possibility seems especially likely, as similar isotopic features to those observed in South Atlantic MORB and OIB are seen in the Paraná flood basalts (Hawkesworth et al 1986).

The younger age peak in the Pb data, around 1.8 b.y., is seen in isotopically extreme form in both the Atlantic (St. Helena) and the South Pacific (Tubuaii), but it is present in virtually all oceanic basalts (except perhaps Indian MORB), including the major hotspots Iceland, the Galapagos, and Hawaii. If ancient subducted and recycled oceanic crust is one of the components seen in modern OIB, then this younger component, especially in its extreme examples like St. Helena and Tubuaii, geochemically and isotopically is the most likely example (Chase 1981, Hofmann & White 1982, Zindler et al 1982, Zindler & Hart 1986). For example, in order to evolve the Nd, Sr, and Pb isotopic compositions of Tubuaii starting 1.8 b.y. ago from a source that will evolve to isotopic compositions like those of modern MORB, parent-daughter ratios of Rb/Sr = 0.012, Sm/Nd = 0.403, $^{238}U/^{204}Pb$ = 18, and Th/U = 3.3 in the Tubuaii source are required. All of these values are well within the range observed for fresh modern MORB. If the 1.8 b.y. Pb model age peak is due to the presence of ancient oceanic crust imperfectly mixed into the mantle, then the time of injection of this crust into the mantle probably is no more than 100 to 200 m.y. before the 1.8–1.9 b.y. age recorded for this component because of the short residence time of oceanic crust at the Earth's surface.

This ∼ 1.8-b.y.-old component appears to be distributed globally and, though preferentially sampled by OIB, also seems to be a component of mid-ocean ridge volcanism. Increasingly detailed study is continually expanding the range of isotopic compositions observed for MORB, even for areas well displaced from possible hotspot influence, and shows some MORB to have isotopic compositions similar to those seen for many ocean island basalts (Carlson et al 1978, Cohen & O'Nions 1982, Dupré & Allègre 1983, Zindler et al 1984, Shirey et al 1987, White et al 1987). Thus, at least some of the incompatible-element-enriched components considered characteristic of OIB also seem to be distributed within the mantle sources of MORB, even at the small length scales sampled along individual ridge segments (Shirey et al 1987, White et al 1987), small seamounts (Zindler et al 1984), and fracture zones (Carlson et al 1978).

The samples that exhibit this 1.8-b.y. age represent the majority of the entire oceanic basalt data set; apparently associated with recycled oceanic crust, this component has all of the characteristics of a globally distributed dispersed heterogeneity occurring within a MORB-S matrix.

In what follows we take the residence time of oceanic plate recycling, $\langle t_r \rangle$, to be 1.7 b.y. This represents the ∼ 1.8-b.y. Pb model age minus 0.1 b.y. to account for the time spent between creation of the oceanic crust at the ridge and its subduction at the trench.

GEOCHEMICAL EVIDENCE FOR A NON-PLATE-TECTONIC COMPONENT

A Non-Plate-Tectonic Component

"Recycled" components produced and transported by plate-tectonic processes appear to dominate the compositional heterogeneity observed in the sources of oceanic basalts. At least one component unrelated to plate-tectonic processing also appears to be present, however. Isotopic data for two major hotspots, the Koolau trend from Hawaii (Stille et al 1983) and the Walvis Ridge (Richardson et al 1982), seem to indicate a component with incompatible-element concentrations equal to those expected for the bulk silicate Earth. Rare-gas data for certain OIB also appear to show a component that is "undegassed", with the implication that it has not been through the plate-tectonic process of oceanic crustal formation. Compared with MORB, the less radiogenic rare-gas isotopic compositions often found for ocean island basalts (Craig & Lupton 1976, Craig et al 1978, Kurz et al 1982), particularly those associated with major hotspots (Kaneoka et al 1983, Kurz et al 1983, 1985, Rison & Craig 1983, Poreda et al 1986), often are considered indicative of a "primordial" component in the mantle (e.g. Kaneoka 1983, Lupton 1983). As discussed by Zindler & Hart (1986) and Hofmann et al (1986), and as we have mentioned previously, chemical and isotopic data for OIB and MORB show that BSE is present only as a very rare component in oceanic basalts, substantially less than that suggested by early models (Schilling 1973, De Paolo & Wasserburg 1976). Nevertheless, both the rare-gas and Sr, Nd, and Pb data sets do lead to the important conclusion that large hotspots are tapping a component in the mantle that has not participated in plate-tectonic recycling.

Size of the Non-Plate-Tectonic Component

The question of how much undifferentiated material remains in the mantle has been approached generally with mass-balance models relating the incompatible-element depletion of the MORB source to the formation of the incompatible-element-rich continental crust (Hurley et al 1962, Armstrong & Hein 1973, Jacobsen & Wasserburg 1979, O'Nions et al 1979, DePaolo 1983, Allègre et al 1983a,b). Even before the inception of the plate-tectonic model, the related processes of subduction and island-arc volcanism have been perceived as one of the mechanisms by which continental crust is formed (Coats 1962). This model suggests that island-arc volcanism transfers incompatible elements from the oceanic slab to the overriding plate. It has thus become customary to treat the MORB source and continental crust as complementary reservoirs. Assuming that the combination of these two reservoirs has incompatible trace-element abundances similar to BSE, roughly one third of the mantle would have to be as depleted as the MORB source to complement the existing volume of continental crust (Jacobsen & Wasserburg 1979, O'Nions et al 1979, Allègre 1983a,b, Turcotte & Kellogg 1986). These calculations have been used to argue that the MORB source occupies the upper mantle above the 650-km seismic discontinuity, with an undepleted lower mantle that has not been involved in the plate-tectonic process (e.g. Wasserburg & DePaolo 1979).

Since the introduction of these mass-balance calculations, however, a number of workers have shown that the uncertainties in the incompatible-element concentrations in the various reservoirs involved allow anywhere from 30 to 90% of the mantle to be as depleted as the MORB source (e.g. Allègre et al 1983a,b, Davies 1984, Zindler & Hart 1986). These volume estimates are most sensitive to the assumed average composition of the continental crust. For example, if the crust is closer to high-Al basalt in average composition rather than the generally assumed andesitic composition (e.g. Taylor & McLennan 1981), the MORB source may occupy significantly *less* than 30% of the mantle, perhaps even less than 10%. Another important factor pushing estimated MORB source volumes to higher percentages is the fact that there appear to be incompatible-element-enriched components in the Earth other than the continental crust. In particular, the subcontinental lithospheric mantle appears to consist of thick sections of geochemically distinct material, often enriched in incompatible elements, that have been isolated from mixing with the main mass of the mantle for time periods that in some cases exceed 3 b.y. (e.g. Leeman 1975, Brooks et al 1976a, Stosch et al 1980, Kramers et al 1983, Cohen et al 1984, Carlson 1984, Hawkesworth et al 1984, Richardson et al 1984, 1985, Boyd & Gurney 1986). Including the subcontinental mantle as an additional and even less well-known enriched reservoir would serve to increase the minimum fraction of depleted mantle (Carlson 1984; reproduced here as Figure 8) as well as the general uncertainty in this kind of mass-balance calculation.

Another difficulty in mass-balance calculations concerns the basic premise that the depletion of the MORB source has occurred through time by

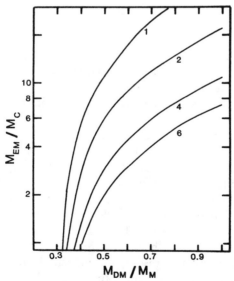

Figure 8   Ratio of mass of depleted mantle (M_{DM}) to total mantle (M_M) required to account for Nd enrichment in a given mass of enriched mantle (M_{EM}) plus continental crust (M_C). Numbers along each curve give the enrichment factor (concentration ratio) of enriched relative to undepleted mantle (from Carlson 1984).

the continual extraction of continental crust. While island-arc volcanism would suggest such a relationship, continuing research on subduction-related volcanism appears to suggest that the slab acts primarily as a fluxing agent, with most of the arc magmas derived instead by melting of the overlying mantle wedge (e.g. Gill 1981). The validity of assuming complementarity between MORB-S and continental crust can be assessed by examining the depletion history of the MORB source provided by isotopic systematics.

Age of Depletion of the MORB Source

Perhaps one of the most important constraints on the age of MORB-S depletion is the radiogenic character of rare gases in modern MORB (e.g. Ozima & Podosek 1983). The moderately high $^4He/^3He$ and very high $^{40}Ar/^{36}Ar$ ratios common to MORB indicate that the MORB source was extensively degassed within about the first billion years of Earth history (Ozima & Kudo 1972, Allègre et al 1983, Sarda et al 1985). Supporting evidence for a very early outgassing of the MORB source comes from the observation of excess ^{129}Xe in MORB (Allègre et al 1983, 1987) attributed to the decay of primordial ^{129}I. Taken at face value, the ^{129}Xe excess suggests that degassing of the MORB source had to occur within a few ^{129}I half-lives (17 m.y.), or only about 50 m.y. of the Earth's formation. However, the Xe isotopic data do not show a signature of the ^{244}Pu fission (half-life = 76 m.y.) that should be present if the excess ^{129}Xe is from the decay of now-extinct ^{129}I. This may imply that the observed excess in ^{129}Xe is not from primordial ^{129}I decay but rather from the decay of ^{129}I produced over Earth history as a product of U fission (Caffee & Hudson 1987). Nevertheless, the general conclusion derived from the rare-gas data set is that the outgassing (and, by inference, the incompatible-element depletion) of the MORB source occurred within the first billion years after the Earth's formation.

An early depletion age is supported further by the initial Nd isotopic composition of ancient mantle-derived rocks now preserved on the continents. As reproduced in Figure 9, initial Nd isotopic compositions for Archean komatiites, tholeiites, and monzodiorites generally yield positive values of ε_{Nd}. Positive initial ε_{Nd} indicates a source with a time-averaged Sm/Nd greater than the bulk-earth, reflecting relative depletion in the light rare-earth elements, and presumably other incompatible elements. Modern MORB has ε_{Nd} generally in the range +8 to +12.

The majority of initial Nd isotopic compositions for Archean mantle-derived rocks fall within, or below, the range of values expected for a source, now similar to the MORB source, that has had a constant Sm/Nd ratio for 4.55 b.y. Data falling below this trend potentially reflect inter-

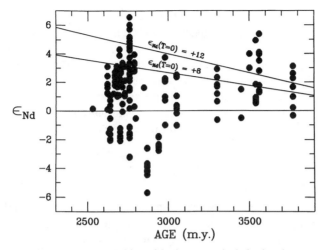

Figure 9 Initial Nd isotopic composition of Archean mantle-derived rocks compared with growth curves expected for the bulk-earth ($\varepsilon_{Nd} \equiv 0$) and for reservoirs with constant Sm/Nd from 4.55 b.y. ago to the present that have present day ε_{Nd} values of $+8$ and $+12$. Here ε_{Nd} is the difference in initial $^{143}Nd/^{144}Nd$ of the sample, expressed in parts in 10,000, relative to that expected for BSE at the time of formation of the sample (modified from Shirey & Hanson 1986).

action between mantle-derived melts and slightly older continental crust or subcontinental mantle. Positive initial ε_{Nd} observed in Archean mantle-derived rocks suggests that the relative incompatible trace-element abundances of the MORB source were established very early in Earth history and have not changed greatly since formation. The data, however, also allow curved (i.e. varying Sm/Nd) evolutionary trajectories for the Nd isotopic composition of the MORB source, consistent with models where the composition of the MORB source changes as a result of continuing crust formation and resubduction into the mantle (e.g. Armstrong 1981, DePaolo 1983).

If the MORB source was in fact depleted by continent formation, the positive ε_{Nd} of even the oldest crustal rocks requires the extraction of significant volumes of continental crust from the mantle during the time period from 3.8 to 4.5 b.y. ago. However, as seen in Figure 9, the general lack of this ancient, negative ε_{Nd} crust as even a highly reworked component in the oldest rocks preserved in any craton is surprising if significant volumes of silicic crust existed prior to 3.8 b.y. ago. Because of its relatively low density and, if wet, its low melting point, continental crust similar to that existing today is likely to stay near the surface either by resisting subduction or by returning as a component of subduction-related volcanism if forced into the mantle.

Thus, both the evidence for early outgassing of the MORB source and the Nd isotopic data presented in Figure 9 imply that the incompatible-element-depleted characteristic of the MORB source was present before 3.8 b.y. ago and therefore may not be related to crust formation. Further, it suggests that the MORB source is a robust mantle component that has supplied surface volcanism throughout Earth history without experiencing significant changes in its defining compositional characteristics. In this case, both MORB-S and the continental crust would have their own complementary reservoirs.

Complementarity Between the Continental Crust and Subcontinental Mantle

An alternative to the conventional assumption of continental crust–MORB source complementarity is that the continental crust is complemented by the subcontinental lithospheric mantle. In their major-element characteristics, garnet lherzolite xenoliths, representing fragments of the subcontinental mantle carried to the surface by kimberlites, suggest that the subcontinental mantle, on average, is depleted in Ca and Al compared with pyrolite (Jordan 1979, Richardson et al 1985, Boyd & Gurney 1986, Boyd & Mertzman 1987). This apparent depletion in "basaltic" elements is in accord with the suggestion that the subcontinental lithosphere forms through melt removal that leaves behind a garnet-clinopyroxene-poor residue less dense than the underlying mantle (Jordan 1981), a process similar to the formation of oceanic crust. If the average composition of these garnet lherzolite xenoliths (e.g. Jordan 1979) represents that of subcontinental mantle, a mixture of average continental crust (e.g. Taylor & McLennan 1981) with 4 to 6 times its volume of subcontinental mantle will have Si, Al, Mg, and Fe contents similar to pyrolite (Ringwood 1975), a major-element model for MORB source and BSE (Table 4). Thus, on a major-element basis alone, a 150- to 200-km-thick subcontinental mantle, similar to the thickness actually observed, could represent the complement to the continental crust, with MORB-S or BSE as the starting composition.

This is not true if incompatible trace-element abundances are taken into account. Given current estimates for the composition of the continental crust (e.g. Taylor & McLennan 1981), an absolute minimum of 10–20% of the mass of the mantle of BSE starting composition is required to provide the mass of incompatible elements in the continental crust. The mass of the subcontinental lithospheric mantle, if assumed to be on average 150 km thick, is only about 3% of the whole mantle. In addition, the available samples of the subcontinental mantle often do not show the incompatible-element depletion that would be expected to accompany their Ca- and Al-depleted character (Stosch et al 1980, Harte 1983, Menzies

Table 4   Mass-balance relations in the continental crust–subcontinental mantle system[a]

	BSE	ACGL	ACC	M_{CL}/M_M (%)	C_{CL}	C_{CC}
SiO_2	46.0	45.6 ± 1.5	58.0	17	43.7	48.1
Al_2O_3	4.06	1.43 ± 0.6	18.0	2.9	1.38	17.8
FeO	7.55	7.61 ± 0.94	7.5	0.4	7.56	7.2
MgO	37.8	43.6 ± 2.0	3.5	3.2	44.4	7.6
CaO	3.21	1.05 ± 0.77	7.5	1.1	2.39	14.5
TiO_2	0.18	0.11 ± 0.06	0.8	4.8	0.06	0.55
Na_2O	0.33	0.14 ± 0.10	3.5	9.0	—	1.32
K_2O	0.032	0.11 ± 0.10	1.5	—	—	—
Cr_2O_3	0.47	0.34 ± 0.14	0	—	0.56	1.15

[a] Abbreviations as follows: BSE, bulk silicate Earth composition (Zindler & Hart 1986); ACGL, average continental garnet lherzolite (Jordan 1979); ACC, average continental crust (Taylor & McLennan 1981); M_{CL}/M_M, mass (expressed in percent) of subcontinental mantle (M_{CL}) relative to total mantle (M_M) required for ACGL and ACC to sum to BSE composition; C_{CL}, composition of a 150-km-thick section of subcontinental mantle from ACC and BSE mass balance; C_{CC}, composition of continental crust produced from BSE to leave behind a 150-km-thick section of composition equal to ACGL. All compositions (BSE, ACGL, ACC, C_{CL}, C_{CC}) expressed as weight percent.

1983, Kramers et al 1983, Menzies & Wass 1983, Cohen et al 1984, Richardson et al 1985, Boyd & Mertzmann 1987). Instead, many reflect interaction with melts or metasomatic fluids that have greatly bolstered their abundances of incompatible elements but left their major-element compositions nearly unchanged (e.g. Menzies & Hawkesworth 1987). Thus, both the continental crust and subcontinental mantle *appear* to be more enriched in incompatible elements than the MORB source.

The apparent complementarity of continental crust and subcontinental mantle in major elements is thus in conflict with the enrichment of both in incompatible elements such as K (Table 4), the REE, and Ba. To reconcile this observation and still maintain strict complementarity between continental crust and subcontinental mantle would require that current estimates of incompatible-element abundances in the crust be high by factors of 2 (light REE) to 10 (Ba) depending on the element used for the calculation, and that the subcontinental mantle is not incompatible-element enriched on average. A continental crust more basaltic than andesitic (Table 4) would still satisfy the major-element mass balance but would be lower in incompatible-element abundances and hence would lessen the mismatch of the mass-balance calculation for major and trace elements. However, even in this case, the subcontinental mantle, on average, would have to be depleted in incompatible elements compared with MORB-S. This is in conflict with the observational evidence for incompatible-element *enrichment* in many xenoliths from the subcontinental mantle, leaving only the possibility that the trace-element signature of these xenoliths is a peculiarity related to the events that caused their capture by host lavas. Though possible, the common occurrence of incompatible-element enrichment in xenoliths from many localities suggests instead that this enrichment is a feature of a substantial portion of the subcontinental mantle. If so, the enrichment requires that the continental lithosphere receive fluxes of incompatible trace-element-rich fluids from some other reservoir, most likely at convergent margins, thereby representing the complementary relation between the continental crust and the MORB source central to the mass-balance models discussed previously. It is important to note, however, that on a major-element basis, the subcontinental mantle *could* be the only portion of the mantle affected by crust extraction, thus implying that the major-element characteristics of the MORB source may have changed little over Earth history. Whether the same can be said of the incompatible trace-element abundances of the MORB source is less clear. The evidence that the MORB source acquired its "depleted" characteristics very early in Earth history, prior to 3.8 b.y. ago, suggests that the major depletion of the MORB source occurred well before the first significant volumes of continental crust were extracted from the mantle. In any case, the questionable complementarity of incompatible-element enrichment in the continental crust and the depletion of the MORB-S in these elements severely weaken any inferences derived from the simple mass-balance models concerning the relative volumes of enriched and depleted reservoirs in the mantle.

Cause of Depletion of the MORB Source

If crustal extraction did not cause the incompatible-element depletion of the MORB source, then what other possibilities exist, and where and how large is the complementary enriched component? At the outset, we emphasize that there is a virtually complete tradeoff between the degree of enrichment of the complement and its size. Small volume but exceedingly incompatible-element-rich materials such as kimberlites have been suggested (Anderson 1982a) in analogy to the KREEP component of the Moon (Meyer 1977), but the isotopic characteristics of at least Type 1 kimberlites show their incompatible-element enrichment to have been produced only shortly before eruption (Smith 1983). Type 2 kimberlites, and many other highly alkaline rock types, show more extreme isotopic compositions suggestive of derivation from an older enriched source, but these in most cases can be reconciled with derivation from an incompatible-element-rich portion of the subcontinental mantle (McCulloch et al 1983,

Smith 1983, Vollmer & Norry 1983a,b, Vollmer et al 1984, Fraser et al 1986, Nelson et al 1986, Dudas et al 1987).

Reservoirs only slightly enriched in incompatible elements compared with undifferentiated BSE need to be much larger in volume and may represent a substantial fraction of the mantle. Given the evidence for ancient depletion of the MORB source, it is conceivable that complementary enriched mantle sources formed in major terrestrial differentiation events early in Earth history, perhaps analogous to the magma ocean processes proposed for early lunar evolution (e.g. Anderson 1982b, Nisbet & Walker 1982, Hofmeister 1983, Herzberg 1984, Ohtani 1985, Anderson & Bass 1986). If so, however, the paucity of negative initial ϵ_{Nd}'s in ancient mantle-derived rocks (Figure 9) strongly suggests that this enriched reservoir did not participate extensively in surface volcanism.

At this time, the significance of major early differentiation processes for later evolution of the Earth is largely conjecture, but the increasing sophistication of models of terrestrial planet formation and early thermal evolution indicate that early extensive processing of the whole Earth is possible, if not likely. With the first data now becoming available for trace-element partitioning for high-pressure phases such as majorite and perovskite (Kato et al 1987) and for the physical properties of silicate liquids at high pressure (Agee & Walker 1987), rigorous tests of several evolutionary scenarios for early Earth history should soon be possible.

We are thus left with the unsolved problem of characterizing the component that is complementary to the MORB source. If it is not the continental crust as in the conventional mass-balance calculations, then this component appears to be only rarely sampled at the Earth's surface. Thus we have a "missing" reservoir that by simple calculation could represent a substantial fraction of the volume of the mantle.

While the two complementary relationships suggested above for MORB-S lead to different views of the mantle, there is one important similarity that should be emphasized—namely, in either case it is suggested that there is a large non-plate-tectonic component in the mantle. The rarity of this component at the Earth's surface may be almost as important as its existence, in that it too implies a component that has not interacted extensively in the plate-tectonic process of creating the oceanic plates.

CONSTRAINTS ON STRATIFICATION FROM SEISMOLOGY/MINERAL PHYSICS

The evidence for a non-plate-tectonic component discussed in the preceding section has been used by many investigators to argue that the mantle is compositionally stratified. While this evidence is based on trace elements, the implied chemical differentiation would be expected to produce major-element stratification as well; the major-element stratification would then serve to preserve trace-element stratification in a convecting mantle.

It is possible to address the question of major-element stratification by comparing the seismically determined shear and compressional velocities (V_S, V_P) and density (ρ) with the theoretical profiles for a homogeneous material, as well as with the elastic parameters of candidate mantle compositions. Such comparisons have had a long history, starting with the pioneering work of Birch (1952). The comparisons of seismically determined V_S and V_P with theoretical profiles in density and squared bulk sound velocity $\phi(= V_P^2 - \frac{4}{3}V_S^2)$ (Birch 1952, Bullen 1965) using the Adams-Williamson equation suggest that the lower mantle is consistent with the self-compression of a homogeneous material, except perhaps at the very bottom; however, there is one region that clearly deviates from this—the transition zone between depths of about 350 and 800 km. More recent seismological observations, including free-oscillation data, have confirmed these conclusions (Dziewonski & Anderson 1981). Some earlier authors used the properties of the transition zone to argue that the mantle was compositionally stratified and either not convecting at all or convecting in separate layers above and below the transition zone (Urey 1952). However, we now know that the mantle *is* convecting and, furthermore, that the transition zone involves several important phase changes in the primary silicates of the mantle. The major question that has been addressed for the last 20 years is whether the properties of this zone can be adequately accounted for solely by a series of phase changes in a homogeneous material or whether a composition change additionally is present. The latter possibility would argue for a chemically distinct upper and lower mantle. The study of the transition zone itself has yielded inconclusive results. There have been many diagnostics used: the locations of seismic discontinuities, the velocity/density contrast across discontinuities, the sharpness of the 650-km discontinuity, and the absolute values of the velocities in the transition between the 400- and 650-km discontinuities. Weidner (1985, 1986) has summarized presently available elastic data for the transition zone and has constructed a model based on pyrolite composition. Comparing this model to the V_P and V_S velocity-depth profiles of Grand & Helmberger (1984) and Walck (1984), he has concluded that within present uncertainties, pyrolite satisfactorily fits the velocity jump at the 400-km discontinuity, the velocity increase between 400 and 650 km, and the velocity between 300- and 400-km depth. This conclusion has received further support from Bina & Wood (1987), who find that both the sharpness and jump across the 400-km discontinuity in bulk sound velocity $(\phi^{1/2})$ is consistent with that computed for these same seismic profiles. This is not to say that a compositional gradient necessarily can be ruled out; several models for a chemically stratified upper mantle have been proposed (see Anderson & Bass 1986) in which it has been argued that phase changes in pyrolite *cannot* match the observed elastic properties. At this stage, however, it does not seem possible to distinguish between these two hypotheses using data from mineral physics and seismology alone (Weidner 1985, 1986).

The sharpness of the 650-km discontinuity has also been used as a diagnostic. Lees et al (1983) have argued that if this boundary represents a phase change, it should be very broad, perhaps on the order of a 30-km depth range, whereas a chemical boundary between separately convecting systems would presumably be sharp (less than about 5 km). Based on the amplitudes of precursors to the body-wave phase $P'P'$, representing bottom-side reflections from this boundary, they argue for a sharp discontinuity of only a few kilometers thickness and thus conclude that the boundary is compositional. However, such an inference is premature, since the expected phase boundary transition may in fact be very sharp (see Ito & Takahashi 1986, Silver et al 1985).

Recently, researchers have begun to address the problem of stratification by comparing the seismic properties of the lower mantle with those of successful candidates for the upper mantle such as pyrolite. Because of the higher pressures, this becomes a more difficult experiment technically. In some ways, however, the lower mantle is easier to study than the transition zone. In the transition zone, many phase changes occur in a relatively small depth interval, which both requires detailed knowledge of the behavior of a variety of phases and places a greater emphasis on the resolution of the seismic velocity models. In contrast, the lower mantle is thought to consist predominantly of one phase, perovskite (with lesser amounts of magnesiowustite), and to be homogeneous in composition; no major phase changes appear to occur throughout the entire lower mantle (Knittle & Jeanloz 1987), at least in (Fe,Mg) perovskite. In this case, radial averages in seismic properties over a few hundred kilometers are sufficient for comparisons with candidate compositions. The frequencies of the Earth's free oscillations, while being of limited utility in the transition zone due to poor resolution, become an important source of information in the lower mantle, especially because they can provide constraints on density, which is the easiest property to measure in the laboratory.

Based on a formal resolution analysis of the free-oscillation data set, Gilbert et al (1973) concluded that the seismically determined density profile $\rho_s(r)$ of the lower mantle is known to better than 1% for averaging lengths of about 300 km. A recent analysis by G. Masters (personal communication) using more precise frequency measurements, incorporating nearly all available digital seismic data, indicates that the uncertainty is now closer to 0.5%.

Thus, if it were possible to make a measurement of pyrolite density or other candidate upper-mantle compositions at lower-mantle conditions, this comparison could be made directly. Because of practical limitations, the measurements are done at atmospheric pressure and at temperatures that are elevated but less than lower-mantle temperatures. As a result, a measurement requires the "decompression" of the seismic density and a determination of the coefficient of thermal expansion, α. Knittle et al (1986) have in fact performed this experiment. They find that perovskite has an unusually large value of α (4×10^{-5} K^{-1}) and conclude that reasonable upper-mantle compositions, such as pyrolite or peridotite, appear to be 2% lighter than $\rho_s(r)$ from PREM (Preliminary Reference Earth Model; Dziewonski & Anderson 1981) in the lower mantle. They suggest that the lower mantle is enriched in iron with a Mg number of 0.84 compared with 0.89 for pyrolite. An increase in silica producing more perovskite and less magnesiowustite could also explain the results. This 2% density difference is much larger than their formal uncertainties or the formal seismic uncertainties and, if true, suggests that the upper and lower mantle are compositionally distinct. The potential importance of this conclusion places a greater emphasis on the uncertainties in both the mineral physics and the seismological procedures. While such issues are fairly technical [for seismology, the potential tradeoffs between density and seismic velocities; for mineral physics, how extrapolations are made between two (P, T) conditions, problems with metastable crystals], a close examination of these uncertainties will be necessary before this result may be taken as highly probable. In principle, one way of reducing the mineral physics uncertainties would be to make the density measurements directly at lower-mantle conditions. While this procedure is not technically feasible at present, current progress with synchrotron X-ray sources suggests that such an experiment will in fact be possible in the next few years (A. Jephcoat, personal communication).

If there is a density difference between the upper and lower mantle, it is possible to discuss some of its characteristics. For example, the observation of lower-mantle slab penetration limits the intrinsic density gradient through the transition zone to about 3% or less. Secondly, if the major seismic discontinuities are primarily due to phase changes, then there is no compelling reason for a compositional change to be associated with a particular discontinuity unless a mechanism can be proposed for the phase and chemical boundaries to track each other in the presence of substantial mass flux through the transition zone. Thus, a gradual gradient above or below 650-km depth would be possible and perhaps even more probable.

AN EXAMINATION OF SOME CANDIDATE MANTLE MODELS

In this section several models for mantle flow are considered primarily on the basis of simple mixing calculations. Initially, we invoke only two constraints: that a substantial fraction of subducted slabs penetrate into the lower mantle, and that the observed mean age or residence time, $\langle t_r \rangle$, of plate-tectonic recycling is 1.7 b.y. These models are then evaluated further on the basis of other geophysical and geochemical constraints. While there are many properties that may be used to classify mantle models, probably the most appropriate for discussing mixing is the mass flux between the upper and lower mantle of material associated with plate-tectonic recycling. In this classification, whole-mantle convection and two-layer convection represent end-member flows of unrestrained flux and zero flux, respectively; both models are examined. Strictly speaking, the two-layer model is inconsistent with lower-mantle slab penetration; instead, a modified two-layer model is considered with minimal flux. Two other models, referred to as boundary-layer convection and penetrative convection, represent intermediate cases, where the plate-tectonic-related flux through the transition zone is approximately equal to that carried by the slab flux.

In mixing calculations, two characteristic times are typically considered: the residence time $\langle t_r \rangle$ and the survival time t_s. They reflect very different properties of the mantle. While $\langle t_r \rangle$ represents a robust, average property of a mixed system, survival time depends on the details of the strain field in the mantle, which are highly uncertain. For this reason, there has been a marked difference in opinion in the literature concerning survivability. While Gurnis & Davies (1986) have indicated that the heterogeneity injected into the mantle by plate subduction can survive billions of years, other investigators have come to the opposite conclusion, namely Allègre & Turcotte (1986), Richter et al (1982), Hoffman & McKenzie (1985), and Olson et al (1984), based on consideration of the role of small-scale boundary-layer instabilities that dramatically increase the rate of mixing. Hoffman & McKenzie (1985) have argued that heterogeneity will survive only about 0.2 b.y. Because t_s is dependent on the details of the flow, $\langle t_r \rangle$ will be our primary constraint on mixing. We thus are not concerned with the "survival" of individual heterogeneities, but rather the contribution of the heterogeneity to the mean properties of the mantle.

The process of plate tectonics is modeled as injecting zero-age heterogeneity into the mantle at subduction zones and randomly sampling it at ridges by the differentiation of oceanic crust. Here $\langle t_r \rangle$ is interpreted as the mean age of either the upper mantle or whole mantle depending on the model considered. Thus it is assumed that this region is well mixed from the point of view of sampling at mid-ocean ridges, and we are not biased by sampling at the top of the reservoir. Finally, undifferentiated material is assigned an age equal to that of the Earth. There have been several recent attempts to construct models that satisfy this residence-time constraint, most notably in a series of papers by Davies and Gurnis (Davies 1984, Gurnis & Davies 1986, Gurnis 1986). We proceed with a development similar to theirs for a single reservoir but extend it to consideration of two-reservoir mixing.

An important quantity in mixing calculations is the flux time for the whole mantle, T_w, which is defined as the mass of the whole mantle M_w divided by the mass flux of chemically differentiated oceanic lithosphere, referred to as f:

$$T_w = M_w / f. \tag{1}$$

This is the time it would take for the entire mantle to be fluxed through the plate-tectonic process of oceanic crustal formation. The importance of this number is that for random sampling models, and for residence times much shorter than the age of the system, the flux time is a good approximation to the residence time (Gurnis & Davies 1986).

We use mass flux in (1) rather than volume flux in order to account for the compression of homogeneous material. The mass flux f of chemically differentiated oceanic lithosphere can be written as

$$f = \dot{A} D \rho_{\text{lith}}, \tag{2}$$

where $\dot{A}$ is the areal plate creation rate, D is the thickness of chemically differentiated material, and ρ_{lith} is the average density of oceanic lithosphere. The present-day plate creation rate can be estimated reasonably well to be $3 \, \text{km}^2 \, \text{yr}^{-1}$ (e.g. Chase 1972). The least well-constrained parameter is D, since it includes not only the 6 km of oceanic crust but also its mantle complement. Here D is not the thickness of the mechanical lithosphere but rather corresponds to the chemically differentiated portion. Estimates of the degree of partial melting to produce ocean-ridge basalts are about 10–30% (Gast 1968, Fuji & Scarfe 1985), which yields $D = 20$–60 km. Larger estimates range up to about $D = 75$ km (e.g. Ringwood 1982). In the calculations that follow, we use $D = 30$ km and $D = 75$ km to represent the reasonable range of values (see Table 5). Taking

Table 5 Mixing parameters[a]

| Region | Volume $(10^{11}\ km^3)$ | Mass $(10^{24}\ kg)$ | Flux time (b.y.) | |
			$D = 30$ km	$D = 75$ km
Upper mantle	3.0	1.0	3.3	1.3
Lower mantle	6.1	3.1	10.3	4.1
Whole mantle	9.1	4.1	13.6	5.4

[a] Values used: areal plate generation rate $= 3\ km^2\ yr^{-1}$; chemically differentiated thickness $D = 30$ km (Gast 1968), $= 75$ km (Ringwood 1982); density of lithosphere $= 3375\ kg\ m^{-3}$; volume flux $= 90\ km^3\ yr^{-1}$ ($D = 30$ km), $= 225\ km^3\ yr^{-1}$ ($D = 75$ km); mass flux ($\times 10^{14}$ $kg\ yr^{-1}$) $= 3.0$ ($D = 30$ km), $= 7.6$ ($D = 75$ km).

$M_w = 4.1 \times 10^{24}$ kg and the density of the lithosphere to be $\rho_{lith} = 3375$ kg m^{-3}, we obtain $f(30) = 3.0 \times 10^{14}\ kg\ yr^{-1}$ and $f(75) = 7.6 \times 10^{14}\ kg\ yr^{-1}$.

The respective flux times from (1) for the whole mantle are $T_w(30) = 13.6$ b.y. and $T_w(75) = 5.4$ b.y. Thus, at the outset it is seen that regardless of the value of D chosen, T_w is substantially greater than $\langle t_r \rangle$, suggesting that much of the mantle may not have been through the process of plate-tectonic chemical differentiation.

Homogeneous Whole-Mantle Convection: Unrestrained Flux

Consider first a mantle that is homogeneous both in composition and rheology and ignore for the present the effects of phase changes. Assuming that the mean ages of the whole mantle τ_w and the material sampled at ridges are the same, and assuming a constant flux, the expression for the rate of change of the mean age of the mantle is

$$\tau'_w = 1 - \tau_w/T_w, \tag{3}$$

where the prime denotes differentiation with respect to time. The first term on the right-hand side represents the change in mean age due simply to the passage of time, while the second term is the result of the flux. The solution to (3) is

$$\tau_w = T_w(1 - e^{-t/T_w}). \tag{4}$$

Taking t to be age of the Earth, 4.55 b.y., the mean ages for the mantle are 3.8 b.y. and 3.1 b.y. for $D = 30$ km and 75 km, respectively. Thus, it may be concluded that the homogeneous whole-mantle convection model yields ages that are much greater than the observed residence time of 1.7 b.y.

From Equation (1) there are two ways to reduce T_w in order to bring the mean age τ_w into agreement with $\langle t_r \rangle$: either reduce the effective mass of the mantle associated with plate recycling or appeal to a mass flux that was much greater in the geologic past [i.e. either choose a constant flux that is 3 to 8 times higher than the present-day estimate or make T_w time dependent in Equation (3)]. Gurnis & Davies (1986), who noted this discrepancy between observed residence time and mantle flux time chose the latter, arguing that plate speeds were much greater in the past owing to the increase in heat flux associated with the greater abundance of radioactive elements. They assume that in time-dependent convection, the plate speed u is proportional to the square of the heat flux q ($u = cq^2$), as predicted for steady state by simple boundary-layer theory (Turcotte & Oxburgh 1967), and that heat flux has been decreasing exponentially over time as $q = q_0 e^{-\gamma t}$. Reasonable values for the thermal decay constant γ predict that plate speeds were roughly an order of magnitude higher in the Archean than in the present. If correct, this model has important implications for the early history of the Earth, as it predicts that all tectonic time scales were shorter in the past by this same factor.

While it is difficult to test this possibility in a definitive manner, there are some observations and dynamical considerations that may provide a start. These results suggest that plate speeds in fact have been more or less constant over the last 2–3 b.y. Numerical experiments by Christensen (1984) on convection with strongly temperature-dependent viscosity indicate a much weaker dependence of heat flux and surface velocity on Rayleigh number than is inferred from heat-transfer laws derived from studies of constant viscosity convection. This suggests that the conventional thermal history calculations based on parameterizations of constant viscosity convection as used by Gurnis & Davies (1986) overestimate the sensitivity of surface variables (heat flow, plate speed, etc) to changes in mantle viscosity, and that the increased viscosity with time due to secular cooling may not have as dramatic an effect as these simple parameterizations predict. The implication is that plate speeds and the characteristic time scales of tectonic processes driven by convection have remained approximately constant through Earth history.

There appears to be observational support for this conclusion as well. Ullrich & Van Der Voo (1981) have computed the minimum plate velocities of four continental plates, two back to the Archean, on the basis of an estimate of latitudinal velocities from apparent polar wander paths. While this does not represent the true velocity, it should reflect the rate of plate-tectonic activity in general and thus show a substantial increase in velocity back to the Archean, assuming that this apparent velocity is not biased, for example, by an increase or decrease in the ratio of latitudinal to longitudinal velocity over time. These authors find that while there are certainly large fluctuations in velocities, there is no evidence that plate velocities have decreased by an order of magnitude since the Archean. It is thus concluded that there is no observational support for significantly faster plate velocities in the past, nor is there necessarily a compelling reason from thermal history considerations.

There is also the possibility of producing a larger flux (with constant plate speeds) by allowing for substantially smaller plates in the past, which can be viewed as increasing the total lengths of ridges and subduction zones by an order of magnitude. This case is not easily evaluated, since there is even less information on this possibility than on past plate speeds. It should be mentioned, however, that this alternative would represent a much more radical departure from the present-day style and geometry of plate tectonics than that implied by greater plate velocities.

In summary, a homogeneous whole-mantle model does not satisfy the observed residence time for mantle recycling unless the plate flux was significantly higher in the past. While such an evolutionary model cannot be firmly rejected, neither can it be taken for granted. It is concluded that the short residence time for material injected into the mantle by plate-tectonic processes may have another explanation. We thus consider the alternative—namely, that the volume associated with plate-tectonic activity is substantially smaller than that of the mantle as a whole.

For model residence times consistent with the observed time of 1.7 b.y., the mass of the mantle associated with plate-tectonic recycling would have to range from one seventh to one third of the whole mantle, depending on the value of D taken. Such models may be referred to as heterogeneous, in that plate-tectonic material has a shorter residence time compared with the average properties of the whole mantle. There are many ways of achieving this goal, although we must at the same time satisfy the constraint on slab penetration. The following models represent some possibilities.

Modified Two-Layer Convection

The first and probably best-known pattern of convection with heterogeneity is two-layer convection, where plate recycling and crustal formation are restricted to the upper mantle and the lower mantle remains an isolated, rarely sampled reservoir. This configuration is usually achieved by interpreting the 650-km seismic discontinuity as a compositional boundary, with an intrinsic density increase that overwhelms the effects of thermal buoyancy. Since the plate-tectonic reservoir is then the size of the upper mantle, its mean age τ_u is in approximate agreement with the Pb model ages because the upper mantle (down to 650 km) is roughly one fourth of the mantle by mass. However, such a model in its strict form is inconsistent with lower-mantle slab penetration. One may attempt to modify this model by proposing a "leaky" upper mantle, where the barrier to mass flux between the upper and lower mantle is not total. Compositional stratification is not the only way to impose this; *homogeneous* mechanisms, such as an endothermic phase change or a pressure-dependent viscosity contrast between the upper and lower mantle, are possibilities.

COMPOSITIONAL BARRIER   Consider the case in which the upper and lower mantle are allowed to have different mean ages. It is assumed that the downward flux from the upper to lower mantle results solely from subducted slabs of zero age; thus entrained flow is ignored. This is probably not a bad assumption in the case of a compositional barrier, since the barrier will serve to minimize the amount of entrained flow. As the upper mantle will be younger than the lower mantle, however, any entrained flow would serve to increase the age of the upper mantle. Ignoring entrained flow thus provides an upper bound on the allowable slab flux. Further, it is assumed that both reservoirs are well mixed, so that the mean age of the upward flux at the surface equals the mean age of the upper mantle τ_u, while the mean age of the upward flux at the transition zone equals the mean age of the lower mantle τ_l. The differential equations describing the time rate of change of the mean ages of the upper and lower mantle are

$$\tau_l'(t) = 1 - \varepsilon\tau_l(t)/T_l, \tag{5}$$

$$\tau_u'(t) = 1 + \varepsilon\tau_l(t)/T_u - \tau_u(t)/T_u, \tag{6}$$

where $T_l = M_l/f$ is the flux time of the lower mantle, and ε is the fraction of the surface flux f that enters the lower mantle. In this case, (5) may be solved for the lower mantle independently of the upper mantle and substituted into (6) with the solution

$$\tau_u = \frac{K+1}{\alpha_2}[1 - e^{-\alpha_2 t}] - \frac{K}{(\alpha_2 - \alpha_1)}[e^{-\alpha_1 t} - e^{-\alpha_2 t}], \tag{7}$$

where $\alpha_1 = \varepsilon T_l^{-1}$, $\alpha_2 = T_u^{-1}$, and $K = M_l/M_u$ is the mass ratio of lower to upper mantle. For $K = 3.1$, $T_u = 1.3$ b.y., $T_l = 4.1$ b.y. ($D = 75$ km), we have plotted τ_u as a function of ε in Figure 10. For $\tau_u = \langle t_r \rangle = 1.7$ b.y., we have $\varepsilon = 0.15$; thus the mass flux across the transition zone into the lower mantle is at most one sixth of the surface flux (a smaller value of ε for lower values of D).

How does a value of $\varepsilon = 0.15$ compare with reasonable estimates of slab flux from seismology? Considering only the western Pacific, where deep slabs have been identified by Creager & Jordan (1984, 1986a), this is about one fourth of the areal flux at the surface of the Earth, corresponding to $\varepsilon = 0.25$. Thus it may be marginally possible for such a small-flux model to satisfy the slab data on the basis of residence-time considerations,

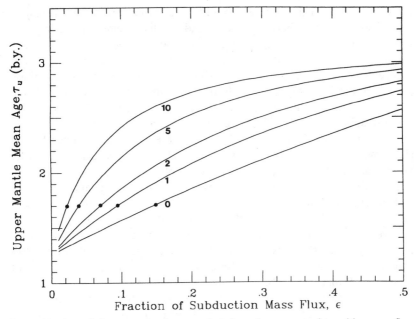

Figure 10 Age of the upper-mantle reservoir (τ_u) vs. the amount of transition-zone flux expressed as a fraction of the surface subduction flux (ε) for the modified two-layer model and considering five levels of entrainment. For no entrainment (curve labelled "0"), the observed residence time of 1.7 b.y. is obtained for $\varepsilon = 0.15$, which represents an upper bound on ε. Even modest amounts of entrainment appropriate for a compositional barrier to flow (entrainment factors of 1 or 2) reduce ε to below 0.1. Entrainment factors more appropriate for a homogeneous barrier to flow (5 or 10) reduce ε to below 0.03. By comparison, a lower bound on ε of 0.25 is suggested by the seismological evidence for lower-mantle slab penetration.

although it should be remembered that any entrainment would serve to reduce the allowable range of ε. If one explicitly allows for entrainment, an entrainment of only a factor of 2 (twice as much entrained material as original material) would put ε below 0.1 (see Figure 10).

HOMOGENEOUS BARRIER   The applicability of this model depends on being able to identify the physical mechanisms responsible for throttling flow through the transition zone. One possibility is the combination of endothermic phase changes plus an increased lower-mantle viscosity. Such a combination could deflect all but the heaviest subducted material and thus limit the permeability of the transition zone considerably. Phase changes in the pressure regime of the lower part of the transition zone are generally thought to be endothermic on thermodynamic grounds (Navrotsky 1980). The principal dynamic effect comes from depression of the phase boundary in the slab, which adds a restoring force proportional to the density difference between phases and the displacement of the boundary. Larger (negative) Clapeyron slopes produce larger phase boundary displacements and are therefore more likely to arrest sinking slabs. Latent heat effects in which the slab absorbs heat across the depressed phase boundary are by comparison only of minor importance (Olson & Yuen 1982). Numerical calculations by Christensen & Yuen (1984) on the resistance to subduction by endothermic phase changes indicate that a phase-change Clapeyron slope of about -20 bar K^{-1}, a likely upper bound (Navrotsky 1980, Liu 1979), is capable of deflecting some slabs, especially young segments of lithosphere. Therefore, phase transitions near 650 km may play an important role in limiting flow into the lower mantle.

The tendency for slabs to be deflected at the transition zone and thus remain within the upper mantle is enhanced by the jump in viscosity between the upper and lower mantle. There is now general agreement that a viscosity increase occurs across the transition zone, although there is still no consensus regarding its magnitude. Rotational data are more consistent with a small increase—about a factor of 3 (Wu & Peltier 1984, Yuen & Sabadini 1985), whereas a factor of 10–100 is required to bring the geoid computed from slab density anomalies into agreement with the geoid measured over subduction zones (Hager 1984). A viscosity increase by a factor of 3 will not alter the behavior of slabs in a significant way compared with their behavior in a uniform mantle, but a factor of 100 increase, especially in conjunction with endothermic phase change, might prevent younger slabs from penetrating much beyond the 650-km discontinuity.

There is a basic distinction in the implied flows of homogeneous and compositional barriers. In the former case, if a slab does in fact penetrate into the lower mantle, a large entrained flow would be anticipated that is perhaps an order of magnitude greater than the negatively (thermally) buoyant portion of the slab. In contrast, compositional stratification acts to restrict the entrained flow, so that while the flux of chemically differentiated material through the transition zone may be the same for the two types of mechanisms, the *total* flux may differ by an order of magnitude. Compared with the homogeneous whole-mantle model, the fluxes for homogeneous and compositional mechanisms would be roughly one order of magnitude and two orders of magnitude smaller, respectively.

This flux difference would be expected to be reflected both in the age of the upper-mantle reservoir and in the temperature profiles. The large return flow implied by the homogeneous barrier model indicates that in fact there will be substantial mixing between the upper and lower mantle, such that the allowable *slab* flux into the lower mantle is reduced from the no-entrainment value of $\varepsilon = 0.15$. With an entrainment factor of 10, ε is reduced to less than 0.03 (see Figure 10). This would make it inconsistent with the observations of lower-mantle slab penetration.

The effect on the geotherm of sharply reduced mass flux between the upper and lower mantle can be analyzed by considering the variation in the Nusselt number (Nu), defined as the spherically averaged ratio of the total to conductive heat flow; it is thus a measure of the efficiency of convection for heat transport. For whole-mantle convection, Nu can be approximated by the ratio of the depth of the convective layer to the thickness of the thermal boundary layer (lithosphere), giving a mantle-average value of about 30 (Olson 1987). The two modified two-layer models predict that in the transition zone, the local Nusselt number would be reduced to about 3 and 1.3 for a restrictive phase boundary and a leaky compositional boundary, respectively. A Nusselt number of 1.3 would effectively correspond to conduction across the transition zone with a temperature profile like that of rigorous two-layer convection. If we assume that the current estimates of lower-mantle and core heat generation are correct, an internal thermal boundary layer with an approximately 1000-K temperature rise is expected near the 650-km discontinuity for an impenetrable barrier (Olson 1981, Spohn & Schubert 1982, Kenyon & Turcotte 1984). A comparable thermal boundary layer would thus be expected for the compositional barrier mechanism. That there is no evidence for a temperature rise of this magnitude is a problem for this model. In the case of Nu = 3, a sharp thermal boundary layer in the transition zone would not be expected.

An additional difficulty for using a compositional mechanism for constraining flow is that even for a slab flux that is one fourth of the surface flux ($\varepsilon = 0.25$) and a plate thickness of 100 km (no additional entrained flow), the flux through the transition zone corresponds to the volume of the entire upper mantle over a time equal to the age of the Earth. If in fact the slab flux into the lower mantle is closer to the surface flux, the time required for emptying the upper mantle is reduced to only 1 b.y. Thus, it appears difficult to maintain significant stratification in the presence of this large amount of flux, unless the upper and lower mantle initially had such distinctly different compositions that mixing still has not been complete.

Regardless of the mechanism used to restrict flow through the transition zone, there is one geochemical consideration that should be raised. With a flux that is only 15% of the surface flux, much of the lower mantle need not be differentiated by plate tectonics, assuming a constant flux. On the other hand, this amount of flux would mean that, over time, the upper mantle would contain a sizable fraction of undifferentiated mantle. If so, it should have distinct and readily observable Nd, Sr, and Pb isotopic compositions, as well as a characteristic "undegassed" rare-gas signature. As discussed previously, these signatures are not found in *any* samples of MORB and only rarely in the oceanic basalts associated with large hotspots. Such samples represent perhaps a fraction of a percent of the total oceanic basalt data set. Thus, this component appears to be conspicuously underrepresented in the oceanic basalt data set compared with what would be expected for the modified two-layer model.

Models With Slab Flux Through the Transition Zone

We now discuss the class of models for which the mass flux of chemically differentiated material across the transition zone is of the same order as the subduction-related surface flux and constant in time (steady-state flux), while the volume of the mantle associated with plate recycling is restricted to roughly one fourth of the mantle. This may be viewed as midway between the two previous cases considered, since homogeneous whole-mantle convection would involve an order of magnitude increase in flux, while the modified two-layer model (compositional barrier) has a flux that is nearly an order of magnitude less. One clear implication of such a model is that the slab material descending into the lower mantle resists mixing and instead returns rapidly to the surface. This is because the flux time for the lower mantle alone is 4.1–10.3 b.y. (Table 5).

Consider the case in which the mean age of the material passing from the lower to upper mantle is allowed to be independent of the mean age of the lower mantle. Modifying (6) results in

$$\tau'_u(t) = 1 + \varepsilon \Delta \tau_l / T_u - \tau_u(t) / T_u, \tag{8}$$

where $\Delta \tau_l$ is the difference in age between the upward and downward flux across the transition zone, which for simplicity is taken to be constant. Since the downward flux is assumed to have zero age, $\Delta \tau_l$ can be interpreted as the average residence time of this material in the lower mantle. Equation (8) yields

$$\tau_u = (\varepsilon \Delta \tau_l + T_u)(1 - e^{-t/T_u}), \tag{9}$$

or

$$\varepsilon \Delta \tau_l = \frac{\tau_u}{1 - e^{-t/T_u}} - T_u. \tag{10}$$

Taking $\tau_u = 1.7$ b.y. and setting ε equal to 1 (transition zone flux equal to surface slab flux), $\Delta \tau_l$ is at most 0.5 b.y. (the value corresponding to $T_u = 1.3$ b.y., or $D = 75$ km), at least 8 times smaller than the lower-mantle flux time. This implies that the return flow associated with descending slabs in fact consists of "recently" subducted slab material.

There are at least two mechanisms that could hasten the return of subducted slabs into the upper mantle. The first, based on thermal buoyancy, is boundary-layer convection with cooling from the Earth's surface and heating from the core-mantle boundary; the other, premised on compositional buoyancy, is referred to as *penetrative* convection.

Boundary-Layer Convection

In this model, illustrated schematically in Figure 11 (after Schubert & Spohn 1981), only the material in thermal boundary layers is involved in plate-tectonic recycling, while the cores of the convection cells are largely unsampled. Schubert & Spohn (1981) first proposed such a model to satisfy the need for isolated geochemical reservoirs in the context of whole-mantle convection. Thus, subducted lithosphere would descend to the CMB, become part of the bottom thermal boundary layer, and return to the surface as a result of its excess thermal buoyancy. The residence time for the subducted material would then be approximately the cycle time around the cell at surface plate speeds, about 0.25 b.y.

While this residence time is much smaller than the observed residence time, there are two ways in which to increase it. For example, the assumed viscosity of the lower mantle may be 30 times greater than that of the upper mantle (Hager 1984), which would reduce convective velocities by

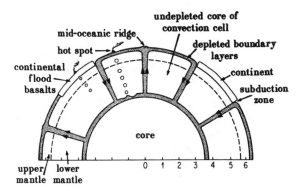

Figure 11   Boundary-layer convection (from Schubert & Spohn 1981). In this model, mid-ocean ridges sample only the boundary-layer products that descend as slabs to the core-mantle boundary, thermally equilibrate, and return to the surface to be resampled. Originally proposed to account for geochemical heterogeneity (cores are chemically distinct from the layers) in the context of whole-mantle convection, this model satisfies the constraint that the volume of the plate-tectonic reservoir is less than the entire mantle and is consistent with the evidence for lower-mantle slab penetration.

about a factor of 10. As a consequence, the thickness of the descending slabs would be increased proportionally, a feature that might be used as a diagnostic. Only recently has the question of lower-mantle slab thickening been addressed, in particular by Fischer et al (1988). They find that both for the Kurils and Mariana subduction zones, an order of magnitude thickening is unlikely; while some thickening might be allowed (factor of 3–5), the undeformed slabs actually provide the best fit to the data. Thus, an order of magnitude decrease in slab velocities is not supported by these data and may reflect the strength of the slab due to its high viscosity, although the thickening might occur at greater depths in the lower mantle once the slab thermally equilibrates.

The other possibility is that the boundary-layer products are stored in a reservoir with a residence time close to $\langle t_r \rangle$. One candidate is the D'' layer above the CMB, similar to that envisioned by Hofmann & White (1982) for subducted oceanic crust. However, given the size of the slab flux at the Earth's surface and for a 300-km-thick layer above the CMB, the residence time of slab material in this reservoir would be much too short.

As mentioned above, a reservoir with about the right flux time is the upper mantle. Thus in such a model, the upper mantle would consist of well-mixed boundary-layer products that have spent only a few hundred million years in the lower mantle, so that the residence time would be only slightly larger than the upper-mantle flux time. This last possibility seems to be the most promising. From a geochemical point of view, the upper mantle appears to be very well mixed in that even with rapid changes in the locations of ridges, indicating unsteady flow, samples of MORB almost invariably appear to tap the same geochemical source. Indeed, one of the major criticisms of this boundary-layer model without a well-mixed upper mantle is that the chemically distinct non-plate-tectonic cores of convection cells would be expected to be sampled whenever a ridge jumps. This clearly is not the case.

In conclusion, this model can be made to be consistent with our two postulates; however, it requires a particular flow regime in which all slabs descend to the core-mantle boundary and then return to an upper mantle that consists of boundary-layer products.

Penetrative Convection

Another mechanism that may reduce the residence time of slabs in the lower mantle is compositional stratification. Consider a lower mantle that is some 2% more dense than the upper mantle (Figure 12) by virtue of increased silica or iron content, as suggested by the recent work of Knittle et al (1986). Christensen & Yuen (1984) showed that the combined effect of the negative thermal buoyancy and high viscosity of the slab produces

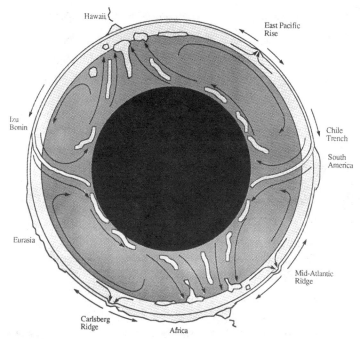

Hawaii

East Pacific
Rise

Izu
Bonin

Chile
Trench

South
America

Eurasia

Mid-Atlantic
Ridge

Carlsberg
Ridge

Africa

Figure 12 Schematic drawing of penetrative convection in the Earth that satisfies both the 1.7-b.y. residence time and the evidence for lower-mantle slab penetration. Because of a ~ 2% density increase from the upper to the lower mantle, slab material will descend into the lower mantle with little entrained flow and rapidly return to the upper mantle as a result of its intrinsic positive buoyancy while in the lower mantle. Once in the upper mantle, this material loses its compositional buoyancy and mixes in, eventually to be sampled at mid-ocean ridges, or occasionally at large hotspots if the rising flow from the lower mantle has sufficient thermal inertia to penetrate through the upper mantle. Slabs represent the only flux through the transition zone, except perhaps for small amounts of entrained lower-mantle material. The model has been superimposed upon the schematic in Figure 6, with the addition of some flow lines (see Figure 6 for description). While the boundary between the upper and lower mantle is taken here to be at 650-km depth, the actual compositional boundary could be either above this depth (throughout the transition zone, for example) or as much as a few hundred kilometers below it.

penetration into the lower layer (see Figure 13) that may continue to the core-mantle boundary. At the same time, the compositional contrast between the upper and lower mantle acts to minimize entrained flow, so that effectively only the negatively thermally buoyant material descends into the lower mantle. When this slab material thermally equilibrates in approximately 0.2 b.y., it will be ~ 2% less dense than its surroundings and have a tendency to ascend. For simple Stokes flow and for a characteristic dimension r of 100 km, the approximate ascent velocity will be about 40

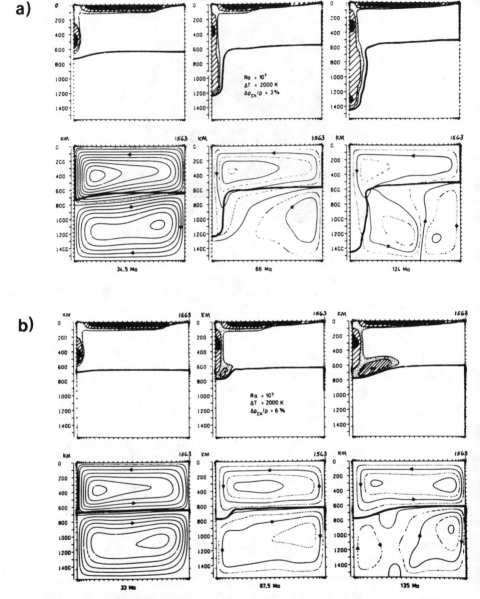

Figure 13 (*a*) Example of penetrative convection for an intrinsic density contrast of 3%. Upper plot is the viscosity distribution, and the lower plot gives the streamlines for three time periods. Slab is able to penetrate into the bottom layer without provoking overturn. (*b*) Example of a density contrast of 6%, showing no penetration. Density contrasts of 1.5% or less appear to provoke overturn (after Christensen & Yuen 1984).

mm yr^{-1}, i.e. on the order of surface plate velocities assuming a viscosity of 10^{21} Pa s. Because of the strong dependence of velocity on r, this model will work only if the slabs are not substantially reduced in size. As in the case of descent, the density contrast will serve to minimize the amount of entrained flow on ascent and maintain the integrity of the boundary. Once the material returns to the upper mantle, its compositional buoyancy is lost and only the thermal buoyancy remains. This would slow its further ascent and provide a natural mechanism for having a well-mixed upper mantle.

This type of circulation falls within the general class of penetrative convective flows. It occurs in a relatively narrow regime of parameter space in which the destabilizing buoyancy force due to temperature differences is comparable to the stabilizing buoyancy force provided by compositional differences. The regime end members are two-layer convection (compositional buoyancy exceeds thermal buoyancy) and overturn leading to single-layer convection (compositional buoyancy is dynamically insignificant). Lying between these two extremes is a continuum of behavior, ranging from convection with minor penetration across the compositional boundary up to the case just described, in which convective elements are able to penetrate the entire depth of the opposite layer. Numerical calculations of penetrative mantle convection have been made by Ellsworth & Schubert (1988) and in the context of slab penetration by Christensen & Yuen (1984). Laboratory simulations by Kincaid & Olson (1987) of highly viscous slabs sinking into a stratified fluid show virtually the same results, namely the ability of slabs to penetrate to the base of the mantle when their negative buoyancy, relative to the upper mantle, is equal to or greater than compositional buoyancy at the transition zone. Since slabs are on average 2–3% denser than the ambient upper mantle, the observed aseismic extension into the lower mantle is consistent with the results of the studies cited above, provided that the compositionally induced density increase across the transition zone is in the neighborhood of 2% or less.

It is important to emphasize that penetrative convection in the mantle does not necessarily imply rapid mixing of geochemical reservoirs, since it may be that only slab material passes between the upper and lower mantle. Thus it is conceivable for this style of convection to operate for billions of years without provoking an overturn of the whole mantle. At the present time it is not possible to make many definite predictions about the consequences of penetrative mantle convection, because the fluid mechanics in this regime have hardly been studied, but in general terms it does appear to present a way to reconcile aseismic slab extension with compositional variations across the transition zone.

Penetrative convection may be contrasted with boundary-layer convection in two important respects. First, since the slabs have an intrinsic positive buoyancy in the lower mantle, they need only thermally equilibrate at some point in the lower mantle to begin their ascent back into the upper mantle. Thus the slabs would serve to cool the lower mantle if they failed to reach the CMB, or would remove heat from the core if they ultimately descended to the base of the mantle. For boundary-layer convection, the slabs would always descend to the CMB. Second, for penetrative convection, slabs lose their intrinsic buoyancy upon reentering the upper mantle, which would provide a natural mechanism for a well-mixed upper mantle. Boundary-layer convection does not provide such a mechanism, and thus some other physical mechanism would have to be invoked.

An advantage of penetrative convection over either the strictly two-layer or the modified two-layer (compositional barrier) models is that it does not produce a strong thermal boundary layer at the base of the transition zone. Rather, the effect of the subducted slabs is to spread this boundary layer over the entire lower mantle (if the slabs extend to the CMB).

Penetrative convection would thus be expected to produce weakly super-adiabatic temperatures throughout the lower mantle. There is, in fact, some evidence for this from seismology. Standard earth models such as PREM generally show lower-mantle density increasing more slowly than would be expected for isentropic compression. The deviation from isentropic stratification is small—the depth-averaged value of Bullen's homogeneity parameter η is about 0.98–0.99 (a value 1.0 corresponds to isentropic stratification)—but this still implies an increase in temperature through the lower mantle of 200–800 K in addition to the adiabatic increase (Shankland & Brown 1985, Jeanloz & Morris 1987). It is known from numerical modeling that the horizontally averaged temperature distribution in a homogeneous convecting fluid is *subadiabatic* except in thermal boundary layers (Weinstein et al 1987, Yuen et al 1987). This is true regardless of whether the fluid is heated internally or at its base. Thus it is difficult to reconcile the apparent superadiabaticity of most of the lower mantle with the structure of homogeneous whole-mantle convection. This would apply to the boundary-layer model as well. A way out of this dilemma is to distribute compositional and thermal gradients over the whole lower mantle by the action of deep penetrative convection in such a way that the lower mantle appears slightly superadiabatic in spherical average. At the present time we do not know enough about penetrative convection to say with certainty that it would produce the observed stratification, but at least superficially it seems possible.

In summary, we have shown that neither of the two standard models—whole-mantle convection in a homogeneous mantle, and two-layer convection—are strictly compatible with lower-mantle slab penetration and a 1.7-b.y. mean age for plate-tectonic recycling. The whole-mantle model predicts a residence time much larger than this, based on the inferred flux time of the whole mantle. This discrepancy can be removed by appealing to greater mass flux in the past, although there is no evidence for this and thermal history considerations do not necessarily require it. The two-layer model, in its strict form, can be rejected on the basis of lower-mantle slab penetration data, since it allows no mass flux through the transition zone. A "leaky" two-layer model with a compositional barrier is marginally acceptable, although it is expected to produce a large thermal boundary layer in the transition zone that is not observed. Also, with the implied slab flux, the integrity of such a boundary appears difficult to maintain over the age of the Earth. The homogeneous barrier mechanism, because of the large entrained flow accompanying any descending slab, appears to be inconsistent with the minimum allowable slab flux through the transition zone. Both two-layer models predict more non-plate-tectonic mantle material to reside in the upper mantle than is inferred from the paucity of samples from such a component.

The other two models considered have the property that slabs descending into the lower mantle spend little time there and avoid being mixed. This can be accomplished either by boundary-layer convection in homogeneous models, where ridges are predominantly sampling boundary-layer products (thermal buoyancy), or by penetrative convection with an intrinsic density difference of $\sim 2\%$ between the upper and lower mantle (compositional buoyancy). The latter mechanism has the advantage of (a) accounting for an apparently well-mixed upper mantle, (b) not requiring that all of the slabs descend to the core-mantle boundary, and (c) producing a slightly superadiabatic thermal gradient in the lower mantle that appears to be consistent with the inferred gradient from the seismically determined lower-mantle density profile.

DISCUSSION

The preceding review of the Earth's mantle provides some insight into the long-standing controversy between advocates of whole-mantle and two-layer convection. If we consider various constraints on mantle flow in isolation and interpret them in the most straightforward way, it appears that the mantle paradoxically has properties of both of these end-member models. Not all constraints, however, are equally compelling. The feature of the mantle that is the least open to alternative interpretations is the coherence of slab-related downwelling from the upper mantle, through the transition zone and into the lower mantle. If this single constraint is accepted, then *all* allowable models are "whole" mantle. As mentioned in the introduction, however, much, if not most, of the debate on convective style really has been over the closely linked question of mantle stratification. While this linkage would appear to argue for an unstratified mantle, it makes the implicit assumption that this downwelling induces mantle-wide overturn.

Compared with the observations of lower-mantle slab penetration, the other constraints discussed are less direct and thus susceptible to multiple interpretations. They are, however, features that most easily would be associated with a stratified mantle with separate flow regimes in the upper and lower mantle: (*a*) a transition zone that appears to constitute a partial barrier to lower-mantle return flow; (*b*) a 1.7-b.y. residence time for oceanic crustal recycling, indicating a plate-tectonic reservoir about the size of the upper mantle; and (*c*) geochemical evidence for a non-plate-tectonic component that is sampled rarely at the Earth's surface and only at large hotspots.

How can these observations be reconciled? There are presently two paths that one may take. The first is to conclude that mantle overturn indeed is occurring and that these less direct arguments for stratification and separate flow regimes can in fact be explained within the context of a homogeneous, unstratified mantle. The other alternative is to challenge the implicit linkage between whole-mantle models and the absence of dynamically significant compositional stratification. To explore these two possibilities further, we discuss two self-consistent scenarios, not only for the present state of the mantle but for its implied evolution as well. For the homogeneous model, we actually consider the three models with homogeneous mechanisms together. For the compositional model, we take penetrative convection. While there is not enough information presently available to constrain more than the general features of these characterizations, this is done as a means of identifying consequences and thus important tests that may be used to evaluate these models.

Homogeneous Convection

For this model we appeal only to homogeneous mechanisms for explaining the data. In the interpretation of the geochemical data, the model of Davies (1984) is followed closely in that only plate-tectonic processes are invoked, and that all heterogeneity is distributed. For this model, there is no dynamically significant compositional stratification in the transition zone, so that the mantle seismic discontinuities are interpreted as phase changes. Mid-ocean ridges represent the loci of mantle upwelling complementary to subduction. The apparent barrier to lower-mantle upwelling in the transition zone is attributed to homogeneous mechanisms, such as an increase in lower-mantle viscosity or an endothermic phase change at the bottom of the transition zone.

Within the context of a homogeneous mechanism, there are two ways of reconciling the 1.7-b.y. residence time for plate-tectonic recycling that, at the present-day value of mass flux, suggests a reservoir the size of the upper mantle. The first is by appealing to much more rapid plate velocities in the past. Alternatively, the structure of convection may be such that the plate-tectonic process only samples boundary-layer products (boundary-layer convection) that move rapidly through the mantle. *All* geochemical heterogeneity is interpreted as distributed and primarily resulting from plate-tectonic recycling. The few samples from large hotspots that argue for a primitive or non-plate-tectonic component are interpreted as trace amounts of distributed heterogeneity that have survived since the formation of the Earth.

In the homogeneous model, hotspots are fundamentally no different than seamounts, except that their source region is much larger but still small compared with the size of the mantle. Alternatively, they may originate from core-mantle boundary instabilities in a way similar to that discussed by Loper (1985). Hotspots represent a secondary and relatively minor component of convection compared with the complementary flow associated with ridges and trenches. There are no scales of convection other than mantle-wide flow and the small-scale flow associated with plumes.

The depletion of the MORB reservoir is due to the plate-tectonic process of creating both continental and oceanic crust. The evidence for the early depletion of the MORB source requires that this was produced by very rapid plate tectonics in the first 0.7 b.y. of the Earth that gave rise to continental crust that has subsequently disappeared and/or to oceanic crust that now resides at the base of the mantle, such as proposed by Hofmann & White (1982), or dispersed throughout the lower half of the lower mantle.

Penetrative Convection

In this model, it is assumed that lower-mantle slab penetration can occur without mantle-wide overturn, which is the most fundamental departure from homogeneous models. This is accomplished by the presence of a $\sim 2\%$ intrinsic density contrast across the transition zone that serves to minimize entrained flow and hasten the return of the penetrating slab material back into the upper mantle. Once it enters the upper mantle, it loses its compositional buoyancy, mixes into the upper mantle, and ultimately is sampled at the ridges. The upper mantle–lower mantle density contrast is not necessarily associated with a seismic discontinuity and may represent a gradient in properties within the transition zone or somewhat below the 650-km discontinuity. The apparent partial barrier to lower-mantle upwelling is due to this intrinsic density contrast; while it allows the localized, coherent downwelling of slabs, it is a barrier to the more diffuse lower-mantle upwelling, with the exception of recently sub-ducted slabs returning from the lower mantle. The flux through the transition zone thus involves only subducted slab material with little entrainment.

This model interprets the features of surface plate tectonics differently from a homogeneous model. In the shallow mantle, ridges are passively drawing material from the upper-mantle reservoir. They are comple-mentary to the trenches in that the rate of spreading must equal the rate of subduction. However, it is hotspots and slabs that reflect complementary upwelling and downwelling in the lower mantle. While this only suggests that hotspots mark regions of large-scale upwelling in the lower mantle, the connection may be taken a bit further. Slabs are argued to be the only component of the lower-mantle return flow that reaches the upper mantle; furthermore, hotspots are statistically associated with lower-mantle upwell-ing and thus somehow communicate with the lower mantle. It is suggested that the largest hotspots actually are sampling unusually thermally buoy-ant slab material (old oceanic crust and upper mantle) that has only recently reentered the upper mantle from the lower mantle, in addition to entrained material intrinsic to the lower mantle.

The penetrative convection model interprets the 1.7-b.y. residence time as indicating that the plate-tectonic reservoir roughly corresponds to the upper mantle (constant plate flux through time) plus the temporary slab extensions into the lower mantle. Nearly all of the geochemical hetero-geneity observed in oceanic basalts is interpreted as distributed and is the result of plate-tectonic recycling. The lower mantle is interpreted as a rarely sampled reservoir that does not participate in plate tectonics. Thus, the few observations of a non-plate-tectonic component are interpreted as the occasional sampling of this lower-mantle reservoir. The roughly 2% greater density of lower-mantle material would explain the relative lack of expression of this component, whereas the slab-related entrainment of lower-mantle material would explain its appearance at the surface, par-ticularly concentrated at large hotspots.

The evidence for the early depletion of the MORB source may be indicative of a large-scale, non-plate-tectonic differentiation process whose complement now resides in the lower mantle. Alternatively, the con-ventional mass-balance complementarity between MORB-S and con-tinental crust would also lead to a lower mantle slightly enriched in incom-patible elements.

Tests

There are six tests that might provide a means of evaluating the mantle models discussed or, more specifically, of choosing between homogeneous and compositionally stratified models.

1. *Observation of an intrinsic density contrast between the upper and lower mantle.* Such an observation would be difficult to reconcile with lower-mantle slab penetration without a mechanism similar to that proposed for the penetrative convection model. The recent evidence provided by Knittle et al (1986) for a 2% density contrast is about the correct size expected in the penetrative convection model; however, this result, because of its importance, will have to be evaluated very carefully. From the seismological point of view, it would be worthwhile to examine the transition zone for the existence of a gradient in material properties that cannot be easily associated with a phase transformation. The change in seismic velocity associated with an intrinsic increase in density is unclear. The two probable ways of increasing lower-mantle density would be to increase Fe, Si, or both, and these two elements are expected to have opposite effects on seismic velocities: Increasing Si alone would raise the *P*-wave velocity by about 2.5–3% (Weidner & Ito 1987), whereas increasing Fe would lower it by about the same amount (Bonczar & Graham 1982). There is an indication from Creager & Jordan (1986a) that the intrinsic *P*-velocity increase between the upper and lower mantle may be no more than about 1%. This would suggest that, if present, the change in composition could involve some mixture of Fe and Si. Similar tests would thus provide constraints on the candidate compositions for the lower mantle.

2. *A demonstration of the dynamic feasibility of penetrative convection.* The assumption that lower-mantle slab penetration provokes mantle-wide overturn is implicit in the interpretation of most convection modeling. The initial work of Christensen & Yuen (1984) and Kincaid & Olson (1987), however, suggests that penetrative convection may be possible, although work in the regime of interest, where thermal and compositional buoyancy are nearly equal, is only beginning. The potential success of this model in reconciling many geochemical and geophysical observations argues strongly for its careful evaluation.

3. *Constraints on the vertical coherence of seismically slow regions within and just below the transition zone.* While the presently existing tomographic images suggest a barrier to lower-mantle upwelling in the transition zone, it will take more detailed vertical resolution to confirm this. A strong indication of obstructed flow would argue for a flux that is close to the slab flux, such as in penetrative convection.

4. *Constraints on past plate speeds.* As mentioned above, one way for homogeneous models to satisfy the 1.7-b.y. residence time for plate-tectonic recycling is to invoke a substantially higher mass flux in the past. We have noted, however, that the few data presently available do not support this contention, nor do the recent attempts to include temperature-dependent viscosity in thermal history calculations. Thus, the proposition that plate tectonics was much faster in the past should not be taken as an unchallenged assumption. The notion that the rates of plate-tectonic processes were once much higher has important consequences in particular for the evolution of the Earth, since it would imply that rates of continent formation, island-arc volcanism, continent-continent collisions, and hotspot volume would all be an order of magnitude greater than at present. A constant rate would suggest that much of the mantle has not been processed through plate-tectonic differentiation. Thus, a more careful examination of this question would be important not only as a check on homogeneous models of mantle convection but also for the Earth's evolution.

5. *Model ages of large hotspots.* The association between hotspots and slabs in the penetrative convection model can be tested. To the extent that most of the time spent by recycled oceanic crust is in the upper-mantle reservoir, one would predict that large hotspots representing recently returning material from the lower mantle would possess significantly younger model ages, whereas small ocean islands and MORB, representing a random sampling of the upper mantle, would yield samples that are older on average. While model age calculations must be done carefully for many of the hotspots, it appears as though one of the trends for Hawaii does in fact indicate a young model age. This trend was pointed out by Tatsumoto (1978) and by Stille et al (1983) with a Pb-Pb secondary isochron of 0.94 b.y. and is suggestive of such a process. A true test, however, would involve a more careful statistical evaluation of the model ages seen for hotspots, other ocean islands, and MORB.

6. *Measurement of Nb/U and Ce/Pb ratios for those particular ocean island basalts that have isotopic compositions most suggestive of a non-plate-tectonic component.* These two chemical fingerprints of distinct geochemical components, when combined with isotopic diagnostics, would further distinguish between recycled and non-plate-tectonic sources in the Earth's interior.

ACKNOWLEDGMENTS

We would like to thank Pat Kenyon, Alan Boss, Steve Shirey, David James, Andy Jephcoat, and Don Wiedner for stimulating discussions during the formative stages in the development of the manuscript. Additionally, we thank Julie Morris, Selwyn Sacks, and Thomas Jordan for thoughtful reviews. Unpublished isotopic data were kindly made available to us by Pat Castillo and Dave Gerlach. We also thank Donna Jurdy for making available her catalogue of hotspot locations. Special thanks go to Janice Dunlap for invaluable help in manuscript preparation.

Literature Cited

Adams, L. H., Williamson, E. D. 1925. The composition of the Earth's interior. *Smithson. Rep. 1923*, pp. 241–60
Agee, C. B., Walker, D. 1987. Static silicate liquid compression to 6 GPa, 2000°C. *Eos, Trans. Am. Geophys. Union* 68: 437 (Abstr.)
Allègre, C. J., Hamelin, B., Provost, A., Dupré, B. 1987. Topology in isotopic multispace and origin of mantle chemical heterogeneities. *Earth Planet Sci. Lett.* 81: 319–37
Allègre, C. J., Hart, S. R., Minster, J.-F. 1983a. Chemical structure and evolution of the mantle and continents determined by inversion of Nd and Sr isotopic data, I. Theoretical methods. *Earth Planet. Sci. Lett.* 66: 177–90
Allègre, C. J., Hart, S. R., Minster, J.-F. 1983b. Chemical structure and evolution of the mantle and continents determined by inversion of Nd and Sr isotopic data, II. Numerical experiments and discussion. *Earth Planet. Sci. Lett.* 66: 191–213
Allègre, C. J., Staudacher, T., Sarda, P., Kurz, M. 1983c. Constraints on evolution of Earth's mantle from rare gas systematics. *Nature* 303: 762–66
Allègre, C. J., Turcotte, D. L. 1986. Implications of a two-component marble-cake mantle. *Nature* 323: 123–27
Anderson, D. L. 1982a. Isotopic evolution of the mantle: the role of magma mixing. *Earth Planet. Sci. Lett.* 57: 1–12
Anderson, D. L. 1982b. Isotopic evolution of the mantle: a model. *Earth Planet. Sci. Lett.* 57: 13–24
Anderson, D. L. 1984. The Earth as a planet: paradigms and paradoxes. *Science* 223: 347–55
Anderson, D. L. 1987. Thermally induced phase changes, lateral heterogeneity of the mantle, continental roots and deep slab anomalies. *J. Geophys. Res.* 92: 13,968–80
Anderson, D. L., Bass, J. D. 1986. Transition region of the Earth's upper mantle. *Nature* 320: 321–28
Armstrong, R. L. 1981. Radiogenic isotopes: the case for crustal recycling on a near-steady-state no-continental-growth Earth. In *The Origin and Evolution of the Earth's Continental Crust*, ed. S. Moorbath, B. F. Windley, pp. 259–87. London: R. Soc.
Armstrong, R. L., Hein, S. M. 1973. Computer simulation of Pb and Sr isotope evolution of the Earth's crust and upper mantle. *Geochim. Cosmochim. Acta* 37: 1–18
Baumgardener, J. R., Schubert, G. 1987. 3-D spherical convection model of the Earth's mantle: geoid and gravity signatures. *Eos, Trans. Am. Geophys. Union* 68: 413
Beck, S. L., Lay, T. 1986. Test of the lower mantle slab penetration hypothesis using broadband *S* waves. *Geophys. Res. Lett.* 13: 1007–10
Bina, C. R., Wood, B. J. 1987. Olivine-spinel transitions: experimental and thermodynamic constraints and implications for the nature of the 400-km seismic discontinuity. *J. Geophys. Res.* 92: 4853–66
Birch, F. 1952. Elasticity and constitution of the Earth's interior. *J. Geophys. Res.* 57: 227–86
Bonczar, L. J., Graham, E. K. 1982. The pressure and temperature dependence of the elastic properties of polycrystal magnesiowustite. *J. Geophys. Res.* 87: 1061–78
Boyd, F. R., Gurney, J. J. 1986. Diamonds and the African lithosphere. *Science* 232: 472–77
Boyd, F. R., Mertzman, S. A. 1981. Composition and structure of the Kaapvaal lithosphere, southern Africa. In *Magmatic Processes: Physiochemical Principles*, ed. B. O. Mysen, pp. 13–24. University Park, Pa: Geochem. Soc.
Brooks, C., Hart, S. R., Hofmann, A., James, D. E. 1976b. Rb-Sr mantle isochrons from oceanic regions. *Earth Planet. Sci. Lett.* 32: 51–61
Brooks, C., James, D. E., Hart, S. R. 1976a. Ancient lithosphere: its role in young continental volcanism. *Science* 193: 1086–94
Bullen, K. E. 1965. *An Introduction to the Theory of Seismology*. Cambridge: Cambridge Univ. Press

Caffee, M. W., Hudson, G. B. 1987. A non-primordial origin for terrestrial ^{129}Xe anomalies. *Lunar Planet. Sci.* 18: 145–46 (Abstr.)

Carlson, R. W. 1984. Isotopic constraints on Columbia River flood basalt genesis and the nature of the subcontinental mantle. *Geochim. Cosmochim. Acta* 48: 2357–72

Carlson, R. W., Macdougall, J. D., Lugmair, G. W. 1978. Differential Sm/Nd evolution in oceanic basalts. *Geophys. Res. Lett.* 5: 229–32

Chase, C. G. 1972. The *N* plate problem of plate tectonics. *Geophys. J. R. Astron. Soc.* 29: 117–22

Chase, C. G. 1979. Subduction, the geoid, and lower mantle convection. *Nature* 282: 464–68

Chase, C. G. 1981. Oceanic island Pb: two-stage histories and mantle evolution. *Earth Planet. Sci. Lett.* 52: 277–84

Chen, C.-Y., Frey, F. A. 1985. Trace element and isotopic geochemistry of lavas from Haleakala Volcano, east Maui, Hawaii: implications for the origin of Hawaiian basalts. *J. Geophys. Res.* 90: 8743–68

Christensen, U. R. 1984. Heat transport by variable viscosity convection and implications for the Earth's thermal evolution. *Phys. Earth Planet. Inter.* 35: 264–82

Christensen, U. R., Yuen, D. A. 1984. The interaction of a subducting lithosphere slab with a chemical or phase boundary. *J. Geophys. Res.* 89: 4389–4402

Coats, R. R. 1962. Magma type and crustal structure in the Aleutian arc in crust of the Pacific basin. *Am. Geophys. Union. Monogr.* 6: 92–109

Cohen, R. S., O'Nions, R. K. 1982. The lead, neodymium and strontium isotopic structure of ocean ridge basalts. *J. Petrol.* 23: 299–324

Cohen, R. S., O'Nions, R. K., Dawson, J. B. 1984. Isotope geochemistry of xenoliths from East Africa: implications for development of mantle reservoirs and their interaction. *Earth Planet. Sci. Lett.* 68: 209–20

Comer, R. P., Clayton, R. W. 1984. Tomographic reconstruction of lateral velocity heterogeneity in the Earth's mantle. *Eos, Trans. Am. Geophys. Union* 65: 236 (Abstr.)

Cormier, V. F. 1987. Slab diffracted *S* waves. *Proc. IUGG. Gen. Assem., 19th, IASPEI Symp.*, 1: 308

Craig, H., Lupton, J. E. 1976. Primordial neon, helium and hydrogen in oceanic basalts. *Earth Planet. Sci. Lett.* 31: 369–85

Craig, H., Lupton, J. E., Welhan, J. A., Poreda, R. 1978. Helium isotope ratios in Yellowstone and Lassen Park volcanic gases. *Geophys. Res. Lett.* 5: 897–900

Creager, K. C., Jordan, T. H. 1984. Slab penetration into the lower mantle. *J. Geophys. Res.* 89: 3031–49

Creager, K. C., Jordan, T. H. 1986a. Slab penetration into the lower mantle beneath the Mariana and other island arcs of the northwest Pacific. *J. Geophys. Res.* 91: 3573–89

Creager, K. C., Jordan, T. H. 1986b. Aspherical structure of the core-mantle boundary from *PKP* travel times. *Geophys. Res. Lett.* 13: 1497–1500

Crough, S. T., Jurdy, D. M. 1980. Subducted lithosphere, hotspots, and the geoid. *Earth Planet. Sci. Lett.* 48: 15–22

Davies, G. F. 1984. Geophysical and isotopic constraints on mantle convection: an interim synthesis. *J. Geophys. Res.* 89: 6017–40

DePaolo, D. J. 1981. Nd in the Colorado Front Range and implications for crust formation and mantle evolution in the Proterozoic. *Nature* 291: 193–96

DePaolo, D. J. 1983. The mean life of continents: estimates of continent recycling rates from Nd and Hf isotopic data and implications for mantle structure. *Geophys. Res. Lett.* 10: 705–8

DePaolo, D. J., Wasserburg, G. J. 1976. Inferences about magma sources and mantle structure from variations of ^{143}Nd/^{144}Nd. *Geophys. Res. Lett.* 3: 743–46

Dudas, F. O., Carlson, R. W., Eggler, D. H. 1987. Regional middle Proterozoic enrichment of the subcontinental mantle source of igneous rocks from central Montana. *Geology* 15: 22–25

Duncan, R. A., McCulloch, M. T., Barsczus, H. G., Nelson, D. R. 1986. Plume versus lithospheric sources for melts at Ua Pou, Marquesas Islands. *Nature* 322: 534–38

Dupré, B., Allègre, C. J. 1983. Pb-Sr isotope variation in Indian Ocean basalts and mixing phenomena. *Nature* 303: 142–46

Dziewonski, A. M. 1984. Mapping the lower mantle: determination of lateral heterogeneity in *P* velocity up to degree and order 6. *J. Geophys. Res.* 89: 5929–52

Dziewonski, A. M., Anderson, D. L. 1981. Preliminary reference earth model. *Phys. Earth Planet. Inter.* 25: 297–356

Dziewonski, A. M., Hager, B. H., O'Connell, R. J. 1977. Large-scale heterogeneities in the lower mantle. *J. Geophys. Res.* 82: 239–55

Ellsworth, K., Schubert, G. 1988. Numerical models of thermally and mechanically coupled two-layer convection of highly viscous fluids. *Geophys. J. R. Astron. Soc.* In press

Engdahl, E. R., Vidale, J. E., Cormier, V. F. 1988. Wave propagation in subducted lithospheric slabs. In *Digital Seismology and Fine Modeling of the Lithosphere*. Sicily, Italy: Plenum. In press

Fischer, K. M., Jordan, T. A., Creager, K. C. 1988. Seismic constraints on the morphology of deep slabs. *J. Geophys. Res.* In press

Fraser, K. J., Hawkesworth, C. J., Erlank, A. J., Mitchell, R. H., Scott-Smith, B. H. Sr. 1986. Nd and Pb isotope and minor element geochemistry of lamproites and kimberlites. *Earth Planet. Sci. Lett.* 76: 57–70

Fuji, T., Scarfe, C. M. 1985. Compositions of liquids coexisting with spinel lherlozite at 10 kbar and the genesis of MORBS. *Contrib. Mineral. Petrol.* 90: 18–28

Gast, P. W. 1968. Trace element fractionation and the origin of tholeiitic and alkaline magma types. *Geochim. Cosmochim. Acta* 32: 1057–86

Gerlach, D. C., Hart, S. R., Morales, V. W. J., Palacios, C. 1986. Mantle heterogeneity beneath the Nazca plate: San Felix and Juan Fernandez Islands. *Nature* 322: 165–69

Giardini, D., Woodhouse, J. H. 1984. Deep seismicity and modes of deformation in Tonga subduction zone. *Nature* 307: 505–9

Gilbert, F., Dziewonski, A., Brune, J. 1973. An informative solution to a seismological inverse problem. *Proc. Natl. Acad. Sci. USA* 70: 1410–13

Gill, J. 1981. *Orogenic Andesites and Plate Tectonics.* Berlin: Springer-Verlag. 390 pp.

Grand, S. 1987. Tomographic inversion for shear velocity beneath the North American plate. *J. Geophys. Res.* 92: 14,065–90

Grand, S., Helmberger, D. 1984. Upper-mantle shear structure of North America. *Geophys. J. R. Astron. Soc.* 76: 399–438

Gudmundsson, O., Clayton, R. W., Anderson, D. L. 1986. CMB topography inferred from ISC PcP travel times. *Eos, Trans. Am. Geophys. Union* 67: 1100 (Abstr.)

Gurnis, M. 1986. The effects of chemical density differences on convective mixing in the Earth's mantle. *J. Geophys. Res.* 91: 11,407–19

Gurnis, M., Davies, G. F. 1986. Mixing in numerical models of mantle convection incorporating plate kinematics. *J. Geophys. Res.* 91: 6375–95

Hager, B. H. 1984. Subducted slabs and the geoid: constraints on mantle rheology and flow. *J. Geophys. Res.* 89: 6003–15

Hager, B. H., Clayton, R. W., Richards, M. A., Comer, R. P., Dziewonski, A. M. 1985. Lower mantle heterogeneity, dynamic topography and the geoid. *Nature* 313: 541–45

Hamelin, B., Dupré, B., Allègre, C. J. 1984. Lead-strontium isotopic variations along the East Pacific Rise and the Mid-Atlantic Ridge: a comparative study. *Earth Planet Sci. Lett.* 67: 340–50

Hamelin, B., Dupré, B., Allègre, C. J. 1985. Pb-Sr-Nd isotopic data of Indian Ocean ridges: new evidence of large-scale mapping of mantle heterogeneities. *Earth Planet. Sci. Lett.* 76: 288–98

Hanan, B. B., Kingsley, R. H., Schilling, J.-G. 1986. Pb isotope evidence in the South Atlantic for migrating ridge-hotspot interactions. *Nature* 322: 137–44

Hart, S. R. 1984. A large-scale isotope anomaly in the Southern Hemisphere mantle. *Nature* 309: 753–57

Hart, S. R., Gerlach, D. C., White, W. M. 1986. A possible new Sr-Nd-Pb mantle array and consequences for mantle mixing. *Geochim. Cosmochim. Acta* 50: 1551–63

Harte, B. 1983. Mantle peridotites and processes—the kimberlite sample. In *Continental Basalts and Mantle Xenoliths*, ed. C. J. Hawkesworth, M. J. Norry, pp. 46–91. Nantwich, Engl: Shiva

Hawkesworth, C. J., Mantovani, M. S. M., Taylor, P. N., Palacz, Z. 1986. Evidence from the Paraná of south Brazil for a continental contribution to Dupal basalts. *Nature* 322: 356–59

Hawkesworth, C. J., Rogers, N. W., van Calsteren, P. W. C., Menzies, M. A. 1984. Mantle enrichment processes. *Nature* 311: 331–35

Herzberg, C. T. 1984. Chemical stratification in the silicate Earth. *Earth Planet. Sci. Lett.* 67: 249–60

Hoffman, N. R. A., McKenzie, D. P. 1985. The destruction of geochemical heterogeneities by differential fluid motions during mantle convection. *Geophys. J. R. Astron. Soc.* 82: 163–206

Hofmann, A. W., Jochum, K. P., Seufert, M., White, W. M. 1986. Nb and Pb in oceanic basalts: new constraints on mantle evolution. *Earth Planet. Sci. Lett.* 79: 33–45

Hofmann, A. W., White, W. M. 1982. Mantle plumes from ancient oceanic crust. *Earth Planet. Sci. Lett.* 57: 421–36

Hofmeister, A. M. 1983. Effect of a Hadean terrestrial magma ocean on crust and mantle evolution. *J. Geophys. Res.* 88: 4963–83

Hurley, P. M., Hughs, H., Faure, G., Fairbairn, H. W., Pinson, W. H. 1962. Radiogenic strontium-87 model of continent formation. *J. Geophys. Res.* 67: 5315–34

Isacks, B., Molnar, P. 1971. Distribution of stresses in the descending lithosphere from a global survey of focal-mechanism solutions of mantle earthquakes. *Rev. Geophys. Space Phys.* 9: 103–74

Isacks, B., Oliver, J., Sykes, L. R. 1968. Seismology and the new global tectonics. *J. Geophys. Res.* 73: 5855–5900

Ito, E., Takahashi, E. 1986. The phase equilibria and the constitution in the deep mantle. In *High Pressure Research in Mineral Physics. Proc. US-Jpn. Semin. High Pressure.* Washington, DC: Am. Geophys. Union

Jacobsen, S. B., Wasserburg, G. J. 1979. The mean age of mantle and crustal reservoirs. *J. Geophys. Res.* 84: 7411–27

Jeanloz, R., Morris, S. 1987. Is the mantle geotherm subadiabatic? *Geophys. Res. Lett.* 14: 335–38

Jeffreys, H. 1939. The times of *P, S* and *SKS. Mon. Not. R. Astron. Soc., Geophys. Suppl.* 4: 498–533

Jordan, T. H. 1977. Lithospheric slab penetration into the lower mantle beneath the Sea of Okhotsk. *J. Geophys.* 43: 473–96

Jordan, T. H. 1978. A procedure for estimating lateral variations from low-frequency eigenspectra data. *Geophys. J. R. Astron. Soc.* 52: 441–55

Jordan, T. H. 1979. Mineralogies, densities and seismic velocities of garnet lherzolite and their geophysical implications. In *The Mantle Sample: Inclusions in Kimberlites and Other Volcanics,* ed. H. O. A. Meyer, F. R. Boyd, pp. 1–14. Washington, DC: Am. Geophys. Union

Jordan, T. H. 1981. Continents as a chemical boundary layer. *Philos. Trans. R. Soc. London Ser. A* 301: 359–73

Jordan, T. H., Lynn, W. S. 1974. A velocity anomaly in the lower mantle. *J. Geophys. Res.* 79: 2679–85

Kaneoka, I. 1983. Noble gas constraints on the layered structure of the mantle. *Nature* 302: 698–700

Kaneoka, I., Takaoka, N., Clague, D. A. 1983. Noble gas systematics for coexisting glass and olivine crystals in basalts and dunite xenoliths from Loihi Seamount. *Earth Planet. Sci. Lett.* 66: 427–37

Kato, T., Irifune, I., Ringwood, A. E. 1987. Experimental constraints on the early differentiation of the Earth's mantle. *Lunar Planet. Sci.* 18: 483–84

Kenyon, P. M., Turcotte, D. L. 1984. Convection in a two-layer mantle with a strongly temperature-dependent viscosity. *J. Geophys. Res.* 88: 6403–14

Kincaid, C., Olson, P. 1987. An experimental study of subduction and slab migration. *J. Geophys. Res.* In press

Knittle, E., Jeanloz, R. 1987. Synthesis and equation of state of $(Mg,Fe)SiO_3$ perovskite to over 100 gigapascals. *Science* 235: 668–70

Knittle, E., Jeanloz, R., Smith, G. L. 1986. Thermal expansion of silicate perovskite

and stratification of the Earth's mantle. *Nature* 319: 214–216

Kramers, J. D., Roddick, J. C. M., Dawson, J. B. 1983. Trace element and isotope studies on veined, metasomatic and "MARID" xenoliths from Bultfontein, South Africa. *Earth Planet. Sci. Lett.* 65: 90–106

Kurz, M. D., Jenkins, W. J., Hart, S. R. 1982. Helium isotopic systematics of oceanic islands and mantle heterogeneity. *Nature* 297: 43–47

Kurz, M. D., Jenkins, W. J., Hart, S. R., Clague, D. 1983. Helium isotopic variations in volcanic rocks from Loihi Seamount and the Island of Hawaii. *Earth Planet. Sci. Lett.* 66: 388–406

Kurz, M. D., Meyer, P. S., Sigurdsson, H. 1985. Helium isotopic systematics within the neovolcanic zones of Iceland. *Earth Planet. Sci. Lett.* 74: 291–305

Lay, T. 1983. Localized velocity anomalies in the lower mantle. *Geophys. J. R. Astron. Soc.* 72: 483–516

Leeman, W. P. 1975. Radiogenic tracers applied to basalt genesis in the Snake River Plain–Yellowstone National Park region—evidence for a 2.7 b.y. old upper mantle keel. *Geol. Soc. Am. Abstr. With Programs* 7: 1165

Lees, A. C., Bukowinski, M. S. T., Jeanloz, R. 1983. Reflection properties of phase transition and compositional change models of the 650-km discontinuity. *J. Geophys. Res.* 88: 8145–59

Liotard, J. M., Barsczus, H. G., Dupuy, C., Dostal, J. 1986. Geochemistry and origin of basaltic lavas from Marquesas Archipelago, French Polynesia. *Contrib. Mineral. Petrol.* 92: 260–68

Liu, L.-G. 1979. On the 650-km seismic discontinuity. *Earth Planet. Sci. Lett.* 42: 202–8

Loper, D. E. 1985. A simple model of whole-mantle convection. *J. Geophys. Res.* 90: 1809–36

Lugmair, G. W., Scheinin, N. B., Marti, K. 1975. Sm-Nd age and history of Apollo 17 basalt 75075: evidence for early differentiation of the lunar exterior. *Proc. Lunar Sci. Conf., 6th,* pp. 1419–29

Lupton, J. E. 1983. Terrestrial gases: isotope tracer studies and clues to primordial components in the mantle. *Ann. Rev. Earth Planet. Sci.* 11: 371–414

Macdougall, J. D., Lugmair, G. W. 1985. Extreme isotopic homogeneity among basalts from the southern East Pacific Rise: mantle or mixing effect? *Nature* 313: 209–11

Macdougall, J. D., Lugmair, G. W. 1986. Sr and Nd isotopes in basalts from the East Pacific Rise: significance for mantle heterogeneity. *Earth Planet. Sci. Lett.* 77: 273–84

Masters, G., Jordan, T. H., Silver, P. G., Gilbert, F. 1982. Aspherical Earth structure from fundamental spheroidal-mode data. *Nature* 298: 609–13

McCulloch, M. T., Jaques, A. L., Nelson, D. R., Lewis, J. D. 1983. Nd and Sr isotopes in kimberlites and lamproites from Western Australia: an enriched mantle origin. *Nature* 302: 400–3

McKenzie, D., O'Nions, R. K. 1983. Mantle reservoirs and ocean island basalts. *Nature* 301: 229–31

Menzies, M. 1983. Mantle ultramafic xenoliths in alkaline magmas: evidence for mantle heterogeneity modified by magmatic activity. In *Continental Basalts and Mantle Xenoliths*, ed. C. J. Hawkesworth, M. J. Norry, pp. 92–110. Nantwich, Engl: Shiva

Menzies, M. A., Hawkesworth, C. J., eds. 1987. *Mantle Metasomatism.* London: Academic. 472 pp.

Menzies, M. A., Wass, S. Y. 1983. CO_2 and LREE-rich mantle below eastern Australia: a REE and isotopic study of alkaline magmas and apatite-rich mantle xenoliths from the Southern Highlands Province, Australia. *Earth Planet. Sci. Lett.* 65: 287–302

Meyer, C. 1977. Petrology, mineralogy and chemistry of KREEP basalt. *Phys. Chem. Earth* 10: 239–60

Morelli, A., Dziewonski, A. M. 1987. Topography of the core-mantle boundary and lateral homogeneity of the liquid core. *Nature* 325: 678–83

Navrotsky, A. 1980. Lower mantle phase transitions may generally have negative pressure-temperature slopes. *Geophys. Res. Lett.* 7: 709–12

Nelson, D. R., McCulloch, M. T., Sun, S.-S. 1986. The origins of ultrapotassic rocks as inferred from Sr, Nd and Pb isotopes. *Geochim. Cosmochim. Acta* 50: 231–45

Nisbet, E. G., Walker, D. 1982. Komatiites and the structure of the Archean mantle. *Earth Planet. Sci. Lett.* 60: 105–13

Ohtani, E. 1985. The primordial terrestrial magma ocean and its implication for stratification of the mantle. *Phys. Earth Planet. Inter.* 38: 70–80

Olson, P. 1981. Mantle convection with spherical effects. *J. Geophys. Res.* 86: 4881–91

Olson, P. 1987. A comparison of heat flow lavas for mantle convection at very high Rayleigh numbers. *Phys. Earth Planet. Inter.* 48: 153–60

Olson, P., Yuen, D. A. 1982. Thermochemical plumes and mantle phase transitions. *J. Geophys. Res.* 87: 3993–4002

Olson, P., Yuen, D. A., Balsiger, D. 1984. Convective mixing and the fine structure of mantle heterogeneity. *Phys. Earth Planet. Inter.* 36: 291–304

O'Nions, R. K., Evensen, N. M., Hamilton, P. J. 1979. Geochemical modeling of mantle differentiation and crustal growth. *J. Geophys. Res.* 84: 6091–6101

Ozima, M., Kudo, K. 1972. Excess argon in submarine basalts and an Earth-atmosphere evolution model. *Nature Phys. Sci.* 239: 23–24

Ozima, M., Podosek, F. A., eds. 1983. *Noble Gas Geochemistry.* Cambridge: Cambridge Univ. Press. 367 pp.

Patchett, P. J., White, W. M., Feldmann, H., Kielinczuk, S., Hofmann, A. W. 1984. Hafnium/rare earth element fractionation in the sedimentary system and crustal recycling into the Earth's mantle. *Earth Planet. Sci. Lett.* 69: 365–78

Poreda, R., Schilling, J.-G., Craig, H. 1986. Helium and hydrogen isotopes in ocean-ridge basalts north and south of Iceland. *Earth Planet. Sci. Lett.* 78: 1–17

Richards, M. A., Hager, B. H. 1984. Geoid anomalies in a dynamic Earth. *J. Geophys. Res.* 89: 5987–6002

Richardson, S. H., Erlank, A. J., Duncan, R. A., Reid, D. L. 1982. Correlated Nd, Sr, and Pb isotope variation in Walvis Ridge basalts and implications for the evolution of their mantle source. *Earth Planet. Sci. Lett.* 59: 327–42

Richardson, S. H., Erlank, A. J., Hart, S. R. 1985. Kimberlite-borne garnet peridotite xenoliths from old enriched subcontinental lithosphere. *Earth Planet. Sci. Lett.* 75: 116–28

Richardson, S. H., Gurney, J. J., Erlank, A. J., Harris, J. W. 1984. Origin of diamonds in old enriched mantle. *Nature* 310: 198–202

Richter, F. M. 1979. Focal mechanisms and seismic energy release of deep and intermediate earthquakes and their bearing on the depth extent of mantle flow. *J. Geophys. Res.* 84: 6783–95

Richter, F. M., Daly, S. F., Nataf, H.-C. 1982. A parameterized model for the evolution of isotopic heterogeneities in a convecting system. *Earth Planet. Sci. Lett.* 60: 178–94

Ringwood, A. E. 1975. *Composition and Petrology of the Mantle.* New York: McGraw-Hill. 618 pp.

Ringwood, A. E. 1982. Phase transformations and differentiation in subducted lithosphere: implications for mantle dynamics, basalt petrogenesis, and crustal evolution. *J. Geol.* 90: 611–43

Rison, W., Craig, H. 1983. Helium isotopes

and mantle volatiles in Loihi Seamount and Hawaiian Island basalts and xenoliths. *Earth Planet. Sci. Lett.* 66: 407–26

Sarda, P., Staudacher, T., Allègre, C. J. 1985. $^{40}Ar/^{36}Ar$ in MORB glasses: constraints on atmosphere and mantle evolution. *Earth Planet. Sci. Lett.* 72: 357–75

Schilling, J.-G. 1973. Iceland mantle plume: geochemical study of Reykjanes ridge. *Nature* 242: 565–71

Schubert, G., Spohn, T. E. 1981. Two-layer mantle convection and the depletion of radioactive elements in the lower mantle. *Geophys. Res. Lett.* 8: 951–54

Shankland, T. T., Brown, J. M. 1985. Homogeneity and temperatures in the lower mantle. *Phys. Earth Planet. Inter.* 38: 51–58

Shirey, S. B., Hanson, G. N. 1986. Mantle heterogeneity and crustal recycling in Archean granite-greenstone belts: evidence from Nd isotopes and trace elements in the Rainy Lake area, Ontario. *Geochim. Cosmochim. Acta* 50: 2631–51

Shirey, S. B., Bender, J. F., Langmuir, C. H. 1987. Three-component isotopic heterogeneity near the Oceanographer transform, Mid-Atlantic Ridge. *Nature* 325: 217–23

Silver, P. G., Carlson, R. W., Bell, P., Olson, P. 1985. Mantle structure and dynamics. *Eos, Trans. Am Geophys. Union* 66: 1193, 1196–98

Silver, P. G., Chan, W. W. 1986. Observations of body wave multipathing from broadband seismograms: evidence for lower mantle slab penetration beneath the Sea of Okhotsk. *J. Geophys. Res.* 91: 13,787–13,802

Silver, P. G., Jordan, T. H. 1981. Fundamental spheroidal mode observations of aspherical heterogeneity. *Geophys. J. R. Astron. Soc.* 64: 605–34

Smith, C. B. 1983. Pb, Sr and Nd isotopic evidence for sources of southern African Cretaceous kimberlites. *Nature* 304: 51–54

Spohn, T., Schubert, G. 1982. Modes of mantle convection and the removal of heat from the Earth's interior. *J. Geophys. Res.* 87: 4682–96

Staudigel, H., Zindler, A., Hart, S. R., Leslie, T., Chen, C.-Y., Clague, D. 1984. The isotope systematics of a juvenile intraplate volcano: Pb, Nd, and Sr isotope ratios of basalts from Loihi Seamount, Hawaii. *Earth Planet. Sci. Lett.* 69: 13–29

Stefanick, M., Jurdy, D. M. 1984. The distribution of hot spots. *J. Geophys. Res.* 89: 9919–25

Stille, P., Unruh, D. M., Tatsumoto, M. 1983. Pb, Sr, Nd and Hf isotopic evidence of multiple sources of Oahu, Hawaii

basalts. *Nature* 304: 25–29

Stosch, H.-G., Carlson, R. W., Lugmair, G. W. 1980. Episodic mantle differentiation: Nd and Sr isotopic evidence. *Earth Planet. Sci. Lett.* 47: 263–71

Sun, S.-S. 1980. Lead isotopic study of young volcanic rocks from mid-ocean ridges, ocean islands and island arcs. *Philos. Trans. R. Soc. London Ser. A.* 297: 400–45

Sun, S.-S., Jahn, B. 1975. Lead and strontium isotopes in post-glacial basalts from Iceland. *Nature* 255: 527–30

Tatsumoto, M. 1978. Isotopic composition of lead in oceanic basalt and its implication to mantle evolution. *Earth Planet. Sci. Lett.* 38: 63–87

Tatsumoto, M., Hedge, C. E., Engel, A. E. J. 1965. Potassium, rubidium, strontium, thorium, uranium, and the ratio of strontium-87 to strontium-86 in oceanic tholeiitic basalt. *Science* 150: 886–88

Taylor, S. R., McLennan, S. M. 1981. The composition and evolution of the continental crust: rare-earth element evidence from sedimentary rocks. *Philos. Trans. R. Soc. London Ser. A* 301: 381–99

Turcotte, D. L., Kellogg, L. H. 1986. Isotopic modeling of the evolution of the mantle and crust. *Rev. Geophys.* 24: 311–28

Turcotte, D. L., Oxburgh, E. E. 1967. Finite amplitude convection cells and continental drift. *J. Fluid Mech.* 28: 29–42

Ullrich, L., Van der Voo, R. 1981. Minimum continental velocities with respect to the pole since the Archean. *Tectonophysics* 74: 17–27

Urey, H., C. 1952. *The Planets, Their Origin and Development.* New Haven, Conn: Yale Univ. Press

Vidal, P., Chauvel, C., Brousse, R. 1984. Large mantle heterogeneity beneath French Polynesia. *Earth Planet. Sci. Lett.* 62: 273–82

Vidale, J. E. 1987. Waveform effects of a high-velocity, subducted slab. *Geophys. Res. Lett.* 14: 542–45

Vollmer, R., Norry, M. J. 1983a. Unusual isotopic variations in Nyiragongo nephelinites. *Nature* 301: 141–43

Vollmer, R., Norry, M. J. 1983b. Possible origin of K-rich volcanic rocks from Virunga, East Africa, by metasomatism of continental crustal material: Pb, Nd and Sr isotopic evidence. *Earth Planet. Sci. Lett.* 64: 374–86

Vollmer, R., Ogden, P., Schilling, J.-G., Kingsley, R. H., Waggoner, D. G. 1984. Nd and Sr isotopes in ultrapotassic volcanic rocks from the Leucite Hills, Wyoming. *Contrib. Mineral. Petrol.* 87: 359–68

Walck, M. C. 1984. The *P*-wave upper mantle structure beneath an active spreading centre: the Gulf of California. *Geophys. J. R. Astron. Soc.* 76: 697–723

Wasserburg, G. J., DePaolo, D. J. 1979. Models of Earth structure inferred from neodymium and strontium isotopic abundances. *Proc. Natl. Acad. Sci. USA* 76: 3594–98

Weaver, B. L., Wood, D. A., Tarney, J., Joron, J. L. 1986. Role of subducted sediments in the genesis of ocean-island basalts: geochemical evidence from South Atlantic ocean islands. *Geology* 14: 275–78

Weidner, D. J. 1985. A mineral physics test of a pyrolite mantle. *Geophys. Res. Lett.* 12: 417–20

Weidner, D. J. 1986. Mantle model based on measured physical properties of minerals. In *Advances in Physical Geochemistry*, Vol. 6, *Chemistry and Physics of Terrestrial Planets*, ed. S. K. Saxena, pp. 251–74. New York: Springer-Verlag

Weidner, D. J., Ito, E. 1987. Mineral physics constraints on a uniform mantle composition. In *High Pressure Research in Mineral Physics. Proc. US-Jpn. Semin. High Pressure*, pp. 117–24. Washington, DC: Am. Geophys. Union

Weinstein, S., Olson, P., Yuen, D. A. 1987. Time-dependent convection and the lower mantle geotherm. *Eos, Trans. Am. Geophys. Union* 68: 413 (Abstr.)

Weis, D. 1983. Pb isotopes in Ascension Island rocks: ocean origin for the gabbroic to granite plutonic xenoliths. *Earth Planet. Sci. Lett.* 62: 273–82

White, W. M. 1985. Sources of oceanic basalts: radiogenic isotopic evidence. *Geology* 13: 115–18

White, W. M., Hofmann, A. W., Puchelt, H. 1987. Isotope geochemistry of Pacific mid-ocean ridge basalt. *J. Geophys. Res.* 92: 4881–93

Witte, D. 1987. Numerical simulations of seismic waves disturbed by subducted slabs. *Eos, Trans. Am. Geophys. Union* 68: 352 (Abstr.)

Woodhouse, J. H., Dziewonski, A. M. 1984. Mapping the upper mantle: three-dimensional modeling of Earth structure by inversion of seismic waveforms. *J. Geophys. Res.* 89: 5953–86

Woodhouse, J. H., Dziewonski, A. M. 1987. Models of the upper and lower mantle from waveforms of mantle waves and body waves. *Eos, Trans. Am. Geophys. Union* 68: 356–57 (Abstr.)

Wu, P., Peltier, W. R. 1984. Pleistocene deglaciation and the Earth's rotation: a new analysis. *Geophys. J. R. Astron. Soc.* 76: 753–91

Yuen, D. A., Quareni, F., Hong, H.-J. 1987. Effects from equation of state and rheology in dissipative heating in compressible mantle convection. *Nature* 326: 67–69

Yuen, D. A., Sabadini, R. 1985. Viscosity stratification of the lower mantle as inferred by the $\dot{J}_2$ observation. *Ann. Geophys.* 3: 647–54

Zindler, A., Hart, S. 1986. Chemical geodynamics. *Ann. Rev. Earth Planet. Sci.* 14: 493–571

Zindler, A., Jagoutz, E., Goldstein, S. 1982. Nd, Sr, and Pb isotopic systematics in a three-component mantle: a new perspective. *Nature* 58: 519–23

Zindler, A., Staudigel, H., Batiza, R. 1984. Isotope and trace element geochemistry of young Pacific seamounts: Implications for the scale of upper mantle heterogeneity. *Earth Planet. Sci. Lett.* 70: 175–95

Ann. Rev. Earth Planet. Sci. 1988. 16: 543–603

UNITED PLATES OF AMERICA, THE BIRTH OF A CRATON: Early Proterozoic Assembly and Growth of Laurentia

Paul F. Hoffman

Lithosphere and Canadian Shield Division, Continental Geoscience and Mineral Resources Branch, Geological Survey of Canada, Ottawa, Ontario K1A 0E4, Canada

> *The fact that two provinces of the Canadian Shield have been together during post-Cambrian time does not necessarily mean that they were formed close together ...*
>
> J. Tuzo Wilson (1962)

INTRODUCTION

Cratons are large areas of continental lithosphere that have remained coherent and relatively rigid since the Precambrian. Laurentia, the North American craton, is one of the oldest and largest. It includes the Precambrian shields of Canada and Greenland, the covered platform and basins of the North American interior, and, in this paper, the reactivated Cordilleran foreland of the southwestern United States.

Laurentia owes its existence to a network of Early Proterozoic[1] orogenic belts (Figure 1). Many of the belts are collison zones preserving only the deformed margins of formerly independent microcontinents composed of Archean crust. Other belts contain accreted Early Proterozoic island arcs and associated intraoceanic deposits. The assembly of Laurentia in the Early Proterozoic may be compared with the assembly of Eurasia in the Phanerozoic.

This review of geological and geochronological data concerns the nature

---

[1] Archean = before 2.5 Ga; Early Proterozoic = 2.5–1.6 Ga; Middle Proterozoic = 1.6–0.9 Ga; Late Proterozoic = 0.9–0.6 Ga; Ga = 10^9 yr before present; Myr = 10^6 yr.

543

0084–6597/88/0515–0543$02.00

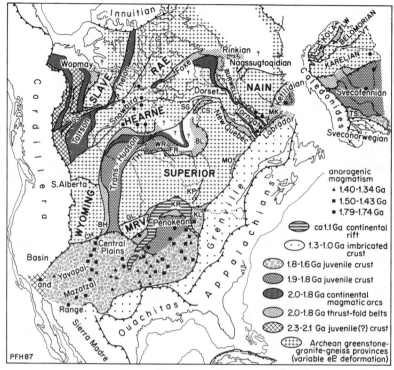

Figure 1 Precambrian tectonic elements of Laurentia. The Baltic shield is shown in a pre-Iapetus reconstruction, and Greenland is restored prior to rifting from North America. Uppercase names are Archean provinces; lowercase names are Proterozoic and Phanerozoic orogens. Abbreviations: BH, Black Hills inlier; BL, Belcher belt; CH, Cheyenne belt; CS, Cape Smith belt; FR, Fox River belt; GL, Great Lakes tectonic zone; GS, Great Slave Lake shear zone; KL, Killarney magmatic zone; KP, Kapuskasing uplift; KR, Keweenawan rift zone; LW, Lapland–White Sea tectonic zone; MK, Makkovik orogen; MO, Mistassini-Otish basins; MRV, Minnesota River Valley terrane; SG, Sugluk terrane; TH, Thompson belt; TS, Transscandinavian (Småland-Värmland) magmatic zone; VT, Vetrenny tectonic zone; WR, Winisk River fault.

and timing of Early Proterozoic orogenic belts in Laurentia. It complements an earlier review of geophysical data on proposed Proterozoic sutures in Canada (Gibb et al 1983). The underlying theme of both reviews is J. Tuzo Wilson's (1962) admonition, quoted above, regarding the implications of mobilism for Precambrian tectonics. (It is revealing that his first paper espousing mobilism concerns the Precambrian. Field work in the Canadian shield was his initiation to earth science, and he, among the founders of plate tectonics, was the one most interested in its long-term implications.)

New impetus for studying the early history of Laurentia comes, above all, from advances in isotopic geochronology. Precise dating of igneous and metamorphic events provides the most effective means of testing and refining dynamic models for Precambrian orogenic belts based on geological or geophysical data. Greatly improved U-Pb dating methods (e.g. Krogh 1973, 1982, Roddick et al 1987) permit igneous and metamorphic ages to be determined with an analytical precision of better than 0.5% (i.e. ± 5 Myr at 2 Ga). Although the actual uncertainties are likely somewhat greater than the formal analytical errors, zircon U-Pb ages are far more accurate and generally much more resistant to isotopic resetting than Rb-Sr or K-Ar ages for Precambrian rocks. *Unless otherwise stated, all ages quoted in this paper are based on U-Pb analyses.* Most are based on multiple age determinations, and the resulting age limits are rounded to the nearest 10 Myr.

Progress has also been spurred by other developments in geochronology, geophysics, and geology. Sm-Nd isotopic systematics, complementing earlier Pb-Pb and Rb-Sr methods, enables crust newly extracted from the Proterozoic mantle ("juvenile crust") to be distinguished from the products of remelting or mechanical reworking of Archean crust (McCulloch & Wasserburg 1978, Patchett & Arndt 1986). Systematic gravity surveys and interpretation of gravity anomalies permitted the early identification of probable collision zones (Gibb et al 1983, and references therein). Digitized high-resolution aeromagnetic surveys now cover much of the continental interior (Committee for the Magnetic Anomaly Map of North America 1987), providing grounds for extrapolating structural trends and magmatic zones (especially the magnetite-bearing rocks of magmatic arcs) across poorly exposed or poorly mapped parts of the shield and beneath the platformal cover (e.g. Dods et al 1985, Hoffman 1987a).

Some evidence of Early Proterozoic relative motions within Laurentia has been obtained from paleomagnetic studies (e.g. McGlynn & Irving 1981, Irving et al 1984). However, the failure to recognize secondary remagnetizations and other problems (Burke et al 1976) led many paleomagnetists (e.g. McElhinny & McWilliams 1977, Embleton & Schmidt 1979, Piper 1983) to question the early proponents of mobilism (e.g. Gibb & Walcott 1971, Fraser et al 1972, Dewey & Burke 1973). Recently recognized criteria for determining the kinematics of shear zones (Berthé et al 1979, White et al 1980, Lister & Snoke 1984, Passchier & Simpson 1986, Hanmer 1986) assist in determining relative motions between adjacent crustal blocks during collisional orogeny, but the lack of information on relative motions prior to collision remains the most serious obstacle for Early Proterozoic paleocontinental reconstructions. Nevertheless, the recent discoveries of obducted slices of oceanic crust and mantle, including "sheeted" dike complexes, within Early Proterozoic orogenic belts (e.g. Kontinen 1987, St-Onge et al 1988) corroborates the now prevailing view of geologists, geophysicists, and geochemists that plate tectonics was in operation during the birth of Laurentia.

ARCHEAN PROVINCES AND PROTEROZOIC OROGENS

The main tectonic elements of Laurentia, stripped of its platformal cover, are shown in Figure 1. The problematic "Churchill province" of previous authors is subdivided into the Archean Hearne and Rae provinces to the northwest and the Early Proterozoic Trans-Hudson orogen to the southeast (Hoffman 1988). The Archean provinces (Slave, Rae, Hearne, Wyoming, Superior, Nain) are clustered in the northern two thirds of the craton and underlie most of the Canadian shield. Each province has an Archean basement complex comprising a granite-greenstone terrain or its high-grade equivalent, overlain by erosional remnants of Early Proterozoic sedimentary cover of platformal facies. Variable deformation and metamorphism of the sedimentary cover indicate degrees of Early Proterozoic reactivation of the Archean provinces. In general, reactivation is related in trend and intensity to the orogenic belts that frame the Archean provinces.

Many of the Early Proterozoic orogenic belts appear to represent collision zones between Archean provinces. In the western shield, the Trans-Hudson orogen welds the Hearne and Wyoming provinces to the Superior province, the Southern Alberta orogen welds the Hearne and Wyoming provinces, the Snowbird orogen welds the Rae and Hearne provinces, and the Thelon orogen welds the Slave and Rae provinces. In the eastern shield, the southeastern branch of the Rae province is welded to the Superior and Nain provinces by the New Quebec ("Labrador Trough") and Torngat orogens, respectively; the northeastern branch of the Rae province is welded to the Nain province by the Rinkian-Nagssugtoqidian orogen; and the Foxe fold belt forms a syntaxis at the junction of the north- and southeastern branches of the Rae province. The collisional orogens are typically asymmetric, with one side characterized by a sedimentary prism overthrust toward the Archean foreland, and the other side characterized by a magmatic arc and by great reactivation of the Archean hinterland. The asymmetry presumably reflects the dominant polarity of subduction and consequent thermal regimes during ocean closure. By implication, the Archean provinces were formerly independent microcontinents, although they may have originated cogenetically as rifted fragments of an earlier continental assembly in the same way that the rifted fragments of Gond-wanaland have been reassembled in Eurasia.

The Early Proterozoic orogenic belts peripheral to the cluster of Archean provinces appear to represent zones of lateral accretion of mainly juvenile Proterozoic crust. Accretion between 2.0 and 1.8 Ga occurred in the Wopmay, Penokean, and Makkovik-Ketilidian orogens, and between 1.8 and 1.6 Ga in the Labradorian, Central Plains, and Yavapai-Mazatzal orogens. In the southwestern United States, the zone of accretion is at least 1200 km wide.

Geochronological data (Table 1) indicate that amalgamation of the Archean provinces occurred between about 2.0 and 1.8 Ga, and that accretion of Laurentia was complete, except for the southeastern part of the Grenville orogen, by about 1.6 Ga. That Laurentia is essentially a product of events occurring between 2.0 and 1.6 Ga is the main conclusion of this review. It gives a mobilistic interpretation to the tectonic episode recognized by Stockwell (1961) as the "Hudsonian orogeny."

The subsequent Proterozoic evolution of Laurentia, including wide-spread Middle Proterozoic "anorogenic" igneous activity (Barager 1977, Emslie 1978, Anderson 1983), the 1.2–1.0 Ga midcontinent (Keweenawan) rift system (Van Schmus & Hinze 1985), and the 1.3–0.9 Ga Grenville orogen (Moore et al 1986, Rivers et al 1988), lies beyond the scope of this review.

Table 1 Comparison of Early Proterozoic orogens in Laurentia

Orogen	Foreland	Hinterland	Obliquity	Magmatism (U-Pb ages)[a]
Thelon	Slave	SW-Rae	right	2.02–1.91 Ga
Wopmay	Slave	arcs	right	1.95–1.84 Ga
Snowbird	Hearne(?)	SW-Rae(?)	(?)	post-1.92/pre-1.85 Ga
S-Alberta	Hearne	Wyoming	(?)	pre-1.85 Ga
Trans-Hudson	Superior	Hearne(+)	left(?)	1.91–1.81 Ga
Cape Smith	Superior	Hearne(?)	nil	1.96–1.84 Ga
New Quebec	Superior	SE-Rae	right	2.14–1.81 Ga
Torngat	Nain	SE-Rae	left	post-2.3/pre-1.65 Ga
Foxe	NE-Rae	SE-Rae	left(?)	1.90–1.82 Ga
Nagssugtoqidian	Nain	NE-Rae	(?)	(?)
Makkovik	Nain	arcs(?)	right	1.86–1.80 Ga
Labrador	Nain(+)	arcs(?)	(?)	1.71–1.63 Ga
Penokean	Superior(+)	arcs	(?)	1.89–1.82 Ga
Yavapai	Wyoming	arcs	(?)	1.79–1.69 Ga
Mazatzal	Wyoming(+)	arcs(+)	(?)	1.71–1.62 Ga

[a] See text for references.

COLLISIONS IN THE NORTH

Thelon Orogen: Rae/Slave Collision

The Slave province is a late Archean (2.7–2.5 Ga) granite-greenstone terrain that served as a foreland for the Thelon orogen (2.02–1.91 Ga) to the east and the Wopmay orogen (1.95–1.84 Ga) to the west (Figure 2). The Thelon orogen resulted from a dextral-oblique collision between the Slave and Rae provinces (Figure 3), followed by indentation of the Rae hinterland by the Slave foreland province (Gibb 1978a, Hoffman 1987a, R. Tirrul & J. P. Grotzinger, in preparation). Indentation was accommodated by dextral-oblique crustal shortening across the Queen Maud uplift, a reactivated Archean(?) granulite-grade domain in the hinterland, and by 600–700 km of right-slip on the Great Slave Lake shear zone, an intracontinental transform structure exposed as a zone of continuous mylonite 25 km wide (Hanmer 1987a, Hoffman 1987a). Crustal wedges were extruded laterally from the indentation zone: The one bounded by the Great Slave Lake and Rutledge-Allen shear zones escaped southwestward; another bounded by the Bathurst and Ellice fault zones escaped north-northwestward. The deformed alluvial-lacustrine Nonacho basin, although undated, is probably related to sinistral wrench faulting associated with tectonic escape from the indentation zone (Aspler & Donaldson 1985).

A cryptic suture zone (Gibb & Thomas 1977), transposed by right-slip and east-dipping dip-slip shear zones (Thompson et al 1986, Henderson et al 1987), is inferred between Archean rocks of the Slave province and a zone to the east characterized by 2.02–1.91 Ga granitic to dioritic plutons (van Breemen et al 1987a,b, Bostock et al 1987). This zone of magnetite-series and subordinate ilmenite-series plutons is expressed as a distinctive belt of magnetic anomalies more than 80 km wide that can be traced for 2550 km from central Alberta to Prince of Wales Island in the central Arctic archipelago (Figure 1). Exposed for 1000 km, the Taltson-Thelon plutonic zone is interpreted as a composite precollisional magmatic arc and postcollisional anatectic batholith.

Foreland thrust-fold belts and autochthonous foreland basins related to the Thelon orogen are preserved in two structural depressions (Figure 2). The Goulburn Supergroup in the northeastern Slave province comprises a basal, eastward-thickening wedge (0–500 m) of shallow-shelf quartzite and carbonate (Kimerot platform) that is overlain by the Bear Creek foredeep (Grotzinger & McCormick 1987). The foredeep was initiated by sudden drowning of the Kimerot platform, followed by deposition of a flysch-molasse wedge that thins from 5.5 km in the east to 1.5 km over a syndepositional flexural arch in the foreland. Turbidity currents flowed

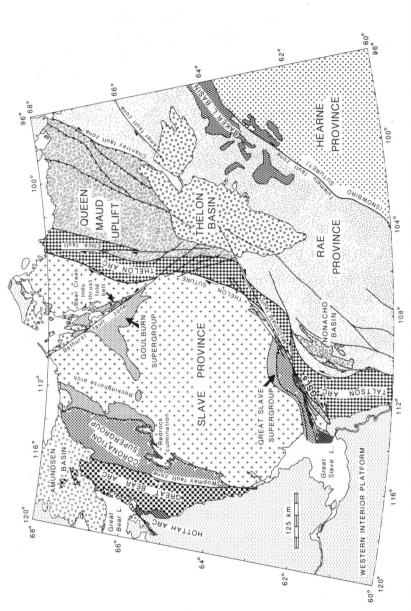

Figure 2   The northwestern Canadian shield, showing the tectonic elements described in the text. The Slave province is bounded by the Thelon orogen on the east and the Wopmay orogen on the west.

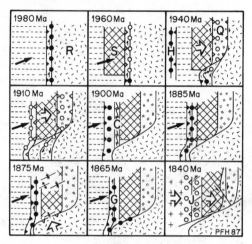

Figure 3   Schematic summary of the plate-tectonic evolution of the Thelon and Wopmay orogens. Solid teeth indicate subduction zones; open teeth indicate intracontinental thrust zones. Solid circles are magmatic arcs above subduction zones; open circles are post-subduction magmatic zones. Horizontal dashed areas are oceanic lithosphere. Abbreviations: C, Coronation Supergroup back-arc basin; G, Great Bear magmatic zone; H, Hottah magmatic zone; J, Johnny Hoe suture(?); Q, Queen Maud uplift; R, Rae province; S, Slave province.

axially in the foredeep, but the overlying fluvial sediments were transported to the west or northwest (Campbell & Cecile 1981, Grotzinger & McCormick 1987). The remnant Bear Creek Hills thrust-fold belt, which is completely exposed in cross section as a result of later folding associated with the Bathurst left-slip fault system, is developed above a WNW-directed ductile sole thrust located near the base of the Kimerot platform (Tirrul 1985, R. Tirrul & J. P. Grotzinger, in preparation). Thin-skinned deformation produced about 60% shortening in the cover transverse to the Thelon orogen. U-Pb zircon dating of a volcanic ash layer near the base of the Bear Creek foredeep establishes a temporal link between foredeep subsidence and the Taltson-Thelon magmatic zone, and provides a preliminary age of about 1.96 Ga for the Rae/Slave collision (S. A. Bowring, personal communication, 1986).

A complex basin in the east arm of Great Slave Lake is located on the southeast margin of the Slave province and is bounded by the Great Slave Lake shear zone (Figure 2). The basin contains three Early Proterozoic sequences, the older two of which are deformed by northwest-directed thrust-nappes and all three by northeast-trending right-slip faults and related folds (Hoffman 1987c). The oldest sequence is the allochthonous Wilson Island Group, a rift-like succession at least 8 km thick in which bimodal subaerial volcanics (1.93 Ga) are overlain by fluvial to marine siliciclastic metasediments intruded by a 1.89-Ga granite (Bowring et al 1984, Johnson 1986). The Great Slave Lake shear zone overlaps the Wilson Island Group in age and contains tentatively correlative metasedimentary protoliths. The 3–7-km-thick Great Slave Supergroup structurally underlies but is apparently younger than the Wilson Island Group. It unconformably overlies the Great Slave Lake shear zone, and its basal conglomerate contains clasts lithologically similar to the Wilson Island Group. Paleocurrents show that the lower clastics (Sosan Group) of the Great Slave Supergroup were shed from the northeast, probably from the Thelon magmatic zone and Queen Maud uplift, and that the upper clastics were derived from the Wopmay orogen to the southwest (Hoffman 1969). The Wilson Island Group and Great Slave Supergroup were telescoped by northwest-directed thrust-nappes and then intruded by calc-alkaline laccoliths (Badham 1981) that are coeval (1.87–1.86 Ga) with the Great Bear magmatic zone of the Wopmay orogen (Bowring et al 1984b). The laccoliths are overlain erosionally by alluvial-fan deposits related to right-slip on the McDonald fault system. The basin was earlier interpreted as an aulacogen related to the Wopmay orogen (Hoffman 1973), but it is now believed to have originated as an Andaman-type basin (Figure 3) resulting from indentation-extrusion tectonics (Tapponnier et al 1982) related to the Rae/Slave collision (Hoffman 1987a).

Wopmay Orogen: Active Western Margin of the Slave Province

The Wopmay orogen evolved on the active western margin of the Slave province (Figure 2). Dividing the orogen is the meridional Wopmay fault zone, a 10-km-wide belt characterized by mylonite having subhorizontal stretching lineations and by younger brittle faults having many kilometers of west-side-down throw (Hildebrand et al 1987a, J. E. King, in preparation). To the west are two continental calc-alkaline volcanic-plutonic belts (Hildebrand 1981, 1984, Hildebrand et al 1987b), in part spatially superimposed but temporally separated by a short-lived episode of rifting evidenced by subsidence and tholeiitic volcanism (Reichenbach 1986, 1987). Based on U-Pb zircon ages (Bowring 1984), the older Hottah magmatic arc was active from at least 1.95 to 1.91 Ga and was deformed and metamorphosed prior to the rifting event at about 1.90 Ga. The younger, more easterly Great Bear magmatic arc was active from about 1.88 to 1.86 Ga, after which dextral shear parallel to the arc produced broad northwest-trending en echelon folds. A 1.86–1.84 Ga suite of anatectic syenogranites postdates the en echelon folds but predates a system of conjugate transcurrent faults that accommodates east-west shortening and north-south extension affecting the entire orogen (Hoffman 1984, Tirrul 1987a).

East of the Wopmay fault zone, rifting marginal to the Slave province at about 1.90 Ga initiated deposition of the Coronation Supergroup, a west-facing passive-margin shelf-rise prism and succeeding foredeep flysch and molasse (Easton 1981, Hoffman & Bowring 1984, Grotzinger 1986, Hoffman 1987b). Soon thereafter, between about 1.89 and 1.88 Ga, the westerly off-shelf facies of the prism was intruded by a suite of gabbroic-dioritic-tonalitic-granitic plutons (Hoffman 1984, Lalonde 1986) and then translated eastward above a ductile sole thrust, producing a thin-skinned foreland thrust-fold belt in shelf and foredeep sediments to the east (Tirrul 1983, King 1986, Hoffman 1987b, Hoffman et al 1987, Tirrul 1987b). Preservation of inverted metamorphic isograds in upward-facing autochthonous strata indicates that thrusting occurred while the plutons and their metamorphic envelope were at least 250°C hotter than the underlying autochthon (St-Onge & King 1987a,b). Microprobe analyses of zoned poikiloblastic garnets document clockwise pressure-temperature paths for various structural levels in the allochthon and autochthon, with the latter having been uplifted from depths exceeding 30 km (St-Onge 1987). Uplift was accompanied by large-scale basement-cover folding coaxial with meridional synmetamorphic stretching ascribed to a dextral component of regional transpression (Hoffman et al 1987, King et al 1987). Deformation of the Coronation Supergroup has been ascribed to oblique collision of an Atlantic-type margin with an exotic island arc (Hoffman 1980, Hildebrand et al 1983, Hoffman & Bowring 1984). Alternatively, the correlation of 1.90-Ga rifting events east and west of the Wopmay fault zone is compatible with short-lived (1.90–1.88 Ga) evolution of the Coronation Supergroup in a back-arc setting (Hildebrand & Roots 1985, Reichenbach 1986, 1987). Rifting of the back-arc basin may have been induced by an arc-continent collision, accounting for the pre-1.90-Ga deformation of the Hottah arc, by a mechanism similar to that postulated for opening of the active Okinawa Trough (Letouzy & Kimura 1986).

Interpretation of magnetic and gravity anomalies west of the exposed part of the Wopmay orogen, constrained by sparse well data, provides a possible explanation for the post-1.84-Ga conjugate transcurrent faulting in the orogen and the correlative McDonald-Bathurst fault system of the eastern Slave province and Thelon orogen (Hoffman 1980). A linear gravity high bounded by an east-facing "escarpment" in the Bouguer anomaly field 180 km west of the Wopmay fault zone occurs between the Hottah/ Great Bear magmatic zones to the east and a parallel 1200-km-long magnetic high to the west, from which a drill core of granodiorite has been dated at about 1.86 Ga (Hildebrand et al 1987b, Hoffman 1987a). The positive magnetic anomaly is tentatively interpreted as a buried magmatic arc developed above a west-dipping subduction zone at the leading edge of a terrane that collided with the Wopmay orogen along a suture delineated by the Bouguer "escarpment." Accordingly, the conjugate transcurrent faulting could be a manifestation of east-west shortening in the foreland of the collision zone.

No exposures of Archean crust are known west of the Wopmay fault zone, and preliminary investigation of Pb and Nd isotopes in the Hottah and Great Bear arcs, as well as U-Pb dating of detrital and xenocrystic zircons, suggests that none was involved in their generation (Bowring & Podosek 1987, S. A. Bowring, personal communication, 1987).

Snowbird Tectonic Zone: Hearne/Rae Collision?

Between the Thelon orogen to the northwest and the Trans-Hudson orogen to the southeast lies a broad region of Archean crust (Figure 4) containing

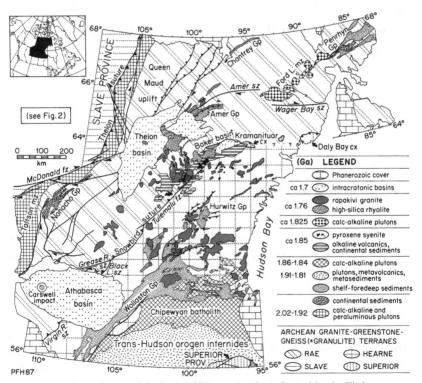

Figure 4 Geology of parts of the Rae and Hearne provinces framed by the Thelon orogen on the west, the Trans-Hudson orogen on the south, the Hudson Bay basin on the east, and the Arctic platform on the north.

scattered outliers of folded sedimentary cover, posttectonic igneous suites, and overlying cratonic basins all of Early Proterozoic age (Davidson 1972, Lewry et al 1985). The region is transected by an anastomosing, northeast-trending crustal break most evident on the horizontal gravity gradient map (Sharpton et al 1987). Segments of the break are recognized geologically as the Virgin River (Lewry & Sibbald 1977), Black Lake (Gilboy 1980, Hanmer 1987b), and Tulemalu (Tella & Eade 1986) fault and/or shear zones. Extending for almost 3000 km from the Rocky Mountains to Hudson Strait (Figure 1), the entire break is here referred to as the Snow-bird tectonic zone. The name is derived from Snowbird Lake in the southeast corner of the District of Mackenzie, where Taylor (1963) de-scribed a zone of mylonite many kilometers wide containing bodies of banded gabbro-anorthosite-pyroxenite, an association now known to occur at many places along the length of zone. The name "Athabasca axis," proposed for the same zone by Darnely (1981), is not retained because its trend is genetically unrelated and almost perpendicular to the depositional and structural axis of the Athabasca basin (Ramaekers 1981).

The initial suggestion that the Snowbird zone might be a suture (Walcott & Boyd 1971, Gibb & Halliday 1974) was based on the associated gravity anomalies. Other evidence comes from magnetic anomalies, which indicate that the Snowbird zone truncates the Taltson magmatic zone of the Thelon orogen at a high angle in the subsurface of central Alberta (Committee for the Magnetic Anomaly Map of North America 1987). Accordingly, the age of suture is less than about 1.92 Ga. It must be older than the 1.85-Ga intrusions and related alkaline volcanics and sediments of the Baker Lake basin (LeCheminant et al 1987a,b), which constitute an overlap assemblage on the central segment of the zone. The Virgin River–Black Lake segment of the Snowbird zone has been interpreted as an intra-continental reactivation structure related to the Trans-Hudson orogen (Lewry & Sibbald 1980), but elsewhere the zone diverges from the trend of the orogen, suggesting an independent origin.

The interpretation of the Snowbird zone as a suture remains hypo-thetical. No magmatic arc related to the Snowbird zone has been clearly identified, although 2.4–2.0 Ga granites occur 150 km northwest of the zone at Lake Athabasca (Van Schmus et al 1986) and ∼1.88-Ga granite occurs about 100 km northwest of the zone near Baker Lake (LeCheminant et al 1987b). There are as yet few kinematic data for the Snowbird zone. Hanmer (1987b) presents preliminary evidence that the Tantato wedge of mafic to felsic granulite-facies mylonite north of the Athabasca basin was driven southwestward, compatible with tectonic "escape" from a zone of convergence between the Hearne and Rae provinces to the northeast. There, granulites of the Rae province are juxtaposed against relatively low-grade Archean rocks of the Hearne province. To the northeast, the Tulemalu fault zone contains fragments of high-pressure (~ 1.1 GPa) garnet-clinopyroxene granulite (Tella & Eade 1986). Farther to the northeast, between the Baker Lake and Hudson Bay basins, layered mafic-felsic granulite-anorthosite complexes appear to be thrust north- or northwestward onto lower-grade rocks of the Rae province (Schau et al 1982, Gordon 1987).

The Hearne and Rae provinces have much in common. Both contain some very old rocks; granites between 3.33 and 2.95 Ga occur widely in the Rae province on the Melville Peninsula (Wanless 1979), near Baker Lake (A. N. LeCheminant, personal communication, 1987), and at Lake Athabasca (Van Schmus et al 1986), and a 3.48-Ga gneiss occurs in the Hearne province 200 km northeast of the Athabasca basin (Wanless 1979, W. D. Loveridge & K. E. Eade, in preparation). Both provinces contain late Archean greenstone belts and associated granites. Those in the Hearne province (Davidson 1970a,b) contain submarine mafic-intermediate-felsic volcanics and associated graywackes and are about 2.7 Ga (Mortensen & Thorpe 1987); those in the Rae province (Schau 1982, Taylor 1985) include ultramafic-mafic-felsic volcanics about 2.8 Ga, associated with cross-bedded quartzarenite, and an epizonal granite-rhyolite ash-flow tuff association about 2.6 Ga (LeCheminant et al 1984, 1987b).

Both provinces preserve erosional remnants of Early Proterozoic platformal sedimentary cover that predate 1.85 Ga. Both the Hurwitz Group of the Hearne province (Bell 1970, Eade & Chandler 1975) and the Amer Group of the Rae province (Tippett & Heywood 1978, Frisch & Patterson 1983, Frisch et al 1985, Patterson 1986) have a prominent orthoquartzite in their lower part; a middle part containing mixed carbonate, pelite, and localized mafic volcanics; and an upper part dominated by fine-grained felspathic sandstone. Deformation of the Hurwitz Group is of two trends, producing a basin-and-dome interference pattern at their intersection. East-striking folds and south-dipping thrusts parallel the Seal River fold belt of the Trans-Hudson orogen in the south; northeast-striking folds and north- to northwest-dipping thrusts parallel the Snowbird zone in the north and west (Eade 1974, 1987). The folds and thrusts involve the Archean basement, and deformation and metamorphic grade increase toward both the Trans-Hudson orogen (Pearson & Lewry 1974, Schledewitz 1978) and the Snowbird zone (Davidson 1970a, Eade 1986, Tella et al 1986). In the Rae province, folds and thrusts affecting the Amer Group and its basement trend northeasterly, but there is no obvious increase in deformation or metamorphism toward the Snowbird zone. Patterson (1986) interprets the thrusts in the Amer Group to have north to northwest vergence, but this is questionable because several of the thrusts ramp down into the basement in that direction, and stratigraphic footwall cutoffs (Tippett & Heywood 1978, Patterson 1986) are more compatible with northeast-directed thrusting.

Extensive posttectonic Early Proterozoic igneous suites straddle the Snowbird zone (Figure 4) and distinguish the Hearne and Rae provinces from the other Archean provinces. A suite of 1.85–1.84 Ga lamprophyre dikes and pyroxene-syenite stocks, associated with alkaline volcanics and continental sediments of the Baker Lake basin (Blake 1980, LeCheminant et al 1981, 1987a,b), overlaps the central part of the Snowbird zone. A younger suite of 1.78–1.75 Ga high-silica rapakivi granites and rhyolite ash-flow tuffs, associated with continental sediments, is widely distributed west of Hudson Bay. The tectonic significance of the magmatism is conjectural, but Hoffman (1980) suggests that the alkaline suite was localized by east-directed "wedging" of the Slave province on the McDonald-Bathurst fault system (Figure 3).

Southern Alberta "Rift": Wyoming/Hearne Collision?

The buried Precambrian basement of southwestern Saskatchewan, southern Alberta, and Montana, lying between the exposed parts of the Hearne and Wyoming provinces, is composed mainly of Archean or reworked Archean crust (Peterman 1981, Frost & Burwash 1986, Peterman & Futa 1987). The Hearne and Wyoming provinces may therefore be mutually continuous. Alternatively, an easterly trending crustal discontinuity that crosses southern Alberta between latitudes 50° and 50°30′N (Thomas et al 1987, Sharpton et al 1987) is a possible suture between the two provinces (J. W. Peirce, personal communication, 1986).

Kanasewich et al (1969) interpreted the southern Alberta structure (Figure 1) as a buried Precambrian rift related to the Middle Proterozoic Belt-Purcell basin on the basis of Bouguer and magnetic anomalies and seismic reflection profiling. Their seismic profile shows that the lower crust dips southward beneath the "rift" but is uplifted sharply at its southern margin. A compatible asymmetry is indicated by the gravity field, which shows a negative anomaly coincident with the "rift," a sharp positive anomaly cresting close to its southern margin, and a broad positive anomaly cresting 30–70 km north of the rift. The seismic evidence for a thickened crust and the nature of the gravity field are not typical of old, thermally mature rifts. The observations are more compatible with a foredeep (foreland basin) and associated flexural bulge to the north, resulting from north-directed thrusting at the south margin of the "rift" (cf Karner & Watts 1983). Contrasting trends of regional magnetic anomalies, having a strong northwest "grain" south of the "rift" but an irregular north to northeast "grain" north of the "rift" (Committee for the Magnetic Anomaly Map of North America 1987), suggest the juxtaposition of different crustal blocks (Green et al 1985b). Accordingly, the Wyoming province may have collided with the Hearne province along a south-dipping suture zone, producing a foredeep consistent with the polarity implied by the seismic and gravity data.

The age of the proposed collision is conjectural, but its presumed eastward extension appears to be truncated by and therefore older than the subsurface part of the Trans-Hudson orogen (Green et al 1985b, Thomas et al 1987). This implies a pre-1.9 Ga collision age but does not preclude reactivation of the suture as a rift arm during deposition of the Belt-Purcell basin (Kanasewich et al 1969) or as a transform structure during Late Proterozoic rifting (Lis & Price 1976) and early Paleozoic subsidence (Bond & Kominz 1984) of the Cordilleran passive margin.

Trans-Hudson Orogen: Hearne-Wyoming/Superior Collision

As exposed in the provinces of Saskatchewan and Manitoba, the 500-km-wide Trans-Hudson orogen forms a dogleg, convex to the northwest, bounded by the Hearne and Superior provinces (Figures 1, 5). Southward, the orogen has been outlined in the subsurface between the Wyoming and Superior provinces as far as South Dakota (Green et al 1985a, Klasner & King 1986, Thomas et al 1987), where it appears to be truncated by the Central Plains orogen (Sims & Peterman 1986). To the northeast, the main part of the orogen passes beneath the Paleozoic Hudson Bay basin (Gibb & Walcott 1971, Coles & Haines 1982, Gibb 1983, Sharpton et al 1987), but its southeastern margin is discontinuously exposed south of Hudson Bay and along the coast and offshore islands of eastern Hudson Bay (Baragar & Scoates 1981, Mukhopadhyay & Gibb 1981). The northwest-dipping Sugluk suture zone in northernmost Quebec (see Figure 8) may represent the pinched extension of the orogen, its internal zone having been obducted southward and preserved as an infolded klippe (the Cape Smith thrust-fold belt) within the Superior province (Hoffman 1985, Doig 1987, St-Onge & Lucas 1988).

Essentially, the orogen comprises an internal zone of intraoceanic rocks flanked by ensialic external belts (Stauffer 1984). The Wathaman-Chipewyan batholith (Figure 5) is interpreted to be an ensialic magmatic arc (Ray & Wanless 1980, Lewry et al 1981, Fumerton et al 1984, Meyer 1987), and its position bordering the Hearne province implies a gross tectonic polarity for the orogen, in which the Superior province is the foreland and the Hearne province is the hinterland.

558   HOFFMAN

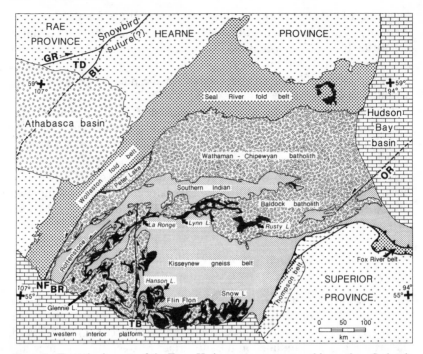

Figure 5 Tectonic elements of the Trans-Hudson orogen as exposed in northern Saskatchewan and Manitoba. Volcanic belts are shown in black. Major shear zones: BL, Black Lake; BR, Birch Rapids; GR, Grease River; NF, Needle Falls; OR, Owl River; TB, Tabbernor; TD, Tantato granulite-facies mylonite domain.

EXTERNAL BELTS   The ensialic belts of the Superior margin are exposed in four segments named, from southwest to northeast, the Thompson, Fox River, Belcher, and Cape Smith belts. The most complete stratigraphic sequence occurs in the Belcher belt, an orogenic salient in eastern Hudson Bay. On the coast, an arcuate west-dipping autochthonous sequence exposes, in oblique cross section, a basal east-trending rift-valley prism, 70 km wide, overlain by postrift shelf strata that onlap the basement north and south of the rift (Chandler 1984, 1987). The islands to the west belong to an arcuate meridional belt of doubly plunging folds that have been transported eastward, relative to the autochthon, on a sole thrust that carries 7–9 km of strata (Jackson 1960). The fold belt exposes at least 3.4 km of shelf carbonates and clastics containing a unit of plateau basalt up to 0.9 km thick (Ricketts & Donaldson 1981). The shelf strata are overlain paraconformably by foredeep deposits (Hoffman 1987b), comprising a transgressive sequence of quartzarenite, ironstone, tholeiitic-komatiitic volcanics, and euxinic shale, followed by a regressive sequence of graywacke turbidites and fluvial arkose (Ricketts & Donaldson 1981, Ricketts et al 1982, Baragar 1984). There are no U-Pb ages for the Belcher belt, but Pb isochrons for flows in the shelf and foredeep sequences are 1.96 ± 0.08 and 1.81 ± 0.03 Ga, respectively (Todt et al 1984), and Sm-Nd isotopic data indicate that both sequences contain flows contaminated by Archean crust (Chauvel et al 1987).

The Cape Smith belt (see Figure 7) is an erosional remnant of a thin-skinned, south-vergent thrust-fold belt preserved as a doubly plunging synclinorium resulting from two episodes of basement-involved refolding (St-Onge et al 1986, St-Onge & Lucas 1988). The volcanic and sedimentary rocks record the evolution of the rifted north margin of the Superior province from continental, through transitional oceanic, to true oceanic crust (St-Onge et al 1988). The continental rift sequence (Povungnituk Group) is represented by basal imbricates of semipelite, ironstone, and proximal to distal arkosic submarine-fan deposits, which pass stratigraphically upward into light-rare-earth-element-enriched basalt, minor rhyolite (1.96 Ga), volcaniclastic deposits, and fault-scarp breccias (Hynes & Francis 1982, St-Onge & Lucas 1988). The transitional crust (Chukotat Group) occurs in structurally higher thrust sheets and is dominated by pillowed basalt distinguished by a komatiitic to midoceanic-ridge-basalt-(MORB)-like tholeiitic chemistry (Francis et al 1983). Both the continental and transitional oceanic crust sequences are intruded by layered peridotite-gabro sills (St-Onge & Lucas 1988). Structurally overlying the Chukotat Group is a metamorphosed ophiolite suite exposed in inverse stratigraphic order (St-Onge et al 1988). From structural bottom to top, the ophiolite consists of thrust sheets composed of (a) laminated graphitic pelite and semipelite; (b) pillowed basalt cut by mafic dikes, "sheeted" dikes, and gabbro cut by dikes; and (c) layered mafic-ultramafic cumulates. Early thrusting occurred in a "piggy-back" sequence and was followed by "out-of-sequence" thrusting, with both forms of thrusting rooted on a sole thrust at the basement-cover contact in the preserved belt (St-Onge & Lucas 1988). Thermal relaxation following the early thrusting resulted in syndeformational metamorphism and is probably responsible for minor tonalite to granite plutons (1.84 Ga) that intrude the ophiolitic thrust sheets (St-Onge & Lucas 1988).

The east-trending Fox River belt (Figure 5) and Sutton inlier south of Hudson Bay are poorly exposed but include steeply north-dipping thrust stacks of relatively thin basal shelf sediments and overlying foredeep flysch, the latter associated with differentiated tholeiitic-komatiitic sills and flows (Bostock 1971, Scoates 1981, Baragar & Scoates 1981). The 2-km-thick Fox River sill and the extensive NNE-trending Molson dike swarm of the northwestern Superior province are coeval at 1883 ± 2 Ma (Scoates &

Macek 1978, Heaman et al 1986), but the tectonic significance of the magmatism is unclear. The north margin of the Fox River belt is a probable north-dipping thrust separating subgreenschist-facies rocks to the south from amphibolite-facies metasedimentary paragneisses similar to those of the Kisseynew belt (Figure 5).

In the northeast-trending Thompson belt, equivalents of the Fox River belt strata are tightly infolded with Archean basement (Weber & Scoates 1978, Peredery et al 1982). Foliations in the basement and cover dip very steeply to the southeast, and stretching lineations in mylonite zones are generally subvertical throughout the Thompson belt (Fueten et al 1986). Kinematic indicators in the western part of the belt show that the Superior province has moved up relative to the Trans-Hudson orogen (W. Bleeker, personal communication, 1987). The bulk of the deformation postdates the 1.88-Ga Molson dikes and predates 1.79–1.77 Ga pegmatites; post-metamorphic cooling is recorded by a concordant titanite age of 1.72 Ga (Machado et al 1987).

The northwest marginal zone of the orogen includes the Wollaston and Seal River fold belts (Figure 5) of Saskatchewan and Manitoba. The cover sequence on the margin of the Hearne province is thought to include an early synrift assemblage, subsequent shelf quartzite, and late arkosic synorogenic deposits (Lewry & Sibbald 1980, Stauffer 1984). However, the stratigraphic order and its interpretation are uncertain because of poor outcrop, complex structure involving upright basement-cover folds super-imposed on polyphase recumbent folds (Lewry & Sibbald 1980), and upper-amphibolite- to granulite-facies metamorphism (Lewry et al 1978, Schledewitz 1978). There are numerous granitic bodies, particularly in the Seal River belt (Schledewitz 1986), but it is difficult in the absence of U-Pb ages to ascertain the relative proportions of granite related to the Wathaman-Chipewyan batholith (1.86–1.85 Ga; Meyer 1987) to the south, the 1.75-Ga rapakivi suite of the Hearne province, and Archean basement inliers.

The Black Hills uplift in South Dakota (Redden & Norton 1975) exposes a fold belt marginal to the Wyoming province near the junction of the Trans-Hudson and Central Plains orogens (Sims & Peterman 1986). According to Redden et al (1987), Archean basement is overlain by synrift clastic sediments that were intruded by a 2.17-Ga mafic sill, then deformed and eroded, prior to deposition of younger rift- and shelf-facies clastic sediments and mafic volcanics. The shelf quartzites, which contain a 1.97-Ga tuff bed, are laterally equivalent to and overlain by thick turbidites capped by 1.88-Ga alkalic tuffs. The entire sequence was recumbently folded about easterly axes, refolded about NNW-trending axes, and intruded by the 1.72-Ga Harney Peak granite.

The Black Hills uplift lies just west of the North American Central Plains electrical conductivity anomaly (Jones & Savage 1986, and references therein), which trends northward from the Cheyenne belt of southeastern Wyoming (Figure 1) to the Canadian shield, where it follows the curvature of the Rottenstone–Southern Indian gneiss belt (Figure 5) and continues eastward to Hudson Bay. The conductivity anomaly is generally regarded as being related to a suture zone within the Trans-Hudson orogen, and it may coincide with the deep subsurface limit of Archean crust contiguous with the Wyoming and Hearne provinces. Thomas et al (1987) question this interpretation because of differences in trend between the conductivity anomaly and horizontal gravity gradients. However, the latter emphasize near-surface density variations and may reflect the structure of Early Proterozoic allochthons that need not parallel the deep-seated limit of autochthonous Archean crust.

INTERNAL BELTS   The internal zone of the Trans-Hudson orogen is a complex of plutonic, metavolcanic, and metasedimentary rocks, varying proportions of which have been used to distinguish different belts (Figure 5). The Baldock and Wathaman-Chipewyan belts are compound calc-alkaline batholiths of granodiorite and granite, with subordinate tonalite, diorite, and gabbro. The La Ronge, Lynn Lake, Rusty Lake, Glennie Lake, Hanson Lake, and Flin Flon belts consist mainly of plutonic, meta-volcanic, and subordinate metasedimentary rocks of greenschist to lower amphibolite grade. The Kisseynew and Rottenstone–Southern Indian belts are composed of upper-amphibolite-grade metasedimentary paragneiss and felsic-to-mafic orthogneiss.

The internal zone was long throught to be a typical "granite-greenstone-gneiss" complex of Archean age (Harrison 1951); its true age of 1.9–1.8 Ga (discounting 1.8–1.7 Ga Rb-Sr isochrons reset during metamorphism) was first indicated by Pb isotopic ratios of syngenetic massive-sulfide mineralization (Sangster 1972, 1978). Recent U-Pb zircon dating has confirmed the Early Proterozoic age (Van Schmus & Schledewitz 1986, Baldwin et al 1987, Gordon et al 1987a,b, Syme et al 1987, Van Schmus et al 1987a), showing that the earliest known volcanism occurred in the Lynn Lake belt at 1.91 Ga, followed by volcanism at 1.89–1.88 Ga in the La Ronge, Rusty Lake, Glennie Lake, Hanson Lake, and Flin Flon belts. Plutonism occurred between 1.88 and 1.83 Ga in all the volcanic-plutonic belts and in the Rottenstone–Southern Indian gneiss belt, contemporaneous with intrusion of the Wathaman-Chipewyan batholith between 1.86 and 1.84 Ga. Peak metamorphism and anatectic plutonism in the Kisseynew gneiss belt occurred at 1.82–1.81 Ga. The only dated Archean rocks occur in the Peter Lake orthogneiss domain (Figure 5) and in small structural culminations within the Hanson Lake (Van Schmus et al 1987a) and Glennie Lake (J. Chiarenzelli, personal communication, 1987) belts. Sm-Nd isotopic data (Chauvel et al 1987, Hegner & Hulbert 1987, Thom et al 1987) indicate that the magmas of the Flin Flon volcanic and plutonic rocks were derived from sources having no perceptible crustal residence time, but that contributions from older crust increase toward the northwestern Archean hinterland. These data indicate that magma generation involved little or no Archean crust in the southeastern part of the internal zone, but they do not rule out subsequent underthrusting by Archean crust (cf Green et al 1985a). The isotopic data strongly support the view that most of the rocks exposed in the internal zone represent juvenile 1.9–1.8 Ga crust of intraoceanic origin (Bailes 1971, Stauffer 1974, 1984, Green et al 1985a) bounded by an Andean-type continental margin to the northwest (Ray & Wanless 1980, Lewry et al 1981, Fumerton et al 1984, Meyer 1987).

The volcanic-plutonic belts contain structurally dismembered piles of mainly submarine, mafic-through-felsic, metavolcanic rocks having the major- and trace-element characteristics of Cenozoic island arcs (Stauffer et al 1975, Gilbert et al 1980, Syme 1985, 1987, Bailes & Syme 1987, Thom et al 1987, Watters et al 1987). They include tholeiitic basalts, high-alumina/low-titania basalts and basaltic andesites, and calc-alkaline andesite-dacite-rhyolite centered complexes. Mafic-to-felsic heterolithic breccias and redeposited volcaniclastic debris are common. The volcanic piles are intruded by synvolcanic gabbro, diorite, and tonalite sheets and stocks, above which extensive mineralized alteration haloes are developed (Baldwin 1980, Barham & Froese 1986, Bailes et al 1987). In the Bear Lake block east of Flin Flon (Bailes & Syme 1987), a thick pile of arc-like mafic flows is overlain by a ferrobasalt-ferrogabbro-rhyolite association succeeded by andesitic turbidites, a sequence possibly related to intraarc rifting. Elsewhere, volcanic piles are intercalated with and grade laterally into volcaniclastic turbidites (Bailes 1980a). The intercalated volcanic and sedimentary rocks (Amisk and Wasekwan groups of the Flin Flon and Lynn Lake belts, respectively) and related plutons are overlain unconformably by alluvial fanglomerate and sandstone (Missi and Sickle groups of the same belts) (Milligan 1960, Mukherjee 1974). Alluvial sedimentation accompanied D1 (discussed in what follows) folding (Stauffer & Mukherjee 1971, Bailes et al 1987), and the 1.83-Ga age of a rhyolite tuff in the Missi Group (Gordon et al 1987a) indicates that this occurred at least 50 Myr after Amisk volcanism.

The transitions into the Kisseynew gneiss belt from the bounding Flin Flon and Lynn Lake volcanic-plutonic belts represent changes in both primary facies, from volcanics to turbidites (Zwanzig 1976, Bailes 1980a,b,

Ashton & Wheatley 1986), and marked increases in metamorphic grade from chlorite-biotite-muscovite schist to K feldspar-cordierite-garnet paragneiss (Bailes & McRitchie 1978, Bailes 1980b, Jackson & Gordon 1985, 1986). It has been suggested that turbidite sedimentation occurred in a zone of high heat flow because of the spatial correlation between turbidite facies and high metamorphic grade, and because of the high-temperature (750°C)/moderate pressure (0.5 GPa) character of the metamorphism. This would favor an immature back-arc basin (Lewry 1981) over a fore-arc setting (Green et al 1985b). However, the current view that turbidite sedimentation was coeval with the 1.88-Ga Amisk volcanism (e.g. Bailes 1980a), rather than with the 1.83-Ga Missi sedimentation (Harrison 1951), implies a 60–70 Myr lapse between sedimentation and peak metamorphism. The enormous volume of metasediment in the Kisseynew belt suggests proximity to a contemporaneous collision zone, consistent with isotopic evidence for a greater contribution from older crust in the Kisseynew belt than in the bounding volcanic-plutonic belts (Chauvel et al 1987).

Although complex in detail, much of the structural evolution of the internal zone can be described in terms of two main phases of deformation, here referred to as D1 and D2 (e.g. Froese & Moore 1980, Stauffer 1984, Bailes et al 1987, Lewry et al 1987). The dominant map-scale structures are broad, upright D2 folds plunging gently to the northeast. They refold systems of stratigraphically distinct thrust-nappes containing originally recumbent, near-isoclinal D1 folds (e.g. Bailes & Syme 1987, Lewry et al 1987). The nappes are bounded by faults and shear zones having north-west-trending stretching lineations. Preliminary analysis of stratigraphic cutoffs and kinematic indicators suggests that both southeast- and north-west-vergent thrusting is present, with the former predominant. For example, the Railway fault system at Flin Flon appears to be a folded east- or southeast-vergent thrust fan (cf Bailes & Syme 1987), whereas the McLeod Road fault at Snow Lake, 120 km to the east, appears to be a northwest-vergent thrust (cf Froese & Moore 1980). Preliminary observations of rotated porphyroclasts in mylonite of the northwest-dipping Stanley shear zone indicate southeast-vergent thrusting betweeen the La Ronge and Glennie Lake belts; however, shear bands with top-side-down sense of movement occur in the northwest-dipping McLennan Lake tectonic zone within the La Ronge belt (cf Lewry & Slimmon 1985).

Although the gross structural evolution (i.e. D1 thrust-nappes deformed by broad upright D2 folds) appears similar throughout the internal zone, the timing of the thermal peak with respect to the structural evolution varies across the zone. In the northwest, strong synmetamorphic fabrics were developed during D1 thrusting; in the southeast, the dominant synmetamorphic fabrics are related to the upright D2 folds. Accordingly, stretching lineations in the northwest were folded during D2 folding and have moderately large rakes in the D1 foliation planes; stretching lineations in the southeast at Snow Lake plunge gently to the northeast, coaxial with the D2 folds (Bailes et al 1987). The latter observation implies that D2 folding was accompanied by hinge-line extension, compatible with deformation involving a significant component of noncoaxial strain.

One possible model for the tectonic evolution of the orogen (Figure 6A) involves meridional convergence and collision between the Superior and Hearne provinces. Convergence was accommodated by prevailing north-dipping subduction of oceanic lithosphere resulting in the development of an accretionary prism(s) in which dominantly south-directed D1 thrusting occurred. During collision, the accretionary complex was subjected to sinistral transpression between the ENE-trending continental margins,

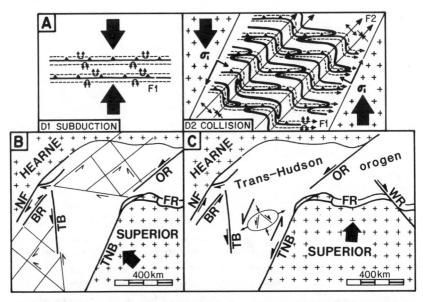

Figure 6 (*A*) Two-stage structural model for the internal zone of the Trans-Hudson orogen discussed in the text. (*B*) Slip-line field model proposed by Gibb (1983), assuming north-westward convergence of the Superior province relative to the Hearne province. (*C*) Model compatible with structural model (*A*), assuming northward convergence of the Superior province relative to the Hearne province. Note the difference in sense of movement on the Needle Falls and Birch Rapids shear zones in the two models (*B*) and (*C*). Abbreviations: BR, Birch Rapids shear zone; FR, Fox River belt; NF, Needle Falls shear zone; OR, Owl River shear zone; TB, Tabbernor fault zone; TNB, Thompson belt; WR, Winisk River shear zone.

causing the D2 folding and counterclockwise rotation of the D1 lineations. The model is consistent with the maximum development of the Wathaman-Chipewyan batholith on the latitudinal margin of the orogen and with the interpretation of the Tabbernor fault (Figure 5) as a meridional transform fault linked to north-dipping subduction beneath the Lynn Lake arc (Lewry 1981). The model predicts a component of left-slip between the Kisseynew and Thompson belts (Green et al 1985a), for which preliminary evidence has been observed (W. Bleeker et al, in preparation). An alternative model (Gibb 1983), based on slip-line field analysis of the transcurrent shear zones, implies late-stage northwest-southeast-oriented convergence between the Superior and Hearne provinces (Figure 6B). This model predicts right-slip motion on the Needle Falls and Birch Rapids shear zones, where preliminary indications of left-slip have been observed (J. F. Lewry, personal communication, 1987). More structurally oriented field work is urgently needed to clarify the tectonic evolution of this key orogenic belt.

Kapuskasing Uplift: Intracratonic Thrust Related to the Trans-Hudson Orogen?

The Kapuskasing uplift (Figure 1) is a composite northeast-trending, 500-km-long zone of granulite and upper-amphibolite-facies rocks that transects the east-west Archean granite-greenstone and metasedimentary-gneiss belts of the central Superior province. The belts on either side of the uplift correlate lithologically and geochronologically, and therefore the uplift is likely intracratonic in origin (Percival & Card 1983, 1985). The deep structure of the uplift has been inferred from paleomagnetism (Ernst & Halls 1984), seismic reflection profiling (Cook 1985), metamorphic geobarometry (Percival & McGrath 1986), and modeling of gravity and magnetic anomalies (Percival & McGrath 1986). The east side of the uplift is bounded by a system of west-dipping, crustal-scale thrusts. In the north, the west side is defined by an east-dipping back-thrust, and the uplift has the geometry of a pop-up; in the central part, the west side is defined by a system of west-dipping normal faults, and the uplift has the geometry of a perched thrust-tip; and in the south the uplift is a simple west-dipping slab (Percival & McGrath 1986). Geometrically, the uplift resembles the Laramide structures of Wyoming and adjacent states. Integrated isotopic studies (U-Pb zircon and sphene; K-Ar and $^{40}Ar/^{39}Ar$ hornblende, biotite, and whole-rock; Rb-Sr biotite) suggest that the high-grade rocks cooled nearly isobarically in the lower crust from about 2.6 Ga until they were brought to the surface by thrusting at about 2.0–1.9 Ga (J. A. Percival, personal communication, 1986). Thrusting must be older than carbonatite and alkaline igneous complexes that intruded the uplift and its boundary faults at 1.87–1.86 Ga (T. E. Krogh, personal communication, 1984; Bell et al 1987). Gibb (1978b, 1983) suggests a genetic link between the Kapuskasing uplift and the parallel Thompson belt of the Trans-Hudson orogen, 1250 km to the northwest. This suggestion is compatible with the possible synchroneity of the Kapuskasing uplift with foredeep(?) magmatism in the Fox River belt (1.88 Ga; Heaman et al 1986) and with deformation in the Thompson belt (post-1.88/pre-1.79 Ga; Machado et al 1987). Geometrically, however, the Gibb (1983) model is difficult to reconcile with right-slip on the Winisk River fault evident from magnetic anomalies (Committee for the Magnetic Anomaly Map of North America 1987).

New Quebec ("Labrador Trough") Orogen: Rae/Superior Collision

The northeast margin of the Superior province and the opposing margin of the Nain province (Figure 1) are bounded by the New Quebec and Torngat orogens, respectively. The orogens share a common 180–280-km-wide Archean hinterland (Ashwal et al 1986, Machado et al 1988), occupied by the George River drainage basin and Ungava Bay (Figure 7). The hinterland appears to represent, together with the Dorset fold belt of southern Baffin Island (Jackson & Taylor 1972), a southeastern extension of the Rae province. The sedimentary-volcanic belt related to the Rae/Slave collision (Kearey 1976) is widely known as the "Labrador Trough," a name that this author has been urged to abandon in order to avoid confusion with the recently named, geologically younger "Labrador orogen" (Thomas et al 1985) and "Trans-Labrador batholith" (Gower & Owen 1984) to the southeast (Figure 7). The proposed name suits because most of the belt lies in New Quebec, not Labrador, and because the belt is a structural depression but not a primary depositional "trough" (e.g. Dimroth 1981). The belt has long been considered as being continuous with the Trans-Hudson orogen (e.g. Baragar & Scoates 1981), but this remains to be demonstrated geochronologically and assumes that the Rae and Hearne provinces were already joined when they collided with the Superior province.

The New Quebec orogen contains an 800-km-long, southwest-vergent, foreland thrust-fold belt (Wardle 1982, Boone & Hynes 1987) that has been eroded at its north end across a transverse flexural culmination parallel to the Cape Smith belt (Hoffman 1985) and redeformed at its south end by northwest-vergent structures of the Grenville orogen (Rivers & Wardle 1985, Rivers & Chown 1986). The belt is bounded to the southwest by the flexurally arched Archean foreland of the Superior

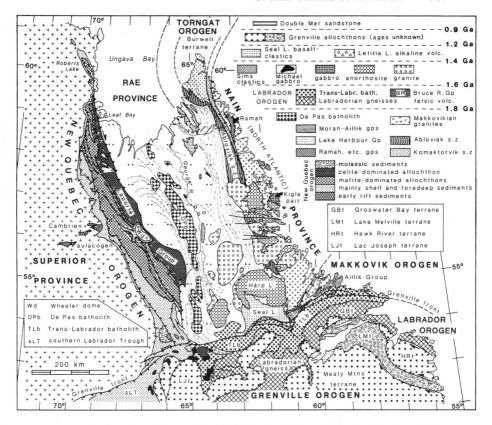

Figure 7   Geology of northeastern Quebec ("New Quebec") and Labrador. The New Quebec ("Labrador Trough") orogen is the Early Proterozoic collision zone between the Archean Superior and Rae provinces. The Torngat orogen is the Early Proterozoic collision zone between the Archean Rae and Nain provinces. The Makkovik and Labrador orogens involved accretion of Early Proterozoic crust onto the Nain province and the Rae-Nain assembly, respectively. Middle Proterozoic anorogenic granite-anorthosite-gabbro suites intruded the Rae and Nain provinces and the Torngat orogen. During the late Middle Proterozoic Grenville orogeny, all of the older crustal elements were telescoped by means of northwest-directed thrusting.

province and to northeast by the allochthonous Archean hinterland of the Rae province. From southwest to northeast, the belt comprises an autochthonous sedimentary zone, a medial zone of igneous and sedimentary allochthons that can be more or less confidently correlated with the autochthonous cover, and a metasedimentary zone (Laporte Group) of uncertain ancestry. The ductile sole thrust of the belt, characterized by northeast-trending stretching lineations and sheath-folds that have been refolded by northwest-trending basement-involved folds, is exposed in the structural culmination on the west side of Ungava Bay (Figure 7).

The autochthonous strata consist of three unconformity-bounded sequences that overstep sequentially onto the Archean foreland (Le Gallais & Lavoie 1982). The oldest sequence is exposed mainly in the medial allochthons but also is observed in the autochthonous Cambrian Lake aulacogen (Clark 1984). From the base, it comprises fluvial redbeds (Seward Subgroup), marine-shelf quartzite and carbonate (Pistolet Subgroup), foredeep black shale and turbidites (Swampy Bay Subgroup), and a regressive carbonate reef-complex (Denault/Abner Formation). The middle sequence begins with a transgressive high-energy quartzarenite (Wishart Formation), overlain by an iron-formation (Sokoman Formation) and a second foredeep flysch (Menihek Formation). The third sequence is a fluvial molasse (Chioak/Tamarack River Formation) containing westerly derived clasts of basement and middle-sequence lithologies. The overall sequence resembles that of the central Appalachians in having two successive foredeeps developed above initial-rift and passive-margin prisms (Hoffman 1987b).

The thrust sheets to the northeast carry sequences in part correlative with those already described but also containing thick piles of mafic flows and sills, locally with small felsic centers (Baragar 1960, Dimroth 1971). Although correlations are uncertain, major tholeiitic suites appear to occur in the Swampy Bay and Menihek foredeeps, as do alkalic(?) suites in the basal rift sequence and in the Sokoman iron-formation. The allochthon structurally beneath the Laporte Group is composed of mafic and ultramafic rocks (Doublet Group) of unknown stratigraphic affinity. The basal rift-related volcanics have not been dated, but rhyolites tentatively included in the Swampy Bay foredeep are 2.14 Ga (T. E. Krogh & S. A. Bowring, personal communications discussed in Hoffman & Grotzinger 1987). Syenite associated with the Sokoman iron-formation and a glomerophyric gabbro sill in the Menihek foredeep are 1.88 Ga (R. Parrish, personal communication, 1987), and a similar sill in the correlative(?) Hellancourt Formation is 1.87 Ga (Machado et al 1988). These are maximum ages for thrusting and metamorphism. Pb isotope ages for galena mineralization in various stratigraphic units (Clark & Thorpe 1987) are in broad agreement with the U-Pb zircon ages. These preliminary geochronological data indicate that while initial rifting in the New Quebec orogen (pre-2.14 Ga) greatly predated rifting in the Cape Smith belt (1.96 Ga), the Wishart-Sokoman-Menihek sequence (1.88 Ga) was deposited during the tectonic evolution of the Cape Smith belt (1.96–1.84 Ga).

The Laporte Group consists of arkosic semipelitic schist that may represent a passive continental-rise prism belonging to the Superior province (Wardle & Bailey 1981) or a fore-arc basin deposited on the leading edge of the Rae province (van der Leeden et al 1988). The northeast boundary of the Laporte zone west of Ungava Bay is a northeast-dipping, post-metamorphic fault having a major component of dextral transcurrent slip (Goulet et al 1987). There, basement just east of the Laporte zone is overlain by coarse grits and metatuffs, intruded by metatonalite, possibly representing a proximal fore-arc environment (Poirier et al 1987). The tectonic significance of the domal Archean basement inliers within the Laporte zone (e.g. Wheeler dome; see Figure 7) is uncertain; they may be either tectonic windows exposing the autochthonous Superior province beneath the allochthonous Laporte zone or structural culminations of allochthonous basement on which the Laporte metasediments were deposited. Metamorphic grade in the Laporte zone increases eastward from greenschist to granulite facies, and the thermal peak was preceded by the pressure acme of 0.8 GPa, suggesting that metamorphism was a consequence of thermal relaxation following thrusting (Perreault et al 1987). Peak metamorphism postdates emplacement of the domal basement inliers, and its age is given by U-Pb monazite dates of 1.79 Ga for the basement inliers and 1.78 Ga for basement on the east margin of the belt, which have U-Pb zircon ages of 2.88 and 2.71 Ga, respectively (Machado et al 1988). As thrusting and metamorphism appear to be continuous throughout the New Quebec orogen, it is inferred that thrusting in response to the Rae/Superior collision occurred after 1.87 Ga (the age of gabbro sills emplaced prior to thrusting) and before 1.79 Ga (Machado et al 1988).

Intruding the Rae province 25–100 km east of the trough is the NNW-trending De Pas batholith (Figure 7), a calc-alkaline plutonic belt coincident with a pre- to synplutonic dextral granulite-grade shear zone involving Archean basement in the north and the Laporte Group in the south (Taylor 1979, van der Leeden et al 1988). Preliminary zircon ages for the batholith are 1.84 and 1.81 Ga (S. A. Bowring & T. E. Krogh, personal communications, 1986), and a monazite age of 1.81 Ga dates the metamorphism of an Archean migmatitic host rock of the batholith (Machado et al 1988). The batholith is interpreted as a continental magmatic arc related to dextral oblique convergence between the Superior and Rae provinces (van der Leeden et al 1988); however, its age relative to the inferred collision suggests that plutonism may be partly or entirely due to postcollisional anatectic melting. The southern part of the batholith and adjacent Archean rocks are thrust westward on structures that parallel the Torngat orogen (Figure 6). This deformation clearly postdates the thin-skinned thrusting in the New Quebec orogen (cf Wardle 1982), suggesting that the Torngat orogen is the younger of the two collision zones.

Torngat Orogen: Rae/Nain Collision

Torngat orogen is the name proposed for the zone of intense Early Proterozoic deformation and metamorphism between the Rae and Nain provinces, best exposed in the Torngat Mountains of northern Labrador (Figure 7). The orogen is thought to be a compound suture zone, bisected in the north by the Burwell terrane (Taylor 1979, Korstgård et al 1987), a possible Archean(?) microcontinent.

The Torngat orogen is a mirror image of the New Quebec orogen, thrusting in the Nain foreland being east-vergent and transcurrent shear in the Rae hinterland being sinistral (Korstgård et al 1987). In the foreland, sedimentary cover of the Ramah Group is preserved in a 110 by 10 km fold belt, which comprises a lower siliclastic shelf sequence overlain by foredeep flysch intruded by mafic sills (Knight & Morgan 1981). Metamorphism of the cover increases from greenschist to upper amphibolite facies toward the western margin of the fold belt, which is truncated by west-dipping basement-rooted thrusts (Morgan 1979, 1981, Mengel 1985).

To the west of the Ramah Group is a crustal-scale shear zone that bifurcates northward into the Komaktorvik and Abloviak shear zones, which bound the eastern and western sides of the Burwell terrane, respectively (Wardle 1983, Korstgård et al 1987). The former is an asymmetric zone, up to 20 km wide, of east-vergent thrusting and sinistral transcurrent shear, in which the latitudinal pre–Ramah Group dikes of the Nain province foreland are progressively transposed. Metamorphic grade increases westward across the zone, where granulite-facies rocks characterized by large bodies of sheared Archean anorthosite occur (Wardle 1983). Farther to the west, the Abloviak shear zone is marked by a belt, 5–45 km wide, of garnetiferous mylonite derived from pelitic diatexite. The mylonite contains a subhorizontal stretching lineation and coincides with a marked negative magnetic anomaly that contrasts with the linear magnetic highs corresponding to the adjacent granulite-facies orthogneisses. The magnetic low is continuous for 900 km along strike, from the Grenville front to Resolution Island off the southeast tip of Baffin Island (Figure 8). A sinistral kink in the Abloviak shear zone occurs near the southern termination of the Burwell terrane (Taylor 1979). The absence of a kink in the adjacent Komaktorvik shear zone suggests that the Abloviak shear zone may be the older of the two shear zones.

Near the Korok River west of the Abloviak shear zone, a fold belt involving granitoid basement and sedimentary cover of probable Archean and Early Proterozoic age, respectively, has been sinistrally sheared under metamorphic conditions decreasing westward from granulite to amphibolite facies (Wardle 1984). The cover contains quartzite, marble, and euxinic

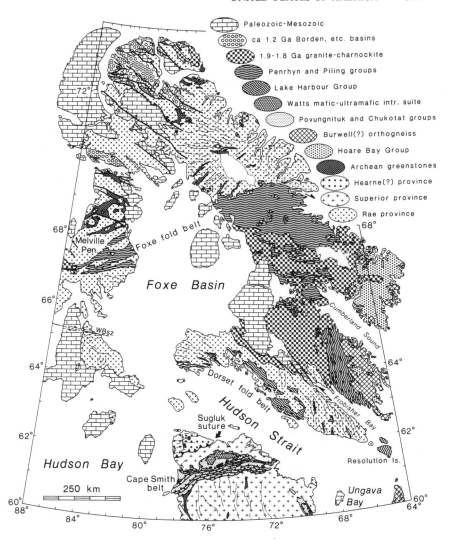

Figure 8 Geology of the Baffin Island, Foxe. Basin, and Hudson Strait area, showing possible extensions of the Archean Rae and Hearne provinces and of the Burwell terrane. Interpretations for southern Baffin Island are extremely tenuous because of the absence of U-Pb geochronology. Abbreviation: WBsz, Wager Bay shear zone.

pelite, resembling the Lake Harbour Group of the Dorset fold belt on southern Baffin Island (Jackson & Taylor 1972, Taylor 1979). The Korok River and Lake Harbour sediments are interpreted as epicontinental deposits of the Rae province, lithologically similar to the Amer Group of the northwestern Rae province and cover strata of the Foxe fold belt (Figure 8) in the northeastern Rae province. The protoliths of the Abloviak shear zone possibly include off-shelf correlatives of the Korok River (epi-Rae) and/or Ramah (epi-Nain) sedimentary cover.

The age of the Torngat orogen is very poorly constrained. It must be younger than the Ramah Group, which unconformably overlies dikes that have a minimum (Rb-Sr) age of about 2.3 Ga, and is older than the 1.65-Ga Trans-Labrador batholith (Figure 7). A single zircon age of about 1.91 Ga has been obtained from a mylonite in the Komaktorvik shear zone southwest of the Kiglapait intrusion (U. Schärer, personal communication, 1987), but the significance of this age is uncertain. If meridional basement-involved thrusts in the southern New Quebec orogen (Figure 7) are cogenetic with the Torngat orogen, as suggested earlier, then the Torngat orogen must be younger than 1.88 Ga, possibly younger than 1.81 Ga.

South Baffin Batholith and Its Bounding Fold Belts

Northern Baffin Island is underlain by a 2.9–2.7 Ga granite-greenstone-gneiss terrane that represents an extension of the Rae province of the northern District of Keewatin (Jackson & Taylor 1972). It includes scattered nonfoliated granites and charnockites of possible Early Proterozoic age (Jackson & Morgan 1978) and the Borden and related basins of Middle Proterozoic age (Jackson & Iannelli 1981). South-central Baffin Island contains a 250,000 km^2 charnockite-granite batholith bounded by three fold belts (Figure 8): the Foxe fold belt to the north, the Dorset fold belt to the southwest, and the Hoare Bay fold belt to the east (Jackson & Taylor 1972). The batholith is 1.90–1.87 Ga (Pidgeon & Howie 1975, Henderson 1985a) and is intrusive into granulite-facies equivalents of the bounding fold belts (Jackson & Morgan 1978).

The Early Proterozoic Foxe fold belt of Baffin Island extends westward onto Melville Peninsula (Figure 8) and eastward, prior to the opening of Baffin Bay, into the Rinkian belt (Escher & Pulvertaft 1976) of central west Greenland (Figure 1). The Early Proterozoic cover strata of the belt comprise a thin lower sequence of quartzite, schist, and marble of platformal facies, and an upper sequence of ferruginous pelite, locally with mafic-ultramafic flows and sills, and a great thickness of graywacke turbidites, resembling foredeep flysch (Morgan et al 1976, Henderson & Tippett 1980, Henderson 1983). The cover sequence is known as the Piling Group on Baffin Island, the Penrhyn Group on Melville Peninsula, and the Karrat Group in west Greenland, all possibly correlative.

On central Baffin Island, the upper sequence has been shortened to produce tight, closely spaced, steeply inclined, ENE-trending folds that predate the metamorphic peak (Morgan 1983). The upright folds pass downward into polyphase recumbent folds possibly associated with a ductile detachment zone separating the cover from the basement. Lobate basement-cored nappes trending SSE occur near the southern margin of the belt (Henderson et al 1979, Henderson 1985a). These early folds are refolded by two sets of upright basement-involved folds (Henderson & Tippett 1980, Henderson 1985a,b). ENE-trending lobate basement-cored antiforms and cuspate synforms (Morgan 1983) were formed during waning metamorphism, which increases in grade outward and structurally downward from greenschist facies in the central part of the belt to upper-amphibolite facies along the northern margin and eastern end of the belt and granulite facies along its southern margin (Jackson & Morgan 1978). A younger set of ESE-trending folds is developed toward the east end of the fold belt and affects the Archean rocks far to the north, producing recumbent folds that are spectacularly exposed in the fiords along the northeast coast of Baffin Island. A 1.81-Ga tonalitic pegmatite is syntectonic with the ESE-trending folds, dating the deformation that has been tentatively attributed to dextral transcurrent shear parallel to the coast (Henderson & Loveridge 1981).

On Melville Peninsula, the Foxe fold belt is characterized by belt-parallel stretching lineations and sheath folds, superimposed on a large-scale basement-cover nappe system, that converge to the southwest, where the cover pinches out (Henderson 1983, 1984). The charnockite-granite batholith of Baffin Island does not appear on Melville Peninsula, but a zone of composite calc-alkaline plutons extends for 350 km southwestward from the west end of the fold belt to the Wager Bay shear zone (Figure 4). The plutons were emplaced at 1.83–1.82 Ga, during waning metamorphism of the fold belt, and slightly predate pegmatites (1.82–1.81 Ga) that are synkinematic with respect to the dextral transcurrent Wager Bay shear zone (Henderson et al 1986, LeCheminant et al 1987c).

The Dorset fold belt appears to be coextensive with the western hinterland of the Torngat orogen (Jackson & Taylor 1972). The fold belt comprises amphibolite- to granulite-grade flyschoid metasediments; subordinate quartzite, marble, and rusty pelite; rare mafic bands; and granitoid orthogneiss. Although no internal stratigraphy, basement/cover relations, or U-Pb dating is available for this superbly exposed 700-km-long belt, the presence of platformal facies suggests that much of the orthogneiss may be basement to the metasediments. The belt is wrenched along a northwest-trending sinistral shear zone occupied by Frobisher Bay (Figure 8). East of the batholith is the Hoare Bay Group, an undated assemblage of mainly semipelitic schist and gneiss, lesser mafic metavolcanic and ultramafic rocks, and granitoid orthogneiss. The more westerly parts of this assemblage have been compared lithologically with rocks of the Foxe and Dorset fold belts, but other parts more closely resemble Archean greenstone belts north of the Foxe fold belt (Jackson & Taylor 1972). The age of the Hoare Bay Group is unknown, but the fold belt east of the batholith may represent an extension of the Burwell terrane of northern Labrador. Prior to the opening of Baffin Bay, the Hoare Bay belt was situated adjacent to the Nagssugtoqidian belt (Figure 1) of west Greenland (Myers 1984).

The tectonic significance of the 1.90-Ga charnockite-granite batholith of southern Baffin Island and its bounding Early Proterozoic fold belts is enigmatic, as are their relations to the Nagssugtoqidian belt of Greenland. A pivotal problem is whether the presumed Archean basement of the Dorset fold belt is an extension of the Rae province (Figure 1). If so, the Trans-Hudson and Snowbird orogens do not extend northeastward beyond Hudson Bay (e.g.. Lewry et al 1985) but must curve around the northeast margin of Superior province and link up with the New Quebec orogen. These important problems may remain unresolved until high-resolution magnetic anomaly data are available for Foxe Basin and Hudson Strait.

ACCRETION IN THE SOUTH

The north-central part of Laurentia was assembled through a rapid succession of microcontinental collisions, from 1.96 to about 1.80 Ga in age, involving the Archean Slave, Rae, Hearne, Wyoming, Superior, and Nain provinces (Figure 1). The remaining one third of the craton consists mostly of juvenile Early Proterozoic crust accreted prior to 1.6 Ga (Patchett & Arndt 1986). It is best to describe the Early Proterozoic accreted terranes south of the Nain, Superior, and Wyoming provinces independently, because they are exposed in widely separated areas and it is not yet certain how much accretion occurred before or after the collisions to the north.

The Early Proterozoic accreted terranes of Laurentia are correlative with belts in the Baltic shield (Figure 1), with which they may have been coextensive prior to opening of the Iapetus paleoocean (Gower & Owen 1984, Gower 1985). There are three sets of coeval accreted belts, decreasing in age from north to south. The Penokean orogen of the Great Lakes area, the Makkovik orogen of Labrador (Gower & Ryan 1986), the Ketilidian orogen of south Greenland (Allaart 1976), and the Svecofennian orogen of the Baltic shield (Gaal & Gorbatschev 1987) are all 1.9–1.8 Ga. The Yavapai orogen extending through northwestern Arizona and Colorado (Karlstrom & Bowring 1987), the Central Plains orogen (Sims & Peterman 1986), and the Killarney belt of Lake Huron (van Breemen & Davidson 1987) are 1.8–1.7 Ga. The Mazatzal orogen of southeastern Arizona and

New Mexico (Karlstrom & Bowring 1987), the Labrador orogen (Thomas et al 1986), and the Trans-Baltic (or Småland-Värmland) belt (Gaal & Gorbatschev 1987) are all 1.7–1.6 Ga. Collectively, an area of new crust up to 1200 km wide and 5000 km long was accreted in less than 300 Myr.

Isotopic data indicate that Early Proterozoic crust underlies the external parts of the Grenville orogen (Ashwal et al 1986, Schärer et al 1986, Schärer & Gower 1987, van Breemen & Davidson 1987, Rivers et al 1988) and the correlative Sveconorwegian orogen (Demaiffe & Michot 1985), but that internal parts of the orogen on both sides of the Atlantic consist mainly of rocks younger than 1.3 Ga.

Accretion to Nain Province: Makkovik and Labrador Orogens

The Nain province is bounded to the southeast by the Ketilidian orogen of south Greenland (Allaart 1976) and the coextensive Makkovik orogen (Gower & Ryan 1986), which occupies a wedge-shaped area of coastal Labrador truncated to the south by the Trans-Labrador batholith (Figure 7). The Ketilidian orogen and the correlative Svecofennian orogen of the Baltic shield have external zones developed on or adjacent to Archean crust; their internal zones are composed of juvenile 1.9–1.8 Ga crust (Patchett & Bridgwater 1984, Kalsbeek & Taylor 1985, Huhma 1986, Patchett & Kouvo 1986). The Makkovik orogen corresponds to the external zones of the Ketilidian orogen, but possible equivalents of the internal zones occur as enclaves within the younger (1.7–1.6 Ga) Labrador orogen to the south.

In the Makkovik orogen (Figure 7), Archean basement continuous with the Nain province is overlain unconformably by the Early Proterozoic Moran Lake Group (Ryan 1984) and structurally by the correlative lower Aillik Group (Marten 1977). Both groups consist of a basal quartzarenite overlain by graywacke-semipelite and mafic metavolcanic rocks, which are in fault contact to the east with the upper Aillik Group, comprising calc-alkaline, dacitic to rhyolitic volcanic and associated clastic sedimentary rocks (Gower et al 1982, Gandhi 1984, Ryan 1984). These rocks were deformed during an early phase of north(?)-directed low-angle thrusting (including basement-cover imbrication), a main phase of upright basement-involved folding of NNE trend, and a late phase of northeast-trending dextral shearing (Marten 1977, Clark 1979, Korstgård & Ermanovics 1984, Gower & Ryan 1986). Metamorphic grade increases southeastward into the orogen from lower-greenschist facies at the Nain province margin to middle-amphibolite facies. Thrusting postdated the 1.86–1.81 Ga upper Aillik volcanics; the main upright folding event occurred during or after the emplacement of 1.81–1.80 Ga granites, and waning metamorphism is dated at 1.79 (U-Pb monazite) and 1.76 (U-Pb titanite) Ga (Schärer et al 1987, Ermanovics et al 1987). In the Ketilidian orogen, synmetamorphic(?) granites are 1.81–1.80 Ga and postmetamorphic rapakivi granites are 1.76–1.74 Ga (van Breemen et al 1974, Gulson & Krogh 1975, Patchett & Bridgwater 1984).

The recently defined Labrador orogen (Thomas et al 1985, 1986, Wardle et al 1986, Schärer et al 1986) is a broad, ENE-trending zone of 1.71–1.63 Ga gneissic, plutonic, and volcanic-sedimentary rocks that make up most of the northeastern Grenville orogen and, locally, extend north of the Grenville front. These rocks were deformed and metamorphosed up to granulite facies during the Labradorian orogeny, which culminated at about 1.65 Ga, before being structurally telescoped and metamorphosed again during the Grenvillian orogeny about 1.03–0.97 Ga. Rocks of the Labrador orogen overlie, intrude, or truncate the Rae and Nain provinces and the New Quebec, Torngat, and Makkovik orogens (Figure 7).

The zonation of the Labrador orogen comprises a northern volcanic-sedimentary belt (Bruce River Group) preserved in the foreland of the Grenville orogen, a Grenvillian parautochthonous zone containing the 1.65-Ga Trans-Labrador batholith flanked to the south by amphibolite-grade Labradorian para- and orthogneisses, and a zone of Grenvillian allochthons (folded about southeast-plunging axes) composed of granulite-grade Labradorian gneisses. The 1.65-Ga Bruce River Group contains a lower assemblage of alluvial fanglomerate, sandstone, and minor bimodal lava flows; a middle sequence of volcaniclastic arenites; and an upper association of potassic calc-alkaline mafic-intermediate-felsic lava flows, pyroclastic flows, related intrusions, and derived volcaniclastic sediments (Baragar 1981, Ryan 1984, Schärer et al 1987). The 500-km-long Trans-Labrador batholith consists mainly of calc-alkaline granite and grano-diorite, along with subordinate early gabbro-diorite bodies, and is syn- to posttectonic with respect to the Labradorian orogeny (Gower & Owen 1984, Thomas et al 1986). The amphibolite- and granulite-grade Lab-radorian metasediments are generally monotonous pelitic to psammitic gneisses, containing strongly to weakly deformed granitic to gabbroic and anorthositic intrusions older than about 1.63 Ga. Isotopically, the Labradorian gneisses and associated intrusions appear not to contain any crustal components older than about 1.71 Ga (Wardle et al 1986, Schärer et al 1986). Thus, the Labradorian orogeny involved major accretion of new continental crust that was later incorporated in the eastern Grenville orogen.

Accretion to the Superior Province: Penokean Orogen

The evolution of the Early Proterozoic southern margin of the Superior province is recorded in the Penokean orogen exposed around Lake

Superior (Figure 9). This 1.90–1.83 Ga orogen forms an embayment on the Superior province margin, also occupied by the younger Keweenawan rift system. Uplift of the Penokean shield south of Lake Superior is of Keweenawan (1.14 Ga) age (Peterman et al 1985) and may be a flexural consequence of loading by rift volcanics, focused inside the U-shaped bend in the rift system (Peterman & Sims 1986). The westward extension of the orogen appears to be truncated in the subsurface by the 1.80–1.63 Ga Central Plains orogen (Sims & Peterman 1986). Its eastward extension is truncated by or incorporated within the Grenville orogen. The Otish and

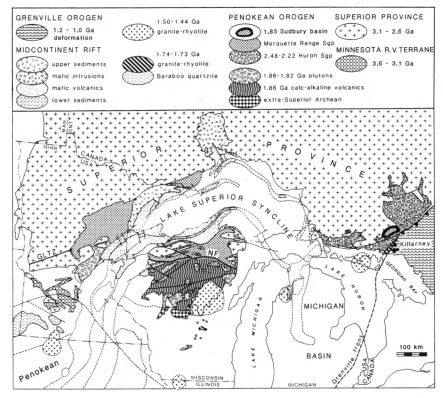

Figure 9 Precambrian tectonic elements of the Great Lakes region, modified from P. K. Sims (personal communication, 1985). The Great Lakes tectonic zone (GLTZ) is a possible suture, occluded by late Archean granitic intrusions, between the Superior province and the Minnesota River Valley terrane. The Niagara Fault zone (NF) separates the ensialic Marquette Range Supergroup from ensimatic magmatic-arc terranes accreted during the ca 1.85 Ga Penokean orogeny. Note the small area of Archean rocks south of the Penokean island-arc terrane, and the post-Penokean (ca 1.75 Ga) granite-rhyolite inliers west of Lake Michigan and in the Killarney area of Georgian Bay.

Mistassini outliers (Chown & Caty 1973, Rivers & Chown 1986) in the foreland of the Grenville orogen in Quebec may also be related to the Penokean orogen.

The Niagara fault zone divides the Penokean orogen into a northern domain of sedimentary and minor tholeiitic volcanic rocks deposited on Archean basement of the Superior province and the Minnesota River Valley terrane, and a southern domain composed mainly of Early Proterozoic volcanic and plutonic rocks resembling island-arc suites (Morey et al 1982, Greenberg & Brown 1983, Schulz 1983, Klasner et al 1985). The absence of Early Proterozoic arc-type igneous rocks north of the Niagara fault zone favors models involving south-dipping subduction terminated by collision of island arc(s) against a south-facing passive continental margin (Cambray 1978, Larue & Sloss 1980, Schulz et al 1984).

The deformed continental margin includes two different depositional prisms: the Huron ("Huronian") and Marquette Range supergroups. East of Lake Superior, the Huron Supergroup is a southward-thickening siliciclastic prism (Roscoe 1968) deposited between 2.48 and 2.22 Ga (Krogh et al 1984, Corfu & Andrews 1986). The prism has an aggregate maximum thickness of 12 km and progressively onlaps the Superior province to the north, from which most of the sediment was derived (Card 1984). Polyphase deformation of its southern part into easterly trending folds and northward-vergent thrusts (Zolnai et al 1984) has been ascribed to the Penokean orogeny (Brocoum & Dalziel 1974). Much of the deformation postdated the 1.850-Ga (Krogh et al 1984) Sudbury impact event (Rousell 1984), and most of it predated the 1.74-Ga Killarney granite (Davidson 1986). However, folding began before intrusion of the 2.22-Ga (Corfu & Andrews 1986) Nipissing diabase, and prograde metamorphism of Huronian strata occurred prior to the Sudbury event (Card 1978).

Conversely, the Marquette Range Supergroup south of Lake Superior and its equivalents to the northwest may be closely related to the Penokean orogeny in age and origin, although this remains to be demonstrated radiometrically. They contain lower assemblages (Mille Lacs and Chocolay groups) in which immature clastics and mafic volcanics are overlain by shelf-type quartzarenite and dolomite (Larue 1983). Overstepping these rocks is a more extensive upper sequence (Animikie Group), in which a basal trangressive littoral quartzarenite (Ojakangas 1983) is overlain successively by an iron-formation, mafic tholeiites, and a thick succession of turbidites (Morey 1983, Cambray 1987). Schulz (1987a) relates the Animikie Group to initial continental rifting, whereas Hoffman (1987b) suggests that the lower assemblages represent rift and passive-margin deposits, and that the Animikie Group filled a foredeep related to the Penokean orogeny. The foredeep model predicts a synorogenic age for

Animikie volcanism and possible derivation of the Animikie turbidites from the Early Proterozoic arc terrane to the south.

During the Penokean orogeny, the Marquette Range Supergroup underwent early thin-skinned, northward-directed thrusting, followed by upright basement-involved folding of easterly trend (Holst 1982, 1984, Sims et al 1987). The upright folds are typically doubly plunging (Morey et al 1982), and the gross meridional alignment of successive culminations, expressed by elliptical basement domes, is suggestive of transverse cross folding. Peak metamorphism postdates thrusting (Klasner 1978) and ranges in grade from anchizonal to the sillimanite+muscovite zone in pelites of the Marquette Range Supergroup and its correlatives (Morey 1978).

The southern limit of the Marquette Range Supergroup is the 0–12-km-wide Niagara fault zone, a broadly arcuate, convex northward system of fault slices composed of strongly flattened, steeply dipping rocks having dominantly downdip stretching lineations (Larue & Ueng 1985, Sedlock & Larue 1985). Although structurally dismembered, ophiolitic crustal components are present, and they appear to floor the calc-alkaline volcanic rocks occurring to the south (Schulz 1987b). Mafic-intermediate-felsic volcanic rocks of calc-alkaline character and related tonalitic-granitic intrusions occupy a 140-km-wide zone south of the suture (Greenberg & Brown 1983, Schulz 1983). The felsic volcanics are about 1.86 Ga, and the intrusions range in age from 1.86 to 1.82 Ga (Van Schmus 1980, Peterman et al 1985). Deformation attributed to mainly subhorizontal compression under metamorphic conditions ranging from lower-greenschist to upper-amphibolite facies occurred within the same time interval (Maass 1983), perhaps culminating prior to 1.84 Ga (Maass et al 1980). The Penokean magmatic arc terrane, 120–140 km wide, separates Archean rocks continuous with the Superior province from a zone of high-grade Archean (2.8 Ga) rocks in central Wisconsin invaded by 1.89–1.82 Ga tonalites (Van Schmus & Anderson 1977, Maass et al 1980, Anderson & Cullers 1987). The central Wisconsin Archean rocks and the arc terrane are believed to have been accreted to the craton during the Penokean orogeny (Schulz et al 1984).

Following a magmatic hiatus of 60 Myr, a suite of epizonal syenogranites and related rhyolites was emplaced within and south of the exposed Penokean orogen at about 1.76 Ga (Van Schmus 1980, Van Schmus & Bickford 1981). Metaluminous and peraluminous granites and rhyolites of this age exposed in south-central Wisconsin comprise an anorogenic igneous suite associated with dikes of tholeiitic basalt and andesite (Anderson et al 1980, Smith 1983). In the Killarney area northwest of Lake Huron, a wedge-shaped complex of 1.74–1.73 Ga epizonal granites and related rhyolites truncates Huronian rocks of the Penokean fold belt to the northwest and is truncated in turn by the Grenville orogenic front to the southeast (Davidson 1986, van Breemen & Davidson 1987). Rocks of the Killarney wedge are massive adjacent to the Penokean belt, but deformation increases toward the southeast, producing a steeply dipping, northeast-striking foliation and associated gently plunging stretching lineation. The deformation predates 1.40-Ga pegmatite dikes and may be causally related to the ~1.63-Ga resetting of Rb-Sr isochrons in the Killarney complex, also observed in the 1.76-Ga granites of southern Wisconsin. If the deformation occurred at ~1.63 Ga, it would be coeval with the important Labradorian and Mazatzal orogenies observed in the extreme east and southwest of Laurentia, respectively.

A 1.50–1.42 Ga epizonal granite-rhyolite suite is exposed in the St. Francois Mountains inlier of southeast Missouri and occurs widely in the subsurface of Illinois, Indiana, Missouri, and Kentucky (Anderson 1983, Bickford et al 1986, Van Schmus et al 1987b). This suite is coeval with isolated intrusions to the north, such as the Wolf River batholith exposed in central Wisconsin and the Manitoulin granite in the subsurface of northern Lake Huron (Van Schmus et al 1975a,b). Nd isotopic ratios (Nelson & De Paolo 1985) indicate that the subsurface granite-rhyolite suite was mostly derived from crust having a model age of crust-mantle separation ~1.9 Ga, indicating little or no involvement of Archean crust. In contrast, Penokean granites and younger granites exposed within the Penokean orogen have Nd model ages of ~2.3–2.1 Ga and were probably derived from crustal sources containing mixed Archean and Early Proterozoic components.

Accretion to the Wyoming Province: Yavapai and Mazatzal Orogens

Most of the southwestern United States is underlain by Early Proterozoic crust accreted to the south margin of the Wyoming province between about 1.8 and 1.6 Ga (Figure 10). The main exposures of Early Proterozoic crust are in the Cordilleran front ranges extending from southern Wyoming through Colorado to northern New Mexico, in the ranges of central and southern New Mexico bordering the Rio Grande rift, in the transition zone between the Colorado Plateau and the Basin and Range province of Arizona and adjacent states, and along the San Andreas fault system of southern California (Condie 1981). The correlative Central Plains orogen (Sims & Peterman 1986) of the midcontinent truncates the respective southerly and southwesterly extensions of the Trans-Hudson and Penokean orogens in the subsurface of Nebraska (Van Schmus & Bickford 1981, Arvidson et al 1984, Bickford et al 1986).

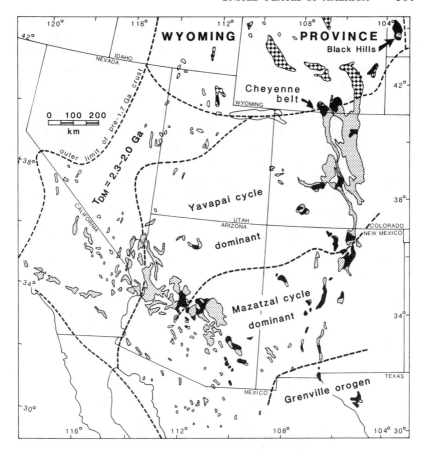

Figure 10 Distribution of Early Proterozoic volcanic and sedimentary rocks (solid) accreted to the Archean Wyoming province, modified from Condie (1986). Plutonic rocks (stippled) include both Early Proterozoic intrusions related to the accreted terranes and Middle Proterozoic postaccretionary intrusions. Rocks of the Yavapai cycle evolved between 1.79 and 1.69 Ga, and those of the Mazatzal cycle between 1.71 and 1.61 Ga. The distribution of crust having "model" (i.e. crust-mantle separation) ages of 2.3–2.0 Ga is from Bennett & DePaolo (1987).

Isotopic ratios of Pb, Nd, and Sr indicate that the crust of Arizona, New Mexico, and Colorado has a mantle-separation age of roughly 1.8 Ga, and that little or no Archean crust is present (Zartman 1974, DePaolo 1981, Condie 1982, Stacey & Hedlund 1983, Nelson & DePaolo 1984, 1985, Wooden et al 1987). In contrast, Mesozoic-Tertiary granites in western Utah, southern Nevada, and southern California have Nd model ages of 2.3–2.0 Ga (Farmer & DePaolo 1984, Bennett & DePaolo 1987), and Pb isotopic data require an Archean component for crustal genesis in this area (Wooden et al 1987). However, the older model-age province, as exposed in southern California, underwent lower-granulite-grade metamorphism at about 1.71 Ga (Wooden et al 1986), essentially coeval with the main orogenic event in Arizona.

Three main stages of Proterozoic crustal development have been recognized from geological relationships and U-Pb zircon dating throughout much of the Arizona-Sonora-New Mexico-Colorado area (Silver 1969, 1984, 1987, Silver et al 1977a). The Yavapai cycle (1.79–1.69 Ga) involved the generation of volcanic-plutonic suites and associated graywacke-pelite facies, interpreted as relics of islands arcs and related sedimentary basins (Anderson 1978, 1987, Condie 1982, 1986). These rocks were consolidated at about 1.70 Ga during an episode of deformation, metamorphism, and plutonism referred to as the Yavapai orogeny (Karlstrom & Bowring 1987). The Mazatzal cycle (1.71–1.62 Ga) is characterized both by subaerial felsic volcanics and shelf-facies arenites that overlie Yavapai-type rocks in central Arizona and northern New Mexico and by turbidites in central Colorado, southeastern Arizona, and southern New Mexico for which no basement is known. These rocks experienced northwest-directed folding and thrusting, followed by plutonism (1.65–1.62 Ga), referred to as the Mazatzal orogeny (Karlstrom & Bowring 1987). The third stage involved the widespread emplacement of "anorogenic"calc-alkaline to alkaline plutons (rapakivi granite, granodiorite, anorthosite, syenite) and related volcanics in two pulses (1.50–1.42 and 1.40–1.34 Ga; Silver et al 1977b, Anderson 1983, Silver 1984, Thomas et al 1984, Bickford et al 1986, Van Schmus et al 1987b). Northwestern Arizona is dominated by rocks of the first stage (Yavapai cycle), whereas southeastern Arizona and southern New Mexico are dominated by rocks of the second stage (Mazatzal cycle). However, the effects of the two orogenic cycles overlap in much of the terrane (e.g. Silver 1984, Karlstrom et al 1987, Karlstrom & Bowring 1987, Reed et al 1987).

Although the general character of the Arizona-New Mexico-Colorado terrane and its boundary with the Wyoming province are reasonably well understood, the specific details of the tectonic evolution remain controversial because of their intrinsic complexity and the discontinuous exposure of the terrane. One scenario involves the initial collision of an island arc with the Wyoming province as a consequence of southeast-dipping subduction, followed by arc-polarity reversal and progressive southward accretion of an Andean margin above a northwest-dipping subduction zone (Hills & Houston 1979, Karlstrom & Houston 1984, Anderson 1986, Condie et al 1987, Reed & Premo 1987). The role of cyclic back-arc basins during accretion is stressed by Condie (1982, 1986), particularly for the sediment-dominated zone in central Colorado (Condie et al 1987, Reed et al 1987). In contrast, the sediment-dominated zone of southeastern Arizona and southern New Mexico is interpreted as a prograded trench-forearc complex (Anderson 1986, Swift 1987). Unlike other modelers, who invoke progressive southward accretion, Karlstrom & Bowring (1987) view central Arizona as having been assembled by nonsystematic juxtaposition of crustal blocks having disparate histories along strike-slip and dip-slip shear zones of various ages.

WYOMING   The Cheyenne belt forms the boundary between the Wyoming province and the accreted terranes to the south, and it is exposed in the Medicine Bow Mountains and Sierra Madre of southern Wyoming (Karlstrom et al 1983, Karlstrom & Houston 1984, Duebendorfer & Houston 1987). North of the boundary, an Early Proterozoic passive-margin sedimentary prism (Snowy Pass Supergroup) is exposed in two south- to southeast-dipping thrust sheets and an underlying autochthon floored by Archean basement. The sediments were intruded at about 2.0 Ga by mafic sills and dikes that predate thrusting. The autochthonous cover sequence comprises quartzite, conglomerate, phyllite, and minor marble of the Deep Lake Group. The allochthonous Libby Creek Group consists of a lower thrust sheet of quartzite, diamictite, and phyllite, and an upper thrust sheet in which stromatolitic dolomite, nearly 2 km thick, is overlain by mafic flows(?) and laminated ferruginous slate. The slate may mark the onset of foredeep sedimentation, related to abortive subduction of the margin beneath the arc terranes to the south (Hills & Houston 1979).

The boundary zone is best exposed in the Medicine Bow Mountains, where it consists of a pair of steeply dipping northeast-trending mylonite zones, between which granitic orthogneiss and amphibolite of unknown ages separate the Snowy Pass Supergroup to the north from an Early Proterozoic metamorphic-plutonic assemblage to the south. The latter includes metamorphosed graywacke, pelite, intermediate-mafic volcanics, peridotite, and 1.78–1.76 Ga dioritic to granitic plutons. The mylonite zones record two components of simple shear: a dominant synmetamorphic southeast-side-up dip-slip component, and a subordinate dextral strike-slip component. The dominant downdip stretching lineation resulted from northwest-directed thrusting on planes that were subsequently steepened. This is consistent with the observed increase in metamorphic grade to the southeast across the boundary, including inverted metamorphic zonation in the southeast-facing upper Snowy Pass Supergroup (Duebendorfer 1987).

COLORADO   South of the Cheyenne belt, metavolcanic rocks are exposed in southern Wyoming (Condie & Shadel 1984) and southern Colorado

(Bickford & Boardman 1984, Boardman 1986, Boardman & Condie 1986), but the intervening area is characterized by turbiditic metasediments and derived paragneiss. In southern Wyoming, felsic volcanism occurred at 1.79 Ga, plutonism at 1.78–1.74 Ga, and the peak of metamorphism and deformation at about 1.77 Ga (Reed et al 1987). In south-central Colorado, calc-alkaline volcanism at 1.77–1.76 Ga and syntectonic plutonism at 1.76–1.75 Ga were followed by felsic volcanism and plutonism at 1.74–1.70 Ga and a major deformation at about 1.71 Ga (Reed et al 1987). In south-western Colorado, 1.79–1.73 Ga volcanic and plutonic rocks were deformed and metamorphosed (Yavapai orogeny) prior to intrusion of 1.69-Ga plutons, collectively forming a basement complex to quartzites and pelites (Uncompahgre Group) that were deformed by NNW-directed thin-skinned thrusting and subsequent ENE-trending thick-skinned folding during the Mazatzal orogeny (Gibson et al 1987). In the sediment-dominated zone of northern and central Colorado, syntectonic plutons range from 1.75 to 1.67 Ga and major deformation occurred at 1.68–1.67 Ga (Reed et al 1987). The current interpretation of these relations (Condie et al 1987, Reed & Premo 1987) is that the northern volcanic belt represents an island arc that collided with the Wyoming province at about 1.77 Ga, that the southern volcanic belt represents a continental-margin arc developed above a northwest-dipping subduction zone immediately following the arc-continent collision, and that the dominantly meta-sedimentary zone represents a composite back-arc basin that closed as a result of collision(s) farther south.

ARIZONA   Both the Yavapai and Mazatzal cycles are represented in central Arizona and northern New Mexico, but relations between them are uncertain. A northeast-trending boundary that angles midway between Flagstaff and Phoenix in Arizona separates a northwestern domain dominated by the older cycle from a southeastern domain dominated by the younger cycle at the present erosion level (Silver & Ludwig 1986, Karlstrom & Conway 1986, Karlstrom et al 1987, Karlstrom & Bowring 1987). The boundary fault is a relatively late high-angle structure (the southeast side having moved down), but Karlstrom et al (1987) propose that the fundamental boundary is a subhorizontal surface on which rocks of the southeastern domain were thrust over unrelated rocks to the northwest during the Mazatzal orogeny. Thin-skinned thrusting is well documented, and their evidence for tectonic juxtaposition of unrelated terranes is that intense ductile compressional deformation and contemporaneous emplacement of peraluminous granite at a depth of at least 8 km were occurring in the northwestern domain at 1.70 Ga (Yavapai orogeny) while, at the same time, alkaline rhyolite was being erupted on little-deformed Yavapai-type rocks in the southeastern domain. Thus, peraluminous magmatism accompanied by strong horizontal compression in the northwest occurred simultaneously with alkaline magmatism unaccompanied by deformation in the southeast. Karlstrom et al (1987) suggest that the one outlier of rhyolite and quartzite representing the Mazatzal cycle overlying deformed Yavapai rocks in the northwestern domain may be a klippe emplaced by thin-skinned thrusting. The interpretation of central Arizona as being composed of "accreted terranes" is extended by Karlstrom & Bowring (1987) to include three composite terranes of contrasting history, superposed along northwest-directed subhorizontal thrusts, and subsequently chopped up by late high-angle strike-slip and dip-slip faults.

NEW MEXICO Northern New Mexico is similar to the southeastern terrane of central Arizona in that mafic-felsic arc-type volcanics and immature sediments, intruded by plutons, of the Yavapai cycle were not tightly folded prior to deposition of shelf quartzite (Ortega Group) of the Mazatzal cycle (Bowring & Condie 1982, Grambling & Codding 1982, Bowring et al 1984a, Silver 1984, Soegaard & Eriksson 1986). Unlike the thin-skinned thrust-fold belt of central Arizona, the Mazatzal orogeny in northern New Mexico was characterized by progressive ductile shearing that produced large-scale polyphase recumbent folds (Grambling & Codding 1982, Holcombe & Callender 1982, Williams & Grambling 1987). Deformation accompanied prograde metamorphism that peaked after folding at regionally uniform conditions near the aluminosilicate triple point and was followed by slow isobaric cooling (Grambling 1986).

In southern New Mexico and southeasternmost Arizona, the Mazatzal cycle is represented by turbiditic metasediments (Pinal schist), possibly correlative with shelf-facies quartzite and conglomerate to the northwest; by minor mafic and felsic metavolcanic rocks; and by many 1.65-Ga and younger plutons (Silver 1978, 1987, Condie & Budding 1979, Bowring et al 1983, Copeland & Condie 1986, Conway & Silver 1987). The turbidites are interpreted as deep-sea or trench deposits incorporated into a southeast-facing accretionary prism, based on sedimentological and structural evidence (Swift 1987), in accord with the arc-trench progradation model of Anderson (1986) for the Mazatzal terrane. However, this model implies that synsedimentary shearing and thrusting should be southeast-directed, in contrast to the northwest-directed deformation generally observed in the Mazatzal orogen.

The southwestern United States and the central Trans-Hudson orogen of Manitoba and Saskatchewan are important as the only parts of Laurentia where juvenile Early Proterozoic crust is extensively exposed through a significant range of crustal depths and metamorphic grades.

They appear to have much in common with the Archean ("granite-green-stone-gneiss") provinces of Laurentia, aside from the obvious differences in age and certain petrochemical details.

DISCUSSION

Archean crustal provinces are clustered in the northern part of Laurentia; its southern part is underlain by juvenile Early Proterozoic crust. The Canadian shield is not representative of the craton as a whole, but rather is biased in favor of Archean crust. The Early Proterozoic crust is preferentially hidden beneath Phanerozoic sedimentary cover. The narrowest part of the ring-shield enclosing the Hudson Bay basin is underlain by Early Proterozoic crust of the Trans-Hudson orogen. As the shield is believed to have been an area of positive relief through Phanerozoic time, the question arises as to why the Archean lithosphere is more buoyant than Early Proterozoic lithosphere. As depleted mantle has a lower density than fertile mantle due to its lower garnet content (Oxburgh & Parmentier 1978), the relative buoyancy of the Archean provinces may reflect a relatively depleted underlying mantle lithosphere, as inferred from isotopic ratios in carbonatites intruding the Superior province (Bell & Blenkinsop 1987). The presumed secular decline in mean temperature of the asthenosphere implies a greater depth and volume of melting accompanying mantle upwelling in the Archean, and consequently a thicker, more depleted mantle lithosphere (Sleep & Windley 1982, Bickle 1986). Various mechanisms have been discussed whereby such a depleted mantle "tectosphere" might develop beneath the Archean cratons (e.g. Oxburgh & Parmentier 1978, McKenzie 1984, Pollack 1986, MacGregor & Manton 1986, Kramers 1987).

Laurentia contains at least six Archean provinces sutured along Early Proterozoic orogenic belts. The pattern of intersecting orogenic belts bounding provinces that behaved as relatively stable platforms during the Early Proterozoic is comparable to composite Phanerozoic continents such as Eurasia, formed by amalgamation of microcontinents (many of common Gondwanaland ancestry) and intervening island arcs. Interpretation of the Early Proterozoic orogenic belts as collision zones resulting from subduction of oceanic lithosphere is consistent with their sedimentary, structural, metamorphic, and magmatic asymmetry. This is well displayed in the Thelon, Wopmay, Trans-Hudson, New Quebec, Torngat, Penokean, and Cheyenne belts, all of which feature foredeep basins, thrust-fold belts, sparse mafic magmatism, and relatively low metamorphic grades in their forelands, contrasted with relatively high metamorphic grades, major calc-alkaline magmatic belts, and complex patterns of ductile deformation in their hinterlands. Their forelands were dominated by thrusting; their hinterlands by transcurrent shearing accompanying arc magmatism, a consequence of oblique subduction (Fitch 1972), and manifestations of "propagating extrusion tectonics" (Tapponnier et al 1982) in response to continent-continent indentation.

Phanerozoic orogens represent a spectrum of cases between "accretionary" and "collisional" end members, and the same is true for the Early Proterozoic orogens of Laurentia. The orogens of southern Laurentia, the Svecofennian orogen, and possibly the Wopmay orogen involved the accretion of broad zones of juvenile Early Proterozoic crust (Patchett & Arndt 1986). In the Trans-Hudson orogen, a 350-km-wide central zone of juvenile Early Proterozoic crust is trapped between opposing Archean provinces. The Thelon, Snowbird, New Quebec, and Torngat orogens are characterized by the "tight" suturing of adjacent Archean provinces, without the preservation of intervening intraoceanic relics. Differences between Early Proterozoic and Phanerozoic orogens probably exist, but generalizations are premature given the great variability in history, style, and erosion level among orogens of either age.

The amalgamation of the Archean provinces of Laurentia took place in only about 150 Myr. The times of collision, based on ages of foredeep sedimentation and/or cessation of arc-type magmatism, are estimated to be about 1.96 Ga for the Thelon orogen, between 1.92 and 1.85 Ga for the Snowbird zone, between about 1.85 and 1.83 Ga for the Trans-Hudson orogen, and between 1.87 and 1.81 Ga for the New Quebec orogen. Accretion in the Wopmay orogen began about 1.91 Ga, in the Penokean orogen about 1.85 Ga, in the Makkovik orogen about 1.81 Ga, and in the Cheyenne belt about 1.75 Ga. Thus, the inter-Archean collision events occurred between about 1.96 and 1.81 Ga, and accretion of intraoceanic terranes is documented between about 1.91 and 1.63 Ga. The limited time span required to assemble most of Laurentia is the single most startling conclusion to emerge from this synthesis.

Many important problems remain. Did the Archean provinces originate independently, or were they products of the breakup of one or more large continents? Why is crust formed between 2.5 and 2.0 Ga apparently of limited distribution (e.g. northwest Africa, northeast South America, North American Cordillera, and possibly western Wopmay orogen)? What is the significance of the apparent progressive southward growth of Laurentia and its Baltic extension? How large did Laurentia ultimately become, and are its extensions, removed by post-1.6-Ga rifting, to be found on other continents? What is the origin and significance of the extensive Middle Proterozoic "anorogenic" magmatism? Was the discordant trend of the 4000-km-long Grenville orogen dictated by Middle Proterozoic rifting? Where is the Grenville hinterland? Is North America a fragment of Middle and/or Late Proterozoic supercontinent(s) rifted at about 0.6 Ga? Such questions demand a global synthesis of the extant Precambrian lithosphere.

ACKNOWLEDGMENTS

This synthesis is an outgrowth of the Decade of North American Geology program. It is dedicated to John C. McGlynn, who introduced me to Proterozoic rocks and who is responsible for the recent rejuvenation of Precambrian geology at the Geological Survey of Canada. Many workers unselfishly offered their latest observations, ideas, and preprints in order to make the presentation as up-to-date as possible. Nick Arndt, Alan Bailes, Sam Bowring, Ken Card, Jeff Chiarenzelli, Ernie Duebendorfer, Ingo Ermanovics, Tony Frith, Dick Gibb, Normand Goulet, Charlie Gower, Alan Green, John Grotzinger, Robert Hildebrand, Steve Jackson, Brad Johnson, Feiko Kalsbeek, Tony LeCheminant, Bill Muehlburger, Nuno Machado, John Percival, Zell Peterman, Chris Pulvertaft, Jack Reed, Klaus Schulz, Paul Sims, the Cape Smith Bros., Mel Stauffer, Eric Syme, Mike Thomas, Otto van Breemen, John van der Leeden, Randy Van Schmus, Cees van Staal, Dick Wardle, and Hank Williams commented in writing on the first draft of the manuscript. However, the author is solely responsible for errors of fact and judgment, and welcomes corrections. Tim West is thanked for drafting some of the figures during the dog days of July. This article is Geological Survey of Canada Contribution No. 32187.

Literature Cited

Allaart, J. H. 1976. Ketilidian mobile belt in south Greenland. In *Geology of Greenland*, ed. A. Escher, W. S. Watt, pp. 120–51. Copenhagen: Geol. Surv. Greenland. 603 pp.

Anderson, J. L. 1983. Proterozoic anorogenic granite plutonism of North America. See Medaris et al 1983, pp. 133–54

Anderson, J. L., Cullers, R. L. 1987. Crust-enriched, mantle-derived tonalites in the Early Proterozoic Penokean orogen of Wisconsin. *J. Geol.* 95: 139–54

Anderson, J. L., Cullers, R. L., Van Schmus, W. R. 1980. Anorogenic metaluminous and peraluminous plutonism in the mid-Proterozoic of Wisconsin, USA. *Contrib. Mineral. Petrol.* 74: 311–28

Anderson, P. 1978. The island arc nature of Precambrian volcanic belts in Arizona. *Geol. Soc. Am. Abstr. With Programs* 10: 156 (Abstr.)

Anderson, P. 1986. Summary of the Proterozoic plate tectonic evolution of Arizona from 1900 to 1600 Ma. *Ariz. Geol. Soc. Dig.* 16: 5–11

Anderson, P. 1987. Proterozoic plate tectonic assembly of the United States. *Geol. Soc. Am. Abstr. With Programs* 19: 572 (Abstr.)

Arvidson, R. E., Bindschadler, D., Bowring, S., Eddy, M., Guinness, E., Leff, C. 1984. Bouguer images of the North American craton and its structural evolution. *Nature* 311: 241–43

Ashton, K. E., Wheatley, K. J. 1986. Preliminary report on the Kisseynew gneisses in the Kisseynew-Wildnest lakes area, Saskatchewan. In *Current Research, Part B. Geol. Surv. Can. Pap. 86-1B*, pp. 305–17

Ashwal, L. D., Wooden, J. L., Emslie, R. F. 1986. Sr, Nd, and Pb isotopes in Proterozoic intrusives astride the Grenville Front in Labrador: implications for crustal contamination and basement mapping. *Geochim. Cosmochim. Acta* 50: 2571–85

Aspler, L. B., Donaldson, J. A. 1985. The Nonacho basin (Early Proterozoic), Northwest Territories, Canada: sedimentation and deformation in a strike-slip setting. In *Strike-Slip Deformation, Basin Formation, and Sedimentation. Soc. Econ. Paleontol. Mineral. Spec. Publ. 37*, ed. K. T. Biddle, N. Christie-Blick, pp. 193–209. Tulsa, Okla: Soc. Econ. Paleontol. Mineral. 386 pp.

Ayres, L. D., Thurston, P. C., Card, K. D., Weber, W., eds. 1985. *Evolution of Archean Supracrustal Sequences. Geol. Assoc. Can. Spec. Pap. 28*. St. John's: Geol. Assoc. Can. 380 pp.

Badham, J. P. N. 1981. Petrochemistry of late Aphebian (ca. 1.8 Ga) calc-alkaline diorites from the East Arm of Great Slave Lake, N.W.T., Canada. *Can. J. Earth Sci.* 18: 1018–28

Bailes, A. H. 1971. Preliminary compilation of the geology of the Snow Lake–Flin Flon–Sherridon area. *Manitoba Mines Branch Geol. Pap. 1/71*. 27 pp.

Bailes, A. H. 1980a. Origin of Early Proterozoic volcaniclastic turbidites, south margin of the Kisseynew sedimentary gneiss belt, File Lake, Manitoba. *Precambrian Res.* 12: 197–225

Bailes, A. H. 1980b. Geology of the File Lake area. *Manitoba Energy Mines Miner. Res. Div. Geol. Rep. 78-1*. 134 pp.

Bailes, A. H., McRitchie, W. D. 1978. The transition from low to high grade metamorphism in the Kisseynew sedimentary gneiss belt, Manitoba. See Fraser & Heywood 1978, pp. 155–78

Bailes, A. H., Syme, E. C. 1987. Geology of the Flin Flon–White Lake area. *Manitoba Energy Mines Miner. Res. Div. Map GR87-1-1*, scale 1 : 20,000.

Bailes, A. H., Syme, E. C., Galley, A., Price, D. P., Skirrow, R., Ziehlke, D. J. 1987. *Early Proterozoic Volcanism, Hydrothermal Activity, and Associated Ore Deposits at Flin Flon and Snow Lake, Manitoba. Geol. Assoc. Can. Meet. Ann. Field Trip 1 Guideb., Saskatoon.* 95 pp.

Baldwin, D. A. 1980. Porphyritic intrusions and related minerlization in the Flin Flon volcanic belt. *Manitoba Energy Mines Miner. Res. Div. Econ. Geol. Rep. ER79-4*. 23 pp.

Baldwin, D. A., Syme, E. C., Zwanzig, H. V., Gordon, T. M., Hunt, P. A., Stevens, R. D. 1987. U-Pb zircon ages from the Lynn Lake and Rusty Lake metavolcanic belts, Manitoba: two ages of Proterozoic magmatism. *Can. J. Earth Sci.* 24: 1053–63

Baragar, W. R. A. 1960. Petrology of basaltic rocks in part of the Labrador Trough. *Geol. Soc. Am. Bull.* 71: 1589–1644

Baragar, W. R. A. 1977. Volcanism of the stable crust. In *Volcanic Regimes in Canada., Geol. Assoc. Can. Spec. Pap. 16*, ed. W. R. A. Baragar, L. C. Coleman, J. M. Hall, pp. 377–406, St. John's: Geol. Assoc. Can. 476 pp.

Baragar, W. R. A. 1981. Tectonic and regional relationships of the Seal Lake and Bruce River magmatic provinces. *Geol. Surv. Can. Bull. 314*. 72 pp.

Baragar, W. R. A. 1984. Pillow formation and layered flows in the Circum-Superior Belt of eastern Hudson Bay. *Can. J. Earth Sci.* 21: 781–92

Baragar, W. R. A., Scoates, R. F. J. 1981. The Circum-Superior belt: a Proterozoic plate margin? See Kröner 1981, pp. 297–330

Barham, B. A., Froese, E. 1986. Geology of the New Fox alteration zone, Laurie Lake, Manitoba. In *Current Research, Part B. Geol. Surv. Can. Pap. 86-1B*, pp. 827–35

Bell, K., Blenkinsop, J. 1987. Archean depleted mantle: evidence from Nd and Sr initial isotopic ratios of carbonatites. *Geochim. Cosmochim. Acta* 51: 291–98

Bell, K., Blenkinsop, J., Kwon, S. T., Tilton, G. R., Sage, R. P. 1987. Age and radiogenic systematics of the Borden carbonatite complex, Ontario, Canada. *Can. J. Earth Sci.* 24: 24–30

Bell, R. T. 1970. The Hurwitz Group: a prototype for deposition on metastable cratons. In *Basins and Geosynclines of the Canadian Shield, Geol. Surv. Can. Pap. 70-40*, ed. A. J. Baer, pp. 159–69. 265 pp.

Bennett, V., DePaolo, D. J. 1987. Proterozoic crustal history of the western United States as determined by neodymium isotopic mapping. *Geol. Soc. Am. Bull.* 99: 674–85

Berthé, D., Choukroune, P., Jegouzo, P. 1979. Orthogneiss, mylonite and noncoaxial deformation of granites: the example of the South American Shear Zone. *J. Struct. Geol.* 1: 31–42

Bickford, M. E., Boardman, S. J. 1984. A Proterozoic volcano-plutonic terrane, Gunnison and Salida areas, Colorado. *J. Geol.* 92: 657–66

Bickford, M. E., Van Schmus, W. R., Zietz, I. 1986. Proterozoic history of the midcontinent region of North America. *Geology* 14: 492–96

Bickle, M. J. 1986. Implications of melting

for stabilisation of the lithosphere and heat loss in the Archaean. *Earth Planet. Sci. Lett.* 80: 314–24

Blake, D. H. 1980. Volcanic rocks of the Paleohelikian Dubawnt Group in the Baker Lake-Angikuni Lake area, District of Keewatin, N.W.T. *Geol. Surv. Can. Bull. 309.* 39 pp.

Boardman, S. J. 1986. Early Proterozoic bimodal volcanic rocks in Central Colorado, U.S.A., Part 1: petrography, stratigraphy and depositional history. *Precambrian Res.* 34: 1–36

Boardman, S. J., Condie, K. C. 1986. Early Proterozoic bimodal volcanic rocks in Central Colorado, U.S.A., Part II: geochemistry, petrogenesis and tectonic setting. *Precambrian Res.* 34: 37–68

Bond, G. C., Kominz, M. A. 1984. Construction of tectonic subsidence curves for the early Paleozoic miogeocline, southern Canadian Rocky Mountains: implications for subsidence mechanisms, age of breakup, and crustal thinning. *Geol. Soc. Am. Bull.* 95: 155–73

Boone, E., Hynes, A. 1987. A structural cross-section of the northern Labrador Trough, New Quebec. *Geol. Assoc. Can. Programs Abstr.* 12: 26 (Abstr.)

Bostock, H. H. 1971. Geological notes on Aquatuk River map-area, Ontario. *Geol. Surv. Can. Pap. 70-42.* 57 pp. with map, scale 1 : 500,000

Bostock, H. H., van Breemen, O., Loveridge, W. D. 1987. Proterozoic geochronology in the Taltson Magmatic Zone, N.W.T. In *Radiogenic Age and Isotopic Studies. Geol. Surv. Can. Pap. 87-2.* In press

Bowring, S. A. 1984. *U-Pb zircon geochronology of Early Proterozoic Wopmay Orogen, N.W.T., Canada: an example of rapid crustal evolution.* PhD. thesis. Univ. Kans., Lawrence. 148 pp.

Bowring, S. A., Condie, K. D. 1982. U-Pb zircon ages from northern and central New Mexico. *Geol. Soc. Am. Abstr. With Programs* 14: 304 (Abstr.)

Bowring, S. A., Podosek, F. A. 1987. Nd isotopic study of two batholithic belts, Wopmay orogen, N.W.T., Canada. *Geol. Soc. Am. Abstr. With Programs* 19: 597 (Abstr.)

Bowring, S. A., Kent, S. C., Sumner, W. 1983. Geology and U-Pb geochronology of Proterozoic rocks in the vicinity of Socorro, New Mexico. *N. Mex. Geol. Soc. Guideb. 34th Field Conf.*, pp. 137–42

Bowring, S. A., Reed, J. C., Condie, K. C. 1984a. U-Pb geochronology of Proterozoic volcanic and plutonic rocks, Sangre de Cristo Mtns., New Mexico. *Geol. Soc. Am. Abstr. With Programs* 16: 216 (Abstr.)

Bowring, S. A., Van Schmus, W. R., Hoffman, P. F. 1984b. U-Pb zircon ages from Athapuscow aulacogen, East Arm of Great Slave Slave, N.W.T., Canada. *Can. J. Earth Sci.* 21: 1315–24

Brocoum, S. J., Dalziel, I. W. D. 1974. The Sudbury Basin, the Southern Province, the Grenville Front, and the Penokean Orogeny. *Geol. Soc. Am. Bull.* 85: 1571–80

Burke, K., Dewey, J. F., Kidd, W. S. F. 1976. Precambrian paleomagnetic results compatible with the Wilson Cycle. *Tectonophysics* 33: 287–99

Cambray, F. W. 1978. Plate tectonics as a model for the environment of deposition and deformation of the early Proterozoic of northern Michigan. *Geol. Soc. Am. Abstr. With Programs* 10: 376 (Abstr.)

Cambray, F. W. 1987. An alternate stratigraphic correlation within the Marquette Range Supergroup, N. Michigan. *Geol. Soc. Am. Abstr. With Programs* 19: 192 (Abstr.)

Campbell, F. H. A., ed. 1981. *Proterozoic Basins of Canada. Geol. Surv. Can. Pap. 81-10.* 444 pp.

Campbell, F. H. A., Cecile, M. P. 1981. Evolution of the Early Proterozoic Kilohigok Basin, Bathurst Inlet-Victoria Island, Northwest Territories. See Campbell 1981, pp. 103–31

Card, K. D. 1978. Metamorphism of the Middle Precambrian (Aphebian) rocks of the eastern Southern Province. See Fraser & Heywood 1978, pp. 269–82

Card, K. D. 1984. Proterozoic geology of the north shore of Lake Huron, Ontario. *GAC-MAC '84 Field Trip Guideb. 15.* London, Ontario: Dept. Geol., Univ. West. Ontario. 49 pp.

Chandler, F. W. 1984. Metallogenesis of an early Proterozoic foreland sequence, eastern Hudson Bay, Canada. *J. Geol. Soc. London* 141: 299–313

Chandler, F. W. 1987. *Geology of the Early Proterozoic Richmond Gulf graben, east coast of Hudson Bay, Canada. Geol. Surv. Can. Bull. 362.* In press

Chauvel, C., Arndt, N. T., Kielinzcuk, S., Thom, A. 1987. Formation of Canadian 1.9 Ga old continental crust. I: Nd isotopic data. *Can. J. Earth Sci.* 24: 396–406

Chown, E. H., Caty, J. L. 1973. Stratigraphy, petrography and paleocurrent analysis of the Aphebian clastic formations of the Mistassini-Otish basins. In *Huronian Stratigraphy and Sedimentation. Geol. Assoc. Can. Spec. Pap. 12*, ed. G. M. Young, pp. 51–71. St. John's: Geol. Assoc. Can. 271 pp.

Clark, A. M. S. 1979. Proterozoic deformation and igneous intrusions in part of the Makkovik Subprovince, Labrador. *Precambrian Res.* 10: 95–114

Clark, T. 1984. Géologie de la région du Lac Cambrien, Territoire du Nouveau Québec. *Québec Dir. Rech. Géol. ET83-02.* 71 pp. with map, scale 1 : 50,000

Clark, T., Thorpe, R., 1987. Lead isotope galena ages from the Labrador Trough. *Geol. Assoc. Can. Programs Abstr.* 12: 33 (Abstr.)

Coles, R. L., Haines, G. V. 1982. Regional patterns in magnetic anomalies over Hudson Bay. *Can. J. Earth Sci.* 19: 1116–21

Committee for the Magnetic Anomaly Map of North America 1987. *Magnetic Anomaly Map of North America*, 4 sheets, scale 1 : 5,000,000. Boulder, Colo: Geol. Soc. Am.

Condie, K. C. 1981. Precambrian rocks of southwestern United States and adjacent areas of Mexico. *N. Mex. Bur. Mines Miner. Res. Resour. Map 13*, scale 1 : 1,500,000

Condie, K. C. 1982. Plate-tectonics model for Proterozoic continental accretion in the southwestern United States. *Geology* 10: 37–42

Condie, K. C. 1986. Geochemistry and tectonic setting of Early Proterozoic supracrustal rocks in the southwestern United States. *J. Geol.* 94: 845–64

Condie, K. C., Budding, A. J. 1979. Geology and geochemistry of Precambrian rocks, central and south-central New Mexico. *N. Mex. Bur. Mines Miner. Res. Mem. 35.* 58 pp.

Condie, K. C., Shadel, G. A. 1984. An early Proterozoic volcanic arc succession in southeastern Wyoming. *Can. J. Earth Sci.* 21: 415–27

Condie, K. C., Bickford, M. E., Van Schmus, W. R. 1987. Accretion and tectonic evolution of North America 1600–1800 Ma ago. *Geol. Assoc. Can. Programs Abstr.* 12: 33 (Abstr.)

Conway, C. M., Silver, L. T. 1987. Early Proterozoic rocks (1710–1615 Ma) in central to southeastern Arizona. In *Geology of Arizona*, ed. S. Reynolds, J. Jenney, Tucson: Ariz. Geol. Soc. In press

Cook, F. A. 1985. Geometry of the Kapuskasing structure from a Lithoprobe pilot reflection survey. *Geology* 13: 368–71

Copeland, P., Condie, K. C. 1986. Geochemistry and tectonic setting of lower Proterozoic supracrustal rocks of the Pinal Schist, southeastern Arizona. *Geol. Soc. Am. Bull.* 97: 1512–20

Corfu, R., Andrews, A. J. 1986. A U-Pb age for mineralized Nipissing diabase, Gowganda, Ontario. *Can. J. Earth Sci.* 23: 107–9

Darnley, A. G. 1981. The relationship between uranium distribution and some major crustal features in Canada. *Mineral. Mag.* 44: 425–36

Davidson, A. 1970a. Precambrian geology, Kaminak Lake map-area, District of Keewatin. *Geol. Surv. Can. Pap. 69-51.* 27 pp. with Map 1285A, scale 1 : 250,000

Davidson, A. 1970b. Eskimo Point and Dawson Inlet map-areas (north halves), District of Keewatin. *Geol. Surv. Can. Pap. 70-27.* 21 pp.

Davidson, A. 1972. The Churchill Province. In *Variations in Tectonic Styles in Canada. Geol. Assoc. Can. Spec. Pap. 11,* ed. R. A. Price, R. J. W. Douglas, pp. 381–433. St. John's: Geol. Assoc. Can. 688 pp.

Davidson, A. 1986. Grenville Front relationships near Killarney, Ontario. See Moore et al 1986, pp. 107–17

Demaiffe, D., Michot, J. 1985. Isotope geochronology of the Proterozoic crustal segment of southern Norway: a review. See Tobi & Touret 1985, pp. 411–33

DePaolo, D. J. 1981. Neodymium isotopes in the Colorado Front Range and crustmantle evolution in the Proterozoic. *Nature* 271: 193–96

Dewey, J. F., Burke, K. C. A. 1973. Tibetan, Variscan, and Precambrian basement reactivation: products of continental collision. *J. Geol.* 81: 683–92

Dimroth, E. 1971. The evolution of the central segment of the Labrador geosyncline, Part II: the ophiolite suite. *Neus. Jahrb. Geol. Paläont. Abh.* 137: 209–48

Dimroth, E. 1981. Labrador geosyncline: type example of early Proterozoic cratonic reactivation. See Kröner 1981, pp. 331–52

Dods, S. D., Teskey, D. J., Hood, P. J. 1985. The new series of 1 : 1,00,000-scale magnetic anomaly maps of the Geological Survey of Canada: compilation techniques and interpretation. See Hinze 1985, pp. 69–87

Doig, R. 1987. Rb-Sr geochronology and metamorphic history of Proterozoic to early Archean rocks north of the Cape Smith Fold Belt, Quebec. *Can. J. Earth Sci.* 24: 813–25

Duebendorfer, E. M. 1987. Evidence for an inverted metamorphic gradient associated with a Precambrian suture, southern Wyoming. *J. Metamorph. Geol.* In press

Duebendorfer, E. M., Houston, R. S. 1987. Proterozoic accretionary tectonics at the southern margin of the Archean Wyoming craton. *Geol. Soc. Am. Bull.* 98: 554–68

Eade, K. E. 1974. Geology of Kognak River area, District of Keewatin. *Geol. Surv. Can. Mem. 377.* 66 pp. with Map 1364A, scale 1 : 250,000

Eade, K. E. 1986. Precambrian geology of the Tulemalu Lake-Yathkyed Lake area,

592 HOFFMAN

District of Keewatin. *Geol. Surv. Can. Pap. 84-11.* 31 pp. with Map 1604A, scale 1:250,000

Eade, K. E. 1987. Dubawnt River and Maguse River map areas, Northwest Territories. *Geol. Surv. Can. Maps,* scale 1:1,000,000. In press

Eade, K. E., Chandler, F. W. 1975. Geology of Watterson Lake (west half) map-area, District of Keewatin. *Geol. Surv. Can. Pap. 74-64.* 10 pp.

Easton, R. M. 1981. Stratigraphy of the Akaitcho Group and development of an early Proterozoic continental margin, Wopmay orogen, Northwest Territories. See Campbell 1981, pp. 79–95

Embleton, B. J. J., Schmidt, P. W. 1979. Recognition of common Precambrian polar wandering. *Nature* 282: 705–7

Emslie, R. F. 1978. Anorthosite massifs, rapakivi granites, and late Proterozoic rifting of North America. *Precambrian Res.* 7: 61–98

Ermanovics, I. F., Loveridge, W. D., Sullivan, R. W. 1987. U-Pb zircon studies of the Maggo gneiss, Kanairiktok plutonic suite and Lake Harbour plutonic suite, Coast of Labrador. In *Radiogenic Age and Isotope Studies. Geol. Surv. Can. Pap. 87-2.* In press

Ernst, R. E., Halls, H. C. 1984. Paleomagnetism of the Hearst dike swarm and implications for the tectonic history of the Kapuskasing structural zone, northern Ontario. *Can. J. Earth Sci.* 21: 1499–1506

Escher, A., Pulvertaft, T. C. R. 1976. Rinkian mobile belt of west Greenland. In *Geology of Greenland,* ed. A. Escher, W. S. Watt, pp. 104–19. Copenhagen: Geol. Surv. Greenland. 603 pp.

Farmer, G. L., DePaolo, D. J. 1984. Origin of Mesozoic and Tertiary granite in the western United States and implications for pre-Mesozoic crustal structure 2. Nd and Sr isotopic studies of unmineralized and Cu- and Mo-mineralized granite in the Precambrian craton. *J. Geophys. Res.* 89: 10,141–60

Fitch, T. J. 1972. Plate convergence, transcurrent faults, and internal deformation adjacent to southeast Asia and the western Pacific. *J. Geophys. Res.* 77: 4432–60

Francis, D., Ludden, J., Hynes, A. 1983. Magma evolution in a Proterozoic rifting environment. *J. Petrol.* 24: 556–82

Fraser, J. A., Heywood, W. W., ed. 1978. *Metamorphism in the Canadian Shield. Geol. Surv. Pap. 78-10.* 367 pp.

Fraser, J. A., Hoffman, P. F., Irvine, T. N., Mursky, G. 1972. The Bear Province. In *Variations in Tectonic Styles in Canada, Geol. Assoc. Can. Spec. Pap. 11,* ed. R. A. Price, R. J. W. Douglas, pp. 453–503. St.

John's: Geol. Assoc. Can. 688 pp.

Frisch, T., Patterson, J. G. 1983. Preliminary account of the geology of the Montresor River area, District of Keewatin. In *Current Research, Part A. Geol. Surv. Can. Pap. 83-1A,* pp. 103–8

Frisch, T., Annesley, I. R., Gittins, C. A. 1985. Geology of the Chantrey belt and its environs, Lower Hayes River and Darby Lake map areas, District of Keewatin. In *Current Research, Part A. Geol. Surv. Can. Pap. 85-1A,* pp. 259–66

Froese, E., Moore, J. M. 1980. Metamorphism in the Snow Lake area, Manitoba. *Geol. Surv. Can. Pap. 78-27.* 16 pp.

Frost, C. D., Burwash, R. A. 1986. Nd evidence for extensive Archean basement in the western Churchill Province, Canada. *Can. J. Earth Sci.* 23: 1433–37

Fueten, F., Robin, P.-Y., Pickering, M. E. 1986. Deformation in the Thompson Belt, central Manitoba: A progress report. In *Current Research, Part B. Geol. Surv. Can. Pap. 86-1B,* pp. 797–809

Fumerton, S. L., Stauffer, M. R., Lewry, J. F. 1984. The Wathaman batholith: largest known Precambrian pluton. *Can. J. Earth Sci.* 21: 1082–97

Gaal, G., Gorbatschev, R. 1987. An outline of the Precambrian evolution of the Baltic Shield. *Precambrian Res.* 35: 15–52

Gandhi, S. S. 1984. Uranium in Early Proterozoic Aillik Group, Labrador. In *Proterozoic Unconformity and Stratabound Uranium Deposits,* ed. J. Ferguson, pp. 35–67. Vienna: Int. At. Energy Agency

Gibb, R. A. 1978a. Slave-Churchill collision tectonics. *Nature* 271: 50–52

Gibb, R. A. 1978b. A gravity survey of James Bay and its bearing on the Kapuskasing Gneiss Belt, Ontario. *Tectonophysics* 45: T7–13

Gibb, R. A. 1983. Model for suturing of Superior and Churchill plates: an example of double indentation tectonics. *Geology* 11: 413–17

Gibb, R. A., Halliday, D. W. 1974. Gravity measurements in southern District of Keewatin and southeastern District of Mackenzie, with maps. *Earth Phys. Branch Gravity Map Ser. No. 124-131,* Ottawa. 36 pp.

Gibb, R. A., Thomas, M. D. 1977. The Thelon Front: a cryptic suture in the Canadian Shield? *Tectonophysics* 38: 211–22

Gibb, R. A., Walcott, R. I. 1971. A Precambrian suture in the Canadian Shield. *Earth Planet. Sci. Lett.* 10: 417–22

Gibb, R. A., Thomas, M. D., Lapointe, P. L., Mukhopadhyay, M. 1983. Geophysics of proposed Proterozoic sutures in Canada. *Precambrian Res.* 19: 349–84

Gibson, R. G., Harris, C. W., Eriksson, K. A., Simpson, C. 1987. Regional implications of Proterozoic deformation and lithostratigraphy in the Needle Mtns., Colorado. *Geol. Soc. Am. Abstr. With Programs* 19: 675 (Abstr.)

Gilbert, H. P., Syme, E. C., Zwanzig, H. V. 1980. Geology of the metavolcanic and volcaniclastic metasedimentary rocks in the Lynn Lake area. *Manitoba Mines Energy Miner Res. Div. Geol. Pap. GP-80-1.* 118 pp.

Gilboy, C. F. 1980. Bedrock compilation geology: Stony Rapids area. *Saskatchewan Geol. Surv. Prelim. Geol. Map,* scale 1: 250,000

Gordon, T. M. 1987. Precambrian geology of the Daly Bay area, District of Keewatin. *Geol. Surv. Can. Mem. 422.* 21 pp. In press

Gordon, T. M., Hunt, P. A., Loveridge, W. D., Bailes, A. H., Syme, E. C. 1987a. U-Pb zircon ages from the Flin Flon and Kisseynew belts, Manitoba: chronology of Early Proterozoic crust formation. *Geol. Assoc. Can. Programs Abstr.* 12: 48 (Abstr.)

Gordon, T. M., Lemkow, D. R., Roddick, J. C. 1987b. Metamorphism of the Kisseynew sedimentary gneiss belt, Manitoba: thermal evolution of an Early Proterozoic interarc basin. *Geol. Assoc. Can. Programs Abstr.* 12: 48 (Abstr.)

Goulet, N., Gariépy, C., Machado, N. 1987. Structure, geochronology, gravity and the tectonic evolution of the northern Labrador Trough. *Geol. Assoc. Can. Programs Abstr.* 12: 48 (Abstr.)

Gower, C. F. 1985. Correlations between the Grenville Province and the Sveconorwegian orogenic belt—implications for Proterozoic evolution of the southern margins of the Canadian and Baltic Shields. See Tobi & Touret 1985, pp. 247–57

Gower, C. F., Owen, V. 1984. Pre-Grenvillian and Grenvillian lithotectonic regions in eastern Labrador—correlations with the Sveconorwegian Orogenic Belt in Sweden. *Can. J. Earth Sci.* 21: 678–93

Gower, C. F., Ryan, A. B. 1986. Proterozoic evolution of the Grenville Province and adjacent Makkovik Province in eastern-central Labrador. See Moore et al 1986, pp. 281–96

Gower, C. F., Flanagan, M. J., Kerr, A., Bailey, D. G. 1982. Geology of the Kaipokok Bay–Big River area, Central Mineral Belt, Labrador. *Newfoundland Miner. Dev. Div. Rep. 82-7.* 77 pp.

Grambling, J. A. 1986. Crustal thickening during Proterozoic metamorphism and deformation in New Mexico. *Geology* 14: 149–52

Grambling, J. A., Codding, D. B. 1982. Stratigraphic and structural relationships of multiply deformed Precambrian metamorphic rocks in the Rio Mora area, New Mexico, *Geol. Soc. Am. Bull.* 93: 127–37

Green, A. G., Hajnal, Z., Weber, W. 1985a. An evolutionary model of the western Churchill province and western margin of the Superior province in Canada and the north-central United States. *Tectonophysics* 116: 281–322

Green, A. G., Weber, W., Hajnal, Z. 1985b. Evolution of Proterozoic terrains beneath the Williston Basin. *Geology* 13: 624–28

Greenberg, J. K., Brown, B. A. 1983. Lower Proterozoic volcanic rocks and their setting in the southern Lake Superior district. See Medaris 1983, pp. 67–84

Grotzinger, J. P. 1986, Evolution of early Proterozoic passive-margin carbonate platform, Rocknest Formation, Wopmay orogen, Northwest Territories, Canada. *J. Sediment. Petrol.* 56: 831–47

Grotzinger, J. P., McCormick, D. S. 1987. Flexure of the early Proterozoic lithosphere and the evolution of Kilohigok basin (1.9 Ga), northwest Canadian shield. In *New Perspectives in Basin Analysis,* ed. K. Kleinspehn, C. Paola. Heidelberg: Springer-Verlag. In press

Gulson, B. L., Krogh, T. E. 1975. Evidence of multiple intrusion, possible resetting of U-Pb ages, and new crystallization of zircons in the post-tectonic intrusions ("Rapakivi granites") and gneisses from South Greenland. *Geochim. Cosmochim. Acta* 39: 65–82

Hanmer, S. 1986. Asymmetrical pull-aparts and foliation fish as kinematic indicators. *J. Struct. Geol.* 8: 111–22

Hanmer, S. 1987a. Great Slave Lake Shear Zone, Canadian Shield: depth-sections through a crustal scale fault zone. *Tectonophysics.* In press

Hanmer, S. K. 1987b. Granulite facies mylonites: a brief structural reconnaissance north of Stony Rapids, northern Saskatchewan. In *Current Research, Part A. Geol. Surv. Can. Pap. 87-1A,* pp. 563–72

Harrison, J. M. 1951. Precambrian correlation and nomenclature, and problems of the Kisseynew gneisses, in Manitoba. *Geol. Surv. Can. Bull. 20.* 53 pp.

Heaman, L. M., Machado, N., Krogh, T. E., Weber, W. 1986. Precise U-Pb zircon ages for the Molson dyke swarm and the Fox River sill: constraints for Early Proterozoic crustal evolution in northeastern Manitoba, Canada. *Contrib. Mineral. Petrol.* 94: 82–89

Hegner, E., Hulbert, L. 1987 Sm-Nd ages and origin of mafic-ultramafic intrusions from the Proterozoic Lynn Lake domain,

Manitoba, Canada. *Geol. Assoc. Can. Programs Abstr.* 12: 54 (Abstr.)

Henderson, J. B., McGrath, P. H., James, D. T., Macfie, R. I. 1987. An integrated geological, gravity and magnetic study of the Artillery Lake area and the Thelon Tectonic Zone, District of Mackenzie. In *Current Research, Part A. Geol. Surv. Can. Pap. 87-1A*, pp. 803–14

Henderson, J. R. 1983. Structure and metamorphism of the Aphebian Penrhyn Group and its Archean basement complex in the Lyon Inlet area, Melville Peninsula, District of Franklin. *Geol. Surv. Can. Bull. 324.* 50 pp. with map 1510A, scale 1: 100,000.

Henderson, J. R. 1984. Description of a virgation in the Foxe Fold Belt, Melville Peninsula, Canada. See Kröner & Greiling 1984, pp. 251–61

Henderson, J. R. 1985a. Geology, Ekalugad Fiord–Home Bay, Northwest Territories. *Geol. Surv. Can. Map 1606A*, scale 1: 250,000.

Henderson, J. R. 1985b. Geology, McBeth Fiord–Cape Henry Kater, Northwest Territories. *Geol. Surv. Can. Map 1605A*, scale 1 : 250,000

Henderson, J. R., Loveridge, W. D. 1981. Age and geological significance of a tonalite pegmatite from east-central Baffin Island. In *Current Research, Part C. Geol. Surv. Can. Pap. 81-1C*, pp. 135–37

Henderson, J. R., Tippett, C. R. 1980. Foxe Fold Belt in eastern Baffin Island, District of Franklin. In *Current Research, Part A. Geol. Surv. Can. Pap. 80-1A*, pp. 147–52

Henderson, J. R., LeCheminant, A. N., Jefferson, C. W., Coe, K., Henderson, M. N. 1986. Preliminary account of the geology around Wager Bay, District of Keewatin. In *Current Research, Part A. Geol. Surv. Can. Pap. 86-1A*, pp. 159–76

Henderson, J. R., Shaw, D., Mazurski, M., Henderson, M., Green, R., Brisbin, D. 1979. Geology of part of Foxe Fold Belt, central Baffin Island, District of Franklin. In *Current Research, Part A. Geol. Surv. Can. Pap. 79-1A*, pp. 95–99

Hildebrand, R. S. 1981. Early Proterozoic LaBine Group of Wopmay Orogen: remnant of a continental volcanic arc developed during oblique convergence. See Campbell 1981, pp. 133–56

Hildebrand, R. S. 1984. Geology of the Rainy Lake–White Eagle Falls area, District of Mackenzie: Early Proterozoic cauldrons, stratovolcanoes and subvolcanic plutons. *Geol. Surv. Can. Pap. 83-20.* 42 pp.

Hildebrand, R. S., Roots, C. F. 1985. Geology of the Riviere Grandin map area (Hottah Terrane and western Great Bear Magmatic Zone), District of Mackenzie.

In *Current Research, Part A. Geol. Surv. Can. Pap. 85-1A*, pp. 373–83

Hildebrand, R. S., Bowring, S. A., Steer, M. E., Van Schmus, W. R. 1983. Geology and U-Pb geochronology of parts of the Leith Peninsula and Riviere Grandin map areas, District of Mackenzie. In *Current Research, Part A. Geol. Surv. Can. Pap. 83-1A*, pp. 329–42

Hildebrand, R. S., Bowring, S. A., Andrew, K. P. E., Gibbins, S. F., Squires, G. C. 1987a. Geological investigations in Calder River map area, central Wopmay Orogen, District of Mackenzie. In *Current Research, Part A. Geol. Surv. Can. Pap. 87-1A*, pp. 699–711

Hildebrand, R. S., Hoffman, P. F., Bowring, S. A. 1987b. Tectono-magmatic evolution of the 1.9-Ga Great Bear magmatic zone, Wopmay orogen, northwestern Canada. *J. Volcanol. Geotherm. Res.* 32: 99–118

Hills, F. A., Houston, R. S. 1979. Early Proterozoic tectonics of the central Rocky Mountains, North America. *Contrib. Geol.* 17: 89–109

Hinze, W. J. 1985. *The Utility of Regional Gravity and Magnetic Anomaly Maps.* Tulsa, Okla: Soc. Explor. Geophys. 454 pp.

Hoffman, P. F. 1969. Proterozoic paleocurrents and depositional history of the East Arm fold belt, Great Slave Lake, Northwest Territories. *Can. J. Earth Sci.* 6: 441–62

Hoffman, P. F. 1973. Evolution of an early Proterozoic continental margin: the Coronation geosyncline and associated aulacogens of the northwestern Canadian shield. *Philos. Trans. R. Soc. London Ser. A* 273: 547–81

Hoffman, P. F. 1980. Wopmay orogen: a Wilson cycle of early Proterozoic age in the northwest of the Canadian shield. In *The Continental Crust and Its Mineral Deposits. Geol. Assoc. Can. Spec. Pap. 20*, ed. D. W. Strangway, pp. 523–49. St. John's: Geol. Assoc. Can. 804 pp.

Hoffman, P. F. 1984. Geology of the northern internides of Wopmay orogen, District of Mackenzie, Northwest Territories. *Geol. Surv. Can. Map 1576A*, scale 1: 250,000

Hoffman, P. F. 1985. Is the Cape Smith belt (northern Quebec) a klippe? *Can. J. Earth Sci.* 22: 1361–69

Hoffman, P. F. 1987a. Continental transform tectonics: Great Slave Lake shear zone (1.9 Ga), northwest Canada. *Geology* 15: 785–88

Hoffman, P. F. 1987b. Early Proterozoic foredeeps, foredeep magmatism, and Superior-type iron-formations of the Canadian shield. See Kröner 1987, pp. 85–98

Hoffman, P. F. 1987c. Geology and tectonics, east arm of Great Slave Lake, Northwest Territories. *Geol. Surv. Can. Map 1628A*, scale 1 : 250,000

Hoffman, P. F. 1988. The extent of Trans-Hudson orogen and subdivision of the Churchill province. See Lewry & Stauffer 1988. In press

Hoffman, P. F., Bowring, S. A. 1984. Short-lived 1.9 Ga continental margin and its destruction, Wompay orogen, northwest Canada. *Geology* 12: 68–72

Hoffman, P. F., Grotzinger, J. P. 1987. Abner/Denault reef complex (2.1 Ga), Labrador Trough, NE Quebec. In *Reef Case Histories*, ed. H. H. Geldsetzer, Calgary: Can. Soc. Pet. Geol. In press

Hoffman, P. F., Tirrul, R., King, J. E., St-Onge, M. R., Lucas, S. B. 1987. Axial projections and modes of crustal thickening, eastern Wopmay orogen, northwest Canadian shield. In *Processes of Continental Lithospheric Deformation. Geol. Soc. Am. Spec. Pap. 218*, ed. S. P. Clark Jr. Boulder: Geol. Soc. Am. In press

Holcombe, R. J., Callender, J. F. 1982. Structural analysis and stratigraphic problems of Precambrian rocks of the Picuris Range, New Mexico. *Geol. Soc. Am. Bull.* 93: 138–49

Holst, T. B. 1982. Evidence for multiple deformation during the Penokean orogeny in the Middle Precambrian Thompson Formation, Minnesota. *Can. J. Earth Sci.* 19: 2043–47

Holst, T. B. 1984. Evidence for nappe development during the early Proterozoic Penokean orogeny, Minnesota. *Geology* 12: 135–38

Huhma, H. 1986. Sm-Nd, U-Pb and Pb-Pb isotopic evidence for the origin of the Early Proterozoic Svecokarelian crust in Finland. *Geol. Surv. Fin. Bull. 337*. 48 pp.

Hynes, A. J., Francis, D. M. 1982. Komatiitic basalts of the Cape Smith foldbelt, New Quebec, Canada. In *Komatiites*, ed. N. T. Arndt, E. G. Nisbet, pp. 159–70. New York: Allen & Unwin. 520 pp.

Irving, E., Davidson, A., McGlynn, J. C. 1984. Paleomagnetism of gabbros of the Early Proterozoic Blachford Lake Intrusive Suite and the Easter Island Dyke, Great Slave Lake, NMT: Possible evidence for the earliest continental drift. *Geophys. Surv.* 7: 1–25

Jackson, G. D. 1960. Belcher Islands, Northwest Territories. *Geol. Surv. Can. Pap. 60-20*. 13 pp. with map, scale 1 : 126,720.

Jackson, G. D., Iannelli, T. R. 1981. Rift-related cyclic sedimentation in the Neohelikian Borden Basin, northern Baffin Island. See Campbell 1981, pp. 269–302

Jackson, G. D., Morgan, W. C. 1978. Pre-cambrian metamorphism on Baffin and Bylot islands. See Fraser & Heywood 1978, pp. 249–67

Jackson, G. D., Taylor, F. C. 1972. Correlation of major Aphebian rock units in the northeastern Canadian Shield. *Can. J. Earth Sci.* 9: 1650–69

Jackson, S. L., Gordon, T. M. 1985. Metamorphism and structure of the Laurie Lake region, Manitoba. In *Current Research, Part A. Geol. Surv. Can. Pap. 85-1A* pp. 753–59

Jackson, S. L., Gordon, T. M. 1986. Metamorphic studies in the transition zone between the Lynn Lake Greenstone Belt and the Kisseynew Gneiss Belt, Laurie Lake, Manitoba. In *Current Research, Part B. Geol. Surv. Can. Pap. 86-1B*, pp. 539–46

Johnson, B. J. 1986. *Geology of the Wilson Island Group, Great Slave Lake, Northwest Territories*. MS thesis. Carleton Univ., Ottawa, Can. 121 pp.

Jones, A. G., Savage, P. J. 1986. North American Central Plains conductivity anomaly goes east. *Geophys. Res. Lett.* 13: 685–88

Kalsbeek, F., Taylor, P. N. 1985. Isotopic and chemical variation in granites across a Proterozoic continental margin—the Ketilidian mobile belt of south Greenland. *Earth Planet. Sci. Lett.* 73: 65–80

Kanasewich, E. R., Clowes, R. M., McLoughan, C. H. 1969. A buried Precambrian rift in western Canada. *Tectonophysics* 8: 513–27

Karlstrom, K. E., Bowring, S. A. 1987. Early Proterozoic assembly of tectonostratigraphic terranes in southwestern North America. *J. Geol.* In press

Karlstrom, K. E., Conway, C. M. 1986. Deformational styles and contrasting lithostratigraphic sequences within an Early Proterozoic orogenic belt, central Arizona. In *Geology of Central and Northern Arizona*, ed. J. D. Nations, C. M. Conway, G. A. Swann, pp. 1–25. Flagstaff, Ariz: Geol. Soc. Am. Rocky Mt. Sect. 176 pp.

Karlstrom, K. E., Houston, R. S. 1984. The Cheyenne belt: analysis of a Proterozoic suture in southern Wyoming. *Precambrian Res.* 25: 415–46

Karlstrom, K. E., Bowring, S. A., Conway, C. M. 1987. Tectonic significance of an Early Proterozoic two-province boundary in central Arizona. *Geol. Soc. Am. Bull.* In Press

Karlstrom, K. E., Flurkey, A. J., Houston, R. S. 1983. Stratigraphy and depositional setting of the Proterozoic Snowy Pass Supergroup, southeastern Wyoming: record of an early Proterozoic Atlantic-type cratonic margin. *Geol. Soc. Am. Bull.* 94: 1257–74

Karner, G. D., Watts, A. B. 1983. Gravity anomalies and flexure of the lithosphere at mountain ranges. *J. Geophys. Res.* 88: 10,449–77

Kearey, P. 1976. A regional structural model of the Labrador Trough, northern Quebec, from gravity studies, and its relevance to continental collision in the Precambrian. *Earth Planet. Sci. Lett.* 28: 371–78

King, J. E. 1986. The metamorphic internal zone of Wopmay orogen, Canada: 30 km of structural relief in a composite section based on plunge projection. *Tectonics* 5: 973–94

King, J. E., Barrette, P. D., Relf, C. D. 1987. Contrasting styles of basement deformation and longitudinal extension in the metamorphic-internal zone of Wopmay Orogen, N.W.T. In *Current Research, Part A. Geol. Surv. Can. Pap. 87-1A,* pp. 515–31

Klasner, J. S. 1978. Penokean deformation and associated metamorphism in the western Marquette Range, northern Michigan. *Geol. Soc. Am. Bull.* 89: 711–22

Klasner, J. S., King, E. R. 1986. Precambrian basement geology of North and South Dakota. *Can. J. Earth Sci.* 23: 1083–1102

Klasner, J. S., King, E. R., Jones, W. J. 1985. Geologic interpretation of gravity and magnetic data for northern Michigan and Wisconsin. See Hinze 1985, pp. 267–86

Knight, I., Morgan, W. C. 1981. The Aphebian Ramah Group, northern Labrador. See Campbell 1981, pp. 313–30

Kontinen, A. 1987. An early Proterozoic ophiolite—the Jormua mafic-ultramafic complex, northeastern Finland. *Precambrian Res.* 35: 313–41

Korstgård, J., Ermanovics, I. 1984. Archaean and early Proterozoic tectonics of the Hopedale Block, Labrador, Canada. See Kröner & Greiling 1984, pp. 295–318

Korstgård, J., Ryan, B., Wardle, R. 1987. The boundary between Proterozoic and Archaean crustal blocks in central west Greenland and northern Labrador. *J. Geol. Soc. London.* In press

Kramers, J. D. 1987. Link between Archaean continent formation and anomalous subcontinental mantle. *Nature* 325: 47–50

Krogh, T. E. 1973. A low-contamination method for hydrothermal decomposition of zircon and extraction of U and Pb for isotopic age determinations. *Geochim. Cosmochim. Acta* 37: 485–94

Krogh, T. E. 1982. Improved accuracy of U-Pb zircon ages by creation of more concordant systems using an air abrasion technique. *Geochim. Cosmochim. Acta* 46: 637–49

Krogh, T. E., Davis, D. W., Corfu, F. 1984. Precise U-Pb zircon and baddeleyite ages for the Sudbury area. See Pye et al 1984, pp. 431–46

Kröner, A., ed. 1981. *Precambrian Plate Tectonics.* Amsterdam: Elsevier. 781 pp.

Kröner, A., ed. 1987. *Proterozoic Lithospheric Evolution. Geodyn. Ser.,* Vol. 17. Washington DC: Am Geophys. Union. 271 pp.

Kröner, A., Greiling, R., eds. 1984. *Precambrian Tectonics Illustrated.* Stuttgart: Schweizerbart'sche. 419 pp.

Lalonde, A. E. 1986. *The intrusive rocks of the Hepburn metamorphic-plutonic zone of central Wopmay orogen, N.W.T.* PhD. thesis. McGill Univ., Montreal, Can. 258 pp.

Larue, D. K. 1983. Early Proterozoic tectonics of the Lake Superior region: tectonostratigraphic terranes near the purported collision zone. See Medaris 1983, pp. 33–47

Larue, D. K., Sloss, L. L. 1980. Early Proterozoic sedimentary basins of the Lake Superior region. *Geol. Soc. Am. Bull.* 91(1): 450–52, 91(2): 1836–74

Larue, D. K., Ueng, W. L. 1985. Florence-Niagara terrane: an early Proterozoic accretionary complex, Lake Superior region, U.S.A. *Geol. Soc. Am. Bull.* 96: 1179–87

LeCheminant, A. N., Iannelli, T. R., Zaitlin, B., Miller, A. R. 1981. Geology of Tebesjuak Lake map area, District of Keewatin: a progress report. In *Current Research, Part B. Geol. Surv. Can. Pap. 81-1B,* pp. 113–28

LeCheminant, A. N., Jackson, M. J., Galley, A. G., Smith, S. L., Donaldson, J. A. 1984. Early Proterozoic Amer Group, Beverly Lake map area, District of Keewatin. In *Current Research, Part B. Geol. Surv. Can. Pap. 84-1B,* pp. 159–72

LeCheminant, A. N., Miller, A. R., LeCheminant, G. M. 1987a. Early Proterozoic alkaline igneous rocks, District of Keewatin, Canada: petrogenesis and mineralization. In *Geochemistry and Mineralization of Proterozoic Volcanic Suites,* ed. T. C. Pharoah, R. D. Beckinsale, D. T. Rickard. Oxford: Blackwell. In press

LeCheminant, A. N., Roddick, J. C., Henderson, J. R. 1987b. Geochronology of Archean and Early Proterozoic magmatism in the Baker Lake–Wager Bay region, N.W.T. *Geol. Assoc. Can. Programs Abstr.* 12: 66 (Abstr.)

LeCheminant, A. N., Roddick, J. C., Tessier, A. C., Bethune, K. M. 1987c. Geology and U-Pb ages of early Proterozoic calc-alkaline plutons northwest of Wager Bay, District of Keewatin. In *Current Research, Part A. Geol. Surv. Can. Pap. 87-1A*, pp. 773–82

Le Gallais, C. J., Lavoie, S. 1982. Basin evolution of the Lower Proterozoic Kaniapiskau Supergroup, central Labrador miogeocline (Trough), Quebec. *Bull. Can. Pet. Geol.* 30: 150–66

Lewry, J. F. 1981. Lower Proterozoic arc-microcontinent collisional tectonics in the western Churchill Province. *Nature* 294: 69–72

Lewry, J. F., Sibbald, T. I. I. 1977. Variation in lithology and tectononmetamorphic relationships in the Precambrian basement of northern Saskatchewan. *Can. J. Earth Sci.* 14: 1453–67

Lewry, J. F., Sibbald, T. I. I. 1980. Thermotectonic evolution of the Churchill Province in northern Saskatchewan. *Tectonophysics* 68: 45–82

Lewry, J. F., Slimmon, W. L. 1985. Compilation bedrock geology, Lac LaRonge, NTS 73P/73I. *Saskatchewan Energy Mines Rep.* 225, 1 : 250,000 scale map with marginal notes.

Lewry, J. F., Stauffer, M. R., eds. 1988. *The Early Proterozoic Trans-Hudson Orogen: lithotectonic correlations and evolution.* St. John's: Geol. Assoc. Can. In press

Lewry, J. F., Sibbald, T. I. I., Rees, C. J. 1978. Metamorphic patterns and their relation to tectonism and plutonism in the Churchill Province in Northern Saskatchewan. See Fraser & Heywood 1978, pp. 139–54

Lewry, J. F., Sibbald, T. I. I., Schledewitz, D. C. P. 1985. Variation in character of Archean rocks in the western Churchill Province and its significance. See Ayres et al 1985, pp. 239–61

Lewry, J. F., Stauffer, M. R., Fumerton, S. 1981. A Cordilleran-type batholithic belt in the Churchill Province in northern Saskatchewan. *Precambrian Res.* 14: 277–313

Lewry, J. F., Thomas, D. J., MacDonald, R., Chiarenzelli, J. 1987. Structural relations in accreted terranes of the Trans-Hudson Orogen, Saskatchewan: telescoping in a collisional regime? *Geol. Assoc. Can. Programs Abstr.* 12: 67 (Abstr.)

Letouzy, J., Kimura, M. 1986. The Okinawa Trough: genesis of a back-arc basin developing along a continental margin. *Tectonophysics* 125: 209–30

Lis, M. G., Price, R. A. 1976. Large-scale block faulting during deposition of the Windermere Supergroup in southeastern British Columbia. In *Current Research, Part A. Geol. Surv. Can. Pap. 76-1A*, pp. 135–36

Lister, G. S., Snoke, A. W. 1984, S-C mylonites. *J. Struct. Geol.* 6: 617–38

Maass, R. S. 1983. Early Proterozoic tectonic style in central Wisconsin. See Medaris 1983, pp. 85–95

Maass, R. S., Medaris, L. G. Jr., Van Schmus, W. R. 1980. Penokean deformation in central Wisconsin. See Morey & Hanson 1980, pp. 147–57

MacGregor, I. D., Manton, W. I. 1986. Roberts Victor eclogites: ancient oceanic crust. *J. Geophys. Res.* 91: 14,063–79

Machado, N., Heaman, L., Krogh, T. E. 1987. U-Pb geochronology of the Thompson mobile belt, Manitoba: preliminary results. *Geol. Assoc. Can. Programs Abstr.* 12: 69 (Abstr.)

Machado, N., Goulet, N., Gariépy, C. 1988. U-Pb ages of reactivated Archean basement and of Hudsonian metamorphism in the northern Labrador Trough. *Can. J. Earth Sci.* In press

Marten, B. E. 1977. *The relationship between the Aillik Group and the Hopedale gneiss, Kaipokok Bay, Labrador.* PhD thesis. Memorial Univ., St. John's, Can. 389 pp.

McCulloch, M. T., Wasserburg, G. J. 1978. Sm-Nd and Rb-Sr chronology of continental crust formation. *Science* 200: 1003–11

McElhinny, M. W., McWilliams, M. O. 1977. Precambrian geodynamics—a palaeomagnetic view. *Tectonophysics* 40: 137–59

McGlynn, J. C., Irving, E. 1981. Horizontal motions and rotations in the Canadian Shield during the Early Proterozoic. See Campbell 1981, pp. 183–90

McKenzie, D. 1984. A possible mechanism for epeirogenic uplift. *Nature* 307: 616–18

Medaris, L. G., ed. 1983. *Early Proterozoic Geology of the Great Lakes Region. Geol. Soc. Am. Mem. 160.* 141 pp.

Medaris, L. G., Byers, C. W., Mickelson, D. M., Shanks, W. C., eds. 1983. *Proterozoic Geology. Geol. Soc. Am. Mem. 161.* Boulder, Colo: Geol. Soc. Am. 315 pp.

Mengel, F. 1985. Nain-Churchill province boundary: a preliminary report on a cross-section through the Hudsonian front in the Saglek Fiord area, northern Labrador. In *Current Research. Newfoundland Miner. Dev. Div. Rep. 85-1*, pp. 33–42

Meyer, M. T. 1987. *Geochronology and geochemistry of the Wathaman Batholith, the remnant of an Early Proterozoic continental-arc in the Trans-Hudson orogen, Saskatchewan, Canada.* MS thesis. Univ. Kans., Lawrence. 107 pp.

Milligan, G. C. 1960. Geology of the Lynn Lake District. *Manitoba Dept. Mines Nat. Resour. Publ. 57-1.* 317 pp.

Moore, J. M. Davidson, A., Baer, A. J., eds. 1986. *The Grenville Province. Geol. Assoc. Can. Spec. Pap. 31.* St. John's: Geol. Assoc. Can. 358 pp.

Morey, G. B. 1978. Metamorphism in the Lake Superior region, U.S.A., and its relation to crustal evolution. See Fraser & Heywood 1978, pp. 283–314

Morey, G. B. 1983. Animikie Basin, Lake Superior region, U.S.A. In *Iron-Formation: Facts and Problems*, ed. A. F. Trendall, R. C. Morris, pp. 13–67. Amsterdam: Elsevier. 558 pp.

Morey, G. B., Hanson, G. N., eds. 1980. *Selected Studies of Archean Gneisses and Lower Proterozoic Rocks, Southern Canadian Shield. Geol. Soc. Am. Spec. Pap 182.* Boulder, Colo: Geol. Soc. Am. 175 pp.

Morey, G. B., Sims, P. K., Cannon, W. F., Mudrey, M. G. Jr., Southwick, D. L. 1982. Geologic map of the Lake Superior region, Minnesota, Wisconsin, and northern Michigan. *Minn. Geol. Surv. State Map Ser. S-13*, scale 1 : 1,000,000

Morgan, W. C. 1979. Geology Nachvak Fiord–Ramah Bay, Newfoundland-Quebec. *Geol. Surv. Can. Map 1469A*, scale 1 : 50,000

Morgan, W. C. 1981. Geology Bears Gut–Saglek Fiord, Newfoundland. *Geol. Surv. Can. Map 1478A*, scale 1 : 50,000

Morgan, W. C. 1983. Geology, Lake Gillian, District of Franklin. *Geol. Surv. Can. Map 1560A*, scale 1 : 250,000

Morgan, W. C., Okulitch, A. V., Thompson, P. H. 1976. Stratigraphy, structure and metamorphism of the west half of the Foxe Fold Belt, Baffin Island. In *Report of Activities, Part A. Geol. Surv. Can. Pap. 76-1A*, pp. 387–91

Mortensen, J. K., Thorpe, R. I. 1987. U-Pb zircon ages of felsic volcanic rocks in the Kaminak Lake area, District of Keewatin. In *Radiogenic Age and Isotopic Studies. Geol. Surv. Can. Pap. 87-2.* In press

Mukherjee, A. C. 1974. Some aspects of sedimentology of the Missi Group and its environment of deposition. *Can. J. Earth Sci.* 11: 1018–19

Mukhopadhyay, M., Gibb, R. A. 1981. Gravity anomalies and deep structure of eastern Hudson Bay. *Tectonophysics* 72: 43–60

Myers, J. S. 1984. The Nagssugtoqidian mobile belt of Greenland. See Kröner & Greiling 1984, pp. 237–50

Nelson, B. K., DePaolo, D. J. 1984. 1,700-Myr greenstone volcanic successions in southwestern North America and isotopic evolution of Proterozoic mantle. *Nature* 312: 143–46

Nelson, B. K., DePaolo, D. J. 1985. Rapid production of continental crust 1.7–1.9 b.y. ago: Nd isotopic evidence from the basement of the North American midcontinent. *Geol. Soc. Am. Bull.* 96: 746–54

Ojakangas, R. W. 1983. Tidal deposits in the early Proterozoic basin of the Lake Superior region—the Palms and the Pokegama Formations: evidence for subtidal-shelf deposition of Superior-type banded iron-formation. See Medaris 1983, pp. 49–66

Oxburgh, E. R., Parmentier, E. M. 1978. Thermal processes in the formation of continental lithosphere. *Philos. Trans. R. Soc. London Ser. A* 288: 415–29

Passchier, C. W., Simpson, C. 1986. Porphyroclast systems as kinematic indicators. *J. Struct. Geol.* 8: 831–43

Patchett, P. J., Arndt, N. T. 1986. Nd isotopes and tectonics of 1.9–1.7 Ga crustal genesis. *Earth Planet. Sci. Lett.* 78: 329–38

Patchett, J., Bridgwater, D. 1984. Origin of continental crust of 1.9–1.7 Ga defined by Nd isotopes in the Ketilidian terrain of south Greenland. *Contrib. Mineral. Petrol.* 87: 311–18

Patchett, J., Kouvo, O. 1986. Origin of continental crust of 1.9–1.7 Ga age: Nd isotopes and U-Pb zircon ages in the Svecokarelian terrain of South Finland. *Contrib. Mineral. Petrol.* 92: 1–12

Patterson, J. G. 1986. The Amer Belt: remnant of an Aphebian foreland fold and thrust belt. *Can. J. Earth Sci.* 23: 2012–23

Pearson, D. E., Lewry, J. F. 1974. Large-scale fold interference structures in the Mudjatik River area of northern Saskatchewan. *Can. J. Earth Sci.* 11: 619–34

Percival, J. A., Card, K. D. 1983. Archean crust as revealed in the Kapuskasing uplift, Superior province, Canada. *Geology* 11: 323–26

Percival, J. A., Card, K. D. 1985. Structure and evolution of Archean crust in central Superior Province, Canada. See Ayres et al 1985, pp. 179–92

Percival, J. A., McGrath, P. H. 1986. Deep crustal structure and tectonic history of the northern Kapuskasing uplift of Ontario: an integrated petrological-geophysical study. *Tectonics* 5: 553–72

Peredery, W. V., and Geological Staff. 1982. Geology and nickel sulphide deposits of the Thompson Belt, Manitoba. In *Precambrian Sulphide Deposits. Geol. Assoc. Can. Spec. Pap. 25*, ed. R. W. Hutchinson, C. D. Spence, J. M. Franklin, pp. 165–209. St. John's: Geol. Assoc. Can.

Perreault, S., Hynes, A., Moorehead, J. 1987. Metamorphism of the eastern flank of the Labrador Trough, Kuujjuaq, Ungava, northern Quebec. *Geol. Assoc. Can. Programs Abstr.* 12: 80 (Abstr.)

Peterman, Z. E. 1981. Dating of Archean basement in northeastern Wyoming and southern Montana. *Geol. Soc. Am. Bull.* 92: 139–46

Peterman, Z. E., Futa, K. 1987. Is the Archean Wyoming province exotic to the Superior craton? Evidence from Sm-Nd model ages of basement cores. *Geol. Soc. Am. Abstr. With Programs* 19: 803 (Abstr.)

Peterman, Z. E., Sims, P. K. 1986. The Goodman Swell: a major off-axis, rift-related uplift of Keweenawan age in northern Wisconsin. *Geol. Soc. Am. Abstr. With Programs* 18: 717 (Abstr.)

Peterman, Z. E., Sims, P. K., Zartman, R. E., Schulz, K. J. 1985. Middle Proterozoic uplift events in the Dunbar dome of northeastern Wisconsin, USA. *Contrib. Mineral. Petrol.* 91: 138–50

Pidgeon, R. T., Howie, R. A. 1975. U-Pb age of zircon from a charnockite granulite from Pangnirtung on the east coast of Baffin Island. *Can. J. Earth Sci.* 12: 1046–47

Piper, J. D. A. 1983. Dynamics of the continental crust in Proterozoic times. See Medaris et al 1983, pp. 11–34

Poirier, G., Perreault, S., Hynes, A. 1987. The nature of the eastern boundary of the Labrador Trough near Kuujjuaq, Quebec. *Geol. Assoc. Can. Programs Abstr.* 12: 81 (Abstr.)

Pollack, H. N. 1986. Cratonization and thermal evolution of the mantle. *Earth Planet. Sci. Lett.* 80: 175–82

Pye, E. G., Naldrett, A. J., Giblin, P. E., eds. 1984. *The Geology and Ore Deposits of the Sudbury Structure. Ontario Geol. Surv. Spec. Vol. 1.* Toronto: Ontario Geol. Surv. 603 pp.

Ramaekers, P. 1981. Hudsonian and Helikian basins of the Athabasca region, northern Saskatchewan. See Campbell 1981, pp. 219–33

Ray, G. E., Wanless, R. K. 1980. The age and geological history of the Wollaston, Peter Lake, and Rottenstone domains in northern Saskatchewan. *Can. J. Earth Sci.* 17: 333–47

Reed, J. C., Premo, W. R. 1987. Accretionary history of the Colorado Province. *Geol. Soc. Am. Abstr. With Programs* 19: 814 (Abstr.)

Reed, J. C., Bickford, M. E., Premo, W. R., Aleinikoff, J. N., Palliser, J. S. 1987. Evolution of the Early Proterozoic Colorado province: constraints from U-Pb geochronology. *Geology* 15: 861–65

Redden, J. A., Norton, J. J. 1975. Precambrian geology of the Black Hills. *S. Dak. Geol. Surv. Bull.* 16: 21–28

Redden, J. A., Peterman, Z. E., Zartman, R. E., DeWitt, E. 1987. U-Th-Pb zircon and monazite ages and preliminary interpretation of Proterozoic tectonism in the Black Hills, South Dakota. *Geol. Assoc. Can. Programs Abstr.* 12: 83 (Abstr.)

Reichenbach, I. G. 1986. *An ensialic marginal basin in Wopmay Orogen, northwestern Canadian Shield.* MS thesis. Carleton Univ., Ottawa, Can. 120 pp.

Reichenbach, I. G. 1987. Geochemistry of ensialic marginal basin basalts, Wopmay orogen, northwestern Canadian shield. In *Proterozoic Geochemistry Abstracts*, pp. 77–78. Lund, Swed: Int. Geol. Correlations Program

Ricketts, B. D., Donaldson, J. A. 1981. Sedimentary history of the Belcher Group of Hudson Bay. See Campbell 1981, pp. 235–54

Ricketts, B. D., Ware, M. J., Donaldson, J. A. 1982. Volcaniclastic rocks and volcaniclastic facies in the Middle Precambrian (Aphebian) Belcher Group, Northwest Territories, Canada. *Can. J. Earth Sci.* 19: 1275–94

Rivers, T., Chown, E. H. 1986. The Grenville orogen in eastern Quebec and western Labrador: definition, identification and tectonometamorphic relationships of autochthonous, parautochtonous and allochthonous terranes. See Moore et al 1986, pp. 31–50

Rivers, T., Wardle, R. J. 1985. Geology of the Lac Virot, Opacopa Lake, Wightman Lake, Evening Lake, and Gabbro Lake areas, Labrador-Quebec. *Newfoundland Miner. Dev. Div. Maps 85-24 through 85-28,* scale 1 : 100,000

Rivers, T., Martignole, J., Gower, C. F., Davidson, A. 1988. A new tectonic division of the Grenville Province. *Tectonics.* In press

Roddick, J. C., Loveridge, W. D., Parrish, R. R. 1987. Precise U/Pb dating at the subnanogram Pb level. *Chem. Geol.* In press

Roscoe, S. M. 1968. Huronian rocks and uraniferous conglomerates in the Canadian Shield. *Geol. Surv. Can. Pap. 68-40.* 205 pp.

Rousell, D. H. 1984. Structural geology of the Sudbury Basin. See Pye et al 1984, pp. 83–95

Ryan, B. 1984. Regional geology of the central part of the Central Mineral Belt, Labrador. *Newfoundland Miner. Dev. Div. Mem. 3.* 185 pp.

Sangster, D. F. 1972. Isotopic studies of ore-leads in the Hanson Lake–Flin Flon–Snow Lake mineral belt, Saskatchewan and Manitoba. *Can. J. Earth Sci.* 9: 500–13

Sangster, D. F. 1978. Isotopic studies of ore-leads of the circum-Kisseynew volcanic belt of Manitoba and Saskatchewan. *Can. J. Earth Sci.* 15: 1112–21

Schärer, U., Gower, C. F. 1987. Crustal evolution in eastern Labrador: constraints from precise U-Pb ages. *Precambrian Res.* In press

Schärer, U., Krogh, T. E., Gower, C. F. 1986. Age and evolution of the Grenville Province in eastern Labrador from U-Pb systematics in accessory minerals. *Contrib. Mineral. Petrol.* 94: 438–51

Schärer, U., Krogh, T. E., Wardle, R. J., Ryan, B., Gandhi, S. S. 1987. U-Pb ages of rhyolites and gneisses, and implications for the evolution of the Makkovik province, eastern Labrador. *Can. J. Earth Sci.* In press

Schau, M. 1982. Geology of the Prince Albert Group in parts of Walker Lake and Laughland Lake map areas, District of Keewatin. *Geol. Surv. Can. Bull. 337.* 62 pp. with map, scale 1 : 250,000

Schau, M., Tremblay, F., Christopher, A. 1982. Geology of Baker Lake map area, District of Keewatin: a progress report. In *Current Research, Part A. Geol. Surv. Can. Pap. 82-1A*, pp. 143–50

Schledewitz, D. C. P. 1978. Patterns of regional metamorphism in the Churchill Province of Manitoba (north of 58°). See Fraser & Heywood 1978, pp. 179–90

Schledewitz, D. C. P. 1986. Geology of the Cochrane and Seal Rivers area. *Manitoba Energy Mines Miner. Res. Div. Geol. Rep. GR80-9.* 139 pp. and map, scale 1 : 500,000

Schulz, K. J. 1983. Geochemistry of volcanic rocks of northern Wisconsin: implications for early Proterozoic tectonism, Lake Superior region. *Ann. Inst. Lake Superior Geol., 29th, Houghton, Mich.*, pp. 39–40 (Abstr.)

Schulz, K. J. 1987a. Early Proterozoic evolution of a rifted continental margin in the Lake Superior region. *Geol. Soc. Am. Abstr. With Programs* 19: 243 (Abstr.)

Schulz, K. J. 1987b. An Early Proterozoic ophiolite in the Penokean Orogen. *Geol. Assoc. Can. Programs Abstr.* 12: 87 (Abstr.)

Schulz, K. J., LaBerge, G. L., Sims, P. K., Peterman, Z. E., Klasner, J. S. 1984. The volcanic-plutonic terrane of northern Wisconsin: implications for Early Proterozoic tectonism, Lake Superior region. *Geol. Assoc. Can. Programs Abstr.* 9: 103 (Abstr.)

Scoates, R. F. J. 1981. Volcanic rocks of the Fox River belt, northeastern Manitoba. *Manitoba Energy Mines Miner. Res. Div. Geol. Rep. GR81-1.* 109 pp. with map, scale 1 : 50,000

Scoates, R. F. J., Macek, J. J. 1978. Molson dyke swarm. *Manitoba Miner. Res. Div. Geol. Pap. 78-1.* 51 pp.

Sedlock, R. L., Larue, D. K. 1985. Fold axes oblique to the regional plunge and Proterozoic terrane accretion in the southern Lake Superior region. *Precambrian Res.* 30: 249–62

Sharpton, V. L., Grieve, R. A. F., Thomas, M. D., Halpenny, J. F. 1987. Horizontal gravity gradient: an aid to the definition of crustal structure in North America. *Geophys. Res. Lett.* 14: 808–11

Silver, L. T. 1969. Precambrian batholiths of Arizona. In *Abstracts for 1968. Geol. Soc. Am. Spec. Pap. 121*, pp. 558–59. Boulder, Colo: Geol. Soc. Am.

Silver, L. T. 1978. Precambrian formations and Precambrian history in Cochise County, southeastern Arizona. In *Land of Cochise, 29th Field Conf. Guideb.*, pp. 157–63. Socorro: N. Mex. Geol. Soc.

Silver, L. T. 1984. Observations on the Precambrian evolution of northern New Mexico and adjacent regions. *Geol. Soc. Am. Abstr. With Programs* 16: 256 (Abstr.)

Silver, L. T. 1987. A Proterozoic history for southwestern North America. *Geol. Soc. Am. Abstr. With Programs* 19: 845 (Abstr.)

Silver, L. T., Ludwig, K. R. 1986. Implications of a precise chronology for early Proterozoic crustal evolution and caldera formation in the Tonto Basin–Mazatzal Mountains region, Arizona. *Geol. Soc. Am. Abstr. With Programs* 18: 413 (Abstr.)

Silver, L. T., Anderson, C. A., Crittenden, M., Robertson, J. M. 1977a. Chronostratigraphic elements of the Precambrian rocks of the southwestern and far western United States. *Geol. Soc. Am. Abstr. With Programs* 9: 1176 (Abstr.)

Silver, L. T., Bickford, M. E., Van Schmus, W. R., Anderson, J. L., Anderson, T. H., Medaris, L. G. 1977b. The 1.4–1.5 b.y. transcontinental anorogenic plutonic perforation of North America. *Geol. Soc. Am. Abstr. With Programs* 9: 1176–77

Sims, P. K., Peterman, Z. E. 1986. Early Proterozoic Central Plains orogen: a major buried structure in the north-central United States. *Geology* 14: 488–91

Sims, P. K., Peterman, Z. E., Klasner, J. S., Cannon, W. F., Schulz, K. J. 1987. Nappe development and thrust faulting in the upper Michigan segment of the Early Proterozoic Penokean orogen. *Geol. Soc. Am. Abstr. With Programs* 19: 246 (Abstr.)

Sleep, N. H., Windley, B. F. 1982. Archaean plate tectonics: constraints and inferences. *J. Geol.* 90: 363–79

Smith, E. I. 1983. Geochemistry and evolution of the early Proterozoic, post-Penokean rhyolites, granites, and related rocks of south-central Wisconsin, U.S.A. See Medaris 1983, pp. 113–28

Soegaard, K., Eriksson, K. A. 1986. Transition from arc volcanism to stable-shelf and subsequent convergent-margin sedimentation in northern New Mexico from 1.76 Ga. J. Geol. 94: 47–66

Stacey, J. S., Hedlund, D. C. 1983. Lead-isotopic compositions of diverse igneous rocks and ore deposits from southwestern New Mexico and their implications for early Proterozoic crustal evolution in the western United States. Geol. Soc. Am. Bull. 94: 43–57

Stauffer, M. R. 1974. Geology of the Flin Flon area: a new look at the sunless city. Geosci. Can. 1: 30–35

Stauffer, M. R. 1984. Manikewan: an Early Proterozoic ocean in central Canada, its igneous history and orogenic closure. Precambrian Res. 25: 257–81

Stauffer, M. R., Mukherjee, A. 1971. Superimposed deformations in the Missi metasedimentary rocks near Flin Flon, Manitoba. Can. J. Earth Sci. 8: 217–42

Stauffer, M. R., Mukherjee, A. C., Koo, J. 1975. The Amisk Group: an Aphebian(?) island arc deposit. Can. J. Earth Sci. 12: 2021–35

Stockwell, C. H. 1961. Structural provinces, orogenies, and time classification of rocks of the Canadian Precambrian Shield. In Age Determinations by the Geological Survey of Canada. Geol. Surv. Can. Pap. 61-17, ed. J. A. Lowden, pp. 108–18.

St-Onge, M. R. 1987. Zoned poikiloblastic garnets: documentation of P-T paths and syn-metamorphic uplift through thirty kilometers of structural depth, Wopmay orogen, Canada. J. Petrol 28: 1–28

St-Onge, M. R., King, J. E. 1987a. Thermotectonic evolution of a metamorphic internal zone documented by axial projections and petrological P-T paths, Wopmay orogen, northwest Canada. Geology 15: 155–58

St-Onge, M. R., King, J. E. 1987b. Evolution of regional metamorphism during back-arc stretching and subsequent crustal shortening in the 1.9 Ga Wopmay Orogen, Canada. Philos. Trans. R. Soc. London Ser. A 321: 199–218

St-Onge, M. R., Lucas, S. B. 1988. Structural and thermal history of the Cape Smith thrust-fold belt, northern Quebec. See Lewry & Stauffer 1988. In press

St-Onge, M. R., Lucas, S. B., Scott, D. J., Bégin, N. J. 1986. Eastern Cape Smith Belt: an early Proterozoic thrust-fold belt and basal shear zone exposed in oblique section, Wakeham Bay and Cratere du Nouveau Québec map areas, northern Quebec. In Current Research, Part A Geol.

Surv. Can. Pap. 86-1A, pp. 1–14

St-Onge, M. R., Lucas, S. B., Scott, D. J., Bégin, N. J. 1988. Thin-skinned imbrication and subsequent thick-skinned folding of rift-fill, transitional-crust, and ophiolite suites in the 1.9 Ga Cape Smith belt, northern Quebec. In Current Research, Part A. Geol. Surv. Can. Pap. 88-1A. In press

Swift, P. N. 1987. Subduction zone setting for Early Proterozoic turbidite deposition and melange deformation, SE Arizona. Geol. Soc. Am. Abstr. With Programs 19: 862 (Abstr.)

Syme, E. C. 1985. Geochemistry of metavolcanic rocks in the Lynn Lake belt. Manitoba Energy Mines Geol. Rep. GR84-1. 84 pp.

Syme, E. C. 1987. Stratigraphy and geochemistry of two Proterozoic arcs: Lynn Lake and Flin Flon metavolcanic belts, Manitoba. Geol. Assoc. Can. Programs Abstr. 12: 94 (Abstr.)

Syme, E. C., Bailes, A. H., Gordon, T. M., Hunt, R. A. 1987. U-Pb zircon geochronology in the Flin Flon belt: age of Amisk volcanism. In Report of Field Activities, 1987, Manitoba Energy Mines Miner. Div. In press

Tapponnier, P., Peltzer, G., Le Dain, A. Y., Armijo, R. 1982. Propagating extrusion tectonics in Asia: new insights from simple experiments with plasticine. Geology 10: 609–88

Taylor, F. C. 1963. Snowbird Lake map-area, District of Mackenzie. Geol. Surv. Can. Mem. 333. 23 pp. with Map 1138A, scale 1 : 253,440

Taylor, F. C. 1979. Reconnaissance geology of a part of the Precambrian Shield, northeastern Quebec, northern Labrador and Northwest Territories. Geol. Surv. Can. Mem. 393. 99 pp. with 17 maps, scale 1 : 250,000

Taylor, F. C. 1985. Precambrian geology of the Half Way Hills area, District of Keewatin. Geol. Surv. Can. Mem. 415. 19 pp. with Map 1602A, scale 1 : 50,000

Tella, S., Eade, K. E. 1986. Occurrence and possible tectonic significance of high-pressure granulite fragments in the Tulemalu fault zone, District of Keewatin, N.W.T., Canada. Can. J. Earth Sci. 23: 1950–62

Tella, S., Annesley, I. R., Borradaile, G. J., Henderson, J. R. 1986. Precambrian geology of parts of Tavani, Marble Island and Chesterfield Inlet map areas, District of Keewatin: a progress report. Geol. Surv. Can. Pap. 86-13. 20 pp.

Thom, A., Arndt, N. T., Chauvel, C. 1987. Volcanic arcs and granitoids in the Reindeer zone, Saskatchewan. Geol. Assoc. Can. Programs Abstr. 12: 95 (Abstr.)

Thomas, A., Nunn, G. A. G., Krogh, T. E. 1986. The Labradorian orogeny: evidence for a newly identified 1600 to 1700 Ma orogenic event in Grenville Province crystalline rocks from central Labrador. See Moore et al 1986, pp. 175–89

Thomas, A., Nunn, G. A. G., Wardle, R. J. 1985. A 1650 Ma orogenic belt within the Grenville Province of northwestern Canada. See Tobi & Touret 1985, pp. 151–61

Thomas, M. D., Sharpton, V. L., Grieve, R. A. F. 1987. Gravity patterns and Precambrian structure in the North American Central Plains. Geology 15: 489–92

Thomas, J. J., Shuster, R. D., Bickford, M. E. 1984. A terrane of 1,350- to 1,400-m.y.-old silicic volcanic and plutonic rocks in the buried Proterozoic of the mid-continent and in the Wet Mountains, Colorado. Geol. Soc. Am. Bull. 95: 1150–57

Thompson, P. H., Culshaw, N., Buchanan, J. R., Manojlovic, P. 1986. Geology of the Slave Province and Thelon Tectonic Zone in the Tinney Hills–Overby Lake (west half) map area, District of Mackenzie. In Current Research, Part A. Geol. Surv. Can. Pap. 86-1A, pp. 275–89

Tippett, C. R., Heywood, W. W. 1978. Stratigraphy and structure of the northern Amer Group (Aphebian), Churchill structural province, District of Keewatin. In Current Research, Part B. Geol. Surv. Can. Pap. 78-1B, pp. 7–11

Tirrul, R. 1983. Structure cross-sections across Asiak foreland thrust and fold belt, Wopmay orogen, District of Mackenzie. In Current Research, Part B. Geol. Surv. Can. Pap. 83-1B, pp. 253–60

Tirrul, R. 1985. Nappes in Kilohigok basin, and their relation to the Thelon tectonic zone, District of Mackenzie. In Current Research, Part A. Geol. Surv. Can. Pap. 85-1A, pp. 407–20

Tirrul, R. 1987a. Precambrian geology, Calliope Lake area, central Asiak foreland thrust and fold belt, Wopmay orogen, Northwest Territories. Geol. Surv. Can. Map 1654A, scale 1 : 50,000. In press

Tirrul, R. 1987b. Precambrian geology, Asiak River area, northern Asiak foreland thrust and fold belt, Wopmay orogen, Northwest Territories. Geol. Surv. Can. Map 1653A, scale 1 : 50,000. In press

Tobi, A. C., Touret, J. L. R., eds. 1985. The Deep Proterozoic Crust in the North Atlantic Provinces. Dordrecht, Neth: Reidel. 603 pp.

Todt, W., Chauvel, C., Arndt, N., Hofmann, A. W. 1984. Pb isotopic composition and age of Proterozoic komatiites and related rocks from Canada. Eos, Trans. Am. Geophys. Union 65: 1129 (Abstr.)

van Breemen, O., Aftalion, M., Allaart, J. H. 1974. Isotopic and geochronologic studies on granites from the Ketilidian mobile belt of South Greenland. Geol. Soc. Am. Bull. 85: 403–12

van Breemen, O., Davidson, A. 1987. Northeast extension of Proterozoic terranes of midcontinental North America. Geol. Soc. Am. Bull. In press

van Breemen, O., Henderson, J. B., Loveridge, W. D., Thompson, P. H. 1987a. U-Pb zircon and monazite geochronology and zircon morphology of granulites and granite from the Thelon Tectonic Zone, Healey Lake and Artillery Lake map areas, N.W.T. In Current Research, Part A. Geol. Surv. Can. Pap. 87-1A, pp. 783–801

van Breemen, O., Thompson, P. H., Hunt, P. A., Culshaw, N. 1987b. U-Pb zircon and monazite geochronology from the northern Thelon Tectonic Zone, District of Mackenzie. In Radiogenic Age and Isotopic Studies. Geol. Surv. Can. Pap. 87-2. In press

van der Leeden, J., Bélanger, M., Danis, D., Girard, R., Martelain, J. 1988. Lithotectonic domains in the high-grade terrain east of the Labrador Trough (Quebec). See Lewry & Stauffer 1988. In press

Van Schmus, W. R. 1980. Chronology of igneous rocks associated with the Penokean orogeny in Wisconsin. See Morey & Hanson 1980, pp. 159–68

Van Schmus, W. R., Anderson, J. L. 1977. Gneiss and migmatite of Archean age in the Precambrian basement of central Wisconsin. Geology 5: 45–48

Van Schmus, W. R., Bickford, M. E. 1981. Proterozoic chronology and evolution of the mid-continent region, North America. See Kröner 1981, pp. 261–96

Van Schmus, W. R., Hinze, W. J. 1985. The midcontinent rift system. Ann. Rev. Earth Planet. Sci. 13: 345–83

Van Schmus, W. R., Schledewitz, D. C. P. 1986. U-Pb zircon geochronology of the Big Sand Lake area, northeastern Manitoba. In Report of Field Activities 1986, pp. 207–10. Winnipeg: Manitoba Miner. Res. Div.

Van Schmus, W. R., Bickford, M. E., Lewry, J. F., Macdonald, R. 1987a. U-Pb geochronology in the Trans-Hudson Orogen, northern Saskatchewan, Canada. Can. J. Earth Sci. 24: 407–24

Van Schmus, W. R., Bickford, M. E., Zietz, I. 1987b. Early and Middle Proterozoic provinces in the central United States. See Kröner 1987, pp. 43–68

Van Schmus, W. R., Card, K. D., Harrower, K. L. 1975a. Geology and ages of buried Precambrian basement rocks, Manitoulin

Island, Ontario. *Can. J. Earth Sci.* 12: 1175–89

Van Schmus, W. R., Medaris, L. G. Jr., Banks, P. O. 1975b. Geology and age of the Wolf River Batholith, Wisconsin. *Geol. Soc. Am. Bull.* 86: 907–14

Van Schmus, W. R., Persons, S. S., Macdonald, R., Sibbald, T. I. I. 1986. Preliminary results from U-Pb zircon geochronology of the Uranium City region, northwest Saskatchewan. In *Summary of Investigations 1986. Saskatchewan Geol. Surv. Misc. Rep. 86-4*, pp. 108–11

Walcott, R. I., Boyd, J. B. 1971. The gravity field of northern Alberta, and part of Northwest Territories and Saskatchewan, with maps. *Earth Phys. Branch Gravity Map Series No. 103-111*, Ottawa. 13 pp.

Wanless, R. K. 1979. Geochronology of Archean rocks of the Churchill Province. *Geol. Assoc. Can. Programs Abstr.* 4: 85 (Abstr.)

Wardle, R. J. 1982. Geology of the south-central Labrador Trough. *Newfoundland Miner. Dev. Div. Maps 82-5 and 82-6*, scale 1 : 100,000 and cross-sections

Wardle, R. J. 1983. Nain-Churchill province cross-section, Nachvak Fiord, northern Labrador. In *Current Research. Newfoundland Miner. Dev. Div. Rep. 83-1*, pp. 68–90

Wardle, R. J. 1984. Nain-Churchill province cross-section: Riviere Baudancourt–Nachvak Lake. In *Current Research. Newfoundland Miner. Dev. Div. Rep. 84-1*, pp. 1–11

Wardle, R. J., Bailey, D. G. 1981. Early Proterozoic sequences in Labrador. See Campbell 1981, pp. 331–58

Wardle, R. J., Rivers, T., Gower, C. F., Nunn, G. A. G., Thomas, A. 1986. The northeastern Grenville Province: new insights. See Moore et al 1986, pp. 13–29

Watters, B. R., Thomas, D. J., Harper, C. T. 1987. Tectonic setting of Early Proterozoic volcanism of the La Ronge domain, northern Saskatchewan. *Geol. Assoc. Can. Programs Abstr.* 12: 100 (Abstr.)

Weber, W., Scoates, R. F. J. 1978. Archean and Proterozoic metamorphism in the northwestern Superior Province and along the Churchill-Superior boundary, Manitoba. See Fraser & Heywood 1978, pp. 5–16

White, S. H., Burrows, S. E., Carreras, J., Shaw, N. D., Humphreys, F. J. 1980. On mylonites in ductile shear zones. *J. Struct. Geol.* 2: 175–87

Williams, M. L., Grambling, J. A. 1987. Mid-crustal exposure of a Proterozoic orogenic belt. *Geol. Soc. Am. Abstr. With Programs* 19: 890–91 (Abstr.)

Wilson, J. T. 1962, The effect of new orogenic theories upon ideas of the tectonics of the Canadian Shield. In *The Tectonics of the Canadian Shield. R. Soc. Can. Spec. Publ. No. 4*, ed. J. S. Stevenson, pp. 174–80. Toronto: Univ. Toronto Press. 180 pp.

Wooden, J., Miller, D., Elliott, G. 1986. Early Proterozoic geology of the northern New York Mountains, SE California. *Geol. Soc. Am. Abstr. With Programs* 18: 424 (Abstr.)

Wooden, J. L., Stacey, J. S., Doe, B. R., Howard, K. A., Miller, D. M. 1987. Pb isotopic evidence for the formation of Proterozoic crust in the southwestern United States. In *Metamorphic and Crustal Evolution of the Western United States. Rubey Vol. 7*, ed. W. G. Ernst. Englewood Cliffs, NJ: Prentice-Hall. In press.

Zartman, R. E. 1974. Lead isotope provinces in the Cordilleran of the western United States and their geologic significance. *Econ. Geol.* 69: 792–805

Zolnai, A. I., Price, R. A., Helmstaedt, H. 1984. Regional cross section of the Southern Province adjacent to Lake Huron, Ontario: implications for the tectonic significance of the Murray Fault Zone. *Can. J. Earth Sci.* 21: 447–56

Zwanzig, H. V. 1976. Laurie area (Fox Lake project). In *Report of Field Activities 1976*, pp. 26–32. Winnipeg: Manitoba Energy Mines Miner. Res. Div.

Ann. Rev. Earth Planet. Sci. 1988. 16: 605–54

CONCEPTS AND METHODS OF HIGH-RESOLUTION EVENT STRATIGRAPHY

Erle G. Kauffman

Department of Geological Sciences, University of Colorado CB-250, Boulder, Colorado 80309

INTRODUCTION

Modern geological analyses involving sedimentary rocks require a high-resolution system of dating, integrating, and correlating diverse stratigraphic, geochemical, and paleontological data that also takes into account the possibility that short-term phenomena (100 kyr or less) exercise a strong control on sedimentation. Such phenomena may be extraterrestrial, tectonic, volcanic, oceanographic, climatic, sedimentologic, and/or biologic in origin. Some reflect unpredictable perturbations of local to global scale. But most short-term phenomena are predictable and may be either autocyclic (locally regulated, with limited stratigraphic continuity) or allocyclic in nature (regionally to globally regulated, with extensive stratigraphic continuity) (Beerbower 1964). A stratigraphic system based on short-term phenomena would ideally be chronostratigraphic, involving the identification and regional tracing of "time lines" (isochronous surfaces or very thin event deposits), and easily integrated with refined, independently derived biostratigraphic and geochronologic systems.

An intensive search for such a new and revolutionary system of stratigraphy has been prompted in many parts of the Phanerozoic by attainment of "plateaus" in stratigraphic resolution utilizing the more standard tools of magnetostratigraphy, biostratigraphy, and geochronology: Most magnetic reversals have been identified; biostratigraphic systems can be no more refined than the maximum evolutionary rates of component taxa and their major structures (known for many favored groups); and geochronology is regulated by the analytical confidence limits of radiometric

605

or other dating techniques. For the Cretaceous of North America, as an example, magnetostratigraphy is only useful for detailed correlation within the Berriasian-Hauterivian and Maastrichtian stages because of the long middle and early Late Cretaceous quiet interval. Average durations of Cretaceous intervals bounded by magnetostratigraphic "surfaces" are 1.6 Myr. The interval of polarity reversal in magnetostratigraphy is estimated at 4–5 kyr (Tarling 1983), but such reversal intervals are too widely spaced in the Cretaceous to be consistently useful in high-resolution correlation. Similarly, even with $^{39}Ar/^{40}Ar$ dating techniques, geochronology cannot consistently resolve Cretaceous ash or igneous rock dates to any higher resolution than 0.5 Myr; such dates can only be obtained from rocks that are relatively scattered through the stratigraphic column. Finally, biostratigraphic zonation in the Cretaceous of the Western Interior Basin of North America, whether utilizing ammonites (e.g. Cobban 1951, 1958, 1969, 1971, 1984, 1985), inoceramid bivalves (Kauffman 1975, 1976), or composite assemblage biozones (Kauffman 1970), has reached a limit of resolution of 0.1–0.5 Myr per biozone, depending upon the stage analyzed (Kauffman 1970, 1977).

As stratigraphic observation in field and subsurface data (aided by new logging techniques) has recently become more detailed and comprehensive, many stratigraphers have recognized that short-term to isochronous event deposits of millimeter to meter scale are far more common than predicted by prevalent uniformitarian philosophy in geology, based on today's highly variable, environmentally resilient, glacially influenced Earth that favors autocyclically dominated sedimentary systems. In fine-grained basinal facies, short-term event deposits may in fact dominate the stratigraphic record; most of them reflect widespread allocyclic forcing mechanisms such as rapid regional tectonic movements, tsunamis, explosive volcanism, extraterrestrial impact, rapid shifts in ocean currents and stratification, giant and prolonged storm events, influence of short-term climate cycles, major climate perturbations, and, among the global biota, mass extinctions and regional short-term evolutionary, migration, productivity, colon-ization, or mass mortality events. Many of these phenomena are regarded as "geologically instantaneous" in terms of their stratigraphic expression, i.e. they represent anywhere from a few hours to 100 kyr in time; they are essentially isochronous or near-isochronous deposits and, as such, comprise a working chronostratigraphy based on real data.

In the last few years, stratigraphic geologists and paleobiologists have become increasingly interested in the use of regionally distributed, short-term sedimentary "event deposits" as a new tool of refined regional and global correlation to supplement existing systems. The publication of *Cyclic and Event Stratification* (Einsele & Seilacher 1982) provided an important compendium of stratigraphic models for diverse event deposits. Widespread application of these data to development of a practical system of chronostratigraphy for sedimentary basins around the world is an important revolution in stratigraphic geology—termed *high-resolution event stratigraphy* (*HIRES*). In principle, HIRES proceeds where other systems of stratigraphy leave off in refinement, at the 100 kyr and less level of resolution and regional correlation. This concept, coupled with the recognition that many of the dynamic forces shaping basin history in marine systems operate primarily on a short-time basis, gives credence to attempts to develop a high-resolution event chronostratigraphy within and between sedimentary basins.

High-resolution event stratigraphy is rapidly evolving as the working tool of chronostratigraphy (North American Commission on Stratigraphic Nomenclature 1983). Its primary purpose is to provide an independent means of regional and interregional correlation based on isochronous to near-isochronous surfaces/strata. Proponents of HIRES predict that, given enough data, correlation based on such surfaces/strata can be consistently resolved to 100 kyr or less intervals of time. This has already been achieved in middle and Late Cretaceous sequences of the Western Interior Basin of North America, where over 1300 volcanic ash (bentonite) beds, hundreds of climate cycle beds, and many other event units allow division of the marine record into 40–50-kyr-long (average) event units (Kauffman 1986a, 1987, Kauffman et al 1987).

The evolution of HIRES as a concept and as a working tool of stratigraphy has been hindered by two historical factors: (*a*) The strict application of uniformitarian philosophy to the interpretation of geological processes and their stratigraphic reflection; and (*b*) the broadly held view, based on application of the modern Earth as a model for the geological past, that localized autocyclic processes dominate in the development of the stratigraphic record. Both concepts argue against an important role for regional to global unpredictable short-term events (perturbations) and the effects of predictable allocyclic forcing mechanisms (e.g. through eustatic sea-level changes, climate cycles, major impacting events, etc) in shaping the stratigraphic record. Uniformitarian concepts predict a stratigraphic record dominated by strata with restricted geographic extent, and thus limited correlation potential; this is the antithesis of HIRES, which depends largely upon allocyclically influenced sedimentation and major environmental perturbations to produce individual cyclic or event strata with significant regional extent. The key question to both uniformitarian and autocyclic arguments is whether or not the present is really the key to the past, or is it the other way around?

The uniformitarian concept of autocyclic dominance on sedimentation may in part be defensible from observation of the modern Earth, with its highly exaggerated diversity and resiliency of environments, high seasonality, comparatively low sea level, cool temperatures, and steep climatic and environmental gradients associated with extensive ice-bound poles. But today's pattern of environments and sedimentation does not make a good model for the 90% or more of geologic time that probably lacked permanent polar ice sheets. The great majority of geological history was characterized instead by significantly higher sea level (Vail et al 1977, Haq et al 1987), essentially ice-free polar areas, an absence of cold climatic zones, lower seasonality, and warmer, more equable, maritime-dominated global climates. Marine systems would have predictably been less variable in terms of temperature, chemistry, and circulation and more intensely stratified, with more sluggish circulation and lower dissolved oxygen levels than those systems of today. Such environments would result in broad expansion, equalization, and stabilization of global depositional environments and thus in more delicately balanced environmental and biological systems, especially in the marine realm. This, in turn, would diminish autocyclic effects on sedimentation.

Because of the less variable, more delicately balanced aspect of many environmental systems during most of geologic time, even moderate-level regional to global perturbations and allocyclic processes (related to, for example, rapid tectonic movements; explosive volcanism; monsoonal storm events; Milankovitch and other climate cycles; abrupt changes in sea level, ocean chemistry, circulation, or stratification; and diverse extraterrestrial phenomena) should have had a much enhanced and more obvious effect on sedimentation during these periods. This is especially true for shallow coastal and epicontinental seas. The net effect of "normal" Precambrian-Phanerozoic environmental conditions would therefore be to favor dominance of allocyclic, short-term sedimentary processes with regional extent over more localized, autocyclic processes. This, in turn, should enhance the possibility of developing high-resolution event-chrono-stratigraphic systems for much of the stratigraphic column.

Tests of this hypothesis, and of the feasibility of HIRES, have been made for the Cretaceous in the Western Interior Basin of North America, the site of one of the largest and most-studied epicontinental seas of the Phanerozoic [see Kauffman (1984b, 1985a, 1986a, 1987) and papers in Pratt et al (1985) and in Kauffman et al (1987)]. Examples are drawn from these works throughout this paper. In this Cretaceous setting, reflecting significantly higher sea level, warmer and more ameliorated, global, maritime-dominated climates, and ice-free polar areas, I estimate that between 75 and 80% of the bedding features in basinal settings, and up to 50% in nearshore settings, reflect short-term perturbations and allocyclic sedimentation processes. These event deposits are dominated by bentonites (volcanic ash falls), Milankovitch climatic cycle deposits (limestone-shale, chalk-marl, etc, bedding rhythms), anoxic to dysaerobic event deposits (dark, laminated organic-carbon-rich shales), concretion horizons, sediment bypass or starvation (lag) surfaces, regional giant storm deposits, and many other event strata. Event-bounded intervals average between 40–50 kyr for much of the section and have regional correlation of hundreds to thousands of square kilometers.

HIRES not only involves an understanding of the major differences between more equable, sensitive ancient environments and the more variable, resilient modern global environments (favoring allocyclic versus autocyclic controls on sedimentation, respectively), but also focuses on a different stratigraphic philosophy. This philosophy recognizes that the past is the key to the present, and that warm, stable, global environmental systems associated with eustatic highstand are much more typical of Earth history than are extensive glacial intervals ("ice ages"), including the present. Widespread, short-term dynamic changes in regional and global systems are more predictable during equable environmental periods. Thus event-stratigraphic units, from large eustatic cyclothems to millimeter-size volcanic, chemical, oceanographic, or biologic event strata, should characterize and even dominate ancient stratigraphic systems. This, in turn, dictates that observational resolution and data-collecting methods in the field should be far more detailed than in normal stratigraphic analyses, and it fosters a philosophy of collecting data—the expectation that one will commonly find regionally extensive, short-term event deposits in a stratigraphic sequence—that enhances development and application of high-resolution event stratigraphy to geological problem solving.

METHODOLOGY

Regional short-term event deposits are characteristically thin (a meter or less and commonly measured in centimeters or millimeters). This reflects both the short-term nature of the event and its sedimentary response and also the sedimentation rates characteristic of fine-grained shelf and basinal facies (0.5–4 cm kyr^{-1} before compaction), which dominate many parts of the Phanerozoic record. Consequently, high-resolution event stratigraphy depends upon observation and collection of key data at the centimeter-or-less scale in even the thickest stratigraphic sequences. It is at times a tedious process, but the rewards of discovery in terms of stratigraphic and paleobiologic phenomena, and the resolution achieved through this process in correlation, are well worth the efforts. To date, research designed to test high-resolution event stratigraphy has been mainly focused on

Mesozoic strata and linked to studies of mass extinction, climatic cycles, and eustatically influenced cyclothems or sequences (Haq et al 1987). In each case, HIRES analysis of numerous stratigraphic sections, cores, and logs within a sedimentary basin has led to definition of abundant, regionally correlative chronostratigraphic units of short (less than 100 kyr) duration.

At each section studied for HIRES data, the entire interval is trenched, with trenches up to a meter wide and as deep as necessary to consistently encounter fresh rock. A precisely machined, centimeter-scale version of the Jacob's staff with a sliding housing for a Brunton compass (the Elder staff, designed by W. P. Elder of the University of Colorado) is used to develop a 1-m flagged grid throughout the stratigraphic section; these flags serve as calibration points for high-resolution stratigraphic description of smaller units. The stratigraphic section is normally measured and described by two or more scientists to provide observational checks and balances. Each is responsible for collecting different data and samples, and ideally a physical stratigrapher, a paleontologist, and a sedimentary geochemist of sedimentologist work the sections simultaneously. Observations are made at the finest scale possible, but certainly at least at the centimeter scale (the average thickness of most bentonites). Individual stratigraphic units described and sampled during HIRES analysis may be as small as a few millimeters, or, if sedimentation has been monotonous and uninterrupted by events over a long interval, as large as several meters. There is no differentiation in technique made between basinal fine-grained rocks and basin margin shoreface and foreshore/delta or strand-plain facies; the latter naturally yield thicker individual descriptive units.

In each measured section, in addition to detailed lithologic description of major units, all potential event-stratigraphic units/surfaces are noted, sampled, and described in great detail. Each surface or unit so documented is regarded as a hypothetical isochronous to short-term event deposit to be tested through correlation (standard or graphic techniques) to numerous other sections. These potential event units fall into three basic categories, as diagramatically shown in Figure 1 within the context of a typical upward-fining transgressive hemicyclothem reflecting eustatic rise (modeled after the Cretaceous Greenhorn Cyclothem of Colorado; see papers in Pratt et al 1985).

Physcial event units (*PE*) include volcanic ash and bentonite deposits (the most trusted event units in concept and practice), volcanic flows and tuffs, regional channelization and scour events, giant storm beds, mass flow deposits, regional sediment bypass/starvation surfaces, rapidly formed transgressive disconformity surfaces (diachronous but regionally short term in many cases), and levels of meteorite impact debris; still to be tested are wind-blown silt deposits and synorogenic deposits (initial surface of deposition) resulting from rapid tectonic movement.

Chemical event units (*CE*) are determined from 10-cm to 1-m spot or channel sampling and analysis of such chemical components as C_{org}, C_{carb}, ^{18}O, and ^{13}C, and, less commonly, Sr and S isotopes, rare elements, and noble metals. Two kinds of chemostratigraphic events emerge from these analyses (see Figure 1, left side): (*a*) regionally correlative short-term excursions, or "spikes," of unusual magnitude in the chemical data [e.g. Pratt's (1985) correlation of the global ^{13}C and regional ^{18}O and C_{org} excursions across the Cenomanian-Turonian mass extinction boundary interval in the Western Interior Basin of North America]; and (*b*) bounding surfaces of longer, geochemically abnormal intervals (for example, anoxic or dysaerobic events that represent rapid regional stratification and destratification of the marine water column). Also included in chemostratigraphic events are chemical precipitate layers, primary chert layers resulting from the formation of silica gels on the seafloor, and similar deposits. I include under chemostratigraphic events various types of concretions and nodules formed early in diagenesis along specific isochronous or near-isochronous horizons, probably reflecting some unusual primary chemical or mineralogical character of specific buried sediment layers. In examples from the Cretaceous of the Western Interior Basin of North America, limestone and limestone-siderite concretion zones especially show remarkable parallelism to discrete bentonite beds when traced laterally. Septarian limestone concretions and siderite/limonite nodule zones are equally good in some cases.

Biological event units (*BE*) are deposits representing ecological, evolutionary, and extinction events, and are commonly discrete from biostratigraphic zones or zone boundaries. Thus, they may be represented by punctuated evolutionary events, by mass mortality surfaces resulting, for example, from rapid onset or overturn of anoxic watermasses, or by large volcanic ash falls. They may represent discrete steps in mass extinction (e.g. Elder & Kirkland 1986, Kauffman 1986a), rapid regional dispersal and colonization events reflecting oxygenation of a stagnant seafloor after a long interval of anoxia, and/or rapid changes in substrate characteristics. Immigration or emigration events associated with rapid watermass movements (Kauffman 1984b), population bursts, and productivity events producing a unique sedimentary deposit (e.g. a foraminiferal ooze) also comprise the kinds of data applied to bioevent stratigraphy (Kauffman 1986a).

Not all event-stratigraphic units/surfaces fall directly into one of these three categories; some are clearly *composite event units* (*CPE*) combining physical, chemical, and/or biological event characteristics in their definition. Simple examples (Figure 4*A*) would be a major volcanic ash fall (physical event) that had a unique elemental composition or chemical effect on sediments and/or watermasses (chemical event), and that caused both a mass mortality below it (by suffocation) and a unique sediment surface for colonization above it due to a change in dominant grain size or the sealing off of toxic pore-water seepage at the sediment-water interface; either of these changes would produce widely traceable bioevents.

The most important composite event units (e.g. in the Cretaceous of the Western Interior Basin of North America), second only to volcanic ashes or bentonites in regional event correlation, are Milankovitch or similar climate cycle deposits (Figure 15). These bedding rhythms represent alternating dry (possibly warmer) and wet (possibly cooler) climates produced by variation in orbital parameters of the Earth-Sun system (Barron et al 1985, Fischer et al 1985, and references therein). Average Phanerozoic durations of these cycles and the sedimentary deposits influenced by them are calculated to be 20–25 kyr (variation in orbital eccentricity/precession), 40–50 kyr (variation in axial obliquity of the Earth), and 100–125 kyr ["bundles" of Fischer et al (1985), reflecting orbital eccentricity]. Fischer et al (1985) provide the most comprehensive assessment of the potential stratigraphic record of these allocycles and their sedimentary response. Barron et al (1985) present a detailed case history supporting climatic forcing of limestone-shale (or chalk-marl) cycles of sedimentation in the Cretaceous Greenhorn Formation of the Western Interior Basin; Kauffman (1986a) developed sedimentary models for the effects of Milankovitch

Figure 1 Schematic model for components and methods of high-resolution event stratigraphy (HIRES), plotted against an upward-fining transgressive (eustatic rise) stratigraphic sequence. Physical (PE), chemical (CE), biological (BE), and composite events (CPE) determined from centimeter-scale description of a stratigraphic column are composited to right into an integrated event stratigraphy (IES) with an average HIRES interval of 50–100 kyr at present. Key (left to right): TOC, total organic carbon (in weight percent) curve; STRAT, stratigraphic column; PE, physical events such as (down-column) Milankovitch climate cycle and/or productivity event limestones, volcanic ash or bentonite deposits (dark bands and X's), early diagenetic concretion zones, storm beds (thin sandstones), regional bypass surfaces or disconformities of short duration (wavy lines), mass flow deposits (graded beds), ferruginous and phosphatic nodule horizons (dark spheres), channelization events, paleosols, and volcanic flows; CE, chemical events from isotopic, carbon, and elemental analyses; BE, biological events such as mass mortalities, immigration and emigration events, colonization and productivity events, and mass extinctions; CPE, composite events such as Milankovitch climate cycle beds, ash falls (mass mortality, recolonization, physical and chemical events), storm beds (mass mortality, recolonization, physical events); IES, integrated event stratigraphy for each column and collectively for many columns in a basin as correlated by graphic techniques. IES can be integrated with both geochronologic and biostratigraphic data, as diagramatically shown on right. Based on actual data from papers in Pratt et al (1985).

climate cyclicity in various facies of an epicontinental seaway, using the Cretaceous of the Western Interior Basin of North America as an example.

Application of climate cycles to HIRES is based on the concept that broad regional climate changes producing alternating wet and dry intervals are amplified by delicately balanced Phanerozoic ocean-climate systems, especially during warm, equable eustatic highstand intervals. Such intervals are further characterized by increased potential for widespread giant

EVENTS

AVERAGE ISOCHRON INTERVAL = 50 - 100 Ka

MODEL FOR HIGH RESOLUTION EVENT STRATIGRAPHY

storm or monsoon events. This exaggerated climate cyclicity should have had a profound effect on weathering, transport, and depositional rates of marine and terrestrial sediments derived from surrounding landmasses; on salinity in the surface waters of epicontinental seas (and thus stratification of watermasses and oxygen content of deeper benthic waters); on turbidity levels; and on the kind and level of biotic productivity in the photic part of the water column. Collectively, these factors should simultaneously affect sedimentation patterns in all facies, from marginal marine and shoreface to deep basinal settings (Figure 15A). This hypothesis has been successfully tested in a basin-to-margin transect of middle Cretaceous strata in the Western Interior Basin of North America (Elder, Gustason & Sageman, in Kauffman et al 1987), in which several individual ± 100 kyr Milankovitch climate cycles were traced relative to specific marker bentonites and bio-stratigraphic zone boundaries from limestone (dry)–calcareous shale (wet) bedding rhythms in the center of the basin, into marine shale (dry)–sandstone (wet) progradational units at the basin's western margin.

Detailed field documentation of short-term event deposits must be supplemented by laboratory analyses to reveal data on such things as small-scale sedimentologic and chemostratigraphic changes, impact events, and many biological event units/surfaces. This further requires a rigorous sampling program. Normally, bulk samples are taken either continuously or at very closely spaced intervals for laboratory analysis in each section. This is regulated by sedimentation rate and the nature of the events under study. Typically, 10–20 cm channel samples for micropaleontological and chemical analyses are taken at 0.5-m average intervals; 10–20 × 50-cm-thick block (bulk) samples are taken for macropaleontologic and sedimentary analyses, either continuously or at 50-cm intervals; and all unusual facies are sampled in great detail, as are any ash beds or bentonites that appear to be thick enough and rich enough in biotite, sanidine, and/or zircons for radiometric and fission-track dating. Subsequent laboratory analyses of these samples are added to the integrated event-stratigraphic data base initially developed in the field for each section.

For each stratigraphic sequence analyzed, high-resolution event-stratigraphic data are composited into an integrated event stratigraphy (IES; see Figure 1), and it is this composite set of chronostratigraphic surfaces and short-term event intervals that is applied to detailed correlation between sections, normally utilizing graphic correlation techniques (Edwards 1984). At present, for the Greenhorn and Niobrara eustatic cyclothems (late Albian–early Campanian) in the Cretaceous of the Western Interior Basin of North America, integrated event stratigraphy has attained levels of correlation between nearby sections averaging 40–50 kyr per event-bounded interval, and averaging 100 kyr or less for more distant correlations spanning hundreds to thousands of kilometers. The expectations for the system were exceeded in preliminary work in this basin, and this work promises higher precision of correlation in sedimentary basins elsewhere in the world and throughout the Phanerozoic. Many workers now recognize that event bedding is a common to dominant component of marine strata everywhere.

SELECTED CRETACEOUS EXAMPLES OF HIGH-RESOLUTION EVENT UNITS, WESTERN INTERIOR BASIN, NORTH AMERICA

During the middle and Late Cretaceous, the great Western Interior foreland basin of North America was broadly flooded by a shallow epicontinental sea, 1000 mi east-west and 3000 mi north-south in dimensions, connecting the cool temperate Circumboreal Ocean and its biotas to the north with an extension of the subtropical to warm temperate Caribbean Sea and Gulf of Mexico and its biotas to the south for over 36 Myr. The deposits of this epicontinental sea reflect highly diverse environmental settings (Figure 2). The tectonically active Cordilleran fold-and-thrust belt lay to the west. Eastward, the seaway covered five major tectonic belts: (a) a rapidly subsiding, coarse, siliciclastic-filled foredeep basin; (b) a north-south linear forebulge zone of active discontinuous uplifts and arches over which sediments were episodically eroded; (c) a broad, deep marine axial basin characterized by fine-grained shales and limestones and restricted marine circulation; (d) a tectonically active eastern hinge zone characterized by small, intermittently active uplifts and depressions; and (e) a broad, shallow eastern cratonic platform characterized by thin, disconformity-shortened sequences of fine-grained siliciclastic and carbonate rocks. Kauffman (1969, 1977, 1984b, 1985a, 1986a, and references therein) has summarized the tectonic, volcanic, sedimentary, and oceanographic history of this basin utilizing detailed studies of many workers.

The Cretaceous of the Western Interior Basin was affected, in its evolution, by episodic large-scale periods of tectonism and volcanism almost wholly correlated with intervals of eustatic rise and regional transgression, between which there were intervals of relatively low tectonic and volcanic activity associated with eustatic fall and regional regression. Major (second- and third-order) eustatic fluctuations caused large-scale changes in the size, shape, depth, and environments of the seaway every 5–10 Myr. The seaway was characteristically slightly brackish, as determined from faunal analyses, and somewhat dysaerobic during much of its history.

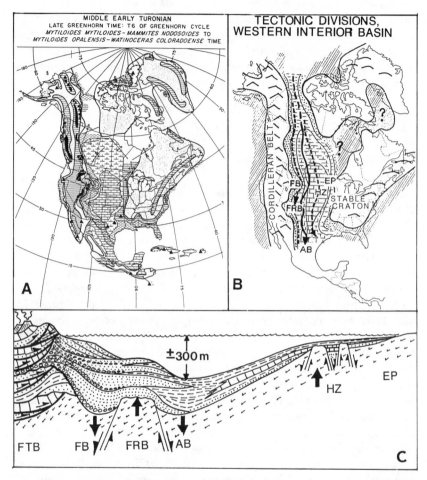

Figure 2   Distribution of facies and tectonic zones in the Cretaceous Western Interior Basin of North America (modified after Kauffman 1984b, 1985a). (*A*) Facies map of the Western Interior of North America at peak mid-lower Turonian eustatic highstand showing size and shape of seaway; HIRES techniques have been largely developed and tested in this area. (*B*) Generalized facies map during eustatic highstand showing active tectonic zones within the basin during the Cretaceous. Key to (*B*) and (*C*) as follows: FTB, Cordilleran fold-and-thrust belt; FB, foredeep basin; FRB, forebulge zone (in this basin a linear series of isolated Precambrian and Paleozoic tectonic blocks reelevated during the Cretaceous); AB, axial or central basin characterized by greatest water depths and fine-grained marine strata; HZ, hinge zone between the stable craton to the east (right) and the axial basin; EP, epicontinental or cratonic platform. Kauffman (1984b, 1986a; Figure 18 herein) has shown that most active tectonism and volcanism in these zones is correlated with eustatic rise events.

Normal marine salinity and circulation/oxygenation were achieved mainly during peak eustatic highstand intervals. Widespread dysaerobic and anaerobic events, reflecting long intervals of density stratification and sluggish basinal circulation of the seaway, were common (especially during eustatic rise and highstand intervals) and rapidly emplaced or broken down. Temperature ranged from subtropical (southern aperture) to cool temperate (northern basin), but this broad horizontal and vertical thermal gradient was frequently punctuated by rapid northward incursions of warm subtropical–warm temperate watermasses associated with mid- to peak eustatic rise, followed by rapid southern retreat of these watermasses during early eustatic fall. Kauffman (1969, 1977, 1984b, 1985a, 1986a) provides detailed analyses of the geological and oceanographic characteristics of the Western Interior Seaway during the Cretaceous.

The Western Interior Seaway of North America was thus large, persistent, environmentally diverse, open at both ends to allow mixing of watermasses, thermally equable and broadly graded. It was characterized by sluggish circulation, lower-than-normal oxygen levels, and frequent intervals of density (salinity, thermal) stratification. It was a delicately perched water body, developed during typical Phanerozoic conditions, that was highly sensitive to perturbations or cyclic environmental changes over broad areas (i.e. to allocyclic forcing of the sedimentary record within short time intervals), conditions favoring high-resolution event stratigraphy. The stratigraphic record of this seaway is probably better studied, with more precise data, than that for any similar epicontinental sea formed during the Phanerozoic. For these reasons, it has been a primary testing ground for concepts and methods of HIRES. In the following pages, examples are drawn from this region to demonstrate the nature and application of HIRES to geological problem solving.

PHYSICAL EVENTS

Volcanic Ash and Bentonite Deposits (Figure 4)

Explosive volcanism associated with major intervals of tectonic deformation in the Western Interior Basin was intermittently very common; bentonite forms 2–10% or more of the marine basinal sedimentary record after compaction (Kauffman 1985a; E. G. Kauffman & D. Beeson, in preparation), and 85% of all volcanic ash/bentonite deposits are associated with tectonically active intervals of basin history (Kauffman 1984b, 1985a), especially along the western Cordilleran fold-and-thrust belt. Over 750 discrete bentonite layers have been described and traced over at least 100 km in the late Albian Kiowa–Skull Creek Cyclothem, the latest Albian–

middle Turonian Greenhorn Cyclothem, and in the late Turonian–early Campanian Niobrara Cyclothem (18 Myr, average of 41.7 persistent ashes Myr^{-1}). Up to 1300 bentonites are projected for the entire Cretaceous marine section once it is studied through HIRES. This suggests a potential regional HIRES correlation interval averaging between 23.8 and 26.5 kyr based on volcanic ash events alone, the most trusted of event markers, for the Western Interior Basin. Ashes and bentonites are essentially isochronous surfaces from which a working chronostratigraphy and an initial graphic axis of real time can be best developed in correlation.

The major problem in HIRES bentonite/ash correlation is the proof of absolute equivalency of each ash from section to section. The ash record of the Cretaceous sequence of the Western Interior Basin was sourced from several places (Idaho-Montana, Arizona–New Mexico, Texas, and possibly the Pacific island arc, the Caribbean island arc, and the Sierra Nevadas). Each area probably had different chemical phases of volcanism. Therefore, sequential elemental analyses of ashes in each stratigraphic sequence would seem to be the primary means of ensuring accurate correlation. But this is an expensive, time-consuming process, and test runs are only now beginning. More normal means of regional ash correlation include unusual color and lithology as determined in the field; recognition of persistent composite ashes with multiple heavy mineral zones; the geometry of bentonite associations or bundles [i.e. persistent doublets, triplets, or larger stratigraphic bundles between which ashes are sparse (Figure 4B)]; the persistent occurrence of certain bentonite beds with other easily recognized event markers or with widespread biostratigraphic zone boundaries (Figure 3); and through graphic correlation (see the subsequent discussion).

Within the Western Interior Basin, an initial focus on larger, easily recognized, more regionally persistent bentonite deposits has provided an important set of event-stratigraphic units against which smaller ashes, other event units, and biostratigraphic data can be compared. Some large beds, like the X-Bentonite marker bed (Hattin 1975, Kauffman 1985b) and the Clayspur Bentonite (Knechtel & Patterson 1962) marking the lower-middle Cenomanian and the Albian-Cenomanian boundary intervals, respectively, extend over thousands of square miles within the Western Interior Basin. Most significant bentonites with thicknesses of at least 1–2 cm have distributions of at least a thousand square miles. Recent examples of high-resolution ash/bentonite correlations are those of Sageman (1985) for the Hartland Shale Member between Arizona and Kansas (680 mi), and of Elder (1985) and Hattin (1985) for the Cenomanian-Turonian boundary interval over the same transect. Figure 3 shows Hattin's (1985) data for the regional correlation of regionally persistent Cen-

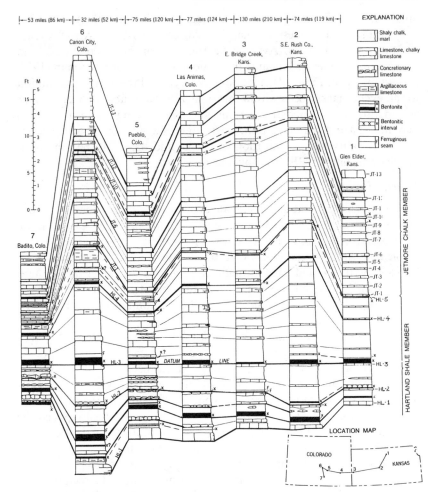

HIRES CORRELATION OF BRIDGE CREEK LIMESTONE, COLORADO TO KANSAS
[MODIFIED FROM D. HATTIN, 1985]

Figure 3   Regional correlation of HIRES units (volcanic ashes/bentonites: bold dark lines and X's) (probable Milankovitch climate cycle beds, especially persistent limestone and limestone concretion units), Hartland and Jetmore members, Greenhorn Limestone, from central Colorado (left) to central Kansas (right; see index map). Figure from work of Hattin (1985; see locality details), as modified to show not only Hattin's marker horizons (HL-1–4, JT-1–13) in bold dark lines, but also probable correlations of other units over this 700+ km transect (thin lines). Average event-bounded interval is about 40 kyr. This is a typical example of the quality and refinement of HIRES correlation in the Cretaceous sequence of the Western Interior Basin of North America, using standard correlation techniques.

omanian-Turonian (Bridge Creek Limestone and equivalents) bentonites vs cyclically bedded limestone-shale bedding couplets (Milankovitch cycles; see Barron et al 1985). The average bentonite event-bounded intervals in this example, compared against the radiometric scale of Kauffman (1977; revised in 1985a, Figure 4), are about 225 kyr, and the average resolution for all event marker beds within this same interval is about 66 kyr, *before* consideration of chemostratigraphic and bioevent units. Large ashfalls may also have a prominent chemostratigraphic and bioevent signature (mass mortality and colonization surfaces), and these falls become composite event units (CPE), as shown in Figure 4*A*.

Storm Beds

Eustatic highstand and global climate amelioration during the Cretaceous and other warm, equable Phanerozoic intervals probably caused more widespread northward and southward migration, and enhancement, of tropical storm belts and monsoonal tracks (Barron & Washington 1984). Seasonally, this would have brought these storms belts over the Western Interior Basin. These factors, as well as the delicately balanced nature of Cretaceous climate/ocean systems, would have combined to enhance the size, regional extent, and sedimentary reflection of storms. Many authors have noted a strong dominance of storm-related sedimentation in Western Interior Basin shoreface and proximal offshore facies belts (Figure 5); field stratigraphers have been able to use storm beds that have not been subsequently destroyed through heavy bioturbation as event marker units across many miles of continuously exposed outcrop. The regional along- and offshore extent of these Cretaceous storm units seems to exceed the predictions based on modern hurricane deposits along the Atlantic Coast, for example. The only real test of storm units for high-resolution event correlation over significant distances, however, is the unpublished work of Glenister (1985), who used graphic correlation to show the regional time equivalency of certain middle Turonian Codell Sandstone Member storm beds (examples in Glenister & Kauffman 1985) for nearly 70 mi along the northern Colorado Front Range in lower shoreface deposits. New data are summarized in Figure 6. Figure 5 illustrates sedimentary structures of a typical Cretaceous storm bed and its effect on the biota.

Regional Bypass and Disconformity Surfaces

Many transgressive disconformities characterize the Cretaceous cyclo-thems of the Western Interior Basin, especially in shoreface sequences and as a result of ravinement (migrating erosional surface cut at normal wave base) processes during early transgression. Except for those formed during exceptionally rapid sea-level rise and/or regional subsidence, most of these

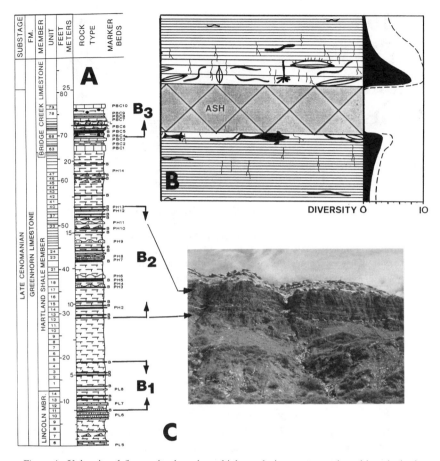

Figure 4 Volcanic ash/bentonite deposits as high-resolution event stratigraphic units in the Cretaceous of the Western Interior Basin of North America. (*A*) Stratigraphic column of the Cenomanian Lincoln, Hartland, and basal Bridge Creek Limestone members, Greenhorn Formation, at Rock Canyon Anticline west of Pueblo, Colorado (after Sageman 1985). Dark bold lines with "B" to right are bentonite deposits arranged in three major clusters, or "swarms" (B_1, B_2, B_3), reflecting intense volcanic activity, with sparser intermediate marker units. These swarms are the first level of regional HIRES correlation, followed by individual ash correlations. (*B*) Model of a single bentonite bed from (*A*) (gray, with X's) showing composite nature of this event having mass mortality bioevent at base (through suffocation), forming prominent physical and chemical event bed (the bentonite), and having regional colonization surface on top as a result of bentonite sealing off toxic interstitial pore-water seepage to sediment-water interface. Diversity reflects benthonic foraminifer species (black) and molluscs (dashed). (*C*) Photograph of (*A*) section, with prominent bentonite swarm represented by white lines across outcrop and bracketed with arrows.

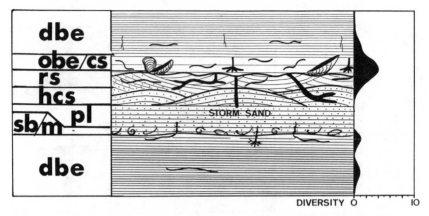

Figure 5 Model of a regional storm bed (lower shoreface expression), taken from the Codell Sandstone near Pueblo, Colorado (middle Turonian) (Kauffman, 1986a), showing the composite event nature of the bed. Note mass mortality surface among burrowing organisms and storm-transported molluscs at base, storm bed as a physical event unit, and colonization surface with increased macrofossil diversity on top (in oxygenated water) before return to laminated shales and dysaerobic benthic conditions (line pattern). Key (base to top): dbe, dysaerobic benthic environment (dark laminated shales); m, mass mortality surface at base of storm bed; sb, basal scour bed of unit; pl, planar laminated, high-flow regime beds of storm unit; hcs, hummocky cross-stratified beds; rs, ripple-laminated beds on upper surface of storm bed; obe/cs, oxygenated benthic environment with colonization surface.

disconformities are regionally diachronous and therefore not suited for high-resolution event stratigraphy. But disconformities and para-conformities that characterize offshore basinal sequences, especially those around eustatic highstands, seem to form regionally within very narrow time intervals and thus to represent bypass and condensation intervals of short duration reflecting sediment starvation. One of the best global examples of this phenomenon is the bypass/starvation/condensation surface associated with peak mid-lower Turonian eustatic highstand (*Watinoceras* and lower *Mammites nodosoides* biozones) in many parts of the world, and as seen in much of the Deep Sea Drilling Project (DSDP) data. In the Western Interior Basin, these strata are partially or completely missing over large parts of the central axial basin, and in western Europe across many somewhat elevated, older massifs. This disconformity surface lies within a 1.0 Myr interval; the diachroneity of bounding depositional surfaces is regionally very slight, probably within 100 kyr from initial tests.

Impact Event Beds

The impact of large extraterrestrial objects on continents and shallow marine shelf areas would predictably throw great quantities of dust-size

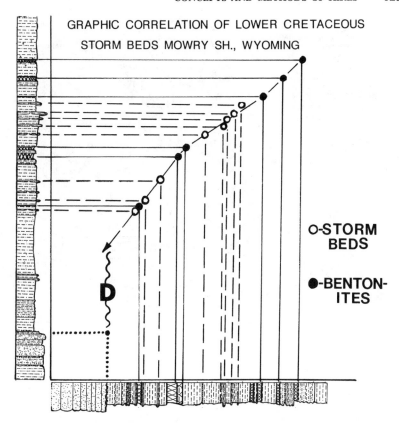

Figure 6   Storm beds as regional event correlation surfaces, as proven by graphic correlation. Graphic plot of two stratigraphic sections of the lower Shell Creek–Mowry Shale units in the western Powder River Basin (Kaycee, Arminto), Wyoming, showing line of isochron correlation drawn through intersection points of same bentonites in two sections (dark dots) and the relationship of distal toes of storm beds (open circles) to this time line between sections. Key: D, major transgressive disconformity on top of Albian Muddy Sandstone, which omits large portion of lower shale section in vertical column.

and larger impact debris into the atmosphere. This would ultimately settle out over the surface of the Earth as a thin, chemically and physically unique deposit within a few years to tens of years. Such a deposit would not only contain a unique mineral/element suite of mixed origin (bolide and target area) but also might contain shock-metamorphic mineral grains, microtekites, ash fragments, and unique concentrations of iridium and other rare elements. The Cretaceous-Tertiary boundary clay, worldwide, represents such a deposit and is the best example of a short-term impact event unit in the Phanerozoic (Alvarez et al 1980, 1984; Figure 7*A* herein).

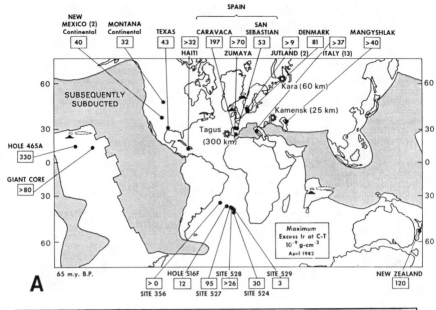

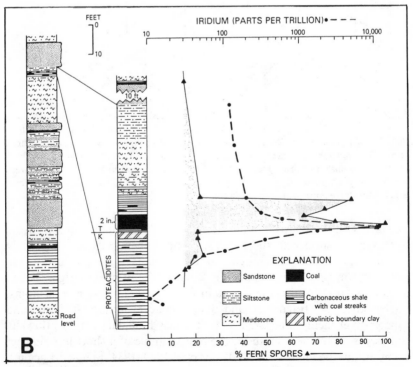

This globally distributed boundary clay is present in the San Juan Basin of New Mexico (Figure 7*B*) at numerous localities (Orth et al 1981, Bohor et al 1984) and contains the iridium-rich chemical signal, shocked quartz and feldspar grains, and chemically unique clay minerals. Recent discovery of two or more iridium peaks just below the Cenomanian-Turonian mass extinction boundary near Pueblo, Colorado (Orth et al 1987), suggests that other "impactites" may be present in the Western Interior Basin as HIRES correlation horizons.

CHEMICAL EVENTS

Closely spaced analyses for organic and carbonate carbon, ^{18}O and ^{13}C isotopic values, and other elements are a normal part of high-resolution stratigraphic research. Major excursions ("spikes") of extraordinary magnitude and short duration (100 kyr or less), or longer intervals characterized by unusual geochemical values (i.e. widespread anoxic or dysaerobic events, especially their boundaries) are the tools of regional Cretaceous chemostratigraphy. The best published examples of these phenomena in the Western Interior Basin of North America are associated with the Cenomanian-Turonian boundary interval (Figure 4); Pratt (1985) has provided the most comprehensive geochemical data. Zelt (1985) has also shown that natural gamma-ray spectrometry can yield data on short-term Th/U spikes that are regionally correlative.

Light-Stable Isotope Chemostratigraphic Events

Pratt (1985) has published an excellent example of light stable isotope event chemostratigraphy in the Western Interior Basin, based on detailed analysis of ^{13}C and ^{18}O values through the latest Albian–middle Turonian Greenhorn Cyclothem. Figure 8*A* shows typical detailed chemostratigraphic data from near Pueblo, Colorado, associated with the Bonarelli global oceanic anoxic event. This major ^{13}C excursion (Figure 8*A*) is a global HIRES chemostratigraphic signal also found widely in DSDP core analyses at the same level (Pratt 1985); in North America, it is precisely correlative with regional ^{18}O and C_{org} excursions (Figure 8). Within this

---

Figure 7   Global event marker bed reflecting chemical and physical fallout from a large meteorite impact at the Cretaceous-Tertiary boundary. (*A*) World map showing dispersion of impact fallout debris (boundary clay) bearing iridium (values for each locality; Alvarez et al 1982). (*B*) Stratigraphic sections of the Cretaceous-Tertiary boundary (K-T) showing kaolinitic boundary clay, iridium spike, and peak in fern spores (ecological generalists and forest floor dwellers in low light situations) reflecting filtered sunlight after impact due to atmospheric dust/debris/smoke cloud (Pilmore & Flores 1987).

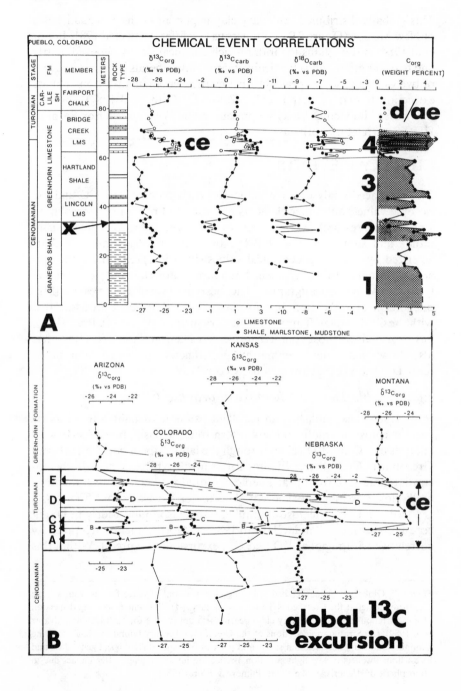

PUEBLO, COLORADO

CHEMICAL EVENT CORRELATIONS

o LIMESTONE
● SHALE, MARLSTONE, MUDSTONE

global ¹³C excursion

global ^{13}C chemoevent marker, at least two regionally correlative positive spikes separated by a short negative excursion (Figure 8*B*) can be defined across the Western Interior Basin. A smaller pair of negative and positive excursions just below and above the Graneros Shale–Lincoln Limestone boundary may be regionally important but remain to be tested. The smaller-scale excursions in these data are of 100 kyr or less magnitude, based on an average radiometric scale (Kauffman 1977, 1985a).

There also exist in these data prominent short-term (50–100 kyr) fluctuations in the $^{18}O_{carb}$ values. The most prominent excursion spans the Cenomanian-Turonian boundary interval, where at least two major positive spikes and one negative short-term ^{18}O spike can be defined regionally as chemoevent units (Pratt 1985). In addition, a rapid set of ^{18}O fluctuations in the upper Graneros Shale and the Lincoln Limestone Member of the Greenhorn Formation (Figure 8*A*) may be of value in HIRES, but these fluctuations remain to be tested regionally.

Organic Carbon Chemostratigraphic Events

Short (25–100 kyr) and long-term density stratification events in the Western Interior Seaway, as well as episodic incursions of oceanic oxygen-minima zones, produced widespread anoxic and dysaerobic intervals that resulted in deposition of regionally correlative, organic-carbon-rich strata. Closely spaced analyses of C_{org} reveal these intervals, which are of particular interest because of their potential as petroleum source rocks.

There are two major aspects of C_{org} chemostratigraphy that are useful in HIRES. The first is short-term spikes in the data representing less than 100 kyr of time and having regional expression. [Figure 8*A* (after Pratt 1985) shows typical data.] Kauffman (1986a) proposed a model whereby many of these could represent the wet phase of wet-dry and possibly cool-warm Milankovitch climate cycles (Figure 15). The effect of this phase (high rainfall and salinity stratification of the seaway, limiting downward mixing of oxygen) would be exaggerated during eustatic highstand, broad

---

Figure 8   High-resolution chemoevent stratigraphy and regional correlation, as shown by organic carbon, ^{13}C, and ^{18}O analyses of the Cenomanian-Turonian boundary interval from Arizona to Montana, and from Colorado to Nebraska (after Pratt 1985). (*A*) Major interval of carbon and stable-isotope disruption around the Cenomanian-Turonian extinction boundary at Pueblo, Colorado. Note position of X-bentonite marker bed (Texas-to-Alberta event bed) (X), major global chemical event interval (ce), and the development of four regionally correlative anoxic/dysaerobic events in the basin (see Figure 10), labeled 1–4 and shaded, as depicted by high C_{org} levels. (*B*) Details of the zone of chemical disruption (ce) shown in (*A*), with data from points throughout the Western Interior Basin. Note regional correlation potential of small-scale isotopic fluctuations (^{13}C) represented by fine lines, with most prominent lines of correlation regionally labeled A–E (modified from Pratt 1985).

global warming, climate amelioration, and sluggish ocean circulation (Figures 10, 15). Second, longer-term anoxic or dysaerobic intervals are represented by relatively thick, regionally correlative, organic-carbon-rich stratigraphic sequences. The apparent rapidity with which these density stratification events became established or broke down, however, as represented in rapid shifts (100 kyr or less) from high to low C_{org} values in strata, indicates that the boundaries of these events may be regionally useful in HIRES. For example, Kauffman (1985b), Sageman (1985), Pratt (1985), and Elder (1985) noted five regionally persistent intervals of high (3% or greater) C_{org} in the latest Albian–early Turonian transgressive phase of the Greenhorn Cyclothem throughout the Western Interior Basin. Four of these are shown by Pratt's (1985) data from Pueblo, Colorado (Figure 8A). When traced regionally relative to individual bentonite or ash beds, C_{org}-enriched intervals appear to consistently occur at the same stratigraphic levels. Their relative synchroneity strongly suggests that the density stratification of the seaway associated with anoxic and dysaerobic intervals became established and broke down very rapidly, usually within 100-kyr intervals or less of strata. L. K. Barlow (in Pratt & Barlow 1985) provided an excellent example of a rapid shift in C_{org}, and in bioturbation levels, associated with onset and breakdown of a major oxygen depletion event during deposition of the Coniacian lower shale unit, Smoky Hill Member, Niobrara Formation, along the Colorado Front Range (Figure 9). Organic carbon levels climb from 1.2 to 5% by weight, and bioturbation levels drop from "severely bioturbated" to "nonbioturbated," within less than 1 m of strata, probably less than 100 kyr in time. After 0.4–0.5 Myr, C_{org} drops from 3.2 to 1.0% and bioturbation abruptly becomes intense again within less than 100 kyr. Both contacts of this regional oxygen depletion event, when traced relative to other event markers (e.g. bentonites), are abrupt and near-isochronous over a wide area in both C_{org} levels and bioturbation. Both contacts are therefore important regional HIRES chemistratigraphic events.

Such sharp changes in benthic and midwater oxygen levels suggest abrupt stratification and destratification of the Western Interior Seaway over wide areas; Kauffman (1986a; Figure 10) has suggested circulation models for these events in which rapid changes in stratification history may be a result of intermittent desalination events during high rainfall and accelerated interior drainage of freshwater runoff; of thermal stratification resulting from overlap of northern and southern watermasses; of rapid incursion of warm tropical/subtropical watermasses from the south as sea level rose sufficiently to breach the southern silled aperture to the Western Interior Basin; and of rapid incursion of oxygen minima zones across the

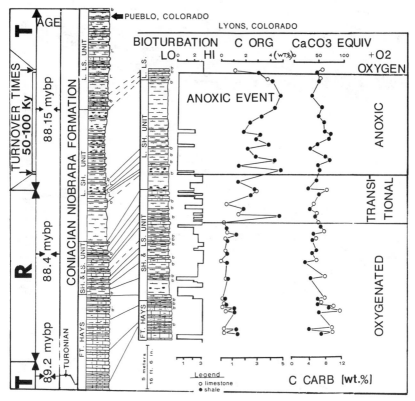

Figure 9 Rapid establishment and overturn of a regional anoxic/dysaerobic event, as reflected in bioturbation levels (left) and C_{org} levels (middle) through the lower Niobrara Formation (Fort Hays and lower Smoky Hill members) at Pueblo and Lyons, Colorado (modified from Barlow & Kauffman 1985). Note that interval of change is 50–100 kyr (one small Milankovitch climate cycle deposit), and that the anoxic event (as elsewhere in section) is correlated with eustatic rise and transgression (T), but not with regression (R).

benthic zone with rising sea level and expanding oceanic oxygen depletion. Figure 10 outlines these models.

Concretion Zones

It has been widely documented that persistent zones of limestone, lime- stone-siderite, siderite, dolomite, and even limonitic concretion and nodu- lar zones regionally retain their relative stratigraphic position when com- pared with persistent bentonite beds. This suggests, therefore, a primary regional (allocyclic) short-term depositional control on their formation.

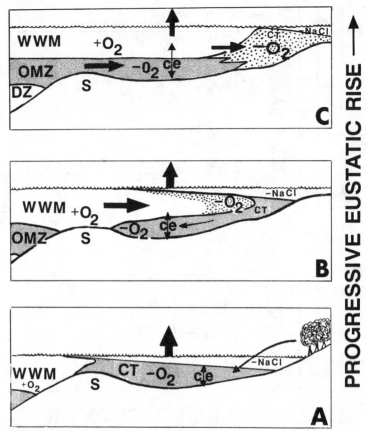

ANOXIC/DYSAEROBIC WATERMASS DEVELOPMENT, CRETACEOUS WESTERN INTERIOR BASIN

Figure 10   Model of water stratification and watermass immigration events associated with rapid development and destruction of anoxic/dysaerobic intervals in the Cretaceous of the Western Interior Seaway of North America (after Kauffman 1986a). (*A*) Early eustatic rise with embayed seas, high freshwater, and terrestrial organic carbon input causes salinity stratification and excessive input of C_{org} to create an early anoxic event (Mowry Shale, Graneros Shale); (*B*) With continued rise of sea level, warm Gulf of Mexico watermasses breach the southern sill of the Western Interior Basin and wedge over cooler, denser northern waters, and below an episodically emplaced subnormal saline wedge, to again stratify the water column and cause anoxic benthic conditions (Hartland Shale, Smoky Hill Shale); (*C*) Near peak transgression, expansion of oceanic oxygen minimum zones causes their incursion into epicontinental seas, already devoid of oxygen (Bridge Creek limestone event, Bonarelli global oceanic anoxic event). Key: WWM, warm water Gulf of Mexico watermass; C-T, cool temperate dense watermass; OMZ, oxygen minimum zone; DZ, dysaerobic deep-ocean watermass; ce, fluctuations of density stratification boundary reflecting Milankovitch climate cycle control.

Many limestone concretion zones laterally grade into persistent limestone beds representing the dry phases of Milankovitch climate cycles. Figure 3 (Hattin 1985) shows this phenomenon with regard to beds of the lower Bridge Creek Limestone Member, Greenhorn Formation, from Kansas to Colorado; Kauffman et al (1987) provide stratigraphic correlation of the same beds to Utah. Similar regional correlations of septarian limestone concretion zones of the upper Blue Hill and lower Codell Members, Carlile Shale, have been noted by Glenister & Kauffman (1985) along the Colorado Front Range outcrop belt. Inasmuch as most concretions seem to represent early diagenetic phenomena rather than primary deposits, it is probable that persistent, near-isochronous concretion zones form in response to some primary regional sedimentary control. Possibly concretions form where there was a concentration of biogenic carbonate or organic material along certain bedding planes, which was then remobilized in such a way as to initiate accretionary carbonate deposition early in diagenesis.

BIOLOGICAL EVENTS

The role of biological events (bioevents) in high-resolution event stratigraphy has recently been discussed in the review by Kauffman (1986a), from which the following synthesis is made. Bioevents include episodes of punctuated evolution, mass mortalities, steps and catastrophes during mass extinction, rapid immigration and emigration events, productivity events and population bursts (acmezones), rapid regional benthic colonization events, and other ecostratigraphic phenomena. In some cases, bioevents mark biostratigraphic boundaries, but the two systems of stratigraphy are not interdependent. Bioevents occur at three scales: local (extending less than 100 mi), regional (basinwide), and intercontinental to global in aspect. The last two are important in HIRES. Many of these types of bioevents are common in the Cretaceous of the Western Interior Basin of North America, from which examples are taken, as well as in other Phanerozoic basins.

Punctuated Evolutionary Events

The rapid origin of new subspecies and species through mechanisms inherent in punctuated evolutionary theory (Eldredge & Gould 1972, and subsequent papers), coupled with rapid, widespread marine dispersal of newly derived taxa through planktotrophic larvae or mobile adult stages (see Kauffman 1975), produces "geologically instantaneous," geographically widespread first appearances of these taxa. These levels comprise not only important biostratigraphic boundaries, but also evolutionary bioevent marker horizons. Kauffman (1986a) gives examples from the Western Interior Basin. The abrupt regional first appearances of new taxa of scaphitid and baculitid ammonites (Cobban 1951, 1962, and other papers) and of inoceramid bivalves (Kauffman 1975, and other papers), within existing lineages of biostratigraphically important Cretaceous molluscs, probably represents to a large degree punctuated evolution coupled with rapid marine dispersal of new taxa. For example, the biostratigraphic zone boundaries associated with the sections shown in Figure 3 are based on the first occurrences of new ammonite and/or inoceramid bivalve species, the abrupt appearance of which, throughout the Western Interior Basin, strongly suggests punctuational evolutionary events. Regionally, these particular zonal boundaries are essentially isochronous, maintaining a nearly constant stratigraphic position relative to bentonite beds and climate cycle-related limestone units (Figure 3). A regional test of this apparent synchroneity, utilizing graphic correlation of zonal ammonite first-appearance data to a line of isochronous correlation (LOIC) constructed from bentonites (Figure 17), shows that first appearances are near-isochronous between sections (punctuational?), whereas last appearances of the same taxa are diachronous (differential extinction rates among subpopulations).

Population Bursts

The onset of favorable environmental conditions for certain taxa may be regionally rapid in delicately balanced, slowly circulating epicontinental seas like that of the Cretaceous Western Interior Basin. Short-lived regional population bursts may result, especially among plankton, leaving a sedimentary bioevent signal. Coccolith-rich varves at 1 cm or less intervals and planktonic foraminifer-rich calcarenites or pelagic limestones are common Cretaceous examples from the Western Interior Basin. A benthic example is the regional (Wyoming to Texas) colonization of strata directly overlying the Cenomanian X-bentonite marker bed (Figure 8A) by abundant calcareous benthic foraminifera and beds or biostromes of the otherwise moderately rare oyster *Ostrea beloiti*. Population bursts among biostratigraphically important ammonite and foraminifer species have been used to define acmezones (epiboles) in the Cretaceous. The persistent short-term (less than 100 kyr) population explosion of several species of calcareous benthic and planktonic foraminifera and diverse molluscs, within a meter of strata encompassing the early middle Cenomanian Thatcher Limestone Member, Graneros Shale (Figure 14), is a regional bioevent marker extending from central Colorado to New Mexico. This interval reflects a short-term immigration event of warm oxygenated water from the Gulf of Mexico region.

Productivity Events

Like population bursts, productivity events represent a short interval of highly favorable conditions for biotic proliferation, but in this case the conditions extend to diverse organisms or whole paleocommunities. In some cases, productivity events are measured by planktonic rain and rates of benthic accumulation of tests; these productivity events commonly produce a major change in lithology toward more calcareous (or siliceous) strata, in contrast to clay-dominated sedimentation in marine basins. In limestone-shale or chalk-marl bedding rhythms associated with Milankovitch climate cycles (Barron et al 1985; Figures 3, 15), the carbonate-enriched portion of the cycle is in part diagenetic, and in part due to increasing primary productivity among calcareous plankton associated with normalization of surface water conditions in the seaway during the dry and potentially warmer climatic intervals (Figure 15). These pelagic biogenic limestones form regional event marker beds second only to volcanic ash/bentonite horizons for regional HIRES correlation across the Cretaceous Western Interior Basin (Hattin 1985, Elder 1985, and references therein).

Ecostratigraphic Events

These represent abrupt widespread changes in community structure as a result of regional (allocyclic) changes in environment (climate, oceanography, paleobathymetry, water chemistry, etc). Detailed bed-by-bed examination of all biotic components in high-resolution stratigraphic analysis is required to see these rapid changes [see Figure 12 for typical data (from Elder 1985)], which maintain their stratigraphic position relative to bentonites and other chronostratigraphic event markers over great distances in the Western Interior Basin. Typical examples are (*a*) the great diversification of warm-water taxa associated with rapid incursion of subtropical watermasses into the Western Interior Seaway during eustatic rise and breaching of the silled southern aperture of the basin (Kauffman 1984b, Elder 1985); the abrupt regional appearance of subtropical and warm temperate molluscan elements of the *Sciponoceras gracile* biozone fauna (Figures 11, 13) represents this event. The previously cited Thatcher Limestone bioevent (Figure 14) is a similar and shorter-term event. (*b*) Rapid changes in benthic faunas associated with the onset and overturn of regional oxygen depletion events, from depauperate inoceramid bivalve–dominated low-oxygen communities to more diverse benthic molluscan communities, has been well documented for the Hartland oxygen depletion event by Sageman (1985, Figures 2, 3), for the recovery from the Cenomanian-Turonian Bonarelli oceanic anoxic event by Elder (1985; and

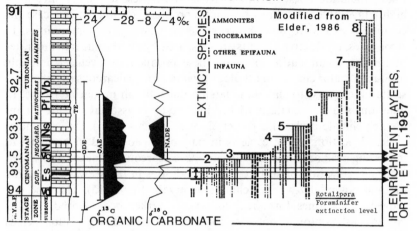

Figure 11   Stepwise mass extinction bioevents in the Cenomanian-Turonian mass extinction interval (modified after Elder 1986). All data compiled here collected at 1–10 cm HIRES intervals. Note stepwise nature of molluscan extinctions, each within 50 kyr or less time (extinction bioevents), the inclusion of most mass extinction events within an interval of global stable isotope (climate, ocean) disruption (ce in Figure 8), and the correlation of lower stepwise extinction events with an interval of iridium enrichment (five small spikes, or chemoevents, marked by thin lines) at Pueblo, Colorado (Orth et al 1987). Letters in column 4 from left refer to ammonite subzones (see Elder 1986). Key: ODE oceanic disruption interval; OAE, oceanic anoxic event (Bonarelli event); NADE, North American desalination event (as indicated by rapid negative excursions of ^{18}O).

Figure 11 herein) and Kauffman (1984a), and for the ichnofauna of the lower Niobrara (lower shale unit) oxygen depletion event by Barlow (in Pratt & Barlow 1985) (Figure 9) in the Cretaceous sequence of the Western Interior Basin.

Immigration-Emigration Bioevents

The silled nature of the southern aperture of the Western Interior Basin, blocked by the central Texas uplift and choked by impinging Precambrian and Paleozoic ranges to the east and west, inhibited influx of warm subtropical waters into the seaway until middle to late stages of eustatic rise during the Cretaceous. Breaching of this sill during eustatic highstand resulted in rapid immigration (and, with lowering sea level, emigration) of subtropical faunal elements into the basin. Kauffman (1984b) documented these rapid changes and estimated their timing as under 0.5 Myr, and in some cases within 100 kyr, over hundreds to thousands of square miles.

ECOSTRATIGRAPHY OF CENOMANIAN–TURONIAN BOUNDARY INTERVAL

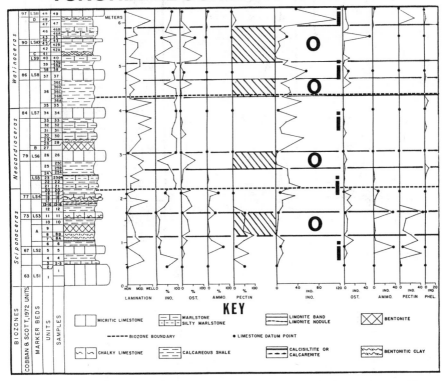

Figure 12 Ecostratigraphy of the Cenomanian-Turonian boundary interval at Pueblo, Colorado (after Elder 1985). Graphs show absolute and relative abundance of major macrofossils; horizontal lines bound taxonomically consistent communities (o, *Ostrea*-dominated; i, *Inoceramus*-dominated) with regional correlation potential (striped lines accentuate *Ostrea* community). Note rapid turnover from one community to another.

The Cenomanian Thatcher Limestone event (Figure 14) lasted about 100 kyr and was characterized by a dramatic change from totally arenaceous to calcareous benthic and planktonic foraminiferal assemblages and by a major increase in molluscan diversity (4 to 36 species, including rudistid bivalves, within 50 cm of rock). This represents an early pulse of an immigration-emigration event associated with a warm-water incursion. The most dramatic of these events in the Western Interior Basin is that associated with the Cenomanian-Turonian boundary interval (Figure 11; Elder 1985, Elder & Kirkland 1986, Kauffman 1984b). Figure 13*A* shows

CENOMANIAN IMMIGRATION EVENTS: WESTERN INTERIOR SEAWAY

A UPPERMOST GRANEROS SH. mid–MIDDLE CENOMANIAN

B MID–HARTLAND SH., MID–LATE CENOMANIAN

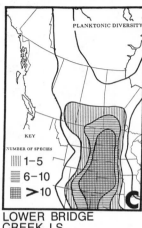

C LOWER BRIDGE CREEK LS LATEST CENOMANIAN

KEY

NUMBER OF SPECIES

||||| 1–5
≡ 6–10
⊞ >10

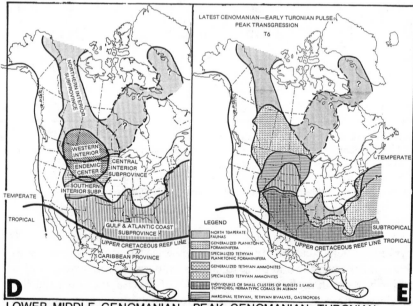

D LOWER-MIDDLE CENOMANIAN PALEOBIOGEOGRAPHIC SUBPROVINCES, NORMAL CRETACEOUS DISTRIBUTION

E PEAK CENOMANIAN–TURONIAN IMMIGRATION OF TROPICAL MOLLUSCS, FORAMINIFERA

MID-CENOMANIAN THATCHER LS IMMIGRATION- EMIGRATION EVENT MODELS

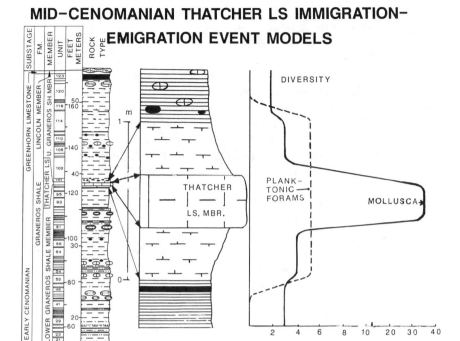

Figure 14 Rapid immigration-emigration event of subtropical molluscs and planktonic foraminifera into the Western Interior Seaway of North America during Thatcher Limestone (middle Cenomanian) time (100 kyr), middle Graneros Shale (Kauffman 1986a). Note rapid increase in carbonate planktonic foraminifers and molluscs within approximately 20 kyr, the short diversity peak, and the equally rapid emigration with a short watermass migration. Background strata are dark organic-carbon-rich clay with mild temperate, low-diversity molluscan communities.

the average paleobiogeographic distribution of temperate subprovinces in the Western Interior Seaway prior to the rapid northward immigration of subtropical and southern warm temperate taxa (Figure 13*B*) during a series of northward pulses of warm watermasses approaching the Cenomanian-

---

Figure 13 Rapid immigration event of tropical-subtropical molluscs and foraminifers into the Western Interior Seaway of North America near the Cenomanian-Turonian boundary, marking peak eustatic rise and transgression. (*A–C*) Maps showing planktonic foraminifer diversity and aerial extent during this 1–2 Myr interval (from Eicher & Diner 1985). (*D, E*) Average paleobiogeographic distribution of Mollusca (predominantly temperate) and (in *E*) distribution of tropical-subtropical faunal elements into the Western Interior Seaway during eustatic highstand (different patterns represent different groups; from Kauffman 1984b). Major immigration time is estimated between 100–500 kyr.

Turonian eustatic high-stand [see Kauffman (1984b) for examples and Kauffman (1986a) for detailed discussion].

Mass Mortality Events

Short-term (100 kyr or less) deterioration of environmental conditions in the Western Interior Seaway linked to mainly allocyclic (oceanographic, climatic) phenomena produced many regional mass mortality surfaces that are useful in high-resolution event stratigraphy. The "fish-scale marker bed" at the Albian-Cenomanian boundary represents one such marker that can be traced from Alberta to Colorado as a single surface or closely spaced series of mass mortalities, probably associated with overturn of the anoxic bottom waters characterizing the upper Albian Mowry Sea (Kauffman 1986a). Sageman (1985) documented similar mass mortalities associated with oxygen depletion events in the upper Cenomanian Hartland Shale Member, Greenhorn Formation, and Barlow (in Pratt & Barlow 1985) (Figure 9) documented onset of one of several Niobrara Formation (Coniacian-Santonian) anoxic events, almost totally decimating the benthic trace and shelly fossil biota, within a meter of strata (100 kyr or less).

The most common type of mass mortality surfaces in the Western Interior Basin are associated with short-term dynamic fluctuations of benthic oxygen within long-term oxygen depletion events. Oxygenated intervals of a few months or years permit settling and initial growth of benthic molluscan larvae, but the entire spat or juvenile crop is abruptly killed off, in situ, during more anoxic intervals. Cretaceous examples are *Entolium* and *Inoceramus* mass mortality surfaces found throughout the upper Cenomanian Hartland Shale Member (Sageman 1985) in the Western Interior Basin. Kauffman (1978, 1981) detailed similar mass mortality surfaces among *Inoceramus*, *Bositra*, and *Ostrea* in the Jurassic Posidonenschiefer and Solnhofen/Nusplingen formations of southern Germany.

Large bentonite falls should suffocate larger organisms and produce regional mass mortality surfaces at their bases (Figure 4, showing a model based on several analyses of the middle Cenomanian Graneros Shale), but initial research shows variable and inconclusive evidence for this from ash to ash. Similarly, some but not all large storm beds have mass mortality surfaces, especially among larger-shelled benthos and trace-making organisms, at their base (Figure 5), as do modern storm deposits spread over a few hundred square miles of sea floor.

Mass Extinction Bioevents

Episodic mass extinctions have broadly affected global biotas and are well documented in the Cretaceous sequence of the Western Interior Basin, e.g.

the Cenomanian-Turonian (C-T) and Cretaceous-Tertiary (K-T) extinctions. The K-T boundary impact-extinction event is already regarded as a global event marker of unusual proportions (Alvarez et al 1984). The C-T extinction is one of the best documented in the Phanerozoic (Kauffman 1984a, 1986a, Elder 1985, Elder & Kirkland 1986) and provides an excellent example of mass extinction bioevent stratigraphy (Kauffman 1986a). In both the K-T and C-T boundary intervals, mass extinction was found to proceed through a series of discrete steps of accelerated to catastrophic extinction among certain ecologically related taxa sets, each step being very short lived (100 kyr or less) and forming a regional to global bioevent surface for correlation [see Elder & Kirkland (1986) and Kauffman (1986a) for summaries]. Individual steps commonly reflect rapid changes in the oceanography (chemistry, stratification) and the climate (temperature, storm cycles) of the basin. These may be expressed as extraordinarily rapid, large-scale shifts in stable isotopes of ^{13}C and ^{18}O (see Pratt 1985; Figure 8 herein), and a single extinction step may be linked to an individual 50–100 kyr geochemical excursion. Apparently, the rate and magnitude of each large oceanographic fluctuation exceeded the adaptive ranges and evolutionary potential of diverse taxa (Kauffman 1986a). The C-T mass extinction data (Elder & Kirkland 1986; Figure 11 herein) show six to seven short-term to near-isochronous steps of molluscan extinction that Elder has traced as short-term events throughout much of the Western Interior Basin, plotting their position against bentonite (ash) marker horizons. New data on iridium enrichment levels (Orth et al 1987) further show low to moderate Ir-enrichment peaks, some of which may indicate meteorite or comet impacts, precisely correlative with the lower four of Elder's molluscan extinction steps and with the large planktonic foraminifer extinction (*Rotalipora* extinction; Eicher & Diner 1985, Leckie 1985) beginning between molluscan steps 1A and 1B. A chronostratigraphic relationship between impact, climate/ocean disruptions, and mass extinction steps is implied across the C-T boundary interval.

COMPOSITE EVENTS

Milankovitch Climate Cycle (M-Cycle) Deposits

Of the diverse kinds of composite event units previously discussed, those representing Milankovitch or similar climate cycle deposits are of particular value in HIRES correlation of Cretaceous strata in the Western Interior Basin; they are second only to ash/bentonite beds in regional correlation potential. Facies interpreted as representing Milankovitch cycles are varied. Kauffman (1986a) proposed facies models for the North American Cretaceous (Figure 15A), involving asymmetrical bedding couplets, or "rhythms," with varying lithologic composition depending upon the depositional setting. Strata representing the wet (possibly cooler) phase of the cycles lie at the base of each couplet model, and different lithologies representing the dry (possibly warmer) phase lie at the top (Figure 15A). The end-member lithotypes may grade or appear to change abruptly (mainly in carbonate facies where the cyclicity has been enhanced by diagenesis; Ricken 1986). In offshore facies of the Western Interior Basin these cycles, normally a meter or less in thickness, are thus expressed as marl (wet; basal)–chalk (dry; top), calcareous shale/shaley chalk–pelagic limestone, clay shale–calcareous shale, and silty clay shale (wet; basal)–clay shale (dry; top) bedding couplets with slight to broad facies gradations between depositional end members. In nearshore facies, Kauffman (1986a) proposed that these cycles might be represented by sandy shale (wet; basal)–silty shale (dry; top) couplets, by bundled distal storm bed (wet)–sandy and silty shale (dry) couplets in distal lower shoreface sequences, and by progradational shoreface sandstone (wet)–silty marine shale (dry) couplets attaining 10 m or more in thickness. Facies are normally graded between dry to wet end members in nearshore settings.

In these Cretaceous models (Kauffman 1986a; Figure 15A herein), variations in basinal M-Cycle couplets were generally attributed to a combination of fluctuations in calcareous plankton productivity (highest during normal marine dry periods, producing pelagic limestone, chalk, and calcareous shale) and/or offshore transport potential of terrigenous clay (highest during wet climate phases as a result of increased erosion and current velocities, coupled with temporary development of a brackish water lens across parts of the seaway that partially restricted calcareous plankton populations). Variations in more nearshore facies were attributed solely to the increased erosion, supply, and fluvial-coastal-offshore transport of relatively coarser siliciclastic material during prolonged wet phases of Milankovitch cycles. This caused near-synchronous, small-scale coastal progradation (a few miles seaward at the most) of strand and delta plains over more normal, marine clay–dominated facies representing dry phases.

Figure 15 (A) Depositional models for Milankovitch climate cycles from the center (left) to the margins of the Cretaceous of the Western Interior Basin of North America (modified from Kauffman 1986a), taken from actual examples in the Cenomanian-Turonian Greenhorn Cyclothem. (B) Generalized model, from several Cenomanian-Coniacian data sets, of a Milankovitch climate cycle deposit in the carbonate facies of the axial basin, Western Interior Seaway (from Kauffman 1986a), showing high organic carbon (reflecting low oxygen), negative ^{18}O values [reflecting subsaline lenses on the Interior seaway during wet seasons causing density stratification (and thus low benthic oxygen)], and low molluscan and trace fossil diversity during wet climate phases, as compared with low carbon preservation in the benthic zone, normal ^{18}O values, and high diveristy during normal warm dry climatic phases. [See Kauffman (1986a) for details].

FACIES MODELS FOR CRETACEOUS MILANKOVITCH CYCLES

OFFSHORE ←——————————————————————————→ ONSHORE

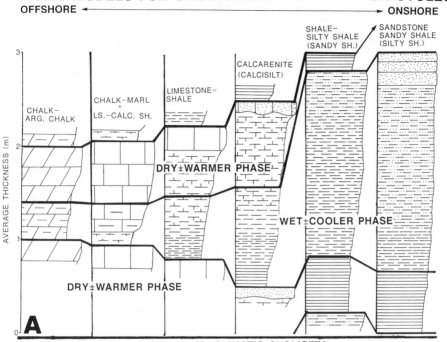

100–125 Ka CLIMATIC CYCLICITY

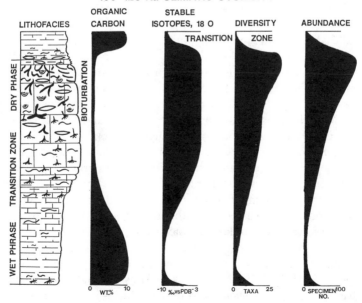

Three tests of these facies models to prove their origins through Milankovitch climate forcing are necessary: (a) They must have durations and stratigraphic repetition compatible with M-cycle periodicity; (b) they must be internally consistent with the predictions of Milankovitch climate effects on sedimentation in terms of their physical, chemical, and biological characteristics; and (c) they must be simultaneously developed in different depositional settings, among different facies suites, as a result of regional to global climate forcing. Preliminary testing of these three criteria in the Cretaceous of the Western Interior Basin of North America has strongly supported the concept of Milankovitch climate forcing for these cycles, as follows:

1. Utilizing K-Ar radiometric dates (average values) clustered around basinal limestone–calcareous shale cycles of the Greenhorn and Niobrara formations of the Western Interior Basin, Kauffman (1977) roughly calculated periodicities of 40–60 kyr for Niobrara bedding rhythms and 60–80 kyr for the most obvious Greenhorn bedding rhythms, both well within the predicted range of M-cycles. Lack of attention to the more subtle carbonate cycles in these sequences, formed during predictable intervals when the dry (warm?) Milankovitch effect might have been damped, or due to diagenetic remobilization of carbonates within them, prevented these early calculations from being more precise. Fischer et al (1985), however, summarized data from more detailed calculations to suggest that these sequences, and similar Mesozoic cycles in Europe, did locally demonstrate a 21-kyr, probable 41-kyr, and 100+-kyr cyclicity compatible with the Milankovitch forcing hypothesis.

2. Selected basinal marl-chalk, shale-limestone, and calcareous shale–clay shale couplets from the Greenhorn Formation (Cenomanian–lower Turonian) of the Western Interior Basin have been tested in detail (centimeter-scale) for physical, chemical, and biological evidence pertaining to the M-cycle hypothesis. Barron et al (1985) and Kauffman (1986a) collectively showed the following changes upsection (from wet to dry phases) in calcareous shale–limestone and marl-chalk bedding couplets of the Bridge Creek Limestone Member, Greenhorn Formation: (a) decreasing clay content; (b) increasing pelagic carbonate content; (c) decreasing organic carbon content, from 3–10% to less than 1% by weight; (d) normalization of ^{18}O values, from abnormally negative at the base (-6 to -11 $^o/_{oo}$ vs the PDB standard; a brackish water signal) to -2 or -3 $^o/_{oo}$ vs PDB at the top (normal Cretaceous marine values); (e) increasing numbers, generations, and diversity of trace fossils upsection, with a correlative change from predominantly detritus-feeding infaunal forms at the base (*Chondrites*, *Planolites*) to mixed detritus- and suspension-feeding

groups at the top; and (f) few or rare epifaunal body fossils in the lower part to larger, more numerous and diverse body fossils (predominantly bivalves) at the top.

Collectively these observations support the hypothesis of Milankovitch climate forcing for these bedding cycles. Wet (basal) phases of sedimentation were characterized by increased rainfall and internal drainage into the basin, producing a brackish water lens on top of the seaway, diminishing calcareous plankton production, establishing a density stratification in the water column, and preventing extensive downward mixing of oxygen to the sea floor. This, in turn, resulted in dysaerobic to anaerobic benthic conditions, preservation of large amounts of C_{org}, and greatly diminished benthic habitation by molluscs and trace-making organisms. Wet cycles were also characterized by increased rainfall, erosion, runoff velocities, and siliciclastic sediment availability to the basin, producing a predominance of clay sedimentation. Dry periods (tops of cycles) were characterized by normal marine surface waters, high calcareous plankton production, breakdown of salinity stratification in the water column, and benthic oxygenation. This resulted in diverse benthic faunas and low levels of C_{org} preservation due to aerobic bacterial decay, organism recycling, and oxygenation of organic carbon. Low rainfall, reduced runoff, and siliciclastic sediment supply also characterized dry phases, allowing pelagic carbonate to dominate over clay sedimentation.

3. Finally, Elder, Gustason & Sageman (in Kauffman et al 1987) carefully documented and traced the largest limestone beds of the Bridge Creek Limestone Member, Greenhorn Formation (late Cenomanian–early Turonian) throughout the central Western Interior Basin (Kansas, Colorado), i.e. those representing the ± 100 kyr Milankovitch cyclicity as defined by Barron et al (1985) and Fischer et al (1985). These events could be identified westward to the shoreline in Utah when traced relative to well-defined regional bentonite (chronostratigraphic) marker horizons. In this transect, they found that the same cycles characterized by calcareous shale (wet)–pelagic limestone (dry), or marl (wet)–chalk (dry) bedding cycles in the central basin, were represented in proximal offshore settings by alternating cycles of clay shale (wet)–limestone concretion horizons (dry) and clay to silty clay shale (wet)–lag calcarenites or shell coquinas (dry). In shoreface settings the M-cycles were represented by progradational sandstone sequences (wet) alternating with silty and sandy clay shale sequences with a basal shell lag (dry) that disconformably overlay the shoreface sandstones. Recognition of Milankovitch effects in shoreface sedimentation and strand plain–delta plain progradation is of great importance to basin analysis inasmuch as many of these progradational events were considered to represent small-scale eustatic

changes. In either case, they are important event units in nearshore correlation.

The most dramatic example of regional HIRES correlation of Milankovitch climate cycle deposits (composite event units) is the regional tracing of individual limestone or chalk beds (dry phase) as isochronous to near-isochronous event deposits retaining a constant position relative to bentonite/ash marker horizons, from central Kansas to central Colorado (Hattin 1971, 1985), into northeastern Arizona [Black Mesa area (Elder & Kirkland 1985)] and south-central Utah (Elder, Gustason & Kirkland in Kauffman et el 1987). Figure 3 shows part of this correlation; compare the relative positions of limestones and volcanic ash deposits across this 440-mi transect.

CORRELATION

Definition and detailed description of potential physical, chemical, and biological event units in individual measured sections of a traverse are only the first steps in HIRES. Each such unit must subsequently be tested and proven to be representative of a short-term isochronous regional event before it can be incorporated into a high-resolution event chronostratigraphy. The proof we seek is our ability to precisely correlate such units among numerous stratigraphic sections within a study area, with each event unit maintaining a relatively constant position in comparison to adjacent event units, some of which must be indisputable chronostratigraphic surfaces, i.e. volcanic ash/bentonite horizons.

Correlation is accomplished by two methods: (a) Standard visual section matching, which may or may not involve lateral tracing of individual beds over essentially continuous outcrop belts between sections; and (b) graphic correlation techniques (Miller 1977, Edwards 1984).

Standard correlation techniques (Figure 3) rely heavily upon the initial recognition of lithologically unique event-marker beds and, independently, upon detailed biostratigraphy to establish a coarse but secure correlation between two sections; this correlation then moves to the examination of smaller-scale event-stratigraphic units and their correlation between sections. The process proceeds from the section with the most data toward the nearest satellite section, and from there toward the next nearest section, and so on. As seen in Figure 3, correlations can be made using this method, relying primarily on event-stratigraphic units, to a high level of resolution (100 kyr or less per event-bounded interval).

Graphic correlation provides an even more detailed and objective means of correlating event units and more dissimilar, distantly spaced sections. Edwards (1984) provides the most up-to-date and easily understood review of the technique. In standard graphic correlation, as normally practiced prior to the development of HIRES, two stratigraphic sections plotted to the same scale are laid out along the X- and Y-axes of the graphic plot (see Figures 16, 17) and adjusted so that some known correlation surface(s) or biozone(s) common to both would produce lines of intersection near the center of the graphic field. First and last occurrences of key biostratigraphic indices in both sections are then plotted as separately marked intersects in the field, and when all such data are plotted, a line of graphic correlation is plotted by eye or (more commonly) by regression analysis. With this line in place, variations in slope of the line across the field of correlation are taken to indicate differences in sedimentation rates between the sections, and breaks in the line are taken to indicate missing sections (i.e. disconformities or faults) on one or both axes. Intervals of the stratigraphic sequence in both sections, and the historical events they represent, are considered to be essentially coeval if their points of intersection in the field of correlation fall on or within a standard deviation of the line of (biostratigraphic) correlation (LOC). After correlation of the first two sections (commonly the most complete available), data from both are composited on the vertical axis, calibrated, and a third section is placed on the horizontal axis for correlation to the composite standard data set [see Edwards (1984) and Miller (1977) for technique]. Figure 17 shows two examples from the Cretaceous of the Western Interior Basin of biostratigraphic vs event-stratigraphic correlations between (A) two Coniacian, lower Niobrara Formation sections (Barlow 1985), and (B) two Cenomanian–lower Turonian Greenhorn cycle transgressions in the Western Interior Basin. In this standard type of graphic system, regional correlation is wholly based on biostratigraphy without consideration of the relative value of first vs last taxa occurrences in the section (although these can be sorted out, and first occurrences give more consistent correlation). Only recently have the users of standard graphic correlation begun to integrate available magnetostratigraphic, geochronologic, and chronostratigraphic data into older, biostratigraphically based systems.

High-resolution event stratigraphy (HIRES) makes possible a new dimension in graphic correlation. It develops both the line of correlation between two sections and the composite standard, based primarily on isochronous or short-term event deposits. The resultant line of correlation becomes a line of isochronous correlation (LOIC), and any two points or intervals that intersect it, or come within a standard deviation of the line, are considered to have occurred at the same absolute point or interval of time. The line can further be calibrated to real time by tying it to trusted radiometric dating. The advantages of this system are (a) it can be based on many more points (event-stratigraphic units); (b) it does not have

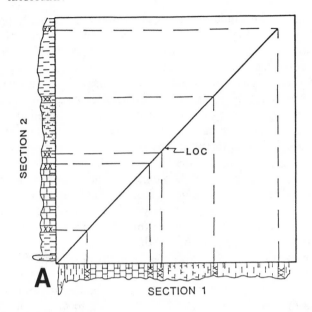

SECTION 2

LOC

A

SECTION 1

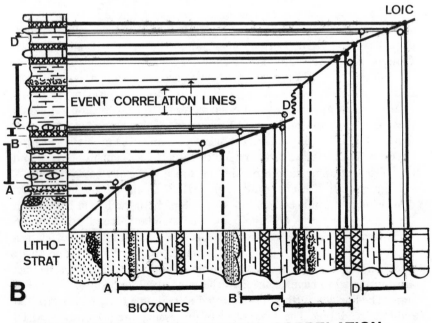

LOIC

D

EVENT CORRELATION LINES

D

C

B

A

LITHO-
STRAT

B

A

BIOZONES

B

C

D

HIRES MODELS FOR GRAPHIC CORRELATION

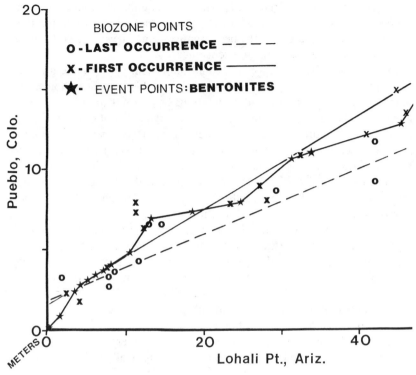

Figure 17 Graphic correlation of HIRES levels (volcanic ash or bentonite beds) between two Cenomanian-Turonian boundary sections developed to test ammonite biostratigraphic zone boundaries. Stars and solid line mark line of isochronous (ash) correlation (LOIC), with changes in slope probably reflecting changes in sedimentation rate between sections. Note that regression line drawn through ammonite biozone first occurrences closely fits the LOIC and is thus nearly isochronous, but that the last occurrence regression line departs markedly from the LOIC. This suggests that only the first occurrence of a biozone marker species is valid as a basis for precise regional correlation (J. I. Kirkland & W. P. Elder, in preparation).

the same potential for error as biostratigraphic correlations in which ecological, evolutionary, and preservational controls can severely distort or modify the field data for any section, and thus the correlations; (c) it provides a real time matrix for basin analysis and for looking at the short-

←

Figure 16   (*A*) Simple model of graphic correlation to develop an isochron event correlation line from volcanic ash beds in two sections [X's and dashed lines; LOC is solid diagonal line (from Barlow 1985)]. (*B*) Detailed model for graphic correlation of high-resolution event stratigraphic data from two sections (after Kauffman 1986) showing development of line of isochron correlation (LOIC) from event units (ash layers and climate cycle beds), near-correlation of mass flow deposits to this line, and position of biozone first and last occurrence points (open circles), which are close but not isochronous.

term dynamics of any system; and (*d*) if the event units are carefully chosen, this system has a more regional, facies-wide potential for correlation than biostratigraphy, where even the best organisms may be subjected to subtle ecological (habitat) controls on occurrence.

Further, the HIRES graphic correlation system can provide an independent check on the accuracy of biostratigraphic, geochronologic, and magnetostratigraphic data and correlations based on them; yet the system is open ended enough to incorporate any type of data for testing or compositing. Figure 16*B* shows a model of the HIRES graphic correlation system, based on a set of a actual correlations between two complex Cretaceous sequences in Arizona and Colorado. Clearly, ash/bentonite correlation units form the basis for construction of the HIRES graphic correlation system, with Milankovitch climate cycle deposits and concretion horizons running a close second in utility.

Figure 17 shows a Cenomanian-Turonian correlation matrix of HIRES event units (volcanic ashes/bentonites) in the Western Interior Basin against which are plotted the first and last occurrence points of key ammonite biostratigraphic indices. Note that the first occurrences of ammonite taxa fall very close to LOIC, but that last occurrences (local population extinctions) in some cases fall considerably beyond the LOIC; this, in turn, suggests that first occurrences of index taxa in biostratigraphy are the most trustworthy in correlation, and, in addition, that dispersal of newly evolved or immigrant taxa can be very rapid and geographically widespread (data from J. I. Kirkland & W. P. Elder, in preparation).

APPLICATIONS OF HIRES

High-resolution event stratigraphy, especially when correlated through graphic correlation, forms a powerful new tool of stratigraphy and basin analysis. It achieves, for the Cretaceous of the Western Interior Basin of North America as an example, a resolution averaging about 50 kyr per HIRES event-bounded interval for basinal settings, and about 100–150 kyr for marginal facies in the basin. At the coarsest level, HIRES equals the finest level of biostratigraphic zonation developed for the basin; HIRES thus creates a new and more precise dimension in regional correlation. As such, and especially when integrated with a large radiometric data set, HIRES is well suited for looking at the rates and patterns of geological, biological, oceanographic, and climatic evolution within sedimentary basins, and for integrating diverse data in basin analysis. Some basic examples of the interpretive potential of HIRES within the Western Interior Basin are given by Kauffman (1987), who demonstrated the following applications (Figure 18):

HIRES REGIONAL CORRELATION OF MARINE AND TECTONIC EVENTS

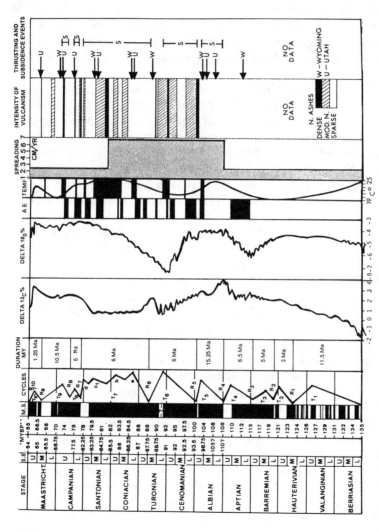

Figure 18 Summary diagram of tectonic and oceanic events for the Cretaceous of the Western Interior Basin of North America (from Kauffman 1984), with correlations calibrated by high-resolution event stratigraphy. Note close correlation of tectonic and volcanic events to global sea-level rise and transgression into epicontinental seas (right-sloping line and T numbers on left side), and correlation of tectonic quiescence with regression and sea-level fall (left-sloping lines), suggesting a dominant plate-tectonic control on both eustatic history and continental tectonic history. Note also correlation of anoxic events (AE) with eustatic rise (CYCLES) and with warm temperature shifts (dark shading under TEMP).

1. Calculation of the timing, volume, and magnitude of volcanism, based on bentonite (ash) volume and distribution within the basin and mapping of the source and distribution of single ash beds within narrow time intervals. From the latter, paleowind directions can be calculated. Most volcanic ash was derived from northern (Idaho, Montana) and south-western (Arizona, New Mexico, Texas) source areas; 80–85% of explosive volcanism was correlated with active thrusting and tectonic movement in the Cordilleran region and with eustatic rise, suggesting a connection to increased spreading and subduction rates along the Pacific margin of North America.

2. Calculation of the timing and relative magnitude of tectonic move-ment within the basin, including thrusting, normal faulting, subsidence, rebound, and regional movement on tectonic blocks. It was concluded that most active tectonism in this region was associated with eustatic rise events, reflecting accelerated Pacific spreading and subduction rates; that tectonic activity was episodic and commonly short term; and that intervals of tectonic and volcanic quiescence were largely associated with eustatic fall intervals and with slowing of rates of ocean spreading and subduction. From this, a two-phase model of tectonics within the basin was developed (Kauffman 1985a, 1986a), linked to plate movements and sea level, and it was determined that active subduction drove regional Cordilleran and basin tectonics without significant time lag.

3. The rates and magnitude of paleoceanographic phenomena were documented in great detail. Second through sixth-order absolute and rela-tive sea-level changes (Kauffman 1985a) were calibrated and correlated within very narrow time intervals to global standards of eustatic history and sequence stratigraphy (e.g. to those in Haq et al 1987). Anoxic and dysaerobic events were identified throughout the Cretaceous and linked in time to eustatic rise and highstand intervals, and rarely to early regression and eustatic fall. It was determined that these events resulted from very rapid (100 kyr or less) stratification and destratification within the marine water column. The effects of Milankovitch climate cycles on stratification and on salinity and oxygen levels within the basin were documented to within at least 40–50 kyr intervals involving cyclic stratification and deoxy-genation events (wet phase), followed by normalization of marine chem-istry and circulation (dry phase). Major changes in watermass charac-teristics (temperature, circulation, chemistry) were shown to be closely related to rapid immigration of warm watermasses from the Gulf of Mexico (Kauffman 1984b); these pulses have been calibrated (intervals of rapid immigration or emigration take under 100 kyr regionally) and correlated to peak eustatic rise events. Biological response to these diverse

oceanographic changes could be documented within 100-kyr intervals or less.

4. Sedimentation rates can be calibrated within 50 kyr stratigraphic intervals for all facies within the basin (see Elder & Kirkland 1985), and thus short-term changes in these rates due to tectonic movement, sediment supply, or biological productivity can easily be plotted. Tracing of individual ash beds across the basin relative to sediment thickness between them allows detailed calibration and identification of bypass and disconformity surfaces associated with eustatic highstand, ravinement development, regional submarine uplift, and sediment starvation in the central basin.

5. HIRES provides a powerful tool in the interpretation of biological events within the basin: rates and patterns of evolution; dispersal rates and patterns; the nature and biogeography of population bursts, productivity events, community and species dispersion, and both population and ecological structure. Biostratigraphic zone boundaries, and individual taxa used in biostratigraphy, can be calibrated and evaluated against HIRES matrices, especially by utilizing graphic correlation. Most biostratigraphic indices for the basin correlate closely to graphic axes of event stratigraphy, suggesting near- (but not absolute) isochronous boundaries. First appearances of biostratigraphically useful taxa fall closer to "time lines" than do last appearances.

SUMMARY

High-resolution event stratigraphy is a new and powerful tool of regional stratigraphy. It can be used to precisely correlate stratigraphic sections and develop theory and tests for basin analysis, paleoceanography, paleoclimatology, evolution, and biostratigraphy. In the Western Interior Basin of North America, the Cretaceous marine sequence is dominantly composed of allocyclic and perturbational event units that are isochronous or short term (100 kyr or shorter) in aspect. These have allowed a very detailed HIRES to be developed, with resolution in correlation between 50 and 100 kyr, and less for some intervals. This exceeds the resolution of any other system of stratigraphy. HIRES has many applications, and it is still in the development and testing stage.

Literature Cited

Alvarez, L. W., Alvarez, W., Asaro, F., Michel, H. V. 1980. Extraterrestrial cause for the Cretaceous-Tertiary extinction. *Science* 208: 1095–1108

Alvarez, W., Alvarez, L., Asaro, F., Michel, H. V. 1982. Current status of the impact theory for the terminal Cretaceous extinction. In *Geological Implications of Impacts of Large Asteroids and Comets on the Earth. Geol. Soc. Am. Spec. Pap. 190,* ed. L. T. Silver, P. H. Schultz, pp. 305–15

Alvarez, W., Kauffman, E. G., Surlyk, F., Alvarez, L. W., Asaro, F., Michel, H. V. 1984. The impact theory of mass extinctions and the marine invertebrate record across the Cretaceous-Tertiary boundary. *Science* 223: 1135–41

Barlow, L. K. 1985. *Event stratigraphy, paleoenvironments, and petroleum source rock potential of the lower Niobrara Formation (Cretaceous), northern Front Range, Colorado.* MS thesis. Univ. Colo., Boulder. 288 pp.

Barlow, L. K., Kauffman, E. G. 1985. Depositional cycles in the Niobrara Formation, Colorado Front Range. See Pratt et al 1985, pp. 199–208

Barron, E. J., Washington, W. M. 1984. The role of geographic variables in explaining paleoclimates—Results from Cretaceous climate model sensitivity studies. *J. Geophys. Res.* 89: 267–79

Barron, E. J., Arthur, M. A., Kauffman, E. G. 1985. Cretaceous rhythmic bedding sequences: a plausible link between orbital variations and climate. *Earth Planet. Sci. Lett.* 72: 327–40

Beerbower, J. R. 1964. *Kans. State Geol. Surv. Bull.* 169: 33–42

Bohor, B. F., Foord, E. E., Modreski, P. J., Triplehorn, D. M. 1984. Mineralogic evidence for an impact event at the Cretaceous-Tertiary boundary. *Science* 224: 867–69

Cobban, W. A. 1951. *US Geol. Surv. Prof. Pap. 239.* 14 pp.

Cobban, W. A. 1958. *Wyo. Geol. Assoc. Guideb. 13th Ann. Field Conf., Powder River Basin,* pp. 114–19

Cobban, W. A. 1962. Baculites from the lower part of the Pierre Shale and equivalent rocks in the Western Interior. *J. Paleontol.* 36(4): 704–18

Cobban, W. A. 1969. *US Geol. Surv. Prof. Pap. 619.* 29 pp.

Cobban, W. A. 1971. *US Geol. Surv. Prof. Pap. 699.* 24 pp.

Cobban, W. A. 1984. *Bull. Geol. Soc. Den.* 33: 71–89

Cobban, W. A. 1985. Ammonite record from the Bridge Creek Member of the Greenhorn Limestone at Pueblo State Recreation Area, Colorado. See Pratt et al 1985, pp. 135–38

Edwards, L. E. 1984. Insights of why graphic correlation (Shaw's method) works. *J. Geol.* 92: 583–97

Eicher, D. L., Diner, R. 1985. Foraminifera as indicators of watermass in the Cretaceous Greenhorn Sea, Western Interior. See Pratt et al 1985, pp. 60–71

Einsele, G., Seilacher, A. 1982. *Cyclic and Event Stratification.* Berlin: Springer-Verlag. 536 pp.

Elder, W. P. 1985. Biotic patterns across the Cenomanian-Turonian extinction boundary near Pueblo, Colorado. See Pratt et al 1985, pp. 157–69

Elder, W. P. 1986. See Elder & Kirkland 1986, Figure 10

Elder, W. P., Kirkland, J. I. 1985. Stratigraphy and depositional environments of the Bridge Creek Limestone Member of the Greenhorn Limestone at Rock Canyon Anticline near Pueblo, Colorado. See Pratt et al 1985, pp. 122–34

Elder, W. P., Kirkland, J. I. 1986. The Bridge Creek Limestone Member of the Greenhorn Limestone at Rock Canyon Anticline near Pueblo, Colorado. See Kauffman 1986b, pp. 91–111

Eldredge, N., Gould, S. J. 1972. Punctuated equilibria: an alternative to phyletic gradualism. In *Models in Paleontology,* ed. T. J. M. Schopf, pp. 82–115. San Francisco: Freeman. 250 pp.

Fischer, A. G., Herbert, T., Premoli-Silva, I. 1985. Carbonate bedding cycles in Cretaceous pelagic and hemipelagic sequences. See Pratt et al 1985, pp. 1–10

Glenister, L. M. 1985. *High-resolution stratigraphy and interpretation of the depositional environments of the Greenhorn Cyclothem regression (Turonian; Cretaceous), Colorado Front Range.* MS thesis. Univ. Colo., Boulder. 184 pp.

Glenister, L. M., Kauffman, E. G. 1985. High-resolution stratigraphy and depositional history of the Greenhorn regressive hemicyclothem, Rock Canyon Anticline, Pueblo, Colorado. See Pratt et al 1985, pp. 170–83

Haq, B. V., Hardenbol, J., Vail, P. R. 1987. Chronology of fluctuating sea-levels since the Triassic. *Science* 235: 1156–67

Hattin, D. E. 1971. *Am. Assoc. Pet. Geol. Bull.* 55: 110–19

Hattin, D. W. 1975. *Kans. State Geol. Surv. Bull. 209.* 128 pp.

Hattin, D. E. 1985. Distribution of widespread, time-parallel pelagic limestone beds in the Greenhorn Limestone (Upper Cretaceous) of the central Great Plains and southern Rocky Mountains. See Pratt et al 1985, pp. 28–37

Kauffman, E. G. 1969. Cretaceous marine cycles of the Western Interior. *Mt. Geol.* 4: 227–45

Kauffman, E. G. 1970. *Proc. North Am. Paleontol. Conv. F,* pp. 612–66. Lawrence, Kans: Allen Press

Kauffman, E. G. 1975. Dispersal and biostratigraphic potential of Cretaceous benthonic Bivalvia in the Western Interior. In *The Cretaceous System in the Western Interior of North America. Geol. Assoc. Can. Spec. Pap. 13*, ed. W. G. E. Caldwell, pp. 163–94. 666 pp.

Kauffman, E. G. 1976. Brittish Middle Cretaceous inoceramid biostratigraphy. In *Événements de la Partie Moyenne du Crétacé (Mid-Cretaceous events), Uppsala-Nice Symp., 1975–76. Ann. Hist. Nat. Nice*, ed. G. Thomel et al, 4: IV-1–11

Kauffman, E. G. 1977. Geological and biological overview: Western Interior Cretaceous Basin. In *Cretaceous Facies, Faunas, and Paleoenvironments across the Western Interior Basin. Mt. Geol.*, ed. E. G. Kauffman, 13(3,4): 75–99

Kauffman, E. G. 1978. Short-lived benthic communities in the Solnhofen and Nusplingen limestones. *Neus Jahrb. Geol. Paläontol. Monatsh.* 12: 714–17

Kauffman, E. G. 1981. Ecological reappraisal of the German Posidonienscheifer (Toarcian) and the stagnant basin model. In *Communities of the Past*, ed. J. Gray, A. J. Boucot, W. B. N. Berry, pp. 311–81. Stroudsburg, Pa: Hutchinson-Ross. 623 pp.

Kauffman, E. G. 1984a. The fabric of Cretaceous marine extinctions. In *Catastrophes and Earth History: The New Uniformitarianism*, ed. W. A. Berggren, J. Van Couvering, pp. 151–246. Princeton, NJ: Princeton Univ. Press. 464 pp.

Kauffman, E. G. 1984b. Paleobiogeography and evolutionary response dynamic in the Cretaceous Western Interior Seaway of North America. In *Jurassic-Cretaceous Biochronology and Paleogeography of North America. Geol. Assoc. Can. Spec. Pap. 27*, ed. G. E. G. Westermann, pp. 273–306. 315 pp.

Kauffman, E. G. 1985a. Cretaceous evolution of the Western Interior Basin of the United States. See Pratt et al 1985, pp. iv–xiii

Kauffman, E. G. 1985b. Depositional history of the Graneros Shale (Cenomanian), Rock Canyon Anticline. See Pratt et al 1985, pp. 90–99

Kauffman, E. G. 1986a. High-resolution event stratigraphy: regional and global bio-events. In *Global Bioevents. Lect. Notes Earth Hist.*, ed. O. H. Walliser, pp. 279–335. Berlin: Springer-Verlag. 442 pp.

Kauffman, E. G., ed. 1986b. *Cretaceous Biofacies of the Central Part of the Western Interior Seaway: A Field Guidebook. North Am. Paleontol. Conv. NAPC IV*. Boulder, Colo: Univ. Colo. 210 pp.

Kauffman, E. G. 1987. High-resolution event stratigraphy: concepts, methods, and Cretaceous examples. See Kauffman et al 1987, pp. 2–34

Kauffman, E. G., Sageman, B. B., Gustason, E. R., Elder, W. P., eds. 1987. *A Field Trip Guidebook: High-resolution Event Stratigraphy, Greenhorn Cyclothem (Cretaceous: Cenomanian-Turonian), Western Interior of Colorado and Utah*. Boulder, Colo: Geol. Soc. Am., Rocky Mt. Sect. 198 pp.

Knechtel, M. M., Patterson, S. H. 1962. *US Geol. Surv. Bull. 1082-M*, pp. 893–1030

Leckie, R. M. 1985. Foraminifera of the Cenomanian-Turonian boundary interval, Greenhorn Formation, Rock Canyon Anticline, Pueblo, Colorado. See Pratt et al 1985, pp. 139–50

Miller, F. X. 1977. The graphic correlation method in biostratigraphy. In *Concepts and Methods of Biostratigraphy*, ed. E. G. Kauffman, J. E. Hazel, pp. 165–86. Stroudsburg, Pa: Dowden, Hutchinson & Ross

North American Commission on Stratigraphic Nomenclature. 1983. North American stratigraphic code. *Am. Assoc. Pet. Geol. Bull.* 67(5): 841–75

Orth, C. J., Gilmore, J. S., Knight, J. D., Pillmore, C. L., Tschudy, R. H., Fassett, J. E. 1981. An iridium anomaly at the palynological Cretaceous-Tertiary boundary in northern New Mexico. *Science* 214: 1341–43

Orth, C. J., Attrep, M., Mao, X., Kauffman, E. G., Diner, R., Elder, W. P. 1987. Iridium abundance maxima at Upper Cenomanian extinction horizons. Submitted for publication

Pillmore, C. L., Flores, R. M. 1987. Stratigraphy and depositional environments of the Cretaceous-Tertiary boundary clay and associated rocks, Raton basin, New Mexico and Colorado. In *The Cretaceous-Tertiary Boundary in the San Juan and Raton Basins, New Mexico and Colorado. Geol. Soc. Am. Spec. Pap. 209*, ed. J. E. Fassett, J. K. Rigby Jr., pp. 111–30

Pratt, L. M. 1985. Isotopic studies of organic matter and carbonate in rocks of the Greenhorn Marine Cycle. See Pratt et al 1985, pp. 38–48

Pratt, L. M., Barlow, L. K. 1985. Isotopic and sedimentological study of the lower Niobrara Formation, Lyons, Colorado. See Pratt et al 1985, pp. 209–14

Pratt, L. M., Kauffman, E. G., Zelt, F. B., eds. 1985. *Fine-Grained Deposits and Biofacies of the Cretaceous Western Interior Seaway: Evidence of Cyclic Sedimentary Processes. Soc. Econ. Paleontol. Mineral. Field Trip Guideb. 4*. 288 pp.

Ricken, W. 1986. *Diagenetic Bedding: A Model for Marl-Limestone Alternations. Lect. Notes Earth Sci.*, Vol. 6. Berlin: Springer-Verlag. 210 pp.

Sageman, B. B. 1985. High-resolution stratigraphy and paleobiology of the Hartland Shale Member: analysis of an oxygen-deficient epicontinental sea. See Pratt et al 1985, pp. 112–21

Tarling, D. H. 1983. *Paleomagnetism.* London: Chapman & Hall. 379 pp.

Vail, P. R., Mitchum, R. M. Jr., Thompson, S. III 1977. Seismic stratigraphy and global changes of sea level, Part 4: Global cycles of relative changes of sea level. In *Seismic Stratigraphy. Am. Assoc. Pet. Geol. Mem. 26,* ed. C. E. Payton, pp. 83–97. 516 pp.

Zelt, F. G. 1985. Paleoceanographic events and lithologic/geochemical facies of the Greenhorn Marine Cycle (Upper Cretaceous) examined using natural gamma-ray spectrometry. See Pratt et al 1985, pp. 49–59

SUBJECT INDEX

A

Accelerator mass spectrometry
(AMS) technique, 357
and cosmogenic isotope studies, 358
Accretion
in southern Laurentia, 574–86
Accretionary wedge(s)
of the Greater Antilles, 216
of the Lesser Antilles, 220–21
in the Scotland district of
Barbados, 220
Acoustic imagery
and subaqueous mass wasting,
116
Acoustic impedance
and seismic reflections, 325,
326–28, 330, 332, 339,
343–44
Aerosol cloud(s)
and flood-basalt volcanism,
92–93
Tambora, 83
from a Toba-sized explosive
eruption, 90–92
Aerosol optical depth(s), 77, 79,
90–92
Aerosols
from El Chichón eruption, 81
stratospheric
and volcanic eruptions, 75–
81, 91–92, 94
Africa
its fit to the Americas, 202–3
plate models of, 210
Agulhas slide
off South Africa, 105
Agung, Mt.
eruption of, 77–78, 80, 82
Albedo(s)
of cometary nuclei, 275–80,
287–88, 290
Aleutians
sedimentary rocks in, 15
Alkali
in granitic rocks, 43, 45–46
Aluminum saturation index
(ASI), 22
Aluminum silicate polymorphs
in granitic rocks, 25
Americium
in spent fuel of nuclear reactors, 174
Amplitude Variations with
Offset (AVO)
in seismic analysis, 344

Andalusite
and AFM phase relations in
peraluminous granitic
rocks, 34–39
in granitic rocks, 25
magmatic, 32
Anorthosite
and titanium, 158
Antarctic ice
fallout from Tambora eruption
in, 83
Antilles
Greater, 213
accretionary wedges of,
216
plate boundary zone in,
217–18
strike-slip motion in, 217
Lesser
accretionary wedge of,
220–21
see also Greater Antillean arc;
Lesser Antillean arc
Appalachian Piedmont
metaluminous rocks in, 43
Archaeomagnetic data, 443,
445–47
Archaeomagnetism, 390, 393,
443, 468
Archean greenstones, 162–63,
168
gold in, 148, 155–56, 169
Archean ore deposits, 161–63
Archean provinces
and Proterozoic orogens, 546–
72, 586–88
Arizona
and the Yavapai and Mazatzal
orogens, 584–85
Artificial intelligence
and integrated seismic
stratigraphic analysis,
349
Asama
eruption of, 82
Asperities
and earthquake ground mo-
tions, 133–36, 138
Asteroid impact
in the Late Cretaceous, 93
Asteroids
Amor-Apollo (AA), 274, 286,
289–90
brightness fluctuations, 282
and comets, 274, 286–90
determining size, 277
formation of, 55

Atlantic
spreading of the Central,
206–7
Atmosphere
and Earth rotation, 461, 463–
64
Atmospheric pressure
and the Chandler wobble, 245
and length-of-day variability,
243–44
Augustine, Mt.
eruption of, 82
Auroral electrojets, 400, 402
Aves Swell, 218, 220
formation of, 215

B

Bahama Banks, 217
Banded iron formation (BIF),
156, 161, 163, 166–67,
169
Algoma-type, 154–55, 162
Superior-type, 154–55
Basalt(s)
oceanic
geochemical heterogeneity
in, 494–501, 532
as a repository for nuclear
waste disposal, 186–88,
193
see also Mid-ocean ridge
basalts (MORB)
Basaltic composition
of the Venusian surface, 298–
99, 300
Basin structures
and earthquake ground mo-
tions, 127–29
Batholith(s)
Cornubian, 32, 35, 37
Idaho, 43–44
in New South Wales, 41, 43
Pioneer, 43
South Baffin, 572–74
Bedding surfaces
and sedimentary facies, 327–
30
Bentonite deposits
and high-resolution event
stratigraphy, 617–21
Bioevents
immigration-emigration
and high-resolution event
stratigraphy, 634–38
Biostratigraphic boundaries, 631

655

CUMULATIVE INDEXES

CONTRIBUTING AUTHORS VOLUMES 1–16

CHAPTER TITLES VOLUMES 1–16

669

Annual Reviews Inc.

A NONPROFIT SCIENTIFIC PUBLISHER

⚏ 4139 El Camino Way
P.O. Box 10139
Palo Alto, CA 94303-0897 • USA

Annual Reviews Inc. publications may be ordered directly from our office by mail or use our Toll Free Telephone line (for orders paid by credit card or purchase order, and customer service calls only); through booksellers and subscription agents, worldwide; and through participating professional societies. Prices subject to change without notice. ARI Federal I.D. #94-1156476

- **Individuals:** Prepayment required on new accounts by check or money order (in U.S. dollars, check drawn on U.S. bank) or charge to credit card — American Express, VISA, MasterCard.
- **Institutional buyers:** Please include purchase order number.
- **Students:** $10.00 discount from retail price, per volume. Prepayment required. Proof of student status must be provided (photocopy of student I.D. or signature of department secretary is acceptable). Students must send orders direct to Annual Reviews. Orders received through bookstores and institutions requesting student rates will be returned. You may order at the Student Rate for a maximum of 3 years.
- **Professional Society Members:** Members of professional societies that have a contractual arrangement with Annual Reviews may order books through their society at a reduced rate. Check with your society for information.
- **Toll Free Telephone orders:** Call 1-800-523-8635 (except from California) for orders paid by credit card or purchase order and customer service calls only. California customers and all other business calls use 415-493-4400 (not toll free). Hours: 8:00 AM to 4:00 PM, Monday-Friday, Pacific Time.

Regular orders: Please list the volumes you wish to order by volume number.
Standing orders: New volume in the series will be sent to you automatically each year upon publication. Cancellation may be made at any time. Please indicate volume number to begin standing order.
Prepublication orders: Volumes not yet published will be shipped in month and year indicated.
California orders: Add applicable sales tax.
Postage paid (4th class bookrate/surface mail) **by Annual Reviews Inc.** Airmail postage or UPS, extra.

ANNUAL REVIEWS SERIES		Prices Postpaid per volume USA & Canada/elsewhere	Regular Order Please send:	Standing Order Begin with:
			Vol. number	Vol. number
Annual Review of ANTHROPOLOGY				
Vols. 1-14	(1972-1985)	$27.00/$30.00		
Vols. 15-16	(1986-1987)	$31.00/$34.00		
Vol. 17	(avail. Oct. 1988)	$35.00/$39.00	Vol(s). _____	Vol. _____
Annual Review of ASTRONOMY AND ASTROPHYSICS				
Vols. 1-2, 4-20	(1963-1964; 1966-1982)	$27.00/$30.00		
Vols. 21-25	(1983-1987)	$44.00/$47.00		
Vol. 26	(avail. Sept. 1988)	$47.00/$51.00	Vol(s). _____	Vol. _____
Annual Review of BIOCHEMISTRY				
Vols. 30-34, 36-54	(1961-1965; 1967-1985)	$29.00/$32.00		
Vols. 55-56	(1986-1987)	$33.00/$36.00		
Vol. 57	(avail. July 1988)	$35.00/$39.00	Vol(s). _____	Vol. _____
Annual Review of BIOPHYSICS AND BIOPHYSICAL CHEMISTRY				
Vols. 1-11	(1972-1982)	$27.00/$30.00		
Vols. 12-16	(1983-1987)	$47.00/$50.00		
Vol. 17	(avail. June 1988)	$49.00/$53.00	Vol(s). _____	Vol. _____
Annual Review of CELL BIOLOGY				
Vol. 1	(1985)	$27.00/$30.00		
Vols. 2-3	(1986-1987)	$31.00/$34.00		
Vol. 4	(avail. Nov. 1988)	$35.00/$39.00	Vol(s). _____	Vol. _____

ANNUAL REVIEWS SERIES	Prices Postpaid per volume USA & Canada/elsewhere	Regular Order Please send:	Standing Order Begin with:
		Vol. number	Vol. number

Annual Review of COMPUTER SCIENCE

Vols. 1-2	(1986-1987)................$39.00/$42.00		
Vol. 3	(avail. Nov. 1988)..............$45.00/$49.00	Vol(s). _____	Vol. _____

Annual Review of EARTH AND PLANETARY SCIENCES

Vols. 1-10	(1973-1982).................$27.00/$30.00		
Vols. 11-15	(1983-1987).................$44.00/$47.00		
Vol. 16	(avail. May 1988)..............$49.00/$53.00	Vol(s). _____	Vol. _____

Annual Review of ECOLOGY AND SYSTEMATICS

Vols. 2-16	(1971-1985).................$27.00/$30.00		
Vols. 17-18	(1986-1987).................$31.00/$34.00		
Vol. 19	(avail. Nov. 1988).............$34.00/$38.00	Vol(s). _____	Vol. _____

Annual Review of ENERGY

Vols. 1-7	(1976-1982).................$27.00/$30.00		
Vols. 8-12	(1983-1987).................$56.00/$59.00		
Vol. 13	(avail. Oct. 1988).............$58.00/$62.00	Vol(s). _____	Vol. _____

Annual Review of ENTOMOLOGY

Vols. 10-16, 18-30	(1965-1971; 1973-1985)........$27.00/$30.00		
Vols. 31-32	(1986-1987).................$31.00/$34.00		
Vol. 33	(avail. Jan. 1988).............$34.00/$38.00	Vol(s). _____	Vol. _____

Annual Review of FLUID MECHANICS

Vols. 1-4, 7-17	(1969-1972, 1975-1985)........$28.00/$31.00		
Vols. 18-19	(1986-1987).................$32.00/$35.00		
Vol. 20	(avail. Jan. 1988).............$34.00/$38.00	Vol(s). _____	Vol. _____

Annual Review of GENETICS

Vols. 1-19	(1967-1985).................$27.00/$30.00		
Vols. 20-21	(1986-1987).................$31.00/$34.00		
Vol. 22	(avail. Dec. 1988).............$34.00/$38.00	Vol(s). _____	Vol. _____

Annual Review of IMMUNOLOGY

Vols. 1-3	(1983-1985).................$27.00/$30.00		
Vols. 4-5	(1986-1987).................$31.00/$34.00		
Vol. 6	(avail. April 1988).............$34.00/$38.00	Vol(s). _____	Vol. _____

Annual Review of MATERIALS SCIENCE

Vols. 1, 3-12	(1971, 1973-1982)............$27.00/$30.00		
Vols. 13-17	(1983-1987).................$64.00/$67.00		
Vol. 18	(avail. August 1988)...........$66.00/$70.00	Vol(s). _____	Vol. _____

Annual Review of MEDICINE

Vols. 1-3, 6, 8-9	(1950-1952, 1955, 1957-1958)		
11-15, 17-36	(1960-1964, 1966-1985)........$27.00/$30.00		
Vols. 37-38	(1986-1987).................$31.00/$34.00		
Vol. 39	(avail. April 1988).............$34.00/$38.00	Vol(s). _____	Vol. _____

Annual Review of MICROBIOLOGY

Vols. 18-39	(1964-1985).................$27.00/$30.00		
Vols. 40-41	(1986-1987).................$31.00/$34.00		
Vol. 42	(avail. Oct. 1988).............$34.00/$38.00	Vol(s). _____	Vol. _____